**Honey**

# Honey: Composition and Health Benefits

*Edited by*

*Md. Ibrahim Khalil*
*Jahangirnagar University*
*Savar, Dhaka*
*Bangaladesh*

*Siew Hua Gan*
*Monash University Malaysia*
*Bandar Sunway*
*Malaysia*

*Bey Hing Goh*
*Monash University Malaysia*
*Bandar Sunway*
*Malaysia*

*Zhejiang University*
*Hangzhou, Zhejiang*
*PR China*

WILEY

This edition first published 2023


*Registered Office(s)*
John Wiley & Sons, Inc., 111 River Street, Hoboken, NJ 07030, USA
John Wiley & Sons Ltd, The Atrium, Southern Gate, Chichester, West Sussex, PO19 8SQ, UK

*Editorial Office*
The Atrium, Southern Gate, Chichester, West Sussex, PO19 8SQ, UK
Boschstr. 12, 69469 Weinheim, Germany

For details of our global editorial offices, customer services, and more information about Wiley products visit us at www.wiley.com.

Wiley also publishes its books in a variety of electronic formats and by print-on-demand. Some content that appears in standard print versions of this book may not be available in other formats.



A catalogue record for this book is available from the Library of Congress

Hardback ISBN: 9781119113294; ePub ISBN: 9781119870616; ePDF ISBN: 9781119113300; oBook ISBN: 9781119113324

Cover image: © Sumiko Scott/Getty Images
Cover design by Wiley

Set in 9.5/12.5pt STIXTwoText by Integra Software Services Pvt. Ltd., Pondicherry, India

Printed and bound by CPI Group (UK ) Ltd, Croydon, CR0 4YY

C9781119113294_170323

# Contents

## List of Contributors

***Rizwana Afroz***
School of Pharmacy
The University of Queensland
Queensland, Australia

***Aimen A. Akbar***
Department of Parasitology
McGill University
Montreal, Quebec, Canada

***Fahmida Alam***
Department of Life Sciences
School of Environment and Life Sciences
Independent University, Bangladesh

***Nur-E-Alam***
Department of Environmental Science
Baylor University
Waco, Texas, USA

***Md. Yousuf Ali***
Department of Biochemistry
Primeasia University
Banani, Dhaka
Bangladesh

***Noor Suryani Mohd Ashari***
Department of Immunology
School of Medical Sciences
Health Campus Universiti Sains Malaysia
Kelantan, Malaysia

***Che Badariah Abd Aziz***
Department of Physiology
School of Medical Sciences
Health Campus Universiti Sains Malaysia
Kelantan, Malaysia

***Ahmad Yasser Hamdi Nor Azlan***
Faculty of Pharmacy and Health Sciences
Royal College of Medicine Perak
Universiti Kuala Lumpur
Ipoh, Perak, Malaysia

***Dhirendra Nath Barman***
Department of Biotechnology and Genetic Engineering
Noakhali Science and Technology University
Noakhali, Bangladesh

***Saziya R. Bidi***
Department of Urology
JN Medical College
KLE Academy of Higher Education & Research
Karnataka, India

***Srinivasa Reddy Bonam***
Institut National de la Santé et de la Recherche Médicale
Centre de Recherche des Cordeliers
Equipe-Immunopathologie et Immunointervention Thérapeutique
Sorbonne Université de Paris
Paris, France

***Chee Keong Chen***
Exercise and Sports Science Programme
School of Health Sciences
Universiti Sains Malaysia
Kelantan, Malaysia

***Siew Liang Ching***
Department of Pharmaceutical Chemistry
Faculty of Pharmacy and Health Sciences
Universiti Kuala Lumpur Royal College of Medicine Perak
Perak, Malaysia

***Nusrat Fatima***
Laboratory of Molecular Medicine
Jahangirnagar University
Dhaka, Bangladesh

***Siew Hua Gan***
Department of Biochemistry and Molecular Biology
Jahangirnagar University
Dhaka, Bangladesh

School of Pharmacy
Monash University Malaysia
Bandar Sunway, Malaysia

***Shridhar C. Ghagane***
Department of Biotechnology
KAHER's Dr. Prabhakar Kore Basic Science Research Center
V.K. Institute of Dental Sciences
Belagavi, India

Department of Urology
JN Medical College
KLE Academy of Higher Education & Research
Karnataka, India

Urinary Biomarkers Research Centre
KLE Academy of Higher Education and Research
Karnataka, India

***Md. Murad Hossain***
Department of Biotechnology and Genetic Engineering
Noakhali Science and Technology University
Noakhali, Bangladesh

***Md. Sakib Hossen***
Laboratory of Preventive and Integrative Biomedicine
Department of Biochemistry and Molecular Biology
Jahangirnagar University
Savar, Dhaka, Bangladesh

***Yim Yee Jin***
Department of Pharmaceutical Chemistry
Faculty of Pharmacy and Health Sciences
Universiti Kuala Lumpur Royal College of Medicine Perak
Perak, Malaysia

***Md. Ibrahim Khalil***
Laboratory of Preventive and Integrative Biomedicine
Department of Biochemistry and Molecular Biology
Jahangirnagar University
Dhaka, Bangladesh

***Shahad Saif Khandker***
Department of Biochemistry and Molecular Biology
Jahangirnagar University
Dhaka, Bangladesh

***Pei Teng Lum***
Department of Pharmaceutical Chemistry
Faculty of Pharmacy and Health Sciences
Universiti Kuala Lumpur Royal College of Medicine Perak
Perak, Malaysia

***Sathiya Maran***
School of Pharmacy
Monash University Malaysia
Bandar Sunway, Malaysia

***Kashif Maroof***
School of Pharmacy
Monash University Malaysia
Bandar Sunway, Malaysia

Department of Pharmaceutical Chemistry
Faculty of Pharmacy and Health Sciences
Universiti Kuala Lumpur Royal College of Medicine Perak
Perak, Malaysia

***Ng Choon Ming***
School of Pharmacy
Monash University Malaysia
Bandar Sunway, Malaysia

***R.B. Nerli***
Department of Urology
JN Medical College
KLE Academy of Higher Education & Research
Karnataka, India

***Foong Kiew Ooi***
Exercise and Sports Science Programme
School of Health Sciences
Universiti Sains Malaysia
Kelantan, Malaysia

***Sridevi I. Puranik***
Department of Zoology
KLES B.K. Arts, Science and Commerce College
Karnataka, India

***Md. Mijanur Rahman***
Department of Biology
University of Alabama at Birmingham
Birmingham, Alabama, USA

***Md. Anisur Rahman***
Department of Biotechnology and Genetic Engineering
Noakhali Science and Technology University
Noakhali, Bangladesh

***Nur Najihah Izzati Mat Rani***
Faculty of Pharmacy and Health Sciences
Royal College of Medicine Perak
Universiti Kuala Lumpur
Ipoh, Perak, Malaysia

***Pasupuleti Visweswara Rao***
Department of Biotechnology
Centre for International Relations and Research Collaborations
Reva University
Karnataka, India

***Mahendran Sekar***
Associate Professor
Department of Pharmaceutical Chemistry
Faculty of Pharmacy and Health Sciences
Universiti Kuala Lumpur Royal College of Medicine Perak
Perak, Malaysia

***Ummay Mahfuza Shapla***
Department of Biochemistry and Molecular Biology
Bangabandhu Sheikh Mujibur Rahman Science and Technology University
Dhaka, Bangladesh

***Md. Solayman***
Institute for Glycomics
Griffith University
Brisbane, Australia

***Usha Sundralingam***
School of Pharmacy
Monash University Malaysia
Bandar Sunway, Malaysia

***Suk Peng Tang***
Department of Pharmacology
School of Medical Sciences
Health Campus Universiti Sains Malaysia
Kelantan, Malaysia

***E.M. Tanvir***
School of Pharmacy
The University of Queensland
Queensland, Australia

Institute of Food and Radiation Biology
Atomic Energy Research Establishment
Bangladesh Atomic Energy Commission
Dhaka, Bangladesh

***Wendy Wai Yeng Yeo***
School of Pharmacy
Monash University Malaysia
Bandar Sunway, Malaysia

***Wan Nazirah Wan Yusuf***
Department of Pharmacology
School of Medical Sciences
Health Campus Universiti Sains Malaysia
Kelantan, Malaysia

***Nurul Amirah Mohd Zaid***
Faculty of Pharmacy and Health Sciences
Royal College of Medicine Perak
Universiti Kuala Lumpur
Ipoh, Perak, Malaysia

***Nur Zulaikha Azwa Zuraini***
Faculty of Pharmacy and Health Sciences
Royal College of Medicine Perak
Universiti Kuala Lumpur
Ipoh, Perak, Malaysia

## Preface

Honey is commonly found in many kitchens as a sweetener and natural food flavoring. Although it has been used since ancient times, the value of both honey and honey products is not fully appreciated. In fact, not many are aware of the unique applications and versatility of honey and its products, including propolis, royal jelly, and bee venom, as well as their economic values.

This book is written by a team of researchers from all over the world who are passionate about natural products, in order to revisit honey and honey products and highlight the scientific research conducted in the hope that the value of honey is more widely appreciated. It also touches on the challenges involved when investigating honey and honey products for various medicinal uses. It unravels the mysteries of the potential of honey and honey products that can be further explored in future studies.

*Md. Ibrahim Khalil*
*Siew Hua Gan*
*Bey Hing Goh*

# 1

# General Introduction

*Pasupuleti Visweswara Rao, Ng Choon Ming, Md. Ibrahim Khalil, and Siew Hua Gan*

## Introduction

Apiculture is a specialized area in science study about beekeeping or maintenance. In Latin, "Apis" means "bee," and "culture" means "keep." In other words, apiculture simply means beekeeping. Although honey is one of the most important products from apiculture, other valuable products, such as pollen, bee wax, royal jelly (RJ), propolis, and bee venom, are also available (Posey 1983). Throughout the years, we could observe the vital role of honey in human lives in various ways due to its highly economic and medicinal values. In fact, the collection of honey has been recognized as one of the major economic areas for rural communities across the world for their livelihood. Honey is produced by honeybees as a result of mixing of the nectar from various flowers and different types of enzymes within their honey sacs, which are then stored in storage cells for a few days to mature (Seeley 2009). At this particular stage, the matured or ripened substance is considered honey.

The honey-ripening process not only involves dehydration of the nectar but also includes different physical and chemical progressions. The constituents of honey tend to fluctuate based on the nectar source and various other factors such as flowering seasons and environmental conditions. Honey has a unique taste because of the combination of the enzymes from the honey sacs of the honeybees and the varying moisture content. In addition, the presence of vital saccharides, sucrose, glucose, and fructose also plays a potential role in its taste (Doner 1977) (Figure 1.1).

### Nectar

Nectar is a liquid substance from various types of flowering plants. It consists of water and sugars (Garcia et al. 2005), which attract the bees. The bees collect the nectar and suck it via their proboscises or long tongues. The honeybees (worker bees) store the nectar in their stomachs for a short duration until it is transferred to the comb with the help of other honeybees (house bees). The nectar and its components play an important role in the taste of honey, which is also influenced by seasonal variations and other environmental factors (Afik et al. 2006).

### Composition of Honey

Honey is a natural product consisting of a combination of sugar, water, and other ingredients. Honey consists of sugar at approximately 76%, and the water content in honey is 18%, with other components making up the remaining 6% (Wedmore 1955). Sugars are the major constituents of honey responsible for honey's sweetness, water content, and several other constituents found in trace amounts that differentiate honey types and may vary in aroma, color, and taste.

### Carbohydrates

Sugars are generally considered saccharides. The saccharides present in honey do not belong to the same category of a single saccharide but are composed of mono- and disaccharides. The monosaccharides present in honey include fructose and glucose, and the disaccharides include sucrose, turanose, maltose, maltulose, and isomaltose (White and Doner 1980). Other constituents, including phenolic compounds, vitamins, amino acids, proteins, and minerals, are also available in

*Honey: Composition and Health Benefits*, First Edition. Edited by Md. Ibrahim Khalil, Gan Siew Hua, and Bey Hing Goh.

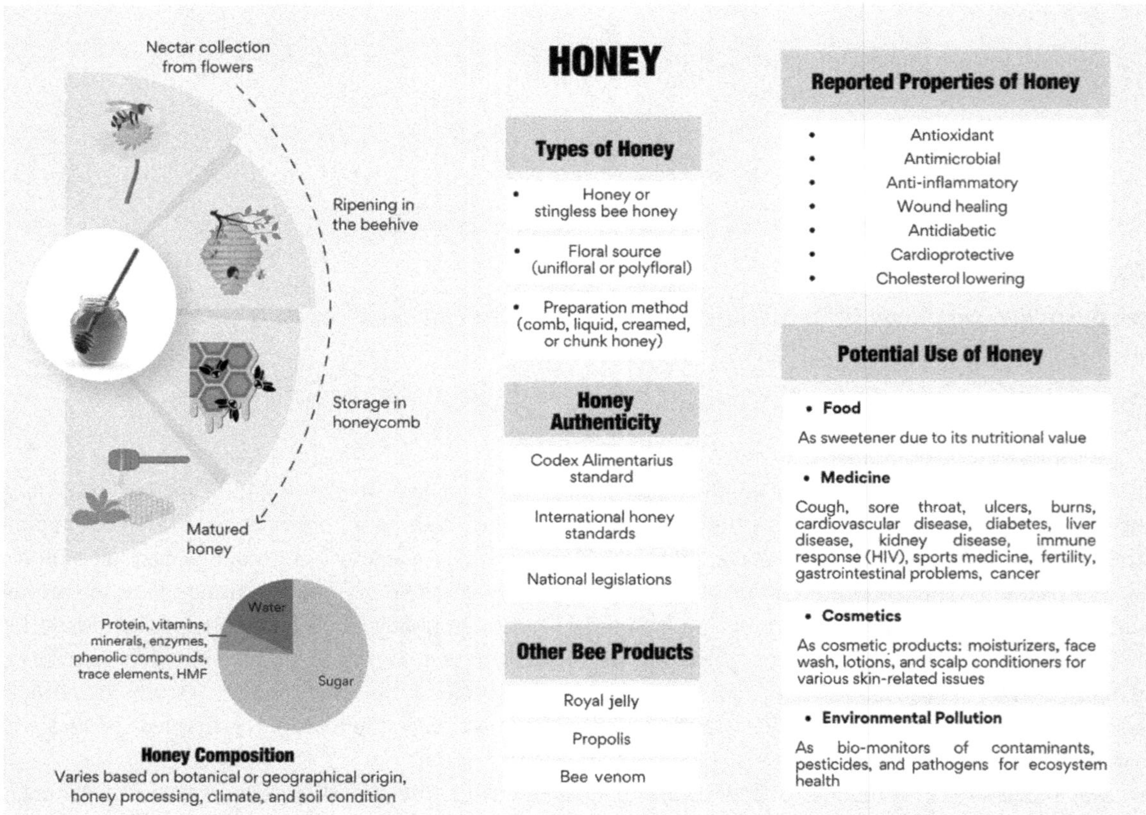


**Figure 1.1** Summary of information about honey. HIV, human immunodeficiency virus; HMF, 5-hydroxymethyl furfural. BillionPhotos.com / Adobe Stock.

honey at various concentrations based on the botanical origin of the honey and the seasons (Huang and Robinson 1995). The available sugars in several types of honey promote the growth of healthy cells and continuous formation of fresh white blood cells. Sucrose generally consists of one fructose molecule linked to glucose through α-1-4 binding and is hydrolyzed by invertase enzyme (Da Silva et al. 2016).

Storage time, heat treatment, and several chemical and physical changes in honey result in changing the darkness of the honey as well as the flavor (Da Silva et al. 2016). Monosaccharide decomposition occurs, thereby resulting in the formation of furans. These furans, composed of furfural and 5-hydroxymethyl furfural (HMF), are derived from pentoses and hexoses, respectively (Anese et al. 2013).

### Minerals

Minerals are imperative and make up 3.68% of the composition of honey, playing a vital role in honey's nutritional value. Various minerals, such as chlorine, phosphorus, potassium, calcium, silicon, sulfur, magnesium, and manganese, have been reported in honey. Potassium is the major mineral found in honey, which makes up approximately one-third of the total mineral content (Bogdanov et al. 2007). Beekeeping practices, honey processing, and conservational effluence have added value to the different types of minerals and their quantities in honey (Pohl et al. 2009). In essence, the wide-ranging mineral profile of honey, present in minute amounts, encourages its nutritional use as food in addition to being part of a healthy diet (Ajibola et al. 2012).

### Proteins

Proteins occupy a minor portion of honey's composition (0.1–0.3 g/100 g) (Anklam 1998). Proteins are available in various honeys in several forms, such as simple or complex structures of amino acids. Generally, proteins are present in low quantities, and hence the nutritional impact is also low. Several researchers have reported that the protein quantity in different types of honey is often lower than 0.5%. The amino acid content depends on the floral sources, geographical regions, and the processing capacity of bees. In honeys, one of the many and important amino acids is proline, which is an indicator of honey's quality and possible adulteration. The proline content should be permissible if the value is below 180 mg/kg (Bogdanov et al. 2002).

### Enzymes

Enzymes are complex structures found in active cells responsible for various reactions and processes in living organisms. Generally, honey consists of small quantities of enzymes, and a large portion is composed of diastase and invertase (White et al. 1961). The enzyme contents and concentration in honey are also dependent on the floral sources and seasonal variations.

One of the key roles of enzymes in honey is to contribute to the functional properties of honey. Several types of enzymes, including oxidases, acid phosphatases, amylases, invertases, catalases, and others, are available in honey. Essentially, the invertase, glucose oxidase, and diastase are considered the key enzymes of honey. Diastase (amylase) converts starch to different carbohydrates such as mono-, di-, and oligosaccharides and dextrins. Invertase, sucrose hydrolase, sucrase, and saccharases are the enzymes that are useful in converting sucrose to glucose and fructose (invert sugar). Glucose oxidase present in honey converts glucose to gluconolactone and is subsequently further processed into gluconic acid and hydrogen peroxide. Subsequently, β-glucosidase-1 transforms β-glucans to oligosaccharides and glucose. Catalase is also one of the major enzymes present in honey that transforms the peroxides into water and oxygen. Proteases are the enzymes that hold vital roles in hydrolyzing the proteins (White and Doner 1980).

### Vitamins

Vitamins are important in determining honey's quality. Ascorbic acid, riboflavin, nicotinic acid, pantothenic acid, and folic acid are some of the vitamins available in honey in minute amounts, to the extent of describing them in parts per millions (Da Silva et al. 2016). Generally, the quantity of vitamins in the food materials is difficult to be determined because they are not stable in various conditions. Over time, foods tend to lose vitamins because of storage and aging processes. Besides, filtration, a process whereby honey is filtered to improve its appearance, diminishes the quantity of the vitamins because pollens containing vitamins are removed during the process (Wilczyńska 2014).

### Trace Elements

The quantity of various types of heavy metals in honey basically relies on the composition of the soil elements and the source of flowers in the region. Honey is not measured as a vital basis of trace elements because the total amounts of elemental quantity or ash amount in nectar honeys and honeydew honeys are typically recorded as below 0.6% and 1.0%, respectively. Generally, the elemental mixture or trace elemental composition depends on the honeydew, nectar, and pollen from the region where the honey was harvested. Bogdanov et al. (2007) has confirmed that botanical aspects have the utmost stimulus on the trace element quantity of honey. The microelement amount was found to be higher than 1.0% in different types of honey. The microelements found in honeys are aluminum, boron, barium, bromine, calcium, chlorine, ferrous, magnesium, manganese, sodium, phosphorus, rubidium, sulfur, strontium, and zinc. The trace elements found to be present in honey are silver, arsenic, cadmium, chromium, copper, lithium, molybdenum, nickel, selenium, and lead (Solayman et al. 2016). Overall, the element composition of honey is useful for assessment of honey's quality to detect adulteration such as honey dilution with water, addition of sugars or syrups, and assessment of the botanical or geographical origins of honey (Sager 2020).

### Hydroxymethylfurfural

Hydroxymethylfurfural (Figure 1.2) is used as an indicator of honey's quality and purity because fresh honey does not include HMF or has very low HMF (0–0.2 mg/kg). HMF is formed as a result of the degradation of glucose and fructose when honey is acidic, and the formation speed usually depends on the temperature (Molan and Allen 1996). The honeys

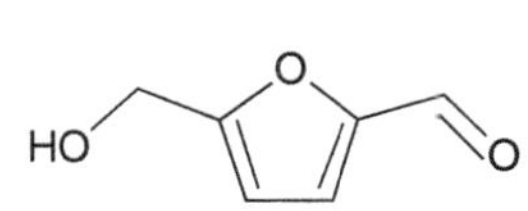


**Figure 1.2** Structure of hydroxymethylfurfural.

containing high HMF signify improper heating and storage. The maximum limits of HMF in honey are 40 mg/kg in normal regions and 80 mg/kg in tropical regions to assure safety for consumption (Bogdanov et al. 2007). It was revealed that HMF has both detrimental and beneficial implications on human health (Shapla et al. 2018). The adverse effects reported include being mutagenic, genotoxic, organotoxic, DNA damaging, and enzyme inhibitory. Conversely, HMF exerts desirable benefits with its antioxidative, anti-allergic, anti-inflammatory, antihypoxic, antisickling, and antihyperuricemic properties. Research has shown that humans can consume between 30 and 150 mg of HMF daily from foods; however, the safe level is not well established yet (Glatt and Sommer 2007).

## Types of Honey

There are broadly two types of honey based on honeybees; these are honey and stingless bee honey (SBH). The culture of the former is generally known as apiculture, and the latter is known as meliponiculture. Stingless bees (*Meliponines*) belong to the genus Apidae, and as opposed to their other counterpart honeybees, SBH is less explored because of its limited production. Some distinctive characteristics of stingless bees include being less vulnerable to diseases, the capability to pollinate small flowers, easy extraction of its product (honey, pollen, propolis), and convenience in maintenance because they do not abandon their hives (Abd Jalil et al. 2017). Recent evidence has highlighted the therapeutic potential of SBH, including its antioxidant properties, which can prevent and manage diseases related to oxidative stress, microbial infections, and inflammatory disorders (Al-Hatamleh et al. 2020).

Honey is further divided into two types based on the floral sources of the nectar. They are monofloral (or unifloral) and polyfloral (or multifloral). Monofloral honeys have a unique flavor from which they originate, which is primarily from the nectar of a single plant species. Because various nutritional, therapeutic, and sensory properties of honey arise based on botanical origin, the distinctive monofloral honeys are generally considered more valuable among consumers compared with polyfloral honeys (Schievano et al. 2016).

Honey can also be categorized into several types based on the preparation. They are comb, liquid, creamed, and chunk honeys (Anklam 1998; Isengard et al. 2001). Comb honey is directly collected from the honeycomb, where the honeybees generally store it. Liquid honey is extracted via cutting of the wax capping and spinning the honeycomb in a specified honey extractor (Abramovič et al. 2008). Creamed honey, also known as granulated honey, is a mixture of finely granulated honey and liquid honey in a 1:9 ratio. Generally, creamed honey is stored at approximately 57°C until it becomes stable and safe. Chunk honey is a combination of comb and liquid honeys. It is prepared in a way that the comb honey floats in the liquid honey in a jar (Chesson et al. 2011).

### Honey as Food

Honey is a solution of sugars, proteins, vitamins, minerals, flavonoids, phenolic compounds, and organic acids. Generally, its composition and nutritional values vary depending on the floral sources and seasonal variations (Gheldof et al. 2002). Nevertheless, honey has been used as food since ancient times because of its nutritional value and medicinal properties, including its wound-healing and antimicrobial and antioxidant capacities. The potential use of honey as food is of great prospect, particularly as an alternate sweetener for sugar. Considerable evidence from animal and human studies has concurred that honey could be a better alternative than sugar for healthy individuals and for those with impaired glucose tolerance, hyperlipidemia, and diabetes and their related comorbidities (Cortés et al. 2011). This in part could be related to the beneficial effect of honey on glycemic regulation and lipid profile. Despite this, the mechanisms of honey in modulating desirable health effects are not well established yet. Long-term randomized controlled clinical trials with sufficient samples and varying amount of honey consumed are much needed to reach a conclusion (Bobiş et al. 2018).

### Honey as Medicine

Since ancient times, humans have been consuming and collecting honey. In fact, approximately 8000 years ago, cave paintings in Valencia, Spain, suggest that humans began hunting honey and honeycomb from a wild bee nest (Nayik et al. 2014). Besides this, there is evidence of honey being kept in earthenware pots in Southern England in approximately 2500 BC (Crane 1999). In addition, 8000 years of evidence exist in the world for which honey is recognized as a precious product by humans (Samarghandian et al. 2017). Historical reports documented that ancient civilizations, including the Egyptians, Greeks, Chinese, Mayans, Romans, and Babylonians, utilized honey for medicinal and nutritional uses (Jones 2009).

To date, several types of biological properties and medicinal properties of honeys have been reported, including antimicrobial, antioxidant, antidiabetic, anticancer, anti-inflammatory, and wound-healing activities; for cataract diseases, fertility, and gastrointestinal problems; and for its cardioprotective and cholesterol-lowering activities (El-Soud and Helmy 2012; Miguel et al. 2017). Additionally, honey has been tested for its organo-protective effects in different disease conditions in several in vivo systems.

Apart from all this, honey is a natural wound-healing agent compared with modern synthetic drugs. Since ancient times, people in various parts of the world, including Egypt, China, Greece, and Romania, have explored diverse types of honey as wound-healing agents for several types of intestinal diseases. Additionally, honey has been mixed with herbs and spices for the treatment of carbuncle infections (Radhakrishnan et al. 2011).

## Honey's Application in Modern Medicine

The use of honey as medicine can be revealed even in ancient written records and has continued into present-day folk medicine. For example, lotus honey is believed to be a remedy for eye ailments in India (Pasupuleti et al. 2017). In Ghana, honey is also used as a remedy for septic leg ulcers in folk medicine, and it is used for earache in Nigeria (Molan 1999). Moreover, honey is a worthy medicine for coughs and sore throats. Honey has also been used in the treatment of gastroenteritis, gastric ulcers, surface wounds, peptic ulcers, and ophthalmology issues (Cooper and Molan 1999). Many researchers also found that honey encourages tissue regeneration by enhancing angiogenesis and the growth of epithelial and fibroblast cells (Nour et al. 2021; Vijaya and Nishteswar 2012). Additionally, honey is used to cure external surface wounds and burns (Bardy et al. 2008).

Promisingly, research has raised the potential value of honey for oncology care, including in radiation-induced mucositis; for skin-related problems in patients undergoing radiotherapy; for dermal problems, especially on the skin of the feet and hands of patients undergoing chemotherapy; and for treatment of the oral cavity. For instance, randomized controlled trials among patients with head and neck cancer elucidated the improvement seen in chemoradiation-induced mucositis with topical application of honey compared with a control group treated with saline solution (Howlader et al. 2019). Among patients with oral carcinoma undergoing radiation therapy, honey limited the severity of mucositis compared with a control group that received the usual treatment gel (Khanal et al. 2010).

## Honey as Cosmetics

Honey is one of the best sources for cosmetics products. Honey from various types of bees is used as several cosmetic products, including moisturizers, face wash lotions, and scalp conditioners, and for other skin-related issues (Ediriweera and Premarathna 2012).

## Use of Honey as an Indicator for Environmental Pollution

Honeybee acts as pollinators and biomonitors of contaminants, pesticides, and pathogens, which is critical for accessing environmental pollution and overall ecosystem health. During foraging, honeybees are exposed to various pollutants and carry these pollutants to the hives. Specifically, bees serve as indicators of environmental pollution by signaling increased mortality rates caused by toxic molecules or by the presence of heavy metals, fungicides, and herbicides in honey, pollen, and larvae (Celli and Maccagnani 2003). As a whole, honeybee colonies are resilient against contaminants, allowing for long-term detection and quantification of pollution in the given studied territory (Cunningham et al. 2022).

## Honey's Authenticity and Quality

Honey's quality and authenticity are based on legislative requirements, set by the Codex Alimentarius standard, international honey standards, and varying national legislations (Codex Alimentarius 2001). Two main aspects of honey authenticity are (1) production and processing without adulteration and (2) authenticity in terms of geographical and botanical origins (Bogdanov 2007). According to the standards set, honey should meet the compositional criteria in terms of sugar content, moisture content, electrical conductivity, and free acid and HMF content. During production or processing by beekeepers or industry, issues can arise, including mislabeling (unlabeled pasteurized honey, harvested in cold), improper filtering, addition of sweeteners, addition of water (resulting in fermentation and spoilage), and harvesting unripe honey (Bogdanov and Martin 2002). In terms of the botanical and geographical origin of honey, misdescription can occur, including the labeling of the wrong botanical or geographical source for a higher price point. The botanical origin of honey can be tested using methods such as sensory analysis, pollen analysis, routine physicochemical parameters (e.g. glucose and fructose content, electrical conductivity), and determination of aroma compounds or other minor components (amino acids, phenolics, trace elements). On the other hand, the geographical origin of honey can be assessed using methods such as pollen analysis, routine parameters (pH, acidity, electrical conductivity, glucose, fructose), and minor components (amino acids, flavonoids, trace elements).

The various types of physicochemical properties, including moisture, ash, pH, HMF content, and other beneficial effects of honey, are discussed in detail in other chapters.

## Other Bee Products

### Royal Jelly

Royal jelly is a creamy substance that is chemically synthesized from plant sources and secreted by the worker *Apis mellifera* (honeybees) from its mandibular and hypopharyngeal glands (Kunugi and Ali 2019). The queen larvae consume RJ throughout their lifetimes, which contributes to their large size, long lifespan, and functioning sexual organs. RJ is mainly composed of water, sugar, proteins, lipids, vitamins, polyphenols, mineral salts, and other unspecified substances present in minor amounts. RJ exhibits antibacterial properties that reduce bacterial motility, exert an inhibitory effect against various numbers of gram-positive and gram-negative bacteria, and synergistically promote antioxidant activities (Cooper et al. 2002; Paul et al. 2007).

Thus far, the potential of RJ in improving health has been widely studied in vivo, in vitro, and in randomized clinical studies. For instance, RJ has displayed antiproliferative and antitumor properties in both cell lines and animal studies (Gismondi et al. 2017; Zhang et al. 2017). Plus, clinical studies have reported the benefits of RJ in ameliorating symptoms of malignancies (Erdem and Güngörmüş 2014), further supporting the prospect of RJ as an anticancer agent. Additionally, the highly nutritious RJ is valuable for health maintenance, longevity, and age-related disorders, particularly in reducing oxidative damage (Inoue et al. 2003), providing protection against the harmful effects of ultraviolet radiation (Zheng et al. 2013), and boosting estrogenic activities (Bălan et al. 2020). Moreover, the beneficial effect on aging extends to optimal neural function, including enhanced memory, thereby suggesting promising therapeutic value on the prevention or treatment of neurodegenerative disorders such as Alzheimer's and Parkinson's diseases (Ali and Kunugi 2020).

Furthermore, there seems to be evidence on the use of RJ for people with diabetes. This is built on research that revealed RJ's use for reducing serum glucose levels, glycosylated hemoglobin, and oxidative stress and increasing insulin concentrations (Mousavi et al. 2017; Pourmoradian et al. 2014). The role of RJ in obesity has also been explored, to which favorable outcomes were shown, including the inhibition of lipid peroxidation; reduction of cholesterol; and a positive effect on satiety, inflammation, and antioxidant capacity (Pan et al. 2018; Petelin et al. 2019; Zahmatkesh et al. 2014). Other benefits reported include RJ's potential effect on skeletal muscle dysfunction, particularly in delaying age-related motor function impairment (Okumura et al. 2018) and on fertility with protective effects on sperm parameters, testosterone levels, and ovarian hormones (Zahmatkesh et al. 2014).

### Propolis

Propolis is a natural bee product retrieved from the flowers, buds, exudates, bark of trees, and plants by honeybees (Maroof and Gan 2020). Specifically, it is composed of different types of material, including resins, beeswax, pollen, balsams, essential oils, and various organic compounds. Propolis contains amino acids, minerals, vitamins, and biochemical compounds such as phenolic acids and flavonoids (Maroof et al. 2020). The medicinal value of propolis has been well recognized since ancient times. First, diverse compounds from propolis are potent antioxidants, including flavonoids, polyphenols, vitamin C, vitamin E, tannins, reducing sugars, caffeic acid phenethyl ester, and chalcones (Tanvir et al. 2018; Turan et al. 2020). These compounds can scavenge free radicals, thereby protecting the cells against lipid peroxidation and reducing oxidative stress (Martinello and Mutinelli 2021). Propolis is also studied for its potential against various types of cancer, with several mechanisms reported, including antiproliferation, the ability to induce apoptosis and to ameliorate the effects of chemotherapy (Catchpole et al. 2015; Kumari et al. 2017; Yilmaz et al. 2016). Apart from this, propolis contains various anti-inflammatory compounds that can inhibit the activation of inflammatory transcription factors, reduce the production of pro-inflammatory cytokines, and alleviate inflammatory responses (Hwang et al. 2018; Jin et al. 2017; Melero-Jerez et al. 2016). Other potential benefits of propolis include its antiprotozoal activity; antibacterial properties, especially toward gram-positive bacteria; and antifungal properties with possible prospect as treatment for onychomycosis as well as various *Candida* yeast strains (Khurshid et al. 2017; Veiga et al. 2018). Plus, propolis is also antiviral against DNA and RNA viruses, demonstrated in vitro and in animal models (Amoros et al. 1992; Nolkemper et al. 2010). Notably, growing evidence suggests the possibility of propolis usage in the prevention or management of chronic diseases such as diabetes and cardiovascular diseases. This is mainly attributed to its antioxidant capacity, anti-inflammation properties, and favorable effects on lipid profile and glycemic level (Chen et al. 2018; Koya-Miyata et al. 2009). Nevertheless, high-quality clinical studies are needed to ascertain the pharmacological potentials of propolis in addition to the exploration of allergens present in propolis for consumer safety.

### Bee Venom

Bee venom is a transparent and odorless liquid containing various pharmacologically active components, including polypeptides, enzymes, sugars, amino acids, minerals, and catecholamines (Wehbe et al. 2019). Bee venom has been extensively studied for the management of various diseases because of its anti-inflammatory, antioxidant, antibacterial, anticancer, analgesic, and anti-atherogenic capacities. For instance, the potentiality of bee venom usage for neurologic disorders such as Alzheimer's disease, Parkinson's disease, amyotrophic lateral sclerosis, and multiple sclerosis has been uncovered in numerous in vivo models. The neuroprotective effect is related to bee venom's ability to enhance cognitive function, reduce inflammatory response, lower oxidative stress, restore apoptotic markers, enhance immune response, and improve motor function (Tanner et al. 2011; Yang et al. 2010; Ye et al. 2016). Additionally, considerable literature corroborated that bee venom could be an alternative therapy to control inflammation and pain and to alleviate the symptoms of arthritis (El-Tedawy et al. 2020; Son et al. 2007).

Another important medicinal value of bee venom emerged based on in vitro cancer cell models, including liver, renal, prostate, ovarian, lung, and melanoma cancer cells, particularly owing to the antitumor, apoptotic, antibacterial, and anti-melanoma activities of bee venom. Furthermore, the antibacterial and anti-inflammatory properties of bee venom have made it a potential agent against inflammatory skin diseases, including atopic dermatitis and acne vulgaris, as reported earlier in in vivo studies. Plus, clinical study has demonstrated the use of bee venom on human aging skin to decrease facial wrinkles in terms of the average depth, total count, and total area of wrinkles (Han et al. 2015). Other medicinal values of bee venom have extended to the treatment of various disease models, such as atherosclerosis, acute kidney injury, and gastric ulceration. Despite the promising therapeutic applications of bee venom, clinical studies are critical to establish the use of bee venom in practice, including its toxicity and further drug development process.

## References


Abd Jalil, M.A., Kasmuri, A.R., and Hadi, H. (2017). Stingless bee honey, the natural wound healer: a review. *Skin Pharmacology and Physiology* 30 (2): 66–75. doi: 10.1159/000458416.

Abramovič, H., Jamnik, M., Burkan, L., et al. (2008). Water activity and water content in Slovenian honeys. *Food Control* 19 (11): 1086–1090. doi: 10.1016/J.FOODCONT.2007.11.008.

Afik, O., Dag, A., Kerem, Z., et al. (2006). Analyses of avocado (Persea americana) nectar properties and their perception by honey bees (Apis mellifera). *Journal of Chemical Ecology* 32 (9): 1949–1963. doi: 10.1007/S10886-006-9120-1.

Ajibola, A., Chamunorwa, J.P., and Erlwanger, K.H. (2012). Nutraceutical values of natural honey and its contribution to human health and wealth. *Nutrition and Metabolism* 9: 61. doi: 10.1186/1743-7075-9-61.

Al-Hatamleh, M.A.I., Boer, J.C., Wilson, K.L., et al. (2020). Antioxidant-based medicinal properties of stingless bee products: recent progress and future directions. *Biomolecules* 10 (6): 923. doi: 10.3390/biom10060923.

Ali, A.M. and Kunugi, H. (2020). Royal jelly as an intelligent anti-aging agent-a focus on cognitive aging and Alzheimer's disease: a review. *Antioxidants* 9 (10): 937. doi: 10.3390/ANTIOX9100937.

Amoros, M., Simões, C.M.O., Girre, L., et al. (1992). Synergistic effect of flavones and flavonols against herpes simplex virus type 1 in cell culture. Comparison with the antiviral activity of propolis. *Journal of Natural Products* 55 (12): 1732–1740. doi: 10.1021/NP50090A003.

Anese, M., Manzocco, L., Calligaris, S., et al. (2013). Industrially applicable strategies for mitigating acrylamide, furan and 5-hydroxymethylfurfural in food. *Journal of Agricultural and Food Chemistry* 61 (43): 10209–10214. doi: 10.1021/JF305085R.

Anklam, E. (1998). A review of the analytical methods to determine the geographical and botanical origin of honey. *Food Chemistry* 63 (4): 549–562. doi: 10.1016/S0308-8146(98)00057-0.

Bălan, A., Moga, M.A., Dima, L., et al. (2020). Royal jelly-a traditional and natural remedy for postmenopausal symptoms and aging-related pathologies. *Molecules* 25 (14): 3291. doi: 10.3390/MOLECULES25143291.

Bardy, J., Slevin, N.J., Mais, K.L., et al. (2008). A systematic review of honey uses and its potential value within oncology care. *Journal of Clinical Nursing* 17 (19): 2604–2623.

Bobiş, O., Dezmirean, D.S., and Moise, A.R. (2018). Honey and diabetes: the importance of natural simple sugars in diet for preventing and treating different type of diabetes. *Oxidative Medicine and Cellular Longevity*. doi: 10.1155/2018/4757893.

Bogdanov, S. (2007). Authenticity of honey and other bee products: state of the art. *Bulletin USAMV_CN* 63: 64.

Bogdanov, S., Haldimann, M., Luginbühl, W., et al. (2007). Minerals in honey: environmental, geographical and botanical aspects. *Journal of Apicultural Research* 46 (4): 269–275. doi: 10.1080/00218839.2007.11101407.

Bogdanov, S. and Martin, P. (2002). Honey authenticity. *Mitteilungen aus Lebensmitteluntersuchung und Hygiene* 93 (3): 232–254.

Bogdanov, S., Martin, P., and Lullmann, C. (2002). Harmonised methods of the international honey commission. *Swiss Bee Research Centre, FAM, Liebefeld* 5: 1–62.

Catchpole, O., Mitchell, K., Bloor, S., et al. (2015). Antiproliferative activity of New Zealand propolis and phenolic compounds vs human colorectal adenocarcinoma cells. *Fitoterapia* 106: 167–174. doi: 10.1016/J.FITOTE.2015.09.004.

Celli, G. and Maccagnani, B. (2003). Honey bees as bioindicators of environmental pollution. *Bulletin of Insectology* 56 (1): 137–139.

Chen, L.H., Chien, Y.W., Chang, M.L., et al. (2018). Taiwanese green propolis ethanol extract delays the progression of type 2 diabetes mellitus in rats treated with streptozotocin/high-fat diet. *Nutrients* 10 (4): 503. doi: 10.3390/NU10040503.

Chesson, L.A., Tipple, B.J., Erkkila, B.R., et al. (2011). B-HIVE: beeswax hydrogen isotopes as validation of environment. Part I: bulk honey and honeycomb stable isotope analysis. *Food Chemistry* 125 (2): 576–581. doi: 10.1016/J.FOODCHEM.2010.09.050.

Codex Alimentarius. (2001). Revised codex standard for honey. *Codex Stan 12*.

Cooper, R. and Molan, P. (1999). The use of honey as an antiseptic in managing Pseudomonas infection. *Journal of Wound Care* 8 (4): 161–164. doi: 10.12968/JOWC.1999.8.4.25867.

Cooper, R.A., Halas, E., and Molan, P.C. (2002). The efficacy of honey in inhibiting strains of Pseudomonas aeruginosa from infected burns. *The Journal of Burn Care & Rehabilitation* 23 (6): 366–370. doi: 10.1097/00004630-200211000-00002.

Cortés, M.E., Vigil, P., and Montenegro, G. (2011). The medicinal value of honey: a review on its benefits to human health, with a special focus on its effects on glycemic regulation. *Ciencia e investigación agraria* 38 (2): 303–317. doi: 10.4067/s0718-16202011000200015.

Crane, E. (1999). *The World History of Beekeeping and Honey Hunting*. Duckworth, London, UK: Routledge.

Cunningham, M.M., Tran, L., McKee, C.G., et al. (2022). Honey bees as biomonitors of environmental contaminants, pathogens and climate change. *Ecological Indicators* 134: 108457. doi: 10.1016/j.ecolind.2021.108457.

Da Silva, P.M., Gauche, C., Gonzaga, L.V., et al. (2016). Honey: chemical composition, stability and authenticity. *Food Chemistry* 196: 309–323. doi: 10.1016/j.foodchem.2015.09.051.

Doner, L.W. (1977). The sugars of honey—a review. *Journal of the Science of Food and Agriculture* 28 (5): 443–456. doi: 10.1002/JSFA.2740280508.

Ediriweera, E.R.H.S.S. and Premarathna, N.Y.S. (2012). Medicinal and cosmetic uses of Bee's Honey – a review. *Ayu (An International Quarterly Journal of Research in Ayurveda)* 33 (2): 182. doi: 10.4103/0974-8520.105233.

El-Soud, N.H.A. and Helmy, N. (2012). Honey between traditional uses and recent medicine. *Macedonian Journal of Medical Sciences* 5 (2): 205–214. doi: 10.3889/MJMS.1857-5773.2012.0213.

El-Tedawy, D.M., Abd-Alhaseeb, M.M., Helmy, M.W., et al. (2020). Systemic bee venom exerts anti-arthritic and anti-inflammatory properties in a rat model of arthritis. *Biomedical Reports* 13 (4): 20. doi: 10.3892/BR.2020.1327.

Erdem, Ö. and Güngörmüş, Z. (2014). The effect of royal jelly on oral mucositis in patients undergoing radiotherapy and chemotherapy. *Holistic Nursing Practice* 28 (4): 242–246. doi: 10.1097/HNP.0000000000000033.

Garcia, J.M., Chambers, I.V.E., Matta, Z., et al. (2005). Viscosity measurements of nectar- and honey-thick liquids: product, liquid and time comparisons. *Dysphagia* 20 (4): 325–335. doi: 10.1007/S00455-005-0034-9.

Gheldof, N., Wang, X., and Engeseth, N. (2002). Identification and quantification of antioxidant components of honeys from various floral sources. *Journal of Agricultural and Food Chemistry* 50 (21): 5870–5877. doi: 10.1021/JF0256135.

Gismondi, A., Trionfera, E., Canuti, L., et al. (2017). Royal jelly lipophilic fraction induces antiproliferative effects on SH-SY5Y human neuroblastoma cells. *Oncology Reports* 38 (3): 1833–1844. doi: 10.3892/OR.2017.5851.

Glatt, H. and Sommer, Y. (2007). Health risks of 5-hydroxymethylfurfural (HMF) and related compounds. In: *Acrylamide and Other Hazardous Compounds in Heat-Treated Foods* (ed. K. Skog and J. Alexander), 328–357. Cambridge, England: Woodhead Publishing Ltd.

Han, S.M., Hong, I.P., Woo, S.O., et al. (2015). The beneficial effects of honeybee-venom serum on facial wrinkles in humans. *Clinical Interventions in Aging* 10: 1592. doi: 10.2147/CIA.S84940.

Howlader, D., Singh, V., Mohammad, S., et al. (2019). Effect of topical application of pure honey in chemo-radiation-induced mucositis and its clinical benefits in improving quality of life in patients of oral squamous cell carcinoma. *Journal of Maxillofacial & Oral Surgery* 18 (1): 79. doi: 10.1007/S12663-017-1077-9.

Huang, Z.Y. and Robinson, G.E. (1995). Seasonal changes in juvenile hormone titers and rates of biosynthesis in honey bees. *Journal of Comparative Physiology B: Biochemical, Systemic and Environmental Physiology* 165 (1): 18–28. doi: 10.1007/BF00264682.

Hwang, S.A., Kim, C.D., and Lee, W.S. (2018). Caffeic acid phenethyl ester protects against photothrombotic cortical ischemic injury in mice. *The Korean Journal of Physiology & Pharmacology* 22 (1): 101–110. doi: 10.4196/KJPP.2018.22.1.101.

Inoue, S.I., Koya-Miyata, S., Ushio, S., et al. (2003). Royal jelly prolongs the life span of C3H/HeJ mice: correlation with reduced DNA damage. *Experimental Gerontology* 38 (9): 965–969. doi: 10.1016/S0531-5565(03)00165-7.

Isengard, H.D., Schultheiß, D., Radović, B., et al. (2001). Alternatives to official analytical methods used for the water determination in honey. *Food Control* 12: 459–466. doi: 10.1016/S0956-7135(01)00044-5.

Jin, X.L., Wang, K., Li, Q.Q., et al. (2017). Antioxidant and anti-inflammatory effects of Chinese propolis during palmitic acid-induced lipotoxicity in cultured hepatocytes. *Journal of Functional Foods* 34: 216–223. doi: 10.1016/J.JFF.2017.04.039.

Jones, R. (2009). Honey and healing through the ages. *Journal of ApiProduct and ApiMedical Science* 1 (1): 2–5. doi: 10.3896/IBRA.4.01.1.02.

Khanal, B., Baliga, M., and Uppal, N. (2010). Effect of topical honey on limitation of radiation-induced oral mucositis: an intervention study. *International Journal of Oral and Maxillofacial Surgery* 39 (12): 1181–1185. doi: 10.1016/j.ijom.2010.05.014.

Khurshid, Z., Naseem, M., Zafar, M.S., et al. (2017). Propolis: a natural biomaterial for dental and oral healthcare. *Journal of Dental Research, Dental Clinics, Dental Prospects* 11 (4): 274. doi: 10.15171/JODDD.2017.046.

Koya-Miyata, S., Arai, N., Mizote, A., et al. (2009). Propolis prevents diet-induced hyperlipidemia and mitigates weight gain in diet-induced obesity in mice. *Biological and Pharmaceutical Bulletin* 32 (12): 2022–2028. doi: 10.1248/BPB.32.2022.

Kumari, S., Nayak, G., Lukose, S.T., et al. (2017). Indian propolis ameliorates the mitomycin C-induced testicular toxicity by reducing DNA damage and elevating the antioxidant activity. *Biomedicine & Pharmacotherapy* 95: 252–263. doi: 10.1016/J.BIOPHA.2017.08.065.

Kunugi, H. and Ali, A.M. (2019). Royal jelly and its components promote healthy aging and longevity: from animal models to humans. *International Journal of Molecular Sciences* 20 (19): 4662. doi: 10.3390/IJMS20194662.

Maroof, K. and Gan, S.H. (2020). A review on chemical compositions, biological activity and formulation techniques of Malaysian honey bee and Meliponine propolis. *Journal of Biologically Active Products from Nature* 10 (6): 507–523. doi: 10.1080/22311866.2020.1856716.

Maroof, K., Lee, R.F.S., Siow, L.F., et al. (2020). Microencapsulation of propolis by spray drying: a review. *Drying Technology*. doi: 10.1080/07373937.2020.1850470.

Martinello, M. and Mutinelli, F. (2021). Antioxidant activity in bee products: a review. *Antioxidants* 10 (1): 71. doi: 10.3390/ANTIOX10010071.

Melero-Jerez, C., Ortega, M.C., Moliné-Velázquez, V., et al. (2016). Myeloid derived suppressor cells in inflammatory conditions of the central nervous system. *Biochimica Et Biophysica Acta* 1862 (3): 368–380. doi: 10.1016/J.BBADIS.2015.10.015.

Miguel, M.G., Antunes, M.D., and Faleiro, M.L. (2017). Honey as a complementary medicine. *Integrative Medicine Insights* 12: 1178633717702869. doi: 10.1177/1178633717702869.

Molan, P.C. (1999). Why honey is effective as a medicine. 1 Its use in modern medicine. *Bee World* 80 (2): 80–92.

Molan, P.C. and Allen, K.L. (1996). The effect of gamma-irradiation on the antibacterial activity of honey. *Journal of Pharmacy and Pharmacology* 48 (11): 1206–1209. doi: 10.1111/J.2042-7158.1996.TB03922.X.

Mousavi, S.N., Jazayeri, S., Khoshpay, B., et al. (2017). Royal jelly decreases blood pressure, serum glucose and interleukin-6 in patients with type 2 diabetes on an iso-caloric diet. *Journal of Nutrition and Food Security* 2 (4): 300–307.

Nayik, G.A., Shah, T.R., Muzaffar, K., et al. (2014). Honey: its history and religious significance: a review. *Universal Journal Pharmacy* 3 (1): 5–8.

Nolkemper, S., Reichling, J., Sensch, K.H., et al. (2010). Mechanism of herpes simplex virus type 2 suppression by propolis extracts. *Phytomedicine* 17 (2): 132–138. doi: 10.1016/J.PHYMED.2009.07.006.

Nour, S., Imani, R., Chaudhry, G.R., et al. (2021). Skin wound healing assisted by angiogenic targeted tissue engineering: a comprehensive review of bioengineered approaches. *Journal of Biomedical Materials Research* 109 (4): 453–478. doi: 10.1002/jbm.a.37105.

Okumura, N., Toda, T., Ozawa, Y., et al. (2018). Royal jelly delays motor functional impairment during aging in genetically heterogeneous male mice. *Nutrients* 10 (9): 1191. doi: 10.3390/NU10091191.

Pan, Y., Xu, J., Chen, C., et al. (2018). Royal jelly reduces cholesterol levels, ameliorates Aβ pathology and enhances neuronal metabolic activities in a rabbit model of Alzheimer's disease. *Frontiers in Aging Neuroscience* 10: 50. doi: 10.3389/FNAGI.2018.00050.

Pasupuleti, V.R., Sammugam, L., Ramesh, N., et al. (2017). Honey, propolis and royal jelly: a comprehensive review of their biological actions and health benefits. *Oxidative Medicine and Cellular Longevity* 2017: 1259510. doi: 10.1155/2017/1259510.

Paul, I.M., Beiler, J., McMonagle, A., et al. (2007). Effect of honey, dextromethorphan and no treatment on nocturnal cough and sleep quality for coughing children and their parents. *Archives of Pediatrics & Adolescent Medicine* 161 (12): 1140–1146. doi: 10.1001/ARCHPEDI.161.12.1140.

Petelin, A., Kenig, S., Kopinč, R., et al. (2019). Effects of royal jelly administration on lipid profile, satiety, inflammation and antioxidant capacity in asymptomatic overweight adults. *Evidence-based Complementary and Alternative Medicine*. doi: 10.1155/2019/4969720.

Pohl, P., Sergiel, I., and Stecka, H. (2009). Determination and fractionation of metals in honey. *Critical Reviews in Analytical Chemistry* 39 (4): 276–288. doi: 10.1080/10408340903001250.

Posey, D.A. (1983). Folk Apiculture of the Kayapo Indians of Brazil. *Biotropica* 15 (2): 158. doi: 10.2307/2387963.

Pourmoradian, S., Mahdavi, R., Mobasseri, M., et al. (2014). Effects of royal jelly supplementation on glycemic control and oxidative stress factors in type 2 diabetic female: a randomized clinical trial. *Chinese Journal of Integrative Medicine* 20 (5): 347–352. doi: 10.1007/S11655-014-1804-8.

Radhakrishnan, N., Kavitha, V., Raja, S., et al. (2011). Embelin-A natural potential cosmetic agent. *Journal of Applied Cosmetology* 29: 99–107.

Sager, M. (2020). Major minerals and trace elements in different honeys. *Bee World* 97 (3): 70–74. doi: 10.1080/0005772X.2020.1785100.

Samarghandian, S., Farkhondeh, T., and Samini, F. (2017). Honey and health: a review of recent clinical research. *Pharmacognosy Research* 9 (2): 127. doi: 10.4103/0974-8490.204647.

Schievano, E., Finotello, C., Uddin, J., et al. (2016). Objective definition of monofloral and polyfloral honeys based on NMR metabolomic profiling. *Journal of Agricultural and Food Chemistry* 64 (18): 3645–3652. doi: 10.1021/acs.jafc.6b00619.

Seeley, T.D. (2009). *The Wisdom of the Hive: The Social Physiology of Honey Bee Colonies*. Cambridge, MA: Harvard University Press. doi: 10.1086/419745.

Shapla, U.M., Solayman, M., Alam, N., et al. (2018). 5-Hydroxymethylfurfural (HMF) levels in honey and other food products: effects on bees and human health. *Chemistry Central Journal* 12 (1): Springer International Publishing. 1–18. doi: 10.1186/s13065-018-0408-3.

Solayman, M., Islam, M.A., Paul, S., et al. (2016). Physicochemical properties, minerals, trace elements and heavy metals in honey of different origins: a comprehensive review. *Comprehensive Reviews in Food Science and Food Safety* 15 (1): 219–233. doi: 10.1111/1541-4337.12182.

Son, D.J., Lee, J.W., Lee, Y.H., et al. (2007). Therapeutic application of anti-arthritis, pain-releasing and anti-cancer effects of bee venom and its constituent compounds. *Pharmacology & Therapeutics* 115 (2): 246–270. doi: 10.1016/J.PHARMTHERA.2007.04.004.

Tanner, C.M., Kame, F., Ross, G.W., et al. (2011). Rotenone, paraquat and Parkinson's disease. *Environmental Health Perspectives* 119 (6): 866–872. doi: 10.1289/EHP.1002839.

Tanvir, E.M., Sakib Hossen, M., Mahfuza Shapla, U., et al. (2018). Antioxidant, brine shrimp lethality and analgesic properties of propolis from Bangladesh. *Journal of Food Biochemistry* 42 (5): e12596. doi: 10.1111/JFBC.12596.

Turan, D., Abdik, H., Sahin, F., et al. (2020). Evaluation of the neuroprotective potential of caffeic acid phenethyl ester in a cellular model of Parkinson's disease. *European Journal of Pharmacology* 883. doi: 10.1016/J.EJPHAR.2020.173342.

Veiga, F.F., Gadelha, M.C., Mrt, D.S., et al. (2018). Propolis extract for onychomycosis topical treatment: from bench to clinic. *Frontiers in Microbiology* 9: 779. doi: 10.3389/FMICB.2018.00779.

Vijaya, K.K. and Nishteswar, K. (2012). Wound healing activity of honey: a pilot study. *Ayu* 33 (3): 377. doi: 10.4103/0974-8520.108827.

Wedmore, E.B. (1955). The accurate determination of the water content of honeys: part I. Introduction and results. *Bee World* 36 (11): 197–206. doi: 10.1080/0005772X.1955.11094904.

Wehbe, R., Frangieh, J., Rima, M., et al. (2019). Bee venom: overview of main compounds and bioactivities for therapeutic interests. *Molecules* 24 (16): 2997. doi: 10.3390/MOLECULES24162997.

White, J. and Doner, L.W. (1980). Honey composition and properties. In: *Beekeeping in the United States* Washington, DC: US Department of Agriculture, Science and Education Administration, 82–91.

White, J.W., Riethof, M.L., and Kushnir, I. (1961). Composition of honey. VI. The effect of storage on carbohydrates, acidity and diastase content. *Journal of Food Science* 26 (1): 63–71. doi: 10.1111/J.1365-2621.1961.TB00042.X.

Wilczyńska, A. (2014). Effect of filtration on colour, antioxidant activity and total phenolics of honey. *LWT – Food Science and Technology* 57 (2): 767–774. doi: 10.1016/J.LWT.2014.01.034.

Yang, E.J., Jiang, J.H., Lee, S.M., et al. (2010). Bee venom attenuates neuroinflammatory events and extends survival in amyotrophic lateral sclerosis models. *Journal of Neuroinflammation* 7 (69). doi: 10.1186/1742-2094-7-69.

Ye, M., Chung, H.S., Lee, C., et al. (2016). Neuroprotective effects of bee venom phospholipase A2 in the 3xTg AD mouse model of Alzheimer's disease. *Journal of Neuroinflammation* 13 (10). doi: 10.1186/s12974-016-0476-z.

Yilmaz, U.C., Bagca, B.G., Karaca, E., et al. (2016). Evaluation of the miRNA profiling and effectiveness of the propolis on B-cell acute lymphoblastic leukemia cell line. *Biomedicine & Pharmacotherapy* 84: 1266–1273. doi: 10.1016/J.BIOPHA.2016.10.056.

Zahmatkesh, E., Najafi, G., Nejati, V., et al. (2014). Protective effect of royal jelly on the sperm parameters and testosterone level and lipid peroxidation in adult mice treated with oxymetholone. *Avicenna Journal of Phytomedicine* 4 (1): 52.

Zhang, S., Shao, Q., Geng, H., et al. (2017). The effect of royal jelly on the growth of breast cancer in mice. *Oncology Letters* 14 (6): 7615. doi: 10.3892/OL.2017.7078.

Zheng, J., Lai, W., Zhu, G., et al. (2013). 10-Hydroxy-2-decenoic acid prevents ultraviolet A-induced damage and matrix metalloproteinases expression in human dermal fibroblasts. *Journal of the European Academy of Dermatology and Venereology* 27 (10): 1269–1277. doi: 10.1111/J.1468-3083.2012.04707.X.

# 2

## Physical Properties of Honey

*Rizwana Afroz, E.M. Tanvir, and Md. Murad Hossain*

### Introduction

Honey is a natural substance with a sweet flavor and viscous consistency (Figure 2.1) that is produced by honeybees, particularly the species *Apis mellifera* (Cortés et al. 2011), from the nectar blossoms or from exudates of trees and plants that produce nectar honeys or honeydews, respectively (Figure 2.2) (Alvarez-Suarez et al. 2010). It is a by-product of flower nectar and the upper aero-digestive tract of honeybees and is concentrated through a dehydration process inside the beehive (Eteraf-Oskouei and Najafi 2013). At least four *Apis* species are native to the Indian subcontinent, that is, *Apis dorsata, Apis cerana, Apis florae*, and *Apis andreniformis. Apis mellifera* bees are imported from Europe and are used for large-scale natural honey production in honey farms on the Indian subcontinent (Bogdanov et al. 2008). Honey is a remarkable, complex natural liquid that has been reported to contain at least 181 substances (Crane 1975). The supersaturated solution consists of fructose (38%) and glucose (31%) as the major constituents, and the rest of the components include minor constituents such as phenolic acids, flavonoids, ascorbic acid, certain antioxidant enzymes (e.g. glucose oxidase and catalase), carotenoid-like substances, organic acids, and Maillard reaction products (Afroz et al. 2016b; El Denshary et al. 2012). In itself, honey is an unique compound because of its highly variable composition, which depends on its floral source, although other factors, such as environment, season, and processing, may also have significant effects on the composition of honey (Afroz et al. 2014; Paul et al. 2017).

The first written reference to honey was on a Sumerian tablet dating back to 2100–2000 BC that mentioned the use of honey as a drug and an ointment. In most ancient cultures, honey was used for both nutritional and medicinal purposes (Alvarez-Suarez et al. 2010). Natural honey has been used as effective medicine around the world since ancient times. It was a valued traditional remedy for centuries. The ancient Egyptians, Assyrians, Chinese, Greeks, and Romans employed honey for wounds and diseases of the gut (Bogdanov et al. 2008). The belief that honey is a nutrient, a drug, and an ointment has persisted to the present time. For centuries in human history, honey was an important source of carbohydrates and the only widely available sweetener until the production of industrial sugar began to replace it after 1800 (Alvarez-Suarez et al. 2010). Honey is a liquid that has been mentioned in all religious books and is accepted by all generations, traditions, and civilizations, both ancient and modern (Ajibola et al. 2012).

**Figure 2.1** Natural honey collected in a jar. Recail / Alamy Stock Photo.

### Brief History

As the only available sweetener, honey was an important food for *Homo sapiens* from our very beginnings. Indeed, the relationship between bees and *H. sapiens* started as early as the stone age (Crane 1983). Honeybees are one of the oldest forms of animal life and have been in existence since the Neolithic age, thus preceding the appearance of

*Honey: Composition and Health Benefits*, First Edition. Edited by Md. Ibrahim Khalil, Gan Siew Hua, and Bey Hing Goh.

humans on earth by 10 to 20 million years. In the course of human history, honey has primarily been used as a sweetener, but it has also been used as a medicine. Honey was mentioned several times in the holy books of ancient India, the Vedas (Crane 2013). In ancient China, honey was mentioned in the book of songs Shi Jing, which was written in the sixth century BC; a honey medicine was mentioned in the "52 Prescription Book" in the third century BC. In ancient Egypt, honey was an important sweetener and was depicted in many wall drawings (Figure 2.3). According to the Ebers papyrus (1550 BC), it is included in 147 prescriptions for external application (Bogdanov 2011).

In ancient Greece, the honeybee, a sacred symbol of Artemis, was an important design on Ephesian coins for almost six centuries (Figure 2.4). Aristoteles first described the production of honey. Hippocrates wrote about the healing virtues of honey. After his death in 323 BC, Alexander the Great was embalmed in a coffin filled with honey. Honey was mentioned many times by the writers Vergil, Varro, and Plinius. During the time of Julius Caesar, honey was used as a substitute for gold to pay taxes (Bogdanov 2011).

In Israel, the land where both honey and milk flow, honey was very important and was mentioned 54 times in the Old Testament. The most famous is the saying of the wise King Solomon, "Eat thou honey because it is good." The Koran recommended honey as a wholesome food and an excellent medicine. In the 16th chapter of the Koran titled "The Bee," we find: "There are proceeded from their bellies a liquor of various colour, wherein is medicine for men." Mohammed pronounced: "Honey is a remedy for all diseases" (Bogdanov 2011). Over the course of human history, honey has not only been a nutrient but also a medicine. A medicine branch, called Apitherapy, has developed in recent years and offers treatments for many diseases using honey and the other bee products (Bogdanov 2011). Therefore, the belief that honey is a nutrient, a drug, and an ointment has persisted to the present day.

**Figure 2.2** Honeybee collecting honey from nectar. dpa/dpa picture alliance archive/Alamy Stock Photo.

**Figure 2.3** A honeybee in an ancient wall drawing. *Source*: Keith Schengili-Roberts / Wikimedia Commons/ CC BY-SA 3.0.

## Composition of Honey

The composition of honey is rather variable and primarily depends on the floral source; however, a number of external factors also play a role, including seasonal and environmental factors and processing (Afroz et al. 2016b; Moniruzzaman et al. 2013). Honey is a sweet and flavorful food that consists of a highly concentrated solution of a complex mixture of sugars. It is a supersaturated solution of sugars, of which fructose (38%) and glucose (31%) are the main contributors (Afroz et al. 2016b; Khalil et al. 2010). Honey also contains small amounts of other constituents, such as minerals, proteins, vitamins, organic acids, flavonoids, phenolic acids, enzymes, and other phytochemicals, which contribute to its antioxidant effects (da Silva et al. 2016). The components in honey that are responsible for its antioxidant effects are flavonoids, phenolic acids, ascorbic acid, catalase, peroxidase, carotenoids, and products of Maillard reactions (Afroz et al. 2016b; Khalil et al. 2011; Paul et al. 2017). The overall composition of natural honey is summarized in Table 2.1.

**Figure 2.4** A coin from Ephesos dated 300 BC, which shows the bee, an emblem of Artemis Ephesia. *Source*: Max Dashu.

**Table 2.1** Average composition of honey (data in g/100 g).

| Component | Average (%) |
|---|---|
| Water | 17.20 |
| Fructose | 38.19 |
| Glucose | 31.28 |
| Disaccharides, calculated as maltose | 7.31 |
| Higher sugars | 1.50 |
| Free gluconic acid | 0.57 |
| Ash | 0.17 |
| Nitrogen | 0.04 |
| Minerals | 0.20 |
| Amino acids, proteins | 0.30 |
| pH value | 3.90 |

Alvarez-Suarez et al. 2010; Bogdanov et al. 2008; Chow 2002; Pérez et al. 2002; Terrab et al. 2003.

### Carbohydrate Profile

Sugar and water are the primary constituents of natural honey. Sugar accounts for 95%–99% of the dry honey matter. The majority of these simple sugars are D-fructose (38.2%) and D-glucose (31.3%), which represent 85%–95% of the total sugars (Aurongzeb and Azim 2011). These six-carbon sugars are immediately digestible by the small intestine. Natural honey samples are rich in both reducing and nonreducing sugars. According to Moniruzzaman et al. (2013), the reducing sugars are the main soluble sugars present in Malaysian honey because the total reducing sugar content in the samples was as high as 61.17%–63.89%. Indian and Bangladeshi honey samples were also reported to contain higher amounts of reducing sugars, ranging from 42.95%–60.31% and from 52.3%–66.5%, respectively (Afroz et al. 2016b; Jahan et al. 2015; Saxena et al. 2010). Tables 2.2 and 2.3 summarize the different di- and trisaccharides reported by Moreira and De Maria (Moreira and Maria 2001). Many of these sugars are not found in nectar but are formed during ripening and storage because of the effects of bee enzymes and the acids in honey. During the process of digestion after honey intake, the principal carbohydrates fructose and glucose are quickly transported into the blood and can be utilized as an energy source by the human body. A daily dose of 20 g of honey will meet approximately 3% of daily energy requirements (Alvarez-Suarez et al. 2010).

### Protein, Enzyme, and Amino Acid Profiles

The presence of proteins and amino acids, as well as carbohydrates, vitamins, and minerals, in natural honey was described many years ago (Aurongzeb and Azim 2011). Honey contains a number of proteins and 18 free amino acids (Mohammed and Azim 2012); the approximate percentage of proteins in natural honey is 0.5% (Won et al. 2008). Nineteen bands of honey proteins have been detected in silver-stained SDS-PAGE (sodium dodecyl sulfate–polyacrylamide gel electrophoresis) gels (Marshall and Williams 1987). Depending on the species of the harvesting honeybees, different proteins of diverse molecular weights are found in natural honey (Won et al. 2008). The protein content of honey from different floral sources has been reported, in which high protein contents were considered to be greater than 1000 μg/g (Azeredo et al. 2003). Nevertheless, the contribution of this fraction to human protein intake is low. Most of the enzymes are added by honeybees during the process of natural honey ripening (Aurongzeb and Azim 2011); the three main honey enzymes are (1) diastase (amylase), which decomposes starch or glycogen into smaller sugar units; (2) invertase, which decomposes sucrose into fructose and glucose; and (3) glucose oxidase, which produces hydrogen peroxide and gluconic acid from glucose (Bogdanov et al. 2008). The proteins in natural honey originate from nectar, pollen, and honeybees. The relative quantity of natural honey proteins is measured as a quality indicator (Aurongzeb and Azim 2011). Amino acids account for 1% (w/w) of honey. The amount of total free amino acids in honey ranges from 10 to 200 mg/100 g, with proline as the main contributor because it corresponds to approximately 50% of the total free amino acids (Iglesias et al. 2004; Kowalski et al. 2017). In addition to proline, there are 26 amino acids in honeys; their

**Table 2.2** Disaccharides reported in different honey samples.

| Trivial Nomenclature | Systematic Nomenclature |
|---|---|
| Cellobiose[a] | O-β-D-glucopyranosyl-(1→4)-D-glucopyranose |
| Gentiobiose[a] | O-β-D-glucopyranosyl-(1→6)-D-glucopyranose |
| Isomaltose[a] | O-α-D-glucopyranosyl-(1→6)-D-glucopyranose |
| Isomaltulose[b] | O-α-D-glucopyranosyl-(1→6)-D-fructofuranose |
| Kojibiose[c] | O-α-D-glucopyranosyl-(1→2)-D-glucopyranose |
| Laminaribiose[d] | O-β-D-glucopyranosyl-(1→3)-D-glucopyranose |
| Leucrose[b] | O-α-D-glucopyranosyl-(1→5)-D-fructofuranose |
| Maltose[c] | O-α-D-glucopyranosyl-(1→4)-D-glucopyranose |
| Maltulose[a] | O-α-D-glucopyranosyl-(1→4)-D-fructose |
| Melibiose[b] | O-α-D-galactopyranosyl-(1→6)-D-glucopyranose |
| Neo-trehalose[d] | O-α-D-glucopyranosyl- β -D- glucopyranoside |
| Nigerose[a] | O-α-D-glucopyranosyl-(1→3)-D-glucopyranose |
| Palatinose[a] | O-α-D-glucopyranosyl-(1→6)-D-fructose |
| Saccharose[c] | O-α-D-glucopyranosyl- β -D- fructofuranoside |
| Turanose[c] | O-α-D-glucopyranosyl-(1→6)-D-fructose |

[a] Minority.
[b] Not confirmed.
[c] Majority
[d] Traces.
Moreira and Maria 2001 / SciELO.

**Table 2.3** Trisaccharides reported in different honey samples.

| Trivial Nomenclature | Systematic Nomenclature |
|---|---|
| Kestose[a] | O-α-D-glucopyranosyl-(1→4)- O-α-D-glucopyranosyl-(1→2)-D-glucopyranose |
| 1-Kestose[a] | O-α-D-glucopyranosyl-(1→2)- β -D- fructofuranosyl-(1→2) – β-D- fructofuranoside |
| Erlose[b] | O-α-D-glucopyranosyl-(1→4)- O-α-D-glucopyranosyl- β-D- fructofuranoside |
| Isomaltotrise[c] | O-α-D-glucopyranosyl-(1→6)- O-α-D-glucopyranosyl-(1→6)-D-glucopyranose |
| Isopanose[c] | O-α-D-glucopyranosyl-(1→4)- O-α-D-glucopyranosyl-(1→6)-D-glucopyranose |
| Laminaritriose[a] | O-β-D-glucopyranosyl-(1→3)- O-β-D-glucopyranosyl-(1→3)-D-glucopyranose |
| Maltotriose[c] | O-α-D-glucopyranosyl-(1→4)- O-α-D-glucopyranosyl-(1→4)-D-glucopyranose |
| Melezitose[c] | O-α-D-glucopyranosyl-(1→3)- O-β-D-fructofuranosyl-(2→1)-D-glucopyranoside |
| Panose[c] | O-α-D-glucopyranosyl-(1→6)- O-α-D-glucopyranosyl-(1→4)-D-glucopyranose |
| Raffinose[c] | O-α-D-glucopyranosyl-(1→6)- O-α-D-glucopyranosyl- β-D- fructofuranoside |
| Teanderosec[c] | O-α-D-glucopyranosyl-(1→6)-α-D-glucopyranosyl- β-D- fructofuranoside |

[a] Not confirmed.
[b] Majority.
[c] Minority.
Moreira and Maria (2001) / SciELO.

relative proportions depend on their origin (nectar or honeydew). Because pollen is the main source of honey's amino acids, the amino acid profile of a type of honey could be a characteristic of its botanical origin (Alvarez-Suarez et al. 2010; Azevedo et al. 2017). The main amino acids identified in honey samples from different botanical and geographical origins are listed in Table 2.4.

**Table 2.4** Free amino acids reported in different honey samples.

| Free Amino Acid | Abbreviation |
|---|---|
| Glutamic acid | Glu |
| Aspartic acid | Asp |
| Asparagine | Asn |
| Serine | Ser |
| Glutamine | Gln |
| Histidine | His |
| Threonine | Thr |
| b-Alanine | b-Ala |
| a-Alanine | a-Ala |
| Tryptophan | Trp |
| Phenylalanine | Phe |
| Lysine | Lys |
| Arginine | Arg |
| Proline | Pro |
| Tyrosine | Tyr |
| Valine | Val |
| Methionine | Met |
| Cysteine | Cys |
| Isoleucine | Ile |
| Leucine | Leu |
| g-Aminobutyric acid | GABA |
| Ornithine | Orn |

Hermosín et al. 2003; Iglesias et al. 2004; Paramás et al. 2006; Pérez et al. 2007.

## Phenolic Composition

Although studies of honeys and honeybees and the basic composition of honeys began 100 years ago, the interest in honey phenolic compounds has only recently increased. Many authors have studied the phenolic and flavonoid contents of honey to determine if they are correlated with their floral origins (Ferreres et al. 1991; Martos et al. 2000a; Roby et al. 2020; Tomás-Barberán et al. 2001). The distribution of three main phenolic families (benzoic and cinnamic acids, as well as flavonoids) shows different profiles in honey from different floral origins, with flavonoids being the most common in floral honeys. Therefore, a characteristic distribution pattern of phenolic compounds should be observed in unifloral honeys sourced from the corresponding plant sources (Estevinho et al. 2008; Gil et al. 1995; Michalkiewicz et al. 2008; Truchado et al. 2008; Vela et al. 2007). The flavonoids in honey and propolis have been identified as flavanones and flavanones or flavanols. In general, the flavonoid concentration in honey is approximately 20 mg/kg (Ferreres et al. 1991; Gil et al. 1995). The polyphenols in honey are mainly flavonoids (e.g. quercetin, luteolin, kaempferol, apigenin, chrysin, and galangin), phenolics, and phenolic acid derivatives (Ferreres et al. 1991; Gil et al. 1995; Michalkiewicz et al. 2008; Truchado et al. 2008; Waheed et al. 2019). The major phenolic acid and flavonoids identified in honey are presented in Table 2.5.

Free radicals and reactive oxygen species (ROS) are involved in processes of cellular dysfunction, the pathogenesis of metabolic and cardiovascular diseases (CVDs) and aging. The consumption of foods and substances rich in antioxidants can protect against these pathological changes and consequently prevent the pathogenesis of these and other chronic ailments (Bouacha et al. 2018). Researchers noted that natural honey contains several important compounds, which include antioxidants (Al-Waili 2003; Schramm et al. 2003). The qualitative and quantitative compositions of honey (including the antioxidant constituents and the other phytochemical substances) are a reflection of the floral source, as well as the variety

**Table 2.5** The phenolic acid and flavonoids identified in honey from different floral sources.

| Phenolic Acids | Flavonoids |
|---|---|
| 4-Dimethylaminobenzoic acid | Apigenin |
| Caffeic acid | Genistein |
| p-Coumaric acid | Pinocembrin |
| Gallic acid | Tricetin |
| Vallinic acid | Chrysin |
| Syringic acid | Luteolin |
| Chlorogenic acid | Quercetin |
| | Quercetin 3-methyl ether |
| | Kaempferol |
| | Galangin |
| | Pinobanksin |
| | Myricetin |

Alvarez-Suarez et al. 2010; Ferreres et al. 1991; Gil et al. 1995; Martos et al. 2000a, 2000b; Tomás-Barberán et al. 2001.

of the particular honey (do Nascimento et al. 2018). The color of the honey also influences its antioxidant content because darker honeys are known to have higher levels of antioxidants than lighter honeys (Frankel et al. 1998; Pauliuc et al. 2020).

### Compositions of Vitamins, Minerals, and Trace Compounds

Usually, natural honey contains a very low concentration of vitamins. Phyllochinon (vitamin K), thiamine (vitamin $B_1$), riboflavin (vitamin $B_2$), pyridoxine (vitamin B6), and niacin (vitamin $B_3$) have been reported in different honey samples. The contribution of honey to the Recommended Dietary Intake of the different trace substance is small (Bogdanov et al. 2008). It is known that the concentrations of different trace and mineral elements in honey depend on its botanical and geological origin (Alvarez-Suarez et al. 2010; Bilandžić et al. 2019; Solayman et al. 2016). Trace elements play a crucial role in the biomedical activities associated with this food because these elements have a multitude of known and unknown biological functions. For this reason, the concentrations of different trace and mineral elements were systematically investigated in botanically and geologically defined honey samples (Alvarez-Suarez et al. 2010; Solayman et al. 2016; Squadrone et al. 2020).

### Profiles of Aromatic Compounds

The aroma profile is one of the most typical features of a food product, both for its organoleptic quality and authenticity (Careri et al. 1993; Rahman et al. 2017). Because of the high number of volatile components, the aroma profile represents a "fingerprint" of the product, which could be used to determine its origin (Anklam and Radovic 2001). In the past few decades, extensive research has been performed on aroma compounds, and more than 500 different volatile compounds have been identified in different types of honey. Indeed, depending on its botanical origin, the levels of most aroma-building compounds vary in the different types of honey (An et al. 2020; Bogdanov et al. 2004). Honey's flavor is an important quality for its application in the food industry and is a selection criterion for the consumer. Aroma compounds are present in honey at very low concentrations as complex mixtures of volatile components of different functionalities with relatively low molecular weights (Cuevas-Glory et al. 2007). An important number of organic compounds have been identified as the volatile components of different types of honeys (An et al. 2020; Rahman et al. 2017). Thus, methyl anthranilate was identified as a compound that was characteristic of citrus honey (Alissandrakis et al. 2005). Other volatile compounds that were suggested to be markers for citrus honey include lilac aldehyde (Alissandrakis et al. 2005, 2007; Piasenzotto et al. 2003), hotrienol (Piasenzotto et al. 2003), and 1-p-menthen-al (Alissandrakis et al. 2005, 2007). Eucalyptus honey was shown to be distinctive because of the content of the volatile compounds nonanol, nonanak, and nonanoic acid. High levels of isophorone (3,5,5-trimethylcyclohexen-2-enone) were found in heather honey (Alissandrakis et al. 2005, 2007; Cuevas-Glory et al. 2007; Piasenzotto et al. 2003).

## Physical Properties of Honey

Honey has several important features in addition to its composition and taste (Deng et al. 2018). Freshly extracted honey is a viscous liquid. Its viscosity depends on large variety of substances and therefore varies with its composition and particularly with its water content. Hygroscopicity is another property of honey and describes the ability of honey to absorb and hold moisture from the environment. Normal honey has a water content of 18.8% or less and absorbs moisture from the air when the relative humidity is greater than 60%. The surface tension of honey varies with the origin of the honey and is likely due to the presence of colloidal substances. Together with high viscosity, it is responsible for the foaming characteristics of honey (Olaitan et al. 2007). The color in liquid honey varies from clear and colorless (like water) to dark amber or black. The various honey colors basically include all shades of yellow and amber. The colors vary with the botanical origin, age, and storage conditions as well as the phenolic and flavonoid contents, but the transparency or clarity depends on the amount of suspended particles, such as pollen (Dżugan et al. 2020; Kulkarni et al. 2020; Oskouei and Najafi 2013). Less common honey colors are bright yellow (sunflower), reddish undertones (chestnut), greyish (eucalyptus), and greenish (honeydew). Once crystallized, honey turns lighter in color because glucose crystals are white. Honey crystallization results from the formation of monohydrate glucose crystals, which vary in their numbers, shapes, dimensions, and quality according to the composition of the honey and its storage conditions. The lower the water and the higher the glucose content of honey, the faster the crystallization (Olaitan et al. 2007). Islam et al. (2012) investigated the color intensity and characteristics (Figure 2.5) of different honey samples from different locations in Bangladesh and showed that they ranged from amber to dark amber colors. According to their study, the color intensity of the honey samples ranged from 254 to 2034 mAU, which is comparable to the values reported by other authors (Bertoncelj et al. 2007; Mendiola et al. 2008; Saxena et al. 2010).

Honey is basically acidic in nature. The pH and acidity levels change depending on the botanical and geographical origin of the honey (Bogdanov et al. 2008; Shamsudin et al. 2019). Natural honey contains minerals and acids that serve as electrolytes and can conduct an electrical current. Electric conductivity (EC) is an indicator of the botanical origin of honey (Roby et al. 2020; Shamsudin et al. 2019). It has been reported that blossom honeys and mixtures of blossom and honeydew honeys should ideally have EC values of less than 0.8 mS/cm according to the European Union (EU Directive 2002). The moisture content of the honey samples is important and contributes to their ability to resist fermentation and granulation during storage (Islam et al. 2012). According to the Codex standard for honey, the maximum limit for the moisture content of honey is below 20% (Codex Alimentarius 2001; Pauliuc et al. 2020).

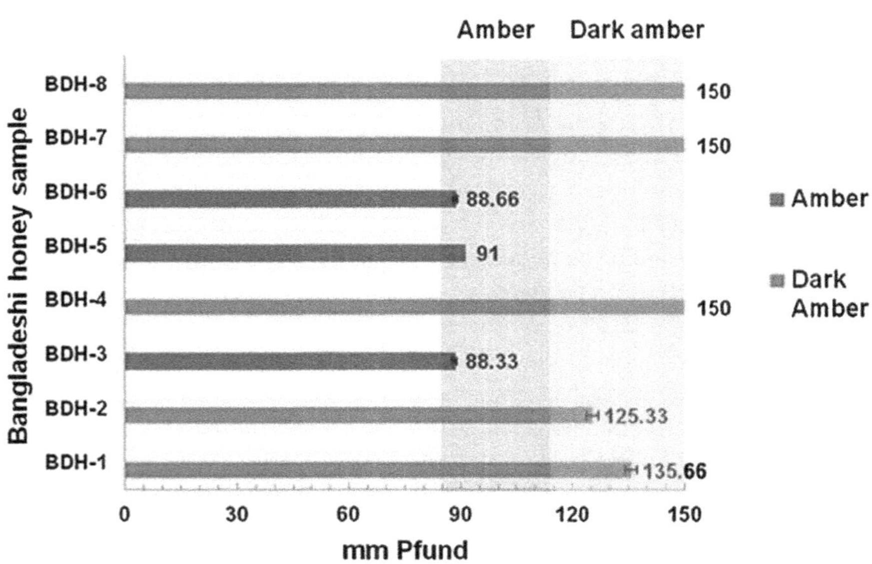


Figure 2.5 Color characteristics of different Bangladeshi honey samples. Islam, Khalil et al. 2012 / Springer Nature / Licensed under CC BY 2.0.

## Chemical Properties of Honey

Honey is mainly composed of sugars and water (Table 2.1). The other chemical constituents of honey are amino acids, antibiotic-rich inhibine, proteins, phenol antioxidants, and micronutrients (da Silva et al. 2016; White and Doner 1980). In addition, it also contains several vitamins and minerals, including vitamin B complex (Table 2.6). The concentration of mineral compounds ranges from 0.1% to 1.0%. Potassium is the major metal followed by calcium, magnesium, sodium, sulphur, and phosphorus. The trace elements include iron, copper, zinc, and manganese (Kumar et al. 2010; Lachman et al. 2007; Solayman et al. 2016).

Organic acids constitute 0.57% of honey and include gluconic acid, which is a by-product of the enzymatic digestion of glucose. The organic acids are responsible for the acidity of honey and largely contribute to its characteristic taste (Olaitan et al. 2007). The characteristic aroma and flavor of honey, which are often associated with the dominant source of pollen, such as "heather honey" in England, "lotus tree honey" in the Arabian Gulf, and "buckwheat honey" in North America (Zhou et al. 2002), are two of the most attractive features of the product, and Castro-Vázquez et al. (2003) identified more than 120 volatile compounds that may contribute to the unique aroma of rosemary honey.

**Table 2.6** Chemical elements found in honey.

| Minerals | Amount (mg/100 g) | Vitamins | Amount (mg/100 g) |
|---|---|---|---|
| Sodium (Na) | 1.600–17.000 | Thiamin (vitamin $B_1$) | 0.000–0.010 |
| Calcium (Ca) | 3.000–31.000 | Riboflavin (vitamin $B_2$) | 0.010–0.020 |
| Potassium (K) | 40.00–3500.00 | Niacin (vitamin $B_3$) | 0.100–0.200 |
| Magnesium (Mg) | 0.700–13.000 | Pantothenic acid (vitamin $B_5$) | 0.020–0.110 |
| Phosphorus (P) | 2.000–15.000 | Pyridoxine (vitamin $B_6$) | 0.010–0.320 |
| Selenium (Se) | 0.002–0.010 | Folic acid (vitamin $B_9$) | 0.002–0.010 |
| Copper (Cu)[a] | 0.020–0.600 | Ascorbic acid (vitamin C) | 2.200–2.500 |
| Iron (Fe)[a] | 0.030–4.000 | Phyllochinon (vitamin K) | 0.025 |
| Manganese (Mn)[a] | 0.020–2.000 | | |
| Chromium (Cr)[a] | 0.010–0.300 | | |
| Zinc (Zn)[a] | 0.050–2.000 | | |
| Aluminium (Al) | 0.010–2.400 | | |
| Arsenic (As)[a] | 0.014–0.026 | | |
| Sulphur (S) | 0.700–26.000 | | |
| Chlorine (Cl) | 0.400–56.000 | | |
| Bromide (Br) | 0.400–1.300 | | |
| Fluorine (F) | 0.400–1.340 | | |
| Iodide (I) | 10.000–100.000 | | |
| Nickel (Ni) | 0.000–0.051 | | |
| Lead (Pb)[a] | 0.001–0.030 | | |
| Boron (B) | 0.050–0.300 | | |
| Cadmium (Cd)[a] | 0–000.001 | | |
| Cobalt (Co)[a] | 0.100–0.350 | | |
| Barium (Ba) | 0.010–0.080 | | |
| Molybdenum (Mo)[a] | 0.000–0.004 | | |
| Silicon (Si) | 0.050–24.000 | | |
| Lithium (Li) | 0.225–1.560 | | |
| Vanadium | 0.000–0.013 | | |

[a] Heavy metals.

Ajibola et al. 2012; Bogdanov et al. 2008; White and Doner 1980.

## Functional and Therapeutic Properties of Honey

Honey is an ancient remedy for the treatment of various health diseases and disorders. Recently, it has been scientifically proven to have functional and biological properties (Figure 2.6). Honey is a sweet and flavorful product that has been consumed over the years for its high nutritional values and beneficial effects on human health. A number of functional properties of natural honey are discussed in the next sections of this chapter:

### Antioxidant Potential

Honey has long been used as a medicine and for domestic needs, but only recently have its antioxidant properties been identified. With increasing demands for antioxidants supplied by food, honey is becoming a popular source of antioxidants because it is rich in phenolic acids, flavonoids, and many other antioxidants (Khalil et al. 2010). The importance of protecting the cell's defense systems against the damage caused by oxygen is well known. Although free radicals of oxygen are a natural metabolic by product within the organism, they cause cellular damage and disrupt the structure of DNA. These processes cause premature aging. Antioxidants bind these dangerous molecules, thus preventing their harmful effects (Jaganathan and Mandal 2009; Karapetsas et al. 2020; Tanvir et al. 2018). Unlike synthetic compounds, honey represents a natural product that does not produce side effects that can be harmful to health. Among the compounds found in honey, phenol compounds, vitamin C, catalase, peroxidase, and glucose oxidase enzymes have antioxidant properties (Gheldof and Engeseth 2002; Tanvir et al. 2015). Honey also contains flavonoids and carotenoids. High levels of these indicators ensure a high level of antioxidants in honey. According to Aljadi and Kamaruddin (2004), the antioxidant capacity of honey is mainly due to the phenolic compounds and flavonoids, and there is a high degree of correlation between these substances and the antioxidant capacity of honey, although a synergistic action between several compounds cannot be discounted (Viuda Martos et al. 2008).

As mentioned previously, the antioxidant activity is primarily due to the presence of phenolic compounds and flavonoids, although the exact mechanism of action is still unknown. Among the proposed mechanisms are free radical sequestration, hydrogen donation, metallic ion chelation, and their ability to act as substrates for radicals, such as superoxide and hydroxyl radicals (Al-Mamary et al. 2002). These biophenols may also interfere with propagation reactions (Russo et al. 2000) or inhibit the enzymatic systems involved in the initiation reactions (You et al. 1999). The more hydroxyl groups that are present in the flavonoids, the more easily they are oxidized (Meyer et al. 1998). It has also been suggested that the organic acids present in honey, such as gluconic, malic, and citric acids, contribute to its antioxidant capacity by chelating metals. Several enzymes, such as glucose oxidase and catalase, also show antioxidant potential through their ability to eliminate oxygen from foods (Viuda Martos et al. 2008). The antioxidant potential of honey is presented in Figure 2.7.

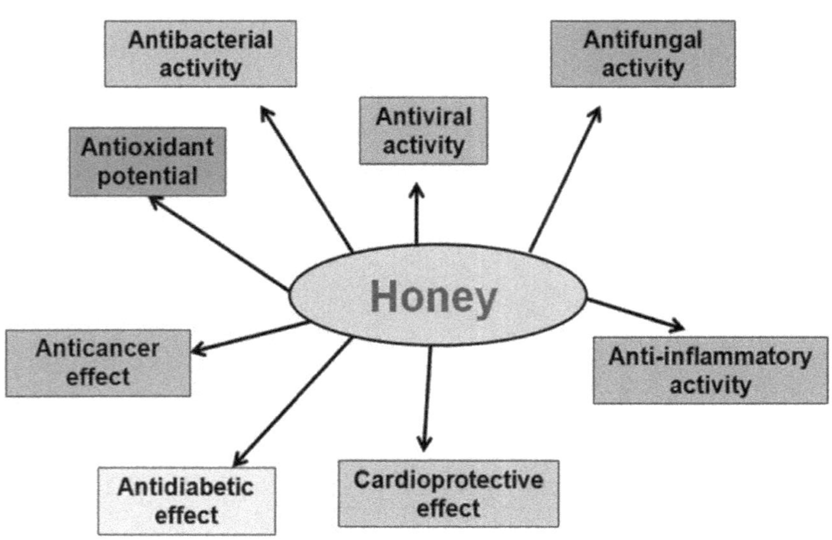


**Figure 2.6** Functional properties of honey.

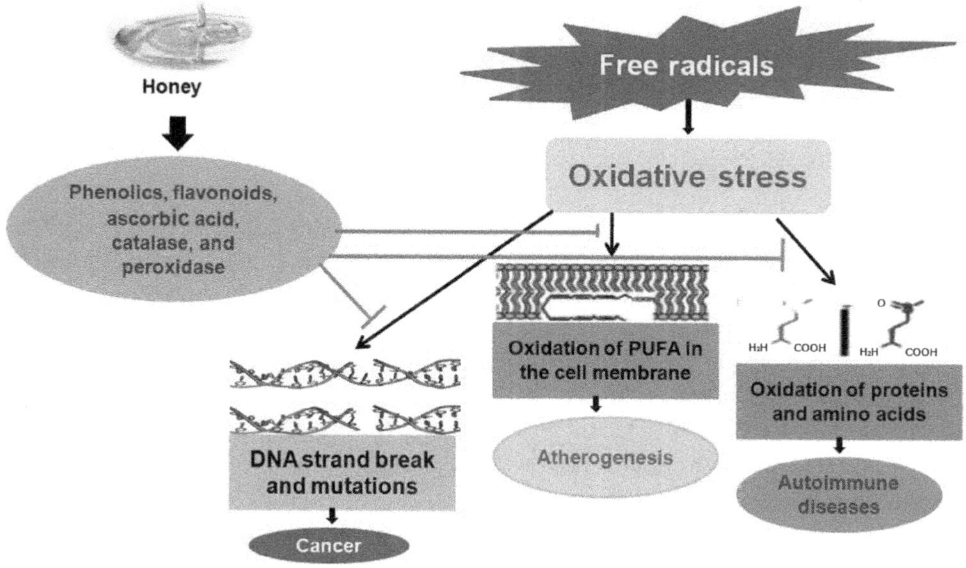


**Figure 2.7** Free radical scavenging activity of honey. PUFA, polyunsaturated fatty acid.

However, the quantity of the antioxidant constituents varies widely, depending on the floral and geographical origin of honey; however, a number of researchers have demonstrated that natural honey samples might be considered as a good source of natural antioxidants (Afroz et al. 2014, 2016c; Aljadi and Kamaruddin 2004; Al-Mamary et al. 2002; Bertoncelj et al. 2007; Islam et al. 2012; Khalil et al. 2010,2011; Saxena et al. 2010; Tanvir et al. 2015).

## Antibacterial Properties

Knowledge of the antibacterial capacity of honey, which was first reported in the 1980s, is currently being revised (Viuda Martos et al. 2008). Two main theories have been proposed to explain this capacity. One is that the antibacterial activity results from the action of the hydrogen peroxide in honey that is produced by glucose oxidase in the presence of light and heat (Dustmann 1979). The other theory is that non-peroxide activity, which is independent of both light and heat, inhibits bacterial growth (Bogdanov 1997). This non-peroxide activity, which remains unaltered, even during long storage times, mainly depends on the floral source (Molan and Russell 1988).

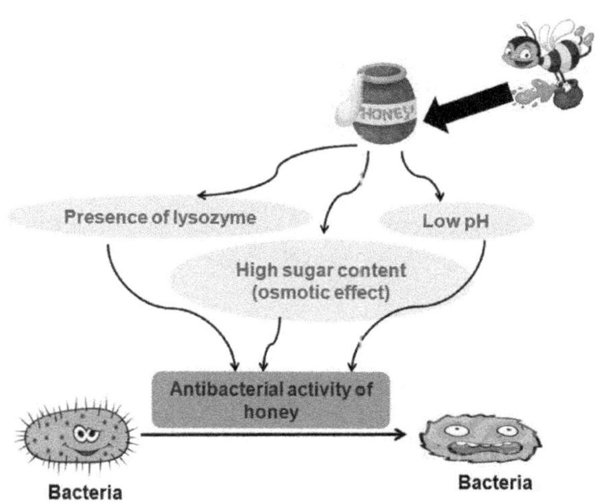


**Figure 2.8** Likely pathways of the antibacterial activity of honey.

The major components of honey are sugars, which themselves possess antibacterial activity because of their osmotic effect (Molan 1992). It is also well known that honey contains lysozyme, a powerful antimicrobial agent (Bogdanov 1997). Other researchers attribute the antibacterial capacity of honey to a combination of properties, such as its low pH and high osmolarity (Yatsunami and Echigo 1984), or to the presence of certain volatile substances, although this has not been studies in great depth (Toth et al. 1987). The probable pathways through which honey exerts its antibacterial activity are illustrated in Figure 2.8.

Honey primarily exerts its antibacterial activity against gram-positive bacteria (Marcucci et al. 2001; Srećković et al. 2019). Burdock (1998) attributed this capacity to the presence of aromatic acids and esters, but Takaisi et al. (1994) suggested that it is due to the action of the flavonone pinocembrin, the flavonol galangin, and caffeic acid phenethyl ester, whose mechanisms of action are based on the inhibition of bacterial RNA polymerase. Cushnie and Lamb (2005) reported that other flavonoids, such as galangin, also exhibit antibacterial action. The mode of action involves the degradation of the bacterial cytoplasmic membrane, which leads to the loss of potassium ions and bacterial cell damage by provoking autolysis. Quercetin, a well-known flavonoid, is also present in a number of honey samples (Afroz et al. 2016b, 2016c; Khalil and Sulaiman 2010) and increases membrane permeability by dissipates its potential, thus preventing the bacteria from synthesizing and transporting adenosine triphosphate (Mirzoeva and Calder 1996; Syed Yaacob et al. 2020). The antibacterial properties of honey have great potential for applications in medicine and the food industry.

#### Antiviral Properties

Natural honey and many other bee products, such as propolis, have the capacity to inhibit viral propagation (Miguel et al. 2017; Viuda Martos et al. 2008). Critchfield et al. (1996) reported that typical honey flavonoids, such as chrysin, acacetin and apigenin, can inhibit the activation of human immunodeficiency virus-1 (HIV-1) in latent models of infection through a mechanism that likely includes the inhibition of viral transcription. The flavonoids present in different types of honeys, namely chrysin and kaempferol (Khalil and Sulaiman 2010), were found to be very active in inhibiting the replication of several herpes viruses, adenoviruses, and rotaviruses (Cheng and Wong 1996). Other studies showed that quercetin and rutin (available flavonoids in honey) exerted antiviral activity against herpes simplex virus (HSV), syncytial virus, poliovirus, and Sindbis virus (Middleton and Kandaswami 1994; Selway 1986; Semprini et al. 2019). These compounds exert their action by inhibiting the viral polymerase and binding to the viral nucleic acids or viral capsid proteins (Selway 1986). Cushnie and Lamb (2005) and Amoros et al. (1992) described the synergistic effect of kaempferol and apigenin on HSV, which may explain why honey exhibits greater antiviral activity than its individual components.

#### Antifungal Properties

Although several in vitro studies have demonstrated the antibacterial properties of honey, only a few have examined its action against fungi (Irish et al. 2006). Recently, the potential antifungal effects of honey have attracted serious attention within the scientific community. Several factors may influence the antifungal activity of honey (Israili 2014). DeMera and Angert (2004) report that honeys from different phytogeographic regions vary in their ability to inhibit the growth of yeasts, suggesting that the botanical origin of the honey plays an important role in its antifungal activity. Like many other biological properties of honey, its antifungal potential is also attributed to its polyphenolic composition (Moussa et al. 2011).

#### Anti-inflammatory Capacity

The inflammatory process is triggered by several chemicals and biological compounds, including pro-inflammatory enzymes and cytokines and low-molecular-weight compounds, such as eicosanoids (Dao et al. 2004; Oryan and Alemzadeh 2017). According to several studies, cyclooxygenase-2 (COX-2), an isoform of COX, is the most important enzyme in the inflammatory process (Cho et al. 2004; Griswold and Adams 1996; Nguyen et al. 2019). This enzyme catalyzes the transformation of arachidonic acid to prostaglandin (Viuda Martos et al. 2008). In the past 30 years, a number of studies noted the anti-inflammatory effects of honey and other bee products (Ali et al. 1991; Mobarok 1994; Nguyen et al. 2019). Flavonoids are primarily responsible for the anti-inflammatory effect of honey. Galangin, a well-known flavonoid found in different honey samples (Khalil and Sulaiman 2010), is capable of inhibiting COX and lipo-oxygenase enzyme activity, limiting the action of polygalacturonase, and reducing the expression of the inducible isoform of COX-2 (Raso et al. 2001; Rossi et al. 2002). Another flavonoid compound in honey, chrysin, also shows strong anti-inflammatory activity (Kim et al. 2002). Chrysin exerts this activity by suppressing the pro-inflammatory activities of COX-2 and inducible nitric oxide synthase (Cho et al. 2004). Furthermore, the ingestion of diluted natural honey can reduce the concentrations of prostaglandins (PGEs), such as $PGE_2$ and $PGF_{2\alpha}$ and thromboxane B2, in the plasma of normal individuals. Recently, a type of Malaysian honey named gelam honey has been demonstrated to decrease the levels of inflammatory mediators, such as COX-2 and tumor necrosis factor-$\alpha$ (TNF-$\alpha$), by attenuating the translocation of nuclear factor–$\kappa$B (NF-$\kappa$B) to the nucleus, thus inhibiting the activation of the NF-$\kappa$B pathway (Al-Waili 2004; Vallianou et al. 2014). It is widely known that the activation of NF-$\kappa$B plays a key role in the pathogenesis of inflammation (Johnston et al. 2005). Although nonsteroidal anti-inflammatory drugs and corticosteroids may have many serious side effects, natural honey has an anti-inflammatory action that is free from any major side effects (Sun et al. 2020; Vallianou et al. 2014).

## Anti-ulcerous Properties

Another functional property of honey is its anti-ulcerous capacity (Ramirez-Acuña et al. 2019). Again, this ability has been attributed to the presence of phenolic compounds, particularly flavonoids (Viuda Martos et al. 2008). Vilegas et al. (1999) described the inhibitory effect of flavonoids on acid secretions, which prevents the formation of peptic ulcers. Young et al. (1999) and Martin et al. (1998) reported that ulcers are associated with ROS and flavonoids protect against ulcers by inhibiting lipid peroxidation, which considerably increases the glutathione peroxidase activity. Many other flavonoids, including quercetin and kaempferol, both of which are present in various honey samples, exhibit protective activity against ulcers (Viuda Martos et al. 2008).

## Antidiabetic Effect

The role of oxidative stress in the pathogenesis and complications of diabetes mellitus is well recognized (Erejuwa et al. 2010; Ramli et al. 2018). Both human and experimental animal models of diabetes exhibit high oxidative stress caused by persistent and chronic hyperglycemia, which depletes the activity of the free radical scavenging enzymes and subsequently promotes free radical generation (Bobiş et al. 2018; Bonnefont-Rousselot et al. 2000; Telci et al. 2000). Oxidative stress has recently been reported to be responsible, to a certain extent, for the β-cell dysfunction caused by glucose toxicity (Evans et al. 2003). Pancreatic β-cells are highly prone to oxidative stress and damage because they exhibit low expression levels and activities of antioxidant enzymes, which are the first line of defense against oxidative insult (Lenzen 2008). Like many of the other functional properties of honey, polyphenolic constituents are a cornerstone of the antidiabetic effect of honey because they protect pancreatic β-cells from oxidative damage by scavenging free radicals.

It is still unknown how honey mediated its hypoglycemic effect in diabetes. Moreover, recent literature assumed that honey may exert this effect through fructose, which is its predominant constituent (Bobiş et al. 2018; Erejuwa et al. 2011a). Fructose does not increase the plasma glucose levels, and its metabolism does not require insulin secretion (Mayes 1993). Dietary fructose is known to activate glucokinase, which is a key enzyme involved in the intracellular metabolism of glucose. It catalyzes the conversion of glucose to glucose-6-phosphate, thereby decreasing the glucose level in the blood (Watford 2002). A previous study also reported that fructose stimulated insulin secretion from an isolated pancreas (Grodsky et al. 1963). However, stronger evidence in support of the role of fructose in mediating the hypoglycemic effect of honey was provided by Curry et al. (1972). These authors found that there was no insulin response to fructose in rat pancreas preparations when glucose was present at very low concentrations or absent from the medium. In contrast, with higher glucose concentrations, an insulin response to fructose was elicited. Furthermore, honey is reported to have a lower glycemic index compared with many other carbohydrates (Abdulrhman et al. 2011).

## Anticancer Effect

Cancers are to the unrestrained growth of cells, which may exhibit malignant behavior. The process of cancer development includes three key stages: initiation, promotion, and progression. Initiation involves irreversible genetic damage and is characterized by the accumulation of mutated DNA (Pitot 1993). This is followed by the promotion stage, which is characterized by the excessive proliferation and growth of the mutated cells, as well as additional genomic alterations in the replicated cells, giving rise to a benign mass of abnormal cells known as a tumor (Tubiana 1997). Then the progression stage occurs, which entails the metastasis of the cancer cells to distant sites (tissues and organs) through the lymphatic or circulatory systems (Pitot 1993; Tubiana 1997). In addition to the limitations of current cancer management strategies (surgery, chemotherapy, and radiotherapy), the available cytotoxic drugs are expensive and are not readily available (particularly in developing countries), and their use is also associated with a number of undesirable adverse side effects (Chidambaram et al. 2011; Wedding 2010). Consequently, a large proportion of the population prefers to patronize complementary and alternative medicine. The phenolic and flavonoid constituents of honey have been shown to exert antioxidant, antiproliferative, antitumor, antimetastatic, and anticancer effects (Aumeeruddy et al. 2019; Erejuwa et al. 2014; Waheed et al. 2019). Therefore, the inhibitory effects of honey on tumorigenesis and carcinogenesis can be attributed to the presence of these flavonoids and phenolic acids (Imtara et al. 2019; Waheed et al. 2019). In fact, honey can suppress all the three steps of cancer development (Figure 2.9).

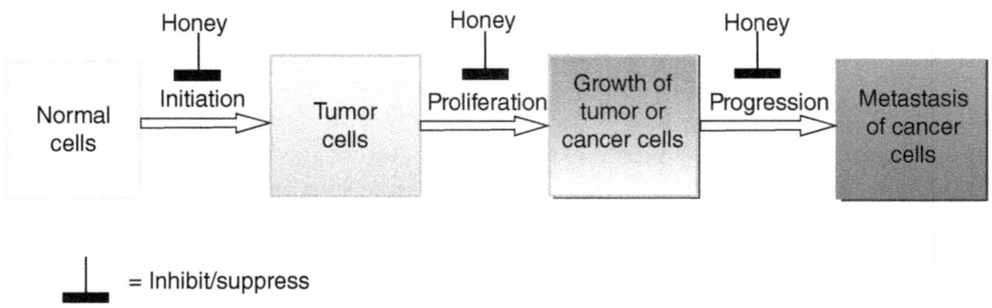


**Figure 2.9** Honey blocks the three stages of carcinogenesis. Erejuwa, Sulaiman et al. 2014 / MPDI / Licensed under CC BY 3.0.

Honey and cancer have a sustainable, inverse relationship. Honey acts a "natural cancer vaccine" because it can reduce chronic inflammatory processes, improve the immune status, and reduce infections by hardy organisms. Some of the simple polyphenols found in honey, namely caffeic acid, chrysin, galangin, quercetin, kaempferol, acacetin, pinocembrin, pinobanksin, and apigenin, have evolved as promising pharmacological agents for the prevention and treatment of cancer (Jaganathan and Mandal 2009; Waheed et al. 2019). Honey may provide the basis for the development of novel therapeutics for patients with cancer and cancer-related tumors. Jungle honey fragments were shown to induce the chemotaxis of neutrophils and inhibit ROS, thus demonstrating its antitumor activity (Fukuda et al. 2010). Honey is rich in flavonoids, and the anticancer properties of flavonoids have created great interest among researchers (Othman 2012). The proposed mechanisms are rather diverse and include various signaling pathways (Woo et al. 2004), such as the stimulation of TNF-α release (Tonks et al. 2001), inhibition of cell proliferation, induction of apoptosis (Jaganathan and Mandal 2010), cell cycle arrest (Pichichero et al. 2010), and inhibition of lipoprotein oxidation (Gheldof and Engeseth 2002).

Although honey has other substances, of which the most predominant are a mixture of sugars (fructose, glucose, maltose, and sucrose) (Aljadi and Kamaruddin 2004) that are carcinogenic (Heuson et al. 1972), it is understandable that some are sceptical of its beneficial effect on cancer. The mechanism by which honey exerts its anticancer effect has recently become an area of great interest. The effects of honeys on hormone-dependent cancers, such as breast, endometrial, and prostate cancer, still remain largely unknown (Othman 2012).

### Cardioprotective Effect

Cardiovascular diseases and their underlying oxidative stress have received global attention and have aroused increased interest in the identification of natural sources of antioxidants that have minimal side effects and can be used as preventive medicines. In recent years, the prevention of CVDs has been linked to the consumption of fresh food items and plants rich in natural antioxidants because they exhibit superior efficacy and safety compared with synthetic products (Topliss et al. 2002). Flavonoids, such as catechin and kaempferol; phenolic acids; ascorbic acid; and proteins are important constitutive antioxidants that have been detected in honey (Khalil et al. 2011; Khalil and Sulaiman 2010; Moniruzzaman et al. 2013). Honey is also reported to be a natural source of antioxidants. All of these compounds can work synergistically to scavenge and eliminate free radicals (Johnston et al. 2005). It is plausible that the presence of these antioxidants may help to protect against oxidative cardiac injury, thus restricting the leakage of cardiac marker enzymes from the myocardium (Afroz et al. 2016a; Khalil et al. 2015; Olas 2020). Lipid peroxidation is an important pathogenic event in CVDs (Rajadurai and Prince 2006). Natural honey has been reported to prevent lipid peroxidation in the myocardium in vivo (Afroz et al. 2016a; Khalil et al. 2015). These synergistic radical scavenging effects of natural honey may be mediated by both the enzymatic and non-enzymatic antioxidants that are involved in the cardiovascular defense mechanisms (Beretta et al. 2007; Bt Hj Idrus et al. 2020; Olas 2020; Rakha et al. 2008).

Honey also boosts the activity of antioxidant enzymes to enable these enzymes to prevent free radical–induced cardiac cell damage (Afroz et al. 2016a; Khalil et al. 2015). The possible mechanisms through which honey supplementation restores the antioxidant enzyme function may include the up-regulation of the activity or expression of Nrf2 (Erejuwa et al. 2011b), a transcription factor that is released from its repressor (Keap1) under oxidative or xenobiotic stress (Kobayashi et al. 2009). The released Nrf2 binds to the antioxidant response element of cytoprotective genes and induces their expression, which subsequently induce the expression of free radical scavenging enzymes to neutralize and eliminate the cytotoxic oxidants (Erejuwa et al. 2012; Kobayashi et al. 2009).

In summary, there is now sizeable evidence that honey is a natural immune booster, natural anti-inflammatory agent, natural antimicrobial agent, natural cancer "vaccine," and natural agent for healing chronic ulcers and wounds, which are some of the risk factors for cancer development.

## References


Abdulrhman, M., El-Hefnawy, M.,Hussein, R., and El-Goud, A.A. (2011). The glycemic and peak incremental indices of honey, sucrose and glucose in patients with type 1 diabetes mellitus: effects on C-peptide level—a pilot study. *Acta Diabetologica* 48 (2): 89–94.

Afroz, R., Tanvir, E., Hossain, M.F., et al. (2014). Protective effect of Sundarban honey against acetaminophen-induced acute hepatonephrotoxicity in rats. *Evidence-Based Complementary and Alternative Medicine* 2014: 143782.

Afroz, R., Tanvir, E., Karim, N., et al. (2016a). Sundarban honey confers protection against isoproterenol-induced myocardial infarction in Wistar rats. *BioMed Research International* 2016: 6437641.

Afroz, R., Tanvir, E., Zheng, W., and Little, P. (2016b). Molecular pharmacology of honey. *Clinical and Experimental Pharmacology* 6 (3).

Afroz, R., Tanvir, E., Paul, S., et al. (2016c). DNA damage inhibition properties of Sundarban honey and its phenolic composition. *Journal of Food Biochemistry* 40: 436–445.

Ajibola, A., Chamunorwa, J.P. and Erlwanger, K.H. (2012). Nutraceutical values of natural honey and its contribution to human health and wealth. *Nutrition and Metabolism (London)* 9: 61.

Ali, A.T.,Chowdhury, M.,and Al Humayyd, M. (1991). Inhibitory effect of natural honey on Helicobacter pylori. *Tropical Gastroenterology* 12 (3): t99.

Alissandrakis, E., Tarantilis, P.A., Harizanis, P.C., and Polissiou, M. (2005). Evaluation of four isolation techniques for honey aroma compounds. *Journal of the Science of Food and Agriculture* 85 (1): 91–97.

Alissandrakis, E., Tarantilis, P.A., Harizanis, P.C., and Polissiou, M. (2007). Aroma investigation of unifloral Greek citrus honey using solid-phase microextraction coupled to gas chromatographic–mass spectrometric analysis. *Food Chemistry* 100 (1): 396–404.

Aljadi, A. and Kamaruddin, M. (2004). Evaluation of the phenolic contents and antioxidant capacities of two Malaysian floral honeys. *Food Chemistry* 85 (4): 513–518.

Al-Mamary, M., Al-Meeri, A. and Al-Habori, M. et al. (2002). Antioxidant activities and total phenolics of different types of honey. *Nutrition Research* 22 (9): 1041–1047.

Alvarez-Suarez, J.M., Tulipani, S., Romandini, S., et al. (2010). Contribution of honey in nutrition and human health: a review. *Mediterranean Journal of Nutrition and Metabolism* 3 (1): 15–23.

Al-Waili, N.S. (2003). Effects of daily consumption of honey solution on hematological indices and blood levels of minerals and enzymes in normal individuals. *Journal of Medicinal Food* 6 (2): 135–140.

Al-Waili, N.S. (2004). Natural honey lowers plasma glucose, C-reactive protein, homocysteine and blood lipids in healthy, diabetic and hyperlipidemic subjects: comparison with dextrose and sucrose. *Journal of Medicinal Food* 7 (1): 100–107.

Amoros, M., Simõs, C., Girre, L., et al. (1992). Synergistic effect of flavones and flavonols against herpes simplex virus type 1 in cell culture. Comparison with the antiviral activity of propolis. *Journal of Natural Products* 55 (12): 1732–1740.

An, N., Cai, W.-J., Zhu, Q.-F., et al. (2020). Metabolic profiling of organic acids in honey by stable isotope labeling assisted liquid chromatography-mass spectrometry. *Journal of Food Composition and Analysis* 87: 103423.

Anklam, E. and Radovic, B. (2001). Suitable analytical methods for determining the origin of European honey. *American Laboratory* 33 (10): 60–62.

Aumeeruddy, M., Aumeeruddy-Elalfi, Z., Neetoo, H., et al. (2019). Pharmacological activities, chemical profile and physicochemical properties of raw and commercial honey. *Biocatalysis and Agricultural Biotechnology* 18: 101005.

Aurongzeb, M. and Azim, M.K. (2011). Antimicrobial properties of natural honey: a review of literature. *Pakistan Journal of Biochemistry and Molecular Biology* 44 (3): 118–124.

Azeredo, L. da C, Azeredo, M., De Souza, S., and Dutra, V. (2003). Protein contents and physicochemical properties in honey samples of Apis mellifera of different floral origins. *Food Chemistry* 80 (2): 249–254.

Azevedo, M.S., Seraglio, S. K. T., Rocha, G., et al. (2017). Free amino acid determination by GC-MS combined with a chemometric approach for geographical classification of bracatinga honeydew honey (Mimosa scabrella Bentham). *Food Control* 78: 383–392.

Beretta, G., Orioli, M., and Facino, R.M. (2007). Antioxidant and radical scavenging activity of honey in endothelial cell cultures (EA.hy926). *Planta Medica* 73 (11): 1182–1189.

Bertoncelj, J., Doberšek, U., Jamnik, M., and Golob, T. (2007). Evaluation of the phenolic content, antioxidant activity and colour of Slovenian honey. *Food Chemistry* 105 (2): 822–828.

Bilandžić, N., Sedak, M., Đokić, M., et al. (2019). Element content in ten Croatian honey types from different geographical regions during three seasons. *Journal of Food Composition and Analysis* 84: 103305.

Bobiş, O., Dezmirean, D. S., and Moise, A R. (2018). Honey and diabetes: the importance of natural simple sugars in diet for preventing and treating different type of diabetes. *Oxidative Medicine and Cellular Longevity* 2018: 4757893.

Bogdanov, S. (1984). Characterisation of antibacterial substances in honey. Lebensmittel-Wissenschaft± Technologie= Food science± technology.

Bogdanov, S. (1997). Nature and origin of the antibacterial substances in honey. *LWT-Food Science and Technology* 30 (7): 748–753.

Bogdanov, S., Ruoff, K., and Persano Oddo, L. (2004). Physico-chemical methods for the characterisation of unifloral honeys: a review. *Apidologie* 35 (Suppl 1): 4–17.

Bogdanov, S., Jurendic, T., Sieber, R., and Gallmann, P. (2008). Honey for nutrition and health: a review. *Journal of the American College of Nutrition* 27 (6): 677–689.

Bogdanov, S. (2011). The bee products: the wonders of the bee hexagon. www.bee.hexagon.net.

Bonnefont-Rousselot, D., Bastard, J., Jaudon, M., and Delattre, J. (2000). Consequences of the diabetic status on the oxidant/antioxidant balance. *Diabetes and Metabolism* 26 (3): 163–177.

Bouacha, M., Ayed, H., and Grara, N. (2018). Honey bee as alternative medicine to treat eleven multidrug-resistant bacteria causing urinary tract infection during pregnancy. *Scientia Pharmaceutica* 86 (2): 14.

Bt Hj Idrus, R., Sainik, N. Q. a. V., Nordin, A., et al. (2020). Cardioprotective effects of honey and its constituent: an evidence-based review of laboratory studies and clinical trials. *International Journal of Environmental Research and Public Health* 17 (10): 3613.

Burdock, G. (1998). Review of the biological properties and toxicity of bee propolis (propolis). *Food and Chemical Toxicology* 36 (4): 347–363.

Careri, M., Mangia, A., Barbieri, G., et al. (1993). Sensory property relationships to chemical data of italian-type dry-cured ham. *Journal of Food Science* 58 (5): 968–972.

Castro-Vázquez, L., Pérez-Coello, M., and Cabezudo, M. (2003). Analysis of volatile compounds of rosemary honey. Comparison of different extraction techniques. *Chromatographia* 57 (3): 227–233.

Cheng, P.C. and Wong, G. (1996). Honey bee propolis: prospects in medicine. *Bee World* 77 (1): 8–15.

Chidambaram, M., Manavalan, R., and Kathiresan, K. (2011). Nanotherapeutics to overcome conventional cancer chemotherapy limitations. *Journal of Pharmacy & Pharmaceutical Sciences* 14 (1): 67–77.

Cho, H., Yun, C.-W., Park, W.-K., et al. (2004). Modulation of the activity of pro-inflammatory enzymes, COX-2 and iNOS, by chrysin derivatives. *Pharmacological Research* 49 (1): 37–43.

Chow, J. (2002). Probiotics and prebiotics: a brief overview. *Journal of Renal Nutrition* 12 (2): 76–86.

Codex Alimentarius, C. (2001). Commission standards. Codex Standards for Honey (1981/revised 1987/revised 2001), Rome: FAO–Rome. 1–7.

Cortés, M.E., Vigil, P., and Montenegro, G. (2011). The medicinal value of honey: a review on its benefits to human health, with a special focus on its effects on glycemic regulation. *Ciencia e Investigacion Agraria* 38: 303–317.

Crane, E. (1975). *Honey. A Comprehensive Survey*. Heinemann [for] the Bee Research Association

Crane, E. (1983). *The Archaeology of Beekeeping*. London: Duckworth.

Crane, E.E. (2013). *The World History of Beekeeping and Honey Hunting*. Oxfordshire, UK: Routledge.

Critchfield, J.W., Butera, S.T., and Folks, T.M. (1996). Inhibition of HIV activation in latently infected cells by flavonoid compounds. *AIDS Research and Human Retroviruses* 12 (1): 39–46.

Cuevas-Glory, L.F., Pino, J.A., Santiago, L.S., and Sauri-Duch, E. (2007). A review of volatile analytical methods for determining the botanical origin of honey. *Food Chemistry* 103 (3): 1032–1043.

Curry, D.L., Curry, K.P., and Gomez, M. (1972). Fructose potentiation of insulin secretion. *Endocrinology* 91 (6): 1493–1498.

Cushnie, T.P. and Lamb, A.J. (2005). Antimicrobial activity of flavonoids. *Int J Antimicrob Agents* 26(5):343–356. Erratum in: *Int J Antimicrob Agents* 2006;27(2):181.

da Silva, P.M., Gauche, C., Gonzaga, L.V., et al. (2016). Honey: chemical composition, stability and authenticity. *Food Chemistry* 196: 309–323.

Dao, T.T., Chi, Y.S., Kim, J., et al. (2004). Synthesis and inhibitory activity against COX-2 catalyzed prostaglandin production of chrysin derivatives. *Bioorganic & Medicinal Chemistry Letters* 14 (5): 1165–1167.

DeMera, J.H. and Angert, E.R. (2004). Comparison of the antimicrobial activity of honey produced by Tetragonisca angustula (Meliponinae) and Apis mellifera from different phytogeographic regions of Costa Rica. *Apidologie* 35 (4): 411–417.

Deng, J., Liu, R., Lu, Q., et al. (2018). Biochemical properties, antibacterial and cellular antioxidant activities of buckwheat honey in comparison to manuka honey. *Food Chemistry* 252: 243–249.

do Nascimento, K.S., Sattler, J.a.G., Macedo, L.F.L., et al. (2018). Phenolic compounds, antioxidant capacity and physicochemical properties of Brazilian Apis mellifera honeys. *LWT* 91: 85–94.

Dustmann, J. (1979). Antibacterial effect of honey. *Apiacta* 14 (1): 7–11.

Dżugan, M., Grabek-Lejko, D., Swacha, S., et al. (2020). Physicochemical quality parameters, antibacterial properties and cellular antioxidant activity of Polish buckwheat honey. *Food Bioscience* 34: 100538.

El Denshary, E.S., Al-Gahazali, M.A., Mannaa, F.A., et al. (2012). Dietary honey and ginseng protect against carbon tetrachloride-induced hepatonephrotoxicity in rats. *Experimental and Toxicologic Pathology* 64 (7): 753–760.

Erejuwa, O.O., Sulaiman, S.A., Wahab, M.S., et al. (2010). Antioxidant protective effect of glibenclamide and metformin in combination with honey in pancreas of streptozotocin-induced diabetic rats. *International Journal of Molecular Sciences* 11 (5): 2056–2066.

Erejuwa, O.O., Sulaiman, S.A., Wahab, M.S., et al. (2011a). Glibenclamide or metformin combined with honey improves glycemic control in streptozotocin-induced diabetic rats. *International Journal of Biological Sciences* 7 (2): 244.

Erejuwa, O.O., Sulaiman, S., Suhaimi, M., et al. (2011b). Impaired Nrf2-ARE pathway contributes to increased oxidative damage in kidney of spontaneously hypertensive rats: effect of antioxidant (honey). *International Journal of Cardiology* 152: S45.

Erejuwa, O.O., Sulaiman, S.A., and Wahab, M.S. (2012). Honey: a novel antioxidant. *Molecules* 17: 4400–4423.

Erejuwa, O.O., Sulaiman, S.A., and Wahab, M.S.A. (2014). Effects of honey and its mechanisms of action on the development and progression of cancer. *Molecules* 19 (2): 2497–2522.

Estevinho, L., Pereira, A. P., Moreira, L., et al. (2008). Antioxidant and antimicrobial effects of phenolic compounds extracts of Northeast Portugal honey. *Food and Chemical Toxicology* 46 (12): 3774–3779.

Eteraf-Oskouei, T. and Najafi, M. (2013). Traditional and modern uses of natural honey in human diseases: a review. *Iranian Journal of Basic Medical Sciences* 16 (6): 731.

EU Directive (2002). Council directive 2001/110/EC relating to honey. *Official Journal of the European Communities* 47–52.

Evans, J.L., Goldfine, I.D., Maddux, B.A., and Grodsky, G.M. (2003). Are oxidative stress–activated signaling pathways mediators of insulin resistance and β-cell dysfunction? *Diabetes* 52 (1): 1–8.

Ferreres, F., Tom as-Barber An F.A., Gil, M.A., and Tom as-Lorente, F. (1991). An HPLC technique for flavonoid analysis in honey. *Journal of the Science of Food and Agriculture* 56 (1): 49–56.

Frankel, S., Robinson, G., and Berenbaum, M. (1998). Antioxidant capacity and correlated characteristics of 14 unifloral honeys. *Journal of Apicultural Research* 37 (1): 27–31.

Fukuda, M., Kobayashi, K., Hirono, Y., et al. (2010). Jungle honey enhances immune function and antitumor activity. *Evidence-Based Complementary and Alternative Medicine* 2011: 908743.

Gheldof, N. and Engeseth, N.J. (2002). Antioxidant capacity of honeys from various floral sources based on the determination of oxygen radical absorbance capacity and inhibition of in vitro lipoprotein oxidation in human serum samples. *Journal of Agricultural and Food Chemistry* 50 (10): 3050–3055.

Gil, M.I., Ferreres, F., Ortiz, A., et al. (1995). Plant phenolic metabolites and floral origin of rosemary honey. *Journal of Agricultural and Food Chemistry* 43 (11): 2833–2838.

Griswold, D.E. and Adams, J.L. (1996). Constitutive cyclooxygenase (COX-1) and inducible cyclooxygenase (COX-2): rationale for selective inhibition and progress to date. *Medicinal Research Reviews* 16 (2): 181–206.

Grodsky, G.M., Batts, A. A., Bennett, L.L., et al. (1963). Effects of carbohydrates on secretion of insulin from isolated rat pancreas. *American Journal of Physiology–Legacy Content* 205 (4): 638–644.

Hermosín, I., Chicón, R.M., and Cabezudo,M.D. (2003). Free amino acid composition and botanical origin of honey. *Food Chemistry* 83 (2): 263–268.

Heuson, J.-C., Legros, N., and Heimann, R. ( (1972). Influence of insulin administration on growth of the 7, 12-dimethylbenz (a) anthracene-induced mammary carcinoma in intact, oophorectomized and hypophysectomized rats. *Cancer Research* 32 (2): 233–238.

Iglesias, M.T., De Lorenzo, C.,Polo, M.D.C.,et al. (2004). Usefulness of amino acid composition to discriminate between honeydew and floral honeys. Application to honeys from a small geographic area. *Journal of Agricultural and Food Chemistry* 52 (1): 84–89.

Imtara, H., Kmail, A., Touzani, S., et al. (2019). Chemical analysis and cytotoxic and cytostatic effects of twelve honey samples collected from different regions in Morocco and Palestine. *Evidence-Based Complementary and Alternative Medicine* 2019: 8768210.

Irish, J., Carter, D.A., Shokohi, T., and Blair, S.E. (2006). Honey has an antifungal effect against Candida species. *Medical Mycology* 44 (3): 289–291.

Islam, A., Khalil, I., Islam, N., et al. (2012). Physicochemical and antioxidant properties of Bangladeshi honeys stored for more than one year. *BMC Complementary and Alternative Medicine* 12 (1): 177.

Israili, Z.H. (2014). Antimicrobial properties of honey. *American Journal of Therapeutics* 21 (4): 304–323.

Jaganathan, S.K. and Mandal, M. (2009). Antiproliferative effects of honey and of its polyphenols: a review. *Journal of Biomedicine and Biotechnology* 2009: 830616.

Jaganathan, S.K. and Mandal, M. (2010). Involvement of non-protein thiols, mitochondrial dysfunction, reactive oxygen species and p53 in honey-induced apoptosis. *Investigational New Drugs* 28 (5): 624–633.

Jahan, N., Islam, M.A., Alam, F., et al. (2015). Prolonged heating of honey increases its antioxidant potential but decreases its antimicrobial activity. *African Journal of Traditional, Complementary and Alternative Medicines* 12 (4): 134–144.

Johnston, J., Sepe, H., Miano, C., et al. (2005). Honey inhibits lipid oxidation in ready-to-eat ground beef patties. *Meat Science* 70 (4): 627–631.

Karapetsas, A., Voulgaridou, G.-P., Iliadi, D., et al. (2020). Honey extracts exhibit cytoprotective properties against UVB-induced photodamage in human experimental skin models. *Antioxidants* 9 (7): 566.

Khalil, M., Sulaiman, S., and Boukraa, L. et al. (2010). Antioxidant properties of honey and its role in preventing health disorder. *The Open Nutraceuticals Journal* 3 (1): 6–16.

Khalil, M., Alam, N., Moniruzzaman, M., et al. (2011). Phenolic acid composition and antioxidant properties of Malaysian honeys. *Journal of Food Science* 76 (6): C921–C928.

Khalil, M. and Sulaiman, S. (2010). The potential role of honey and its polyphenols in preventing heart disease: a review. *African Journal of Traditional, Complementary and Alternative Medicines* 7 (4): 315–321.

Khalil, M.I., Tanvir, E., Afroz, R., et al. (2015). Cardioprotective effects of Tualang honey: amelioration of cholesterol and cardiac enzymes levels. *BioMed Research International* 2015: 286051.

Kim, E.J., Kwon, K.J., Park, J.-Y., et al. (2002). Effects of peroxisome proliferator-activated receptor agonists on LPS-induced neuronal death in mixed cortical neurons: associated with iNOS and COX-2. *Brain Research* 941 (1): 1–10.

Kobayashi, M., Li, L., Iwamoto, N., et al. (2009). The antioxidant defense system Keap1-Nrf2 comprises a multiple sensing mechanism for responding to a wide range of chemical compounds. *Molecular and Cellular Biology* 29: 493–502.

Kowalski, S., Kopuncová, M., Ciesarová, Z., and Kukurová, K. (2017). Free amino acids profile of Polish and Slovak honeys based on LC–MS/MS method without the prior derivatisation. *Journal of Food Science and Technology* 54 (11): 3716–3723.

Kulkarni, S.S., Patil, S.B., Mishra, S., et al. (2020). Physicochemical characteristics, heavy metal analysis and antioxidant potential of Jamun Honey (Syzygium Cumini L.) from Western Ghats, India. *Indian Journal of Public Health Research & Development* 11 (2): 17–22.

Kumar, K., Bhowmik, D., Biswajit, C., and Chandira, M. (2010). Medicinal uses and health benefits of honey: an overview. *Journal of Chemical and Pharmaceutical Research* 2 (1): 385–395.

Lachman, J., Kolihova, D., Miholova, D., et al. (2007). Analysis of minority honey components: possible use for the evaluation of honey quality. *Food Chemistry* 101 (3): 973–979.

Lenzen, S. (2008). Oxidative stress: the vulnerable beta-cell. *Biochemical Society Transactions* 36 (3): 343.

Marcucci, M.C., Ferreres, F., Garcıa-Viguera, C., et al. (2001). Phenolic compounds from Brazilian propolis with pharmacological activities. *Journal of Ethnopharmacology* 74 (2): 105–112.

Marshall, T. and Williams, K.M. (1987). Electrophoresis of honey: characterization of trace proteins from a complex biological matrix by silver staining. *Analytical Biochemistry* 167 (2): 301–303.

Martin, M., La-Casa, C., Alarcon-De-La-Lastra, C., et al. (1998). Anti-oxidant mechanisms involved in gastroprotective effects of quercetin. *Zeitschrift für Naturforschung C* 53 (1–2): 82–88.

Martos, I., Ferreres, F., and Tomás-Barberán, F.A. (2000a). Identification of flavonoid markers for the botanical origin of Eucalyptus honey. *Journal of Agricultural and Food Chemistry* 48 (5): 1498–1502.

Martos, I., Ferreres, F., Yao, L., et al. (2000b). Flavonoids in monospecific Eucalyptus honeys from Australia. *Journal of Agricultural and Food Chemistry* 48 (10): 4744–4748.

Mayes, P.A. (1993). Intermediary metabolism of fructose. *The American Journal of Clinical Nutrition* 58 (5): 754S–765S.

Mendiola, J., Marin, F., Seňoráns, R., et al. (2008). Profiling of different bioactive compounds in functional drinks by high performance liquid chromatography. *Journal of Chromatography* 1188: 234–241.

Meyer, A.S., Donovan, J. L., Pearson, D.A., et al. (1998). Fruit hydroxycinnamic acids inhibit human low-density lipoprotein oxidation in vitro. *Journal of Agricultural and Food Chemistry* 46 (5): 1783–1787.

Michalkiewicz, A., Biesaga, M., and Pyrzynska, K. (2008). Solid-phase extraction procedure for determination of phenolic acids and some flavonols in honey. *Journal of Chromatography A* 1187 (1): 18–24.

Middleton Jr, E. and Kandaswami, C. (1994). The impact of plant flavonoids on mammalian biology: implications for immunity, inflammation and cancer. In: Harborne, J.B., ed., *The Flavonoids: Advances in Research Since 1986*. London: Chapman and Hall.

Miguel, M., Antunes, M.D., and Faleiro, M.L. (2017). Honey as a complementary medicine. *Integrative Medicine Insights* 12: 1178633717702869.

Mirzoeva, O. and Calder, P. (1996). The effect of propolis and its components on eicosanoid production during the inflammatory response. *Prostaglandins, Leukotrienes and Essential Fatty Acids* 55 (6): 441–449.

Mobarok, A.A. (1994). Natural honey exerts its protective effects against ethanol-induced gastric lesions in rats by preventing depletion of glandular nonprotein sulfhydryls. *Tropical Gastroenterology* 16 (1): 18–26.

Mohammed, S.E.A. and Azim, M.K. (2012). Characterisation of natural honey proteins: implications for the floral and geographical origin of honey. *International Journal of Food Science & Technology* 47 (2): 362–368.

Molan, P. and Russell, K. (1988). Non-peroxide antibacterial activity in some New Zealand honeys. *Journal of Apicultural Research* 27 (1): 62–67.

Molan, P.C. (1992). The antibacterial activity of honey: 1. The nature of the antibacterial activity. *Bee World* 73 (1): 5–28.

Moniruzzaman, M., Khalil, M.I., Sulaiman, S.A., and Gan, S.H. (2013). Physicochemical and antioxidant properties of Malaysian honeys produced by Apis cerana, Apis dorsata and Apis mellifera. *BMC Complementary and Alternative Medicine* 13 (1): 43.

Moreira, R.F.A. and Maria, C. (2001). Glicídios no mel. *Química Nova* 24 (4): 516–525.

Moussa, A., Noureddine, D., Saad, A., et al. (2011). Antifungal activity of four honeys of different types from Algeria against pathogenic yeast: Candida albicans and Rhodotorula sp. *International Journal of Microbiology Research* 2 (3): 276–279.

Nguyen, H.T.L., Panyoyai, N., Kasapis, S., et al. (2019). Honey and its role in relieving multiple facets of atherosclerosis. *Nutrients* 11 (1): 167.

Olaitan, P.B., Adeleke, E., and Ola, O. (2007). Honey: a reservoir for microorganisms and an inhibitory agent for microbes. *African Health Sciences* 7 (3): 159–165.

Olas, B. (2020). Honey and its phenolic compounds as an effective natural medicine for cardiovascular diseases in humans? *Nutrients* 12 (2): 283.

Oryan, A. and Alemzadeh, E. (2017). Potential mechanisms and application of Honeybee products in wound management: wound healing by apitherapy. In: Shiffman, M.A. and Low, M., eds., *Burns, Infections and Wound Management*, 267–284. Cham, Switzerland: Springer.

Oskouei, T. and Najafi, M. (2013). Traditional and modern uses of natural honey in human diseases: a review. *Iranian Journal of Basic Medical Sciences* 16: 731–742.

Othman, N.H. (2012). Honey and cancer: sustainable inverse relationship particularly for developing nations—a review. *Evidence-Based Complementary and Alternative Medicine* 2012: 410406.

Paramás, A.M.G., Bárez, J.a.G., Marcos, C.C., et al. (2006). HPLC-fluorimetric method for analysis of amino acids in products of the hive (honey and bee-pollen). *Food Chemistry* 95 (1): 148–156.

Paul, S., Hossen, M., Tanvir, E., et al. (2017). Minerals, toxic heavy metals and antioxidant properties of honeys from Bangladesh. *Journal of Chemistry* 2017: 6101793.

Pauliuc, D., Dranca, F., and Oroian, M. (2020). Antioxidant activity, total phenolic content, individual phenolics and physicochemical parameters suitability for Romanian honey authentication. *Foods* 9 (3): 306.

Pérez, R.A., Sánchez-Brunete, C., Calvo, R.M., and Tadeo, J.L. (2002). Analysis of volatiles from Spanish honeys by solid-phase microextraction and gas chromatography-mass spectrometry. *Journal of Agricultural and Food Chemistry* 50 (9): 2633–2637.

Pérez, R.A., Iglesias, M. T., Pueyo, E., González, M., and De Lorenzo, C. (2007). Amino acid composition and antioxidant capacity of Spanish honeys. *Journal of Agricultural and Food Chemistry* 55 (2): 360–365.

Piasenzotto, L., Gracco, L., and Conte, L. (2003). Solid phase microextraction (SPME) applied to honey quality control. *Journal of the Science of Food and Agriculture* 83 (10): 1037–1044.

Pichichero, E., Cicconi, R., Mattei, M., et al. (2010). Acacia honey and chrysin reduce proliferation of melanoma cells through alterations in cell cycle progression. *International Journal of Oncology* 37 (4): 973–981.

Pitot, H.C. (1993). The molecular biology of carcinogenesis. *Cancer* 72 (S3): 962–970.

Rahman, M.M., Alam, M.N., Fatima, N., et al. (2017). Chemical composition and biological properties of aromatic compounds in honey: an overview. *Journal of Food Biochemistry* 41 (6): e12405.

Rajadurai, M. and Prince, P.S.M. (2006). Preventive effect of naringin on lipid peroxides and antioxidants in isoproterenol-induced cardiotoxicity in Wistar rats: biochemical and histopathological evidences. *Toxicology* 228: 259–268.

Rakha, M.K., Nabil, Z.I., and Hussein, A.A. (2008). Cardioactive and vasoactive effects of natural wild honey against cardiac malperformance induced by hyperadrenergic activity. *Journal of Medicinal Food* 11 (1): 91–98.

Ramirez-Acuña, J.M., Cardenas-Cadena, S.A., Marquez-Salas, P.A., et al (2019). Diabetic foot ulcers: current advances in antimicrobial therapies and emerging treatments. *Antibiotics* 8 (4): 193.

Ramli, N.Z., Chin, K.-Y., Zarkasi, K.A., and Ahmad, F. (2018). A review on the protective effects of honey against metabolic syndrome. *Nutrients* 10 (8): 1009.

Raso, G.M., Meli, R., Di Carlo, G., et al. (2001). Inhibition of inducible nitric oxide synthase and cyclooxygenase-2 expression by flavonoids in macrophage J774A. 1. *Life Sciences* 68 (8): 921–931.

Roby, M.H., Abdelaliem, Y.F., Esmail, A.-H.M., et al. (2020). Evaluation of Egyptian honeys and their floral origins: phenolic compounds, antioxidant activities and antimicrobial characteristics. *Environmental Science and Pollution Research* 27: 1–9.

Rossi, A., Ligresti, A., Longo, R., et al. (2002). The inhibitory effect of propolis and caffeic acid phenethyl ester on cyclooxygenase activity in J774 macrophages. *Phytomedicine* 9 (6): 530–535.

Russo, A., Acquaviva, R., Campisi, A., et al. (2000). Bioflavonoids as antiradicals, antioxidants and DNA cleavage protectors. *Cell Biology and Toxicology* 16 (2): 91–98.

Saxena, S., Gautam, S., and Sharma, A. (2010). Physical, biochemical and antioxidant properties of some Indian honeys. *Food Chemistry* 118 (2): 391–397.

Schramm, D.D., Karim, M., Schrader, H.R., et al. (2003). Honey with high levels of antioxidants can provide protection to healthy human subjects. *Journal of Agricultural and Food Chemistry* 51 (6): 1732–1735.

Selway, J.T. (1986). Antiviral activity of flavones and flavans. *Progress in Clinical and Biological Research* 213: 521.

Semprini, A., Singer, J., Braithwaite, I., et al. (2019). Kanuka honey versus aciclovir for the topical treatment of herpes simplex labialis: a randomised controlled trial. *BMJ open* 9 (5): e026201.

Shamsudin, S., Selamat, J., Sanny, M., et al. (2019). Influence of origins and bee species on physicochemical, antioxidant properties and botanical discrimination of stingless bee honey. *International Journal of Food Properties* 22 (1): 239–264.

Solayman, M., Islam, M. A., Paul, S., et al. (2016). Physicochemical properties, minerals, trace elements and heavy metals in honey of different origins: a comprehensive review. *Comprehensive Reviews in Food Science and Food Safety* 15 (1): 219–233.

Squadrone, S., Brizio, P., Stella, C., et al. (2020). Trace and rare earth elements in monofloral and multifloral honeys from Northwestern Italy; a first attempt of characterization by a multi-elemental profile. *Journal of Trace Elements in Medicine and Biology* 61: 126556.

Srećković, N.Z., Mihailović, V.B., and Katanić-Stanković, J.S. (2019). Physico-chemical, antioxidant and antimicrobial properties of three different types of honey from Central Serbia. *Kragujevac Journal of Science* 41: 53–68.

Sun, L.-P., Shi, F.-F., Zhang, W.-W., et al. (2020). Antioxidant and anti-inflammatory activities of Safflower (Carthamus tinctorius L.) honey extract. *Foods* 9 (8): 1039.

Syed Yaacob, S.N., Wahab, R.A., Huyop, F., et al. (2020). Morphological alterations in gram-positive and gram-negative bacteria exposed to minimal inhibitory and bactericidal concentration of raw Malaysian stingless bee honey. *Biotechnology & Biotechnological Equipment* 34 (1): 575–586.

Takaisi-Kikuni, N.B. and Schilcher, H. (1994). Electron microscopic and microcalorimetric investigations of the possible mechanism of the antibacterial action of a defined propolis provenance. *Planta medica* 60 (3): 222–227.

Tanvir, E., Afroz, R., Chowdhury, M.a.Z., et al. (2015). Honey has a protective effect against chlorpyrifos-induced toxicity on lipid peroxidation, diagnostic markers and hepatic histoarchitecture. *European Journal of Integrative Medicine* 7 (5): 525–533.

Tanvir, E., Sakib Hossen, M., Mahfuza Shapla, U., et al. (2018). Antioxidant, brine shrimp lethality and analgesic properties of propolis from Bangladesh. *Journal of Food Biochemistry* 42 (5): e12596.

Telci, A., Çakatay, U., Salman, S., et al. (2000). Oxidative protein damage in early stage Type 1 diabetic patients. *Diabetes Research and Clinical Practice* 50 (3): 213–223.

Terrab, A., González, A.G., Díez, M.J., and Heredia, F.J. (2003). Characterisation of Moroccan unifloral honeys using multivariate analysis. *European Food Research and Technology* 218 (1): 88–95.

Tomás-Barberán, F.A., Martos, I., Ferreres, F., et al. (2001). HPLC flavonoid profiles as markers for the botanical origin of European unifloral honeys. *Journal of the Science of Food and Agriculture* 81 (5): 485–496.

Tonks, A., Cooper, R.A., Price, A., et al. (2001). Stimulation of TNF-α release in monocytes by honey. *Cytokine* 14 (4): 240–242.

Topliss, J., Clark, A., Ernst, E., et al. (2002). Natural and synthetic substances related to human health (IUPAC Technical Report). *Pure and Applied Chemistry* 74 (10): 1957–1985.

Toth, G., Lemberkovics, E, and Kutasi-Szabo, J. (1987). The volatile components of some Hungarian honeys and their antimicrobial effects. *American Bee Journal (USA)* 127 (7): 496–497.

Truchado, P., Ferreres, F., Bortolotti, L., et al. (2008). Nectar flavonol rhamnosides are floral markers of acacia (Robinia pseudacacia) honey. *Journal of Agricultural and Food Chemistry* 56 (19): 8815–8824.

Tubiana, M. (1997). [Carcinogenesis: from epidemiology to molecular biology]. *Bulletin de l'Academie Nationale de Medecine* 182 (1): 19–29. discussion 29-31.

Vallianou, N.G., Gounari, P., Skourtis, A., et al. (2014). Honey and its anti-inflammatory, anti-bacterial and anti-oxidant properties. *General Medicine* 2 (2).

Vela, L., De Lorenzo, C., and Perez, R.A. (2007). Antioxidant capacity of Spanish honeys and its correlation with polyphenol content and other physicochemical properties. *Journal of the Science of Food and Agriculture* 87 (6): 1069–1075.

Vilegas, W., Sanommiya, M., Rastrelli, L., and Pizza, C. (1999). Isolation and structure elucidation of two new flavonoid glycosides from the infusion of Maytenus aquifolium leaves. Evaluation of the antiulcer activity of the infusion. *Journal of Agricultural and Food Chemistry* 47 (2): 403–406.

Viuda Martos, M., Ruiz-Navajas, Y., Fernández-López, J., and Pérez-Álvarez, J. (2008). Functional properties of honey, propolis and royal jelly. *Journal of Food Science* 73 (9): R117–R124.

Waheed, M., Hussain, M. B., Javed, A., et al. (2019). Honey and cancer: a mechanistic review. *Clinical Nutrition* 38 (6): 2499–2503.

Watford, M. (2002). Small amounts of dietary fructose dramatically increase hepatic glucose uptake through a novel mechanism of glucokinase activation. *Nutrition Reviews* 60 (8): 253.

Wedding, U. (2010). Chances and limitations of chemotherapy in elderly patients. *Der Internist* 51 (3): 402, 404–406, 408–409.

White, J. and Doner, L.W. (1980). Honey composition and properties. *Beekeeping in the United States Agriculture Handbook* 335: 82–91.

Won, S.-R., Lee, D.-C., Ko, S.H., et al. (2008). Honey major protein characterization and its application to adulteration detection. *Food Research International* 41 (10): 952–956.

Woo, K.J., Jeong, Y.-J., Park, J.-W., and Kwon, T.K. (2004). Chrysin-induced apoptosis is mediated through caspase activation and Akt inactivation in U937 leukemia cells. *Biochemical and Biophysical Research Communications* 325 (4): 1215–1222.

Yatsunami, K. and Echigo, E.T. (1997). Antibacterial activity of honey and royal jelly. *Honey Bee Science* 5: 125–130.

You, K.M., Jong, H.-G., and Kim, H.P. (1999). Inhibition of cyclooxygenase/lipoxygenase from human platelets by polyhydroxylated/methoxylated flavonoids isolated from medicinal plants. *Archives of Pharmacal Research* 22 (1): 18–24.

Young, J.F., Nielsen, S. E., Haraldsdóttir, J., et al. (1999). Effect of fruit juice intake on urinary quercetin excretion and biomarkers of antioxidative status. *The American Journal of Clinical Nutrition* 69 (1): 87–94.

Zhou, Q., Li, P., Cheng, N., et al. (2002). Protective effects of buckwheat honey on DNA damage induced by hydroxyl radicals. *Food and Chemical Toxicology* 50 (7): 2766–2773.

# 3

# Carbohydrates in Honey

*Md. Murad Hossain, Dhirendra Nath Barman, Md. Anisur Rahman, and Shahad Saif Khandker*

## Introduction

Honey is a natural sweet food and a complex mixture of approximately 180 different compounds. Honey is highly rich in sugars, and more than 95% of the solids of honey are carbohydrate in nature. Honeybees produce honey sugars from nectar sucrose, which is transformed through the action of enzymes such as α- and β-glucosidase, α- and β-amylase, and β-fructosidase (Da Silva et al. 2016). Honey sugars represent about 75% monosaccharides and 10%–15% disaccharides, and the remainder of the sugar consists of trisaccharides and a few higher oligosaccharides (Da Silva et al. 2016; De la Fuente et al. 2006). The major compositions of these oligosaccharides are glucose and fructose, which are linked by glycosidic bond. Mono- and oligosaccharide profiles can help in discriminating different honeys according to their botanical and geographical origin, floral characteristics, and inter-annual variability (Escuredo et al. 2014; Tedesco et al. 2020). These oligosaccharides also contribute significantly to the high nutritional and medicinal value as a potential "prebiotic" property of honey by balancing the growth of intestinal microflora in animal and human intestines, controlling the gastrointestinal peristalsis, and reducing the incidence of serious illness such as colon cancer and diarrhea (Ouchemoukh et al. 2010; Zhou et al. 2016). The sugar composition depends mainly on the honey's botanical origin (the types of flowers used by the bees) and geographical origin and is affected by climate, environmental and seasonal conditions, processing, and storage, and the processes and transformations occur in bees (Buba et al. 2013). Fructose and glucose are the predominant monosaccharides that represent about 65%–85% of total soluble solids in honey (Tedesco et al. 2020). The concentrations of fructose and glucose, as well as the ratio between them, are useful indicators of honey's quality and for the classification of monofloral honey (Kaškonienė et al. 2010). Sugars present in honey are responsible for properties such as hygroscopy, viscosity, granulation, and energy value (Kamal and Klein 2011). Honey is used as an ingredient in hundreds of manufactured foods. Honey oligosaccharides present potential prebiotic activity (prebiotic index values between 3.38 and 4.24), increasing the populations of *Bifidobacterium* and *Lactobacillus* (Sanz et al. 2005).

## Carbohydrate Profile of Honey

The carbohydrate profile of honey has been studied by scientists throughout the world. Carbohydrates in honey are represented by monosaccharides such as glucose and fructose followed by disaccharides such as sucrose, maltose, turanose, isomaltose, maltulose, trehalose, nigerose, and kojibiose and trisaccharides such as maltotriose and melezitose. Honey has also been reported to contain numerous oligosaccharides (Meo et al. 2017; Mohan et al. 2017).

Glucose and fructose are present in the greatest percentage in carbohydrate composition of honey. They make up the invert sugar in honey that accounts for about 80%–85% of the honey solids. In some sources, they are referred to as dextrose or grape sugar (which stands for *glucose*) and levulose or fruit sugar (for *fructose*). Honey crystallization processes are greatly dependent on the proportions of those monosaccharides. The greater the percentage of glucose and fructose (invert sugar), the better the quality of honey.

In almost all types of honey, fructose is the carbohydrate in greatest proportion, except in some honeys such as rapeseed honey (*Brassica napus*) and dandelion honey (*Taraxacum officinale*), wherein the fraction of glucose may be higher than the fraction of fructose (Escuredo et al. 2014); consequently, these honeys generally have a rapid crystallization. Fructose

*Honey: Composition and Health Benefits*, First Edition. Edited by Md. Ibrahim Khalil, Gan Siew Hua, and Bey Hing Goh.

is sweeter than sucrose (known as table sugar). This explains why honey is sweeter than sugar. Anyone who has tasted rapeseed honey and dandelion honey knows that they are not as sweet as other honey varieties. This is due to the predominance of glucose over fructose in the composition (Oddo et al. 2004).

Monosaccharides are the structural elements (the building blocks) of complex sugars – di-, tri-, and oligosaccharides. Complex sugars in the body should decompose to simple ones to be assimilated. Honey contains a number of disaccharides such as sucrose, maltose, isomaltose, maltulose, isomaltulose, gentabiose, laminaribiose, trigalose, and turanose. The presence of sucrose in nectar honey reaches up to 1%–6% and in honeydew honey up to 10% (Chua and Adnan 2014). Sucrose consists of one molecule of fructose linked with glucose through α-1,4 binding. It is hydrolyzed by the enzyme invertase, yielding an equimolar mixture of hexoses (Kamal and Klein 2011). The amount of sucrose gradually decreases when stored in normal conditions because of its long enzymatic decomposition. The invertase enzyme remains active even after honey has been extracted from the combs and stored. Regardless of the constant effect of invertase, the level of sucrose in honey can never reach zero. The increased percentage of sucrose is therefore a sign of poor quality and adulteration of honey. Maltose affects the speed of honey crystallization. If the level of maltose reaches 6%–9% in honey (acacia), the honey crystallizes slowly. If the level of maltose is 2%–3% in honey (sunflower, rapeseed, and sainfoin), crystallization occurs faster (Chua and Adnan 2014).

Honey contains some oligosaccharides that have more than two monosaccharides in their molecules. The majority of the oligosaccharides present in honey are trisaccharides such as centose, erlose, maltotriose, isomaltotriose, and kestose. Trisaccharides are hydrolyzed enzymatically to monosaccharides. For example, maltotriose consists of three glucose units (α-1,4 glycosidic bonds), which are hydrolyzed by enzymes to maltose. Maltose is then hydrolyzed by enzymes, but in this case, the enzyme is aglucosidase, resulting in two glucose molecules (Soldatkin et al. 2013). Some studies have suggested that raffinose is a minor sugar in honey, but if so, galactose would also be expected. However, galactose has never been observed by paper chromatography or by gas chromatography of honey sugar hydrolysates (Doner 1977). It should be mentioned that gluconic acid (in equilibrium with its lactone) was found in honey by Stinson et al. in 1960 (Stinson et al. 1960). Honey has also been found to contain tetrasaccharides (e.g., maltotetraose, nystose, stachyose), pentasaccharides (isomaltopentaose), and hexasaccharides.

Different chromatographic techniques such as high-pressure liquid chromatography (HPLC), paper chromatography, thin-layer chromatography, high-pressure anion exchange chromatography, and gas chromatography–mass spectroscopy have been used for sugar analysis (Dumté 2010; Ouchemoukh et al. 2010). These methods are validated by the International Honey Commission (Bogdanov et al. 2004). High-performance anion-exchange chromatography with pulsed amperometric detection is one of the most useful techniques for oligosaccharide determination. Size exclusion chromatography coupled with matrix-assisted laser desorption/ionization time-of-flight mass spectrometry is also useful for the analysis of oligosaccharides. In a study, Molan (1996) reported the presence of 27 oligosaccharides in honey. Using capillary gas chromatography, Low and Sporns (1988) found 16 sugars in honey, including 11 disaccharides (maltose, turanose, kojibiose, sucrose, palatinose, laminaribiose, gentiobiose, cellobiose, isomaltose, neotrehalose, nigerose) and 5 trisaccharides (erlose, isopanose, panose, theanderose, maltotriose). The presence of four tetrasaccharides, one pentasaccharide, and one hexasaccharide was found in a New Zealand honeydew honey (Sanz et al. 2005). The names and formulas of the di-, tri-, and oligosaccharides found in honey are shown in Tables 3.1 to 3.3. Many of these sugars are not found in nectar but are formed during the ripening and storage effects of bee enzymes and the acids of honey. In the process of digestion after honey intake, the principal carbohydrates, fructose and glucose, are quickly transported into the blood and can be used for energy requirements by the human body.

## Floral Honey Carbohydrates

Generally, honey is classified by the floral source of the nectar from which it was made. Monofloral honey is made primarily from the nectar of one type of flower. Different types of monofloral honey have a distinctive flavors, colors, and carbohydrate contents because of differences between their principal nectar sources. Polyfloral honey, also known as wildflower honey, is derived from the nectar of many types of flowers. In the early years of honey research, honey was believed to be a simple mixture of dextrose (glucose), levulose (fructose), and sucrose, with an undefined carbohydrate material called "honey dextrin," believed to be analogous to starch dextrin. Over the years, improvements in analytical and separation procedures have revealed honey to be a highly complex mixture of sugars of which glucose and fructose account for 85% of the honey solids (Doner 1977; White 1992). White (1992) analyzed 490 samples of floral honey from the United States, and the results are summarized in Table 3.4.

Siddiqui and colleagues (Siddiqui and Furgala 1967; Slddiqui and Purgala 1968) analyzed the oligosaccharide content of honey produced by bees foraging on alfalfa and red clover in Canada. The results of their analysis are shown in Table 3.5.

**Table 3.1** Names and formulas of disaccharides found in honey.

| Trivial Name | Systematic Name | References |
|---|---|---|
| **Disaccharides** | **$C_{12}H_{22}O_{11}$** | |
| Cellobiose | β-D-glucopyranosyl-(1→4)-D-glucopyranose | Horvath and Molnár-Perl 1997; Low and Sporns 1988 |
| Gentiobiose | β-D-glucopyranosyl-(1→6)-D-glucopyranose | Horvath and Molnár-Perl 1997; Low and Sporns 1988; Siddiqui and Furgala 1967; Swallow and Low 1990 |
| Inulobiose | β-D-fructofuranosyl-(2→1)-D-fructose | Ruiz-Matute et al. 2007 |
| Isomaltose | α-D-glucopyranosyl-(1→6)-D-glucopyranose | Horvath and Molnár-Perl 1997; Low and Sporns 1988; Siddiqui and Furgala 1967; Swallow and Low 1990; Watanabe and Aso 1960; White and Hoban 1959 |
| Isomaltulose (palatinose) | α-D-glucopyranosyl-(1→6)-D-fructose | Horvath and Molnár-Perl 1997; Low and Sporns 1988; Siddiqui and Furgala 1967; Swallow and Low 1990 |
| Kojibiose | α-D-glucopyranosyl-(1→2)-D-glucopyranose | Low and Sporns 1988; Siddiqui and Furgala 1967; Swallow and Low 1990; Watanabe and Aso 1960 |
| Laminaribiose | β-D-glucopyranosyl-(1→3)-D-glucopyranose | Low and Sporns 1988; Siddiqui and Furgala 1967; Swallow and Low 1990 |
| Leucrose | α-D-glucopyranosyl-(1→5)-D-fructopyranose | Sanz et al. 2004; Watanabe and Aso 1960 |
| Maltose | α-D-glucopyranosyl-(1→4)-D-glucopyranose | Horvath and Molnár-Perl 1997; Low and Sporns 1988; Siddiqui and Furgala 1967; Swallow and Low 1990; Watanabe and Aso 1960; White and Hoban 1959 |
| Maltulose | α-D-glucopyranosyl-(1→4)-D-fructose | Siddiqui and Furgala 1967; Swallow and Low 1990; White and Hoban 1959 |
| Melibiose | α-D-galactopyranosyl-(1→6)-D-glucopyranose | Horvath and Molnár-Perl 1997 |
| Neo-trehalose (α,β-trehalose) | α-D-glucopyranosyl-β-D-glucopyranoside | Low and Sporns 1988; Siddiqui and Furgala 1967; Swallow and Low 1990 |
| Nigerose (sakebiose) | α-D-glucopyranosyl-(1→3)-D-glucopyranose | Low and Sporns 1988; Siddiqui and Furgala 1967; Swallow and Low 1990; Watanabe and Aso 1960; White and Hoban 1959 |
| Sophorose | β-D-glucopyranosyl-(1→2)-D-glucopyranose | De la Fuente et al. 2007 |
| Sucrose | β-D-fructofuranosyl-(2→1)-α-D-glucopyranoside | Horvath and Molnár-Perl 1997; Low and Sporns 1988; Siddiqui and Furgala 1967; Swallow and Low 1990; White and Hoban 1959 |
| Trehalose (α,α-trehalose) | α-D-glucopyranosyl-(1→1)-α-D-glucopyranoside | Horvath and Molnár-Perl 1997 |
| Trehalulose | α-D-glucopyranosyl-(1→1)-α-D-fructofuranose | Ruiz-Matute et al. 2007 |
| Turanose | α-D-glucopyranosyl-(1→3)-α-D-fructofuranose | Horvath and Molnár-Perl 1997; Low and Sporns 1988; Siddiqui and Furgala 1967; Swallow and Low 1990; White and Hoban 1959 |

**Table 3.2** Names and formulas of trisaccharides found in honey.

| Trivial Name | Systematic Name | References |
|---|---|---|
| **Trisaccharides** | **$C_{18}H_{32}O_{16}$** | |
| Centose | α-D-glucopyranosyl-(1→4)-α-D-glucopyranosyl-(1→2)-D-glucopyranose | Slddiqui and Purgala 1968 |
| Erlose | α-D-glucopyranosyl-(1→4)-α-D-glucopyranosyl-β-D-fructofuranoside | Horvath and Molnár-Perl 1997; Low and Sporns 1988; Ruiz-Matute et al. 2010; Slddiqui and Purgala 1968; Swallow and Low 1990 |

**Table 3.2** (Continued)

| Trivial Name | Systematic Name | References |
|---|---|---|
| **Trisaccharides** | $C_{18}H_{32}O_{16}$ | |
| Isomaltotriose | α-D-glucopyranosyl-(1→6)-α-D-glucopyranosyl-(1→6)-D-glucopyranose | Horvath and Molnár-Perl 1997; Ruiz-Matute et al. 2010; Slddiqui and Purgala 1968; Swallow and Low 1990 |
| Isomelezitose | α-D-glucopyranosyl-(1→6)-β-D-fructofuranosyl-(2→1)-α-D-glucopyranose | Rittig 2001 |
| Isopanose | α-D-glucopyranosyl-(1→4)-α-D-glucopyranosyl-(1→6)-D-glucopyranose | Low and Sporns 1988; Siddiqui and Purgala 1968; Swallow and Low 1990 |
| Kestose | α-D-glucopyranosyl-(1→4)-α-D-glucopyranosyl-(1→2)-D-glucopyranose | Moreira and De Maria 2001 |
| 1-Kestose | β-D-fructofuranosyl-(2→1)-β-D-fructofuranosyl-(2↔1)-α-D- D-glucopyranose | Ruiz-Matute et al. 2010; Slddiqui and Purgala 1968; Swallow and Low 1990 |
| 6-Kestose | β-D-fructofuranosyl-(2→6)-β-D-fructofuranosyl-(2↔1)-α-D- D-glucopyranose | Ruiz-Matute et al. 2010 |
| Laminaritriose | β-D-glucopyranosyl-(1→3)-β-D-glucopyranosyl-(1→3)-D-glucopyranose | Swallow and Low 1990 |
| Maltotriose | α-D-glucopyranosyl-(1→4)-α-D-glucopyranosyl-(1→4)-D-glucopyranose | Horvath and Molnár-Perl 1997; Low and Sporns 1988; Ruiz-Matute et al. 2010; Slddiqui and Purgala 1968; Swallow and Low 1990 |
| Melezitose | α-D-glucopyranosyl-(1→3)-β-D-fructofuranosyl-(2→1)-D-glucopyranoside | Horvath and Molnár-Perl 1997; Ruiz-Matute et al. 2010; Slddiqui and Purgala 1968; Swallow and Low 1990 |
| Neokestose | β-D-fructofuranosyl-(2→6)-α-D-glucopyranosyl-(1↔2)-β-D-fructofuranose | Ruiz-Matute et al. 2010 |
| Panose | α-D-glucopyranosyl-(1→6)-α-D-glucopyranosyl-(1→4)-D-glucopyranose | Horvath and Molnár-Perl 1997; Low and Sporns 1988; Ruiz-Matute et al. 2010; Slddiqui and Purgala 1968; Swallow and Low 1990 |
| Planteose | α-D-galactopyranosyl-(1→6)-β-D-fructofuranosyl-(2↔1)-α-D-glucopyranose | Ruiz-Matute et al. 2010 |
| Raffinose | α-D-galactopyranosyl-(1→6)-α-D-glucopyranosyl-(1↔2)-β-D-fructofuranoside | Horvath and Molnár-Perl 1997; Ruiz-Matute et al. 2010 |
| Theanderose | α-D-glucopyranosyl-(1→6)-α-D-glucopyranosyl-(1→2)-β-D-fructofuranoside | Low and Sporns 1988; Ruiz-Matute et al. 2010; Slddiqui and Purgala 1968; Swallow and Low 1990 |
| 3-α-Isomaltosylglucose[a] | α-D-glucopyranosyl-(1→6)-α-D-glucopyranosyl-(1→3)-D-glucopyranose | Slddiqui and Purgala 1968 |
| 4-α-Gentiobiosylglucose[a] | β-D-glucopyranosyl-(1→6)-α-D-glucopyranosyl-(1→4)-D-glucopyranose | Slddiqui and Purgala 1968 |

[a] No trivial name exists.

**Table 3.3** Names and formulas of oligosaccharides found in honey.

| Trivial Name | Systematic Name | References |
|---|---|---|
| **Tetrasaccharides** | $C_{24}H_{42}O_{21}$ | |
| α-4′-Glucosyl erlose | α-D-glucopyranosyl-(1→4)-α-D-glucopyranosyl-(1→4)-α-D-glucopyranosyl-(1↔2)-β-D-fructofuranoside | Astwood et al. 1998 |
| α-6′-Glucosyl erlose | α-D-glucopyranosyl-(1→6)-α-D-glucopyranosyl-(1→4)-α-D-glucopyranosyl-(1↔2)-β-D-fructofuranoside | Astwood et al. 1998 |

*(Continued)*

**Table 3.3** (Continued)

| Trivial Name | Systematic Name | References |
|---|---|---|
| **Tetrasaccharides** | $C_{24}H_{42}O_{21}$ | |
| Isomaltotetraose | α-D-glucopyranosyl-(1→6)-α-D-glucopyranosyl-(1→6)-α-D-glucopyranosyl -(1→6)-D-glucopyranose | Slddiqui and Purgala 1968 |
| Maltotetraose | α-D-glucopyranosyl-(1→4)-α-D-glucopyranosyl-(1→4)-α-D-glucopyranosyl -(1→4)-D- glucopyranose | Astwood et al. 1998 |
| Nystose | β-D-fructofuranosyl-(2→1)-β-D-fructofuranosyl-(2→1)-β-D-fructofuranosyl-(2↔1)-α-D-glucopyranose | Rittig 2001; Ruiz-Matute et al. 2010 |
| Stachyose | α-D-galactopyranosyl-(1→6)-α-D-galactopyranosyl-(1→6)-α-D-glucopyranosyl-(1↔2)-β-D-fructofuranoside | Rittig 2001 |
| **Pentasaccharides** | $C_{30}H_{52}O_{26}$ | |
| Isomaltopentaose | α-D-glucopyranosyl-(1→6)-α-D-glucopyranosyl-(1→6)-α-D-glucopyranosyl-(1→6)-α-D-glucopyranosyl-(1→6)-D- glucopyranose | Slddiqui and Purgala 1968 |
| [a] | α-D-glucopyranosyl-(1→4)-α-D-glucopyranosyl-(1→4)-α-D-glucopyranosyl-(1→4)-α-D-glucopyranosyl-(1↔2)-β-D-fructofuranoside | Astwood et al. 1998 |
| [a] | α-D-glucopyranosyl-(1→6)-α-D-glucopyranosyl-(1→4)-α-D-glucopyranosyl-(1→4)-α-D-glucopyranosyl-(1↔2)-β-D-fructofuranoside | Astwood et al. 1998 |
| **Hexasaccharides** | $C_{36}H_{62}O_{31}$ | |
| [a] | α-D-glucopyranosyl-(1→6)-α-D-glucopyranosyl-(1→4)-α-D-glucopyranosyl-(1→4)-α-D-glucopyranosyl-(1→4)-α-D-glucopyranosyl-(1↔2)-β-D-fructofuranoside | Astwood et al. 1998 |

[a] No trivial name exists.

**Table 3.4** Average carbohydrate composition of 490 samples of U.S. floral honey.

| Component | Average (%) | Standard Deviation | Range (%) |
|---|---|---|---|
| Fructose | 38.4 | 1.8 | 30.9–44.3 |
| Glucose | 30.3 | 3.0 | 22.9–40.7 |
| Sucrose | 1.3 | 0.9 | 0.2–7.6 |
| Reducing disaccharides | 7.3 | 2.1 | 2.7–16.0 |
| Higher sugars | 1.4 | 1.1 | 0.1–3.8 |

White J.W., 1992 / Dadant & Sons, Inc.

**Table 3.5** Yields of the principal sugars in the oligosaccharide fraction (3.65%) of floral honey.

| Disaccharides | % | Trisaccharides | % | Higher Oligosaccharides | % |
|---|---|---|---|---|---|
| Maltose | 29.4 | Erlose | 4.5 | Isomaltotetraose | 0.33 |
| Kojibiose | 8.2 | Theanderose | 2.7 | Isomaltopentaose | 0.16 |
| Turanose | 4.7 | Panose | 2.5 | | |
| Isomaltose | 4.4 | Maltotriose | 1.9 | | |
| Sucrose | 3.9 | 1-Kestose | 0.9 | | |
| Maltulose and isomaltulose | 3.1 | Isomaltotriose | 0.6 | | |
| Nigerose | 1.7 | Melezitose | 0.3 | | |
| α,β-Trehalose | 1.1 | Isopanose | 0.24 | | |

**Table 3.5** (Continued)

| Disaccharides | % | Trisaccharides | % | Higher Oligosaccharides | % |
|---|---|---|---|---|---|
| Gentiobiose | 0.4 | Centose | 0.05 | | |
| Laminaribiose | 0.09 | 3-α-Isomaltosylglucose | Trace | | |
| Total[a] | 56.99 | 13.69 | 0.49 | | |

[a] Quantitative recoveries were calculated after allowing for loss of material during separation; they are therefore approximate.
Data from Siddiqui and Furgala 1967; Slddiqui and Purgala 1968.

A similar analysis of the oligosaccharide content (~3%) of an Alsike honey from Canada was performed (Low and Sporns 1988). More recently, Ruiz-Matute et al. (2007) analyzed 35 honey samples purchased in Spain but from various origins and identified a new disaccharide from honey, inulobiose, ranging in concentration from 0.93 to 6.14 mg/g. This new disaccharide, inulobiose, could be formed by transfructosylation.

## Honeydew Honey Carbohydrates

Honeydew passes through the digestive system of insects and in the process is altered, and honeydew honey (also called forest honey) is subsequently produced by the bees (Siddiqui 1970). On average, compared with floral honey, honeydew honey is lower in glucose by 5.2% and lower in fructose by 6.4% but higher in reducing disaccharides and higher sugars (Doner 1977). An average composition of honeydew honey was determined by White et al. (1962) and is given in Table 3.6.

White et al. (1992) pointed out that these averages compare well with those obtained for 38 Swiss honeydew honey types by Bogdanov and Baumann (1988). White et al. (1992) also stated that there are at least two types of honeydew honey, containing erlose or melezitose or mixtures of both, depending on the insect(s) involved. The melezitose type can granulate rapidly (frequently in the comb itself), and the erlose type does not granulate (Siddiqui 1970). New Zealand honeydew honey was analyzed by Astwood et al. (1998), and their results are summarized in Table 3.7. This study noted that compared with American honeydew honey, New Zealand honeydew honey showed smaller amounts of sucrose and significantly smaller amounts of maltose. Conversely, the total percentage of higher sugars (without maltose and sucrose) was greater in the New Zealand honeydew honey. An argument for the differences included a difference in the enzymatic activity of the scale insect or the bee or to seasonal or atmospheric changes. A distinctive feature of honeydew honey compared with floral honey is its optical rotation. Honeydew honey is dextrorotatory, while floral honey is invariably levorotatory (Doner 1977).

## Carbohydrate Profile of Asian Honey

The carbohydrate profiles of honey were investigated in many Asian countries such as China, India, Vietnam, Japan, South Korea, Indonesia, and Malaysia. An average carbohydrate composition for honey originating from Asia is presented in Table 3.8. This average composition agrees with the limits set for honey by the Codex Alimentarius

**Table 3.6** Average carbohydrate composition of 14 honeydew honey samples.

| Component | Average | Standard Deviation | Range |
|---|---|---|---|
| Fructose (%) | 31.8 | 4.2 | 23.9–38.1 |
| Glucose (%) | 26.0 | 3.0 | 19.2–31.9 |
| Sucrose (%) | 0.8 | 0.2 | 0.4–1.1 |
| Maltose (%) | 8.8 | 2.5 | 5.1–12.5 |
| Melezitose (%) | 2.3 | 4.6 | 0.0–13.4 |
| Higher sugars (%) | 4.7 | 1.0 | 1.3–11.5 |

Adapted from White et al, 1962.

**Table 3.7** Components of the oligosaccharide fraction of New Zealand honeydew honey.

| Component | Mean % of Honey Solids[a] | Standard Deviation | Range |
|---|---|---|---|
| Sucrose | 0.55 | 0.12 | 0.40–0.77 |
| Trehalose | 0.044 | 0.0036 | 0.042–0.05 |
| Cellobiose | 0.33 | 0.046 | 0.25–0.37 |
| Turanose | 1.5 | 0.26 | 1.2–1.8 |
| Nigerose | 1.1 | 0.13 | 0.94–1.3 |
| Maltose | 2.2 | 0.20 | 1.9–2.6 |
| Gentiobiose | 0.85 | 0.26 | 0.45–1.3 |
| Palatinose | 1.3 | 0.34 | 0.66–1.8 |
| Isomaltose | 0.32 | 0.072 | 0.24–0.42 |
| Erlose | 1.2 | 0.33 | 0.87–1.8 |
| Melezitose | 0.085 | 0.0080 | 0.069–0.09 |
| Maltotriose | 0.54 | 0.19 | 0.32–0.84 |
| Panose | 0.51 | 0.16 | 0.27–0.73 |
| Maltotetraose | 0.51 | 0.23 | 0.18–0.85 |
| α-D-glucopyranosyl-(1→4)-α-D-glucopyranosyl-(1→4)-α-D-glucopyranosyl-(1↔2)-β-D-fructofuranoside | 3.9 | 1.2 | 1.8–5.5 |
| α-D-glucopyranosyl-(1→6)-α-D-glucopyranosyl-(1→4)-α-D-glucopyranosyl-(1↔2)-β-D-fructofuranoside | 0.46 | 0.13 | 0.21–0.63 |
| α-D-glucopyranosyl-(1→4)-α-D- glucopyranosyl-(1→4)-α-D- glucopyranosyl-(1→4)-α-D-glucopyranosyl-(1↔2)-β-D- fructofuranoside | 1.1 | 0.89 | 0.11–2.9 |
| α-D-glucopyranosyl-(1→6)-α-D- glucopyranosyl-(1→4)-α-D, glucopyranosyl-(1→4)-α-D- glucopyranosyl-(1↔2)-β-D- fructofuranoside | 0.33 | 0.15 | 0.08–0.60 |
| α-D-glucopyranosyl-(1→6)-α-D-glucopyranosyl-(1→4)-α-D, glucopyranosyl-(1→4)-α-D-glucopyranosyl-(1→4)-α-D, glucopyranosyl-(1↔2)-β-D-fructofuranoside | 0.28 | 0.13 | 0.07–0.53 |
| Higher sugars (except maltose and sucrose) | 14.31 | 3.5 | 8.71–20.57 |

[a] Mean of duplicate gas chromatography or triplicate liquid chromatography samples of six different honeydew honey samples.

Astwood et al., 1998 / American Chemical Society.

**Table 3.8** Average Asian honey carbohydrate composition.

| Sugar | China (n = 8) | India (n = 7) | Vietnam (n = 3) | Japan (n = 3) | Malaysia (n = 2) | South Korea (n = 1) | Indonesia (n = 1) | Mean | Standard Deviation | Range |
|---|---|---|---|---|---|---|---|---|---|---|
| | % (w/w) | | | | | | | | | |
| Monosaccharides | 65.56 | 69.35 | 62.11 | 73.30 | 66.92 | 54.91 | 68.30 | 65.78 | 5.90 | 54.91–73.30 |
| Sucrose | 0.70 | 0.09 | 0.10 | 0.18 | 0.06 | 0.03 | 0.27 | 0.20 | 0.23 | 0.03–0.27 |
| Trehalose | 0.10 | 0.12 | 0.06 | 0.09 | 0.10 | 0.04 | ND | 0.08 | 0.03 | 0.04–0.12 |
| Cellobiose | 0.33 | 0.48 | 0.44 | 0.38 | 0.47 | 0.47 | 0.49 | 0.44 | 0.06 | 0.33–0.49 |
| Laminaribiose | 0.32 | 0.27 | 0.19 | 0.34 | 0.60 | 0.11 | 0.17 | 0.29 | 0.16 | 0.11–0.60 |
| Nigerose + turanose | 1.03 | 1.41 | 1.24 | 1.05 | 0.52 | 1.69 | 0.83 | 1.11 | 0.38 | 0.52–1.41 |
| Turanose + maltulose | 0.60 | 0.77 | 0.52 | 0.58 | 0.28 | 1.07 | 0.14 | 0.57 | 0.31 | 0.14–1.07 |
| Maltulose + maltose | 1.31 | 1.55 | 1.01 | 1.57 | 1.31 | 1.83 | 1.04 | 1.37 | 0.30 | 1.01–1.83 |
| Kojibiose | 0.28 | 0.48 | 0.40 | 0.20 | 0.21 | 0.63 | 0.09 | 0.33 | 0.19 | 0.09–0.63 |

**Table 3.8** (Continued)

| Sugar | China (n = 8) | India (n = 7) | Vietnam (n = 3) | Japan (n = 3) | Malaysia (n = 2) | South Korea (n = 1) | Indonesia (n = 1) | Mean | Standard Deviation | Range |
|---|---|---|---|---|---|---|---|---|---|---|
| | % (w/w) | | | | | | | | | |
| Melibiose | 0.35 | 0.70 | 0.45 | 0.18 | 0.33 | 1.12 | 0.20 | 0.47 | 0.33 | 0.18–1.12 |
| Gentiobiose | 0.17 | 0.27 | 0.22 | 0.07 | 0.28 | 0.44 | 0.13 | 0.23 | 0.12 | 0.07–0.44 |
| Palatinose | 0.48 | 1.12 | 0.68 | 0.29 | 0.51 | 1.06 | 0.25 | 0.63 | 0.35 | 0.25–1.06 |
| Isomaltose | 0.37 | ND | 0.15 | 0.56 | 0.77 | ND | ND | 0.47 | 0.27 | 0.15–0.77 |
| Raffinose | 0.03 | 0.02 | 0.02 | ND | ND | ND | ND | 0.02 | 0.01 | 0.02–0.03 |
| 1-Kestose | 0.12 | 0.10 | 0.04 | 0.07 | 0.02 | 0.18 | 0.01 | 0.08 | 0.06 | 0.01–0.18 |
| Erlose | 0.44 | 0.12 | 0.02 | 0.17 | 0.02 | 0.20 | 0.01 | 0.14 | 0.15 | 0.01–0.44 |
| Melezitose | 0.04 | 0.03 | 0.01 | 0.01 | 0.02 | 0.06 | ND | 0.03 | 0.02 | 0.01–0.06 |
| Maltotriose | 0.10 | 0.13 | 0.08 | 0.07 | 0.12 | 0.15 | 0.04 | 0.10 | 0.04 | 0.04–0.15 |
| Panose | 0.07 | 0.17 | 0.12 | 0.11 | 0.16 | 0.28 | 0.01 | 0.13 | 0.09 | 0.01–0.28 |
| Isomaltotriose | 0.01 | 0.03 | 0.01 | 0.01 | 0.03 | 0.07 | ND | 0.03 | 0.02 | 0.01–0.07 |

ND = not detected.

Dumté, 2010 / University of Waikato.

Commission (CAC) standard for honey (1981). The CAC standard (1981) for honey states that honey must not contain less than 60 g/100 g of fructose and glucose (honeydew honey or blends with nectar honey must not be less than 45 g/100g). The sucrose content must not be more than 5 g/100 g of honey; for alfalfa (*Medicago sativa*), *Citrus* spp., false acacia (*Robinia pseudoacacia*), French honey uckle (*Hedysarum*), Menzies banksia (*Banksia menziesii*), red gum (*Eucalyptus camaldulensis*), leatherwood (*Eucryphia lucida*), and *Eucryphia milligani*, the sucrose content must not be more than 10 g/100 g of honey; and for lavender (*Lavandula spp*) and borage (*Borago officinalis*) honey, it must not be more than 15 g/100 g.

The Asian honey showed a carbohydrate composition within the limits set for honey by the CAC and had a disaccharide profile similar to honey from elsewhere in the world. Kojibiose, turanose–nigerose, and turanose–maltulose were present in all the Asian honey analyzed and could be potential markers for the presence of Asian honey in products because these disaccharides are not common in foods, and therefore their presence indicates that the product contains honey.

## Honey Carbohydrates and the Specific Rotation ($[\alpha]_D^{20}$) of Linear Polarized Light

Honey carbohydrates have the ability to rotate linear polarized light. The specific rotation power depends on the amount and quality of the sugars present in honey, including oligosaccharides. The direction and degree of rotation are specific for each carbohydrate. It was shown by Battaglini and Bosi (1972) that the nature of specific rotation of honey linked to the fructose to glucose ratio and the percentage of di- and trisaccharides. Honey containing high levels of fructose and glucose along with low di- and trisaccharide levels is levorotatory. Conversely, hone with low fructose and glucose levels with large quantities of di- and higher saccharides (as in honeydew honey) is dextrorotatory (Doner 1977). D-Glucose is dextrorotatory, whereas D-fructose is levorotatory, and therefore the specific rotation of honey depends on their relative proportions. Therefore, the overall optical rotation of honey depends on the contents of different carbohydrates present in it. Fructose, which is the major sugar in nectar honey, has a high negative optical rotation ($[\alpha]_D^{20} = -92.3$); hence, all nectar honey types have the negative specific rotation. In contrast, honeydew honey types have positive specific rotations due to the lower content of fructose and higher contents of glucose, disaccharides, and oligosaccharides that have positive specific rotations (e.g., glucose, +52.7 degrees; maltose, +137.0 degrees; melezitose, +88.2 degrees; erlose, +121.8 degrees; raffinose, +104.1 degrees). The differences in honey's specific rotation resulting from different carbohydrate profiles are primarily used for differentiation between nectar and honeydew honey but can also contribute to honey's characterization (Bogdanov et al. 1999; Oddo et al. 1995). In addition, specific rotation can be a useful parameter for unifloral honey differentiation even though a notable overlapping occurs with different honey types (Bogdanov 2009; Bogdanov et al. 2004). Recently, Primorac et al. (2011) studied the specific rotation property of three types of honey samples, namely, chestnut honey, black locust honey, and sage honey from Croatia, and the authors found that all three types of honey samples showed negative values of specific rotations. The highest negative value was revealed by chestnut honey, with a mean value of −21 degrees, which is in compliance with the data obtained from Italian and European chestnut honey (Oddo et al. 1995, 2004). According to Primorac et al. (2011), sage honey showed a specific rotation from −21 to −11 degrees, which were in agreement with those reported by Kenjerić et al. (2006). Šarić et al. (2008) found considerably lower negative values for specific rotation of chestnut honey as well as for sage honey. Black locust honey, on the other hand, showed the lowest negative specific rotations from −16 to −9 degrees with a mean of −13 degrees (Primorac et al. 2011). Krpan et al. (2009) showed a higher negative value of specific rotation of acacia honey (mean value, −13.93 degrees ), whereas Šarić et al. (2008) reported significantly lower values for specific rotation with a mean of −2.7 degrees. All honeydew honey harvested from different geographical locations showed positive values for specific rotation. Honeydew honey collected from Italy showed higher positive values of specific rotation (17.0 and 14.0 degrees, respectively, for metcalfa and abies honeydew honey) (Oddo et al. 1995). Honeydew honey from Czech showed a moderate value (10.5 degrees), whereas this honey from Croatia showed the lowest positive value (2.4 degrees) of specific rotation (Přidal and Vorlova 2002; Šarić et al. 2008). The specific rotations of different honey are presented in Table 3.9.

**Table 3.9** Specific rotation $[\alpha]_D^{20}$ of different honey types.[a]

| Honey type | Origin | Specific Rotation (degrees) | Reference |
|---|---|---|---|
| Acacia (n = 45) | Croatia | −3.6 to (−1.9)/−2.7 ± 0.41 | Šarić et al. 2008 |
| Acacia (n = 30) | Croatia | −17.5 to (−8.0)/−13.93 ± 2.21 | Krpan et al. 2009 |
| Arbutus (n = 50) | Italy | −16.0 to (−8.2)/−13.0 ± 1.8 | Oddo et al. 1995 |
| Black locust (*n* = 17) | Croatia | −16 to (−9)/−13 ± 2 | Primorac et al. 2011 |
| Chestnut (*n* = 17) | Croatia | −25 to (−16)/−21 ± 3 | Primorac et al. 2011 |
| Chestnut (*n* = 495) | Europe | −16.7 ± 3.4[b] | Oddo et al. 2004 |
| Chestnut (*n* = 180) | Italy | −24.9 to (−10.0)/−17 ± 3.5 | Oddo et al. 1995 |
| Chestnut (*n* = 7) | Croatia | −3.3 to (−2.7)/−3.0 ± 0.19 | Šarić et al. 2008 |
| Citrus (n = 6) | Croatia | −3.0 to (−1.5)/−2.2 ± 0.69 | Šarić et al. 2008 |
| Citrus (n = 105) | Italy | −17.7 to (−9.3)/−14.0 ± 2.0 | Oddo et al. 1995 |
| Erica (n = 31) | Italy | −17.1 to (−11.8)/−14.5 ± 1.5 | Oddo et al. 1995 |
| Eucalyptus (n = 86) | Italy | −19.5 to (10.7)/−14.0 ± 2.1 | Oddo et al. 1995 |
| Floral (n = 7) | Croatia | −2.9 to (−1.8)/−2.5 ± 0.43 | Šarić et al. 2008 |
| Hedysarum (n = 65) | Italy | −15.2 to (−5.4)/−11.0 ± 2.5 | Oddo et al. 1995 |
| Helianthus (n = 58) | Italy | −19.8 to (−15.4)/−18.0 ± 1.2 | Oddo et al. 1995 |
| Honeydew (n = 5) | Croatia | 0.9 to (3.9)/2.4 ± 1.32 | Šarić et al. 2008 |
| Honeydew – abies (n = 52) | Italy | 6.0 to (29.7)/14.0 ± 5.0 | Oddo et al. 1995 |
| Honeydew – metcalfa (n = 78) | Italy | 2.5 to (30.0)/17.0 ± 7.4 | Oddo et al. 1995 |
| Honeydew (n = 6) | Czech | −1.0 to (20.4)/10.5 ± 7.8 | Přidal and Vorlova 2002 |
| Meadow (n = 17) | Croatia | −3.3 to (−0.9)/−2.4 ± 0.59 | Šarić et al. 2008 |
| Rhododendron (n = 42) | Italy | −10.7 to (−2.3)/−6.0 ± 2.4 | Oddo et al. 1995 |
| Robinia (n = 176) | Italy | −23.4 to (−10.9)/−17.0 ± 2.7 | Oddo et al. 1995 |
| Robinia (n = 5) | Czech | −16.9 to (−13.9)/−15.6 ± 1.1 | Přidal and Vorlova 2002 |
| Sage (n = 7) | Croatia | −3.1 to (−0.6)/−2.2 ± 1.01 | Šarić et al. 2008 |
| Sage (n = 41) | Croatia | −21 to (−11)/−15 ± 3 | Primorac et al. 2011 |
| Sage (n = 23) | Croatia | −20.6 to (−11.6)/−14.9 ± 2.9 | Kenjerić et al. 2006 |
| Taraxacum (n = 23) | Italy | −14.8 to (−5.5)/−10.0 ± 2.4 | Oddo et al. 1995 |
| Thymus (n = 54) | Italy | −24.5 to (−17.0)/−20.0 ± 2.4 | Oddo et al. 1995 |
| Tilia (n = 40) | Italy | −18.0 to (−8.0)/−12.5 ± 2.1 | Oddo et al. 1995 |

[a] Data are presented as (min to max/mean ± standard deviation [SD]).
[b] Only mean and SD values are available.

## Effect of Storage on the Carbohydrate Composition of Honey

Sugars and other components of honey may change during storage. In a recent study, the carbohydrate compositions of both stabilized (at a temperature of 100°C for 15 minutes – enzyme inactivated) and nonstabilized honey stored for 24 weeks were analyzed (Rybak-Chmielewska 2007). Sucrose concentrations in honey that were not stabilized decreased 14% in honey stored at 4°C and 79% in honey stored at room temperature (20°C). For other sugars, such as trehalose and isomaltose, percentages showed no significant changes during prolonged storage. The fructose content increased 4%, and glucose content increased 1.1% compared with their initial value at 4°C; however, at a temperature of 20°C, the fructose content increased 7%, and the glucose content increased 8.8%. In stabilized honey, the percentages of sugars varied 0.1% higher in sucrose, trehalose, and melezitose + erlose or 0.1% lower in turanose. Comparing the honey at different temperatures of 4° and 20°C, the variation in levels of individual sugars was 0.2%. Over time, in addition to the chemical changes, there are physical changes in honey, such as a darker color and change in flavor (Da Silva et al. 2016).

A rapid decline of sucrose content was observed during the storage of acacia honey. Gontarski (1960) drew attention to adulterated honey, which, despite its high initial sucrose content, showed enzymatic activity high enough to break down that sugar down to a 5% level (i.e. to a level that complied with the standard for that parameter). The breakdown of sucrose in stored honey occurs rapidly when conditions favored enzymatic activity. In a mixture of honey with winter stores processed by bees from sucrose (at a ratio of 1:1), it was found that sucrose content conform to the honey standard's requirements (did not exceed 5%) as early as after 7 days of incubation at 36°C (Rybak-Chmielewska 2007). Similarly, in the study of Rybak and Achremowicz (1986), sucrose content declined from a dozen or so percent in the fresh material to a few percentage points after two months of storage at 20°C. Carbohydrate contents remain unchanged for half a year by storing the honey samples at 4°C ± 2°C. While storing for half a year at room temperature, the sucrose contents of honey samples greatly changed, which was recorded to drop by as much as 79% compared with its initial value (Rybak-Chmielewska 2007).

When honey is heated or stored for a long time, pentoses and hexoses decompose in a slow enolization and a fast β-elimination of three molecules of water to form undesirable compounds such as furans (Chernetsova and Morlock 2012). The main furans formed as furfural, which is derived from pentoses, and 5-hydroxymethylfurfural (5-HMF), derived from hexoses such as glucose and fructose (Moreira et al. 2010). These are the main degradation products of sugars, and their occurrence in foods is usually related to non-enzymatic browning reactions (i.e. Maillard reaction, sugar degradation in an acidic medium, and caramelization). Actually, these furans have been used as markers for the heat treatment of food (Moreira et al. 2010). The furfuryl alcohol is also an indicator of thermal treatment and storage conditions. Thus, these compounds are not considered to be good markers of floral honey, although they may indicate a possible loss of freshness caused by exposure to high temperatures or prolonged storage (Barra et al. 2010; Castro-Vázquez et al. 2007).

In addition to the previously mentioned compounds, other products of sugar degradation, such as 2-acetylfuran (Wang et al. 2009), isomaltol (Ota et al. 2006), 3,5-dihydroxy-2-methyl-5,6-diidropiran-4-one, and maltol (Jelen 2011), are formed when submitted to heat in the presence of amino acids, contributing to the change in color, taste, and odor of honey.

## Effect of Gastrointestinal Digestion on Carbohydrates

Different types of carbohydrate compounds can be found in honey, and all of them can contribute to the physicochemical, nutritional, or biological properties of honey, and the digestive system can also influence them (Seraglio et al. 2021). A recent study showed that there was no difference in the sugar profile observed by proton nuclear magnetic resonance analysis in manuka honey after static *in vitro* digestion composed by gastric and duodenal steps (Mannina et al. 2016). Similar results were reported by Parkar et al. (2017) in the investigation of sugars by high-performance anion-exchange chromatography also in manuka honey submitted to static *in vitro* digestion consisting of gastric and duodenal steps followed by dialysis and colonic fermentation. The content of fructose and glucose, the main sugars of honey, remained stable after gastric and duodenal digestion but decreased in the dialysis retentate and after fecal fermentation, which was expected because these sugars are the substrates for fermentation. The dialysis data indicated a passive diffusion of sugars throughout the intestine and potential bioavailability. However, these results are not well aligned with the *in vivo* condition, in which the absorption of the sugars through the intestinal wall is by active diffusion of sugars and the dialysis membrane was chosen to prevent the diffusion of α-cyclodextrin used in other experiment performed in the study. In this sense, the results concerning a possible bioavailability should be interpreted with caution. Besides, in both studies, the small signal of lactose was associated with its presence in the digestive enzyme mixtures and not as a product from the honey digestion. It is also important to note that the addition of α-cyclodextrin to honey partially changed the behavior of honey sugars during the digestive process probably by complexation, which seems to promote protection to honey compounds (Parkar et al. 2017). The data presented suggest that the ingestion of other kinds of foods mixed with honey probably affects the *in vitro* bioaccessibility of sugars and other compounds of honey and other foods. Indeed, Helal et al. (2014) proposed that the addition of honey to cinnamon beverages seems to increase the *in vitro* bioaccessibility of the beverages' polyphenols, and they attributed this action to the sugars of honey that can decrease the interaction between tannins and pepsin or increase their complexes solubility.

## Fructose/Glucose Ratio and Glucose/Water Ratios of Honey

The concentration of fructose and glucose as well as their ratio (F/G ratio) and glucose/water (G/W) ratio are useful indicators of honey's quality (Buba et al. 2013; El Sohaimy et al. 2015; Oddo et al. 2004). The F/G ratio indicates the ability of honey to crystallize because glucose is less soluble in water than fructose (Amir et al. 2010). Honey crystallization is faster when the F/G ratio is below 1.0, and it slows when this ratio is more than 1.0 (Draiaia et al. 2015). In a recent investigation, the F/G ratios for Kashmiri, Yemeni, Egyptian, and Saudi honey samples were found to be $0.42 \pm 0.02$, $1.52 \pm 0.04$, $1.63 \pm 0.05$, and $2.35 \pm 0.02$, respectively. Accordingly, Kashmiri honey was crystallized faster than other types of honey, and Saudi honey was the lowest. In nearly all honey types, fructose predominates, but a few types of honey appeared to contain more glucose than fructose. Honey, which contains less glucose than fructose, has the ability to be a fluid (Ouchemoukh et al. 2007). Furthermore, honey crystallization depending on other factors, such as other sugar contains (e.g. sucrose, maltose), insoluble substance (e.g. dextrin, colloids, pollen), and storage temperature, that can influence the crystallization process (Buba et al. 2013; EL-Metwally 2015). The G/W ratio is considered an appropriate indicator rather than the F/G ratio for the prediction of honey crystallization. The least ability of honey crystallization is obtained when the G/W ratio is less than 1.0; it is faster or completely crystallizes when that ratio is more than 2.0 (Amir et al. 2010; Manikis and Thrasivoulou 2001). The G/W ratios were reported $0.72 \pm 0.025$, $1.38 \pm 0.025$, $1.45 \pm 0.025$, and $1.56 \pm 0.025$ for Kashmiri, Saudi, Egyptian, and Yemeni honey, respectively. These results indicated that Kashmiri honey has the lowest ability to crystallize, but the rest of the honey types were moderate. Thus, moisture levels in honey play a crucial role in honey crystallization. According to Buba et al. (2013), the F/G and G/W ratios could be used to predict and control granulation tendencies in honey.

## Carbohydrates: The Parameter of Identity and Quality Determination of Honey

The authenticity of a type of honey is defined internationally by the CAC, which establishes the identity and the essential quality requirements of honey intended for direct human consumption. These standards are applied to honey produced by bees and cover all styles of honey presentations that are processed and ultimately intended for human consumption (Codex Alimentarius Commission 1981). The purpose of these laws is to establish the identity and minimum quality requirements for honey. Among other factors, these regulations take into account the sensory and physicochemical properties of honey by setting the color and the minimum or maximum amounts related to maturity, purity, and deterioration parameters for honey. The sugar content of honey is suggested as one of the parameters of honey maturity by the CAC on sugars (Codex Alimentarius Commission 1981).

As discussed previously, honey is rich in sugars, and the monosaccharides, especially fructose and glucose are the predominant sugars present in it. The average F/G ratio is 1.2:1, but this ratio depends largely on the source of the nectar from which the honey was extracted. This ratio is used to evaluate the crystallization of the honey because of glucose's lower solubility in water compared with fructose (De la Fuente et al. 2011; Escuredo et al. 2014; Tornuk et al. 2013).

According to the standards of the CAC on sugars (Codex Alimentarius Commission 1981), the minimum amount of reducing sugars present is 60 g 100 $g^{-1}$ for floral honey. In general, the sugar composition of honey is affected by the types of flowers used by the bees, as well as regions and climate conditions (Tornuk et al. 2013). Besides the reducing sugar analysis, the amount of sucrose is a very important parameter in evaluating the maturity of honey. The sucrose content in honey is analyzed with the purpose of identifying any improper manipulation of honey, and high levels may indicate a variety of adulterations, such as adding cheap sweeteners such as cane sugar or refined beet sugar; early harvest, indicating that the sucrose was not completely transformed into glucose and fructose; or prolonged artificial feeding of honeybees with sucrose syrups, resulting in high commercial profits (Escuredo et al. 2013; Puscas et al. 2013; Tornuk et al. 2013). Because of these factors, the CAC on sugars stipulates a maximum value of 5 g of total sugar in 100 g of floral honey.

Camiña et al. (2012), in a study focused on the geographical and botanical classification of honey around the world, also provided a great overview of different analytical techniques combined with multivariate analysis to verify authenticity of honey through sugar evaluation. More recently, the adulteration of honey was monitored by $^{13}C/^{12}C$ analysis using an isotope ratio mass spectrometer in combination with an elemental analyzer (Tosun 2013). This technique was useful in detecting the adulteration of honey by the addition of corn-sugar cane syrups but fails to detect the adulteration of beet sugar syrups. To rapidly and efficiently detect the presence of adulterants, such as sugar syrups in honey, Chen et al. (2014) applied three-dimensional fluorescence spectra technology, aiming to replace the previous techniques. It was considered time consuming and expensive and required a high degree of technical knowledge for data interpretation. Multivariate analysis combined with this technique proved to be a potential tool to detect adulterated honey in rice syrup. HPLC can also be used as a fast technique to detect adulteration in honey as presented by Wang et al. (2015), without the requirement of multivariate analysis. HLPC chromatograms indicated a separate peak at 15.25 minutes of retention time indicative of the presence of syrup in honey samples, with detectable syrup content near 2.5%.

Besides sugar content determination, water content is also an effective parameter to evaluate the quality of honey. The authentication of honey from different countries was evaluated by multivariate analysis in commercial, adulterated, and artificial honey. The sum or ratio of sugars (glucose and fructose) and the G/W ratio were found to be more specific and better indicators of honey's quality than any other parameter evaluated (Kukurova et al. 2008).

## References

Amir, Y., Yesli, A., Bengana, M., et al. (2010). Physico-chemical and microbiological assessment of honey from Algeria. *Electronic Journal of Environmental, Agricultural and Food Chemistry* 9: 1485–1494.

Astwood, K., Lee, B., and Manley-Harris, M. (1998). Oligosaccharides in New Zealand honeydew honey. *Journal of Agricultural and Food Chemistry* 46: 4958–4962.

Barra, M.P.G., Ponce-Díaz, M.C., and Venegas-Gallegos, C. (2010). Volatile compounds in honey produced in the central valley of Ñuble province, Chile. *Chilean Journal of Agricultural Research* 70: 75–84.

Battaglini, M. and Bosi, G. (1972). Determination of Glucides by GLC and its possibilities for honey qualification. International Apicultural Congress, 23rd, Moscow, 1971 (Proceedings).

Bogdanov, S. (2009). *Harmonized methods of the International Honey Commission.* International Honey Commission. 1–61.

Bogdanov, S. and Baumann, E. (1988). Bestimmung von Honigzucker mit HPLC. *Mitteilungen aus dem Gebiete der Lebensmittel-untersuchung und Hygiene* 79: 198–206.

Bogdanov, S., Lüllmann, C., and Martin, P. (1999). Honey quality, methods of analysis and international regulatory standards: review of the work of the International Honey Commission. *Mitteilungen aus Lebensmitteluntersuchung und Hygiene* 90: 108–125.

Bogdanov, S., Ruoff, K., and Oddo, L.P. (2004). Physico-chemical methods for the characterisation of unifloral honeys: a review. *Apidologie* 35: S4–S17.

Buba, F., Gidado, A., and Shugaba, A. (2013). Analysis of biochemical composition of honey samples from North-East Nigeria. *Analytical Biochemistry* 2: 139.

Camiña, J.M., Pellerano, R.G., and Marchevsky, E.J. (2012). Geographical and botanical classification of honeys and apicultural products by chemometric methods. A review. *Current Analytical Chemistry* 8: 408–425.

Castro-Vázquez, L., Díaz-Maroto, M., and Pérez-Coello, M. (2007). Aroma composition and new chemical markers of Spanish citrus honeys. *Food Chemistry* 103: 601–606.

Chen, Q., Qi, S., Li, H., et al. (2014). Determination of rice syrup adulterant concentration in honey using three-dimensional fluorescence spectra and multivariate calibrations. *Spectrochimica Acta Part A: Molecular and Biomolecular Spectroscopy* 131: 177–182.

Chernetsova, E.S. and Morlock, G.E. (2012). Assessing the capabilities of direct analysis in real time mass spectrometry for 5-hydroxymethylfurfural quantitation in honey. *International Journal of Mass Spectrometry* 314: 22–32.

Chua, L.S. and Adnan, N.A. (2014). Biochemical and nutritional components of selected honey samples *Acta Scientiarum Polonorum, Technologia Alimentaria* 13: 169–179.

Codex Alimentarius Commission. (1981). Revised codex standard for honey codex stan 12-1981, Rev. 1 (1987), Rev. 2 (2001). Codex Standard. 12: 1–7.

Da Silva, P.M., Gauche, C., Gonzaga, L.V., et al. (2016). Honey: chemical composition, stability and authenticity. *Food Chemistry* 196: 309–323.

De la Fuente, E., Ruiz-Matute, A., Valencia-Barrera, R., et al. (2011). Carbohydrate composition of Spanish unifloral honeys. *Food Chemistry* 129: 1483–1489.

De la Fuente, E., Sanz, M., Martínez-Castro, I., and Sanz, J. (2006). Development of a robust method for the quantitative determination of disaccharides in honey by gas chromatography. *Journal of Chromatography A* 1135: 212–218.

De la Fuente, E., Sanz, M., Martínez-Castro, I., et al. (2007). Volatile and carbohydrate composition of rare unifloral honeys from Spain. *Food Chemistry* 105: 84–93.

Doner, L.W. (1977). The sugars of honey—a review. *Journal of the Science of Food and Agriculture* 28: 443–456.

Draiaia, R., Dainese, N., Borin, A., et al. (2015). Physicochemical parameters and antibiotics residuals in Algerian honey. *African Journal of Biotechnology* 14: 1242–1251.

Dumté, M.E.J. (2010). Development of a method for the quantitative detection of honey in imported products. Hamilton, New Zealand: University of Waikato.

El Sohaimy, S., Masry, S., and Shehata, M. (2015). Physicochemical characteristics of honey from different origins. *Annals of Agricultural Sciences* 60: 279–287.

EL-Metwally, A. (2015). Factors affecting the physical and chemical characteristics of Egyptian Beehoney. Cairo: Faculty of Agriculture, Cairo University.

Escuredo, O., Dobre, I., Fernández-González, M., and Seijo, M.C. (2014). Contribution of botanical origin and sugar composition of honeys on the crystallization phenomenon. *Food Chemistry* 149: 84–90.

Escuredo, O., Míguez, M., Fernández-González, M., and Seijo, M.C. (2013). Nutritional value and antioxidant activity of honeys produced in a European Atlantic area. *Food Chemistry* 138: 851–856.

Gontarski, H. (1960). Fermentbiologische Studien an Honigen. *Z. Bienenforsch.* 5: 30–39.

Helal, A., Tagliazucchi, D., Verzelloni, E., and Conte, A. (2014). Bioaccessibility of polyphenols and cinnamaldehyde in cinnamon beverages subjected to in vitro gastro-pancreatic digestion. *Journal of Functional Foods* 7: 506–516.

Horvath, K. and Molnár-Perl, I. (1997). Simultaneous quantitation of mono-, di-and trisaccharides by GC-MS of their TMS ether oxime derivatives: II. In honey. *Chromatographia* 45: 328–335.

Jelen, H. (2011). *Food Flavors: Chemical, Sensory and Technological Properties*. Boca Raton, FL: CRC Press.

Kamal, M.A. and Klein, P. (2011). Determination of sugars in honey by liquid chromatography. *Saudi Journal of Biological Sciences* 18: 17–21.

Kaškonienė, V., Venskutonis, P., and Čeksterytė, V. (2010). Carbohydrate composition and electrical conductivity of different origin honeys from Lithuania. *LWT-Food Science and Technology* 43: 801–807.

Kenjerić, D., Primorac, L., Mandić, M.L., et al. (2006). Dalmatian sage (Salvia officinalis L.) honey characterization. *Deutsche Lebensmittel-Rundschau* 102: 479–484.

Krpan, M., Marković, K., Šarić, G., et al. (2009). Antioxidant activities and total phenolics of acacia honey. *Czech Journal of Food Sciences* 27: S245–S247.

Kukurova, K., Karovieova, J., Kohajdova, Z., and Bilikova, K. (2008). Authentication of honey by multivariate analysis of its physico–chemical parameters. *Journal of Food and Nutrition Research* 47: 170–180.

Low, N.I.H. and Sporns, P. (1988). Analysis and quantitation of minor di-and trisaccharides in honey, using capillary gas chromatography. *Journal of Food Science* 53: 558–561.

Manikis, I. and Thrasivoulou, A. (2001). The relation of physicochemical characteristics of honey and the crystallization sensitive parameters. *Apiacta* 36: 106–112.

Mannina, L., Sobolev, A.P., Coppo, E., et al. (2016). Antistaphylococcal activity and metabolite profiling of manuka honey (Leptospermum scoparium L.) after in vitro simulated digestion. *Food and Function* 7: 1664–1670.

Meo, S.A., Al-Asiri, S.A., Mahesar, A.L., and Ansari, M.J. (2017). Role of honey in modern medicine. *Saudi Journal of Biological Sciences* 24: 975–978.

Mohan, A., Quek, S.-Y., Gutierrez-Maddox, N., et al. (2017). Effect of honey in improving the gut microbial balance. *Food Quality and Safety* 1: 107–115.

Molan, P.C. (1996). Authenticity of honey. In: P.R. Ashurst, M.J. Dennis(eds) *Food Authentication*, 259–303. New York: Springer.

Moreira, R.F., De Maria, C.A., Pietroluongo, M., and Trugo, L.C. (2010). Chemical changes in the volatile fractions of Brazilian honeys during storage under tropical conditions. *Food Chemistry* 121: 697–704.

Moreira, R.F.A. and De Maria, C.A.B. (2001). Glicídios no mel. *Química Nova* 24: 516–525.

Oddo, L.P., Piazza, M., Sabatini, A., and Accorti, M. (1995). Characterization of unifloral honeys. *Apidologie* 26: 453–465.

Oddo, L.P., Piro, R., Bruneau, É., et al. (2004). Main European unifloral honeys: descriptive sheets. *Apidologie* 35: S38–S81.

Ota, M., Kohmura, M., and Kawaguchi, H. (2006). Characterization of a new Maillard type reaction product generated by heating 1-deoxymaltulosyl-glycine in the presence of cysteine. *Journal of Agricultural and Food Chemistry* 54: 5127–5131.

Ouchemoukh, S., Louaileche, H., and Schweitzer, P. (2007). Physicochemical characteristics and pollen spectrum of some Algerian honeys. *Food Control* 18: 52–58.

Ouchemoukh, S., Schweitzer, P., Bey, M.B., et al. (2010). HPLC sugar profiles of Algerian honeys. *Food Chemistry* 121: 561–568.

Parkar, S.G., Jobsis, C.M., Herath, T.D., et al. (2017). Metabolic and microbial responses to the complexation of manuka honey with α-cyclodextrin after simulated gastrointestinal digestion and fermentation. *Journal of Functional Foods* 31: 266–273.

Přidal, A. and Vorlova, L. (2002). Honey and its physical parameters. *Czech Journal of Animal Science* 47: 439–444.

Primorac, L., Flanjak, I., kenjerić, D., et al. (2011). Specific rotation and carbohydrate profile of Croatian unifloral honeys. *Czech Journal of Food Sciences* 29: 515–519.

Puscas, A., Hosu, A., and Cimpoiu, C. (2013). Application of a newly developed and validated high-performance thin-layer chromatographic method to control honey adulteration. *Journal of Chromatography A* 1272: 132–135.

Rittig, F.T. (2001). Chemische und enzymatische Herstellung von Oligosacchariden, deren Isolierung, Charakterisierung und Prüfung auf funktionelle Eigenschaften: dissertation. FT Rittig.

Ruiz-Matute, A., Brokl, M., Soria, A., et al. (2010). Gas chromatographic–mass spectrometric characterisation of tri-and tetrasaccharides in honey. *Food Chemistry* 120: 637–642.

Ruiz-Matute, A., Sanz, M., and Martinez-Castro, I. (2007). Use of gas chromatography–mass spectrometry for identification of a new disaccharide in honey. *Journal of Chromatography A* 1157: 480–483.

Rybak, H. and Achremowicz, B. (1986). Zmiany w składzie chemicznym miodów naturalnych i zafałszowanych inwertowaną przez pszczoły sacharozą, zachodzące podczas przechowywania. *Pszczelnicze Zeszyty Naukowe* 30: 19–35.

Rybak-Chmielewska, H. (2007). Changes in the carbohydrate composition of honey undergoing during storage. *Journal of Apicultural Science* 51: 39–47.

Sanz, M., Sanz, J., and Martinez-Castro, I. (2004). Gas chromatographic–mass spectrometric method for the qualitative and quantitative determination of disaccharides and trisaccharides in honey. *Journal of Chromatography A* 1059: 143–148.

Sanz, M.L., Polemis, N., Morales, V., et al. (2005). *In vitro* investigation into the potential prebiotic activity of honey oligosaccharides. *Journal of Agricultural and Food Chemistry* 53: 2914–2921.

Šarić, G., Matković, D., Hruškar, M., and Vahčić, N. (2008). Characterisation and classification of Croatian honey by physicochemical parameters. *Food Technology and Biotechnology* 46: 355–367.

Seraglio, S.K.T., Schulz, M., Gonzaga, L.V., et al. (2021). Current status of the gastrointestinal digestion effects on honey: a comprehensive review. *Food Chemistry* 357: 129807.

Siddiqui, I. (1970). The sugars of honey. *Advances in Carbohydrate Chemistry and Biochemistry* 25: 285–309.

Siddiqui, I. and Furgala, B. (1967). Isolation and characterization of oligosaccharides from honey. Part I. Disaccharides. *Journal of Apicultural Research* 6: 139–145.

Slddiqui, I. and Purgala, B. (1968). Isolation and characterization of oligosaccharides from honey. Part II. Trisaccharides. *Journal of Apicultural Research* 7: 51–59.

Soldatkin, O., Peshkova, V., Saiapina, O., et al. (2013). Development of conductometric biosensor array for simultaneous determination of maltose, lactose, sucrose and glucose. *Talanta* 115: 200–207.

Stinson, E.E., Subers, M.H., Petty, J., and White, J.W. (1960). The composition of honey. V. Separation and identification of the organic acids. *Archives of Biochemistry and Biophysics* 89: 6–12.

Swallow, K.W. and Low, N.H. (1990). Analysis and quantitation of the carbohydrates in honey using high-performance liquid chromatography. *Journal of Agricultural and Food Chemistry* 38: 1828–1832.

Tedesco, R., Barbaro, E., Zangrando, R., et al. (2020). Carbohydrate determination in honey samples by ion chromatography–mass spectrometry (HPAEC-MS). *Analytical and Bioanalytical Chemistry* 412: 5217–5227.

Tornuk, F., Karaman, S., Ozturk, I., et al. (2013). Quality characterization of artisanal and retail Turkish blossom honeys: determination of physicochemical, microbiological, bioactive properties and aroma profile. *Industrial Crops and Products* 46: 124–131.

Tosun, M. (2013). Detection of adulteration in honey samples added various sugar syrups with 13 C/12 C isotope ratio analysis method. *Food Chemistry* 138: 1629–1632.

Wang, S., Guo, Q., Wang, L., et al. (2015). Detection of honey adulteration with starch syrup by high performance liquid chromatography. *Food Chemistry* 172: 669–674.

Wang, Y., Juliani, H.R., Simon, J.E., and Ho, C.-T. (2009). Amino acid-dependent formation pathways of 2-acetylfuran and 2, 5-dimethyl-4-hydroxy-3 [2H]-furanone in the Maillard reaction. *Food Chemistry* 115: 233–237.

Watanabe, T. and Aso, K. (1960). Studies on honey II. Isolation of kojibiose, nigerose, maltose and isomaltose from honey. *Tohoku Journal of Agricultural Research* 11: 109–115.

White, J.W. (1962). Composition of American honeys. Washington, DC: US Department of Agriculture.

White, J.W., Jr. (1992). Honey. In: *The Hive and the Honey Bee* (ed. J.M. Graham), 869–925. Hamilton, IL: Dadant & Sons.

White, J.W. and Hoban, N. (1959). Composition of honey. IV. Identification of the disaccharides. *Archives of Biochemistry and Biophysics* 80: 386–392.

Zhou, Y., Xu, D.-S., Liu, L., et al. (2016). A LC–MS/MS method for the determination of stachyose in rat plasma and its application to a pharmacokinetic study. *Journal of Pharmaceutical and Biomedical Analysis* 123: 24–30.

# 4

# Lipid and Fatty Acids in Honey

*Dhirendra Nath Barman, Md. Anisur Rahman, and Md. Murad Hossain*

## Introduction

Natural honey is an attractive source of nutritional and medicinal substances. Honey is the natural sweet substance produced by honeybees, *Apis mellifera*, from the nectar of plants (blossom) or from the secretion of living parts of plants or excretions of plant sucking insects. This nectar or secretion is collected by honeybees and transformed by combining with specific substances from the bees. They deposit, dehydrate, store, and leave the transformed substances in the honeycomb to ripen and mature (Codex Alimentarius Commission, 2001a, 2001b). Honey with intrinsic properties of a natural antibiotic can be kept for long periods of time without becoming spoiled. It has a high osmotic pressure (~2000 mOsM) responsible for resistance to spoilage by microorganisms (White 1975). Honey contains acids (White 1978) that contribute to resistance against bacterial spoilage, enabling lowering the pH from 3.4 to 6.1, averaging at about 3.9. Moreover, glucose oxidase enzyme in honey confers antibacterial activity (Cocker 1951; White et al. 1958, 1963).

## Chemical Composition of Honey

Essentially, natural honey is a sticky and viscous solution with a content of 80%–85% carbohydrate (mainly glucose and fructose); 15%–17% water; 0.1%–0.4% protein; 0.2% ash; and minor quantities of amino acids, enzymes, and vitamins as well as other substances such as phenolic antioxidants (Gheldof and Engeseth 2002; James et al. 2009; Jeffrey and Echazarreta 1996; National Honey Board 2003; White and Doner 1980). Although the major constituents of honey are nearly the same in all honey samples, the precise chemical composition and physical properties of natural honeys differ according to the plant species on which the bees forage (Cantarelli et al. 2008; Ciappini et al. 2008; Ebenezer and Olubenga 2010; James et al. 2009; Omafuvbe and Akanbi 2009).

## Trace Lipids in Honey

In a study, n-Hexane (Skellysolve B) and diethyl ether were used to extract the trace lipids in cotton honey. Gas liquid chromatography was used to determine the relative concentration of purified methyl esters. Smith and Mc Caughey (1966) have shown the presence of methyl esters in cotton honey by gas chromatography using the standard methyl esters such as methyl laurate, myristoleate, palmitate, stearate, oleate, and linoleate. The equivalent chain length of 8.6 in the purified methyl esters might be the unsaturated analog of caprylic and octenoic acids. The purified methyl esters contain either methyl arachidate or linolenate. Methyl myristate and palmitoleate may be present also, but observations have not been enough to be certain. Table 4.1 indicates some examples of the isolated fatty acids from cotton honey (Smith and Mc Caughey 1966).

Smith (1963) reported that the highest fatty acid level in cotton honey was oleic acid (18:1). To get the free fatty acids, the crude extract of honey was subjected to saponification. After saponification, the fatty acids of crude extract were subjected to reverse-phase chromatography using the indicators hydrogen sulfide ($H_2S$) and iodine ($I_2$). The fatty acids had the same *Rf* as that of palmitic and oleic acids (Table 4.2), indicating that both palmitic acid and oleic acids are present in honey (Smith and Mc Caughey 1966).

*Honey: Composition and Health Benefits*, First Edition. Edited by Md. Ibrahim Khalil, Gan Siew Hua, and Bey Hing Goh.

**Table 4.1** Gas chromatography of fatty acids isolated from cotton honey.

| | Equivalent Chain Length | |
|---|---|---|
| | **Methyl Ester Standard** | **Honey Methyl Esters** |
| Caprylate | 8.1 ± 0.3 (3)[a] | 8.6[b] ± 0.21[b] (6)[a] |
| Caprace | 10.0 ± 0.10 (3) | |
| Laurate | 11.9 ± 0.10 (4) | 11.5.45 (5) |
| Myristate | 14.0 ± 0.09 (7) | 13.70[c] (1) |
| Myristoleate | 14.7 ± 0.07 (7) | 14.8 ± 0.10 (2) |
| Palmitate | 16.0 ± 0.08 (7) | 16.0 ± 0.20 (7) |
| Palmitoleate | 16.6 ± 0.11 (7) | 16.7 (1) |
| Stearate | 18.0 ± 0.11 (7) | 18.0 ± 0.25 (4) |
| Oleate | 18.6 ± 0.11 (7) | 18.6 ± 0.25 (7) |
| Linoleate | 19.2 ± 0.15 (7) | 19.1 ± 0.18 (5) |
| Arachidate | 20.0 ± 0.15 (5) | 20.0 ± 0.29 (5) |
| Linolenate | 20.0 ± 0.17 (3) | |
| Vaccinate | 18.3 (1) | |
| Behenate | 21.9 (1) | |

[a] Number of observations.
[b] Standard deviation.
[c] Significantly different from the corresponding methyl ester at the 5% level (Smith and Mc Caughey 1966).
Smith and Mc Caughey, 1966 / John Wiley & Sons.

**Table 4.2** Paper chromatography of honey fatty acids.

| **Sample** | $R_f$ |
|---|---|
| | Indicator: $H_2S$ |
| Honey fatty acids | 0.40 (0.35–0.45) |
| Stearic acid | 0.29 (0.26–0.32) |
| Palmitic acid | 0.43 (0.41–0.46) |
| Oleic acid | 0.40 (0.38–0.42) |
| Palmitic and oleic acid | 0.40 (0.360.44) |
| | Indicator: $I_2$ |
| Honey fatty acids | 0.37 (0.43–0.39) and 0.55 (0.52–0.58) |
| Palmitic and oleic acid | 0.37 (0.33–0.42) |

Smith and Mc Caughey, 1966 / John Wiley & Sons.

The biochemical composition analysis of samples from Northeast Nigeria revealed that the fat contents falls within the range of 0.1 to 0.5 g/100 g, supporting the literature (Singh and Kuar 1997; Tan et al. 1988) and indicating that honey used in this study contains little or no fat. The presence of fatty acids such as palmitic (16:0), lignoceric (24:1), oleic (18:1), and α-linolenic acids (18:1) has been reported in white clover honey, and the fatty acids ranges from the $C_8$–$C_{28}$ (Tan et al. 1988). Khalil et al. (2001) reported total fat contents in the range of 0.134 to 0.146 g/100 g after biochemical analysis of five different brands of unifloral honey available in the northern region of Bangladesh.

In another study, the constituent lipids of Greek honey were isolated and studied by an initial simple extraction procedure (comparable to that of counter-current distribution) and consequent chromatographic separation on a silicic acid column.

The fractions collected were subjected to (1) qualitative thin-layer chromatography before or after saponification and (2) gas-chromatographic analysis of methyl esters. In addition to the fatty acids mentioned by other investigators, honey has been found to contain a number of neutral lipids, albeit in small amounts, including hydrocarbons, waxes, cholesterol esters, fatty acid esters, fatty acids, fatty alcohols, sterols, and dihydroxy and trihydroxy compounds, as well as some esters of polyols (Kapoulas et al. 1977). Scientists were also interested in the lipid content in beeswax. They found that beeswax also contains lipids, and Canadian beeswax samples showed the highest ratio of fatty acid to be lignoceric acid (24:0) (Tulloch and Hoffman 1972). Unsaturated fatty acids of the same chain length were not assessed separately in this study. However, lipid in honey appears to be different from beeswax lipids with regard to structure and composition, which supports that the biological origin of lipids of honey is different from that of beeswax lipids (Kapoulas et al. 1977).

A qualitative analysis of fatty acids in honey collected from Lithuania was performed by Jarukas et al. (2021) and was found that the total number of fatty acids in honey was eight, whereas total number of saturated fatty acids was six, and the total number of unsaturated fatty acids was two.

There have been very few studies done on honey lipid that indicate that honey does not contain lipid and is, in fact, free of fat. Moreover, most of the studies on the lipid content of honey were performed in the early 1950s–1970s. Therefore, further studies using advanced technologies are needed to determine the lipid content of honey, which may provide potential social, environmental, global climate change, and urbanization effect(s) (if any) on the lipid content of honey of this current era.

## References


Cantarelli, M.A., Pellerano, R.G., Marchevsky, E.J., and Camina, J.M. (2008). Quality of honey from Argentina: study of chemical composition and trace elements. *Argentine Chemical Society* 96: 33–41.

Ciappini, M.C., Gatti, M.B., Di Vito, M.V., et al. (2008). Characterization of different floral origins honey samples from Santa Fe (Argentina) by palynological, physicochemical and sensory data. *Apiacta* 43: 25–36.

Cocker, L. (1951). The enzymic production of acid in honey. *Journal of the Science of Food and Agriculture* 2: 411–414.

Codex Alimentarius Commission. (2001a). Codex standard for honey, FAO, Rome. *Alinorm* 1: 19–26.

Codex Alimentarius Commission. (2001b). Codex standard 12, revised codex standard for honey, standards and standard methods 11.

Ebenezer, I.O. and Olubenga, M.T. (2010). Pollen characterization of honey samples from North Central Nigeria. *Journal of Biological Sciences* 10: 43–47.

Gheldof, N. and Engeseth, N.J. (2002). Antioxidant capacities of honeys from various floral sources based on the determination of oxygen radical absorbance capacity and inhibition of in vitro lipoprotein oxidation in human serum samples. *Journal of Agricultural and Food Chemistry* 50: 3050–3055.

James, O.O., Mesubi, M.A., Usman, L.A., et al. (2009). Physical characterization of some honey samples from North-Central Nigeria. *International Journal of Physical Sciences* 4: 464–470.

Jarukas, L., Kuraite, G., Baranauskaite, J., et al. (2021). Optimization and validation of the GC/FID method for the quantification of fatty acids in bee products. *Applied Sciences* 11(1): 83. https://doi.org/10.3390/app11010083.

Jeffrey, A.E. and Echazarreta, C.M. (1996). Medical uses of honey. *Revista Biomedica* 7: 43–49.

Kapoulas, V.M., Mastronicolis, S.K., and Galanos, D.S. (1977). Identification of the lipid components of honey. *Zeitschrift für Lebensmittel-Untersuchung und Forschung* 163: 96–99.

Khalil, M.I., Motallib, M.A., Anisuzzaman, A.S.M., et al. (2001). Biochemical analysis of different brands of unifloral honey available at the Northern region of Bangladesh. *The Sciences* 1: 385–388.

National Honey Board. 2003. Honey: health and therapeutic qualities. Firestone, CO: National Honey Board.

Omafuvbe, B.O. and Akanbi, O.O. (2009). Microbiological and physico-chemical properties of some commercial Nigerian honey. *African Journal of Microbiology Research* 3: 891–896.

Singh, N. and Bath, P.K. (1997). Quality evaluation of different types of Indian honey. *Food Chemistry* 58: 129–133.

Smith, M.R. (1963). Chromatographic investigation of trace lipids in honey. Ph.D. Dissertation, University of Arizona.

Smith, M.R. and Mc Caughey, W.F. (1966). Identification of some trace lipids in honey. *Journal of Food Science* 31: 902–905. doi: 10.1111/j.1365-2621.1966.tb03268.x.

Tan, S.T., Holand, P.T., Wilkins, A.L., and Molan, P.C. (1988). Extractives from New Zealand honeys. 1. White clover, manuka and kanuka unifloral honeys. *Journal of Agricultural and Food Chemistry* 36: 453–460.

Tulloch, M.R. and Hoffman, L.L. (1972). Analytical values and composition of hydrocarbons, free acids and long chains esters. *Journal of the American Oil Chemists Society* 49: 696–699.

White, J.W., Jr (1975). Composition of honey. In: *Honey: A Comprehensive Survey* (ed. E. Crane), 157–205. London: Heinemann in cooperation with International Bee Research Association.

White, J.W., Jr (1978). Honey. *Advances in Food Research* 24: 287–374.

White, J.W. and Doner, L.W. 1980. Honey composition and properties: beekeeping in the United States. Agriculture Handbook No. 335, Revised October 82-91.

White, J.W., Petty, J., and Hager, R.B. (1958). The composition of honey. II. lactone content. *Journal of Association of Official Agricultural Chemists* 41: 194–197.

White, J.W., Jr, Subers, M.H., and Schepartz, A.I. (1963). The identification of inhibine, the antibacterial factor in honey, as hydrogen peroxide and its origin in a honey glucose-oxidase system. *Biochimica et Biophysica Acta* 73: 57–70.

# 5

# Amino Acids, Proteins, and Enzymes

*Md. Murad Hossain, Dhirendra Nath Barman, and Md. Anisur Rahman*

## Introduction

Honey is a natural complex nutritious food product produced by honeybees (*Apis mellifera* L.) from flower nectar or honeydew. Honey is a rich source of various carbohydrates (e.g. glucose, fructose, sucrose, and maltose), thus making it widely useful as a natural sweetener. It is also an important source of other minor constituents, which are more related to its biological properties such as minerals, vitamins, polyphenols, carotenoids, free amino acids, proteins, and enzymes (da Silva et al. 2016; Machado De-Melo et al. 2018). The presence of flavonoids and phenolic compounds implies the role of honey, along with fruits and vegetables, as a potent source of natural antioxidants accountable for protecting human health (Gheldof et al. 2002). However, the amounts and proportions of honey composition are affected by various factors such as the floral source, bee species, geographic region, and storage period (Schievano et al. 2013).

Honey contains a relatively small amount of amino acids (20–300 mg/g of dry matter). Nonetheless, almost all of the physiologically essential amino acids are present in honey (Cotte et al. 2004; Hermosín et al. 2003; Kıvrak 2017). The divergence in the type and concentration of amino acids in honey is related to the amino acid content of the honeybees' diet, which in turn depends on the source of interest, geographical origin, and season of the year (Hermosín et al. 2003). Proline is the most abundant amino acid in honey with 50%–85% of the total amino acids (Del Campo et al. 2016). Proline has been implied to possess the potential for honey maturity, which is associated with proline content in honey (Paramás et al. 2006). Proline in honey is also used as a criterion to estimate the antioxidant quality and activity of honey as well as the characterization of the botanical and geographical origins of honey (Truzzi et al. 2014). Recently, the presence of D-amino acids in honey has been suggested as a way to distinguish adulterations (Del Campo et al. 2016).

Honey also contains a very low amount of proteins (0.1%–0.5% of total honey mass) (Chua et al. 2013, 2015). The variation in protein content of different honeys may be due to differences in honeybee origin as well as botanical sources used by them (Schievano et al. 2013). The most abundant honey proteins are major royal jelly proteins (MRJPs), which are responsible for the antimicrobial, antitumor, anti-inflammatory, and antioxidant activities of honey (Chua et al. 2015; Guo et al. 2009). The proteins present in honey are widely used as markers of honey's authenticity and adulteration (Bilikova et al. 2015; Won et al. 2008) and as a quality indicator (Bilikova et al. 2015; Chua et al. 2013). Moreover, honey contains several enzymes, which constitute a major part of honey protein, such as α- and β-glucosidase (invertase), α- and β-amylase (diastase), glucose oxidase, catalase, and protease (Machado De-Melo et al. 2018). These enzymes in honey play an important role in metabolic processes as well as in determining its overall medicinal and functional properties. Nevertheless, honey's proteome, including honey proteins and enzymes, is still very poorly researched, mostly because of the low amount of protein present in honey. This chapter includes a discussion of the amino acids, proteins, and enzymes profile of honey; their abundance in honeys from different botanical and different geographical origins; their usefulness in honey authentication and adulteration determination; and their importance in honey's quality.

*Honey: Composition and Health Benefits*, First Edition. Edited by Md. Ibrahim Khalil, Gan Siew Hua, and Bey Hing Goh.

## Amino Acids in Honey

Amino acids are the building block of peptides and proteins. Amino acids content in honey ranges between 20 to 300 mg/100 g of dry matter (Biluca et al. 2019). The source of amino acids in honey is attributed to both animal (bee secretions) and vegetal (nectar, honeydew, and mainly pollen) origins (da Silva et al. 2016). Disparity in the type and concentration of amino acids in honey is related to the amino acid content in the bees' diet, which in turn depends on the floral source and season of the year. During spring when the plants sprout and in autumn when leaves change color, the concentration of amino acids and nitrogenous compounds become very high in the phloem sap; resultantly, it increases in nectar and finally enters the honey (Crane 1975). The amino acids present in honey are nutritionally important. The majority of amino acids present in honey are in the bound form, and the free amino acid content is approximately one-fifth of the total amino acids (Paramás et al. 2006). Free amino acids are present in honey as a minor component that contributes to approximately 1% (w/w) of honey's total weight (Adnan et al. 2014; Kowalski et al. 2017). Although honey contains a relatively small amount of amino acids, almost all of the physiologically essential amino acids are reported to be present (Cotte et al. 2004; Hermosín et al. 2003; Kıvrak 2017). The most abundant amino acid in honey is proline (50%–85%), a secondary amino acid originating mainly from hemolymph of bees (Del Campo et al. 2016). However, a large variation of proline content in different unifloral honeys makes it impossible to classify unifloral honey on the basis of this parameter only. The analysis of the amino acid content has been implicated as a good indicator of both the botanical and geographical origin of honey (Azevedo et al. 2017; Sun et al. 2017). Pollen serves as the main source of amino acids in honey. Hence, amino acid profiling of pollen could be useful to characterize the botanical origin of honey. However, free amino acids are also added by honeybees themselves, which leads to a high variability of the amino acid content within honeys from the same botanical source (Bogdanov and Martin 2002). Recent studies have also suggested that amino acid content, especially the presence of D-amino acids, could be a criterion for distinguishing adulterations (Del Campo et al. 2016). Pätzold and Brückner (2006) demonstrated that certain D-amino acids are naturally found in honey. Relative quantities and kinds of D-amino acids detectable therein depend on the honey sample. Moreover, the amount of D-enantiomers increases by heating. Therefore, the presence and relative content of D-amino acids found in honeys could be used as a test for the long-term storage and the nature of the processing of the honey.

### Distribution of Free Amino Acids in Different Types of Honeys

Free amino acid content varies among different botanical origin as well as different geographical origins of honeys. Free amino acid profiling of acacia, lime, rape, multifloral, and forest honeydew types of honey from Poland and Slovakia was conducted, and the presence of all 20 amino acids was detected with a limit of detection ranging from 3.0 ng/mL for valine to 13.0 ng/mL for hydroxyproline based on a liquid chromatography tandem mass spectrometry method (Kowalski et al. 2017). The total contents of free amino acids for Polish honeys were determined to be about 441.35 ± 102.70 μg/g, 453.64 ± 152.36 μg/g, 621.21 ± 123.63 μg/g, 662.79 ± 248.76 μg/g, and 773.94 ± 102.09 μg/g, respectively, for rape, acacia, multifloral, lime, and forest honeydew honey. The total free amino acid contents of rape, acacia, multifloral, lime, and forest honeydew honey from Slovakia were 393.94 ± 21.96 μg/g, 741.31 ± 94.65 μg/g, 555.49 ± 21.07 μg/g, 616.67 ± 2.98 μg/g, and 749.15 ± 70.46 μg/g, respectively (Kowalski et al. 2017). A similar characteristic for Polish honeys (390.63 ± 105.34 μg/g for rape honey and 398.41 ± 135.52 μg/g for honeydew honey) was observed by Janiszewska et al. (2012), who identified 25 free amino acids from Polish honeys of different botanical origin (Janiszewska et al. 2012). The content of proline should ideally be at least 180 μg/g (Kowalski et al. 2017), and a lower value is indicative of immaturity or adulteration of honey with sugar (Bogdanov et al. 1999). Among the honeys of five botanical origin (rape, acacia, multifloral, lime, and forest honeydew honey) obtained from Poland and Slovakia, the lowest proline content was recorded to be 140.14 μg/g (rape honey from Slovakia), and the proline content was found to be 389.66 μg/g for forest honeydew honey from Poland (Kowalski et al. 2017). These values seem to be characteristic of both Polish honeys (Janiszewska et al. 2012) and some European honeys (Kečkeš et al. 2013). However, in the case of Chinese honeys, the proline content was found higher (an average of ~317 μg/g and a maximum above 600 μg/g) in rape honeys (Chen et al. 2017). In another study, the proline content was measured to vary between 139.2 and 631.5 mg/kg among 33 Chinese unifloral honeys. The highest proline content was observed in a Eucalyptus honey (a dark-colored honey), and the lowest proline content was observed in a Vetch honey (a light-colored honey) (Dong et al. 2013). Among the studied unifloral honey species, the Eucalyptus honeys exhibited the highest proline content (average, 574.7 mg/kg) followed by red date honeys (average, 529.1 mg/kg), Coptidis honeys (average, 423.1 mg/kg), Linden honeys (average, 421.7 mg/kg), Codonopsis honeys (average, 346.8 mg/kg),

Eriobotrya honeys (average, 285.4 mg/kg), Astragalus honeys (average, 272.4 mg/kg), acacia honeys (average, 247.9 mg/kg), and Vetch honeys (average, 144.9 mg/kg) (Dong et al. 2013). Like proline content, the total amino acid content was also observed the highest in a Eucalyptus honey and the lowest in a Vetch honey. The total amino acid contents among these Chinese unifloral honeys varied from 240.7 to 812.6 mg/kg (Dong et al. 2013). Hermosín et al. (2003) suggested that the proline content in honey would be more than 200 mg/kg. Besides proline, the other most common amino acids are glutamic acid, alanine, phenylalanine, tyrosine, leucine, and isoleucine (White and Landis 1980).

The amino acid compositions and concentration in honey may vary because of several factors, including soil or climatic conditions, although they originate from similar botanical origin. There is a wide variability of the type of amino acid present in honeys collected from similar botanical origin albeit from different geographical locations (Table 5.1). As expected, the concentration of amino acid present in honeys is greatly affected by the floral origin. A high concentration of proline was detected in sunflower honey (645.0 mg/kg), while a very low content of proline was found in acacia honey (16.35 mg/kg) (Carratù et al. 2011; Chua and Adnan 2014). This finding was also supported by some other studies that indicated that sunflower honey contained slightly higher proline levels than rapeseed and acacia honeys (Kečkeš et al. 2013; Oddo et al. 2004). Similarly, a great variation in concentration of glutamic acid, threonine, and tyrosine was found in honeys from different floral origins. Amino acid concentrations also vary largely because of the climate changes in different geographical locations. For example, the glutamic acid contents were measured as 185.0 mg/kg, 9.12 mg/kg, and 0.145 mg/kg from acacia honey obtained from Italy, Poland, and Malaysia, respectively (Carratù et al. 2011; Chua and Adnan 2014; Janiszewska et al. 2012). A very high concentration of tyrosine was found in Italian acacia honey (326 mg/kg), while, in contrast, it was detected

**Table 5.1** Amino acid compositions of different honeys (mg/kg) obtained from different locations.

| | Acacia | | | Sunflower | | Heather | | Honeydew | |
|---|---|---|---|---|---|---|---|---|---|
| **Amino acid** | **Italy[a]** | **Poland[b]** | **Malaysia[c]** | **Italy[a]** | **Pakistan[d]** | **Italy[a]** | **Poland[b]** | **Italy[a]** | **Poland[b]** |
| Alanine | 5.0 | 7.26 | na | 40.0 | 9.1 | 9.0 | 9.75 | 26.0 | 9.16 |
| Asparagine | 7.0 | 7.16 | na | 8.0 | na | 4.0 | 5.80 | 36.0 | 3.23 |
| Aspartic acid | 8.0 | 7.77 | na | 65.0 | na | 4.0 | 6.39 | 108.0 | 12.16 |
| Arginine | 22.0 | 1.90 | na | 18.0 | na | 12.0 | 3.55 | 43.0 | 1.90 |
| Cysteine | nd | na | 0.044 | nd | na | nd | na | nd | na |
| Glutamine | 11.0 | 18.58 | na | 32.0 | na | 5.0 | 7.56 | 90.0 | 12.53 |
| Glutamic acid | 185.0 | 9.12 | 0.145 | 242.0 | 29.0 | 423.0 | 9.10 | 243.0 | 11.40 |
| Glycine | 7.0 | 2.79 | na | 5.0 | na | 6.0 | 4.08 | 7.0 | 2.27 |
| Histidine | nd | 1.45 | 4.109 | 3.0 | na | nd | 2.57 | nd | 0.86 |
| Isoleucine | 2.0 | 2.83 | 0.273 | 6.0 | na | 2.0 | 4.09 | 13.0 | 7.46 |
| Leucine | 9.0 | 2.41 | 1.214 | 11.0 | na | 11.0 | 5.26 | 10.0 | 9.26 |
| Lysine | 10.0 | 6.51 | na | 9.0 | 13.0 | 3.0 | 8.19 | 6.0 | 2.91 |
| Methionine | nd | 1.58 | 0.253 | nd | na | nd | 1.03 | nd | 3.32 |
| Phenylalanine | 5.0 | 6.44 | na | 13.0 | na | 6.0 | 16.01 | 30.0 | 5.77 |
| Proline | 161.0 | 225.74 | 16.35 | 645.0 | 356.0 | 267.0 | 284.09 | 478.0 | 263.36 |
| Serine | 10.0 | 6.20 | na | 24.0 | na | 9.0 | 6.72 | 26.0 | 9.58 |
| Threonine | 41.0 | 3.33 | 0.743 | 49.0 | na | 66.0 | 5.51 | 49.0 | 4.04 |
| Tryptophan | nd | na | na | 3.0 | 19.0 | nd | na | nd | na |
| Tyrosine | 326.0 | 51.01 | 0.146 | 60.0 | 2.1 | 309.0 | 4.97 | 49.0 | 9.39 |
| Valine | 7.0 | 5.67 | na | 23.0 | na | 7.0 | 6.83 | 14.0 | 9.42 |

[a] Carratù et al. (2011))
[b] Janiszewska et al. (2012).
[c] Chua and Adnan (2014).
[d] Qamer et al. (2007).

na, not available; nd, not detectable (<0.10 mg/100 g is the quantification limit of amino acid determination method).

extremely low in Malaysian acacia honey (0.146 mg/kg) (Carratù et al. 2011; Chua and Adnan 2014). Similarly, a massive difference was also found in the concentration of the glutamic acid, threonine, and tyrosine content of sunflower honey (Italian vs Pakistani), heather honey (Italian vs Poland), and honeydew honey (Italian vs Poland) (Carratù et al. 2011; Janiszewska et al. 2012; Qamer et al. 2007). It has been found that the methods applied for the extraction and detection of amino acids from honeys were not similar in these studies. Therefore, it is assumed that the difference in amino acid concentration in honey is not only attributable to the floral origin and geographical climatic conditions but also to the extraction and detection methods applied. Distinction of origin is not an easy task because honey composition is affected by many factors, including climatic and soil conditions as well as storage, aging, and processing techniques (Hermosín et al. 2003).

### Free Amino Acids as the Determinant of the Botanical Origin of Honey

Determination of the botanical origin of honey is a hard task because of the complex content of the honey matrix. Moreover, the compositions of the honey also vary on several factors, including soil and climatic conditions as well as aging, storage, and processing techniques. For identification of the floral origin of honey, one of the commonly used methods is melissopalynology, which is the microscopic analysis of pollen present in honey after filtration or centrifugation of honey (Bodor et al. 2021). However, this method may not be effective in case of filtered processed honey because of the scarcity of pollen present after filtration and in case of many commercial honeys, which are often a mixture of honeys from different botanical origins. Many other studies have been performed to characterize the floral origin of honey on the basis of sugar profile, flavonoid pattern, non-flavonoid phenolics, aroma compounds, organic acids, isotopic relations, and protein and amino acid compositions as the potential analytical markers. Among these, free amino acid composition present in honey is more likely to determine the botanical origin of a honey.

Amino acid compositions present in honeys are originated both from honeybees and plant sources, the major of these being pollen of the plant. Because pollen is the main source of amino acids in honey, the amino acid profile of pollen present in honey could be useful to characterize the botanical origin of honey. Therefore, identification of pollen is assumed as a key analytical marker for the determination of the botanical or floral origin of honey. However, this method has some limitations. For example, honeybees contribute to the free amino acid composition of honey in addition to pollen, which leads to a high variation of the amino acid content within honeys even they originated from the same botanical source (Bogdanov and Martin 2002). The variations in the amino acid content in honey depend on the content of the amino acids in the bees' diet, which may vary on the plant type and the season of the year (Paramás et al. 2006).

### Thermal Treatment Effect on Free Amino Acids in Honey Samples

After harvesting raw honey, it is subjected to strain and filter to remove suspended materials, including pollen and beeswax. This honey is then thermally treated before packaging, which eases the bottling process by reducing the viscosity of honey. Thermal treatment of honey also exerts some other beneficial effects; for example, it prevents honey fermentation by reducing the water content of honey (Subramanian et al. 2007), homogenizes the color of the honey for consumers' preference (Abu-Jdayil et al. 2002), prevents sugar crystallization (Escriche et al. 2009; Turhan et al. 2008), and increases the shelf life of honey by destroying the sugar tolerant osmophilic yeasts (Guo et al. 2011). Therefore, thermal treatment is a key step in honey processing that directly affects the quality of honey. However, there is no standard and well-received general guideline available to date for the determination of heating temperature and heating time for any particular type of honey. Honey is usually treated with heating temperatures ranging from 30° to 140°C for a few seconds up to several hours by honey producers worldwide, with the aim to reduce the water content in honey below 20% for shelf life prolongation (Guo et al. 2011; Turhan et al. 2008; Wakhle et al. 1996). The optimum heating condition of a particular honey is believed to depend on the geographical and botanical origins of the honey.

Thermal treatment not only reduces the water content but also affects the content of nutritional composition in honey. Upon heating, the protein in honey is degraded to small polypeptides and free amino acids, thus reducing the protein content and consequently increasing the total free amino acid content. Certainly, there is a change in levels of free amino acids after thermal treatment at 90°C for 30 minutes in Tualang, Gelam, and acacia honey samples (Adnan et al. 2014). The heat exposure causes a significant increase in the level of phenylalanine, tyrosine, valine, and lysine content. At the same time, the content of isoleucine, leucine, proline, and threonine was found in a reduced level in the Tualang honey (Adnan et al. 2014). Similar effects were also found in Gelam and acacia honey. The reduction in the content of some amino acids in honey samples could probably be due to denaturation process that is associated with the unfolding or aggregation of protein at high temperature. As the protein in honey unfolds at high temperature, increased number of amino acids becomes exposed to denaturation. The decrease in free amino acids concentration in honey also accounts for the reaction between the free amino acids and the carbonyl group of reducing sugar in honey during thermal treatment (Adnan et al. 2014).

## Proteins in Honey

Proteins are highly complex biomolecules that are present in all living organisms. Proteins are polymer chains made up of amino acids, which are linked together by peptide bonds. They are of great nutritional value and are directly involved in the biochemical processes essential for life, including catalyzing various metabolic reactions by means of producing enzymes, DNA replication, providing structure to cells and organisms, responding to various stimuli, and transporting various molecules from one location to another. Proteins differ from one another primarily because of differences in their sequence of amino acids, and they usually fold into a specific three-dimensional structure that determines their activity. Honey contains a very low amount of proteins, approximately 0.1%–0.5% of total honey mass (Chua et al. 2013, 2015). The proteins and amino acids present in honey come from plant (nectar, honeydew, and mainly pollen) sources as well as animal (bee secretions) sources. Although honey proteins do not have considerable nutritional value, they have significant pharmacological activities. The protein content in honey varies on the honeybees' origin as well as the botanical sources. Honeys obtained from *A. mellifera* bees contain lower amount of proteins compared with those produced by the *Apis cerana* bee species (Lee et al. 1998; Won et al. 2009). Honey produced by *A. cerana* contains 0.1%–3.3% protein, while *A. mellifera* honey contains 0.2%–1.6% protein (Lee et al. 1998; Won et al. 2009). Honey enzymes, which constitute a major part of honey protein, play an important role, including in the production of hydrogen peroxide ($H_2O_2$) that drives the antibacterial activity of honey. Besides enzymes, honey contains several proteins that can exert antimicrobial (Gomes et al. 2010; Taormina et al. 2001), anti-inflammatory (Tonks et al. 2003), and anticancer (Bardy et al. 2012; Tsiapara et al. 2009) activity.

Royal jelly proteins produced by honeybees have been reported to exhibit antimicrobial, antitumor, anti-inflammatory, and antioxidant activities (Chua et al. 2015; Guo et al. 2009). MRJPs are also present in honey at a low concentration. Among the honey proteins, the honey MRJP-1 is the most prominent and functionally active protein. Recent studies showed that MRJP-1 is involved in honey's antibacterial as well as its immunological properties. MRJP-1 has also been suggested as a suitable marker for authentication of honeybee origins (Won et al. 2009). In a recent study, the presence of several MRJPs such as MRJP-1, MRJP-2, MRJP-5, and MRJP-7, as well as a few uncharacterized proteins from *A. mellifera* was revealed by employing a mass spectroscopic method from Malaysian acacia, Tualang, and Gelam honey (Chua et al. 2015). Among these honey samples, acacia honey was reported to contain the maximum amount of MRJPs. Honey MRJPs exhibit significant antioxidant activity in the form of free radical scavengers and reducing agents. Acacia honey was also shown to display the highest antioxidant properties compared with Tualang and Gelam honey, largely because of the presence of high concentrations of MRJPs (Chua et al. 2015). Honey peptides derived from gastrointestinal digestion of honey proteins, specifically MRJPs, showed potent antioxidant activities, which enabled eukaryotic cells to exhibit enhanced tolerance toward oxidative stress (Ibrahim et al. 2021) (Figure 5.1).

Honey proteins are extensively used as markers for the determination of honey's authenticity and adulteration (Bilikova et al. 2015; Won et al. 2008) as well as a quality indicator (Bilikova et al. 2015). Nevertheless, honey proteome still remains very poorly researched, mostly because of the low protein content in honey. This is why analyses of the honey proteins remain a challenge. A number of trendy methods are used for isolation of honey proteins, including dialysis against distilled water and lyophilization of dialysate (Chua et al. 2013; Won et al. 2008) and precipitation with several chemicals, namely, sodium tungstate, ammonium sulfate, saturated phenols, and acetone (Bocian et al. 2019; Borutinskaite et al. 2018; Mohammed and Azim 2012; Steinhorn et al. 2011). A combination approach of protein dialysis and chemical precipitation has also been reported (Chua et al. 2015).

Honey authentication also remains a major challenge. The classical approach in honey authentication studies is used for determining its botanical origin. Monofloral honey origins are determined by sensory and physicochemical analyses, whereas melissopalynological analysis is commonly used to identify floral pollen grains present in honey by microscopic examination (Anklam 1998). However, because of the limitations of the classical sensory, physicochemical, and melissopalynological authentication techniques, more reliable molecular techniques have been developed for authentication of the botanical and geographical origins of honey. Honey contains proteins of different molecular weights (MWs) depending on its bee and floral origin. Thus, protein- and DNA-based techniques such as sodium dodecyl sulphate polyacrylamide gel electrophoresis (SDS-PAGE), Western blot, enzyme-linked immunosorbent assay, and real-time polymerase chain reaction

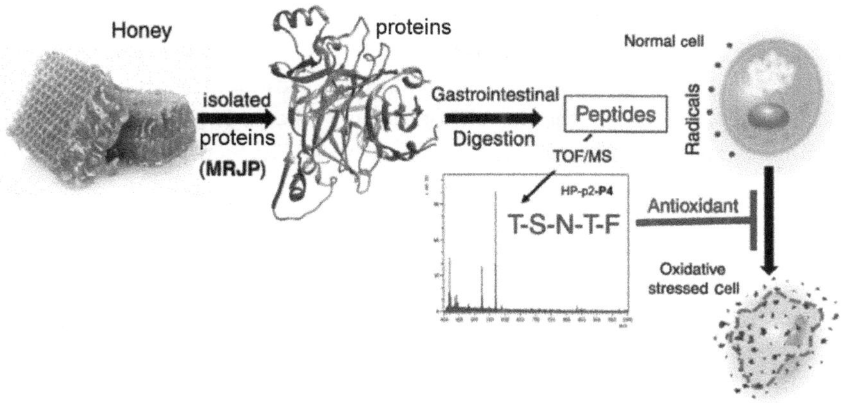


**Figure 5.1** Potent antioxidant peptides derived from honey major protein enhance tolerance of eukaryotic cells toward oxidative stress. MRJP, major royal jelly protein. Ibrahim et al, From Springer Nature / Public domain / CC BY 4.0.

are used to identify, authenticate, and classify honey samples (Chin and Sowndhararajan 2020). Honey contains various proteins as judged by SDS-PAGE of honey samples, producing many protein bands from honeys of different honeybee or floral origin (Lee et al. 1998; Marshall and Williams 1987).

## Bactericidal Activity of Honey Proteins

Honey has been well documented for its antibacterial activity. However, the active component of honey for its bactericidal activity is yet to be determined. Recently, Brudzynski and Sjaarda (2015) isolated active glycoproteins from honey that have the ability to disrupt bacterial cell walls and to exhibit strong growth-inhibitory and antibacterial properties. Sequence analysis revealed that the glycoproteins are identical to the MRJP-1 precursor that harbors three antimicrobial peptides: Jelleins 1, 2, and 4. In this study, it was shown that honey glycoproteins specifically bind to bacterial cells and form visible agglutination of both gram-positive *Bacillus subtilis* and gram-negative *Escherichia coli*. The glycoproteins were also shown to increase membrane permeabilization and to lose membrane integrity of both bacterial cells and erythrocytes. The isolated glycoproteins induced morphological changes in the shape of both *E. coli* and *B. subtilis* cells, which include filamentation, cell wall lysis, and spheroplast formation. The emergence of filaments and spheroplasts indicates the growth inhibition and bactericidal effects of honey glycoproteins, respectively. Deglycosylation of the glycoproteins produced a significant decrease or a complete loss of the antibacterial activity of the glycoproteins. This study explains the mechanism by which honey MRJP-1 produces bactericidal effects. The presence of high-mannose structures explains the lectin-like activity of MRJP-1, and the presence of Jelleins in MRJP-1 may explain cell wall disruptions (Brudzynski and Sjaarda 2015).

In another study, Revamil medical-grade honey, produced under standardized conditions in greenhouses, was shown to exert potent antibacterial activity against laboratory strains *B. subtilis* ATCC6633, *Staphylococcus aureus* 42D, *E. coli* ML-35p, and *Pseudomonas aeruginosa* PAO-1 (ATCC 15692) and against clinical isolates of methicillin-resistant *S. aureus* (MRSA), vancomycin-resistant *Enterococcus faecium*, extended-spectrum β-lactamase–producing *E. coli*, and ciprofloxacin-resistant *P. aeruginosa* (Kwakman et al. 2008). Kwakman et al. (2010) identified all the elements that contribute to the bactericidal activity of this honey using a novel approach of successive neutralization of individual honey bactericidal factors combined with activity-guided identification of unknown factors. After enzymatic neutralization of two potent bactericidal compounds – $H_2O_2$ and methylglyoxal (MGO) – this honey was found to retain substantial antibacterial activity, which is attributable to the presence of honeybee protein defensin-1. Neutralization of $H_2O_2$, MGO, and bee defensin-1 and subsequent adjustment of the pH to 7.0 significantly reduced the antibacterial activity of this honey, and the remaining was accountable solely for the sugar components present. This study suggested that the presence of $H_2O_2$, MGO, and honeybee defensin-1 is required for the broad-spectrum bactericidal activity of honey (Kwakman et al. 2010).

## Immunological Characterization of Honey Protein

Natural honey exhibits immunomodulatory potential, which is attributable to the presence of glycoproteins and glycopeptides, especially the MRJP-1 family proteins, in honey. MRJP-1 family proteins present in honey in both monomeric (55 kDa) as well as oligomeric form (450 kDa) (Simúth 2001). At least four proteins of MRJP-1 family protein were identified in honey, and they were shown to possess the ability to bind immunoglobulin (Ig) E in which the N-linked glycosylation in these proteins is specifically important for IgE–MRJP-1 binding (Hayashi et al. 2011; Won et al. 2009). Natural acacia honey–derived MRJP-1 protein exerts immunomodulatory effects by stimulating the production of tumor necrosis factor-α (TNF-α) in murine peritoneal macrophages, whereas deproteinized acacia honey exerts no effect on TNF-α secretion (Majtán et al. 2006). The authors suggested that honey protein, specially MRJP-1 – the dominant protein of honey – might be responsible for the immunomodulatory effects of honey.

Recently, Ziziphus honey was fractionated to isolate the glycoproteins and glycopeptides, which were then subjected to investigate for their immunomodulatory properties to identify the active honey protein responsible for immunomodulation. Ziziphus honey was fractionated into five proteins and peptides with molecular masses ranging from 2 to 450 kDa (Mesaik et al. 2014). These proteins and peptides showed potent inhibitory effects on the production of reactive oxygen species in zymosan-activated human neutrophils ($IC_{50}$ = 6–14 ng/mL [$IC_{50}$ is the concentration of a drug or compound needed to inhibit a biological process or response by 50%]) and murine macrophages ($IC_{50}$ = 2–9 ng/mL), and the inhibition was found to be dependent on protein concentration. The fractionated honey proteins can also suppress the production of nitric oxide significantly by lipopolysaccharide (LPS)-activated murine macrophages ($IC_{50} = 96 - 450$ ng/mL). Phagocytosis is the primary defense mechanism of the innate immune system to fight against invading microorganisms. Phagocytes (macrophages, dendritic cells [DCs]) trap the bacterial molecules with pattern recognition receptors present on their surface, which then lead them to engulf the pathogen and to wipe out the invaders through the professional killing mechanisms. However, the fractionated proteins obtained from Ziziphus honey showed significant inhibitory effect, roughly 55%–70%, on receptor-mediated phagocytosis of unopsonized latex beads by J774.2 mouse macrophage cells (Mesaik et al. 2014). Hence, honey proteins exert anti-inflammatory potential by interfering with phagocytotic process.

Pro-inflammatory cytokines such as TNF-α and interleukin-1β (IL-1β) play key roles in the initiation of the inflammation-arbitrated defense system. Honey stimulates the production of pro-inflammatory cytokines such as TNF-α, IL-1β, transforming growth factor-β (TGF-β), and matrix metalloproteinases (MMPs) by human monocytes (Tonks et al. 2003). Interestingly, this stimulation of pro-inflammatory cytokine production was shown to be mediated by the protein part of the honey glycoproteins (Majtán et al. 2006). However, Ziziphus honey glycoproteins did not affect IL-1β production by THP-1 human leukemia monocytic cells (Mesaik et al. 2014). Moreover, these glycoproteins significantly suppressed TNF-α production at higher concentration (2000 ng/mL) and stimulated it at lower concentration (80 ng/mL) (Mesaik et al. 2014). These studies suggest that honey glycoproteins and glycopeptides significantly interfere with molecules of the innate immune system.

Honey proteins have also been shown to bind specifically to IgE of the sera of honey-allergic patients. In a previous study, IgE binding honey proteins of molecular masses 54, 46, 17, and 16 kDa were recognized from the sera of three honey-allergic patients (De la Torre et al. 1996). At least five proteins (72, 60, 54, 33, and 30 kDa) from sunflower honey have also been shown capable of binding IgE (Bauer et al. 1996). Hayashi et al. (2011) identified four glycoproteins of varying molecular masses from each of three different honeys – (1) homemade honey composed of honeydew from several flowers, (2) dandelion (*Taraxacum officinale*) honey, and (3) Chinese milk vetch (*Astragalus sinicus*) honey that can specifically bind to IgE of serum from honey-allergic patients. Deglycosylation of these proteins with peptide N-glycosidase F enzyme leads to a drastic reduction in IgE binding activity of at least three IgE binding glycoproteins (54-, 110-, and 138-kDa proteins from homemade honey and 60-, 116-, and 138-kDa proteins from dandelion honey) (Hayashi et al. 2011). This finding suggests that N-linked oligosaccharide is a strong mediator of honey proteins to bind to IgE that is responsible for the immune activity of honey proteins. N-terminal amino acid sequence analysis revealed that the 60-kDa IgE-binding glycoprotein of dandelion honey is originally the MRJP-1.

Apalbumin 1 (Apa1), a 55-kDa glycoprotein of royal jelly produced by honeybees (Schmitzova et al. 1998), is found to be present as a regular protein component of honey (Šimúth et al. 2004). Initially, it was thought that Apa1 was important for the nutrition and development of queen and larval honeybees (Ohashi et al. 1997). Moreover, Apa1 has been shown to enhance hepatocyte proliferation (Kamakura et al. 2001). Peptides derived from enzymatic hydrolysis of Apa1 in the gastrointestinal tract show potent antihypertensive activity (Matsui et al. 2002). It is assumed that the antimicrobial peptides, Jelleines, could be derived from the C-terminal of Apa1 (Fontana et al. 2004). The oligomeric form of Apa1 (MW, 420 kDa) and the monomeric form (MW, 55 kDa) of Apa1 have the ability to stimulate TNF-α release from murine macrophages (Šimúth et al. 2004). Interestingly, the oligomeric form of Apa1 showed a lower stimulatory effect compared with the pronounced effect of the monomeric form. This could be due to the blockage of the active amino acid sequence motifs, which are responsible for the induction of TNF-α release, during oligomer formation by means of protein–protein interactions

(Majtán et al. 2006). In another study, the specific proteinous moiety of Apa1 glycoprotein responsible for the stimulation of macrophages and subsequent production of TNF-α was analyzed. The native honey showed the ability to stimulate TNF-α release from murine peritoneal macrophages, whereas the deproteinized honey had no effect on the release of TNF-α. This result suggests that the immunostimulatory effect of honey is based on its royal jelly protein content, primarily on its dominant protein Apa1 (Majtán et al. 2006). Moreover, the recombinant full-length Apa1 (rApa1) and recombinant protein fragments of four distinct moieties of Apa1, derived on the basis of partial tryptic hydrolysis, were prepared by heterologous expression in *E. coli*. The immunostimulatory effect of these rApa1 proteins (full length and fragments) and the native Apa1 glycoprotein was investigated, revealing that the rApa1 fragments (derived by partial tryptic digestion of Apa1) possess significantly enhanced immunostimulatory effect (Majtán et al. 2006). Furthermore, the recombinant N-terminal fragment of Apa1 was found to be the most prominent stimulator of TNF-α release compared with three other protein fragments of Apa1, as well as with the native Apa1 and rApa1 (Majtán et al. 2006).

Since ancient times, honey has been used as a potential antidote in wound healing. Randomized clinical trials also suggest that honey might speed up wound healing (Ingle et al. 2006; Yaghoobi and Kazerouni 2013). However, no study has been performed on the influence of honey on human skin cells (epidermal keratinocytes and dermal fibroblast). Keratinocytes are the epidermal cells that migrate to and proliferate in damaged regions to cover the denuded dermal surface in wound-healing processes in skin (Wawersik et al. 2001). Honey and its dominant protein MRJP-1 are shown to stimulate and induce elevated expression of inflammatory mediators, including cytokines (such as TNF-α, IL-1β, and TGF-β) and MMP-9 in human keratinocytes (Majtan et al. 2010). Moreover, the increased level of MMP-9 in the epidermis leads to degradation of type IV collagen in the basement membrane (Majtan et al. 2010).

A variety of honey types have been shown to stimulate human immune cells, especially monocytic cells to produce inflammatory cytokines such as TNF-α and IL-6, which play important roles in resolution of infection, wound healing, and tissue repair (Tonks et al. 2003). However, the honey components involved in immune cell stimulation and immune modulation are still unknown. Previously, it was thought that the presence of microorganisms or their cellular components in honey were responsible for the immune-stimulatory activity of honey (Tonks et al. 2007). Currently, MRJP-1 family proteins are largely thought to be involved in immune cell stimulation and cytokine production. However, the mechanism of honey protein–mediated immune cell stimulation is still not clearly understood. In a recent study, Tonks et al. (2007) proposed that a 5.8-kDa protein component isolated from manuka honey (but not bacterial contamination or the presence of bacterial cellular component in honey) stimulates human monocytes via interactions with Toll-like receptor 4 and thereby induces cytokine production. Initially, the authors investigated the possibility that honey-stimulated cytokine induction may be initiated by the presence of microbes or their components present as contaminants in honey. Honey is generally an unsuitable medium for bacterial growth because of its low water activity and acidic nature (Henriques et al. 2006); nevertheless, a number of reports have described bacterial and fungal contamination of honey (Nakano and Sakagucki 1991; Silva et al. 2017). However, bacterial spores of *Bacillus* species were isolated from the honey samples, which raises the possibility that honey samples may be contaminated with bacterial components, including LPS, and this LPS may be responsible for the cytokine-inducing activity of the honey. Because LPS induces inflammatory cytokine production in monocytic cells, the stimulatory activity of the honey samples and an LPS dose response were assessed in Mono Mac 6 (MM6) cells. When MM6 cells were stimulated with LPS at several-fold higher concentrations than that detected in the honey samples, cytokine production was considerably lower than that observed for honey. Statistical analysis revealed that the cytokine-stimulatory activity of honey did not correlate with LPS levels or with bacterial spore content. This study thus confirms that cytokine production by monocytes is not a consequence of bacterial contamination or the presence of bacterial cellular components (e.g. LPS) in honey (Tonks et al. 2007).

## Antigiardial Activity of Honey Protein

Giardiasis is an intestinal disease caused by the protozoan *Giardia lamblia* (also called *Giardia intestinalis* and *Giardia duodenalis*). *Giardia* organisms inhabit the digestive tract and have been the most common pathogenic parasitic infection in humans. In 2013, nearly 280 million people worldwide had symptomatic giardiasis (Esch and Petersen 2013). Besides the antibacterial and antifungal activities, honey has also been reported for its anti-amoebic (Waw and Alvieno 2012) and nematocidal (Azim and Sajid 2009) activities. However, the active component that accounts for honey's antiparasitic activity has not yet been well characterized. Recently, Mohammed et al. (2015) showed that Ziziphus honey proteins and peptides with molecular masses in the range of 2 to greater than 200 kDa contain active components of antigiardial activity, which exhibited *in vitro* activities against *G. lamblia* with $IC_{50}$ values of 25μg/mL or less. It was also shown that Ziziphus honey proteins possessed much stronger bioactivity than an antiprotozoal agent such as metronidazole, a common therapy for the treatment of patients with giardiasis (Mohammed et al. 2015).

### Functional Characterization of Honey's Arabinogalactan Proteins

Highly glycosylated arabinogalactan proteins (AGPs) with molecular masses of approximately 110 kDa have been isolated and characterized from New Zealand kanuka honey (Steinhorn et al. 2011). AGPs are proteoglycans that contain type II arabinogalactans and 2%–10% covalently attached protein that is typically rich in hydroxyproline and serine. AGPs are compact molecules composed of a protein core to which highly branched polysaccharide chains are attached, predominantly through O-linkages to hydroxyproline (Showalter 2001). AGPs are found in the female sexual tissues of flowers and thus could be a potential source of AGPs in honey. AGPs and type II arabinogalactans from different sources have been shown to exert immunomodulatory properties (Grønhaug et al. 2010). Recently, type II arabinogalactans and AGPs isolated from leaves of *Opilia celtidifolia* tree have been shown to be biologically active compounds that possess the ability to activate rat macrophages and DCs as well as to possess complement-fixating ability and to induce B-cell proliferation (Grønhaug et al. 2010). It was also observed that these bioactive elements can stimulate the production and secretion of pro-inflammatory cytokines from purified B cells and DCs (Grønhaug et al. 2010). These findings suggest that the arabinogalactans and AGP polymers present in *O. celtidifolia* leaves may have potential use in developing medical devices to regulate inflammatory processes.

### Major Honey Proteins in Detection of Adulteration

Honey proteins can be used for adulteration detection because the major proteins in honey differ in their MWs depending on the species of honeybee that produced the honey. Two major proteins having MWs of 56 and 59 kDa were identified and isolated from honey produced by *A. cerana* and *A. mellifera*, respectively (Won et al. 2008). Although their MWs are slightly different, they share a common primary structure. For further clarification, these purified honey major proteins were then injected repeatedly into rats to raise anti–honey protein antibodies (Won et al. 2009). The antibodies obtained from the rats were verified with Western blot analysis, which showed that each antibody could react specifically to its particular antigen with high affinity. These results indicated that the surface structures of major proteins in honey derived from *A. cerana* and from *A. mellifera* are different. Hence, the measurement of molecular mass of honey major proteins and the immune specificity of the honey major protein antigen–antibody reactions can be an ideal tool to discriminate honey from different honeybee species and subsequently can be useful for detection of adulteration.

### Honey Allergens

Honey produced by *Trigona laeviceps* (stingless bees) collected from Thailand was shown to possess many distinctive proteins. Among these proteins, a 29-kDa protein was judged to be pollen allergen Lol p VA precursor *Lolium perenne* (perennial ryegrass) (Chanchao 2009). This result is further supported by the findings of Gunduz et al. (2008) and Finola et al. (2007), which imply that honey could be contaminated with pollen. The presence of pollen and pollen allergen proteins (antigens) may cause intoxication of honey. This may raise the issue of the risk to sensitive (hypoallergenic) individuals already sensitized and allergic to pollen antigens from consuming honey.

## Enzymes in Honey

Enzymes are a class of functional proteins that typically catalyze specific chemical reactions. Honey usually contains a number of different types of active enzymes in small quantities, which makes it unique and makes the difference from all other sweetening agents. Enzymes in honey play an important role in determining its overall medicinal and functional properties. Diastase (amylase) and invertase (saccharase or sucrase) are the main enzymes present in honey (Ruth 2006). These enzymes are introduced into honey by honeybees. Glucose oxidase is another important enzyme that goes into honey during nectar processing. Honey also contains small quantities of phosphatase and catalase (Hive & Honey Apiary 2013). The presence of honey's enzymes also vary because of several factors, including the floral origin, nectar composition, and concentration and the age of the bees. Because the enzymes present in honey are heat sensitive and visible and ultraviolet light sensitive, it is recommended to store honey away from sunlight and to avoid extreme heating to preserve its natural properties.

## Honey Enzymes and Their Functions

Honey contains several enzymes that play important role in metabolic processes as well as in its antimicrobial activities. Table 5.2 summarizes the enzymes present in honey and their functions.

### Diastase

Diastase, most commonly known as amylase, is the enzyme that converts starch to simpler sugars. Diastase is added in the nectar by honeybees during the collection and ripening of nectar into honey (Oddo et al. 1990). Nearly all honeys contain diastase in varying amounts, and the floral origin of honey is one of the factors that greatly influences the diastase content. For example, honeys of citrus and clover origin are likely to contain lower amounts of diastase. Other factors that may affect diastase values include the natural difference in pH among honey, nectar flow, and the bees' foraging patterns. Diastase is a fairly stable enzyme in honey. Long storage of honey at moderate temperatures and exposure to high temperatures cause inactivation of diastase enzymes present in honey (Babacan and Rand 2005). The half-lives of diastase are 1480 days, 200 days, and 1.2 hours at 20°C, 30°C, and 80°C, respectively (Airborne 1999). Therefore, the level of diastase present in honey could be used as an indicator of the freshness of honey because the enzyme activity gradually decreases in old or heated honey.

### Invertase

Invertase in honey is the enzyme that catalyzes the hydrolysis of sucrose in nectar into its constituent simple sugars, glucose and fructose. Invertase is also called sucrase. The catalytic activity of this enzyme is a key step of ripening nectar to honey. The ripest honey contains a very small amount of sucrose (usually <5%) because of the catalytic activity of invertase. It is added to the nectar by honeybees. Honey contains a small amount of invertase, which is inactivated by heating. The half-lives of invertase are 820 days, 83 days, and 8.6 minutes at 20°C, 30°C, and 80°C, respectively (Airborne 1999). Similar to diastase, invertase content and activity are also greatly influenced by the floral origin of honey. For example, honeydew honey is likely to possess the highest level of invertase activity compared with the rest of the honey varieties.

### Glucose Oxidase

Glucose oxidase is another interesting enzyme in honey that is important for its relation to the antibacterial properties of honey. It oxidizes the glucose in honey to form gluconic acid and $H_2O_2$; the latter is the key agent responsible for the antibacterial activity of most honeys. On the other hand, gluconic acid is the major acid found in honey that predominantly contributes to most of a honey's acidity. Glucose oxidase is a functionally active enzyme in nectar and remains almost inactive in ripe honey because of the low-pH conditions. In fact, the production of gluconic acid and $H_2O_2$ mostly takes place during the honey-ripening process. Although the enzyme remains in ripe honey, the formation of gluconic acid and $H_2O_2$ from glucose is very slow in ripe honey. Glucose oxidase activity speeds up again if the honey is diluted. When honey is diluted, glucose oxidase is activated and acts on endogenous glucose to produce $H_2O_2$. In fact, the maximum level of

**Table 5.2** Enzymes present in honey and their functions.

| Enzymes | Catalyzed Chemical Reactions |
|---|---|
| Diastase, amylase | Converts starch into simpler carbohydrates such as oligo-, di-, and monosaccharides |
| Invertase, sucrase | Degrades sucrose to glucose and fructose |
| Glucose oxidase | Produces gluconolactone (which in turn converts to gluconic acid) and hydrogen peroxide |
| Catalase | Decomposes hydrogen peroxide to produce water and oxygen |
| Acid phosphatase | Removes phosphate groups from organic phosphates |
| Esterase | Breaks down ester bonds |
| β-Glucosidase | Hydrolyzes the terminal non-reducing residue in β-D-glucosides and release glucose |
| Protease | Catalyzes the cleavage of peptide bonds in protein |

$H_2O_2$ can be achieved by diluting honey by 30%–50%, typically in the range of 5–100 μg $H_2O_2$/g honey (equivalent to ~0.146–2.93 mM) (Almasaudi 2021). Therefore, the concentration of honey is an important factor in determining the bactericidal activity of honey.

The presence of glucose oxidase was detected in the pharyngeal glands of honeybees. Hence, honeybees are thought to be the source of glucose oxidase in honey. Glucose oxidase activity also greatly differs among the honeys depending on the honey sources. Glucose oxidase activity is sensitive to light and heat. Some honeys lose their glucose oxidase activity upon exposure to small amounts of visible light, but others are not affected even if exposed to strong sunlight (Airborne 1999).

#### Catalase

Catalase is the enzyme that decomposes $H_2O_2$. Hence, the presence of catalase reduces the antibacterial activity of honey produced from glucose oxidase activity. Catalase is found in some honeys and is thought to be present in the honey from the nectar of the plant source. The variability in the apparent $H_2O_2$ production in different honeys could be due to the absence or presence of catalase in the nectar from a particular plant species (Airborne 1999).

#### Protease

Honeybees are well known to use the exopeptidase leucine aminopeptidase and three endopeptidases – trypsin, chymotrypsin, and elastase – to digest dietary proteins (Burgess et al. 1996). However, the presence of proteolytic enzymes in honey has only recently been described (Rossano et al. 2012).

In a recent study, honey samples were analyzed to detect the presence of proteolytic enzymes using a bidimensional zymography (2-DZ) technique (Rossano et al. 2012). 2-DZ allows for the detection of the entire proteolytic network present in a biological sample, including isoforms and post-translational variants. This study demonstrated that honey contains proteases, mostly serine proteases, because the proteolytic activities of honey were almost completely inhibited by the serine protease–specific inhibitor phenylmethylsulfonyl fluoride. The origin of honey proteases may be the nectar, the pollen, or the secretions from cephalic glands of honeybees. In honeybees, proteases, particularly serine proteases, not only take part in digestive process but also play an important role in development, immune response, and defense against pathogens (Bogaerts et al. 2009; Zou et al. 2006).

### Honey Enzymes in Relation to Honey Fermentation

Acid phosphatase, one of the enzymes present in honey, has been found to be involved in the fermentation of honey. Acid phosphatase is mainly present in pollen; however, it is also found in nectar. Acid phosphatase shows a positive correlation with honey fermentation. Honeys that have higher acid phosphatase activities become fermented more easily, and the unfermented honeys have relatively lower amounts of acid phosphatase activities. The pH of honey strongly influences its acid phosphatase activity – the higher the pH, the lower the decrease in acid phosphatase activity (Alonso-Torre et al. 2006).

### Honey Enzymes Linked to the Antibacterial Properties of Honey

Traditionally and since ancient times, honey has been used as a substance in wound healing even when the elements associated with healing and antimicrobial activity had not yet been discovered. Now honey has been shown to possess significant antibacterial properties. During the process of honey formation from nectar, honeybees secrete invertase (sucrase) enzymes from their enzyme-rich saliva, which turns the sucrose into glucose and fructose. These sugars then bind to the water in nectar, leaving a concentrated mixture in which bacteria cannot survive (Little Tree Oriental Healing Arts 2001). The antibacterial activity of honey was thought to be due to the high osmotic properties of honey. However, some antibacterial activity of honey remains persistent, suggesting that honey contains some additional factors that are responsible for its antibacterial activity.

Honey is known to contain $H_2O_2$, which accounts for the antibacterial activity of honey. The formation of $H_2O_2$ might occur through the catalytic activity of honey's glucose oxidase enzyme, which converts the glucose into gluconolactone and in turn yields gluconic acid and $H_2O_2$ (Weston 2000). The latter is a deadly agent to microorganisms, including *E. coli*, *Helicobacter pylori* (associated with stomach ulcers), *Salmonella* spp., and antibiotic-resistant bacteria (including hospital superbug MRSA) ( Little Tree Oriental Healing Arts 2001). $H_2O_2$ also stimulates proteolytic enzymes that digest dead tissues and promotes growth of blood vessels, which deliver oxygen, nutrients, and fibroblasts and aid in forming new connective tissue (Little Tree Oriental Healing Arts 2001).

## Discussion

Honey contains a fairly good amount of carbohydrates as well as other minor nutrients, including minerals, vitamins, free amino acids, proteins, and enzymes (da Silva et al. 2016; Machado De-Melo et al. 2018). Moreover, the occurrence of flavonoids, phenolic compounds, and various catalytic enzymes in honey contributes to its role as a natural antioxidant and antimicrobial (Gheldof et al. 2002). Thus, regular use of honey could significantly contribute to protecting human health. To ensure the maximum benefit of the medicinal applications of honey, it is important to control the quality as well as the authenticity of honey. Maintaining honey's quality is particularly important for two major reasons. First, the composition of honey changes during that storage period, resulting from a number of chemical reactions, including fermentation, oxidation, and thermal processing (Moreira et al. 2010). Second, because honey is a relatively expensive food and the market demand is also fairly high for its nutritional and medicinal value, honey has become an object of numerous adulterations. This is why the testing and maintaining of honey parameters may be useful for detecting possible adulterations, as well as for confirming the proper conditions for the manipulation and the storage of honey. Honey quality control could be achieved based on guidelines of the Codex Alimentarius, which describes ash content, electrical conductivity, and insoluble solids in water as purity control parameters and sugar and 5-hydroxymethylfurfural content, diastase activity, acidity, and moisture as the maturity indicators (Codex Alimentarius 2001). However, recently, free amino acid content, protein content, and their variability have also been used as effective parameters for determining the quality, origin, and authenticity of honey (Bilikova et al. 2015; Mohammed and Azim 2012; Wang et al. 2009; Won et al. 2008).

Proline is the most abundant free amino acid (50%–85%) in honey (Del Campo et al. 2016), which could be used as a marker to determine the botanical origin of honey. However, it became impractical to classify unifloral honey on the basis of this parameter only because of the large variation of proline content in different unifloral honeys. Total amino acid profile has been indicated as a good indicator of both the botanical and geographical origins of honey (Azevedo et al. 2017; Sun et al. 2017). Moreover, because pollen is the key source of amino acids in honey, profiling of pollen's amino acids could also be an effective parameter for determining the botanical origin of honey. Honey contains a very low amount of protein, and the content varies with the type of honeybee as well as the botanical sources. Honey contains several proteins that can exert antimicrobial, anti-inflammatory, antitumor, and anticancer activity (Bardy et al. 2012; Gomes et al. 2010; Taormina et al. 2001; Tonks et al. 2003; Tsiapara et al. 2009). These biological functions of honey are mostly attributed to the most functionally active honey proteins, namely, the MRJPs (Chua et al. 2015; Guo et al. 2009). Honey enzymes also contribute to the production of $H_2O_2$, which drives the antibacterial activity of honey (Weston 2000). Honey proteins are extensively used as markers for the determination of honey's authenticity and adulteration as well as a quality indicator (Bilikova et al. 2015; Won et al. 2008). MRJP-1 has been suggested as a suitable marker for authentication of honeybee origins (Won et al. 2009).

Honey is widely consumed for its biological and medicinal value, although it can also be used as a nutritional supplement. Even though the concentrations of amino acids, proteins, and enzymes in honey vary on several factors, including the types of honeybee, season, and storage time as well as the botanical origin of the honey, a complete profiling of these components and their combinatorial analysis could be a novel and effective approach for authentication and quality control of honey.

## References

Abu-Jdayil, B., Ghzawi, A.A.-M., Al-Malah, K.I., and Zaitoun, S. (2002). Heat effect on rheology of light-and dark-colored honey. *Journal of Food Engineering* 51: 33–38.

Adnan, N.A., Chua, L.S., and Sarmidi, M.R. (2014). Thermal treatment effect on free amino acids in honey samples. *Jurnal Teknologi* 69.

Airborne. Technical information: enzymes. http://www.airborne.co.nz/enzymes.shtml.

Almasaudi, S. (2021). The antibacterial activities of honey. *Saudi Journal of Biological Sciences* 28: 2188–2196.

Alonso-Torre, S., Cavia, M., Fernández-Muiño, M., et al. (2006). Evolution of acid phosphatase activity of honeys from different climates. *Food Chemistry* 97: 750–755.

Anklam, E. (1998). A review of the analytical methods to determine the geographical and botanical origin of honey. *Food Chemistry* 63: 549–562.

Azevedo, M.S., Seraglio, S.K.T., Rocha, G., et al. (2017). Free amino acid determination by GC-MS combined with a chemometric approach for geographical classification of bracatinga honeydew honey (Mimosa scabrella Bentham). *Food Control* 78: 383–392.

Azim, M.K. and Sajid, M. (2009). Evaluation of nematocidal activity in natural honey. *Pakistan Journal of Botany* 41: 3261–3264.

Babacan, S. and Rand, A.G. (2005). Purification of amylase from honey. *Journal of Food Science* 70: c413–c8.

Bardy, J., Molassiotis, A., Ryder, W.D., et al. (2012). A double-blind, placebo-controlled, randomised trial of active manuka honey and standard oral care for radiation-induced oral mucositis. *British Journal of Oral and Maxillofacial Surgery* 50: 221–226.

Bauer, L., Kohlich, A., Hirschwehr, R., et al. (1996). Food allergy to honey: pollen or bee products?: characterization of allergenic proteins in honey by means of immunoblotting. *Journal of Allergy and Clinical Immunology* 97: 65–73.

Bilikova, K., Kristof Krakova, T., Yamaguchi, K., and Yamaguchi, Y. (2015). Major royal jelly proteins as markers of authenticity and quality of honey. *Arhiv za higijenu rada i toksikologiju* 66: 259–267.

Biluca, F.C., Bernal, J., Valverde, S., et al. (2019). Determination of free amino acids in stingless bee (Meliponinae) honey. *Food Analytical Methods* 12: 902–907.

Bocian, A., Buczkowicz, J., Jaromin, M., et al. (2019). An effective method of isolating honey proteins. *Molecules* 24: 2399.

Bodor, Z., Kovacs, Z., Benedek, C., et al. (2021). Origin Identification of Hungarian Honey Using Melissopalynology, Physicochemical Analysis, and Near Infrared Spectroscopy. *Molecules* 26: 7274.

Bogaerts, A., Baggerman, G., Vierstraete, E., et al. (2009). The hemolymph proteome of the honeybee: gel-based or gel-free? *Proteomics* 9: 3201–3208.

Bogdanov, S., Lüllmann, C., Martin, P., et al. (1999). Honey quality and international regulatory standards: review by the International Honey Commission. *Bee World* 80: 61–69.

Bogdanov, S. and Martin, P. (2002). Honey authenticity. *Mitteilungen aus Lebensmitteluntersuchung und Hygiene* 93: 232–254.

Borutinskaite, V., Treigyte, G., Čeksteryte, V., et al. (2018). Proteomic identification and enzymatic activity of buckwheat (Fagopyrum esculentum) honey based on different assays. *Journal of Food & Nutrition Research* 57: 57–69.

Brudzynski, K. and Sjaarda, C. (2015). Honey glycoproteins containing antimicrobial peptides, Jelleins of the Major Royal Jelly Protein 1, are responsible for the cell wall lytic and bactericidal activities of honey.

Burgess, E.P., Malone, L.A., and Christeller, J.T. (1996). Effects of two proteinase inhibitors on the digestive enzymes and survival of honey bees (Apis mellifera). *Journal of Insect Physiology* 42: 823–828.

Carratù, B., Ciarrocchi, M., Mosca, M., and Sanzini, E. (2011). Free amino acids, oxalate and sulphate for honey characterization. *Journal of Api Product & Api Medical Science* 3: 81–88.

Chanchao, C. (2009). Antimicrobial activity by Trigona laeviceps (stingless bee) honey from Thailand. *Pakistan Journal of Medical Sciences* 25: 364–369.

Chen, H., Jin, L., Chang, Q., et al. (2017). Discrimination of botanical origins for Chinese honey according to free amino acids content by high-performance liquid chromatography with fluorescence detection with chemometric approaches. *Journal of the Science of Food and Agriculture* 97: 2042–2049.

Chin, N.L. and Sowndhararajan, K. (2020). A review on analytical methods for honey classification, identification and authentication. *Honey Analysis—New Advances and Challenges* 1–33.

Chua, L.S. and Adnan, N.A. (2014). Biochemical and nutritional components of selected honey samples. *Acta Scientiarum Polonorum, Technologia Alimentaria* 13: 169–179.

Chua, L.S., Lee, J.Y., and Chan, G.F. (2013). Honey protein extraction and determination by mass spectrometry. *Analytical and Bioanalytical Chemistry* 405: 3063–3074.

Chua, L.S., Lee, J.Y., and Chan, G.F. (2015). Characterization of the Proteins in Honey. *Analytical Letters* 48: 697–709.

Codex Alimentarius. (2001). Revised standard for honey. Codex Standard 12-1981. Rev 1 (1987), Rev 2 (2001), Rome: FAO

Cotte, J.-F., Casabianca, H., Giroud, B., et al. (2004). Characterization of honey amino acid profiles using high-pressure liquid chromatography to control authenticity. *Analytical and Bioanalytical Chemistry* 378: 1342–1350.

Crane, E. Honey: a comprehensive survey; 1975.

da Silva, P.M., Gauche, C., Gonzaga, L.V., et al. (2016). Honey: chemical composition, stability and authenticity. *Food Chemistry* 196: 309–323.

De la Torre, F., Garcia, J., Martinez, A., et al. (1996). IgE binding proteins in honey: discussion on their origin. *Journal of Investigational Allergology & Clinical Immunology* 7: 83–89.

Del Campo, G., Zuriarrain, J., Zuriarrain, A., and Berregi, I. (2016). Quantitative determination of carboxylic acids, amino acids, carbohydrates, ethanol and hydroxymethylfurfural in honey by 1H NMR. *Food Chemistry* 196: 1031–1039.

Dong, R., Zheng, Y., and Xu, B. (2013). Floral origin identification and amino acid profiles of chinese unifloral honeys. *International Journal of Food Properties* 16: 1860–1870.

Esch, K.J. and Petersen, C.A. (2013). Transmission and epidemiology of zoonotic protozoal diseases of companion animals. *Clinical Microbiology Reviews* 26: 58–85.

Escriche, I., Visquert, M., Juan-Borrás, M., and Fito, P. (2009). Influence of simulated industrial thermal treatments on the volatile fractions of different varieties of honey. *Food Chemistry* 112: 329–338.

Finola, M.S., Lasagno, M.C., and Marioli, J.M. (2007). Microbiological and chemical characterization of honeys from central Argentina. *Food Chemistry* 100: 1649–1653.

Fontana, R., Mendes, M.A., De Souza, B.M., et al. (2004). Jelleines: a family of antimicrobial peptides from the Royal Jelly of honeybees (Apis mellifera). *Peptides* 25: 919–928.

Gheldof, N., Wang, X.-H., and Engeseth, N.J. (2002). Identification and quantification of antioxidant components of honeys from various floral sources. *Journal of Agricultural and Food Chemistry* 50: 5870–5877.

Gomes, S., Dias, L.G., Moreira, L.L., et al. (2010). Physicochemical, microbiological and antimicrobial properties of commercial honeys from Portugal. *Food and Chemical Toxicology* 48: 544–548.

Grønhaug, T.E., Ghildyal, P., Barsett, H., et al. (2010). Bioactive arabinogalactans from the leaves of Opilia celtidifolia Endl. ex Walp. (Opiliaceae). *Glycobiology*. 20: 1654–1664.

Gunduz, A., Turedi, S., Russell, R.M., and Ayaz, F.A. (2008). Clinical review of grayanotoxin/mad honey poisoning past and present. *Clinical Toxicology* 46: 437–442.

Guo, H., Kouzuma, Y., and Yonekura, M. (2009). Structures and properties of antioxidative peptides derived from royal jelly protein. *Food Chemistry* 113: 238–245.

Guo, W., Liu, Y., Zhu, X., and Wang, S. (2011). Temperature-dependent dielectric properties of honey associated with dielectric heating. *Journal of Food Engineering* 102: 209–216.

Hayashi, T., Takamatsu, N., Nakashima, T., and Arita, T. (2011). Immunological characterization of honey proteins and identification of MRJP 1 as an IgE-binding protein. *Bioscience, Biotechnology, and Biochemistry* 75: 556–560.

Henriques, A., Jackson, S., Cooper, R., and Burton, N. (2006). Free radical production and quenching in honeys with wound healing potential. *Journal of Antimicrobial Chemotherapy* 58: 773–777.

Hermosín, I., Chicón, R.M., and Cabezudo, M.D. (2003). Free amino acid composition and botanical origin of honey. *Food Chemistry* 83: 263–268.

Hive & Honey Apiary. (2013). Important nutritional facts about honey. http://www.hiveandhoneyapiary.com/NutritionalFactsaboutHoney.html.

Ibrahim, H.R., Nanbu, F., and Miyata, T. (2021). Potent antioxidant peptides derived from honey major protein enhance tolerance of eukaryotic cells toward oxidative stress. *Food Production, Processing and Nutrition* 3: 1–10.

Ingle, R., Levin, J., and Polinder, K. (2006). Wound healing with honey-a randomised controlled trial. *South African Medical Journal* 96: 831–835.

Janiszewska, K., Aniołowska, M., and Nowakowski, P. (2012). Free amino acids content of honeys from Poland. *Polish Journal of Food and Nutrition Sciences* 62: 85–89.

Kamakura, M., Suenobu, N., and Fukushima, M. (2001). Fifty-seven-kDa protein in royal jelly enhances proliferation of primary cultured rat hepatocytes and increases albumin production in the absence of serum. *Biochemical and Biophysical Research Communications* 282: 865–874.

Kečkeš, J., Trifković, J., Andrić, F., et al. (2013). Amino acids profile of Serbian unifloral honeys. *Journal of the Science of Food and Agriculture* 93: 3368–3376.

Kıvrak, Ş. (2017). Analysis of amino acid and phenolic content in honey by UPLC-ESI-MS/MS. *Honey Analysis* 79: 93.

Kowalski, S., Kopuncová, M., Ciesarová, Z., and Kukurová, K. (2017). Free amino acids profile of Polish and Slovak honeys based on LC–MS/MS method without the prior derivatisation. *Journal of Food Science and Technology* 54: 3716–3723.

Kwakman, P.H., te Velde, A.A., de Boer, L., et al. (2010). How honey kills bacteria. *The FASEB Journal* 24: 2576–2582.

Kwakman, P.H., Van den Akker, J.P., Güçlü, A., et al. (2008). Medical-grade honey kills antibiotic-resistant bacteria in vitro and eradicates skin colonization. *Clinical Infectious Diseases* 46: 1677–1682.

Lee, D.-C., Lee, S.-Y., Cha, S.-H., et al. (1998). Discrimination of native bee-honey and foreign bee-honey by SDS-PAGE. *Korean Journal of Food Science and Technology* 30: 1–5.

Little Tree Oriental Healing Arts. (2001). Bee healthy – why honey is the bee's knees when it comes to healing. http://www.littletree.com.au/manuka.htm.

Machado De-Melo, A.A., Almeida-Muradian, L., Sancho, M.T., and Pascual-Maté, A. (2018). Composition and properties of Apis mellifera honey: a review. *Journal of Apicultural Research* 57: 5–37.

Majtán, J., Kováčová, E., Bíliková, K., and Šimúth, J. (2006). The immunostimulatory effect of the recombinant apalbumin 1–major honeybee royal jelly protein–on TNFα release. *International Immunopharmacology* 6: 269–278.

Majtan, J., Kumar, P., Majtan, T., et al. (2010). Effect of honey and its major royal jelly protein 1 on cytokine and MMP-9 mRNA transcripts in human keratinocytes. *Experimental Dermatology* 19: e73–e9.

Marshall, T. and Williams, K.M. (1987). Electrophoresis of honey: characterization of trace proteins from a complex biological matrix by silver staining. *Analytical Biochemistry* 167: 301–303.

Matsui, T., Yukiyoshi, A., Doi, S., et al. (2002). Gastrointestinal enzyme production of bioactive peptides from royal jelly protein and their antihypertensive ability in SHR. *The Journal of Nutritional Biochemistry* 13: 80–86.

Mesaik, M.A., Dastagir, N., Uddin, N., et al. (2014). Characterization of Immunomodulatory Activities of Honey Glycoproteins and Glycopeptides. *Journal of Agricultural and Food Chemistry* 63: 177–184.

Mohammed, S.E.A. and Azim, M.K. (2012). Characterisation of natural honey proteins: implications for the floral and geographical origin of honey. *International Journal of Food Science & Technology* 47: 362–368.

Mohammed, S.E.A., Kabashi, A.S., Koko, W.S., and Azim, M.K. (2015). Antigiardial activity of glycoproteins and glycopeptides from Ziziphus honey. *Natural Product Research* 29: 2100–2102.

Moreira, R.F., De Maria, C.A., Pietroluongo, M., and Trugo, L.C. (2010). Chemical changes in the volatile fractions of Brazilian honeys during storage under tropical conditions. *Food Chemistry* 121: 697–704.

Nakano, H. and Sakagucki, G. (1991). An unusually heavy contamination of honey products by Clostridium botulinum type F and Bacillus alvei. *FEMS Microbiology Letters* 79: 171–178.

Oddo, L.P., Baldi, E., and Accorti, M. (1990). Diastatic activity in some unifloral honeys. *Apidologie* 21: 17–24.

Oddo, L.P., Piro, R., Bruneau, É., et al. (2004). Main European unifloral honeys: descriptive sheets. *Apidologie* 35: S38–S81.

Ohashi, K., Natori, S., and Kubo, T. (1997). Change in the mode of gene expression of the hypopharyngeal gland cells with an age-dependent role change of the worker honeybee apis mellifera L. *European Journal of Biochemistry* 249: 797–802.

Paramás, A.M.G., Bárez, J.A.G., Marcos, C.C., et al. (2006). HPLC-fluorimetric method for analysis of amino acids in products of the hive (honey and bee-pollen). *Food Chemistry* 95: 148–156.

Pätzold, R. and Brückner, H. (2006). Gas chromatographic detection of D-amino acids in natural and thermally treated bee honeys and studies on the mechanism of their formation as result of the Maillard reaction. *European Food Research and Technology* 223: 347–354.

Qamer, S., Ehsan, M., Nadeem, S., and Shakoori, A.R. (2007). Free amino acids content of Pakistani unifloral honey produced by Apis mellifera. *Pakistan Journal of Zoology* 39: 99.

Rossano, R., Larocca, M., Polito, T., et al. (2012). What are the proteolytic enzymes of honey and what they do tell us? A fingerprint analysis by 2-D zymography of unifloral honeys.

Ruth, T. (2006). Benefits of honey: why digestive enzymes are so vital to us. http://www.benefits-of-honey.com/digestive-enzymes.html.

Schievano, E., Morelato, E., Facchin, C., and Mammi, S. (2013). Characterization of markers of botanical origin and other compounds extracted from unifloral honeys. *Journal of Agricultural and Food Chemistry* 61: 1747–1755.

Schmitzova, J., Klaudiny, J., Albert, Š., et al. (1998). A family of major royal jelly proteins of the honeybee Apis mellifera L. *Cellular and Molecular Life Sciences CMLS* 54: 1020–1030.

Showalter, A. (2001). Arabinogalactan-proteins: structure, expression and function. *Cellular and Molecular Life Sciences CMLS* 58: 1399–1417.

Silva, M.S., Rabadzhiev, Y., Eller, M.R., et al. (2017). Microorganisms in honey. *Honey Analysis* 500: 233–257.

Simúth, J. (2001). Some properties of the main protein of honeybee (Apis mellifera) royal jelly. *Apidologie* 32: 69–80.

Šimúth, J., Bíliková, K., Kováčová, E., et al. (2004). Immunochemical approach to detection of adulteration in honey: physiologically active royal jelly protein stimulating TNF-α release is a regular component of honey. *Journal of Agricultural and Food Chemistry* 52: 2154–2158.

Steinhorn, G., Sims, I.M., Carnachan, S.M., et al. (2011). Isolation and characterisation of arabinogalactan-proteins from New Zealand kanuka honey. *Food Chemistry* 128: 949–956.

Subramanian, R., Umesh Hebbar, H., and Rastogi, N. (2007). Processing of honey: a review. *International Journal of Food Properties* 10: 127–143.

Sun, Z., Zhao, L., Cheng, N., et al. (2017). Identification of botanical origin of Chinese unifloral honeys by free amino acid profiles and chemometric methods. *Journal of Pharmaceutical Analysis* 7: 317–323.

Taormina, P.J., Niemira, B.A., and Beuchat, L.R. (2001). Inhibitory activity of honey against foodborne pathogens as influenced by the presence of hydrogen peroxide and level of antioxidant power. *International Journal of Food Microbiology* 69: 217–225.

Tonks, A., Cooper, R., Jones, K., et al. (2003). Honey stimulates inflammatory cytokine production from monocytes. *Cytokine* 21: 242–247.

Tonks, A.J., Dudley, E., Porter, N., et al. (2007). A 5.8-kDa component of manuka honey stimulates immune cells via TLR4. *Journal of Leukocyte Biology* 82: 1147–1155.

Truzzi, C., Annibaldi, A., Illuminati, S., et al. (2014). Determination of proline in honey: comparison between official methods, optimization and validation of the analytical methodology. *Food Chemistry* 150: 477–481.

Tsiapara, A.V., Jaakkola, M., Chinou, I., et al. (2009). Bioactivity of Greek honey extracts on breast cancer (MCF-7), prostate cancer (PC-3) and endometrial cancer (Ishikawa) cells: profile analysis of extracts. *Food Chemistry* 116: 702–708.

Turhan, I., Tetik, N., Karhan, M., et al. (2008). Quality of honeys influenced by thermal treatment. *LWT-Food Science and Technology* 41: 1396–1399.

Wakhle, D., Phadke, R., Pais, D., and Shakuntala Nair, K. (1996). Design for honey processing unit–Part II. *Indian Bee Journal* 58: 5–9.

Wang, J., Kliks, M.M., Qu, W., et al. (2009). Rapid determination of the geographical origin of honey based on protein fingerprinting and barcoding using MALDI TOF MS. *Journal of Agricultural and Food Chemistry* 57: 10081–10088.

Waw, W. and Alvieno, S. (2012). In vitro study on anti-amebic activity of Tualang honey against Entamoeba histolytica trophozoite. *Health and Environmental Journal* 3: 92–97.

Wawersik, M.J., Mazzalupo, S., Nguyen, D., and Coulombe, P.A. (2001). Increased levels of keratin 16 alter epithelialization potential of mouse skin keratinocytes in vivo and ex vivo. *Molecular Biology of the Cell* 12: 3439–3450.

Weston, R.J. (2000). The contribution of catalase and other natural products to the antibacterial activity of honey: a review. *Food Chemistry* 71: 235–239.

White, J. and Landis, W. (1980). Honey composition and properties. *Beekeeping in the United States Agriculture Handbook* 335: 82–91. https://www.beesource.com/threads/honey-composition-and-properties.365910/

Won, S.-R., Lee, D.-C., Ko, S.H., et al. (2008). Honey major protein characterization and its application to adulteration detection. *Food Research International* 41: 952–956.

Won, S.-R., Li, C.-Y., Kim, J.-W., and Rhee, H.-I. (2009). Immunological characterization of honey major protein and its application. *Food Chemistry* 113: 1334–1338.

Yaghoobi, R. and Kazerouni, A. (2013). Evidence for clinical use of honey in wound healing as an anti-bacterial, anti-inflammatory anti-oxidant and anti-viral agent: a review. *Jundishapur Journal of Natural Pharmaceutical Products* 8: 100.

Zou, Z., Lopez, D.L., Kanost, M.R., et al. (2006). Comparative analysis of serine protease-related genes in the honey bee genome: possible involvement in embryonic development and innate immunity. *Insect Molecular Biology* 15: 603–614.

# 6

# Vitamins

*Ng Choon Ming, Md. Ibrahim Khalil, and Siew Hua Gan*

## Introduction

Honey is a natural bee product, valued for millennia (Rao et al. 2016). In ancient times, the sweet taste of honey was considered a gift from the gods. Honey is produced from the nectar of flowers by bees, stored in the hive or honeycomb, and further processed into its final form. Today, raw honey is still popular because of its wide-ranging health, nutritional, and medicinal value. In terms of its nutritional content, honey is largely made up of sugar and water, with trace amounts of vitamins, minerals, and other minor compounds. Given that honey is a source of food for bees and basically originates from flowers, much interest has been piqued about the vitamin content of this natural product, particularly regarding human nutrition.

Generally, vitamins are vital organic compounds needed by humans in varying amounts throughout the life course for normal body function and health promotion (Gironés-Vilaplana et al. 2017). Because vitamins cannot be synthesized in the human body or are present in an insufficient quantity, food sources of these essential vitamins become crucial (Godswill et al. 2020). The specific vitamins present in honey include water-soluble vitamins (vitamins C and B complex) and fat-soluble vitamins (vitamins A, D, E, and K). These vitamins are present in a small but detectable amounts, to the extent of describing it in parts per million (Table 6.1).

## Vitamin Content of Honey

From the early days, the determination and quantification of vitamins present in honey have made notable progress (Figure 6.1). First, the early exploration of the vitamin content of honey through animal feeding tests was mostly futile; vitamins were present in a negligible amount that was considered insufficient to support growth or protect against diseases in the animals studied (Dutcher 1918; Hawk et al. 1920).

In the early 1940s, the analysis of honey based on chemical and microbiological methods showed the presence of various vitamins in small but detectable amounts (Haydak et al. 1942; Kitzes et al. 1943). By then, it seemed evident that there is a considerable variation in the vitamin content of honey produced from different sources in the United States (31 honeys from different plants and regions) and different countries (7 countries) (Haydak et al. 1942), seen in Table 6.2.

More recently, biological assay and chemical analysis methods were used and validated to assess the vitamins present in different honeys (Ciulu et al. 2011). The water-soluble vitamins in honey were evaluated by chromatographic, titrimetric, spectrometric, and amperometric procedures (Adenekan et al. 2015; Castro et al. 2001; Da Silva et al. 2012; Gheldof et al. 2002; León-Ruiz et al. 2011). Additionally, ascorbic acid, riboflavin, and its vitamers were analyzed using liquid chromatography (Bonta et al. 2013; Viñas et al. 2004). Vitamins C, A, and E have been determined using gas chromatography–mass spectrometry analysis as well (Muhammad et al. 2015). Regardless of the analysis method, the vitamin content in honey seems to be consistently low. Using reverse-phase high-performance liquid chromatography (RP-HPLC gradient mode) analysis, the water-soluble B vitamin content in honey in Italy was similar and consistently below 40 mg/kg, as seen in Table 6.3 (Ciulu et al. 2011). However, other water-soluble vitamins typically present in honey, such as vitamins $B_1$ and $B_6$, were not analyzed. Later, comprehensive analysis of water-soluble vitamins using chromatography analysis (RP-HPLC isocratic mode) of six monoflorals and on honeydew honey detected a minimal amount of the studied vitamins (Table 6.4).

*Honey: Composition and Health Benefits*, First Edition. Edited by Md. Ibrahim Khalil, Gan Siew Hua, and Bey Hing Goh.

**Table 6.1** General vitamin content of honey.

| Vitamins | Amount (mg/100 g of honey) | Parts per million (ppm) |
|---|---|---|
| Retinol (vitamin A) | 0.49[a] | – |
| Thiamine (vitamin $B_1$) | 0.00–0.01 | 0.06 |
| Riboflavin (vitamin $B_2$) | 0.01–0.02 | 0.63 |
| Niacin (vitamin $B_3$) | 0.10–0.20 | 3.2 |
| Pantothenic acid (vitamin $B_5$) | 0.02–0.11 | 0.96 |
| Pyridoxine (vitamin $B_6$) | 0.01–0.32 | 3.2 |
| Biotin (vitamin $B_7$) | – | 0.21 |
| Folic acid (vitamin $B_9$) | 0.002–0.01 | – |
| Ascorbic acid (vitamin C) | 2.20–2.50 | 22 |
| Tocopherol (vitamin E) | 0.74[b] | – |
| Phyllochinon (vitamin K) | 0.025[a] | – |

[a] Measured in mg%.
[b] Measured in mg/kg.
(-), No available data.
Data from Adenekan et al. (2015); Ajibola et al. (2012); Al-Waili (2003); Ball (2007); Bogdanov et al. (2008); Muhammad et al. (2015).

Similarly, vitamin C determination is based on (1) amperometry with gold electrodes altered with the palladium flow injection technique with the enzyme ascorbate oxidase and (2) titration methods demonstrated minor amounts of

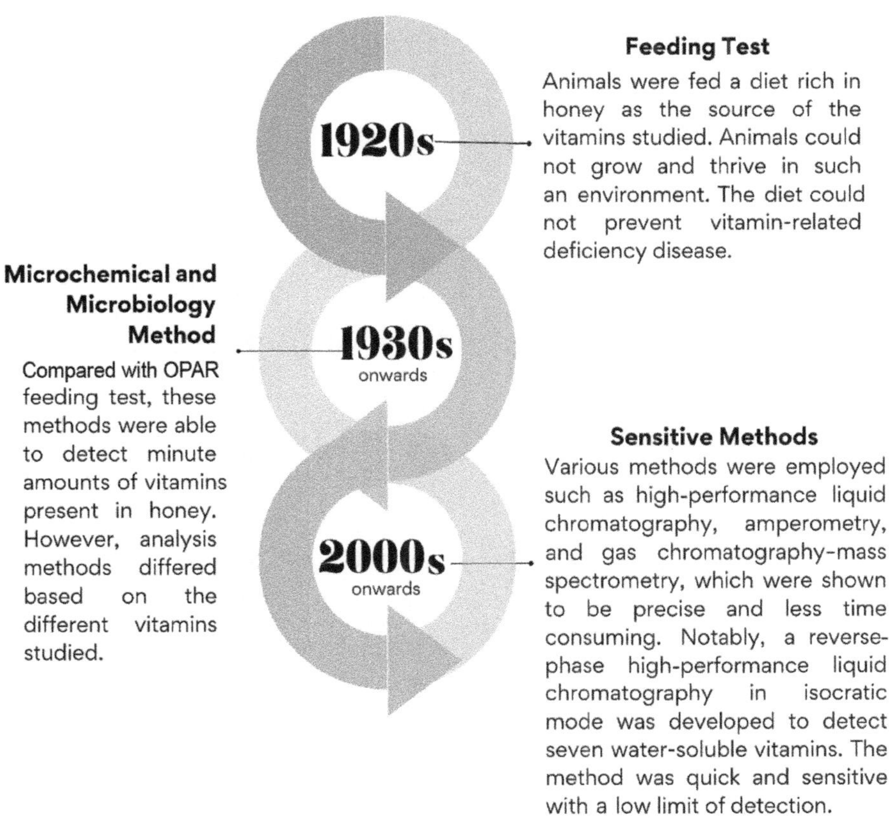


**Figure 6.1** Vitamin detection methods in honey.

**Table 6.2** Vitamin content of honey (per 100 g) in territorial regions of the United States and across countries.

| Origin | Nectar Source | $B_1$ | $B_2$ | C | $B_6$ | $B_5$ | $B_3$ |
|---|---|---|---|---|---|---|---|
| | | (μg) | (μg) | (mg) | (μg) | (μg) | (mg) |
| **United States** | | | | | | | |
| Washington | Mixed | 6.4 | 73 | 0.6 | 300 | 87 | 78 |
| Washington | Clover and alfalfa | 6.8 | 67 | 1.5 | 227 | 96 | 24 |
| Washington | Black locust | 7.4 | 68 | 1.4 | 233 | 100 | 47 |
| Washington | Fireweed | 8.2 | 81 | 4.1 | 397 | 56 | 13 |
| Oregon | Locust | 4.3 | 35 | 0.5 | 260 | 103 | 4 |
| Oregon | Alfalfa-sweet clover | 4.3 | 36 | 1.3 | 430 | 175 | 92 |
| Oregon | Fireweed | 2.2 | 62 | 1.4 | 260 | 87 | 84 |
| Oregon | Wild buckwheat | 4.3 | 56 | 2.8 | 250 | 180 | 16 |
| California | Star thistle | 8.6 | 137 | 6.5 | 410 | 90 | 11 |
| California | Orange | 8.6 | 35 | 2.5 | 210 | 150 | 16 |
| California | Orange | 4.3 | 42 | 1.9 | 310 | 63 | 13 |
| California | Sage | 3.0 | 36 | 5.4 | 320 | 56 | 4 |
| California | Star thistle | 3.0 | 58 | 1.3 | 420 | 58 | 24 |
| California | Manzanita | 7.1 | – | 2.5 | 280 | – | 8 |
| Texas | Cotton | 2.1 | 58 | 0.6 | 350 | 103 | 17 |
| Texas | Rattan | 6.5 | 87 | 2.3 | 440 | 190 | 23 |
| Long Island | Mixed | 6.5 | 46 | 2.0 | 240 | 155 | 26 |
| Florida | Tupelo | – | 67 | 1.1 | 260 | 25 | 24 |
| Florida | Tupelo | 4.3 | 58 | 2.1 | 250 | 118 | 44 |
| New York | Buckwheat | 8.6 | 62 | 1.3 | 250 | 47 | 13 |
| Tennessee | Poplar | 6.5 | – | 1.2 | 240 | – | 14 |
| Tennessee | Crimson clover | 8.6 | – | 2.3 | 400 | – | 56 |
| Hawaii | Algaroba | 8.6 | 46 | 2.3 | 250 | 50 | 32 |
| Idaho | Dandelion | 6.4 | 87 | 2.5 | 267 | 192 | 11 |
| Montana | Clover | 3.3 | 77 | 3.2 | 416 | 141 | 18 |
| Montana | Milkwort | 9.1 | – | 2.9 | 250 | – | 17 |
| Minnesota | Mellon | 7.1 | 36 | 2.6 | 400 | 156 | 58 |
| Minnesota | Mixed | 8.6 | 46 | 2.8 | 480 | 31 | 86 |
| Minnesota | Mixed | 7.1 | 53 | 3.6 | 260 | 25 | 23 |
| Minnesota | Mixed | 9.1 | 40 | .8 | 470 | 103 | 16 |
| Minnesota | Mixed | 6.5 | – | 1.9 | 310 | – | 11 |
| **Other Countries** | | | | | | | |
| Greece | Unknown | 8.6 | 47 | 2.8 | 460 | 58 | 16 |
| France | Lavender | 6.4 | 145 | 2.5 | 410 | 112 | 82 |
| Czechia | Linden | 9.1 | 62 | 2.5 | 240 | 56 | 4 |
| Azores | Unknown | 2.2 | 81 | 2.6 | 250 | 150 | 33 |
| Haiti | Logwood | 3.0 | 65 | 1.5 | 280 | 50 | 15 |
| Cuba | Campanilla | 2.2 | 58 | 1.3 | 420 | 56 | 67 |
| Guatemala | Coffee plant | 7.1 | 62 | 2.5 | 250 | 60 | 94 |

Haydak et al., 1942 / Oxford University Press.

**Table 6.3** Water-soluble vitamins in Sardinian honey samples.

| Sample | Vitamin Content in Honey (mg kg$^{-1}$ ± SD[a]) | | | | |
|---|---|---|---|---|---|
| | $B_2$ | $B_3$ | $B_5$ | $B_9$ | C |
| Eucalyptus | <0.25[a] | <0.75[b] | <1.75[b] | 6.6 ± 0.3 | 5.8 ± 0.5 |
| Eucalyptus | 1.49 ± 0.03 | 2.2 ± 0.7 | <1.75[b] | 5.4 ± 0.2 | 2.0 ± 0.3 |
| Eucalyptus | <0.75[b] | 2.86 ± 0.06 | 5.8 ± 0.7 | 4.32 ± 0.08 | 3.0 ± 0.4 |
| Eucalyptus | 2.90 ± 0.08 | 2.1 ± 0.3 | 4.4 ± 0.4 | 4.54 ± 0.02 | 2.85 ± 0.04 |
| Eucalyptus | 1.90 ± 0.02 | 3.4 ± 0.3 | 4.73 ± 0.02 | 6.92 ± 0.01 | 2.3 ± 0.1 |
| Sulla | <0.25[a] | 4.9 ± 0.6 | 4.73 ± 0.09 | <0.50[b] | 1.5 ± 0.4 |
| Sulla | <0.25[a] | 5.8 ± 0.8 | 6.5 ± 0.4 | <0.15[a] | 0.84 ± 0.09 |
| Sulla | <0.75[b] | 3.8 ± 0.2 | 4.4 ± 0.6 | <0.50[b] | 1.5 ± 0.7 |
| Citrus | 2.0 ± 0.1 | 27 ± 2 | 7 ± 1 | <0.50[b] | 2.0 ± 0.4 |
| Citrus | 2.6 ± 0.2 | 27.80 ± 0.04 | 8.09 ± 0.07 | <0.50[b] | 1.6 ± 0.4 |
| Citrus | 1.86 ± 0.05 | 22 ± 1 | <1.75[b] | <0.15[a] | 3 ± 2 |
| Acacia | <0.25[a] | 6 ± 1 | <1.75[b] | <0.50[b] | 1.015 ± 0.005 |
| Acacia | <0.25[a] | 4.29 ± 0.02 | <1.75[b] | <0.15[a] | 1.3 ± 0.2 |
| Asphodel | 4.8 ± 0.2 | 6.91 ± 0.02 | 18 ± 1 | 3 ± 1 | 4 ± 2 |
| Asphodel | 4.6 ± 0.2 | 6.71 ± 0.05 | 12 ± 1 | <0.15[a] | 2 ± 1 |
| Asphodel | 1.7 ± 0.2 | 3.7 ± 0.1 | 19 ± 6 | <0.15[a] | 1.0 ± 0.1 |
| Thistle | 3.03 ± 0.01 | 6.6 ± 0.7 | <1.75[b] | 1.54 ± 0.08 | 1.4 ± 0.2 |
| Thistle | 9.2 ± 0.2 | 6.8 ± 0.3 | <1.75[b] | 2.3 ± 0.2 | 1.9 ± 0.1 |
| Thistle | <0.25[a] | 12.3 ± 0.2 | <1.75[b] | <0.50[b] | 3.5 ± 0.2 |
| Strawberry tree | 2.1 ± 0.6 | 12.4 ± 0.6 | <1.75[b] | 0.52 ± 0.03 | 3.08 ± 0.02 |
| Strawberry tree | <0.25[a] | <0.75[b] | 28 ± 6 | <0.50[b] | 4.4 ± 0.3 |
| Strawberry tree | <0.25[a] | <0.75[b] | <0.58[a] | <0.15[a] | 4 ± 1 |
| Lavender | 1.92 ± 0.01 | <0.75[b] | <0.58[a] | <0.15[a] | 2.2 ± 0.3 |
| Lavender | 6 ± 1 | 5.5 ± 0.7 | <0.58[a] | 3 ± 2 | 2.3 ± 0.2 |
| Heather | <0.25[a] | 5.92 ± 0.01 | <0.58[a] | <0.50[b] | 2.7 ± 0.9 |
| Rosemary | <0.25[a] | <0.75[b] | <0.58[a] | 1.7 ± 0.2 | 1.5 ± 0.2 |
| Linden | <0.25[a] | 7.0 ± 0.3 | <0.58[a] | 1.28 ± 0.05 | <0.10[a] |
| Multifloral | 1.1 ± 0.5 | 8 ± 1 | <0.58[a] | 1.8 ± 0.3 | <0.10[a] |

[a] Below the detection limit.
[b] Below the quantification limit.
SD, standard deviation.
Ciulu et al. 2011 / With permission of Elsevier.

vitamin C, ranging from 1.5 to 6.2 mg/100 g of honey (Table 6.5). Both methods showed a high correlation with one another ($r^2 = 0.9998$) (Da Silva et al. 2012).

Likewise, biochemical testing of natural, unprocessed honey of multifloral origin in the United Arab Emirates revealed the presence of vitamin A (0.49 mg%), vitamin C (2.3 mg%), and vitamin E (0.74 mg%) in marginal amounts (Al-Waili 2003). Spectrophotometry analysis of honey produced from *Acacia modesta* flowers in Pakistan showed low levels of vitamins A and E (<5 mg/100 mL) (Muhammad et al. 2015). Vitamin determination of honey from Nigeria based on titrimetry analysis reported small but definite values (Adenekan et al. 2015). Significant differences were observed in the vitamin content of honey from various agroecological zones (Table 6.6).

However, there are several exceptions in which the vitamin content is much higher than usual, with variations observed across the origin of specific honey. For instance, the vitamin C content of honey samples from Spain, determined via the

**Table 6.4** Vitamin content in various types of honey in Central Spain (n = 4).

| Honey | B1 | B2[a] | B3[b] | B3[c] | B5 | B6 | C |
|---|---|---|---|---|---|---|---|
| Rosemary | <0.005 | <0.43 | <0.22 | <0.05 | <0.04 | <0.06 | 1.90 (0.01) |
| Thyme | 3.2 (0.2) | <0.43 | <0.06 | 0.192 (0.001) | 0.27 (0.06) | 0.022 (0.002) | 149 (10) |
| Thyme | 2.3 (0.2) | <0.43 | <0.06 | <0.05 | 0.23 (0.05) | 0.022 (0.001) | 109 (4) |
| Echium | 0.26 (0.02) | <0.43 | 0.02 (0.01) | 0.05 (0.01) | <0.04 | 0.044 (0.004) | 2.2 (0.1) |
| Lavender | 0.03 (0.01) | <0.43 | <0.06 | 0.31 (0.02) | <0.04 | <0.06 | 0.98 (0.04) |
| Honeydew | 0.4 (0.1) | 16.52 (0.06) | 0.013 (0.003) | 0.11 (0.04) | <0.04 | <0.21 | 1.93 (0.06) |
| Chestnut | 0.97 (0.02) | <1.42 | 0.07 (0.01) | 0.04 (0.03) | <0.04 | <0.06 | 3.4 (0.2) |

[a] Measured in μ/100 g.
[b] In the form of nicotinamide.
[c] In the form of nicotinic acid.
León-Ruiz et al., 2013 / Springer Nature.

**Table 6.5** Vitamin C content in honey samples in Brazil.

| Flower | Amperometry (mg/100 g) | Iodometry (mg/100 g) |
|---|---|---|
| *Citrus aurantium* var. *sinensis* | 3.10 (0.01) | 3.00 (0.45) |
| *Eucalyptus tereticornis* | 6.20 (0.06) | 5.50 (0.40) |
| *Vernonia polyanthes* | 1.50 (0.02) | 1.90 (0.15) |
| *Cissus rhombifolia* | 1.95 (0.00) | 2.00 (0.10) |
| *Croton campestris* A. *St.-Hill* | 3.20 (0.00) | 3.40 (0.43) |
| *Julocroton triqueter* wild | 1.50 (0.01) | 1.70 (0.10) |
| *Euphobia milii* var. *splendens* | 5.00 (0.01) | 4.30 (0.11) |

Da Silva et al., 2012 / With permission of Elsevier.

**Table 6.6** Vitamin content of honey obtained from various agroecological zones in Nigeria (n = 36).[a]

| Agro-ecological Zone | Biotin (ppm) | Riboflavin (ppm) | Thiamine (ppm) | Ascorbic Acid (ppm) | Folic Acid (ppm) |
|---|---|---|---|---|---|
| Mangrove forest | $0.20 \pm 0.01^{c}$ | $0.22 \pm 0.01^{c}$ | $0.21 \pm 0.01^{c}$ | $0.24 \pm 0.01^{b}$ | $0.12 \pm 0.01^{c}$ |
| Tropical rainforest | $0.25 \pm 0.01^{d}$ | $0.30 \pm 0.02^{a}$ | $0.23 \pm 0.01^{c}$ | $0.85 \pm 0.02^{a}$ | $0.28 \pm 0.02^{a}$ |
| Derived savanna | $0.12 \pm 0.01^{d}$ | $0.26 \pm 0.02^{b}$ | $0.12 \pm 0.01^{d}$ | $0.25 \pm 0.01^{b}$ | $0.21 \pm 0.01^{b}$ |
| Southern Guinea savanna | $0.34 \pm 0.02^{a}$ | $0.31 \pm 0.02^{a}$ | $0.32 \pm 0.02^{a}$ | $0.18 \pm 0.01^{d}$ | $0.28 \pm 0.02^{a}$ |
| Northern Guinea savanna | $0.18 \pm 0.01^{c}$ | $0.23 \pm 0.02^{c}$ | $0.26 \pm 0.02^{b}$ | $0.14 \pm 0.01^{c}$ | $0.22 \pm 0.01^{b}$ |
| Sudan savanna | $0.16 \pm 0.01^{c}$ | $0.28 \pm 0.02^{b}$ | $0.27 \pm 0.02^{b}$ | $0.21 \pm 0.01$ | $0.25 \pm 0.02^{a}$ |
| Mean | 0.21 | 0.22 | 0.26 | 0.31 | 0.23 |

[a] Values with the same letters in the same column indicate insignificant differences based on the Duncan multiple range test.
Adenekan et al., 2015 /Adenekan.

RP-HPLC method (León-Ruiz et al. 2011), demonstrated that vitamin C is exceptionally high in thyme samples compared with other honey types, in certain instances reaching up to 170 mg/100 g. Nevertheless, the high standard deviation values should be noted, indicating a great range of values in thyme samples (Table 6.7).

In addition, honey samples (Tualang, Gelam, and acacia) from Malaysia revealed higher vitamin content than usual, specifically for vitamins $B_3$ and C (Chua et al. 2013) (Table 6.8). No doubt, there is a high variation in the quality and composition of honey, differing based on the geographical origin, climate conditions, and botanical origin of honey (nectar source), which is unsurprising considering the different environments where honey grows (Rao et al. 2016).

**Table 6.7** Vitamin C content in 82 unifloral honeys from the region of Castilla-La Mancha, Spain.

| Honey Type | Mean (SD) for Vitamin C (mg/100 g) |
|---|---|
| Chestnut (n = 11) | 10.66 (5.67) |
| Lavender (n = 9) | 2.11 (1.08) |
| Honeydew (n = 5) | 0.77 (9.89) |
| Rosemary (n = 11) | 5.08 (8.84) |
| Thyme (n = 16) | 57.15 (38.77) |
| Echium (n = 12) | 2.18 (2.87) |

SD, standard deviation.
León-Ruiz et al., 2013 / Springer Nature.

**Table 6.8** Vitamin content of honey plants in Malaysia (mg/kg).

| Vitamin | Tualang | Gelam | Acacia |
|---|---|---|---|
| Vitamin $B_1$ | <LOQ | 13.85 (0.24) | <LOQ |
| Vitamin $B_2$ | <LOD | <LOQ | 11.85 (3.90) |
| Vitamin $B_3$ | 170.38 (9.60) | 355.38 (56.11) | 134.67 (3.53) |
| Vitamin $B_9$ | <LOQ | <LOD | <LOD |
| Vitamin $B_{12}$ | <LOD | <LOQ | <LOD |
| Vitamin C | 52.20 (3.29) | 67.36 (10.48) | 62.80 (8.60) |

LOD, limit of detection; LOQ, limit of quantification.
Adapted from Chua et al., 2013.

## Aging, Storage, and Processing: Influence on Vitamin Content

Aging, storage, and processing conditions may lead to vitamin losses. External factors such as temperature, light, oxygen, moisture, time, and pH, particularly related to post-harvest handling and processing procedures, can affect the vitamin content in foods (Khaneghah et al. 2019). Generally, ascorbic acid, thiamin, biotin, niacin, and riboflavin are retained in food with low-pH foods such as honey (Bonté and Desmoulière 2013). Niacin and biotin are also less affected by oxygen and heat (Lešková et al. 2006). Folate loss during storage is expected, but it is less affected by heat and oxidation (Khaneghah et al. 2019). Thiamin is not light sensitive, but heat can accelerate the rate of thiamin destruction (Lešková et al. 2006). Like thiamin, riboflavin can be mildly affected by heat, but the difference is that riboflavin is extremely sensitive to light (Sheraz et al. 2014). On the other hand, pyridoxine is generally stable with heat application, and losses are largely due to storage conditions and oxygen exposure (Parra et al. 2018). In addition, pantothenic acid is not affected by air but rather is greatly influenced by dry heating (Khaneghah et al. 2019).

Ascorbic acid is one of the most unstable vitamins present in honey. Ascorbic acid is labile, vulnerable to chemical and enzymatic oxidation (World Health Organization and Food and Agriculture Organization of the United Nations 2004). Ascorbic acid is also heavily influenced by prolonged heat, light exposure, oxygen, storage, and enzyme action (León-Ruiz et al. 2013). For instance, it was demonstrated that ascorbic acid could be oxidized by hydrogen peroxide produced by glucose oxidase, thereby reducing the vitamin levels in honey (White and Crane 1975).

As opposed to water-soluble vitamins, fat-soluble vitamins are typically more stable to processing and storage (Devi 2015). Vitamin A is relatively unaffected in an inert atmosphere; however, it mainly loses its properties under the presence of both oxygen and heat (Lešková et al. 2006). Whereas vitamin E is susceptible to oxygen, temperature, and light (Pignitter et al. 2019), vitamin K is generally stable with heat, only influenced by light and alkalis (Lešková et al. 2006).

Apart from these external factors, both the B vitamins (derived principally from pollen) and ascorbic acid are influenced by commercial and industrial processes, such as filtration (Ciulu et al. 2011). The appearance of honey is originally cloudy, and beekeepers may use the clarification process to enhance the clarity and appearance of honey. This process involves thoroughly mixing rapidly filtering diatomaceous earth with honey, heated around 140°–160°F. Analysis of the clarification process with untreated honey in comparison revealed that the processing noticeably lowers the vitamin content for about one-third to approximately half of the initial content, with thiamine decreasing from 8.4 to 5.4 μg, riboflavin from 93 to 64 μg, pyridoxine from 402 to 316 μg, pantothenic acid from 176 to 97 μg, ascorbic acid from 2.7 to 1.9 mg, and nicotinic acid from 87 to 46 mg per 100 g of honey (Haydak et al. 1942). Similarly, commercial filtration lowers vitamin content when pollen is removed almost completely (Da Silva et al. 2016). For instance, using layers of cheesecloth for filtering traps wax, debris, and pollen grains, which are typically discarded (Bryant 2017).

Another point of interest is to study the effect of storage and time on the vitamin content of honey, given that honey can last for a long time if stored properly. Thus, a research study dated back to the 1940s (Kitzes et al. 1943) compared the vitamin content of old honey from 1935–1939 (stored in a dark closet at room temperature) with newer honey at the present year of 1940–1942 (Table 6.9).

Among all the B vitamins studied, pantothenic acid showed the greatest loss over time, with approximately a 50% reduction. One possible explanation could be the properties of pantothenic acid, which are unstable in an acidic condition. Although honey is slightly acidic with a pH of 4 to 5, the destruction of vitamins could occur slowly over the years. A more recent study analyzed the ascorbic acid content of honey harvested in Romania in different years, 2011, 2012, and 2013, using liquid chromatography (Bonta et al. 2013).

The study concluded that the storage period has little effect on the lack of vitamin C content in honey because vitamin C was absent or below the detection limit in a similar number of honey samples based on the 3 years studied (Table 6.10).

**Table 6.9** Comparison of B vitamin content (mg/100 g) in aged and new honey.

| Honey (Year) | Riboflavin | Pantothenic acid | Niacin | Thiamine | Pyridoxine |
|---|---|---|---|---|---|
| 1935–1939 | 21.8 (3.0) | 20.4 (2.4) | 124.4 (11.6) | 3.5 (0.25) | 7.6 (0.66) |
| 1940–1942 | 26.3 (2.1) | 54.4 (3.6) | 108.5 (9.1) | 4.4 (0.50) | 10.0 (1.1) |

**Table 6.10** Vitamin C in various honey samples based on different harvest years.

| 2011 | | 2012 | | 2013 | |
|---|---|---|---|---|---|
| Origin | Vitamin C (mg/kg) | Origin | Vitamin C (mg/kg) | Origin | Vitamin C (mg/kg) |
| Honeydew | 29.50 | Mint | 37.00 | Rape | 3.80 |
| Honeydew | 12.39 | Raspberry | 20.22 | Rape | <0.10[a] |
| Raspberry | 24.70 | Raspberry | 13.30 | Rape | <0.10[a] |
| Raspberry | 6.65 | Honeydew | 18.50 | Rape | <0.10[a] |
| Raspberry | 4.18 | Honeydew | 7.00 | Rape | <0.10[a] |
| Mint | 4.12 | Honeydew | 6.18 | Multifloral | 2.46 |
| Linden | 2.37 | Sunflower | 4.61 | Multifloral | 1.16 |
| Linden | <0.10[a] | Sunflower | 3.03 | Multifloral | <0.10[a] |
| Thyme | 1.90 | Sunflower | 1.66 | Multifloral | <0.10[a] |
| Thyme | <0.10[a] | Sunflower (n = 2) | <0.10[a] | Sunflower | 1.50 |
| Heather | 1.60 | Multiflower | 1.72 | Sunflower | 0.72 |
| Sunflower | 0.76 | Multiflower | <0.10[a] | Sunflower | <0.10[a] |
| Acacia | <0.10[a] | Heather | 0.93 | Sunflower | <0.10[a] |
| Rape | <0.10[a] | Acacia | <0.10[a] | Sunflower | <0.10[a] |
| Multiflower | <0.10[a] | Rape (n = 5) | <0.10[a] | Acacia | 0.71 |

[a] Below the detection limit.

Bonta et al., 2013 / Bonta V.

Nonetheless, keeping the same types of honey in similar conditions and conducting analysis over the years are recommended for better comparison. In essence, it is imperative to note that the stability of vitamin is widely affected by the food matrix itself. In the case of honey, there remains a limited study that explores vitamins losses by specific processing and storage conditions.

## Nutritional and Health Benefits of Vitamins Present in Honey

Generally, the vitamins present in honey are needed for optimal body function and good health (Figure 6.2).

These vitamins must be obtained from the diet because the human body cannot synthesize them (Gironés-Vilaplana et al. 2017). Specifically, each vitamin has its own role, and deficiencies can lead to adverse health conditions (Table 6.11). The B vitamins primarily function as fundamental catalytic coenzymes in vital metabolic and biochemical processes (Khaneghah et al. 2019). The B vitamins are also essential for testosterone production. Research has shown a positive correlation between honey consumption and testosterone concentration, suggesting the possibility of honey use in fertility (Meo et al. 2017). Another vitamin belonging to the water-soluble group is vitamin C, widely studied for its value in human health. Vitamin C is needed to develop and maintain connective tissue and gingiva, bone formation, and wound healing (Chambial et al. 2013). Besides, vitamin C contributes to a healthy immune system, lessens the severity of allergies, and protects the body against infections (Carr and Maggini 2017). Unlike fat-soluble vitamins, vitamins B and C should be constantly supplied through the diet because this vitamin group is not soluble in fat and cannot be stored in the human body for long (Khaneghah et al. 2019).

The fat-soluble vitamins present in honey are vitamins A, E, and K. Vitamin A is needed for optimal vision, bone growth, reproduction, cell division, immunity, and differentiation of the epithelial tissues (Huang et al. 2018). Vitamin E can prevent oxidative stress, protect cell membranes, and help regulate platelet aggregation (Rizvi et al. 2014). Finally, vitamin K functions mainly to activate several proteins involved in blood clotting (Gironés-Vilaplana et al. 2017).

There is no doubt that honey is a source of several essential vitamins. However, the vitamins in honey are present in minute amounts so that the contribution to the recommended daily intake is marginal (Bogdanov et al. 2008). Although vitamins are micronutrients needed only in small amounts for optimal human physiology and metabolism, the vitamins present per 100 g of honey are still considered minimal according to the recommended nutrient intake (RNI). Initially, the RNI values for vitamins are established to determine the basic nutritional needs for mostly all healthy populations (>97% of the population). Over the years, the RNI values have evolved based on the understanding of the optimal nutrient intake to prevent a range of diseases, especially with longer life expectancies and sedentary lifestyles (World Health Organization

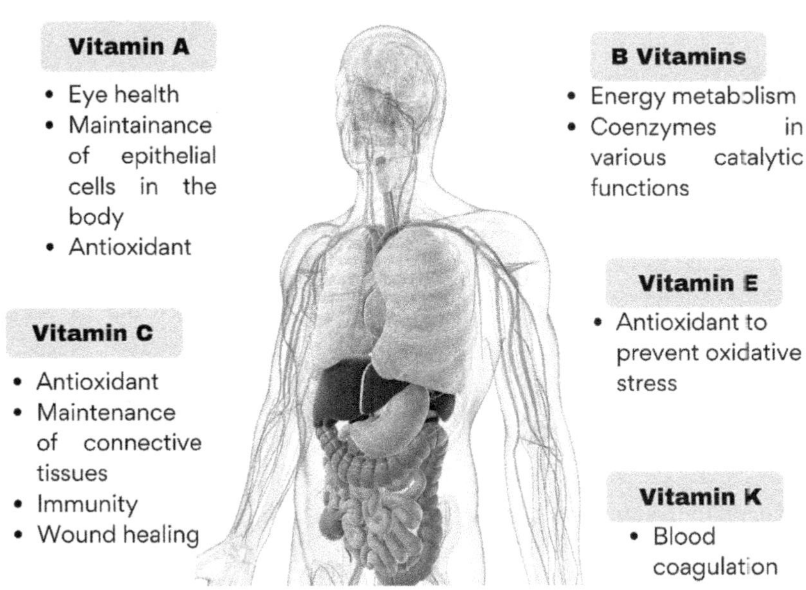


**Figure 6.2** Essential vitamins present in honey. ciePro / Adobe Stock.

**Table 6.11** Importance of vitamins present in honey.

| Vitamin | Functions | Possible Effect of Deficiency |
|---|---|---|
| A | Optimum visual system in the retina of the eyes<br>Maintenance of epithelial cell integrity in all body tissues for growth, development, immunity, and reproduction | Night blindness, impaired growth, increased severity of infection, impaired iron metabolism if vitamin A deficiency coexists with iron deficiency, particularly for children: diarrhea, measles, and increased risk of death |
| $B_1$ | Coenzyme functions in metabolism of carbohydrates and branched-chain amino acids<br>Facilitates energy use | Reduction in the formation of acetylcholine for normal neural function; other condition: beriberi |
| $B_2$ | Coenzyme functions in oxidation and reduction reactions in metabolic pathways and energy production | Hypo- or ariboflavinosis, with sore throat; hyperemia, edema of the pharyngeal and oral mucous membranes, and dermatitis (with complications of other B vitamin deficiency) |
| $B_3$ | Coenzyme function for hydrogen transfer with dehydrogenases | Pellagra with diarrhea (from inflammation of the intestinal mucous surfaces), dermatitis, insomnia, mental impairment, and dementia |
| $B_5$ | Component of coenzyme A and phosphopantetheine involved in fatty acid metabolism | Fatigue, nausea, impaired coordination, and poor sleep quality |
| $B_6$ | Coenzyme functions in protein, glycogen, and sphingoid bases metabolism<br>Maintains optimal level of homocysteine<br>Involved in immunity function<br>Involved in hemoglobin formation | Hypovitaminosis of vitamin $B_6$ may occur with riboflavin deficiency; can adversely affect immune function; other condition: microcytic anemia |
| $B_9$ | Important role in DNA and methylation cycle | Reduction in DNA biosynthesis, particularly in cells that rapidly divide, such as red blood cells; greater vulnerability to infection, lower blood coagulation, intestinal malabsorption, and can negatively affect cognitive function |
| $B_7$ | Coenzyme functions in bicarbonate-dependent carboxylation<br>Important for cell signaling and gene regulations | Fatigue, depression, hallucination, seizures, nausea, dermatitis, conjunctivitis, and muscular pain |
| C | Protective role in human health because of its antioxidant nature<br>Promotes iron absorption<br>Enzymatic function for body processes such as carnitine biosynthesis, collagen hydroxylation, and amidation of peptide hormones and for tyrosine metabolism | Scurvy and other possible conditions such as anemia (often present with vitamin C deficiency); epidemiologic evidence showed low plasma vitamin C in those with diabetes and smokers; inadequate vitamin C intake seems to increase cancer risk |
| E | A part of cell antioxidant system that protects polyunsaturated fatty acids and other components of cell membranes from oxidation stress caused by free radicals | Lack of clinical evidence to suggest deficiency disease associated with vitamin E, possibly because it is naturally present in plant-based food and animal products; however, epidemiologic data suggest that vitamin E can reduce the risk of cardiovascular diseases |
| K | Maintenance of normal blood coagulation | Excessive bleeding |

Data from Khaneghah et al. (2019); World Health Organization and Food and Agriculture Organization of the United Nations (2004).

and Food and Agriculture Organization of the United Nations 2004). Table 6.12 presents the average amount of vitamins in 100 g of honey compared with the daily RNI values for each specific age group. It seems that the contribution of vitamins to RNI values is insignificant (Bogdanov 2012).

Although there are instances when the vitamin content in honey is much higher than usual (Chua et al. 2013; León-Ruiz et al. 2011), the consumption of 100 g of honey per day (~5 tablespoons) is questionable from a nutritional point of view. In both adults and children, the World Health Organization's guideline recommends the intake of free sugars (including natural sugar present in honey) to less than 10% of total energy intake, with a further reduction to 5% for better health

**Table 6.12** Comparison between vitamin content in 100 g of honey with the daily recommended nutrient intake.

| | Recommended Nutrient Intake for Specific Vitamins (mg/day) | | | | | | | | | | |
|---|---|---|---|---|---|---|---|---|---|---|---|
| | A[a] | $B_1$ | $B_2$ | $B_3$[b] | $B_5$ | $B_6$ | $B_7$[c] | B9[c] | C | E[d] | K[c] |
| **Average Amount Present in Honey (mg/100 g)** | 0.49[e] | 0.00–0.01 | 0.01–0.02 | 0.10–0.20 | 0.02–0.11 | 0.01–0.32 | 0.21[f] | 2–10 | 2.20–2.50 | 0.74[e] | 2.5 |
| **Children** | | | | | | | | | | | |
| 1–3 yr | 400 | 0.5 | 0.5 | 6.0 | 2.0 | 0.5 | 8 | 150 | 30 | 5 | 15 |
| 4–6 yr | 450 | 0.6 | 0.6 | 8.0 | 3.0 | 0.6 | 12 | 200 | 30 | 5 | 20 |
| 7–9 yr | 500 | 0.9 | 0.9 | 12.0 | 4.0 | 1.0 | 20 | 300 | 35 | 7 | 25 |
| **Boys** | | | | | | | | | | | |
| 10–18 yr | 600 | 1.2 | 1.3 | 16.0 | 5.0 | 1.3 | 25 | 400 | 40 | 10 | 35–55 |
| **Girls** | | | | | | | | | | | |
| 10–18 yr | 600 | 1.1 | 1.0 | 16.0 | 5.0 | 1.2 | 25 | 400 | 40 | 7.5 | 35–55 |
| **Men** | | | | | | | | | | | |
| 19–65 yr | 600 | 1.2 | 1.3 | 16.0 | 5.0 | 1.3[g] | 30 | 400 | 45 | 10 | 65 |
| >65 yr | 600 | 1.2 | 1.3 | 16.0 | 5.0 | 1.7[h] | 30 | 400 | 45 | 10 | 65 |
| **Women** | | | | | | | | | | | |
| 19–65 yr | 500 | 1.1 | 1.1 | 14.0 | 5.0 | 1.3[g] | 30 | 400 | 45 | 7.5 | 55 |
| >65 yr | 600 | 1.1 | 1.1 | 14.0 | 5.0 | 1.5[h] | 30 | 400 | 45 | 7.5 | 55 |
| **Pregnancy** | 800 | 1.4 | 1.4 | 18.0 | 6.0 | 1.9 | 30 | 600 | 55 | 7.5 | 55 |
| **Lactation** | 850 | 1.5 | 1.6 | 17.0 | 7.0 | 2.0 | 35 | 500 | 70 | 7.5 | 55 |

The average amount of vitamins present in honey is based on data from Adenekan et al. (2015); Ajibola et al. (2012); Al-Waili (2003); Ball (2007); Bogdanov et al. (2008); and Muhammad et al. (2015).

[a] Reported in safe intake of retinol equivalents µg RE/day.
[b] Recommended nutrient intake in niacin equivalents mg/day.
[c] Reported in µg/day.
[d] Reported in acceptable intake of mg α-tocopherol/day.
[e] Reported in mg%.
[f] Reported in mg/kg.
[g] Adults aged 19–50 years old.
[h] Adults aged 51 years and older.

Adapted from World Health Organization and Food and Agriculture Organization of the United Nations, 2004.

outcomes (World Health Organization 2015). For example, 5% of total energy intake from a 2000-calorie diet is equivalent to approximately five teaspoons or 25 g of free sugar. It should be noted that apart from honey, one may also consume other forms of free sugar such as syrups, fruit juices, fruit juice concentrates, and sugar added during cooking or added by manufacturers in ultra-processed foods. With that being said, it is also noteworthy that food value is not necessarily in the proportion to the quantity of the vitamins that it contains but rather the fact that honey, indeed, is a source of vitamins, which is an added value to its myriad benefits (Stoney Archer 1929).

Another interesting health contribution of the vitamins present in honey is the antioxidant profile, contributed by vitamins C, E, A, and B. Oxidation is a biochemical reaction that produces free radicals to chain reactions that can potentially harm cells; tissues; and, subsequently, body functions (Meo et al. 2017). Oxidative stress has been implicated in the onset and development of numerous chronic diseases, but foods containing vitamins can have protective health benefits because of their antioxidant properties (Ferreira et al. 2009). The role of honey as an antioxidant to reduce oxidative stress is well-established, particularly by stimulating cells through carbohydrates, proteins, and nucleic acids to produce reactive oxygen species–reducing oxidative stress (Ahmed et al. 2018).

The antioxidant properties of honey have been determined using various methods, including quantifying antioxidant agents such as vitamin C and β-carotene (provitamin A carotenoids, which are converted to vitamin A in the body) using

**Table 6.13** Vitamins with antioxidant properties found in Portuguese honey.

| Honey Sample | Ascorbic Acid (mg/kg) | β-Carotene (mg/kg) |
|---|---|---|
| Light honey (rosemary) | 140.01 (0.05) | 9.32 (0.01) |
| Ambar honey (viper's bugloss) | 143.86 (1.16) | 8.64 (0.06) |
| Dark honey (heather) | 145.80 (0.02) | 9.49 (0.15) |

Adapted from Ferreira et al., 2009.

spectrophotometric absorbance measurement (Ferreira et al. 2009). For instance, analysis of three unifloral Portuguese honey from Northeast Portugal demonstrated that dark-type honey contains significantly higher vitamins with antioxidant activity than light and amber honey (Table 6.13). On the contrary, it was reported that commercial honey, including fireweed, tupelo, Hawaiian Christmas berry, buckwheat, clover, acacia, and soybean, did not contain any detectable antioxidant vitamin C, possibly because of processing and storage conditions (Gheldof et al. 2002).

Generally, limited studies have analyzed the correlation between vitamins in honey and honey's antioxidant capacity, with most studies focusing on water-soluble vitamins. For example, it was shown that the total water-soluble vitamin content of honey samples in Malaysia correlated ($r = 0.8226$) with the antioxidant scavenging activity based on 1,1-diphenyl-2-picrylhydrazyl (DPPH) assay, demonstrating the potential of these vitamins, which include vitamins $B_1$, $B_2$, $B_3$, $B_9$, $B_{12}$, and C, as a radical scavenger to reduce oxidation (Chua et al. 2013). Despite this, no correlations were found for ferric-reducing antioxidant power (FRAP) assay and β-carotene–bleaching inhibition. Another study reported that the antioxidant capacity of Algerian honey was attributed to vitamin C, as seen in the significant correlations with both DPPH and FRAP assay (Khalil et al. 2012). However, commercial honey of Portuguese origin (strawberry tree, cardoon, carob, citrus, heather, eucalyptus, sunflower, lavender, pennyroyal, multifloral, thyme) showed no significant correlation between vitamin C and antioxidant properties (Aazza et al. 2013). It seems that the antioxidant properties of honey contributed by vitamins can vary based on the botanical origins and processing conditions, highlighting the need for more research to explore the role of various vitamins as antioxidants in different honey samples.

Furthermore, 10 healthy participants (mean age, 31.2 years old) consumed a regular diet 2 with 1.2 g/kg body weight of honey dissolved in 250 mL of water every day for 2 weeks in a pre-post experimental design (Al-Waili 2003). Alongside other antioxidant agents, serum vitamin C concentration increased by almost double (47%), from $0.821 \pm 0.496$ mg/dL (baseline) to $1.209 \pm 0.769$ mg/dL (post-test). The findings suggest the ability of honey to promote health by increasing antioxidant agents such as vitamin C in the body. However, the results did not reach significance ($p = 0.1558$). Other limitations include the low sample size, lack of randomization, and considerably short duration, limiting the long-term effect. Besides, the amount of honey consumed is not small; for example, for a person weighing 60 kg, the required amount of honey to be consumed is 72 g/day for the observed effect.

Research also uncovered the association between reduced cardiovascular diseases risks and the consumption of foods containing antioxidant vitamins, also present in honey (Cianciosi et al. 2018). Moreover, it was demonstrated that honey has protective effects against cardiovascular risks in both healthy subjects and those with higher risk factors (Rao et al. 2016). Antioxidative agents such as vitamins C and E (α-tocopherol) can terminate the chain reactions to defend the body from damaging free radicals (Meo et al. 2017), thereby protecting against chronic diseases. Although it is likely that the various active components of honey act synergistically in augmenting the antioxidant activity (Samarghandian et al. 2017), it is possible that these vitamins could contribute to disease prevention to a certain degree (Viuda-Martos et al. 2008).

## Importance of Vitamin Quantification in Honey

The marked dependence of certain vitamin in concentrations (i.e. vitamins $B_3$, $B_5$, and C) on the botanical origin presents an important use for future identification of honey (Bonta et al. 2013; Ciulu et al. 2011). For example, vitamin C content, along with conductivity properties and sugar content, led to the accurate classification of the six unifloral honeys studied (León-Ruiz et al. 2011). The findings suggest that although the vitamin content in honey may be low to contribute to significant daily micronutrients, the quantification of vitamins in honey samples is useful for determining honey's authenticity or certification and verifying possible adulteration for consumers. Essentially, honey should not be heated or processed to such an extent that its composition and quality are adversely affected (Fakhlaei et al. 2020). Particularly, adulteration of honey can take place during preparation, whereby excessive heating can lead to inadvertent losses of heat-sensitive vitamins, thereby reducing the nutritional value (Islam et al. 2014). Additionally, adulteration of honey can also occur by altering the botanical and geographical origin (Damto 2019), to which the vitamin profiles of different honeys can potentially serve as a comparison indicator.

## Conclusion

Progressively, various methods are used for the determination of vitamins in honey, which are validated in terms of precision and accuracy and shown to be advantageous with a low detection limit and low quantification limit and are less time consuming. Overall, honey is a source of vitamins A, B complex, C, E, and K, all of which are required for the maintenance of optimal body functions, disease prevention, and health promotion. However, these vitamins are present in minute amounts so that the contribution to the daily recommended nutrient intake is minimal. Additionally, the vitamin content in different honey types varies according to geographical origin, botanical origin, climate, processing, and storage conditions. The potential relationship between the vitamins present in honey samples and their respective botanical origins poses an interesting perspective to further explore the means for honey identification in the future.

## References


Aazza, S., Lyoussi, B., Antunes, D., et al. (2013). Physicochemical characterization and antioxidant activity of commercial Portuguese honeys. *Journal of Food Science* 78 (8): C1159–C1165. DOI: 10.1111/1750-3841.12201.

Adenekan, M.O., Amusa, N.A., Owolade, E.O., et al. (2015). Physicochemical and vitamin constituents of honey samples obtained from different agro-ecological zones of Nigeria. *Journal of Multidisciplinary Engineering Science Studies* 7 (2): 36–41.

Ahmed, S., Sulaiman, S.A., Baig, A.A., et al. (2018). Honey as a potential natural antioxidant medicine: an insight into its molecular mechanisms of action. *Oxidative Medicine and Cellular Longevity* 2018: 8367846. DOI: 10.1155/2018/8367846.

Ajibola, A., Chamunorwa, J.P., and Erlwanger, K.H. (2012). Nutraceutical values of natural honey and its contribution to human health and wealth. *Nutrition and Metabolism* 9: 61. DOI: 10.1186/1743-7075-9-61.

Al-Waili, N.S. (2003). Effects of daily consumption of honey solution on hematological indices and blood levels of minerals and enzymes in normal individuals. *Journal of Medicinal Food* 6 (2): 135–140. DOI: 10.1089/109662003322233549.

Ball, D.W. (2007). The chemical composition of honey. *Journal of Chemical Education* 84 (10): 1643–1646. DOI: 10.1021/ed084p1643.

Bogdanov, S. (2012). Honey as nutrient and functional food. *Proteins* 1100: 1400–2700.

Bogdanov, S., Jurendic, T., Sieber, R., et al. (2008). Honey for nutrition and health: a review. *Journal of the American College of Nutrition*. 7 (6): 677–689. DOI: 10.1080/07315724.2008.10719745.

Bonta, V., Mărghitaş, L.A., Bobiş, O., et al. (2013). Quantitative determination of vitamin C in honey and bee pollen using HPLC-DAD method. *UASVM Journal of Animal Science and Biotechnology* 70: 31–36. DOI: 10.15835/buasvmcn-asb:70:1:.

Bonté, F. and Desmoulière, A. (2013). Honey: origin and composition. *Actualités Pharmaceutiques* 52 (531): 18–21. DOI: 10.1016/J.ACTPHA.2013.10.004.

Bryant, V.M. (2017). Filtering honey almost every filter removes. *Bee Culture*. 45 (9): 75–76.

Carr, A.C. and Maggini, S. (2017). Vitamin C and immune function. *Nutrients* 9 (11): 1211. DOI: 10.3390/NU9111211.

Castro, R.N., Azeredo, L.C., Azeredo, M.A.A., et al. (2001). HPLC assay for the determination of ascorbic acid in honey samples. *Journal of Liquid Chromatography & Related Technologies* 24 (7): 1015–1020. DOI: 10.1081/JLC-100103427.

Chambial, S., Dwivedi, S., Shukla, K.K., et al. (2013). Vitamin C in disease prevention and cure: an overview. *Indian Journal of Clinical Biochemistry* 28 (4): 314–328. DOI: 10.1007/S12291-013-0375-3.

Chua, L.S., Rahaman, N.L.A., Adnan, N.A., et al. (2013). Antioxidant activity of three honey samples in relation with their biochemical components. *Journal of Analytical Methods in Chemistry* 2013: 313798. DOI: 10.1155/2013/313798.

Cianciosi, D., Forbes-Hernández, T.Y., Afrin, S., et al. 2018. Phenolic compounds in honey and their associated health benefits: a review. *Molecules*, 23(9), p. 2322.

Ciulu, M., Solinas, S., Floris, I., et al. (2011). RP-HPLC determination of water-soluble vitamins in honey. *Talanta* 83 (3). 924–929. DOI: 10.1016/j.talanta.2010.10.059.

Da Silva, P.M., Gauche, C., Gonzaga, L.V., et al. (2016). Honey: chemical composition, stability and authenticity. *Food Chemistry* 196: 196:309-323. DOI: 10.1016/j.foodchem.2015.09.051.

Da Silva, V.L., Cerqueira, M.R.F., Lowinsohn, D., et al. (2012). Amperometric detection of ascorbic acid in honey using ascorbate oxidase immobilised on amberlite IRA-743. *Food Chemistry* 133 (3): 1050–1054. DOI: 10.1016/j.foodchem.2012.01.066.

Damto, T. (2019). A review on effect of adulteration on honey properties. *SSRN*. DOI: 10.2139/ssrn.3359494.

Devi, R. (2015). Food processing and impact on nutrition. *Scholars Journal of Agriculture and Veterinary Sciences* 2 (4A): 304–311.

Dutcher, R.A. (1918). Vitamine studies. III Observations on the curative properties of honey, nectar, and corn pollen in avian polyneuritis. *Journal of Biological Chemistry* 36 (3): 551–555.

Fakhlaei, R., Selamat, J., Khatib, A., et al. (2020). The toxic impact of honey adulteration: a review. *Foods* 9 (11): 1538. DOI: 10.3390/foods9111538.

Ferreira, I.C.F.R., Aires, E., Barreira, J.C.M., et al. (2009). Antioxidant activity of Portuguese honey samples: different contributions of the entire honey and phenolic extract. *Food Chemistry* 114 (4): 1438–1443. DOI: 10.1016/j.foodchem.2008.11.028.

Gheldof, N., Wang, X., and Engeseth, N. (2002). Identification and quantification of antioxidant components of honeys from various floral sources. *Journal of Agricultural and Food Chemistry* 50 (21). 5870–5877. . DOI: 10.1021/JF0256135.

Gironés-Vilaplana, A., Villanõ, D., Marhuenda, J., et al. (2017). Vitamins. In: *Nutraceutical and Functional Food Components: Effects of Innovative Processing Techniques* (ed. C. Galanakis), 159–201. Cambridge, MA: Academic Press. DOI: 10.1016/B978-0-12-805257-0.00006-5.

Godswill, A.G., Somtochukwu, I.V., Ikechukwu, A.O., et al. (2020). Health benefits of micronutrients (vitamins and minerals) and their associated deficiency diseases: a systematic review. *International Journal of Food Sciences* 3 (1): 1–32. DOI: 10.47604/IJF.1024.

Hawk, P.B., Smith, C.A., and Bergeim, O. (1920). The vitamine content of honey and honeycomb. *The American Journal of Physiology* 55 (3): 339–348. DOI: 10.3181/00379727-18-32.

Haydak, M.H., Palmer, L.S., Tanquary, M.C., et al. (1942). Vitamin content of honeys. *The Journal of Nutrition* 23 (6): 581–588. DOI: 10.1093/jn/23.6.581.

Huang, Z., Liu, Y., Qi, G., et al. (2018). Role of Vitamin A in the immune system. *Journal of Clinical Medicine* 7 (9): 258. DOI: 10.3390/JCM7090258.

Islam, M.N., Khalil, M.I., Islam, M.A., et al. (2014). Toxic compounds in honey. *Journal of Applied Toxicology* 34 (7): 733–742. DOI: 10.1002/jat.2952.

Khalil, M.I., Moniruzzaman, M., Boukraâ, L., et al. (2012). Physicochemical and antioxidant properties of Algerian honey. *Molecules* 17 (9): 11199–11215. DOI: 10.3390/molecules170911199.

Khaneghah, A.M., Hashemi, S.M.B., Es, I., et al. (2019). Water-soluble vitamins. In: *Innovative Thermal and Non-Thermal Processing, Bioaccessibility and Bioavailability of Nutrients and Bioactive Compounds* (ed. F.J. Barba, J.M.A. Saraiva, and G. Cravottoet al.), 241–266. Duxford, UK: Woodhead Publishing. DOI: 10.1016/B978-0-12-814174-8.00008-1.

Kitzes, G., Schuette, H.A., and Elvehjem, C.A. (1943). The B Vitamins in honey. *The Journal of Nutrition* 26 (3): 241–250. DOI: 10.1093/JN/26.3.241.

León-Ruiz, V., Vera, S., González-Porto, A.V., et al. (2011). Vitamin C and sugar levels as simple markers for discriminating Spanish honey sources. *Journal of Food Science* 76 (3): C356– C361. DOI: 10.1111/j.1750-3841.2011.02041.x.

León-Ruiz, V., Vera, S., González-Porto, A.V., et al. (2013). Analysis of water-soluble vitamins in honey by isocratic RP-HPLC. *Food Analytical Methods* 6 (2): 488–496. DOI: 10.1007/s12161-012-9477-4.

Lešková, E., Kubíková, J., Kováčiková, E., et al. (2006). Vitamin losses: retention during heat treatment and continual changes expressed by mathematical models. *Journal of Food Composition and Analysis* 19 (4): 252–276. DOI: 10.1016/j.jfca.2005.04.014.

Meo, S.A., Al-Asiri, S.A., Mahesar, A.L., et al. (2017). Role of honey in modern medicine. *Saudi Journal of Biological Sciences* 24 (5): 975–978. DOI: 10.1016/j.sjbs.2016.12.010.

Muhammad, A., Odunola, O.A., Gbadegesin, M.A., et al. (2015). Inhibitory effects of sodium arsenite and acacia honey on acetylcholinesterase in rats. *International Journal of Alzheimer's Disease*. 2015: 903603. DOI: 10.1155/2015/903603.

Parra, M., Stahl, S., and Hellmann, H. (2018). Vitamin B6 and its role in cell metabolism and physiology. *Cells* 7 (7): 84. DOI: 10.3390/cells7070084.

Pignitter, M., Grosshagauer, S., and Somoza, V. (2019). Stability of Vitamin E in foods. In: *Vitamin E in Human Health* (ed. P. Weber, M. Birringer, J. Blumberg, et al.), 215–232. Cham, Switzerland: Humana Press. DOI: 10.1007/978-3-030-05315-4_16.

Rao, P.V., Krishnan, K.T., Salleh, N., et al. (2016). Biological and therapeutic effects of honey produced by honey bees and stingless bees: a comparative review. *Revista Brasileira de Farmacognosia* 26: 657–664. DOI: 10.1016/j.bjp.2016.01.012.

Rizvi, S., Raza, S.T., Ahmed, F., et al. (2014). The Role of Vitamin E in human health and some diseases. *Sultan Qaboos University Medical Journal* 14 (2): e165.

Samarghandian, S., Farkhondeh, T., and Samini, F. (2017). Honey and health: a review of recent clinical research. *Pharmacognosy Research* 9 (2): 127. DOI: 10.4103/0974-8490.204647.

Sheraz, M.A., Kazi, S.H., Ahmed, S., et al. (2014). Photo, thermal and chemical degradation of riboflavin. *Beilstein Journal of Organic Chemistry* 10: 1999–2012. DOI: 10.3762/BJOC.10.208.

Stoney Archer, G.J. (1929). Vitamins in Honey. *Bee World* 10 (7): 101–102. DOI: 10.1080/0005772x.1929.11092834.

Viñas, P., Balsalobre, N., López-Erroz, C., et al. (2004). Liquid chromatographic analysis of riboflavin vitamers in foods using fluorescence detection. *Journal of Agricultural and Food Chemistry* 52 (7): 1789–1794. DOI: 10.1021/JF030756S.

Viuda-Martos, M., Ruiz-Navajas, Y., Fernández-López, J., et al. (2008). Functional properties of honey, propolis, and royal jelly. *Journal of Food Science* 73 (9): R117–R124. DOI: 10.1111/J.1750-3841.2008.00966.X.

White, J.W. and Crane, E., 1975. Honey a comprehensive survey. *by Eva Crane, Morrison and Gibb Ltd*, pp.194-206.

World Health Organization. (2015). *Guideline: Sugars Intake for Adults and Children*. Geneva: World Health Organization.

World Health Organization and Food and Agriculture Organization of the United Nations. (2004). *Vitamin and Mineral Requirements in Human Nutrition*, 2e. Geneva, Switzerland : World Health Organization. DOI: 92 4 154612 3.

# 7

# Minerals and Trace Elements

*Md. Solayman*

## Minerals in Honey

From a nutritional standpoint, minerals are naturally occurring inorganic solid substances in the biosphere (including all foods) found following the degradation of plant and animal tissues (Belitz et al. 2009). They are formed by geological processes (Krivovichev et al. 2018) and are essential for the regulation of metabolic pathways in the living body (Costa-Pinto and Gantner 2020). Minerals are divided into three groups on the basis of body requirements: (1) major elements, (2) trace elements, and (3) ultra-trace elements. Major elements – sodium (Na), potassium (K), calcium (Ca), magnesium (Mg), phosphorus (P), sulfur (S), and chlorine (Cl) – should be present at greater than 50 mg/d, whereas trace elements – iron (Fe), iodine (I), fluorine (F), zinc (Zn), selenium (Se), cupper (Cu), manganese (Kropf et al. 2008), chromium (Cr), molybdenum (Mo), cobalt (Co), and nickel (Ni) – are required in concentrations of less than <50 mg/d in human beings (Belitz et al. 2009). The calculated dietary requirements for ultra-trace elements – aluminum (Al), arsenic (As), barium (Ba), bismuth (Bi), boron (B), bromine (Br), cadmium (Cd), cesium (Cs), germanium (Ge), mercury (Hg), lithium (Li), lead (Pb), rubidium (Rb), antimony (Sb), silicon (Si), samarium (Sm), tin (Sn), strontium (Sr), thallium (Tl), titanium (Ti), and tungsten (W) – are usually less than 1 μg/d and often present at less than 50 ng/g in the dry diet (Aliasgharpour 2020; Belitz et al. 2009). Micro- or trace minerals are useful for good health, especially if they originate from an organic or plant source. On the other hand, if they originate from an inorganic or metallic source, they will have at least five times the specific gravity of water and can become toxic. In this case, they are then known as heavy metals (Ajibola et al. 2012). The scientific literatures indicate that Pb, Cd, Hg, Cr, Cu, Mn, Ni, Zn, and Ag are the most important heavy metals (Singh et al. 2011). Heavy metals are toxic or poisonous at low concentrations because of their tendency to accumulate in living organisms (Mahurpawar 2015).

Honey also contains different kinds of minerals and heavy metal elements. The types and amounts of them depend on the botanical and geographical origins (Oroian et al. 2016; Solayman et al. 2016). There are 54 minerals reported to be present in honey from all over the world (Solayman et al. 2016), which include both major (Na, K, Ca, Mg, P, S, Cl) and minor (Al, Cu, Pb, Zn, Mn, Cd, Tl, Co, Ni, Rb, Ba, Be, Bi, U, V, Fe, Pt, Pd, Te, Hf, Mo, Sn, Sb, La, I, Sm, Tb, Dy, Sd, Th, Pr, Nd, Tm, Yb, Lu, Gd, Ho, Er, Ce, Cr, As, B, Br, Cd, Hg, Se, Sr) elements (Adugna et al. 2020; Batista et al. 2012; Nkansah et al. 2018; Ruschioni et al. 2013; Taha et al. 2010; Vanhanen et al. 2011; Vincēviča-gaile 2010; Vit et al. 2010).

The total mineral content in honey is relatively low and generally accounts for 0.1%–0.2% of the composition of nectar honey but can exceed 1% in other types of honey such as honeydew honey (Hernández et al. 2005). Some mineral types cannot be detected in some types of honey collected from certain areas but are nevertheless available in other regions. For example, the trace elements Ni and Cr cannot be detected in Egypt (Taha et al. 2010) but are reported to be present in honey samples from Pakistan (Farooq Khan and Maqbool 2008), Turkey (Silici et al. 2008; Tuzen et al. 2007), New Zealand (Vanhanen et al. 2011), Romania (Matei et al. 2004), and Spain (González Paramás et al. 2000; Terrab et al. 2004b), which may be attributed to differences soil composition, floral type, and floral density as well as nectar and pollen (Fernández-Torres et al. 2005; Santos et al. 2008; Yarsan et al. 2007). Therefore, the composition of honey can actually be used to identify its geographical and botanical sources of honey (Bogdanov et al. 2007). Other trace elements (Pb, Cd, Hg, Cu, Mn, Zn, Ag), which are also heavy metals, play important roles as bio-indicators for environmental pollution (Roman et al. 2007; Vit et al. 2010). Overall, the individual amount of minerals and metals may differ from one honey to another. Updated information on the amount of different major elements and minor elements as well as heavy metals found in honey from all over the world is showed with their means, standard deviations, and minimum and maximum levels (Table 7.1).

*Honey: Composition and Health Benefits*, First Edition. Edited by Md. Ibrahim Khalil, Gan Siew Hua, and Bey Hing Goh.

**Table 7.1** Major elements, trace elements, and heavy metals in honey.

| Element | Units | Mean | Standard Deviation | Range | References |
|---|---|---|---|---|---|
| Ag[a,b] | µg/kg | 299.61 | 419.17 | 3.21–596.00 | Agbagwa et al. 2011; Akram 2008; Alvarez-Suarez et al. 2012; Atanassova et al. 2012; Berinde and Michnea 2013; Corredera et al. 2014; Dag et al. 2006; de Alda-garcilope et al. 2012; de Andrade et al. 2014; Dourado et al. 2019; Eissa et al. 2014; Farooq Khan and Maqbool 2008; Fredes and Montenegro 2006; Frías et al. 2008; Hungerford et al. 2020; Mondragón-Cortez et al. 2013; Moniruzzaman et al. 2014; Omode and Ademukola 2008; Perna et al. 2012; Roman et al. 2007; Silici et al. 2013; Solayman et al. 2016; Taha et al. 2010; Tutun et al. 2019; Tuzen et al. 2007; Tuzen and Soylak 2005; Uršulin-Trstenjak et al. 2015; Vanhanen et al. 2011; Yarsan et al. 2007, Zugravu et al. 2009 |
| Al[a] | mg/kg | 5.12 | 3.61 | 1.39–11.36 | |
| As[a] | mg/kg | 0.05 | 0.04 | ND–0.10 | |
| B[a] | mg/kg | 5.36 | 1.33 | 4.42–6.30 | |
| Be[a] | µg/kg | 9.93 | 0.05 | 9.83–10.00 | |
| Ca[c] | mg/kg | 84.36 | 68.34 | 4.85–218.00 | |
| Cd[a,b] | µg/kg | 89.69 | 129.75 | 0.17–373.00 | |
| Cl[c] | mg/kg | 302.63 | 83.81 | 181.30–427.2 | |
| Co[a] | µg/kg | 171.78 | 291.72 | 0.01–800.00 | |
| Cr[a,b] | µg/kg | 152.84 | 139.74 | ND–370.00 | |
| Cu[a,b] | mg/kg | 1.99 | 4.31 | 0.05–17.30 | |
| Fe[a] | mg/kg | 30.34 | 64.99 | 0.41–224.00 | |
| Ga[a,b] | µg/kg | 170.00 | 0.17 | ND–1600.00 | |
| Hg[a,b] | µg/kg | 5.09 | 6.82 | 0.27–9.91 | |
| K[c] | mg/kg | 742.43 | 453.88 | 39.66–1349.34 | |
| Li[a] | µg/kg | 310.00 | 0.07 | 240.00–680.00 | |
| Mg[c] | mg/kg | 74.31 | 163.05 | 2.18–563.72 | |
| Mn[a,b] | mg/kg | 1.42 | 1.47 | 0.00–4.35 | |
| Mo[a] | mg/kg | 0.23 | 0.30 | 0.01–0.44 | |
| Na[c] | mg/kg | 96.48 | 80.58 | 3.23–236.80 | |
| Ni[a, b] | mg/kg | 1.24 | 2.62 | ND–9.00 | |
| P[c] | mg/kg | 84.10 | 48.36 | 28.80–118.50 | |
| Pb[a, b] | µg/kg | 424.57 | 839.65 | 0.63–3232.00 | |
| S[c] | mg/kg | 35.27 | 19.86 | 15.39–72.20 | |
| Se[a] | mg/kg | 0.01 | 0.00 | 0.01–0.01 | |
| Si[a] | mg/kg | 23.52 | 16.13 | 8.70–40.70 | |
| Sr[a] | mg/kg | 1.63 | 1.03 | 0.90–2.36 | |
| Ti[a] | mg/kg | 43.40 | 17.30 | 25.00–71.00 | |
| V[a] | mg/kg | 0.03 | 0.03 | 0.01–0.05 | |
| Zn[a, b] | mg/kg | 9.33 | 19.55 | 0.23–73.60 | |

[a]Trace element.
[b]Heavy metal.
[c]Major element.
ND, not detected.

## Environmental Effect on Mineral Content in Honey

The amount of different minerals and heavy metals in honey is largely dependent on the soil composition as well as various types of floral plants because minerals are transported into plants through the roots and are passed to the nectar and finally into the honey produced from it (Machado De-Melo et al. 2018). In addition to these factors, beekeeping practices, environmental pollution, and honey processing also contribute to the diversified mineral content found to be present in honey (Pohl 2009). Different studies have indicated that almost all macrominerals were commonly found in honey from all countries, with the exception of Cl, which is only detected in honey samples from Spain (González Paramás et al. 2000).

Among the major and minor elements reported to be present in honey, K is reported to be present in the highest concentrations (Dag et al. 2006; Farooq Khan and Maqbool 2008; Fernández-Torres et al. 2005; Terrab et al. 2004b) followed by Na

(Daniel et al. 2011; Mbiri et al. 2011; Moniruzzaman et al. 2014; Nanda et al. 2003; Santos et al. 2008), and Mg, Ca, Fe, Zn, and Cu are present in intermediate amounts (Atanassova et al. 2012; Belouali et al. 2008; Dag et al. 2006; Mondragón-Cortez et al. 2013; Sager 2020; Tuzen and Soylak 2005; Vit et al. 2010). Although the stated amount of minerals found in honey is shown in an apparently low percentage compared with other honey constituents, some precise elements are found in similar or even higher amounts than in some common fruits and vegetables. For instance, from 100 g of honey, 0.4 mg of iron is achievable, which is higher than that of apples (0.2 mg), cranberries (0.2 mg), grapefruits (0.2 mg), kiwis (0.3 mg), mangoes (0.2 mg), papayas (0.2 mg), and pineapple (0.3 mg) (United States Department of Agriculture 2015). The floral types (botanical origins) characterize not only the differences between the mineral and toxic element contents but also the composition, taste, color, and flavor of honey (Bogdanov et al. 2008; Silva et al. 2009).

Honey may be miofloral or the more common polyfloral type (Silici et al. 2008). Monofloral honey originates from areas in the temperate climatic zone (especially from species such as heather, linden, buckwheat, or clover) to the tropical climatic zone (from species such as orange acacia, chestnut, rhododendron, eucalyptus, and rosemary) (Devillers et al. 2002; Fernández-Torres et al. 2005; Silici et al. 2008). Toxic trace elements or heavy metals have been reported to be found in higher concentrations in honey with close proximity to industrial areas (Matin et al. 2016). Similarly, traffic-related pollution and chemical-intensive agriculture, which contaminate the air, water, and soil, can also contribute to the increasing levels of these elements in honey (Silici et al. 2013; Street et al. 2009). Higher Pb concentrations (up to 2370 mg/kg) were detected in honey samples collected from contaminated areas in Italy (D'ambrosio and Marchesini 1982). In another study, the highest concentrations of Pb, Cd, and Cr reported to be present in honey were 45.0, 630, and 102 mg/kg, respectively, which were higher than the tolerable intake levels (Bartha et al. 2020; Conti and Botrè 2001).

Initially, Batista et al. (2012) reported that honey samples from Brazil contained 42 chemical elements (toxic and essential elements) in which some unique components, such as Pd, U, TI, Te, Hf, Sn, Sb, La, I, Sm, Tb, Dy, Sd, Th, Pr, Nd, Tm, Yb, Lu, Gd, Ho, Er, Ce, and C, were reported to be present. Another study, however, indicated that 32 chemical elements were reported to be found in honey from several European countries (Vincēviča-gaile 2010). The summarized major and minor elements, as well as heavy metals commonly found in different countries on several continents, are as reported in Tables 7.2 and 7.3.

**Table 7.2** Major elements reported in honey samples from different countries.

| Country | Na(mg/kg) | K(mg/kg) | Ca(mg/kg) | Mg(mg/kg) | P(mg/kg) | S(mg/kg) | Cl(mg/kg) | References |
|---|---|---|---|---|---|---|---|---|
| **Asia** | | | | | | | | |
| India | 163.30–304.31 | 489.52–932.56 | 32.60–84.63 | | – | – | – | Nanda et al. 2003 |
| Israel | 26.60–132.50 | 189.00–3768.30 | 58.40–137.10 | 18.50–204.60 | 47.00–651.00 | 22.40–188.30 | | Dag et al. 2006 |
| Malaysia | 83.17–732.16 | 413.63–4026.40 | 65.80–567.27 | 21.83–199.33 | – | – | – | Moniruzzaman et al. 2014 |
| Pakistan | 77.00–200.00 | 225.00–1480.00 | 46.1.00–98.10 | 18.00–71.80 | – | – | – | Farooq Khan and Maqbool 2008; Lanjwani and Channa 2019 |
| Palestine (West Bank) | 41.80–306.30 | 42.80–585.00 | 44.50–150.70 | 12.30–46.70 | 39.80–118.10 | – | – | Abdulkhaliq and Swaileh 2017 |
| Saudi Arabia | – | – | – | 18.39–23.21 | – | – | – | Osman et al. 2007 |
| Turkey | 70.64–613.20 | 146.72–7030.00 | 15.20–232.00 | 9.28–117.16 | – | – | – | Silici et al. 2008; Tutun et al. 2019 |
| **Europe** | | | | | | | | |
| Bulgaria | 7.22–16.30 | 105.00–1628.00 | 32.00–110.00 | 6.00–97.00 | 24.00–124.00 | 12.00–41.00 | – | Atanassova et al. 2012 |

**Table 7.2** (Continued)

| Country | Na(mg/kg) | K(mg/kg) | Ca(mg/kg) | Mg(mg/kg) | P(mg/kg) | S(mg/kg) | Cl(mg/kg) | References |
|---|---|---|---|---|---|---|---|---|
| France | – | – | 8.90–130.90 | 3.62–68.78 | 84.39–354.45 | 9.61–118.10 | – | Devillers et al. 2002 |
| Italy | 6.10–62.50 | 205.00–2639.00 | 9.10–402.60 | 3.90–65.00 | – | – | – | Conti et al. 2007 |
| Poland | 6.80–131.00 | 12.00–736.00 | 27.60–61.10 | 2.80–145.00 | 35.70–571.00 | – | – | Grembecka and Szefer 2013 |
| Serbia | 6.75–160.00 | 6.35–466.69 | – | 5.71–63.60 | – | – | – | Mračević et al. 2020 |
| Spain | 55.80–151.65 | 183.30–5570.73 | 15.14–202.30 | 23.90–1078.95 | 67.70–205.00 | 27.80–53.80 | 181.30–427.2 | de Alda-garcilope et al. 2012; González Paramás et al. 2000 |
| **Africa** | | | | | | | | |
| Kenya | 98.04–269.10 | 172.83–564.13 | 19.33–70.17 | 12.64–41.88 | – | – | – | Maiyo et al. 2014; Mbiri et al. 2011 |
| Morocco | – | 30.65–5097.91 | 22.32–228.57 | 5.02–62.80 | – | – | – | Belouali et al. 2008 |
| Nigeria | 289.00 | 28770.00 | 144.00–270.00 | 2660.00 | – | – | – | Daniel et al. 2011; Omode and Ademukola 2008 |
| Sudan | 14.11–28.24 | 17.60–74.66 | 35.63–82.92 | 23.67–177.15 | 28.10–204.60 | 46.10–131.50 | – | Moniruzzaman et al. 2014 |
| **North America** | | | | | | | | |
| Mexico | 20.50–302.00 | 276.70–1760.00 | 38.00–127.30 | 13.10–61.30 | 47.80–104.80 | 13.10–62.90 | – | Mondragón-Cortez et al. 2013 |
| **South America** | | | | | | | | |
| Argentina | 6.10–89.98 | 90.92–1955.75 | 18.60–136.14 | 6.01–46.57 | 1.17–100.66 | – | – | Cantarelli et al. 2008 |
| Brazil | 113.08–205.70 | 489.43–555.94 | 32.48–49.94 | 13.86–16.88 | – | – | – | Santos et al. 2008 |
| Venezuela | – | – | 98.26–214.01 | 12.59–79.00 | – | – | – | Vit et al. 2010 |
| **Australia** | | | | | | | | |
| Australia | 3.70–382.70 | 202.00–4600.00 | 21.00–270.00 | 6.00–190.00 | 12.00–920.00 | – | – | Hungerford et al. 2020 |
| New Zealand | 1.10–110.00 | 34.80–3640.00 | 7.21–94.30 | 7.52–86.30 | 29.50–255.00 | 13.40–93.90 | – | Vanhanen et al. 2011 |

## Role of Minerals And Heavy Metals in Honey

Honey as food and medicine has many nutritional and medicinal benefits (Ajibola et al. 2012; Jahangir et al. 2020). Because honey is a source of nutrients in low quantities, it would be advisable for adults to take honey in larger quantities (70–95 g/d) to obtain the complete desired nutritional and health benefits (Hasam et al. 2020; Münstedt et al. 2009; Yaghoobi et al. 2008). In a comparative study on the importance of honey relative to sucrose in children's nutrition, honey ingestion led to the development of hematological profiles and calcium uptake, lighter and thinner bowel contents, better skin color, and

**Table 7.3** Trace elements and heavy metal levels reported in honey samples from different countries.

| Country | Fe (mg/kg) | Cu (mg/kg) | Zn (mg/kg) | Cd (μg/kg) | Pb (μg/kg) | As (mg/kg) | Ag (μg/kg) | Si (mg/kg) | B (mg/kg) | Co (μg/kg) |
|---|---|---|---|---|---|---|---|---|---|---|
| **Asia** | | | | | | | | | | |
| China | | 0.04–0.31 | 0.59–22.85 | BDL–4.20 | 7.15–85.78 | BDL–0.08 | – | – | – | – |
| India | 8.86–13.25 | 1.74–2.90 | 2.50–16.77 | | | – | – | – | – | – |
| Iran | 2.13–9.24 | 0.05–2.70 | 0.31–7.75 | 330.00–530.00 | 60.00–340.00 | 0.08–0.40 | – | – | – | – |
| Israel | 0.90–9.30 | 9.00–3.18 | 0.80–11.50 | – | 150.00–8220.00 | – | – | 4.20–18.00 | 0.17–12.60 | – |
| Malaysia | 55.83–233.00 | N.D.–2.93 | 4.7–173.77 | ND–1030.00 | ND–1017.00 | 0.03–0.13 | – | – | – | ND–107.00 |
| Pakistan | 2.98–16.2 | 0.08–1.94 | 1.11–4.10 | 10.00–380.00 | 10.00–140.00 | – | – | – | – | 10–1120.00 |
| Palestine West Bank | 2.00–10.80 | ND–7.60 | 1.00–19.90 | – | – | – | – | – | 0.23–6.20 | – |
| Saudi Arabia | 0.31–3.19 | 0.24–0.39 | 0.21–0.75 | 2.00–37.00 | 38.00–80.00 | 0.00–0.53 | – | – | – | 0.60–27.5 |
| Turkey | 0.57–16.36 | 0.04–1.52 | 0.32–6.00 | 0.28–240.00 | 1.54–1240.00 | – | – | – | – | 1.25–30.00 |
| **Europe** | | | | | | | | | | |
| Bulgaria | 0.35–4.37 | <0.01–0.45 | 0.08–1.17 | <10.00 | <80.00–310.00 | < 0.1–0.320 | – | – | – | <10.00 |
| Croatia | – | 124–1272 | – | 1000.00–4000.00 | 4130.00–21590.00 | 10.0–499.00 | – | – | – | – |
| Czech Republic | – | – | – | 0.95–32.35 | 22.80–117.85 | <.01–0.04 | – | – | – | – |
| France | 0.56–86.76 | 0.06–1.71 | 0.17–6.42 | 80.00–250.00 | 280.00–1080.00 | – | 90.00–160.00 | – | – | 100.00–230.00 |
| Italy | 2.00–35.10 | 0.19–2.98 | 0.178–8.120 | 12.00–274.00 | 100.00–1533.00 | – | – | – | – | – |
| Poland | 0.60–5.70 | 0.10–0.90 | 0.30–8.40 | ND | ND | – | – | – | – | ND–300.00 |
| Romania | 2.04–2.91 | 0.01–33.00 | 15.00–36.40 | 10.00–1600.00 | 20.00–6000.00 | – | – | – | – | – |
| Serbia | 0.91–5.99 | N.D.–1.6 | 0.36–20.36 | N.D.–150.00 | N.D.–0.42 | BDL | – | 1.21–7.84 | – | – |
| Spain | ND–7.59 | 0.36–1.36 | 1.19–2.76 | 2.70–54.20 | 31.50–46.32 | – | – | 0.01–0.04 | – | BDL–136.2 |

| Ni (mg/kg) | Hg (μg/kg) | Cr (μg/kg) | Mn (mg/kg) | Se (mg/kg) | Al(mg/ kg) | Mo (mg/kg) | Sr (mg/kg) | V (μg/kg) | Ti (mg/kg) | Be (μg/kg) | References |
|---|---|---|---|---|---|---|---|---|---|---|---|
| – | 0.80–4.00 | – | – | – | – | – | – | – | – | – | Ru et al. 2013 |
| – | – | – | – | – | – | – | – | – | – | – | Nanda et al. 2003 |
| – | 2370.00–3510.00 | – | 0.15–0.62 | 0.06–0.26 | 1.85–16.12 | – | – | – | – | – | Samimi et al. 2001 |
| – | – | – | – | – | – | – | – | – | – | – | Dag et al. 2006 |
| – | – | – | – | – | – | – | – | – | – | – | Moniruzzaman et al. 2014 |
| 0.06–1.48 | – | 0.01–0.10 | 0.12–0.95 | – | – | – | – | – | – | – | Farooq Khan and Maqbool 2008; Lanjwani and Channa 2019 |
| ND–0.51 | – | ND–740.00 | 0.11–0.99 | – | – | – | – | – | – | – | Abdulkhaliq and Swaileh 2017; Swaileh and Abdulkhaliq 2013 |
| 0.00–0.19 | – | – | 2.50–102.30 | | | – | – | – | – | – | Bazeyad et al. 2019; Osman et al. 2007 |
| 1.35–1.44 | – | 1.24–1040.00 | 0.02–1.56 | 0.01–0.45 | 0.18–25.00 | – | 0.20–0.80 | – | – | – | Leblebici and Aksoy 2008; Silici et al. 2008; Silici et al. 2013; Tutun et al. 2019; Yarsan et al. 2007 |
| <0.01–1.00 | – | <10.00–20.00 | 0.06–12.70 | – | 0.24–1.58 | – | 0.12–0.40 | <50 | – | – | Atanassova et al. 2012; Jivan et al. 2008 |
| – | 130.00–900.00 | – | – | – | – | – | – | – | – | – | Bilandžić et al. 2012 |
| – | 3.24–11.31 | – | – | – | – | – | – | – | – | – | Batelková et al. 2012 |
| 0.09–0.34 | NA | 80.00–360.00 | 0.11–42.81 | – | 0.18–9.72 | 0.15–0.33 | – | – | – | – | Devillers et al. 2002 |
| – | – | 33.30–60.00 | 0.08–1.17 | – | – | – | – | – | – | – | Conti et al. 2007; Conti and Botrè 2001; Sanna et al. 2000 |
| ND–0.50 | – | ND–500.00 | 0.10–8.00 | – | – | – | – | – | – | – | Grembecka and Szefer 2013 |
| 0.90–2.50 | – | 30.00–800.00 | 1.75–6.51 | – | – | – | – | – | – | – | Dobrinas et al. 2006; Matei et al. 2004; Purcarea et al. 2017 |
| N.D.–0.08 | – | ND | – | ND | N.D.–4.56 | – | – | – | – | – | Mračević et al. 2020 |
| 0.05–0.75 | – | 50.10–843.30 | 0.98–10.9 | – | 3.04–8.31 | – | – | – | – | – | Frías et al. 2008; de Alda-garcilope et al. 2012; González Paramás et al. 2000 |

*(Continued)*

**Table 7.3** (Continued)

| Country | Fe (mg/kg) | Cu (mg/kg) | Zn (mg/kg) | Cd (μg/kg) | Pb (μg/kg) | As (mg/kg) | Ag (μg/kg) | Si (mg/kg) | B (mg/kg) | Co (μg/kg) |
|---|---|---|---|---|---|---|---|---|---|---|
| **Africa** | | | | | | | | | | |
| Egypt | 21.50–38.82 | BLD | 1.63–2.57 | BLD | 860.00–1880.00 | – | – | – | – | – |
| Ethiopia | 6.33–13.61 | 0.37–13.99 | 0.370–1.124 | 22.00–29.00 | – | – | – | – | – | 0.00–249.00 |
| Kenya | 0.08–1.12 | 0.02–0.05 | 0.05–0.35 | BLD–30.00 | BLD–280.00 | BLD–0.03 | – | – | – | – |
| Libya | | 0.80–10.40 | – | N. D.–5.00 | N. D.–60.00 | – | – | – | – | – |
| Morocco | 0.88–207.65 | 0.51–4.75 | 0.04–2.74 | 1.30–24.9 | 36.00–1880.00 | – | – | – | – | – |
| Nigeria | 0.12–3.42 | 0.01–0.02 | 0.04–0.11 | – | ND–0.90 | – | – | – | – | – |
| Sudan | 2.05–33.65 | 2.94–58.12 | 4.86–9.61 | ND–100.00 | <450.00 | – | – | – | – | 5.00–1172.00 |
| **North America** | | | | | | | | | | |
| Cuba | 4.04–10.24 | 0.04–0.59 | <0.20–1.71 | <0.10–0.50 | 76.70–102.00 | <.0.01 | – | – | – | – |
| Mexico | 0.82–4.72 | – | 1.51–6.80 | – | – | – | – | 23.7–90.9 | – | – |
| United State | <983.30–<999.60 | 0.08–0.13 | 0.31–1.27 | 704.20–1585.40 | <983.30–1534.80 | – | <2.46–6.14 | – | – | – |
| **South America** | | | | | | | | | | |
| Argentina | 1.13–10.32 | 0.05–0.68 | 0.14–3.87 | – | – | – | – | – | – | – |
| Brazil | 3.98–21.99 | ND–2.10 | 0.25–0.79 | <LOQ–8.00 | 141.00–228.00 | – | – | – | – | – |
| Chile | 0.78–7.66 | 0.09–4.32 | 0.19–4.93 | ND–10.00 | ND–100.00 | – | – | – | – | 10.00–620.00 |
| Venezuela | 1.09–4.59 | 0.00–0.97 | 0.60–2.52 | – | – | – | – | – | – | – |
| **Australia** | | | | | | | | | | |
| Australia | 0.20–99.00 | 0.05–4.80 | 0.16–120.00 | 0.00–0.05 | 0.00–0.69 | 0.00–0.01 | 0.01–0.01 | – | 1.00–19.00 | 0.01–0.48 |
| New Zealand | 0.67–3.39 | 0.70–0.09 | 0.20–2.46 | 10.00–450.00 | 10.00–40.00 | 0.04–0.17 | – | – | 0.05–0.49 | – |

BDL, below detection limit; LOQ, limit of quantification; ND, not detected.

| Ni (mg/kg) | Hg (µg/kg) | Cr (µg/kg) | Mn (mg/kg) | Se (mg/kg) | Al(mg/kg) | Mo (mg/kg) | Sr (mg/kg) | V (µg/kg) | Ti (mg/kg) | Be (µg/kg) | References |
|---|---|---|---|---|---|---|---|---|---|---|---|
| BLD | – | BLD | 0.46–1.17 | – | – | – | – | – | – | – | Taha and Ali 2012 |
| 1.99–2.61 | – | 493.00–624.00 | 0.99–1.49 | – | – | – | – | – | – | – | Nigussie et al. 2012 |
| – | – | ND–0.15 | – | – | – | – | – | – | – | – | Maiyo et al. 2014; Mbiri et al. 2011 |
| – | – | – | – | – | – | – | – | – | – | – | Ahmida et al. 2013 |
| – | – | – | 0.08–9.76 | 0.26–17.39 | – | – | – | – | – | – | Belouali et al. 2008 |
| 0.01–0.02 | – | – | – | – | – | – | – | – | 25.0–71.0 | – | Daniel et al. 2011; Omode and Ademukola 2008 |
| ND–4.06 | – | <10.00 | 0.12–1.02 | – | – | – | – | – | – | – | Moniruzzaman et al. 2014 |
| 0.09–0.19 | <10 | 28.2–56.9 | <0.10–1.51 | <10 | – | – | – | <10 | – | – | Alvarez-Suarez et al. 2012 |
| – | – | – | – | – | – | – | – | – | – | – | Mondragón-Cortez et al. 2013 |
| <0.01–0.13 | – | <9.83–<10.00 | – | – | – | – | – | – | – | <9.83–<10.00 | Birge and Price 2001 |
| – | – | – | 0.07–0.68 | – | 0.02–13.04 | – | – | – | – | – | Cantarelli et al. 2008 |
| – | – | 83.00–94.00 | 0.90–3.27 | – | – | – | – | – | – | – | de Andrade et al. 2014; Santos et al. 2008 |
| 0.01–1.48 | – | 10.00–1980.00 | 0.02–6.97 | – | 0.04–22.06 | – | 0.04–22.06 | – | – | – | Fredes and Montenegro 2006 |
| – | – | – | 0.50–10.70 | – | – | – | – | – | – | – | Vit et al. 2010 |
| 0.01–1.40 | 0.00–0.08 | 0.01–0.06 | 0.38–38.00 | 0.01–0.03 | 0.05–14.00 | 0.01–0.29 | 0.08–2.30 | 0.01–0.02 | – | – | Hungerford et al. 2020 |
| 0.02–0.65 | – | 120.00–550.00 | 0.18–4.75 | – | 0.21–21.3 | ND–0.01 | – | – | – | – | Vanhanen et al. 2011 |

improved digestion. It also contributed to a steady weight gain and reduced disease susceptibility (Alvarez-Suarez et al. 2010). Analysis of honey composition indicates that minerals and heavy metals are important constituents and play specific roles in human health. A study observed that the growth of rodents following honey feeding was significantly high because of bone growth and mineralization (Chepulis and Starkey 2008). Ca is an essential mineral because of its important contribution to several biological functions in the cardiac, nervous, and musculoskeletal systems, including contributing to the formation of bone and teeth. Furthermore, Ca is involved in mineral homeostasis and physiological performance as well as acting as a cofactor for many enzymes (Huskisson et al. 2007; Morgan 2008; Theobald 2005).

Another study showed that honey is beneficial to individuals suffering from anemia because it helps to improve hemoglobin concentration, increase erythrocyte count, and elevate hematocrit level. The researchers also attributed this ability to the presence of Fe in honey as well as the increased immunity found in honey eaters (Ajibola et al. 2007). In an electron microscopy and micro-hardness study, honey consumption was safer and provided a better environment for oral health than drinking fruit juice (Grobler et al. 1994). It was reported that drinking fruit juice erodes the tooth enamel in just 10 minutes after consumption, while erosion of teeth was not prominent or visible even up to half an hour after honey consumption. It is plausible that honey constituents such as Ca, F, P, and other colloidal honey components confer a protective role, making honey less cariogenic. Another important point is that Na plays a vital role in the maintenance of optimum blood pressure, kidney function, and nerve and muscle functions (Hall 2003; Sobotka et al. 2008). The presence of K needs to be in balance with the fluid system like Na and also assists in nerve functions, heart activity, and muscle contraction (Ko et al. 2008; Lambert et al. 2008; Sobotka et al. 2008). Another key mineral in honey is Mg, which acts as cofactor for up to 300 enzymes, most of which are related to antioxidant reactions. Mg deficiency contributes to aging and age-related disorders (Barbagallo and Dominguez 2018; Huskisson et al. 2007).

Trace elements are helpful when present in optimum concentrations, but at higher concentrations trace elements (heavy metals) contribute to toxicity in humans. The toxicity occurs because of the inability of the heavy metal to be metabolized by the body, leading to accumulation in human or animal soft tissues without being fully degraded or destroyed (Ajibola et al. 2012). Health problems caused by heavy metals include headache, metabolic abnormalities, respiratory disorders, nausea, and vomiting. For instance, Pb can cause damage to the brain, kidney, nervous system, and red blood cells (Ara and Usmani 2015). According to the Agency for Toxic Substances and Disease Registry, the most hazardous heavy metals include As, Pb, and Cd. As and Cd poisonings are comparatively less common with honey because of their limited use, but contamination with Pb is frequently reported (Bogdanov 2006). Other heavy metals in honey include Cr, Zn, Hg, Mn, and Ag, all of which are toxic to human health but play important roles as bio-indicators of environmental pollution (Celli and Maccagnani 2003; Okeola et al. 2020; Vit et al. 2010).

## Technique Used to Determine Minerals in Honey

A number of techniques are used to determine chemical elements, such as minerals, metals, and nonmetals, at low concentrations in supplied honey samples. Different studies on detecting minerals and heavy metals in honey samples have utilized spectrometry and many other instrumental methods. These analyses are important to determine the nutritional value and the potential effect of honey on human well-being, and they ensure the general safety and the purity of honey. However, analysis of minerals in honey is challenging because of the complex organic matter of the matrix (Ajtony et al. 2007). The commonly used techniques are flame emission spectrometry, flame atomic absorption spectrometry (FAAS), inductively coupled plasma optical emission spectrometry, and inductively coupled plasma mass spectrometry. In some cases, ethylenediaminetetraacetic acid titration or precipitation titrations are used to identify Ca and Ag, respectively (Table 7.4).

There are also some reported marginal uses of ion chromatography (Buldini et al. 2001), total reflection X-ray fluorescence spectrometry (Golob et al. 2005; Ribeiro et al. 2014), potentiometric stripping analysis (Muñoz and Palmero 2006), and anodic stripping voltammetry techniques (Khammas et al. 2012). However, the application of these diverse processes is dependent on their availability in the laboratory. Although different processes can determine the aforementioned elements, they differ in speed, precision, sensitivity, specificity, and accuracy (see Table 7.4). Acid digestion is required in these analyses for honey sample preparation. Then the absorption of radiation is measured at specific wavelengths corresponding to the mineral of interest. The instruments must be calibrated and standardized using relevant working standards. The concentrations of the individual elements are then measured after the instrument has been accurately calibrated.

**Table 7.4** Determination of different elements using various techniques.

| Methods | Elements | References |
|---|---|---|
| Flame emission photometry or spectrometry (FES) | Na, K, Ca, Li | Mbiri et al. 2011 |
| Inductively coupled plasma optical emission spectrometry (ICP–OES) | Na, K, Ca, Li, Mg, Fe, Zn, Mn, Cu, Al, Cd, Co, Cr, Ni, Pb | Akbari et al. 2012; Akram 2008; Belouali et al. 2008; Cantarelli et al. 2008; Demirezen and Aksoy 2010; Devillers et al. 2002; Dobrinas et al. 2006; Fernández-Torres et al. 2005; Leblebici and Aksoy 2008; Ruschioni et al. 2013 |
| Inductively coupled plasma mass spectrometry (ICP-MS) | Al, Cu, Pb, Zn, Mn, Cd, Tl, Co, Ni, Rb, Ba, Be, Bi, U, V, Fe, Pt, Pd, Te, Hf, Mo, Sn, Sb, P, La, Mg, I, Sm, Tb, Dy, Sd, Th, Pr, Nd,Tm, Yb, Lu, Gd,Ho, Er, Ce, Cr | Batista et al. 2012; Bogdanov et al. 2007; Musgrove et al. 2010; Voica et al. 2020 |
| Flame atomic absorption spectrometry (FAAS) | Na, K, Ca, Li, Mg, Fe, Zn, Hg, Mn, Cu, Cd, Co, Cr, Ni, Pb | Agbagwa et al. 2011; Ahmida et al. 2013; Akhtar et al. 2014; Altunay et al. 2019; Farooq Khan and Maqbool 2008; Mbiri et al. 2011; Moniruzzaman et al. 2014; Nigussie et al. 2012; Omode and Ademukola 2008; Osman et al. 2007; Ru et al. 2013; Saghaei et al. 2012; Taha and Ali 2012 |
| Electrothermal atomic absorption spectrometry (ET-AAS) | Fe, Zn, Mn, Cu, Al, Cd, Co, Cr, Ni, Pb | de Andrade et al. 2014; Fiorentini et al. 2019; Fiorentini et al. 2020 |
| Graphite furnace atomic absorption spectrometry (GF-AAS) | Cd, Pb | Ru et al. 2013; Silici et al. 2013 |
| Hydride generation–atomic fluorescence spectrometry (HG-AAS) | As, Hg | Mbiri et al. 2011; Ru et al. 2013 |
| Ion chromatography | Fe, Cu, Zn, Pb, Cd, Co | Buldini et al. 2001; Idoko et al. 2018 |
| Ethylenediaminetetraacetic acid (EDTA) titration | Ca, Mg | Farooq Khan and Maqbool 2008 |

## Correlation Among Minerals, Color, and Taste of Honey

The appearance of food is one of the factors that defines its quality and is actually consumers' first impression. As a part of the appearance, the color has to meet consumers' expectations and be within a selected range. Honey varies in color, ranging from white or pale yellow to dark red or even black. According to the U.S. Department of Agriculture (1985), honey samples with Pfund values of less than 8 mm are categorized as "water white," between 9 and 17 mm as "extra white," between 18 and 34 mm as "white," between 35 and 50 mm as "extra light amber," between 51 and 85 mm as "light amber," between 86 and 114 mm as "amber," and greater than 114 mm as "dark amber." Honey's color mainly differs based on its floral origin, industrial processing methods, temperature, and storage duration (Eshete and Eshete 2019; Šarić et al. 2020). Some studies have considered that in addition to the mineral content, the morphology and color of pollen grains (Terrab et al. 2004a) in honey also have some effects on honey color and taste (González-Miret et al. 2005).

The mineral composition of honey also correlates with its color (Downey et al. 2005; González-Miret et al. 2005; Pohl 2009). Dark and amber honey (avocado, chestnut, honeydew, heather) contain higher amounts of certain major, minor, and trace metals (e.g. Al, Ca, Cd, Cu, Fe, K, Mg, Mn, Na, Ni, Zn) than pale-colored honey (González-Miret et al. 2005; Nalda et al. 2005; Osman et al. 2007; Pisani et al. 2008). Amber light honey (eucalyptus and thyme honey) show rather intermediate quantities of total minerals. Interestingly, the darker colored honeys have been reported to be mostly associated with higher concentrations of Cd, Fe, and Pb, whereas light- and brown-colored honeys are related to Al and Mg concentrations (González-Miret et al. 2005). Several studies have indicated that darker honey (e.g. avocado, chestnut, honeydew, and heather) contain higher levels of minerals (between 1340 and 1879 mg/kg), which correlates with other studies performed

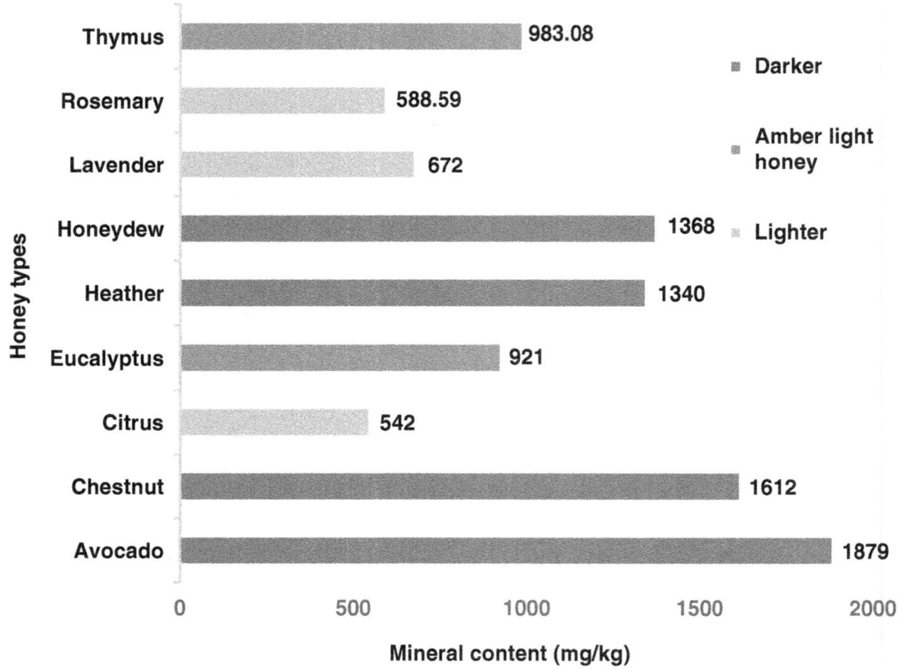


**Figure 7.1** Reported concentrations of minerals and metals in darker honeys (avocado, chestnut, honeydew, and heather honey) compared with amber light honeys (eucalyptus and thyme honey) or lighter honeys (citrus, rosemary, and lavender honey).

on Spanish, Italian, and Moroccan honey (Taleuzzaman et al. 2020; Terrab et al. 2003). Lighter honeys (citrus, rosemary, and lavender) have lower levels of minerals (542, 588.5, and 672 mg/kg, respectively), and amber light honeys (eucalyptus and thyme) have medium quantities of total minerals (921 and 983 mg/kg, respectively) (Figure 7.1).

## Effect of Minerals of Honey on Electrical Conductivity

Electrical conductivity (EC) is the measure of the amount of electrical current that a material can carry. Honey is a mixture of water, minerals, acids, and organic materials and is measured for its EC, which indicates the amount of minerals and acid content in a honey sample (Kropf et al. 2008). There is a relationship between both the ash (minerals) and acid content of honey and its EC. The higher the concentration of minerals, the greater the EC values (Bogdanov et al. 2007; Kropf et al. 2008; Pohl 2009; Silva et al. 2009). Kropf et al. (2008) demonstrated a linear regression model for the relationship between ash and EC for Slovenian honey using 290 samples (including acacia, lime, chestnut, spruce, fir, multifloral, and mixed forest honeydew honey).

Another study performed by Bogdanov et al. (2007) also revealed significant positive correlations between the EC and most of the mineral elements (with the exception of Pb and Cr) on the basis of an examination of 95 honey samples from production areas in Switzerland. Because the mineral content determines the botanical origins and the color of honey, individual mineral measurements can be replaced by EC. For example, lighter honeys usually contain fewer elements than darker ones; consequently, the value of K is lower in light honeys than in dark ones. In addition, EC, along with measurement of the trace elements Ni, Fe, Mn, and Cd, could be used for the classification and isolation of unifloral honey from polyfloral honey (Bogdanov et al. 2004). Currently, EC is one of the most beneficial quality parameters and has been included in the new international standards for honey (Alimentarius 2001; Cahyono 2020).

## Effect of Honeycombs on Mineral Content of Honey

The mineral contents in honey depend on not only botanical origin (Bogdanov et al. 2007; Fernández-Torres et al. 2005; Nanda et al. 2003), geographical origin, and environment conditions (Bogdanov et al. 2007; Bratu and Georgescu 2005; González Paramás et al. 2000; Kropf et al. 2010) but also on the length of storage (Seraglio et al. 2020) and the age of the honeycombs (Taha et al. 2010). Honeycombs are masses of hexagonal wax cells built by honeybees to store honey and pollen and contain their larvae. Taha et al. (2010) showed the levels of K, Na, Ca, Mg, Fe, Zn, Pb, and Mn in clover honey significantly depended on the age of combs. They observed significant ($P < 0.01$) positive correlations between all determined mineral elements in clover honey with the age of honeycombs. They found the highest mineral values in honey and wax from combs more than 4 years old. Ni was not detected in younger combs, but in combs that were more than 4 years old, the content was 1.00 mg/kg. Taha and El-Sanat (2007) found the color intensity and the EC of honey significantly correlated to the age of the comb. The color intensity and the EC of honey increase with a rise in mineral level (González-Miret et al. 2005; Kropf et al. 2008). These results are similar to those of another study performed by Bogdanov et al. (2007), who established significant positive correlations between chemical elements (Fe, Zn, Mn, and Pb) in honey and the age of the honeycomb.

## Effect of Minerals on the Antioxidant Properties of Honey

Free radicals are atoms, molecules, or ions having one or more unpaired electrons and are very reactive and toxic to cells (Surai 2003). The harmful effects of free radicals are lessened by the body's antioxidant defenses. These defense mechanisms against free radical damage usually contain vitamin C, vitamin E, and beta-carotene (and other carotenoids) as the main vitamin antioxidant sources. Furthermore, several metal-containing enzymes, which include glutathione peroxidase (Se), catalase (Fe), and superoxide dismutase (Zn, Cu and Mn), are also critical in defending the internal cellular components from oxidative damage (McDowell et al. 2007). The dietary and tissue balances of all these nutrients (vitamins and trace elements) are important in protecting tissues against free radical damage as well as participating in immune functions. However, when there is a lack of equilibrium in the organism between the production of free radicals and the antioxidant protective activity, then an "oxidative stress" condition develops (Bogdanov 2011).

A varied range of elements is present in honey, and many of these elements are known to have antioxidant properties. The naturally occurring antioxidants present in honey are flavonoids, phenolic acids, some enzymes (e.g. glucose oxidase, catalase), ascorbic acid, carotenoid-like substances, organic acids, Maillard reaction products, amino acids, and proteins (Alam et al. 2014; Aljadi and Kamaruddin 2004; Ibrahimi and Hajdari 2020; Khalil et al. 2011; Nagai et al. 2006; Pérez et al. 2007; Vela et al. 2007). Perna et al. (2012) showed that there was a statistically significant correlation ($P < 0.05$) between metal content and both total phenolic and antioxidant activities based on 78 samples of southern Italian honey from five different floral origins (chestnut, eucalyptus, citrus, multifloral, and sulla). They observed correlations between total phenolic and Co, Cr, Fe, Pb, and Zn contents ($r = 0.34, 0.48, 0.58, 0.37$, and $0.38$, respectively). They also showed that the correlation coefficients between total flavonoid and metal content were low and not significant except for the correlations between total flavonoid and both Co and Fe contents, with $r = 0.22$ ($P < 0.05$) and $0.42$ ($P < 0.001$), respectively. Another study by Ferreira et al. (2009) concluded that dark honey had better antioxidant activity and phenolic content than the other studied honey (amber and light) samples. From these studies, it can be concluded that with an increase in the metal content of honey (color intensity is positively correlated with metal content in honey), there is an increase in its antioxidant properties. There is a significant relationship between the mineral contents and antioxidant properties of honey.

## Honey Minerals and the Gastrointestinal Digestive System

Minerals are important compounds present in honey (Pita-Calvo and Vázquez 2017; Seraglio et al. 2019), and many of these elements are essential and play crucial roles in biological activities in the human body (Pohl et al. 2012; Seraglio et al. 2017). Their presence in honey depends on the soil's mineral composition and the plant's absorption capacity. Therefore, besides the presence of essential elements on honey composition, toxic elements such as Hg, Pb, and Cd can also be found in this product from polluted areas or improper handling (Bergamo et al. 2018; Deng et al. 2018; Seraglio et al. 2019), which justifies the need to investigate the bioaccessibility and bioavailability of various elements in honey. Three different groups of species of these minerals (hydrophobic, cationic, and residual) were evaluated in acacia, buckwheat, multiflower, goldenrod, goldenrod-buckwheat, heather, honeydew, linden, and rape honey using FAAS in conjunction with tandem column solid-phase extraction. The cationic fraction was predominant (up to 96.3% for Ca and 97.0% for Mg) followed by hydrophobic (up to 20.1% for Ca and 23.0% for Mg) and residual (up to 6.5% for Ca and 2.1% for Mg) fractions. The cationic and residual fractions can be considered those of better bioavailability because the first is related to simple ions or linked to low-molecular-weight organic compounds (e.g. organic acids and amino acids). At the same time, the second includes neutral and/or anionic complexes stable also with low-molecular-weight compounds. On the other hand, the hydrophobic fraction includes mainly Ca and Mg bound with phenolic compounds, decreasing the absorption capacity of these minerals.

Following that study, the influence of the digestive system on honey minerals was reported by Pohl et al. (2012) for the first time. Dark honey of buckwheat, heather, and honeydew was chosen to assess the in vitro bioaccessibility of Ca, Cu, Fe, Mg, Mn, and Zn because of their high concentrations of these minerals. In this study, the honey minerals were determined by FAAS in the soluble and permeable fraction obtained after being submitted to static in vitro digestion composed by gastric and duodenal phases followed by dialysis. Considering that the cationic fraction of minerals was proposed as the most accessible class found in honey, the authors used the semi-permeable membrane to separate only these elements and test this hypothesis. Indeed, the results indicated high average values of in vitro bioaccessibility regardless of the honey and mineral evaluated: 87.2 to 100.0% for Ca, 85.4 to 100.0% for Cu, 73.0 to 100.0% for Fe, 82.0 to 96.5% for Mg, 85.8 to 100.0% for Mn, and 71.9 to 98.8% for Zn. Therefore, the high in vitro bioaccessibility of these elements was linked to their presence in honey commonly as non-complex structures, indicating potential bioavailability.

In another study conducted with dark honey, Seraglio et al. (2017) investigated K, Ca, Mg, and Mn in the soluble digested fraction of bracatinga honeydew honey by capillary electrophoresis coupled to diode array detector. In this study, the honey samples were evaluated after both gastric and duodenal phases of static in vitro digestion. As a result, high average values of in vitro bioaccessibility were found for all quantified minerals (Mn was not detected in any sample) after gastric digestion: 94.8% to 102.0% for K, 106.2% to 108.4% for Ca, and 93.1% to 98.6% for Mg. This tendency was maintained after duodenal digestion for K (94.0%–112.7%) and Mg (99.7%–108.9%), but a significant increase in the in vitro bioaccessibility of Ca was verified (164.0%–220.5%). Thus, these data complement the previous findings of Pohl et al. (2012), indicating that honey minerals are also highly in vitro bioaccessibility after gastric digestion. Besides, the presence of minerals, not just in simple forms but also complex ones with organic compounds such as phenolic compounds and proteins, was proposed and associated with the increase in the Ca content after duodenal digestion.

Although the in vitro digestion protocol applied in each study showed differences mainly regarding the concentration of enzymes and other reagents used, the punctual conflicting findings between these studies seem to be more associated with the different chemical compositions of each honey. Besides, both studies evaluated independent replicates and procedural blanks, which minimize analytical errors. In both studies mentioned, the elements investigated were minerals commonly found in honey in high or intermediate concentrations. Besides, they were evaluated only in dark honey from *Apis* bees, which have physical, chemical, and biological properties distinct to those found in light blossom honey from *Apis* and *Meliponinae* bees. In this sense, the study conducted by Biluca et al. (2017) brings new information about the in vitro bioaccessibility of major minerals, trace elements, and heavy metals in blossom honey from stingless bees. Sixteen elements (Al, Ba, Cd, Co, Cu, Fe, Li, Mg, Mn, Na, Ni, Pb, K, Ca, Rh, and Zn) were monitored using inductively coupled plasma mass spectrometer and atomic absorption spectrometer in the samples after being submitted to static in vitro digestion involving the gastric and duodenal phases of digestion. The results found indicated that stingless bee honey present high average values of in vitro bioaccessibility for all minerals quantified (Na not determined): 90.9%–106.8% for K, 85.0%–106.3% for Ca, 98.9%–99.9% for Mg, 94.9%–99.5% for Mn, 84.6%–99.3% for Al, 73.6%–107.6% for Fe, 99.2%–99.8% for Co, 94.%7–100.0% for Cu, 99.4%–104.95% for Zn, 94.9%–100.0% for Ba, 95.3%–99.4% for Ni, and 99.3% for Li. Therefore, these findings reinforce the proposal of high bioaccessibility of honey minerals and highlight the importance of the absence or low concentration of minerals harmful to health in honey because, if present, they show a high absorption potential. On the other hand, Oliveira et al. (2019) reported different results in the determination of Ca, Fe, K, Mg, and Na by inductively coupled plasma optical emission spectrometry after in vitro gastric digestion in blossom honey from *Apis* bees. The results found suggested low in vitro bioaccessibility for Ca (12.9%–21.4%), Fe (14.3%–21.8%), K (9.4%–20.0%), Mg (15.0%–25.0%), and Na (4.0%–20.2%). The authors justified it as the result of the non-execution of the duodenal and oral phases of digestion, in which enzymes with action on carbohydrates are found, such as amylase. However, the authors did not discuss or estimate the effect of the absence of pepsin in the gastric digestion step on their results. In this sense, these data are important to complement discussions on the topic. However, they should be used with caution because estimators of in vitro bioaccessibility in the gastric digestion because they were contrary to the data found by Seraglio et al. (2017), which determined the in vitro bioaccessibility after the gastric phase of digestion using pepsin.

Considering all these reports, it is evident that there is a scarcity of studies on the bioaccessibility and mainly bioavailability of honey minerals. Besides, the available studies suffered from several drawbacks such as the evaluation of a low number of honey samples, the low diversity of honey types, the use of in vitro digestion protocols, and conflicting outcomes among the studies. All these factors made it difficult for robust conclusions to be made about the bioaccessibility and mainly the bioavailability of minerals in honey and their interactions with other compounds present in the food matrix, such as proteins and phenolic compounds, or in the digestive system itself, such as enzymes. Therefore, it is necessary to improve our understanding of the possible interactions between honey minerals and other constituents through parallel investigation of phenolic compounds, proteins, and enzymes, for example. This information is not clearly known yet, but it has a direct impact on the bioaccessibility and bioavailability of minerals, highlighting the importance of its investigation in honey (Seraglio et al. 2017).

## Honey as a Bio-indicator of Environmental Pollution

Honey's composition, flavor, and aroma are derived from the plants used by bees, as well as regional and climatic conditions. Therefore, the specific composition of a type of honey and the possible presence of contaminants also depend on the crops surrounding the beehives (Aliferis et al. 2010; Kujawski and Namieśnik 2008). Bees are considered biological indicators because of their important morphological, ecological, and behavioral characteristics. They act as detectors of environmental pollution in two ways: (1) high mortality rates in the presence of toxic molecules and (2) the presence of residues in honey, pollen, and larvae because of existence of heavy metals, fungicides, and herbicides that are normally harmless to bees (Celli and Maccagnani 2003; Samimi et al. 2001). Therefore, honey can be considered as a bio-indicator for assessing the possible presence of environmental pollutants such as heavy metals (Okeola et al. 2020; Przybyłowski and Wilczyńska 2001).

The heavy metals present in honey may be of environmental origin (soil, water, and air) or result from various anthropogenic activities such as agricultural practices, industries nearby, and waste dumps (Figure 7.2) (Pohl 2009; Przybyłowski and Wilczyńska 2001; Stankovska et al. 2008; Vincēviča-gaile 2010). Many studies have shown that there is a direct relationship between the heavy metal contents in honey samples and the industry-related environmental pollution in the area from which the honey is harvested. For example, the city of Baia Mare and its surroundings are known as one of the most polluted areas with heavy metals both in Romania and worldwide, and Cu, Zn, Cd, and Pb are the main industrial pollutants. Berinde and Michnea (2013) investigated the honey samples harvested from the area and revealed a positive correlation between the metal contents in honey and those in the air, settling dust, and soils from that region. Wieczorek et al. (2020) showed the effects of volcanic and radioactivity on honey, while Smith et al. (2019) successfully claimed the potential use of honey as a bio-monitor of changing world.

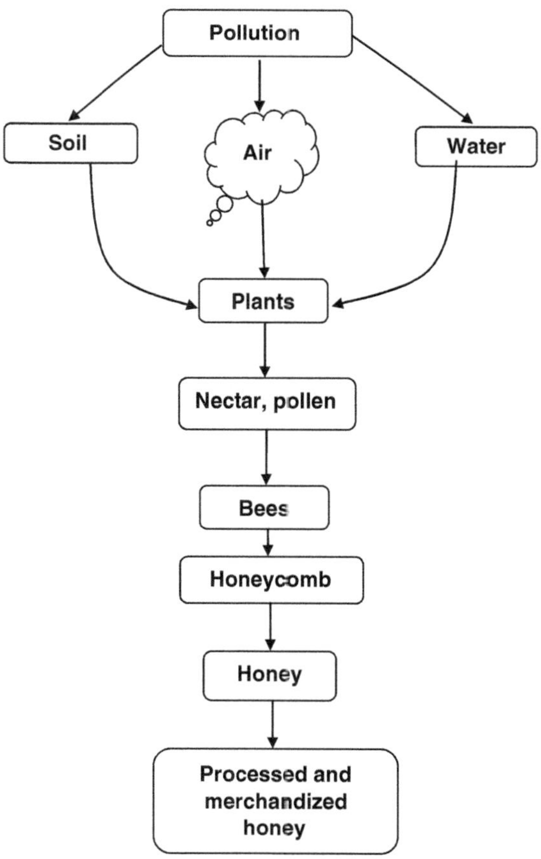


Figure 7.2 Schematic overview of natural and anthropogenic sources of heavy metals in honey, which can influence the composition of honey.

In fact, chemical wastes and exhaust fumes emitted or produced by the mines and steelworks, industrial and urban areas, and highway vehicles near the bees' foraging area increase the concentrations of certain metals in honey (Cd, Al, Ca, Ba, Cu, Mn, Ni, Pb, Zn, or Pd) (Bratu and Georgescu 2005; Przybyłowski and Wilczyńska 2001; Stankovska et al. 2008; Tuzen et al. 2007) depending on the type of flowers visited by the bees. For instance, honey produced from the nectar of pungent plants is categorized by high concentrations of heavy metals because these plants tend to concentrate pollutants more easily than herbaceous plants. In addition, honey derived from caduceus trees usually contains fewer pollutant than honey of other botanical origins (Devillers et al. 2002). Honey can also be contaminated during its processing by the beekeepers, by equipment and tools used, and by the process itself. Materials such as aluminum, stainless steel, and galvanized steel used in tools and equipment for the processing of honey may release some polluting metals (including Al, Cd, Co, Cr, Cu, Fe, Pb, Ni, and Zn) into honey (Santos et al. 2008; Stankovska et al. 2008; Tuzen and Soylak 2009).

## Conclusion

The analysis of minerals and heavy metals in honey gives an indication of its geographical and botanical origins, types, source, and degree of contamination and an overall measure of honey's purity. Such information can help consumers make informed decisions when purchasing honey and help beekeepers to avoid possible contamination during honey processing. In addition, honey can be used as a monitor of environmental pollution. Heavy metals in honey indicate the level of pollution and the type of toxic metals, which can also assist regulatory bodies in taking proper measures for environmental protection. Moreover, the minerals in honey have medicinal and nutritional value that correlates with antioxidants and antioxidant activities. Such information is important to guide honey production in areas distant from highways, railways, and volcanic areas, where the environment is clean. This information will also help in selecting the equipment used in the processing and storing of honey after harvesting because this equipment can also be a possible source of honey contamination with heavy metals. In addition, it can assist in the development of a quality control protocol for maintaining the honey trade in local and international markets. This would also help to prevent health problems associated with honey toxicity, especially poisoning with As, Cd, and Pb. Furthermore, minerals in honey correlate with color, EC, storage time, and even antioxidant activities. Therefore, determination of the mineral contents in honey will assist consumers in choosing a honey of interest based on its quality. It can also encourage native consumption as well as provide more income to the producing country.

## References


Abdulkhaliq, A. and Swaileh, K.M. (2017). Physico-chemical properties of multi-floral honey from the West Bank, Palestine. *International Journal of Food Properties* 20 (2): 447–454.

Adugna, E., Hymete, A., Birhanu, G., and Ashenef, A. (2020). Determination of some heavy metals in honey from different regions of Ethiopia. *Cogent Food & Agriculture* 6 (1): 1764182.

Agbagwa, O., Otokunefor, T., and Frank-Peterside, N. (2011). Quality assessment of Nigeria honey and manuka honey. *Journal of Microbiology & Biotechnology Research* 1.

Ahmida, M.H., Elwerfali, S., Agha, A., et al. (2013). Physicochemical, heavy metals and phenolic compounds analysis of Libyan honey samples collected from Benghazi during 2009-2010. *Food and Nutrition Sciences* 4: 33–40.

Ajibola, A., Chamunorwa, J.P., and Erlwanger, K.H. (2012). Nutraceutical values of natural honey and its contribution to human health and wealth. *Nutrition & Metabolism* 9: 61.

Ajibola, A., Idowu, G., Amballi, A., et al. (2007). Improvement of some haematological parameters in albino rats with pure natural honey. *Journal of Biological Sciences Research* 2: 67–69.

Ajtony, Z., Bencs, L., Haraszi, R., et al. (2007). Study on the simultaneous determination of some essential and toxic trace elements in honey by multi-element graphite furnace atomic absorption spectrometry. *Talanta* 71 (2): 683–690.

Akbari, B., Gharanfoli, F., Khayyat, M.H., et al. (2012). Determination of heavy metals in different honey brands from Iranian markets. *Food Additives and Contaminants: Part B* 5 (2): 105–111.

Akhtar, S., Ali, J., Javed, B., et al. (2014). Comparative physiochemical analysis of imported and locally produced Khyber Pakhtunkhwa honey. *Global Journal of Biotechnology & Biochemistry* 9 (3): 55–59.

Akram, S. (2008). Characterization of Saudi Arabian floral honeys by their physicochemical characteristics and heavy metal contents. *Fresenius Environmental Bulletin* 17 (7b): 877–881.

Alam, F., Islam, M.A., Gan, S.H., and Khalil, M.I. (2014). Honey: a potential therapeutic agent for managing diabetic wounds. *Evidence-Based Complementary and Alternative Medicine*, 2014: 169130.

Aliasgharpour, M. (2020). Trace elements in human nutrition (II)–an update. *International Journal of Preventive Medicine* 1 (1): 11–12.

Aliferis, K.A., Tarantilis, P.A., Harizanis, P.C., and Alissandrakis, E. (2010). Botanical discrimination and classification of honey samples applying gas chromatography/mass spectrometry fingerprinting of headspace volatile compounds. *Food Chemistry* 121 (3): 856–862.

Alimentarius, C. (2001). Codex standard 12. Revised Codex Standard for Honey, Standards and Standard Methods. 11.

Aljadi, A. and Kamaruddin, M. (2004). Evaluation of the phenolic contents and antioxidant capacities of two Malaysian floral honeys. *Food Chemistry* 85 (4): 513–518.

Altunay, N., Elik, A., and Gürkan, R. (2019). Monitoring of some trace metals in honeys by flame atomic absorption spectrometry after ultrasound assisted-dispersive liquid liquid microextraction using natural deep eutectic solvent. *Microchemical Journal* 147: 49–59.

Alvarez-Suarez, J.M., Giampieri, F., Damiani, E., et al. (2012). Radical-scavenging activity, protective effect against lipid peroxidation and mineral contents of monofloral Cuban honeys. *Plant Foods for Human Nutrition* 67 (1): 31–38.

Alvarez-Suarez, J.M., Tulipani, S., Romandini, S., et al. (2010). Contribution of honey in nutrition and human health: a review. *Mediterranean Journal of Nutrition and Metabolism* 3 (1): 15–23.

Ara, A. and Usmani, J.A. (2015). Lead toxicity: a review. *Interdisciplinary Toxicology* 8 (2): 55–64.

Atanassova, J., Yurukova, L., and Lazarova, M. (2012). Pollen and inorganic characteristics of Bulgarian unifloral honeys. *Czech Journal of Food Sciences* 30 (6): 520–526.

Barbagallo, M. and Dominguez, L.J. (2018). Magnesium role in health and longevity. In: *Trace Elements and Minerals in Health and Longevity*. (ed. M.Malavolta and E. Mocchegiani), 235–264. Cham, Switzerland:Springer.

Bartha, S., Taut, I., Goji, G., et al. (2020). Heavy metal content in polyfloral honey and potential health risk. A case study of Copşa Mică, Romania. *International Journal of Environmental Research and Public Health* 17 (5): 1507.

Batelková, P., Borkovcová, I., Čelechovská, O., et al. (2012). Polycyclic aromatic hydrocarbons and risk elements in honey from the South Moravian region (Czech Republic). *Acta Veterinaria Brno* 81 (2): 169–174.

Batista, B., Da Silva, L., Rocha, B., et al. (2012). Multi-element determination in Brazilian honey samples by inductively coupled plasma mass spectrometry and estimation of geographic origin with data mining techniques. *Food Research International* 49 (1): 209–215.

Bazeyad, A.Y., Al-Sarar, A.S., Rushdi, A.I., et al. (2019). Levels of heavy metals in a multifloral Saudi honey. *Environmental Science and Pollution Research* 26 (4): 3946–3953.

Belitz, H., Grosch, W., and Schieberle, P. ed. (2009). *Food Chemistry*, 4th revised and extended edn. Heidelberg, Germany:Springer.

Belouali, H., Bouaka, M., and Hakkou, A. (2008). Determination of some major and minor elements in the east of Morocco honeys through inductively coupled plasma optical emission spectrometry. *Apiacta* 43: 17–24.

Bergamo, G., Seraglio, S.K.T., Gonzaga, L.V., et al. (2018). Mineral profile as a potential parameter for verifying the authenticity of bracatinga honeydew honeys. *LWT* 97: 390–395.

Berinde, Z.M. and Michnea, A.M. (2013). A comparative study on the evolution of environmental and honey pollution with heavy metals. *Journal of Science and Arts* 2 (23): 173–180.

Bilandžić, N., Đokić, M., Sedak, M., et al. (2012). Content of five trace elements in different honey types from Koprivnica-Kri Evci County. *Slovenian Veterinary Research* 49 (4): 167.

Biluca, F.C., de Gois, J.S., Schulz, M., et al. (2017). Phenolic compounds, antioxidant capacity and bioaccessibility of minerals of stingless bee honey (Meliponinae). *Journal of Food Composition and Analysis* 63: 89–97.

Birge, W. and Price, D. (2001). Analysis of metals and polychlorinated biphenyl (PCB) residues in honey bees, honey and pollen samples collected from the Paducah gaseous diffusion plant and other areas. Frankfort, KY: Division of Waste Management, Kentucky Department for Environmental Protection.

Bogdanov, S. (2006). Contaminants of bee products. *Apidologie* 37: 1–18.

Bogdanov, S. (2011). Honey as nutrient and functional food: a review. *Bee Product Science* 1–31.

Bogdanov, S., Haldimann, M., Luginbühl, W., and Gallmann, P. (2007). Minerals in honey: environmental, geographical and botanical aspects. *Journal of Apicultural Research* 46 (4): 269–275.

Bogdanov, S., Jurendic, T., Sieber, R., and Gallmann, P. (2008). Honey for nutrition and health: a review. *Journal of the American College of Nutrition* 27 (6): 677–689.

Bogdanov, S., Ruoff, K., and Persano Oddo, L. (2004). Physico-chemical methods for the characterisation of unifloral honeys: a review. *Apidologie* 35 (Suppl 1): S4–S17.

Bratu, I. and Georgescu, C. (2005). Chemical contamination of bee honey–identifying sensor of the environment pollution. *Journal of Central European Agriculture* 6 (1): 95–98.

Buldini, P.L., Cavalli, S., Mevoli, A., and Sharma, J.L. (2001). Ion chromatographic and voltammetric determination of heavy and transition metals in honey. *Food Chemistry* 73 (4): 487–495.

Cahyono, B.E. (2020). Measurement of optical parameters and electrical and conductivity for determining the quality of Besuki local honey using spectrophotometer.

Cantarelli, M., Pellerano, R., Marchevsky, E., and Camiña, J. (2008). Quality of honey from Argentina: study of chemical composition and trace elements. *The Journal of Argentine Chemical Society* 96 (1–2): 33–41.

Celli, G. and Maccagnani, B. (2003). Honey bees as bioindicators of environmental pollution. *Bulletin of Insectology* 56 (1): 137–139.

Chepulis, L. and Starkey, N. (2008). The long-term effects of feeding honey compared with sucrose and a sugar-free diet on weight gain, lipid profiles, and DEXA measurements in rats. *Journal of Food Science* 73 (1): H1–H7.

Conti, M.E. and Botrè, F. (2001). Honeybees and their products as potential bioindicators of heavy metals contamination. *Environmental Monitoring and Assessment* 69 (3): 267–282.

Conti, M.E., Stripeikis, J., Campanella, L., et al. (2007). Characterization of Italian honeys (Marche Region) on the basis of their mineral content and some typical quality parameters. *Chemistry Central Journal* 1 (14): 14.

Corredera, L., Bayarri, S., Pérez-Arquillué, C., et al. (2014). Evaluation of heavy metals and polycyclic aromatic hydrocarbons in honeys from different origins. *Journal of Food Protection* 77 (3): 504–509.

Costa-Pinto, R. and Gantner, D. (2020). Macronutrients, minerals, vitamins and energy. *Anaesthesia & Intensive Care Medicine* 21 (3): 157–161.

Dag, A., Afik, O., Yeselson, Y., et al. (2006). Physical, chemical and palynological characterization of avocado (Persea americana Mill.) honey in Israel. *International Journal of Food Science & Technology* 41 (4): 387–394.

D'ambrosio, M. and Marchesini, A. (1982). Research on contamination by heavy metals in honey sample. *Atti della Societa Italiana di Scienze Naturali e del Museo di Storia Naturale* 123 (213): 342–348.

Daniel, V., Daniang, I., and Nimyel, N. (2011). Phytochemical analysis and mineral elements composition of Ocimum basilicum obtained in JOS METROPOLIS, Plateau state Nigeria. *IJET-IJENS* 11 (6): 161–165.

de Alda-garcilope, C., Gallego-Picó, A., Bravo-Yagüe, J., et al. (2012). Characterization of Spanish honeys with protected designation of origin "Miel de Granada" according to their mineral content. *Food Chemistry* 135 (3): 1785–1788.

de Andrade, C.K., Dos Anjos, V.E., Felsner, M.L., et al. (2014). Direct determination of Cd, Pb and Cr in honey by slurry sampling electrothermal atomic absorption spectrometry. *Food Chemistry* 146: 166–173.

Demirezen, D. and Aksoy, A. (2010). Determination of heavy metals in bee honey using by inductively coupled plasma optical emission spectrometry (ICP-OES). *Gazi University Journal of Science* 18 (4): 569–575.

Deng, J., Liu, R., Lu, Q., et al. (2018). Biochemical properties, antibacterial and cellular antioxidant activities of buckwheat honey in comparison to manuka honey. *Food Chemistry* 252: 243–249.

Devillers, J., Dore, J., Marenco, M., et al. (2002). Chemometrical analysis of 18 metallic and nonmetallic elements found in honeys sold in France. *Journal of Agricultural and Food Chemistry* 50 (21): 5998–6007.

Dobrinas, S., Matei, N., Soceanu, A., et al. (2006). Estimation of vitamin C and Cd, Cu, Pb content in honey and Propolis. *Scientific Study & Research* VII (4).

Dourado, G.S., Gomes, V.V., Maia, M.T.V., et al. (2019). Determination of macro and trace element levels in honey from the lower amazonian region, Brazil. *Brazilian Journal of Analytical Chemistry* 6 (23): 29–44.

Downey, G., Hussey, K., Kelly, J.D., et al. (2005). Preliminary contribution to the characterisation of artisanal honey produced on the island of Ireland by palynological and physico-chemical data. *Food Chemistry* 91 (2): 347–354.

Eissa, A.A., Hassan, A.S., and El Rahman, T.A.A. (2014). Determination of total aflatoxins and carbamate pesticide residues in some bee honey samples using QuEChERS method and high performance liquid chromatography. *Food and Public Health* 4 (5): 209–213.

Eshete, Y. and Eshete, T. (2019). A review on the effect of processing temperature and time duration on commercial honey quality. *Madridge Journal of Food Technology* 4 (1): 159–163.

Farooq Khan, Z. and Maqbool, T. (2008). Physical and spectroscopic characterization of Pakistani honey. *Ciencia e investigación agraria* 35 (2): 199–204.

Fernández-Torres, R., Pérez-Bernal, J.L., Bello-López, M.Á., et al. (2005). Mineral content and botanical origin of Spanish honeys. *Talanta* 65 (3): 686–691.

Ferreira, I.C., Aires, E., Barreira, J.C., and Estevinho, L.M. (2009). Antioxidant activity of Portuguese honey samples: different contributions of the entire honey and phenolic extract. *Food Chemistry* 114 (4): 1438–1443.

Fiorentini, E.F., Canizo, B.V., and Wuilloud, R.G. (2019). Determination of As in honey samples by magnetic ionic liquid-based dispersive liquid-liquid microextraction and electrothermal atomic absorption spectrometry. *Talanta* 198: 146–153.

Fiorentini, E.F., Oviedo, M.N., and Wuilloud, R.G. (2020). Ultra-trace Cr preconcentration in honey samples by magnetic ionic liquid dispersive liquid-liquid microextraction and electrothermal atomic absorption spectrometry. *Spectrochimica Acta Part B: Atomic Spectroscopy* 169: 105879.

Fredes, C. and Montenegro, G. (2006). Heavy metal and other trace elements contents in honey bee in Chile. *Ciencia e investigación agraria* 33 (1): 50–58.

Frías, I., Rubio, C., González-Iglesias, T., et al. (2008). Metals in fresh honeys from Tenerife Island, Spain. *Bulletin of Environmental Contamination and Toxicology* 80 (1): 30–33.

Golob, T., Doberšek, U., Kump, P., and Nečemer, M. (2005). Determination of trace and minor elements in Slovenian honey by total reflection X-ray fluorescence spectroscopy. *Food Chemistry* 91 (4): 593–600.

González Paramás, A.M., Gómez Bárez, J.A., Garcia-Villanova, R.J., et al. (2000). Geographical discrimination of honeys by using mineral composition and common chemical quality parameters. *Journal of the Science of Food and Agriculture* 80 (1): 157–165.

González-Miret, M.L., Terrab, A., Hernanz, D., et al. (2005). Multivariate correlation between color and mineral composition of honeys and by their botanical origin. *Journal of Agricultural and Food Chemistry* 53 (7): 2574–2580.

Grembecka, M. and Szefer, P. (2013). Evaluation of honeys and bee products quality based on their mineral composition using multivariate techniques. *Environmental Monitoring and Assessment* 185 (5): 4033–4047.

Grobler, S., Du Toit, I., and Basson, N. (1994). The effect of honey on human tooth enamel in vitro observed by electron microscopy and microhardness measurements. *Archives of Oral Biology* 39 (2): 147–153.

Hall, J.E. (2003). The kidney, hypertension, and obesity. *Hypertension* 41 (3): 625–633.

Hasam, S., Qarizada, D., and Azizi, M. (2020). A review: honey and its nutritional composition. *Asian Journal of Research in Biochemistry* 7: 34–43.

Hernández, O., Fraga, J., Jimenez, A., et al. (2005). Characterization of honey from the Canary Islands: determination of the mineral content by atomic absorption spectrophotometry. *Food Chemistry* 93 (3): 449–458.

Hungerford, N.L., Tinggi, U., Tan, B.L., et al. (2020). Mineral and trace element analysis of Australian/Queensland Apis mellifera honey. *International Journal of Environmental Research and Public Health* 17 (17): 6304.

Huskisson, E., Maggini, S., and Ruf, M. (2007). The role of vitamins and minerals in energy metabolism and well-being. *Journal of International Medical Research* 35 (3): 277–289.

Ibrahimi, H. and Hajdari, A. (2020). Phenolic and flavonoid content, and antioxidant activity of honey from Kosovo. *Journal of Apicultural Research* 59: 452–457.

Idoko, J., Haruna, B., Tifwa, P., and Musa, W. (2018). Evaluation of heavy metals in honey from Brinin-Gwari (Nigeria). *Journal of Chemical Society of Nigeria* 43 (1).

Jahangir, M.A., Muheem, A., Anand, C., and Imam, S.S. (2020). Traditional and modern applications of honey: an insight. In: *Therapeutic Applications of Honey and Its Phytochemicals,* 151–169. Singapore : Springer.

Jivan, A., Pătruică, S., and Popescu, G. (2008). Researches concerning the heavy metal content of the rape honey originating from the Banat area in the years 2006-2007. *Scientific Papers Animal Science and Biotechnologies* 41 (2): 302–308.

Khalil, M., Alam, N., Moniruzzaman, M., et al. (2011). Phenolic acid composition and antioxidant properties of Malaysian honeys. *Journal of Food Science* 76 (6): C921–C928.

Khammas, Z.A., Ghali, A.A., and Kadhim, K.H. (2012). Combined cloud-point extraction and spectrophotometric detection of lead and cadmium in honey samples using a new ligand. *International Journal of Chemical Sciences* 10: 1185–1204.

Ko, E.A., Han, J., Jung, I.D., and Park, W.S. (2008). Physiological roles of K+ channels in vascular smooth muscle cells. *Journal of Smooth Muscle Research* 44 (2): 65–81.

Krivovichev, V.G., Charykova, M.V., and Krivovichev, S.V. (2018). The concept of mineral systems and its application to the study of mineral diversity and evolution. *European Journal of Mineralogy* 30 (2): 219–230.

Kropf, U., Jamnik, M., Bertoncelj, J., and Golob, T. (2008). Linear regression model of the ash mass fraction and electrical conductivity for Slovenian honey. *Food Technology and Biotechnology* 46 (3): 335–340.

Kropf, U., Korošec, M., Bertoncelj, J., et al. (2010). Determination of the geographical origin of Slovenian black locust, lime and chestnut honey. *Food Chemistry* 121 (3): 839–846.

Kujawski, M. and Namieśnik, J. (2008). Challenges in preparing honey samples for chromatographic determination of contaminants and trace residues. *TrAC Trends in Analytical Chemistry* 27 (9): 785–793.

Lambert, I., Hoffmann, E., and Pedersen, S. (2008). Cell volume regulation: physiology and pathophysiology. *Acta physiologica* 194 (4): 255–282.

Lanjwani, M.F. and Channa, F.A. (2019). Minerals content in different types of local and branded honey in Sindh, Pakistan. *Heliyon* 5 (7): e02042.

Leblebici, Z. and Aksoy, A. (2008). Determination of heavy metals in honey samples from Central Anatolia using plasma optical emission spectrofotometry (ICP-OES). *Polish Journal of Environmental Studies* 17 (4): 549–555.

Machado De-Melo, A.A., Almeida-Muradian, L.B.D., Sancho, M.T., and Pascual-Maté, A. (2018). Composition and properties of Apis mellifera honey: a review. *Journal of Apicultural Research* 57 (1): 5–37.

Mahurpawar, M. (2015). Effects of heavy metals on human health. *International Journal of Research-Granthaalayah* 2394–3629.

Maiyo, W., Kituyi, J., Mitei, Y., and Kagwanja, S. (2014). Heavy metal contamination in raw honey, soil and flower samples obtained from Baringo and Keiyo Counties, Kenya. *International Journal of Emerging Science and Engineering* 2 (7): 5–9.

Matei, N., Birghila, S., Dobrinas, S., and Capota, P. (2004). Determination of C vitamin and some essential trace. *Acta Chimica Slovenica* 51: 169–175.

Matin, G., Kargar, N., and Buyukisik, H.B. (2016). Bio-monitoring of cadmium, lead, arsenic and mercury in industrial districts of Izmir, Turkey by using honey bees, propolis and pine tree leaves. *Ecological Engineering* 90: 331–335.

Mbiri, A., Onditi, A., Oyaro, N., and Murago, E. (2011). Determination of essential and heavy metals in Kenyan Honey by atomic absorption and emission spectroscopy. *Journal of Agriculture, Science and Technology* 13 (1).

McDowell, L.R., Wilkinson, N., Madison, R., and Felix, T. (2007). Vitamins and minerals functioning as antioxidants with supplementation considerations. The proceedings of the Florida Ruminant Nutrition Symposium. Best Western Gateway Grand. Gainesville, FL.

Mondragón-Cortez, P., Ulloa, J., Rosas-Ulloa, P., et al. (2013). Physicochemical characterization of honey from the West region of México. *CyTA-Journal of Food* 11 (1): 7–13.

Moniruzzaman, M., Chowdhury, M.A.Z., Rahman, M.A., et al. (2014). Determination of mineral, trace element, and pesticide levels in honey samples originating from different regions of Malaysia compared to manuka honey. *BioMed Research International* 2014: 1–10.

Morgan, K.T. (2008). Nutritional determinants of bone health. *Journal of Nutrition for the Elderly* 27 (1–2): 3–27.

Mračević, S.Đ., Krstić, M., Lolić, A., and Ražić, S. (2020). Comparative study of the chemical composition and biological potential of honey from different regions of Serbia. *Microchemical Journal* 152: 104420.

Muñoz, E. and Palmero, S. (2006). Determination of heavy metals in honey by potentiometric stripping analysis and using a continuous flow methodology. *Food Chemistry* 94 (3): 478–483.

Münstedt, K., Hoffmann, S., Hauenschild, A., et al. (2009). Effect of honey on serum cholesterol and lipid values. *Journal of Medicinal Food* 12 (3): 624–628.

Musgrove, M., Stern, L., and Banner, J. (2010). Springwater geochemistry at Honey Creek State Natural Area, central Texas: implications for surface water and groundwater interaction in a karst aquifer. *Journal of Hydrology* 388 (1): 144–156.

Nagai, T., Inoue, R., Kanamori, N., et al. (2006). Characterization of honey from different floral sources. Its functional properties and effects of honey species on storage of meat. *Food Chemistry* 97 (2): 256–262.

Nalda, M.N., Yagüe, J.B., Calva, J.D., and Gómez, M.M. (2005). Classifying honeys from the Soria Province of Spain via multivariate analysis. *Analytical and Bioanalytical Chemistry* 382 (2): 311–319.

Nanda, V., Sarkar, B., Sharma, H., and Bawa, A. (2003). Physico-chemical properties and estimation of mineral content in honey produced from different plants in Northern India. *Journal of Food Composition and Analysis* 16 (5): 613–619.

Nigussie, K., Subramanian, P., and Mebrahtu, G. (2012). Physicochemical analysis of Tigray honey: an attempt to determine major quality markers of honey. *Bulletin of the Chemical Society of Ethiopia* 26 (1).

Nkansah, M.A., Shamsu–Deen, M., and Opoku, F. (2018). Phytocompounds, heavy metal and mineral contents in honey samples from selected markets in the Kumasi metropolis. *Emerging Science Journal* 2 (5): 287–294.

Okeola, F., Oluade, O., and Liad, M. (2020). Stingless bee honey as bio-indicator of heavy metals pollution in and around the University of Ilorin Environ Kwara State, Nigeria. *Journal of Applied Sciences and Environmental Management* 24 (5): 773–778.

Oliveira, S.S., Alves, C.N., Morte, E.S.B., et al. (2019). Determination of essential and potentially toxic elements and their estimation of bioaccessibility in honeys. *Microchemical Journal* 151: 104221.

Omode, P. and Ademukola, S. (2008). Determination of trace metals in Southern Nigerian honey by use of atomic absorption spectroscopy. *Spectroscopy Letters* 41 (7): 328–331.

Oroian, M., Prisacaru, A., Hretcanu, E.C., et al. (2016). Heavy metals profile in honey as a potential indicator of botanical and geographical origin. *International Journal of Food Properties* 19 (8): 1825–1836.

Osman, K.A., Al-Doghairi, M.A., Al-Rehiayani, S., and Helal, M.I. (2007). Mineral contents and physicochemical properties of natural honey produced in Al-Qassim region, Saudi Arabia. *Journal of Food Agriculture and Environment* 5 (3/4): 142.

Pérez, R.A., Iglesias, M.T., Pueyo, E., et al. (2007). Amino acid composition and antioxidant capacity of Spanish honeys. *Journal of Agricultural and Food Chemistry* 55 (2): 360–365.

Perna, A., Simonetti, A., Intaglietta, I., et al. (2012). Metal content of southern Italy honey of different botanical origins and its correlation with polyphenol content and antioxidant activity. *International Journal of Food Science & Technology* 47 (9): 1909–1917.

Pisani, A., Protano, G., and Riccobono, F. (2008). Minor and trace elements in different honey types produced in Siena County (Italy). *Food Chemistry* 107 (4): 1553–1560.

Pita-Calvo, C. and Vázquez, M. (2017). Differences between honeydew and blossom honeys: a review. *Trends in Food Science & Technology* 59: 79–87.

Pohl, P. (2009). Determination of metal content in honey by atomic absorption and emission spectrometries. *TrAC Trends in Analytical Chemistry* 28 (1): 117–128.

Pohl, P., Stecka, H., Greda, K., and Jamroz, P. (2012). Bioaccessibility of Ca, Cu, Fe, Mg, Mn and Zn from commercial bee honeys. *Food Chemistry* 134 (1): 392–396.

Przybyłowski, P. and Wilczyńska, A. (2001). Honey as an environmental marker. *Food Chemistry* 74 (3): 289–291.

Purcarea, C., Dzugan, M., Wesolowska, M., et al. (2017). A comparative study of metal content in selected polish and Romanian honey samples. *Revista de Chimie* 68: 1163–1169.

Ribeiro, R.D.O.R., Mársico, E.T., de Jesus, E.F.O., et al. (2014). Determination of trace elements in honey from different regions in Rio de Janeiro State (Brazil) by total reflection X-ray fluorescence. *Journal of Food Science* 79 (4): T738–T742.

Roman, A., Bartkowiak, A., Reginia, M., and Aland, A. (2007). The cumulation of selected chemical elements of toxic properties in bee honey originating from the industrial and rural-forest areas. Proceedings of the 13th International Congress in Animal Hygiene, Tartu, Estonia, 17–21 June, 2007. Volume 2. Estonian University of Life Sciences, Jõgeva Plant Breeding Institute, Estonian Research Institute of Agriculture.

Ru, Q.-M., Feng, Q., and He, J.-Z. (2013). Risk assessment of heavy metals in honey consumed in Zhejiang province, southeastern China. *Food and Chemical Toxicology* 53: 256–262.

Ruschioni, S., Riolo, P., Minuz, R.L., et al. (2013). Biomonitoring with honeybees of heavy metals and pesticides in nature reserves of the Marche Region (Italy). *Biological Trace Element Research* 154 (2): 226–233.

Sager, M. (2020). Major minerals and trace elements in different honeys. *Bee World* 97 (3): 70–74.

Saghaei, S., Ekici, H., Demirbas, M., et al. (2012). Determination of the metal contents of honey samples from Orumieh in Iran. *Kafkas Universitesi Veteriner Fakultesi Dergisi* 18: 281–284.

Samimi, A., Maymand, O.E., and Mehrtabatabaei, M. (2001). Determination of Cadmium and Arsenic pollution by bee honey based on the study on Ja'far abad area from Saveh city from Iran. *Water and Geoscience* 199–192.

Sanna, G., Pilo, M.I., Piu, P.C., et al. (2000). Determination of heavy metals in honey by anodic stripping voltammetry at microelectrodes. *Analytica Chimica Acta* 415 (1): 165–173.

Santos, J.S.D., Santos, N.S.D., Santos, M.L.P.D., et al. (2008). Honey classification from semi-arid, Atlantic and Transitional Forest zones in Bahia, Brazil. *Journal of the Brazilian Chemical Society* 19 (3): 502–508.

Šarić, G., Vahčić, N., Bursać Kovačević, D., and Putnik, P. (2020). The changes of flavonoids in honey during storage. *Processes* 8 (8): 943.

Seraglio, S.K.T., Bergamo, G., Molognoni, L., et al. (2021). Quality changes during long-term storage of a peculiar Brazilian honeydew honey:"Bracatinga". *Journal of Food Composition and Analysis* 97: 103769.

Seraglio, S.K.T., Silva, B., Bergamo, G., et al. (2019). An overview of physicochemical characteristics and health-promoting properties of honeydew honey. *Food Research International* 119: 44–66.

Seraglio, S.K.T., Valese, A.C., Daguer, H., et al. (2017). Effect of in vitro gastrointestinal digestion on the bioaccessibility of phenolic compounds, minerals, and antioxidant capacity of Mimosa scabrella Bentham honeydew honeys. *Food Research International* 99: 670–678.

Silici, S., Uluozlu, O.D., Tuzen, M., and Soylak, M. (2008). Assessment of trace element levels in Rhododendron honeys of Black Sea Region, Turkey. *Journal of Hazardous Materials* 156 (1): 612–618.

Silici, S., Uluozlu, O.D., Tuzen, M., and Soylak, M. (2013). Honeybee and honey as monitors for heavy metal contamination near the thermal power plants in Mugla, Turkey. *Toxicology and Industrial Health* 32 (3): 507–516.

Silva, L.R., Videira, R., Monteiro, A.P., et al. (2009). Honey from Luso region (Portugal): physicochemical characteristics and mineral contents. *Microchemical Journal* 93 (1): 73–77.

Singh, R., Gautam, N., Mishra, A., and Gupta, R. (2011). Heavy metals and living systems: an overview. *Indian Journal of Pharmacology* 43 (3): 246.

Smith, K.E., Weis, D., Amini, M., et al. (2019). Honey as a biomonitor for a changing world. *Nature Sustainability* 2 (3): 223–232.

Sobotka, L., Allison, S., and Stanga, Z. (2008). Basics in clinical nutrition: water and electrolytes in health and disease. *e-SPEN, the European e-Journal of Clinical Nutrition and Metabolism* 3 (6): e259–e266.

Solayman, M., Islam, M.A., Paul, S., et al. (2016). Physicochemical properties, minerals, trace elements, and heavy metals in honey of different origins: a comprehensive review. *Comprehensive Reviews in Food Science and Food Safety* 15 (1): 219–233.

Stankovska, E., Stafilov, T., and Šajn, R. (2008). Monitoring of trace elements in honey from the Republic of Macedonia by atomic absorption spectrometry. *Environmental Monitoring and Assessment* 142 (1–3): 117–126.

Street, R., Kulkarni, M., Stirk, W., et al. (2009). Effect of cadmium uptake and accumulation on growth and antibacterial activity of Merwilla plumbea—An extensively used medicinal plant in South Africa. *South African Journal of Botany* 75 (3): 611–616.

Surai, P.F. (2003). Selenium-vitamin E interactions: does 1+ 1 equal more than 2. In: *Nutritional Biotechnology in the Feed and Food Industries* (ed. T.P. Lyons and K.A. Jacques), 51–58. Nottingham, UK: Nottingham University Press.

Swaileh, K.M. and Abdulkhaliq, A. (2013). Analysis of aflatoxins, caffeine, nicotine and heavy metals in Palestinian multifloral honey from different geographic regions. *Journal of the Science of Food and Agriculture* 93 (9): 2116–2120.

Taha, E. and Ali, M. (2012). Determination of heavy metals content in cotton honey in Kafr Elshiekh Province, Egypt. *Journal of Plant Protection and Pathology* 3 (11): 1211–1219.

Taha, E.A. and El-Sanat, S.Y. (2007). Effect of combs age on honey production and its physical and chemical properties. Proceedings of the 2nd International Conference of the Entomological Society of Egypt.

Taha, E.-K.A., Manosur, H.M., and Shawer, M.B. (2010). The relationship between comb age and the amounts of mineral elements in honey and wax. *Journal of Apicultural Research* 49 (2): 202–207.

Taleuzzaman, M., Alam, M.J., Kala, C., and Rahat, I. (2020). *Honey of Authenticity: An Analytical Approach. Therapeutic Applications of Honey and Its Phytochemicals*, 101–120. Singapore: Springer.

Terrab, A., Escudero, M.L., González-Miret, M.L., and Heredia, F.J. (2004a). Colour characteristics of honeys as influenced by pollen grain content: a multivariate study. *Journal of the Science of Food and Agriculture* 84 (4): 380–386.

Terrab, A., Gonzalez, A.G., Díez, M.J., and Heredia, F.J. (2003). Mineral content and electrical conductivity of the honeys produced in Northwest Morocco and their contribution to the characterisation of unifloral honeys. *Journal of the Science of Food and Agriculture* 83 (7): 637–643.

Terrab, A., Hernanz, D., and Heredia, F.J. (2004b). Inductively coupled plasma optical emission spectrometric determination of minerals in thyme honeys and their contribution to geographical discrimination. *Journal of Agricultural and Food Chemistry* 52 (11): 3441–3445.

Theobald, H. (2005). Dietary calcium and health. *Nutrition Bulletin* 30 (3): 237–277.

Tutun, H., Kahraman, H.A., Aluc, Y., et al. (2019). Investigation of some metals in honey samples from West Mediterranean region of Turkey. Veterinary Research Forum, Faculty of Veterinary Medicine, Urmia University, Urmia, Iran.

Tuzen, M., Silici, S., Mendil, D., and Soylak, M. (2007). Trace element levels in honeys from different regions of Turkey. *Food Chemistry* 103 (2): 325–330.

Tuzen, M. and Soylak, M. (2005). Trace heavy metal levels in microwave digested honey samples from middle Anatolia, Turkey. *Journal of Food and Drug Analysis* 13 (4): 343.

Tuzen, M. and Soylak, M. (2009). Column solid-phase extraction of nickel and silver in environmental samples prior to their flame atomic absorption spectrometric determinations. *Journal of Hazardous Materials* 164 (2): 1428–1432.

Uršulin-Trstenjak, N., Levanić, D., Primorac, L., et al. (2015). Mineral profile of Croatian honey and differences due to its geographical origin. *Czech Journal of Food Sciences* 33 (2): 156–164.

U.S. Department of Agriculture (1985). United States Standards for Grades of Extracted Honey. May 23.Washington, DC: Agricultural Marketing Service.

U.S. Department of Agriculture. 2015. United States National Nutrient Database for Standard Reference, Release 28. U.S. Dept. of Agriculture, Agricultural Research Service; September 2015. Nutrient Data Laboratory. http://www.ars.usda.gov/ba/bhnrc/ndl.

Vanhanen, L.P., Emmertz, A., and Savage, G.P. (2011). Mineral analysis of mono-floral New Zealand honey. *Food Chemistry* 128 (1): 236–240.

Vela, L., de Lorenzo, C., and Perez, R.A. (2007). Antioxidant capacity of Spanish honeys and its correlation with polyphenol content and other physicochemical properties. *Journal of the Science of Food and Agriculture* 87 (6): 1059–1075.

Vincēviča-gaile, Z. (2010). Macro-and trace elements in honey. *Latvijas Lauksaimniecības Universitāte-Raksti* 25 (320): 54–66.

Vit, P., Rodríguez-Malaver, A., Rondón, C., et al. (2010). Bioactive indicators related to bioelements of eight unifloral honeys. *Archivos Latinoamericanos de Nutricion* 60 (4): 405–410.

Voica, C., Iordache, A.M., and Ionete, R.E. (2020). Multielemental characterization of honey using inductively coupled plasma mass spectrometry fused with chemometrics. *Journal of Mass Spectrometry* 55 (7): e4512.

Wieczorek, J., Kaczor, M., Romańczyk, G., et al. (2020). Radioactivity of honey in central and southern Poland. *Journal of Environmental Radioactivity* 222: 106376.

Yaghoobi, N., Al-Waili, N., Ghayour-Mobarhan, M., et al. (2008). Natural honey and cardiovascular risk factors; effects on blood glucose, cholesterol, triacylglycerole, CRP, and body weight compared with sucrose. *The Scientific World Journal* 8: 463–469.

Yarsan, E., Karacal, F., Ibrahim, I., et al. (2007). Contents of some metals in honeys from different regions in Turkey. *Bulletin of Environmental Contamination and Toxicology* 79 (3): 255–258.

Zugravu, C.-A., Parvu, M., Patrascu, D., and Stoian, A. (2009). Correlations between lead and cadmium pollution of honey and environmental heavy metal presence in two Romanian counties. *Bulletin of University of Agricultural Sciences and Veterinary Medicine Cluj-Napoca Agriculture* 66 (2).

# 8

# Organic Acids in Honey

*Md. Anisur Rahman, Md. Murad Hossain, and Dhirendra Nath Barman*

## Introduction

Honey is a sweet, viscous natural product produced mainly by bees and some related insects from floral nectar and other sugary secretions (like aphid honeydew) by some insect species (Crane 1991). Mention of honey in ancient Greek religion and its use in folk medicine imply that humans have been familiar with honey since the early ages of civilization (Bogdanov et al. 2008; Henrichs 1980). In modern times, its application in the treatment of people with burns, infected wounds, gastrointestinal disorders, asthma, and skin ulcers has been re-investigated (Tezcan et al. 2011). Honey mainly consists of carbohydrates (predominantly glucose and fructose), with more than 95% of the solids of honey being carbohydrate in nature and some other compounds in minor amounts such as phenolics, proteins, enzymes, amino acids, minerals, vitamins, and organic acids (da Silva et al. 2016; Tezcan et al. 2011). Although organic acids make up less than 0.5% of the total contents, they significantly contribute to honey's organoleptic properties, such as color and flavor, and to its physical and chemical properties, such as pH, acidity, and electrical conductivity (Mato et al. 2003). Among the 30 nonaromatic organic acids that are said to be present or probably present in honey, many of them are directly involved in antibacterial activity, the antioxidant activity of honey, and the characterization of botanical or geographical origin of honey (Mato et al. 2003). This chapter discusses the organic acid profile of honey, their abundance in honey from different regions, different analytical methods for their determination, and their significance in honey quality with good details.

## Organic Acid Profile of Honey

Research on determining the organic acids in honey has been carried out since the first decade of twentieth century. To the authors' knowledge, malic acid was first identified in 1904 followed by formic acid in 1908 (Farnsteiner 1908; Hilger 1904; White 1979). Since then, several different researchers worldwide have made their contributions to identify other organic acids present in honey. This scientific endeavor followed a chronological development of efforts and successes to determine the organic acids present in honey. In the 1930s, citric or malic acids were thought to be the primary organic acids in honey (Nelson and Mottern 1931), but in 1960, gluconic acid was first reported as the principal acid of honey (Stinson et al. 1960). Gluconic acid, together with gluconolactone, is present in all types of honey at much higher concentrations – concentrations of grams per kilogram in contrast to milligrams per kilogram (White 1978). A work in 1990 claimed to identify 19 organic acids in honey (Crane 1990). Later, other research conducted in the following decade added more names to the list, and now a total of 30 non-aromatic organic acids altogether (Mato et al. 2003) have been reported to be present or probably present in honey (Table 8.1).

Organic acids are produced in honey as a result of aerobic and anerobic fermentation and are responsible for the special flavors of honey (Tezcan et al. 2011). For example, gluconic acid in honey derives from two main sources: (1) the action of bee glucose-oxidase on nectar glucose (Figure 8.1) and (2) the metabolic activity of certain *Gluconobacter* spp. bacteria present in the bees' gut and during the ripening of honey. The latter produces large quantities of gluconic acids under aerobic conditions in presence of high glucose concentration, thus establishing the fact that gluconic acid of honey originates from bacteria to some extent (Ruiz-Argüeso and Rodriguez-Navarro 1973).

*Honey: Composition and Health Benefits*, First Edition. Edited by Md. Ibrahim Khalil, Gan Siew Hua, and Bey Hing Goh.

$$\text{D-glucose} + H_2O + O_2 \xrightarrow{\text{Glucose Oxidase}} \text{D-gluconic acid} + H_2O_2$$

**Figure 8.1** D-gluconic acid synthesis in honey.

**Table 8.1** Nonaromatic organic acids in honey with their corresponding years of discovery.

| Status | Acid | Year of Discovery |
|---|---|---|
| Known to be present | Acetic | 1960 |
| | Butyric | 1960 |
| | Citric | 1931 |
| | Formic | 1908 |
| | Fumaric | 1974 |
| | Gluconic | 1960 |
| | a-Ketoglutaric | 1974 |
| | Lactic | 1960 |
| | Maleic | 1955 |
| | Malic | 1904 |
| | Oxalic | 1952 |
| | Pyroglutamic | 1960 |
| | Pyruvic | 1994 |
| | Succinic | 1931 |
| | Tartaric | 1974 |
| Probably present | α- or β-Glycerophosphate | 1966 |
| | Glycolic | 1962 |
| | Glucose-6-phosphate | 1966 |
| | 2- or 3-Phosphoglyceric | 1966 |
| Other acids identified and quantified | Abscisic | 1996 |
| | *cis*-Aconitic | 1974 |
| | Citramalic | 1988 |
| | Galacturonic | 1998 |
| | 4-Hydroxy-3methyl-*trans*-2-pentenedioic | 1995 |
| | Isocitric | 1998 |
| | Malonic | 1974 |
| | 2-Methoxybutanedioic | 1995 |
| | Quinic | 1998 |
| | Shikimic | 1998 |

Mato et al., 2003 / International Association for Food Protection.

The gluconic acid level is proportional to the time needed for transforming nectar into honey because a longer time allows a larger amount of glucose oxidase to be added by the bees; ultimately, synthesis of larger amount of gluconic acid occurs. Other intercorrelated factors are the strength of the bee colonies and the quality and quantity of the nectar that arrives at the beehives (White 1979).

Evidence shows that organic acid profiles help to distinguish between honey of different origins (El Mohandes 2011). For example, the citric acid concentration acts as a reliable parameter for the differentiation of two main types of honey, floral and honeydew honey (Bogdanov 1997). Organic acid profiles also contribute to the acidity of honey and can be used to detect honey's spoilage (El Mohandes 2011). The acidity is mainly caused by maleic acid even though it is a non-major organic acid of honey (El Mohandes 2011). The average pH of honey is 3.9 (with a typical range of 3.4–6.1). The low pH of honey prevents the presence and growth of microorganisms (El Mohandes 2011). Artificial honey can also be produced from carbohydrate sources enriched with a glucose–fructose composition that is close to their natural counterparts, but they lack medicinal and nutritional value because of the absence of minor constitutes organic acid and some others phyto-constituents found in natural honey (El Mohandes 2011).

## Organic Acid Profiles of Different Regional Honey Types

### Egyptian Honey

There are no extensive literature sources of organic acid profiles of Egyptian honey. In a small-scale study, four types of Egyptian honey were screened and analyzed to determine their organic acid profiles (El Mohandes 2011). The honey types and their corresponding organic acids profile are presented in Table 8.2.

Eight organic acids – formic, malonic, tartaric, shikimic, maleic, citric, succinic, and propionic acids – were detected in the four types of honey investigated. Only three organic acids – formic, malonic, and maleic – were commonly found in all four types of honey. Citrus honey was the richest in terms of organic acid content because it possessed all eight observed acids. Interestingly, artificial honey contained only four of them – formic, malonic, maleic, and succinic acids – because nectar and pollen grains are absent in artificial diet (sugar syrup). Table 8.2 also indicates that formic acid is the principal acid both in terms of its presence in all four honey types and is its highest component compared with other acids. Naturally, the formic acid content in honey varies from 5–600 mg/kg based on their botanical and geographical origin (Capolongo et al. 1996).

Tartaric acid and shikimic acid could be used to differentiate between the artificial honey and three other natural honey types because they are not found in artificial honey. Among the four honey types analyzed, citric acid was detected only in citrus honey, which differs from other observations that claim citric acid to be present in all floral honeys (Nelson and Mottern 1931). This difference of organic acids profile in Egyptian honey from other investigated honey types may be due to variations in their botanical and geographical origins.

### Turkish Honey

Turkey, which is the fourth largest honey-producing country in the world, has a rare mix of suitable conditions for honey production, including climate, topographical structure, and richness of plant flora (Tezcan et al. 2011). The Black Sea and West Anatolian regions are home to authentic botanical sources of honey, including both floral and honeydew honey (Tezcan et al. 2011). In a study conducted in 2011, a total of 10 groups of honey were constructed from different botanical origins and regions (Tezcan et al. 2011). These included chestnut, rhododendron, lime, acacia, cherry, laurel, and Anzer honey from the Black Sea region, among which Anzer honey is a heterofloral blossom honey containing wildflowers from

**Table 8.2** Organic acids (mg/100 g) in four types of Egyptian honey.

| | Types of Honey | | | | |
|---|---|---|---|---|---|
| **Organic Acid Type** | **Citrus** | **Clover** | **Cotton** | **Artificial** | **LSD** |
| Formic | 1.57 | 1.37 | 1.34 | 0.57 | 0.54 |
| Malonic | 0.07 | 0.27 | 0.80 | 0.20 | 0.23 |
| Tartaric | 0.38 | 0.03 | 0.05 | 0.00 | 0.15 |
| Shikimic | 0.23 | 0.18 | 0.62 | 0.00 | 0.20 |
| Maleic | 0.14 | 0.11 | 0.57 | 0.13 | 0.19 |
| Citric | 0.50 | 0.00 | 0.00 | 0.00 | 0.00 |
| Succinic | 0.16 | 0.11 | 0.00 | 0.07 | 0.08 |
| Propionic | 0.06 | 0.04 | 0.00 | 0.00 | NS |

LSD, least significant difference; NS, not significant.

El Mohandes, 2011 / NotionWave Inc.

the Anzer Plateau near Ikizdere and Rize in the East Black Sea region, Turkey. Being exported to many countries, this is one of the most famous honey types in Turkey (Tezcan et al. 2011). Another plant, *Rhododendron ponticum*, endemic to the Black Sea region produces a honey locally known as "mad" or wild honey (Küçük et al. 2007). The only source of honeydew honey used in the previously mentioned study is pine honey that is produced by honeybees from the honeydew excreted by the insect *Marchalina hellenica* that lives by sucking the sap of pine trees (Tezcan et al. 2011). Different organic acids found in these various groups of Turkish honey are listed in Table 8.3.

Of the six major organic acids screened, only gluconic acid was found in all groups of honey. Oxalic acid could neither be identified nor quantified in any sample. Formic, malic, citric, and succinic acids were quantified in milligrams per kilogram units whether gluconic acid was quantified in grams per kilogram units, denoting it as the principal organic acid in honey. This finding is in accordance with a previous study (Mato et al. 2006a) that found gluconic acid levels from 10 different botanical origins were between 8 and 12.3 g/kg.

Another study (Haroun et al. 2012) analyzed eight different honey samples among which pine and oak honey were of the honeydew type and the rests were of the floral type (cotton, chestnut, sunflower, yayla, multifloral, and citrus). The researchers could detect and quantify seven organic acids in these honey samples, including tartaric, malic, ascorbic, maleic, citric, succinic, and fumaric acids. Citric acid dominates in both honeydew and pine honey (87.22–139.4 mg/100 g) as well as in oak honey (142.28 mg/100 g), thus supporting the previous finding of Talpay (1988). However, the other floral honey samples have much smaller amounts of citric acid, showing that citric acid is surely a reliable parameter to

**Table 8.3** Organic acid contents of some investigated Turkish honey.

| Botanical Origins and Regions | Oxalic Acid[a] (mg/kg) | Formic Acid[a] (mg/kg) | Malic Acid[a] (mg/kg) | Citric Acid[a] (mg/kg) | Succinic Acid[a] (mg/kg) | Gluconic Acid[b] (g/kg) |
|---|---|---|---|---|---|---|
| Chestnut (Zonguldak, West Black Sea) | NQ | 1276.00 ± 3.00 | 476.00 ± 13.00 | 211.00 ± 6.00 | 141.00 ± 12.00 | 8.90 ± 0.22 |
| Chestnut (Eregli, West Black Sea) | NQ | 940.00 ± 47.00 | 878.00 ± 5.00 | 465.00 ± 7.00 | 232.00 ± 3.40 | 10.93 ± 0.17 |
| Chestnut (Görele, East Black Sea) | NQ | 929.00 ± 11.00 | 105.00 ± 7.00 | 78.90 ± 6.70 | 39.00 ± 2.50 | 13.78 ± 0.53 |
| Rhododendron (Eregli, West Black Sea) | ND | 4.70 ± 0.60 | 164.00 ± 1.00 | NQ | 41.70 ± 0.30 | 6.51 ± 0.35 |
| Rhododendron (Görele, East Black Sea) | ND | 40.00 ± 2.90 | 162.00 ± 1.00 | NQ | NQ | 5.79 ± 0.06 |
| Anzer flowers (Anzer plateau, East Black sea) | ND | 17.30 ± 0.50 | ND | ND | ND | 5.52 ± 0.09 |
| Anzer flowers (Anzer plateau, East Black sea) | ND | ND | ND | ND | ND | 4.22 ± 0.02 |
| Flowers (Gümüshane, East Black Sea) | ND | ND | ND | ND | ND | 2.01 ± 0.06 |
| Acacia (Ordu, Middle Black Sea) | ND | ND | ND | ND | ND | 1.50 ± 0.05 |
| Pine (Mugla, Aegean) | NQ | 89.30 ± 5.90 | 257.00 ± 32.00 | 124.00 ± 15.00 | NQ | 3.47 ± 0.30 |

[a] Values are expressed in milligrams of oxalic acid, milligrams of formic acid, milligrams of malic acid, milligrams of citric acid, and milligrams of succinic acid per 1 kg of honey.
[b] Values are expressed in g gluconic acid per 1 kg of honey.
ND, not detectable; NQ, not quantifiable.
Tezcan et al., 2011 / VÚP Food Research Institute.

differentiate between floral and honeydew honey. The highest concentration of malic acid was determined in cotton honey (41.51–180.64 mg/100 g), and therefore it can be used as a floral marker for cotton honey (Haroun et al. 2012). Nevertheless, the study has some unusual and unexpected findings such as it could not identify the most predominant organic acid in honey – gluconic acid – and it failed to detect citric acid in citrus honey samples. The anomaly might have been caused by the method used. The first study (Tezcan et al. 2011) used a capillary electrophoresis method to analyze individual organic acid, whereas the second one (Haroun et al. 2012) used a high-performance liquid chromatography (HPLC) method. The difference in the screening technique possibly has resulted in such abnormal outcomes. We will further discuss this in the next section of this chapter.

## Analytical Methods for the Determination of Organic Acids in Honey

Because the frequency of detecting organic acids in honey increases day by day, scientists have applied several methods to analyze and quantify them. Their attempts veered from simple to complex techniques gradually to identify and quantify most the important organic acids in honey, both individually or as a group simultaneously. This section briefly explores the methods implemented by several researchers.

### Enzymatic Method

Basically, all enzymatic methods are based on the measurement of the increase or decrease in absorbance of the co-enzymes NADH (nicotinamide adenine dinucleotide, reduced form) or NADPH (nicotinamide adenine dinucleotide phosphate, reduced form), which absorb light in the long wavelength region, generally at 340 nm (Mato et al. 2006b). Enzymatic methods have been mainly used for the quantification of citric, malic, formic, D- and L-lactic, oxalic, and total D-gluconic acids in honey. However, it would be possible to determine other acids such as acetic, L-ascorbic, and succinic acids (Bòhringer Mannheim GmBH 1995).Researchers determined D-(-) and L-(+)-lactic acid contents using an enzymatic method both in treated (residual contents of these acids after treatment) and untreated (natural content) honey against *Varroa* infestation (Stoya et al. 1986). The mean recovery percentage for D-(-) lactic acid was of 83 ± 13%, and for L-(+) lactic acid, it was of 96 ± 6%. Regarding formic acid, there are important differences between light and dark honey in enzymatic methods if the honey solution is previously clarified with activated charcoal to mitigate the influence of honey's and color and to remove certain instabilities of the spectrophotometric measurement (Sabatini et al. 1994). An enzymatic micro-method enabled determination of total D-gluconic acid in honey, including D-gluconic acid and D-glucono-δ-lactone, because it could not be quantified separately (Mato et al. 1997). To ensure complete decomposition of D-glucono-δ-lactone, the honey solution was initially taken to a pH of 10.5 for 10 minutes at room temperature and then dropped to 7.8, which is the optimum pH for enzymatic activity. The method combines precision (relative standard deviation [RSD]% $\leq$ 0.30%), good recovery (99.80 ± 0.40%), specificity, and simplicity, as well as low cost.

The main advantages of enzymatic methods include high specificity that even allows the determination of D/L isomers of some organic acids. In addition, the equipment and materials necessary for enzymatic analysis are very simple and are available in the majority of honey quality control laboratories. Enzymatic procedures could also be considered excellent as reference methods (Mato et al. 2006b). The main disadvantage of the enzymatic test is its incapability of determining more than one organic acid at a time, so constructing a profile of organic acids of a given honey would be tedious and time consuming. Another downside of this method is the instability of the kits, requiring the assays to be carried out within a few days (Mato et al. 2006b).

### Chromatographic Methods

Several chromatographic techniques are available to determine organic acid in honey samples, including paper and on-column ion-exchange chromatography, gas chromatography (GC), reverse-phase chromatography, and liquid chromatography. Some chromatographic techniques such as ionic and ion-exchange chromatography and-phase chromatography are not very exclusive nowadays to determine the organic acids in honey because they are routinely exploited in liquid chromatography, especially in HPLC.

### Ion-Exchange Chromatography

Ion-exchange chromatography was a method of choice in the mid-twentieth century before the emergence of more sophisticated technology and contributed to identifying several organic acids in honey. Ion-exchange resins were separated organic acids from honey and then often fractionated them on silicic acid and ion-exchange columns (Stinson et al. 1960). The use of these two columns was based on different principles and gave varying results that helped to compare them. For example, the silicic acid column could hardly elute gluconic acid, whereas the ion-exchange column did this conveniently.

In ion-exchange chromatography, the honey solution was passed through two columns coupled in series. The second column was intended to retain all breakthrough acidity, though none was found. The first column was rinsed with distilled water until a negative carbohydrate test result was obtained, and the acids were removed from it by eluting with a gradient to 4N formic acid. After elution, acid fractions of 5 mL were collected at a flow rate of 70–100 mL/h and then located by paper chromatography using washed Whatman no. 1 paper and an n-butanol–acetic acid–water solvent. Finally, the acid fractions were combined and concentrated to 5 mL by evaporation in a filtered air stream at less than 40°C (Stinson et al. 1960). This method successfully identified gluconic acid for the first time as the principal acid of honey.

### Gas Chromatography

The most common form of GC is gas–liquid chromatography (GLC) in which the mobile phase is an inert carrier gas, and the stationary phase is a high-molecular-weight non-volatile liquid that is coated either on the surface of finely divided particles or on the walls of a long capillary tubing. Samples are mixed with the carrier gas through the injection port at a temperature high enough to vaporize all components. Then the samples along with inert gas pass through the stationary phase and are separated into different components based on their interaction with stationary phase liquid because of their two main properties – volatility and polarity. A detector directly attached to the column exit detects individual sample components as they are eluted from the column (Clement and Karasek 1988). Recently, GC has been combinedly used with mass spectrometry (MS) and jointly termed the GC-MS technique. In this system, GC is coupled with MS that was two separate processes – ionization (electron or chemical) and separation – and recording of the ions formed. The GC-MS system exploits the knowledge of carrier gas flow of GC and the vacuum principles of MS. The connecting lines throughout the GC-MS system contain valves, tubings, and orifices, each with individual effects on gaseous flow behavior (Clement and Karasek 1988).

Detection of some aromatic and aliphatic organic acids by GC involves three steps: extraction, derivatization, and chromatographic analysis. In three separate studies conducted by the same researchers, 24 aromatic organic acids were determined as free acids in honey from different floral sources after extraction with ethyl acetate and then analyzed as their trimethylsilyl derivatives by a GC-MS method (Steeg and Montag 1987, 1988a, 1988b). All samples contained benzoic, β-phenylactic, 4-hydroxybenzoic, and 4-hydroxyphenylacetic acids. Another GC-MS procedure simultaneously determined and quantified five aliphatic carboxylic acids (malic, citric, iso-citric, succinic, and quinic acids) and four aromatic acids (benzoic, 4-hydroxibenzoic, 2,4-dihydroxibenzoic, and 2,6-dihydroxibenzoic acids) (Horváth and Molnár-Perl 1998). GC is mostly applicable when the analytes are volatile (e.g. aromatic acids). But in case of aliphatic acids, which are mostly non-volatile, GC appears to be a less suitable method because these acids need to be derivatized previously (Mato et al. 2006b).

### High-Performance Liquid Chromatography

Whereas traditional liquid chromatography simply relies on gravitational forces to pass the mobile phase through the column, HPLC mechanically pumps the liquid mobile phase using high pressure through a column containing the solid stationary phase (Weston and Brown 1997). Traditional HPLC uses a polar stationary phase with a non-polar mobile phase. Because organic acids are polar in nature, several scientists have frequently used reverse-phase HPLC (RP-HPLC), which uses non-polar stationary phase and an aqueous, moderately polar mobile phase in which the sample organic acids can easily be injected to be carried throughout the column. The retention time is less, and the polar organic acids readily elute to be analyzed.

Scientists have quantified six organic acids (gluconic, pyruvic, malic, citric, succinic, and fumaric acids) in 48 floral honey samples using an HPLC method with ultraviolet (UV) detection (Cherchi et al. 1994). The samples were previously purified by solid-phase extraction with an ionic exchange cartridge followed by chromatographic separation by implementing two reverse-phase C18 columns connected in series. Using sulfuric acid as the mobile phase, this method gave results in 60 minutes with a detection limit, and average recovery percentages ranged between 0.002 to 3 ppm. (w/w) and 89% to 104%, respectively (Cherchi et al. 1994).

Another study quantified gluconic acid and nine other minority organic acids – citric, pyruvic, galacturonic, malic, citramalic, quinic, succinic, fumaric, and formic acids (Del Nozal et al. 1998). Like the previous study, this one used anionic exchange cartridges in solid-phase extraction to purify the samples. Here, two HPLC procedures were used; the first one used two reverse-phase C18 columns connected in series and quantified three organic acids – galacturonic, pyruvic, and citric. The second one used one anionic exchange column and quantified four acids – malic, citramalic, quinic, and formic acids. Ammonium dihydrogen phosphate and 5-mM sulfuric acid were used in the mobile phase in the two chromatographic procedures, respectively. The remaining two minority organic acids (succinic and fumaric) were quantified using both chromatographic methods. The total analysis time was approximately 30 minutes (Del Nozal et al. 1998).

A distinct group of scientists (Suárez-Luque et al. 2002) devised a similar HPLC method as the previous ones with a subtle difference in the single C18 reverse-phase column that uses Milli-Q water acidified with 4.5% m-phosphoric acid as mobile phase instead of sulfuric acid or ammonium dihydrogen phosphate. This method allowed the identification and quantification of malic, maleic, citric, succinic, and fumaric acids in honey. Maleic acid was determined and quantified for the first time. Interestingly, the most predominant, gluconic acid, could not be quantified probably because of the saturation of anion-exchange cartridge. Nevertheless, organic acid concentrations in honey are very low, and there are possible interferences of unwanted compounds, such as sugar, in the quantification of organic acids. As a result, pretreatment of the samples is needed, which makes the method laborious and time consuming (Suárez-Luque et al. 2002).

### Capillary Zone Electrophoresis

To avoid potential matrix interferences faced in HPLC and other liquid chromatography techniques, capillary zone electrophoresis is becoming more popularized to separate low-molecular-weight organic acids because of its good resolution, automation, short analysis time, and reduced sample preparation (Heiger 1992). Basically, capillary zone electrophoresis is a special mode of capillary electrophoresis. In capillary zone electrophoresis, the analytes separate because of their varying electrophoretic mobilities, resulting in different migration velocities in electrophoretic buffer enclosed in fine capillaries (Li 1992). Although the organic acids profiles of several food items have been analyzed by capillary zone electrophoresis, very few articles have used this method to determine the organic acids in honey samples.

The first known electrophoretic separation method was carried out in 8 minutes at 22°C with indirect UV detection at 232 nm using salicylic acid, tris[hydroxymethyl]aminomethane (TRIS), Ca(OH)2, and dodecyl-trimethylammonium hydroxide as the electrolyte. The method could only detect citric acid (no quantification) together with inorganic anions chloride and phosphate (Boden et al. 2000).

A second capillary zone electrophoresis method that was applied with UV detection at 210 nm using sodium phosphate, 2-propanol, and hexadimethrine bromide as running buffer determined oxalic, fumaric, maleic, and malic acids in honey within 14 minutes (Navarrete et al. 2005). Recovery percentages ranged between 97.62% and 101.81%, but this experiment did not provide any evidence of repeatability and reproducibility. As a result, several predominant organic acids, such as gluconic, citric, acetic, formic, and lactic acids, were not determined.

To increase the detection range of organic acids and to decrease the detection time, a simple capillary electrophoresis method with direct UV detection at 185 nm was developed (Mato et al. 2006a), implying repeatability and reproducibility studies (RSD $\leq$4.6% and $\leq$10.0%, respectively). This method used phosphate as the carrier buffer, tetradecyltrimethylammonium hydroxide (TTAOH) as the electroosmotic flow modifier, and $CaCl_2$ as the selectivity modifier. All these electrolytes were adjusted at a constant pH of 6.40 and controlled temperature of 25°C. Just a simple pretreatment of dilution and filtration of the sample was done in this method. After the procedure, this method could finally determine oxalic, formic, malic, succinic, pyruvic, acetic, lactic, citric, and gluconic acids with a recovery percentage ranging between 89.4% and 104.6% (Mato et al. 2006a). To the authors' knowledge, this method quantified acetic acid in honey for the first time ever.

Although the capillary electrophoresis technique is simple, rapid, and more convenient to handle and avoids sugar interferences, its drawbacks are lower reproducibility and lower sensitivity (generally higher detection and quantification limits than enzymatic and chromatographic procedures) compared with other methods (Mato et al. 2006b).

## Significance of Non-aromatic Organic Acids in Honey

Although organic acids constitute a very minute quantity (<0.5%) of honey's total ingredients, several scientific explorations on non-aromatic organic acids in honey divulged many of its magnitudes that can be codified into five main arenas: (1) the antibacterial activities of these acids, (2) the antioxidant activities of these acids, (3) the use of these acids as possible indicators of fermentation, (4) use for treatment of *Varroa* infestation, and (5) the use as factors to characterize the botanical and geographical origins of honey.

### Antibacterial Activity

Hydrogen peroxide that is produced during the synthesis of gluconic acid (see Figure 8.1) was previously considered the main antibacterial component of honey and was named *inhibine* (White et al. 1963). Subsequently, other factors with antibacterial activity, including pH and acidity, osmolarity, and plant constituents, were collectively categorized as "non-peroxide antibacterial activity" (Molan 2006). These activities were considered more important from two points of view. First, glucose oxidase in ripe honey is inactive, and the peroxide concentration is too low to inhibit bacterial growth (Bogdanov 1984). Second, non-peroxide antibacterial activity is insensitive to heat and light and remains perfect after a long period of storage (Molan 2006). Therefore, research has been carried out to determine whether the antibacterial activity can be credited to acidity or pH. It was found that antibacterial activity was significantly correlated with both free and total acidity but not with pH (Bogdanov 1984). Again, studies found no relationship between pH and acidity type (Esti et al., 1997), establishing the fact that honey acid's components were responsible for its antibacterial properties. Identified components are mainly aromatic organic acids such as caffeic and ferulic acids (Wahdan 1998) and benzoic acid derivatives (Weston et al. 1999). Interestingly, there has been no study on the potential influence of non-aromatic organic acids on antibacterial activity. So, this is still an unexplored area for future research.

### Antioxidant Activity

The antioxidant activity of honey is measured by the capacity of oxygen radical absorbance (Gheldof and Engeseth 2002). This capacity results from the combined activity of a variety of compounds, including phenolics (e.g. phenolic acids and flavonoids that have strong antioxidant capacity), peptides and amino acids, organic acids, enzymes (glucose oxidase, catalase, and others), and possibly other Maillard reaction products. Non-aromatic organic acids such as gluconic, citric, and malic acids may contribute to the antioxidant capacity of honey. These organic acids act as chelating agents against metals and jointly accelerate the action of other antioxidants, such as phenolics (Gheldof et al. 2002).

### Indicators of Fermentation

Although acetic acid is generally found in most honey, a high concentration of this acid may indicate possible fermentation of honey (Mato et al. 2003). Yeast varieties that produce ethanol under certain moisture and optimum temperature conditions are responsible for honey's fermentation. If the chemical reaction proceeds further, acetic acid will be produced by oxidation of ethanol (Mato et al. 2003).

### Treatment of *Varroa* Infestation

Beehives are destroyed by infestation of parasitic mite *Varroa jacobsoni* if it goes untreated. Non-aromatic organic acids such as formic, lactic, and oxalic acids have been applied to date against this invasion (Milani 2001). The reason behind this use is the hydrophilic nature of organic acids that does not let them accumulate in wax (Imdorf et al. 1996). It is notable that these acids are also naturally present in honey; hence, it is important to determine whether a given concentration of these acids is due to natural content or to a residual occurrence after introducing *Varroa* infestation control. This can be done by assessing the time needed to mitigate the presence of acids residues in honey as they degrade over time. Based on this principle, the formic acid concentration in honey from treated beehives was determined at two different times – (1) immediately after treatment in beehives that revealed 54–82 times more formic acid concentration than in honey from untreated beehives and (2) 5–8 months after the treatment that reported formic acid levels were more or less similar in honey from both treated and untreated beehives (Stoya et al. 1986).

But the two enantiomers of the chiral molecule lactic acid showed disparate results. When beehives were treated with a racemic mixture of lactic acid, L(+)-lactic acid levels escalated 10-fold compared with natural content, but D(-)-lactic acid levels remained the same in honey from both treated and untreated beehives. Eventually, 7 or 8 weeks after treatment, L(+)-lactic acid levels had dropped asymptotically to normal levels (Stoya et al. 1987).

In terms of oxalic acid, the presence of eggs, larvae, and pupae of honeybees restrains the acaricidal effects of this acid, but oxalic acid is highly effective in the absence of these broods (Milani 2001). Nevertheless, irrespective of the situation, the natural oxalic acid levels of honey remained the same after treating the beehives with this acid (Bogdanov et al. 2002).

### Factors for Characterizing the Botanical and Geographical Origins of Honey

Non-aromatic organic acids can be used as a potential marker to determine the botanical origin of honey (Anklam 1998). For example, the citric acid concentration was found remarkably lower in floral honey than in honeydew honey but was similar in floral and sugar-fed honey (Talpay 1988). Moreover, 2-methoxybutanedioic acid and 4-hydroxy-3-methyl-trans-2-pentenedioic acid have been identified with high proportions only in honey from *Knightea excelsa* pollen, and cis, trans-abscisic acid and trans, trans-abscisic acid were found only in Portuguese heather (*Erica* spp.) honey, suggesting that these acids can be exploited as efficient markers for the respective honey type (Ferreres et al. 1996; Wilkins and Lu 1995). *Erica* spp. honey could also be differentiated from others by the presence of a high quinic acid content (Del Nozal et al. 1998). Galacturonic acid was found only in multifloral honey (Mato et al. 2003). In brief, citric, pyruvic, gluconic, malic, and quinic acids are most reliable to distinguish between honeys of different botanical origins (Del Nozal et al. 1998).

## References


Anklam, E. (1998). A review of the analytical methods to determine the geographical and botanical origin of honey. *Food Chemistry* 63: 549–562.

Boden, J., Haumann, I., and Mainka, A. (2000). Schwerpunkt Biotechnologie -Anwendung der Kapillarelektrophorese in der Lebensmittelanalytik. *GIT Labor-Fachzeitschrift* 44: 924–928.

Bogdanov, S. (1984). Characterization of antibacterial substances in honey. *Lebensmittel Wissenschaft und Technologie* 17: 74–76.

Bogdanov, S. (1997). Nature and origin of the antibacterial substances in honey. *LWT - Food Science and Technology* 30: 748–753.

Bogdanov, S., Charrière, J.D., Imdorf, A., et al. (2002). Determination of residues in honey after treatments with formic and oxalic acid under field conditions. *Apidologie* 33: 399–409.

Bogdanov, S., Jurendic, T., Sieber, R., and Gallmann, P. (2008). Honey for nutrition and health: a review. *Journal of the American College of Nutrition* 27: 677–689.

Bòhringer Mannheim GmBH, B. (1995). *Methods of Enzymatic Bioanalysis and Food Analysis Using Test-Combinations.* Mannheim, Germany: Bòhringer Mannheim GmBH.

Capolongo, F., Baggio, A., Piro, R., et al. (1996). Trattamento della varroasi con acido formico: accumulo nel miele e influenza sulle sue caratteristiche. *L'Ape nostra Amica* 18: 4–11.

Cherchi, A., Spanedda, L., Tuberoso, C., and Cabras, P. (1994). Solid-phase extraction and high-performance liquid chromatographic determination of organic acids in honey. *Journal of Chromatography A* 669: 59–64.

Clement, R.E. and Karasek, F.W. (1988). *Basic Gas Chromatography-Mass Spectrometry: Principles and Techniques.* Amsterdam: Elsevier.

Crane, E. (1990). Bees and beekeeping: science, practice and world resources.

Crane, E. (1991). Honey from honeybees and other insects. *Ethology Ecology & Evolution* 3: 100–105.

da Silva, P.M., Gauche, C., Gonzaga, L.V., et al. (2016). Honey: chemical composition, stability and authenticity. *Food Chemistry* 196: 309–323.

Del Nozal, M.J., Bernal, J.L., Marinero, P., et al. (1998). High performance liquid chromatographic determination of organic acids in honeys from different botanical origin. *Journal of Liquid Chromatography and Related Technologies* 21: 3197–3214.

El Mohandes, S. (2011). Organic acids in different types of Egyptian honey. *Journal of Plant Protection and Pathology* 2: 865–872.

Esti, M., Panfili, G., Marconi, E., and Trivisonno, M.C. (1997). Valorization of the honeys from the Molise region through physico-chemical, organoleptic and nutritional assessment. *Food Chemistry* 58: 125–128.

Farnsteiner, K. (1908). Der Ameisensäuregehalt des Honigs. *Zeitschrift für Untersuchung der Nahrungs und Genußmittel, sowie der Gebrauchsgegenstände* 15: 598–604.

Ferreres, F., Andrade, P., and Tomás-Barberán, F.A. (1996). Natural occurrence of abscisic acid in heather honey and floral nectar. *Journal of Agricultural and Food Chemistry* 44: 2053–2056.

Gheldof, N. and Engeseth, N.J. (2002). Antioxidant capacity of honeys from various floral sources based on the determination of oxygen radical absorbance capacity and inhibition of in vitro lipoprotein oxidation in human serum samples. *Journal of Agricultural and Food Chemistry* 50: 3050–3055.

Gheldof, N., Wang, X.-H., and Engeseth, N.J. (2002). Identification and quantification of antioxidant components of honeys from various floral sources. *Journal of Agricultural and Food Chemistry* 50: 5870–5877.

Haroun, M.I., Konar, N., Poyrazoglu, E.S., et al. (2012). Organic acid profiles and contents of Turkish honeydew and floral honeys. In: International Conference on Environment, Agriculture and Food Sciences (ICEAFS'2012). 55–58.

Heiger, D. (1992). *High Performance Capillary Electrophoresis—An Introduction*. Santa, Clara, CA: Agilent Technologies.

Henrichs, A. (1980). *Harvard Studies in Classical Philology*. Cambridge, MA: Harvard University Press.

Hilger, A. (1904). Zur Kenntnis der im rechtsdrehenden Koniferenhonig vorkommenden Dextrine. *Zeitschrift für Untersuchung der Nahrungs- und Genußmittel, sowie der Gebrauchsgegenstände* 8: 110–127.

Horváth, K. and Molnár-Perl, I. (1998). Simultaneous GC-MS quantitation of o-phosphoric, aliphatic and aromatic carboxylic acids, proline, hydroxymethylfurfurol and sugars as their TMS derivatives: in honeys. *Chromatographia* 48: 120–126.

Imdorf, A., Charriere, J.D., Maqueln, C., et al. (1996). Alternative varroa control. *American Bee Journal* 136: 189–194.

Küçük, M., Kolaylı, S., Karaoğlu, Ş., et al. (2007). Biological activities and chemical composition of three honeys of different types from Anatolia. *Food Chemistry* 100: 526–534.

Li, S.F.Y. (1992). *Capillary Electrophoresis: Principles, Practice and Applications*. Amsterdam: Elsevier.

Mato, I., Huidobro, J.F., Sánchez, M.P., et al. (1997). Enzymatic determination of total d-gluconic acid in honey. *Journal of Agricultural and Food Chemistry* 45: 3550–3553.

Mato, I., Huidobro, J.F., Simal-Lozano, J., and Sancho, M.T. (2003). Significance of nonaromatic organic acids in honey. *Journal of Food Protection* 66: 2371–2376.

Mato, I., Huidobro, J.F., Simal-Lozano, J., and Sancho, M.T. (2006a). Rapid determination of nonaromatic organic acids in honey by capillary zone electrophoresis with direct ultraviolet detection. *Journal of Agricultural and Food Chemistry* 54: 1541–1550.

Mato, I., Huidobro, J.F., Simal-Lozano, J., and Sancho, M.T. (2006b). Analytical methods for the determination of organic acids in honey. *Critical Reviews in Analytical Chemistry* 36: 3–11.

Milani, N. (2001). Activity of oxalic and citric acids on the mite Varroa destructor in laboratory assays. *Apidologie* 32: 127–138.

Molan, P.C. (2006). The antibacterial activity of honey: 1. The nature of the antibacterial activity. *Bee World* 73: 5–28.

Navarrete, M., Casado, S., Minelli, M., et al. (2005). Direct determination of aliphatic acids in honey by coelectroosmotic capillary zone electrophoresis. *Journal of Apicultural Research* 44: 65–70.

Nelson, E.K. and Mottern, H.H. (1931). Some organic acids in honey. *Industrial & Engineering Chemistry* 23: 335–336.

Ruiz-Argüeso, T. and Rodriguez-Navarro, A. (1973). Gluconic acid-producing bacteria from honey bees and ripening honey. *Microbiology* 76: 211–216.

Sabatini, A.G., Marcazzan, G.L., Colombo, R., and Garagnani, M. (1994). Applicazione di un metodo enzimatico per la determinazione dell'acido formico e dell'acido lattico presenti nel miele. *Apicoltura* 9: 135–145.

Steeg, E. and Montag, A. (1987). Nachweis aromatischer Carbonsäuren in Honig-Aromatic carbonic acids of honey. *Zeitschrift für Lebensmittel-Untersuchung und Forschung* 184: 17–19.

Steeg, E. and Montag, A. (1988a). Quantitative Bestimmung aromatischer Carbonsäuren in Honig (Quantitative determination of aromatic carbonic acids in honey). *Zeitschrift für Lebensmittel-Untersuchung und Forschung* 187: 115–120.

Steeg, E. and Montag, A. (1988b). Minorbestandteile des honigs mit aroma-relevanz. Aromatische carbonsauren und ester aromatischer carbonsauren. *Deutsch Lebensittel-Rundschau* 84: 103–108.

Stinson, E.E., Subers, M.H., Petty, J., and White, J.W. (1960). The composition of honey. V. Separation and identification of the organic acids. *Archives of Biochemistry and Biophysics* 89: 6–12.

Stoya, W., Wachendorfer, G., Kary, I., et al. (1986). Formic-acid as a therapeutics to fight varroatose and its effect on the honey. *Deutsche Lebensmittel-Rundschau* 82: 217–221.

Stoya, W., Wachendorfer, G., Kary, I., et al. (1987). Lactic-acid as a therapeutics against varroatose and the effect it has on the honey. *Deutsche Lebensmittel-Rundschau* 83: 283–286.

Suárez-Luque, S., Mato, I., Huidobro, J.F., et al. (2002). Rapid determination of minority organic acids in honey by high-performance liquid chromatography. *Journal of Chromatography A* 955: 207–214.

Talpay, B. (1988). Inhaltsstoffe des Honigs-Citronensa¨ure (Citrat). *Deutsche Lebensmittel-Rundschau* 84: 41–44.
Tezcan, F., Kolayli, S., Ulusoy, H.S.E., and Erim, F.B. (2011). Evaluation of organic acid, saccharide composition and antioxidant properties of some authentic Turkish honeys. *Journal of Food and Nutrition Research* 50: 33–40.
Wahdan, H.A. (1998). Causes of the antimicrobial activity of honey. *Infection* 26: 26–31.
Weston, A. and Brown, P.R. (1997). *High Performance Liquid Chromatography and Capillary Electrophoresis: Principles and Practices*. Amsterdam: Elsevier.
Weston, R.J., Mitchell, K.R., and Allen, K.L. (1999). Antibacterial phenolic components of New Zealand manuka honey. *Food Chemistry* 64: 295–301.
White, J.W., Jr. (1978). Honey. *Advances in Food Research* 24: 287–374.
White, J.W. (1979). Composition of honey. In: *Honey A Comprehensive Survey* (ed. E. Crane), 157–206. Portsmouth, NH: Heinemann.
White, J.W., Subers, M.H., and Schepartz, A.I. (1963). The identification of inhibine, the antibacterial factor in honey, as hydrogen peroxide and its origin in a honey glucose-oxidase system. *Biochimica et Biophysica Acta (BBA) - Specialized Section on Enzymological Subjects* 73: 57–70.
Wilkins, A.L. and Lu, Y. (1995). Extractives from New Zealand honeys. 5. Aliphatic dicarboxylic acids in New Zealand Rewarewa (Knightea excelsa) honey. *Journal of Agricultural and Food Chemistry* 43: 3021–3025.

# 9

# Polyphenols and Antioxidants

*Md. Sakib Hossen and Md. Yousuf Ali*

## Introduction

Honey has been used as a sweetener for thousands of years. The potential health benefits of consuming honey were mentioned in early Greek, Roman, Vedic, and Islamic texts, and the healing qualities of honey were referred to as early as ancient times by philosophers and scientists such as Aristotle (384–322 BC) and Aristoxenus (320 BC) (Samarghandian et al. 2017). Since the early 1970s, researchers from different scientific fields have investigated the chemical and biological properties of honey, including its antibacterial, bacteriostatic, anti-inflammatory, and wound- and sunburn-healing activities (Gośliński et al. 2020; M Alvarez-Suarez et al. 2013; Stagos 2018). Recent views not only propose that honey is a health-promoting dietary supplement but also shed some light on its antioxidant and nonperoxide-dependent properties. This makes honey more than just a food supplement with high nutritional properties; it is also a valuable dietary source of antioxidants (Abdel-Moneim and Ghafeer 2007).

Honey has been both exported and imported globally because of its medicinal value. As with all other foods, the health benefits of honey are mainly dependent on its quality. The quality of honey varies significantly because the pollen collected from the bees' bodies as they move from plant to plant reflects the diversity of the plants they visit. In the United States, there are more than 300 different types of honey, each with a unique flavor and color. Currently, researchers have established that the main factor that contributes to the diverse color, flavor, and functional properties of honey is mainly its phenolic composition rather than variation in other components, including carbohydrates and proteins (Halagarda 2020; Yordi 2012).

The term "phenolic" or "polyphenol" is chemically defined as a substance that possesses an aromatic ring bearing one or more hydroxyl substituents, including functional derivatives (esters, methyl esters, and glycosides). These bioactive substances are widely distributed in the plant kingdom and are closely associated with the sensory and nutritional quality of fresh and processed plant foods (Yordi 2012). Some phenolic compounds are extremely widespread, but others are specific to certain plant families or are found only in certain plant parts or at specific development stages (Cheynier 2012). It is now confirmed that among all the major groups of polyphenols, only flavonoids and phenolic acids are present in honey. Specifically, among the flavonoids, only the flavonols (e.g. myricetin, kaempferol, 8-methoxy kaempferol, quercetin, isorhamnetin, quercetin-3-methyl ether, quercetin-3, 7-dimethyl ether, pinobanksin, rutin, and galangin), flavones (e.g. genkwanin, luteolin, apigenin, tricetin, and chrysin), and flavanones (e.g. pinocembrin and pinostrobin) sub-classes are present in honey.

Among the phenolic acid group, the hydroxybenzoic acids (e.g. methyl syringate, gallic, ellagic, protocatechuic, syringic, benzoic, and 4-hydroxybenzoic acids), hydroxycinnamic acids (e.g. chlorogenic, vanillic, caffeic, p-coumaric and ferulic acids), and hydroxy-phenylacetic acids (e.g. homogentisic and phenylacetic acids) sub-classes are detected in various honey samples. Several in vivo studies have strongly suggested that long-term consumption of diets rich in these types of polyphenols significantly ameliorates the adverse effects of several liver-, heart-, kidney-, brain-, and pancreas-associated diseases as well as those of genetic disorders, including tumors and cancer (Arts and Hollman 2005; Cianciosi 2020; Graf et al. 2005; Samarghandian et al. 2017; Zhao 2018). In this chapter, we report the various types of polyphenols present in several types of honey samples along with their classifications, sources, and functions. We also provide a review of several of the most potential natural polyphenols that exert an effect on several pathogenic pathways of major diseases, including cardiovascular diseases, diabetes, cancer, and neurodegenerative diseases. It is hoped that this review will be a new

*Honey: Composition and Health Benefits,* First Edition. Edited by Md. Ibrahim Khalil, Gan Siew Hua, and Bey Hing Goh.

comprehensive source in the research field, allowing a better understanding of honey polyphenol's mode of action in ameliorating various chronic and acute diseases as well as an emerging source of drug discovery and development.

## Honey's Compositions and Antioxidant Activity

Because of deviations in the botanical origin, different honeys also differ in appearance, sensory perception, and composition (Figure 9.1) (Pauliuc et al. 2020). However, certain external factors may play a role as well, such as seasonal environmental factors and processing (Guo 2020). Previous studies indicate that the compositional chemistry of honey contains at least 181 substances (Chow 2002). Honey is mainly composed of carbohydrates and water. The carbohydrates are the main constituents, making up approximately 95% of its dry weight basis. In addition to carbohydrates, honey contains numerous compounds such as organic acids, proteins, amino acids, certain enzymes (glucose oxidase, catalase), minerals, polyphenols, vitamins, and aroma compounds (Dimitrova et al. 2007; Jaganathan and Mandal 2009), all of which account for 2.1% of honey's composition (Jaganathan and Mandal 2009; Todd and Vansell 1942).

The antioxidants that occur naturally in honey are flavonoids, phenolic acids, enzymes (e.g. glucose oxidase, catalase), ascorbic acid, carotenoid-like substances, organic acids, Maillard reaction products, amino acids, and proteins (Estevinho 2008; Nagai 2006). A prior investigation indicated that the total antioxidant activity of honey is primarily provided by its phenolic composition rather than vitamin C and other components (Kenjerić 2007). Polyphenols mainly exert their antioxidant activity by neutralizing free radicals, which is accomplished by donation of an electron or hydrogen atom (Rice-Evans et al. 1996). For this study, we discuss and highlight potential natural polyphenols found in honey that exert effects in the pathogenic pathways of various acute and chronic diseases because many of these diseases are related to oxidative stress caused by reactive oxygen species (ROS) and nitrogen species (Battino 2020; Duthie and Brown 1994; Milner 1994) (Table 9.1).

## Honey Polyphenols

Polyphenols are an important group of compounds with respect to the appearance and functional properties of honey. The characterization of these compounds is also a promising way of investigating the floral and geographical origins of honeys (Cianciosi 2018; Gašić et al. 2019; Pyrzynska and Biesaga 2009; She 2019). There are different types of techniques available to identify individual honey polyphenols. To analyze polyphenol contents in honey, high-performance liquid chromatography (HPLC) is the main technique applied. Moreover, a newly developed form of HPLC, called ultra-performance liquid chromatography, has also been employed with better resolution, sensitivity, and speed of analysis (Swartz 2005). Gas chromatography has not been employed frequently because most of these compounds are not volatile (Pyrzynska and Biesaga

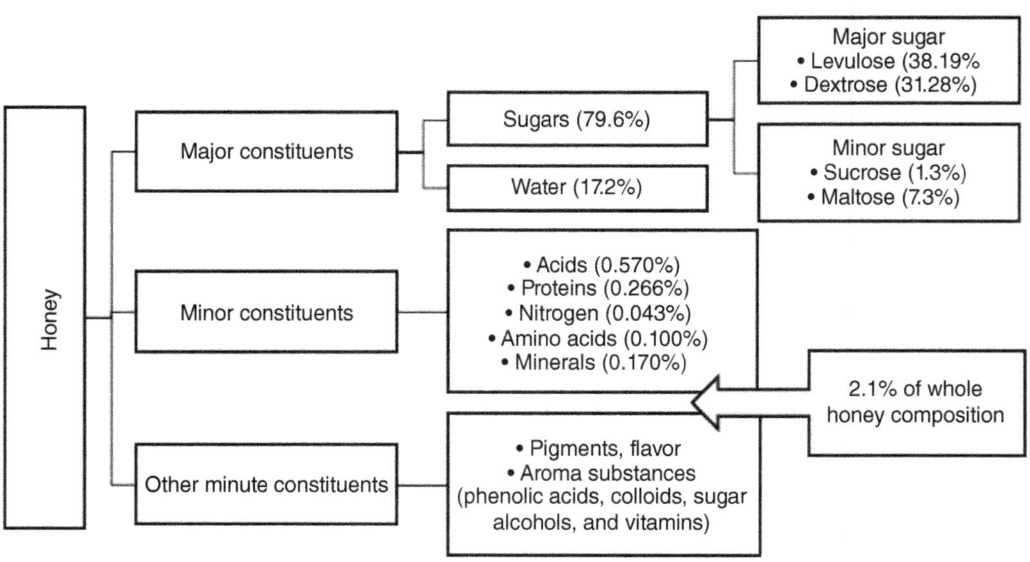


**Figure 9.1** The composition of honey.

**Table 9.1** Honey polyphenols.

| Polyphenol Name | Synonyms | Chemical Formulas | Honey Sources | Other Sources | Investigated Roles | References |
|---|---|---|---|---|---|---|
| Myricetin | 3,5,7,3′,4′,5′ Hexahydroxyflavone | $C_{15}H_{10}O_8$ | Eucalyptus honey, New Zealand and Australian *Leptospermum* honeys, Australian honeys from different botanical species | Grape wines, berry wines, herbs | Protects against oxidative stress induced by tert-butyl hydroperoxide (t-BHP) | Jandera 2005; Pandey et al. 2009 |
| Kaempferol | 3,5,7,4′-Tetrahydroxyflavone | $C_{15}H_{10}O_6$ | Eucalyptus honey, New Zealand and Australian *Leptospermum* honeys, Tunisian honeys, Linden, heather honeys, Australian honeys from different botanical species | Wines, fruit juices, tea infusions, spices, nuts, pulses, fruit vegetables, pod vegetables | Blocks oxidative stress, hepato-protective effect, induces apoptosis, blocks atherosclerosis plaque formation, promotes interleukin-1b and tumor necrosis factor gene expression | Huang 2010; Shakya 2014 |
| 8-Methoxy kaempferol | 8-Methylherbacetin, herbacetin8-methyl ether | $C_{16}H_{12}O_7$ | Honey from Tunisia | Not found | Not found | Pyrzynska and Biesaga 2009 |
| Quercetin | 3,5,7,3′,4′-Pentahydroxyflavone | $C_{15}H_{10}O_7$ | Eucalyptus honey, New Zealand and Australian *Leptospermum* honeys, Tunisian honeys, Australian honeys from different botanical species, Linden, heather honeys | Berry wines, fortified wines, grape wines, sparkling wines, cereals, chocolate, fruit juices, tea infusions, herbs, spices, nuts, fruit vegetables, onion-family vegetable | Hepatoprotective effect, modulates inflammatory cell function, blocks oxidative stress, anti-inflammatory effect, modulates cardiomyocyte gene expression profile, alternative treatment for obesity, prevents nephrotoxicity, promotes growth inhibition and apoptosis, prevents imbalances of calcium, sodium, and potassium in galactosemic cataract | Nabavi 2015; Uzun and Kalender 2013 |
| Isorhamnetin | 3,5,7,4′-Tetrahydroxy-3′-methoxyflavone;3′-methylquercetin | $C_{16}H_{12}O_7$ | Eucalyptus honey, Tunisian honeys | Grape wines, berry juices, nuts, onion-family vegetables | Anti-inflammatory effect, inhibits proliferation and induces apoptosis, prevents endothelial cell injuries, protects against oxidative stress, cardioprotective effects | Yao 2004 |
| Quercetin-3-methyl ether | 3′,4′,5,6,7 Pentahydroxy-3-methoxyflavone;quercetin-3methyl ether;quercetagetin-3-methyl ether | $C_{16}H_{12}O_8$ | Eucalyptus, Tunisian and Australian honeys from different botanical species | Not found | Not found | Naczk and Shahidi 2004 |

*(Continued)*

**Table 9.1** (Continued)

| Polyphenol Name | Synonyms | Chemical Formulas | Honey Sources | Other Sources | Investigated Roles | References |
|---|---|---|---|---|---|---|
| Quercetin 3,7-dimethyl ether | 3,7-Di-O-methylquercetin | $C_{17}H_{14}O_7$ | Tunisian honeys, Australian honeys from different botanical species | Not found | Vasorelaxant activity | Guerrero 2002 |
| Pinobanksin | Pinobanksin; (2R,3R)-3,5,7-trihydroxy-2 phenyl-2,3-dihydrochromen-4-one | $C_{15}H_{12}O_5$ | Sunflower honey, Eucalyptus honey, Tunisian honeys, Australian honeys from different botanical species | Not found | Inhibits peroxidation of low-density lipoprotein, improves cognition, reduces oxidative stress, anti-mutagenic effect, antiproliferative effect | Jaganathan and Mandal 2009; Liu 2015 |
| Rutin | Rutoside; quercetin 3-rutinoside | $C_{27}H_{30}O_{16}$ | Linden, heather honeys | Invasive plant species (*Carpobrotus edulis*) | Improves glucose homeostasis, antidiabetic effects, antitumor effect, protects neural damage, cardioprotective effect, reduces oxidative stress, antidepressant-like effect, prevents renal inflammation and apoptosis, antiplatelet activity, reduces gastric lesions, inhibits β-amyloid aggregation and cytotoxicity, attenuates oxidative stress, decreases the production of nitric oxide and pro-inflammatory cytokines, prevents testicular pathology | Aziza et al. 2014; Chen 2013 |
| Galangin | 3,5,7-Trihydroxyflavone | $C_{15}H_{10}O_5$ | Sunflower honey, honey from Tunisia | Herbs (Mexican oregano, dried) | Anticancer activity, antiproliferative effect, induces apoptosis, increases cytotoxic activity, antineoplastic effects, anti-clastogenic effects, inhibits osteoclastic bone destruction and osteoclastogenesis | Kamaruzzaman 2019; Kim et al. 2013; Zhang 2013 |
| Genkwanin | Apigenin 7-methyl ether; 5-hydroxy-2 – (4-hydroxypenyl)-7 methoxychromen-4-one; 4′,5-dihydroxy-7-methoxyflavone | $C_{16}H_{12}O_5$ | Australian honeys from different botanical species | Not found | Inhibits pro-inflammatory mediators, confers antitumor and immune modulatory activities | Gao 2014 |

| | | | | | | |
|---|---|---|---|---|---|---|
| Luteolin | 5,7,3′,4′-Tetrahydroxyflavone 5,7,3′,4′-Tetrahydroxyflavone | $C_{15}H_{10}O_6$ | Eucalyptus honey, New Zealand and Australian *Leptospermum* honeys, Tunisian honeys, Australian honeys from different botanical species | Fruit vegetable oils, herbs, nuts, lentils, fruit vegetables, shoot vegetables | Ameliorates acute kidney injury; induces apoptosis; provides neuroprotective; intestinal anti-inflammatory, antidiabetic, hepatoprotective, cardioprotective and anticancer effects | Fu 2014; Nishitani 2013 |
| Apigenin | 5,7,4′-Trihydroxyflavone | $C_{15}H_{10}O_5$ | Tunisian honeys | Beers, fruit, vegetable oils, herbs, nuts | Promotes interleukin 1b and tumor necrosis factor gene expression | Carrasco-Pancorbo 2006; Jandera 2005 |
| Tricetin | 5,7,3′,4′,5′-Pentahydroxyflavone Hieracin | $C_{15}H_{10}O_7$ | Eucalyptus honey, New Zealand and Australian *Leptospermum* honeys | Not found | Antiproliferative and anticancer activities | Hsu 2009; Hung 2015 |
| Chrysin | 5,7-Dihydroxyflavone | $C_{15}H_{10}O_4$ | Eucalyptus honey, sunflower honey, Tunisian honeys | Fruit (citrus) juices | Induces apoptosis, anticancer activity, hepatoprotective effect, antihypercholesterolemic effect, promotes angiogenesis, confers cardioprotective effect, and attenuates inflammation | Feng 2014; Mantawy 2014; Saeed 2020; Yang 2018 |
| Pinocembrin | 5,7 Dihydroxyflavanone; dihydrochrysin | $C_{15}H_{12}O_4$ | Eucalyptus, sunflower, New Zealand and Australian *Leptospermum*, Tunisian and Australian honeys from different botanical species | Herbs (Mexican oregano, dried) | Neuroprotective effect, attenuates blood–brain barrier injury, improves memory impairment, induces apoptosis, inhibits inflammatory mediators, inhibits atherosclerosis, reduces cardiac arrhythmia and infarct size, attenuates nephrotoxicity | Lungkaphin 2015; Promsan 2016; Zhou 2015 |
| Pinostrobin | Pinostrobin;2-phenyl-5-hydroxy-7-methoxy-2, 3-dihydro-4H-1-benzopyran-4-one | $C_{16}H_{14}O_4$ | Not found | Not found | Has antiproliferative effect and prevents neurotoxicity | Siekmann 2013; Xian 2012 |
| Methyl syringate | Methyl 4-hydroxy-3,5-dimethoxybenzoate Methyl 3,5-dimethoxy-4-hydroxybenzoate | $C_{10}H_{12}O_5$ | Acacia, Chinese milk vetch, buckwheat, manuka honeys | Not found | Serves as an activator of the chemosensory ion channel TRPA1, an efficient phenolic mediator for bacterial and fungal laccases | Rosado 2012; Son 2012 |

*(Continued)*

**Table 9.1** (Continued)

| Polyphenol Name | Synonyms | Chemical Formulas | Honey Sources | Other Sources | Investigated Roles | References |
|---|---|---|---|---|---|---|
| Gallic acid | 3,4,5- Trihydroxybenzoic acid | $C_7H_6O_5$ | Australian Eucalyptus honeys, New Zealand and Australian *Leptospermum* honeys, acacia, eucalyptus, lime, chestnut, heather, lavender, rosemary, orange, sunflower, rapeseed honeys, Linden, heather honeys | Beers, nut liquors, grape wines, rum, whisky, fortified wines, brandy, sparkling wines, dried fruits, tropical fruits, citrus fruits, pome jams, berry juices, pome juices, tea infusions, soy and soy products, fruits and vegetables (especially the leafy type) | Prevents bacterial cytotoxicity; induces apoptosis; prevents oxidative stress; confers cardioprotective, antihyperglycemic, ant-lipid peroxidative, antimelanogenic, antimicrobial, hepatoprotective, pro-inflammatory, and gastroprotective effects | Nabavi 2013; Umadevi 2012 |
| Ellagic acid | Benzoic acid; lagistase | $C_{14}H_6O_8$ | Australian Eucalyptus honeys, New Zealand and Australian *Leptospermum* and Tunisian honeys | Nut liquors, brandy, rum, whisky, berries, berry jams, berry juices, tropical fruit juices, nuts | Attenuates high carbohydrate, high-fat diet-induced metabolic syndrome; confers anticancer effect; prevents nephrotoxicity; potent inhibitor of protein kinase CK2.; attenuates cisplatin-induced injuries to sperm quality; oxidant–antioxidant system and the histologic structure of the rat testis;antiproliferative, cardioprotective, hepatoprotective, and gastroprotective effects | Panchal et al. 2013; Türk 2008 |
| Protocatechuic acid | 3,4-Dihydroxybenzoicacid | $C_7H_6O_4$ | Acacia, eucalyptus, lime, chestnut, heather, lavender, rosemary, orange, sunflower, rapeseed honeys | Beers, nut liquors, fortified wines, grape wines, sparkling wines, cereals, dried berries, dried fruits, citrus fruits, berry juices, pome juices, tropical fruit juices, fruit vegetable oils, spices, nuts, lentils, cabbages, fruit vegetables, leafy vegetables, onion-family vegetables | Protects against oxidative damage; confers hepatoprotective, antitumor, anti-inflammatory, and antidiabetic effects | Min et al. 2010; Scazzocchio 2011 |

| | | | | | | |
|---|---|---|---|---|---|---|
| Syringic acid | 3,5-Dimethoxy-4- hydroxybenzoic acid | $C_9H_{10}O_5$ | New Zealand and Australian *Leptospermum* honeys, acacia, eucalyptus, lime, chestnut, heather, lavender, rosemary, orange, sunflower, rapeseed honeys, Linden, heather honeys | Beers, nut liquors, brandy, rum, whisky, fortified wines, grape wines, cereals, dried fruits, other fruits, pomes, berry juices, fruit vegetable oils, other seed oils, herbs, other seasonings, spices, nuts, cabbages, fruit vegetables | Hepatoprotective effect, antihyperglycemic effect, antisteatotic and anti-inflammatory effect, antimicrobial activity and fungitoxicity | Francesca 2014; Ham 2016 |
| Benzoic acid | Benzoic acid | $C_7H_6O_2$ | Sunflower honey, acacia, eucalyptus, lime, chestnut, heather, lavender, rosemary, orange and sunflower rapeseed honeys | Cocoa powders and berries | Significant effect on skin diseases, serves as an expectorant, analgesic, and antiseptic | Beale et al. 2010; Dimitrova et al. 2007 |
| 4-Hydroxy benzoic acid | p-Hydroxybenzoic acid | $C_7H_6O_3$ | Acacia, eucalyptus, lime, chestnut, heather, lavender, rosemary, orange, sunflower, rapeseed honeys | Beers, fortified wines, grape wines, sparkling wines, cereals, berries, citrus fruits, other fruits, tropical fruits, fruit vegetable oils, other seed oils, other seasonings, nuts, lentils, fruit vegetables, root vegetables | Used as preservatives, confers antimicrobial activity, has fungi toxicity | Francesca 2014; Londesborough 2015 |
| Chlorogenic acid | 3-Caffeoylquinic and neochlorogenic acid | $C_{16}H_{18}O_9$ | Australian Eucalyptus honeys, New Zealand and Australian *Leptospermum* honeys | Grape wines, dried drupes, pomes, tropical fruits, pome jams, robusta coffee beverages, unknown coffee beverages, tea infusions, cabbages, root vegetables | Involved in glucose and lipid metabolisms; anti-obesity properties; antitumor activity; prevents chromosomal damage; hepatoprotective effect; inhibitor of hepatic glucose 6-phosphate translocase; inhibits DNA methylation; anti-inflammatory effect; attenuates hypertension; modifies plasma and liver concentrations of cholesterol, triacylglycerol, and minerals; wound-healing effect | Hwang 2014; Xu 2014 |

*(Continued)*

**Table 9.1** (Continued)

| Polyphenol Name | Synonyms | Chemical Formulas | Honey Sources | Other Sources | Investigated Roles | References |
|---|---|---|---|---|---|---|
| Vanillic acid | 4-Hydroxy-3-methoxy benzoic acid and p-vanillic acid | $C_8H_8O_4$ | Acacia, eucalyptus, lime, chestnut, heather, lavender, rosemary, orange, sunflower, rapeseed, Linden, heather honeys | Beers, brandy, rum, whisky, fortified wines, grape wines, cereal products, dried fruits, berries, citrus fruits, other seed oils, herbs, other seasonings, nuts, fruit vegetables | Hepatoprotective and antihypertensive effects, prevents oxidative stress | Francesca 2014; Kumar et al. 2011 |
| Caffeic acid | 3,4-Dihydroxycinnamic acid | $C_9H_8O_4$ | Australian Eucalyptus honeys, New Zealand and Australian *Leptospermum* honeys, acacia, eucalyptus, lime, chestnut, heather, lavender, rosemary, orange, sunflower, rapeseed, Linden, heather honeys | Beers, fortified wines, grape wines, sparkling wines, cereal products, cereals, dried drupes, dried other fruits, berries, citrus fruits, drupes, other fruits, pomes, unknown coffee beverages, berry juices, drupe juices, pome juices, tropical fruit juices, fruit vegetable oils, other seed oils, herbs, other seasonings, spices, cabbages, fruit vegetables, leaf vegetables, root vegetables, tubers | Prevents acute renal toxicity, inhibits low-density lipoprotein oxidation, hepatoprotective effect, inhibits activation of nuclear transcription factor nuclear factor κB, antitumor effect, antidiabetic effect, prevents myocardial oxidative stress, inhibits DNA methylation, antimicrobial activity, fungitoxicity | Chong et al. 2009; Nolan et al. 2019 |
| p-Coumaric acid | p-Hydroxycinnamic acid | $C_9H_8O_3$ | Australian Eucalyptus honeys, New Zealand and Australian *Leptospermum* honeys, acacia, eucalyptus, lime, chestnut, heather, lavender, rosemary, orange, sunflower, rapeseed honeys | Beers, nut liquors, fortified wines, grape wines, sparkling wines, dried drupes, dried other fruits, berries, citrus fruits, other fruits, pomes, berry juices, pome juices, tropical fruit juices, fruit vegetable oils, other seed oils, herbs, other seasonings, spices, nuts, lentils, fruit vegetables | Inhibits platelet activity, confers cardioprotective effect, antioxidant protection on low-density lipoprotein cholesterol oxidation, detoxifies certain pesticides in honeybees | Jakopic 2007; Mao et al. 2013 |

| | | | | | | |
|---|---|---|---|---|---|---|
| Ferulic acid | Transferulicacid;4-hydroxy-3-methoxycinnamic acid | $C_{10}H_{10}O_4$ | Sunflower, Australian Eucalyptus, New Zealand, Australian *Leptospermum* honeys, acacia, eucalyptus, lime, chestnut, heather, lavender, rosemary, orange, sunflower, rapeseed honeys | Beers, fortified wines, grape wines, sparkling wines, cereal products, cereals, chocolate, dried other fruits, berries, citrus fruits, other fruits, pomes, berry juices, pome juices, tropical fruit juices, fruit vegetable oils, other seed oils, herbs, nuts, beans, lentils, soy and soy products, cabbages, fruit vegetables | Hepatoprotective effect, ameliorates high-fat diet-induced stress, inhibitory effect on tumor promotion, cardioprotective effect, prevents oxidative stress, prevents DNA damage, alleviates lipid peroxidation, antidiabetic effect, augments angiogenesis, induces circulatory pro-oxidants, prevents colon carcinogenesis, effect on age-related disease, prevents neurotoxicity, anti-inflammatory effect | Das 2014; Kesh 2013 |
| Cinnamic acid | 2-Propenoic acid | $C_9H_8O_2$ | Not in honey; specially propolis extract | Berries, citrus juices, fruit vegetable oils, spices, fruit vegetables | Gastroprotective effect | Boskou 2006; Nanjundaiah et al. 2011 |
| Homogentisic acid | 3,4-Dihydroxyphenylacetic acid | $C_8H_8O_4$ | Strawberry-tree honey | Fruit vegetable oils, fruit vegetables | Serves as a skin pigment | Bianco 2003; Solano 2014 |
| Phenyl acetic acid | Benzyl zinc bromide<br>Phenylactic acid<br>Benzylcarboxylic acid | $C_8H_8O_2$ | Acacia, eucalyptus, lime, chestnut, heather, lavender, rosemary, orange, sunflower, rapeseed honeys | Predominantly in fruits | Anti-inflammatory, analgesic, ulcerogenic, and lipid peroxidation activities | Amir and Shikha 2004; Dimitrova et al. 2007 |

2009). Another new separation technique, called capillary electrophoresis, is gaining popularity and represents an alternative method to analyze phytochemicals. The technique is mainly accomplished by following two types of modes, capillary-zone electrophoresis and micellar electro kinetic chromatography (MEKC). Of these two types of modes, MEKC is superior because electrophoretic behavior is affected by many factors (Wang 2004).

Using these techniques, more than 8000 phenolic structures have been identified to date with more than 4000 flavonoids reported (Bravo 1998; Cheynier 2005; Harborne and Williams 2000). The database of polyphenols, called phenol-explorer, classifies 501 polyphenols into six different classes (flavonoids, lignans, non-phenolic metabolites, "other polyphenols," phenolic acids and stilbenes) and 31 sub-classes (Figure 9.2). The polyphenols found in honey are mainly flavonoids, phenolic acids, and phenolic acid derivatives (Tomás-Barberán 2001). More specifically, the main polyphenols are flavonoids and their content can vary between 60 and 460 μg/100 g of honey, in which the total polyphenol content is approximately 56–500 mg/kg (Al-Mamary et al. 2002; Kenjerić 2007). According to the United States Department of Agriculture database, mixed varieties of honey samples from different countries contain apigenin (0.03 mg/100), luteolin (0.28 mg/100), isorhamnetin (0.06 mg/100), kaempferol (0.06 mg/100), myricetin (0.36 mg/100), and quercetin (0.31 mg/100) (see Table 9.1).

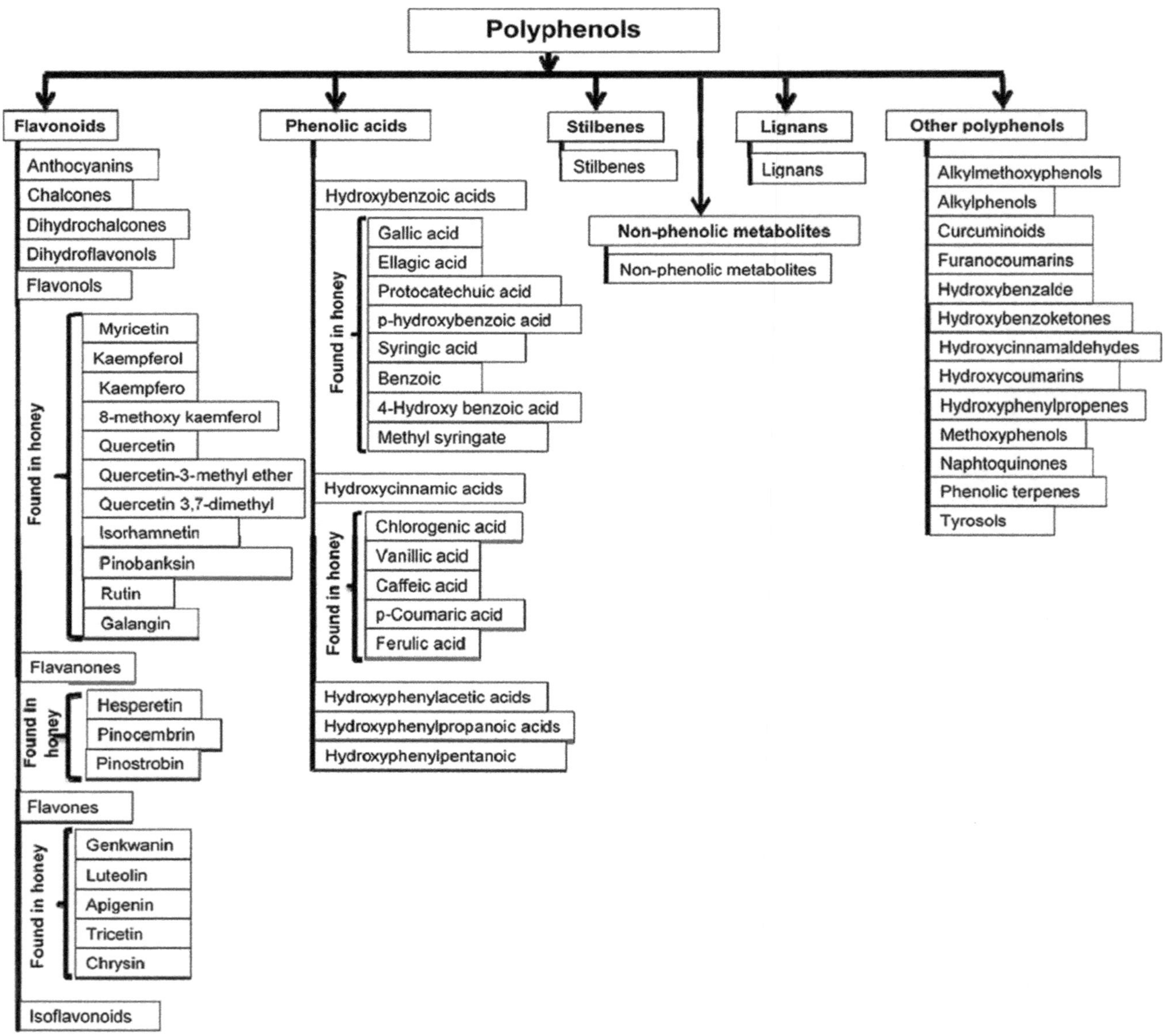


**Figure 9.2** Classification of polyphenols.

## Honey Polyphenols and Cardiovascular Diseases

Cardiovascular diseases (CVDs) are a group of diseases affecting the heart and blood vessels. These include coronary heart disease, stroke, angina, and congenital heart disease, which are the leading causes of death worldwide (World Health Organization and UNAIDS 2007). The most common risk factors for this multi-factorial disease are age, gender, hypertension, hypercholesterolemia, diabetes, obesity, and an unhealthy lifestyle such as tobacco use, physical inactivity, excessive alcohol consumption, or an unhealthy diet (Epstein and Ross 1999; Finks 2012; Kelly and Fuster 2010; Terzo et al. 2020). Various epidemiologic, clinical, and experimental studies demonstrated that higher polyphenol intake from the diet is associated with amelioration of CVD (Ulbricht and Southgate 1991). Polyphenols ameliorate CVD through various mechanisms (Figure 9.3). Most confer antioxidant protection (Rein 2000; Stein 1999), inhibit platelet aggregation (Rein 2000; Stein 1999), lower blood pressure (Desch 2010; Erlund 2008; Taubert 2007), improve endothelial function (Heiss 2003; Heiss 2005), reduce inflammatory responses (Mao 2002; Schramm 2003), decrease oxidative stress (Solayman 2015; Urquiaga and Leighton 2000), improve coronary vasodilatation (Benavente-Garcia and Castillo 2008), and reduce low density lipoprotein (LDL) oxidation (Mathur 2002; Wan 2001).

Antioxidants neutralize ROS and minimize oxidative stress. Dietary polyphenols act as antioxidants. Therefore, the quercetin, caffeic acid, kaempferol, and apigenin found in honey might play a role in the treatment of patients with CVD. These compounds reduce oxidative stress in various ways. Quercetin reduces oxidative stress by scavenging free radicals, chelating metal ions, and inhibiting xanthine oxidase and lipid peroxidation (Hanasaki et al. 1994; Larocca 1995; Namsi 2018). It also reduces oxidative stress in streptozotocin-induced diabetes in rats and in ethanol-induced in mouse livers (Mahesh and Menon 2004; Molina 2003). Caffeic acid inhibits oxidative stress in iron-overloaded rats by reducing lipid peroxidation and increasing vitamin E levels in the plasma (Lafay 2005). Caffeic acid phenethyl ester isolated from the honeybee propolis induced myocardial oxidative stress in rats by scavenging free radicals (Ozguner 2005). Kaempferol is a flavonoid that can reduce oxidative stress caused by glutamate in the mouse hippocampal cell line HT-22 by blocking ROS generation (Ishige et al. 2001). Kaempferol also blocks oxidative stress in granule cells during low potassium–induced apoptosis (Samhan-Arias et al. 2004). Apigenin is a common flavonoid found in honey that is able to reduce oxidative stress and protects endothelium-dependent vasorelaxation of the aorta in male Sprague-Dawley rats (Jin 2009).

Platelet aggregation is another crucial event in the pathogenesis of many types of CVDs such as atherosclerosis, stroke, diabetes mellitus, and coronary heart disease (Angiolillo 2007; O'malley 1995; Ruggeri 2002). Platelet aggregation is the clumping together of platelets in the blood. It is the part of the sequence of events leading to the formation of a thrombus or clot. Experimental studies suggest that dietary polyphenols reduce platelet aggregation and prevent CVD.

Honey is a rich source of polyphenols. Honey polyphenols such as quercetin, apigenin, luteolin, catechin, and rutin inhibit platelet aggregation in vitro through binding to the thromboxane A2 receptor (Guerrero 2005). Consumption of quercetin inhibits collagen-mediated signaling in platelets and reduces platelet aggregation (Hubbard 2004). Lowering blood pressure, improving endothelial function, reducing inflammatory responses, improving coronary vasodilatation, and reducing LDL oxidation are other target steps towards preventing CVDs. Polyphenols contained in honey prevent CVDs by these mechanisms. Quercetin, the main polyphenol in honey, reduces systolic blood pressure and plasma oxidized LDL concentrations in obese people (Egert 2009). Quercetin reduces LDL oxidation by inhibition of myeloperoxidase (Kostyuk 2011). Quercetin also reduces macrophage stimulated LDL oxidation by preventing lipid hydroperoxide generation (Al-Awwadi 2005). Quercetin stimulates coronary vasodilatation by reducing endothelin-1 mRNA expression, and it improves endothelial function (Romero 2009). In another study, quercetin enhanced endothelial nitric oxide synthase activity and decreased NADPH oxidase-mediated superoxide anion ($O^{\cdot}_2$) generation by downregulating the expression of neutrophil cytosol factor 1, also known as $p47^{phox}$, and preventing endothelial dysfunction in hypertensive rats (Sanchez 2006). Daily administration of quercetin, rutin, and hesperidin at 80 mg/kg reduced both acute and chronic phases of inflammation in an adjuvant arthritis rat model. Therefore, quercetin, rutin and hesperidin may exert some anti-inflammatory activities (Guardia 2001).

Apigenin possesses anti-inflammatory activity that is conducted through the following mechanisms: the inhibition of lipopolysaccharide-induced cyclooxygenase-2 expression; suppression of nitric oxide (NO) production in RAW-264.7 macrophage cells; reduction of tumor necrosis factor-α (TNF-α)–induced adhesion of monocytes to a human umbilical vein endothelial cells monolayer; and suppression of TNF-α–stimulated upregulation of vascular cellular adhesion molecule-1 (VCAM-1), intracellular adhesion molecule-1 (ICAM-1), and E-selectin mRNA to basal levels (Lee 2007). Kaempferol, a polyphenol frequently found in honey, exerts anti-inflammatory effects in a high cholesterol–fed rabbits model by decreasing the levels of TNF-α, interleukin-1β, and malondialdehyde (MDA), increasing the activity of superoxide dismutase (SOD) and decreasing the gene and protein expression of inflammatory molecules such as E-selectin, ICAM-1, VCAM-1, and monocyte chemotactic protein-1 (Kong 2013).

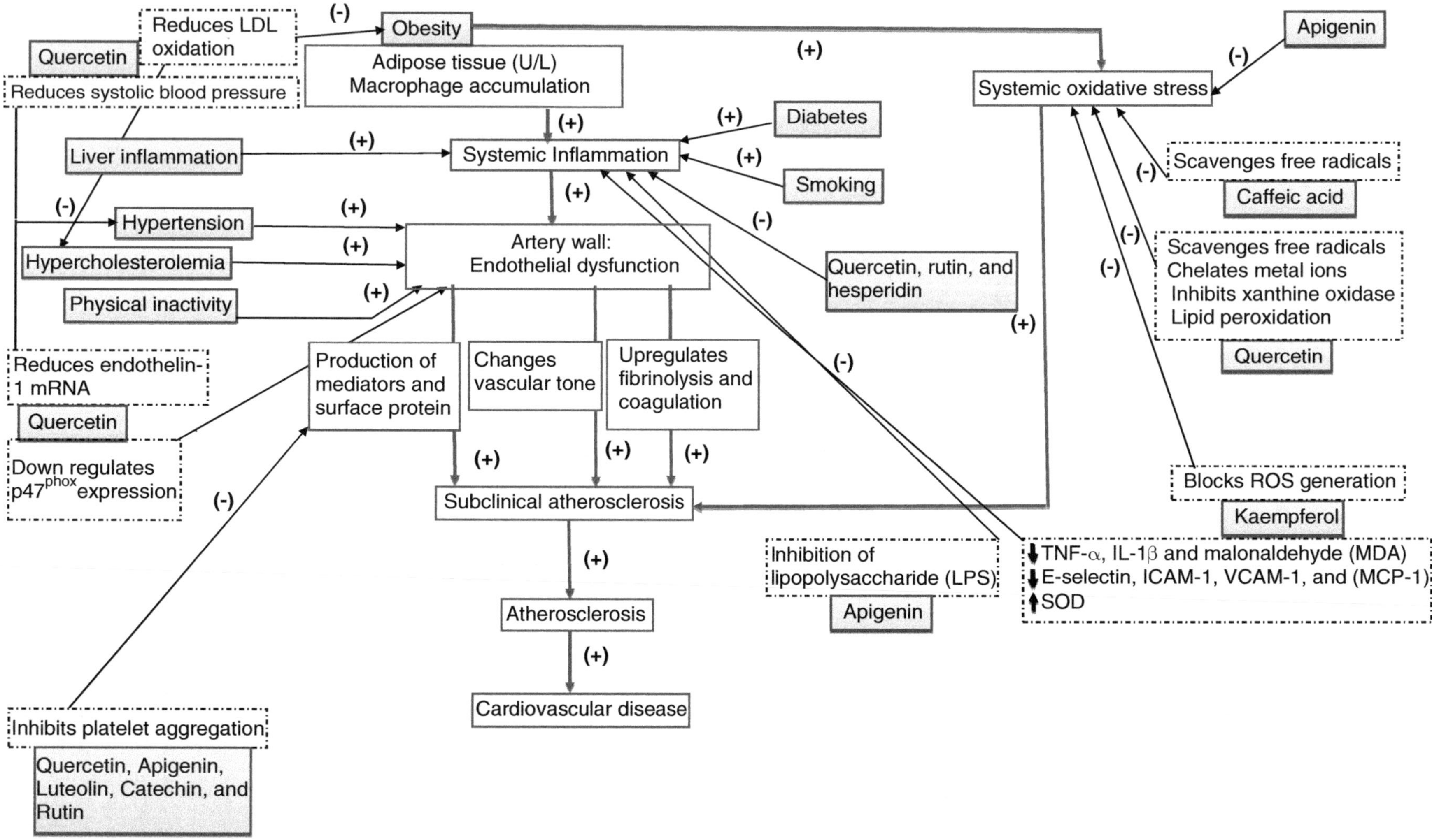


**Figure 9.3** Sequential events of cardiovascular disease and the potential roles of honey polyphenols.

## Honey Polyphenols and Diabetes

Diabetes is one of the most serious endocrine metabolic disorders in the present age. To treat patients with diabetes, several antidiabetic drugs are used. However, because of side effects and the economic burden of these antidiabetic drugs, scientists are eager to discover alternative medicines to treat these patients. Medicinal plants were traditionally and are still used in the treatment of patients with diabetes (Bailey and Day 1989; Patel 2012), with many of the active constituents in these plants reported to be polyphenols (Gray and Flatt 1997; Marles and Farnsworth 1995). There is much experimental evidence that dietary polyphenols can be used in the treatment of diabetes. Although honey contains diverse types of polyphenols, only a few, such as quercetin, apigenin, luteolin, catechin, rutin, and kaempferol, have antidiabetic roles (Figure 9.4), and these act via several mechanisms to reduce blood glucose levels. Some of the mechanisms include inhibition of α-amylase and α-glucosidase (McDougall 2005; McDougall and Stewart 2005), inhibition of gluconeogenic enzymes (Jung 2004), enhancement of pancreatic β-cell protection (Rouse et al. 2014), reduction of oxidative stress (Sabu et al. 2002), enhancement of glucose uptake (Eid 2010), enhancement of insulin secretion and activity (Hii and Howell 1985), production of insulin receptor and glucose transporter type 4 (GLUT4) (Cao et al. 2007), and inhibition of the enzyme aldose reductase (Varma and Kinoshita 1976).

The key enzymes involved in blood glucose elevation through the breakdown of carbohydrates are α-amylase and α-glucosidase. Therefore, inhibition of these two enzymes is considered an effective therapeutic tool to reduce blood glucose levels in patients with diabetes (Tundis et al. 2010). To date, several studies have suggested the potential of honey

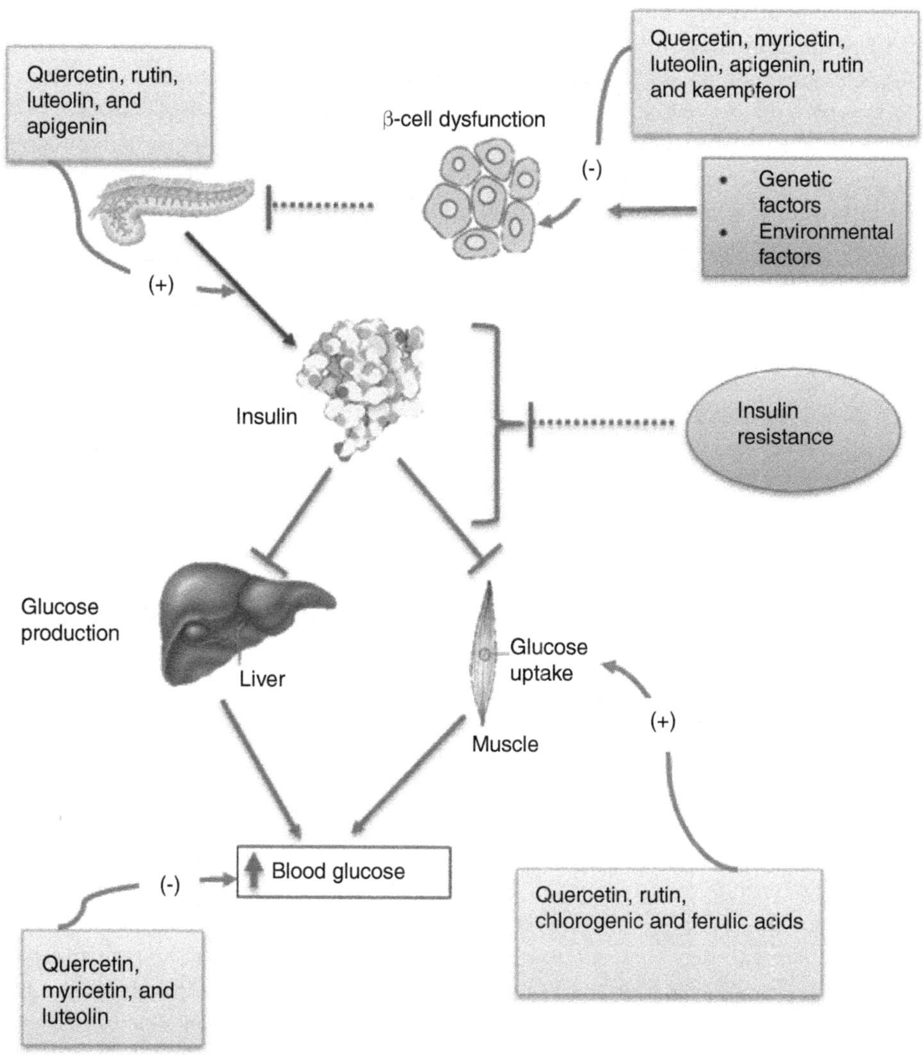


**Figure 9.4** Sequential events of diabetes and the potential roles of honey polyphenols.

polyphenols as well as other polyphenols to inhibit the carbohydrate breakdown enzymes α-amylase and α-glucosidase. In patients with type 2 diabetes, quercetin inhibits the intestinal enzyme α-glucosidase and reduces maltose-induced postprandial hyperglycemia (Hussain et al. 2012). In another in vitro study, quercetin, myricetin, and luteolin at 0.5 mM showed inhibitory effects on porcine pancreatic α-amylase activity with minimum inhibitory concentration values lower than 500, 380, and 360 μM, respectively (Tadera 2006). In another experiment, luteolin and luteolin-7-O-glycoside showed inhibitory effects on α-amylase and α-glucosidase (Kim et al. 2000).

Pancreatic β-cells are responsible for controlling blood glucose level in an animal's body, and disruption of this process leads to diabetes. Therefore, pancreatic β-cell protection is the key target for controlling diabetes. The polyphenols present in honey have protected pancreatic β-cells in many experimental studies. In streptozotocin-induced diabetes in rats, administration of quercetin at a dose of 10–15 mg/kg for 10 days increased the number of pancreatic β-cells (Vessal et al. 2003). In another study, quercetin protected and preserved the pancreatic β-cell structure and integrity while increasing insulin secretion in streptozotocin-induced diabetes in rats (Adewole et al. 2007). Apigenin and rutin protected pancreatic β-cells by increasing antioxidant enzyme levels and insulin levels in streptozotocin-induced diabetes in rats (Esmaeili and Sadeghi 2009). Kaempferol, a flavonol present in honey, was shown to prevent pancreatic β-cell dysfunction and improve peripheral insulin sensitivity in middle-aged obese mice with diabetes (Alkhalidy 2015).

Another approach to treatment of patients with diabetes is the reduction of oxidative stress, which is a primary complication of diabetes (Giacco and Brownlee 2010). Experimental and clinical studies suggest that polyphenols in honey as well as other sources reduce oxidative stress in diabetic animals. Quercetin, present in honey, reduced oxidative stress in streptozotocin-induced diabetes in rats by decreasing the blood glucose level and lipid peroxidation and increasing the levels of vitamins C and E as well as tissue antioxidant activity (Coskun 2005; Mahesh and Menon 2004). Myricetin, another polyphenol of honey, effectively removed ROS and decreased production of free radicals during cell swelling after ischemic injury (Gordon and Roedig-Penman 1998; Panickar et al. 2009). Luteolin is a flavone that has been reported to be present in honey, and it ameliorated cardiac failure by reducing oxidative stress in type 1 diabetic cardiomyopathy (Wang 2012).

In diabetes, glucose utilization is decreased in both insulin- and non–insulin-sensitive tissues. Dietary polyphenols reported in honey increase peripheral glucose uptake in these tissues. Quercetin and rutin are important components of honey and can increase glucose uptake in insulin-resistant tissues (Lee 2012). Chlorogenic and ferulic acids are two other plant phenolic compounds also found in honey that can enhance glucose uptake in myotubes and are reported to be more effective than the two hypoglycemic drugs metformin and 2,4-thiozolodinedione (THZ) (Ding 2019; Prabhakar and Doble 2009). Chlorogenic acid stimulates glucose uptake by expressing the GLUT4 and peroxisome proliferator–activated receptor γ (PPAR-γ) genes via P13K-independent pathways. On the other hand, ferulic acid stimulates glucose uptake by expressing GLUT4 and P13K transcripts via P13K-dependent pathways (Prabhakar and Doble 2009).

Enhancing insulin secretion activity is the most important aim in treating patients with diabetes. Polyphenols such as quercetin, rutin, luteolin, and apigenin are present in honey and contribute to enhanced insulin secretion and activity in many experimental studies. Daily administration of rutin at 25–100 mg/kg for 45 days can increase insulin secretion and decrease fasting blood glucose levels in streptozotocin-induced diabetes in rats (Kamalakkannan and Prince 2006). Apigenin decreased blood glucose levels by increasing insulin secretion after administration at a dosage of 50 mg/kg daily in diabetic rats (Cazarolli 2009). Quercetin can protect pancreatic β-cells against hydrogen peroxide dysfunction and stimulate insulin secretion in insulin secreting cell lines induced by glucose and glibenclamide by the activation of the extracellular signal–regulated kinases pathway (Youl 2010). Luteolin, a flavonoid, reduced hyperglycemia by increasing the secretion of insulin in a diabetic animal model (Zarzuelo 1996). Luteolin also prevented insulin resistance in patients with type 2 diabetes through the production of adipokines (cytokines in adipocytes) by activating the PPAR-γ pathway (Ding et al. 2010).

## Honey Polyphenols and Cancer

Cancer is a group of multifactorial diseases associated with uncontrolled proliferation of cells (Hanahan and Weinberg 2000). Both environmental (90%–95%) and genetic (5%–10%) factors contribute to cancer formation in animals. Among environmental factors, smoke, carcinogenic chemical exposure, radiation, a sedentary lifestyle, infection, and stress are most common. In addition to environmental and genetic factors, hormone status, immune condition, and mutations contribute to cancer development (Anand 2008). Nevertheless, experimental studies suggest that regular consumption of fruits, vegetables, herbs, honey, and other polyphenol-rich foods is associated with a decreased cancer incidence (Block et al. 1992; Swellam 2003). Polyphenols prevent cancer through various mechanisms (Figure 9.5), for example, by the

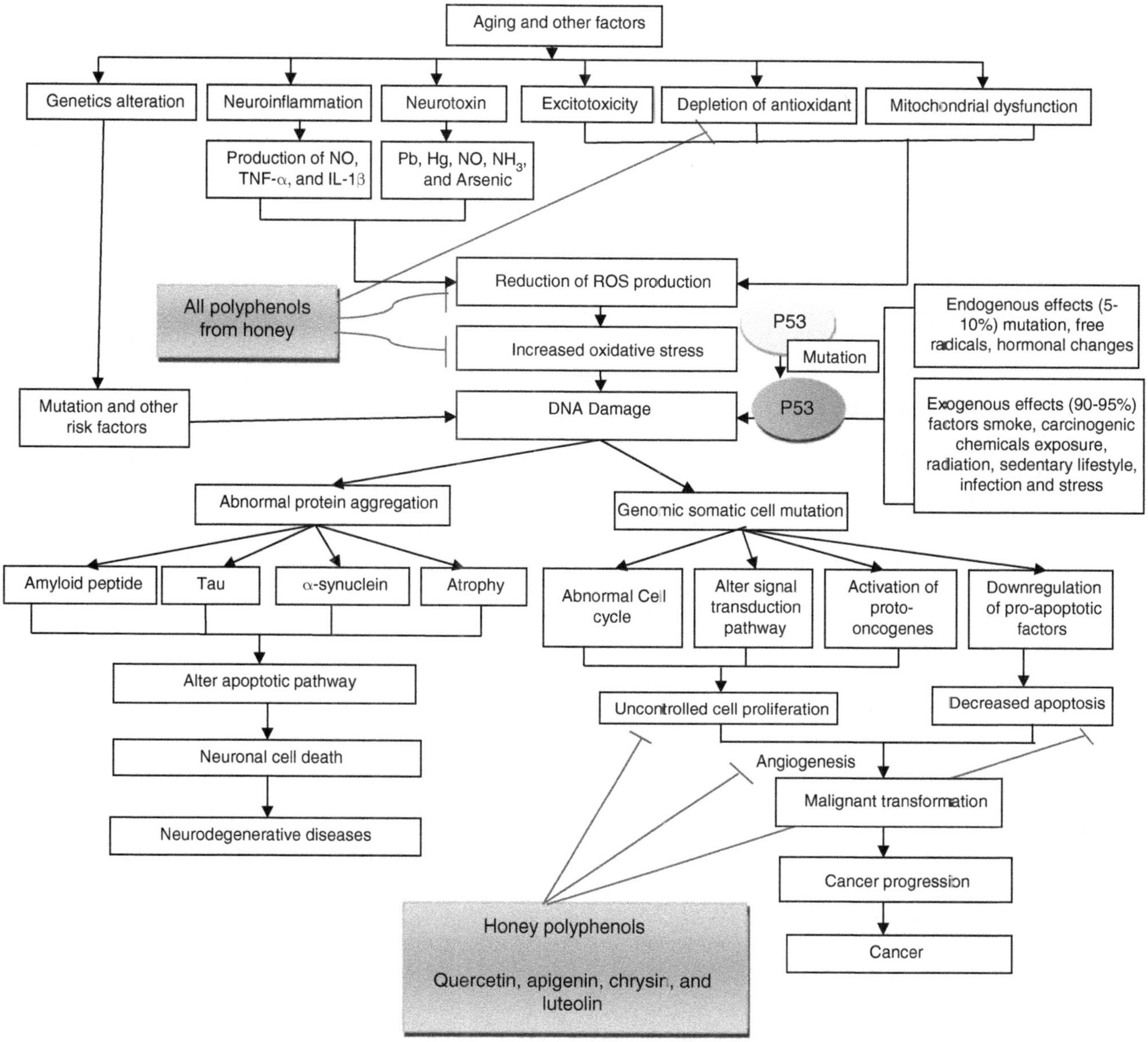


**Figure 9.5** Sequential events of neurodegenerative diseases and cancer: the potential roles of honey polyphenols. IL, interleukin; Hg, mercury; Pb, lead; TNF, tumor necrosis factor.

inhibition of cell proliferation (Kuntz et al. 1999), modulation of cancer cell signaling pathways (Corona 2007; Khan 2006), induction of tumor cell apoptosis (Ishikawa and Kitamura 2000; Mantena et al. 2006), modification of enzymatic activity (Adams and Chen 2008), prevention of oxidative stress via antioxidant effects (Katiyar 2001), and inhibition of DNA damage (Feng 2002).

In cancer, there is uncontrolled cell proliferation. Therefore, inhibition of uncontrolled cell proliferation is the main target goal of cancer prevention. Polyphenols are natural antioxidants, which can prevent cell proliferation in various ways. Quercetin is an antioxidant and anticancer agent found in honey that can inhibit prostate cancer cell proliferation. The mechanism by which it does so is mainly by decreasing the expression of cell cycle regulator cdc2/cdk-1, cyclin B1, phosphorylated pRB, or apoptosis markers Bcl-2, Bcl-$X_L$ and/or by increasing levels of p21, Bax, and caspase-3 (Cianciosi 2020; Vijayababu 2005). Apigenin, a flavonoid present in honey, prevents pancreatic cell proliferation by reducing the levels of cyclin A, cyclin B, and the phosphorylated forms of cdc2 and cdc25, thereby arresting the G2/M phase of the cell cycle (Ujiki 2006). Chrysin is another honey flavone that controls cell proliferation by activating p38-MAPK (mitogen-activated protein kinase) via accumulation of $p21^{Wafi/Cip1}$ in C6 glioma cells of rats (Weng et al. 2005). In another study, chrysin prevented cell proliferation in the prostate cancer cell line PC-3 (Samarghandian et al. 2011). Luteolin, which occurs in honey, inhibits proliferation and DNA synthesis induced by platelet-derived growth factor-BB in rat aortic vascular smooth muscle cells by blocking the phosphorylation of PDGF-β receptor (Kim 2005).

Modulation of cancer cell signaling pathways is another target step in cancer prevention. Experimental evidence suggests that polyphenols prevent cancer through modulation of cancer cell signaling pathways (Afrin 2017). Quercetin prevents colon cancer by inhibiting β-catenin/Tcf signaling in SW480 cell lines and reducing β-catenin/Tcf transcriptional activity (Park 2005). In another study, quercetin exerted an anticancer effect through inhibition of the PI3K-Akt/PKB pathway (Gulati 2006).

## Honey Polyphenols and Neurodegenerative Disease

Neurological disorders such as Alzheimer's disease, Parkinson's disease, Huntington's disease, multiple sclerosis, and stroke cause cognitive impairment in older individuals. These diseases occur because of dysfunction or death of nerve cells as a result of increasing oxidative stress caused by the depletion of antioxidants, neuroinflammation, prions, protein and mitochondrial dysfunction, glutamatergic excitotoxicity, and genetic alterations (Barzilai and Melamed 2003; Jellinger 2001; Spires and Hannan 2005). Dietary polyphenols found in honey as well as other food and plant materials can prevent neurodegenerative disease in various ways (see Figure 9.5). These include oxidative protection of neurons (Inanami 1998), enhancement of neuronal function and regeneration (Youdim and Joseph 2001), protection of neurons from Aβ-induced neuronal injury and neurotoxicity (Luo 2002), protection of hippocampal cells against NO-induced toxicity (Bastianetto et al. 2000), and modulation of neuronal and glial cell signaling pathways (Vauzour 2007).

Luteolin is a flavonoid found in honey that shows neuroprotective activity against microglia-induced neuronal cell death and enhances the spatial working memory via prevention of microglia associated inflammation in the hippocampus of aged rats (Jang et al. 2010). In another study, luteolin protected synaptic function and restored memory in neurodegenerative disorders by enhancing basal synaptic transmission while facilitating the induction of long-term potentiation by high-frequency stimulation in the dental gyrus of the rat hippocampus through the activation of cyclic adenosine monophosphate response element–binding protein (Xu 2010). Quercetin shows neuroprotective activity against chronic lead exposure and colchicine-induced cognitive impairments in rat models (Hu 2008; Kumar 2008). Kaempferol exhibits a neuroprotective effect in Parkinson's disease mice models against the neurotoxin 1-methyl-4-phenyl-1,2,3,6-tetrahydropyridine (MPTP). Induction of MPTP leads to behavioral and biochemical alterations such as behavioral deficits, depletion of dopamine and its metabolites, reduction in SOD and glutathione peroxidase (GSH-PX) activities, and elevation of MDA levels in the substantia nigra of mice. Oral administration of kaempferol to mice every 24 hours for 14 consecutive days improved all the behavioral and biochemical alterations. Its neuroprotective effect was further confirmed by the histochemical findings of the same experiment, in which kaempferol was found to prevent the loss of tyrosine hydroxylase–positive neurons induced by MPTP (Li and Pu 2011).

Ferulic acid, a well-known polyphenol contained in honey, exerts a neuroprotective effect through middle cerebral artery occlusion injury–induced decreases of phospho-PDK1, phospho-Akt, and phospho-Bad levels while preventing injury-induced increases in caspase-3 levels (Koh 2012). In another study, ferulic acid displayed a neuroprotective effect against oxidative stress–associated apoptosis through inhibition of ICAM-1 mRNA expression and by decreasing the number of microglia and macrophages after cerebral ischemia–reperfusion injury in rats (Cheng 2008).

Chlorogenic acid present in honey exerts a neuroprotective effect against methyl mercury–induced apoptosis in pheochromocytoma-12 cell lines. In an experimental study, chlorogenic acid provided a neuroprotective effect by preventing the generation of ROS, suppressing the decreasing action of glutathione peroxidase (GPx) and GSH, and attenuating apoptosis by the activation of caspase-3 (Li 2008). Kwon et al. (2010) reported that chlorogenic acid shows a neuroprotective effect on scopolamine-induced learning and memory impairment by inhibiting the activity of acetylcholine esterase and MDA in the hippocampus as well as in the frontal cortex in mice (Kwon et al. 2010).

## Conclusion

Honey is a rich source of polyphenols, and the amount of each polyphenol varies significantly. These polyphenols prevent the development of various diseases via several specific mechanisms by regulating a particular gene's expression or by promoting or blocking a specific step in a metabolic pathway. Because honey samples are different and may not contain all the polyphenols described and because not all polyphenols exert the same protective effects, consumption of a wide variety of honey samples at any one time is recommended. Finally, this review can be used as a source of information when choosing a specific honey sample for treatment of a specific disease and as a resource for drug discovery and development.

## References


Abdel-Moneim, W.M. and Ghafeer, H.H. (2007). The potential protective effect of natural honey against cadmium-induced hepatotoxicity and nephrotoxicity. *Mansoura Journal of Forensic Medicine & Clinical Toxicology XV* 2 (15): 75–92.

Adams, L.S. and Chen, S. (2008). Phytochemicals for breast cancer prevention by targeting aromatase. *Frontiers in Bioscience (Landmark Edition)* 14: 3846–3863.

Adewole, S.O., Caxton-Martins, E.A., and Ojewole, J.A. (2007). Protective effect of quercetin on the morphology of pancreatic β-cells of streptozotocin-treated diabetic rats. *African Journal of Traditional, Complementary and Alternative Medicines* 4 (1): 64–74.

Afrin, Sadia, Tamara Y., et al. (2017). Strawberry-tree honey induces growth inhibition of human colon cancer cells and increases ROS generation: a comparison with Manuka Honey. *International Journal of Molecular Sciences* 18 (3): 613.

Al-Awwadi, N.A., et al. (2005). Extracts enriched in different polyphenolic families normalize increased cardiac NADPH oxidase expression while having differential effects on insulin resistance, hypertension, and cardiac hypertrophy in high-fructose-fed rats. *Journal of Agricultural and Food Chemistry* 53 (1): 151–157.

Alkhalidy, H., et al. (2015). Small molecule kaempferol promotes insulin sensitivity and preserved pancreatic β-cell mass in middle-aged obese diabetic mice. *Journal of Diabetes Research* 2015.

Al-Mamary, M., Al-Meeri, A., and Al-Habori, M. (2002). Antioxidant activities and total phenolics of different types of honey. *Nutrition Research* 22 (9): 1041–1047.

Amir, M. and Shikha, K. (2004). Synthesis and anti-inflammatory, analgesic, ulcerogenic and lipid peroxidation activities of some new 2-[(2, 6-dichloroanilino) phenyl] acetic acid derivatives. *European Journal of Medicinal Chemistry* 39 (6): 535–545.

Anand, P., et al. (2008). Cancer is a preventable disease that requires major lifestyle changes. *Pharmaceutical Research* 25 (9): 2097–2116.

Angiolillo, D.J., et al. (2007). Impact of platelet reactivity on cardiovascular outcomes in patients with type 2 diabetes mellitus and coronary artery disease. *Journal of the American College of Cardiology* 50 (16): 1541–1547.

Arts, I.C. and Hollman, P.C. (2005). Polyphenols and disease risk in epidemiologic studies. *The American Journal of Clinical Nutrition* 81 (1): 317S–325S.

Aziza, S.A.H., Azab, M.E.-S., and El-Shall, S.K. (2014). Ameliorating role of rutin on oxidative stress induced by iron overload in hepatic tissue of rats. *Pakistan Journal of Biological Sciences* 17 (8): 964.

Bailey, C.J. and Day, C. (1989). Traditional plant medicines as treatments for diabetes. *Diabetes Care* 12 (8): 553–564.

Barzilai, A. and Melamed, E. (2003). Molecular mechanisms of selective dopaminergic neuronal death in Parkinson's disease. *Trends in Molecular Medicine* 9 (3): 126–132.

Bastianetto, S., Zheng, W.H., and Quirion, R. (2000). The ginkgo biloba extract (EGb 761) protects and rescues hippocampal cells against nitric oxide-induced toxicity. *Journal of Neurochemistry* 74 (6): 2268–2277.

Battino, M., et al. (2020). The roles of strawberry and honey phytochemicals on human health: a possible clue on the molecular mechanisms involved in the prevention of oxidative stress and inflammation. *Phytomedicine* 86, 153170.

Beale, J.M., Block, J., and Hill, R. (2010). *Organic Medicinal and Pharmaceutical Chemistry*. Philadelphia: Lippincott Williams & Wilkins.

Benavente-Garcia, O. and Castillo, J. (2008). Update on uses and properties of citrus flavonoids: new findings in anticancer, cardiovascular, and anti-inflammatory activity. *Journal of Agricultural and Food Chemistry* 56 (15): 6185–6205.

Bianco, A., et al. (2003). Analysis by liquid chromatography-tandem mass spectrometry of biophenolic compounds in virgin olive oil, Part II. *Journal of Separation Science* 26 (5): 417–424.

Block, G., Patterson, B., and Subar, A. (1992). Fruit, vegetables, and cancer prevention: a review of the epidemiological evidence. *Nutrition and Cancer* 18 (1): 1–29.

Boskou, G., et al. (2006). Antioxidant capacity and phenolic profile of table olives from the Greek market. *Food Chemistry* 94 (4): 558–564.

Bravo, L. (1998). Polyphenols: chemistry, dietary sources, metabolism, and nutritional significance. *Nutrition Reviews* 56 (11): 317–333.

Cao, H., Polansky, M.M., and Anderson, R.A. (2007). Cinnamon extract and polyphenols affect the expression of tristetraprolin, insulin receptor, and glucose transporter 4 in mouse 3T3-L1 adipocytes. *Archives of Biochemistry and Biophysics* 459 (2): 214–222.

Carrasco-Pancorbo, A., et al. (2006). Rapid quantification of the phenolic fraction of Spanish virgin olive oils by capillary electrophoresis with UV detection. *Journal of Agricultural and Food Chemistry* 54 (21): 7984–7991.

Cazarolli, L.H., et al. (2009). Mechanism of action of the stimulatory effect of apigenin-6-C-(2 ″-O-α-L-rhamnopyranosyl)-β-L-fucopyranoside on 14 C-glucose uptake. *Chemico-biological Interactions* 179 (2): 407–412.

Chen, H., et al. (2013). Anti-tumor effect of rutin on human neuroblastoma cell lines through inducing G2/M cell cycle arrest and promoting apoptosis. *The Scientific World Journal* 2013.

Cheng, C.-Y., et al. (2008). Ferulic acid provides neuroprotection against oxidative stress-related apoptosis after cerebral ischemia/reperfusion injury by inhibiting ICAM-1 mRNA expression in rats. *Brain Research* 1209: 136–150.

Cheynier, V. (2005). Polyphenols in foods are more complex than often thought. *The American Journal of Clinical Nutrition* 81 (1): 223S–229S.

Cheynier, V. (2012). Phenolic compounds: from plants to foods. *Phytochemistry Reviews* 11 (2–3): 153–177.

Chong, K.P., Rossall, S., and Atong, M. (2009). In vitro antimicrobial activity and fungitoxicity of syringic acid, caffeic acid and 4-hydroxybenzoic acid against Ganoderma Boninense. *Journal of Agricultural Science* 1 (2): 15.

Chow, J. (2002). Probiotics and prebiotics: a brief overview. *Journal of Renal Nutrition* 12 (2): 76–86.

Cianciosi, D., et al. (2018). Phenolic compounds in honey and their associated health benefits: a review. *Molecules (Basel, Switzerland)* 23 (9): 2322.

Cianciosi, D., et al. (2020). Phenolic compounds from Mediterranean foods as nutraceutical tools for the prevention of cancer: the effect of honey polyphenols on colorectal cancer stem-like cells from spheroids.Food chemistry, 325, 126881.

Cianciosi, D., et al. (2020). The influence of in vitro gastrointestinal digestion on the anticancer activity of manuka honey. *Antioxidants* 9 (1): 64.

Corona, G., et al. (2007). Inhibition of p38/CREB phosphorylation and COX-2 expression by olive oil polyphenols underlies their anti-proliferative effects. *Biochemical and Biophysical Research Communications* 362 (3): 606–611.

Coskun, O., et al. (2005). Quercetin, a flavonoid antioxidant, prevents and protects streptozotocin-induced oxidative stress and β-cell damage in rat pancreas. *Pharmacological Research* 51 (2): 117–123.

Das, U., et al. (2014). Role of ferulic acid in the amelioration of ionizing radiation induced inflammation: a murine model. *PloS One* 9 (5): e97599.

Desch, S., et al. (2010). Effect of cocoa products on blood pressure: systematic review and meta-analysis. *American Journal of Hypertension* 23 (1): 97–103.

Dimitrova, B., Gevrenova, R., and Anklam, E. (2007). Analysis of phenolic acids in honeys of different floral origin by solid-pase extraction and high-performance liquid chromatography. *Phytochemical Analysis* 18 (1): 24–32.

Ding, L., Jin, D., and Chen, X. (2010). Luteolin enhances insulin sensitivity via activation of PPARγ transcriptional activity in adipocytes. *The Journal of Nutritional Biochemistry* 21 (10): 941–947.

Ding, Y., et al. (2019). Combination of honey with metformin enhances glucose metabolism and ameliorates hepatic and nephritic dysfunction in STZ-induced diabetic mice. *Food & Function* 10 (11): 7576–7587.

Duthie, G.G. and Brown, K.M. (1994). *Reducing the Risk of Cardiovascular Disease, in Functional Foods.* Springer, Boston, MA, 1994. 19–38.

Egert, S., et al. (2009). Quercetin reduces systolic blood pressure and plasma oxidised low-density lipoprotein concentrations in overweight subjects with a high-cardiovascular disease risk phenotype: a double-blinded, placebo-controlled cross-over study. *British Journal of Nutrition* 102 (7): 1065.

Eid, H.M., et al. (2010). Stimulation of AMP-activated protein kinase and enhancement of basal glucose uptake in muscle cells by quercetin and quercetin glycosides, active principles of the antidiabetic medicinal plant Vaccinium vitis-idaea. *Molecular Nutrition & Food Research* 54 (7): 991–1003.

Epstein, F.H. and Ross, R. (1999). Atherosclerosis—an inflammatory disease. *New England Journal of Medicine* 340 (2): 115–126.

Erlund, I., et al. (2008). Favorable effects of berry consumption on platelet function, blood pressure, and HDL cholesterol. *The American Journal of Clinical Nutrition* 87 (2): 323–331.

Esmaeili, M.A. and Sadeghi, H. (2009). Pancreatic B-cell protective effect of rutin and apigenin isolated from Teucrium polium. *Pharmacology Online* 2: 341–353.

Estevinho, L., et al. (2008). Antioxidant and antimicrobial effects of phenolic compounds extracts of Northeast Portugal honey. *Food and Chemical Toxicology* 46 (12): 3774–3779.

Feng, Q., et al. (2002). Black tea polyphenols, theaflavins, prevent cellular DNA damage by inhibiting oxidative stress and suppressing cytochrome P450 1A1 in cell cultures. *Journal of Agricultural and Food Chemistry* 50 (1): 213–220.

Feng, X., et al. (2014). Chrysin attenuates inflammation by regulating M1/M2 status via activating PPARγ. *Biochemical Pharmacology* 89 (4): 503–514.

Finks, S.W., et al. (2012). Key articles of dietary interventions that influence cardiovascular mortality. *Pharmacotherapy: The Journal of Human Pharmacology and Drug Therapy* 32 (4): e54–e87.

Francesca, N., et al. (2014). Evolution of microbiological and chemical parameters during red wine making with extended post-fermentation maceration. *International Journal of Food Microbiology* 171: 84–93.

Fu, X., et al. (2014). Protective role of luteolin against cognitive dysfunction induced by chronic cerebral hypoperfusion in rats. *Pharmacology Biochemistry and Behavior* 126: 122–130.

Gao, Y., et al (2014). Genkwanin inhibits proinflammatory mediators mainly through the regulation of miR-101/MKP-1/MAPK pathway in LPS-activated macrophages. *PloS One* 9 (5): e96741.

Gašić, U.M., Milojković-Opsenica, D.M., and Tešić, Ž.L. (2019). Polyphenols as possible markers of botanical origin of honey. *Journal of AOAC International* 100 (4): 852–861.

Giacco, F. and Brownlee, M. (2010). Oxidative stress and diabetic complications. *Circulation Research* 107 (9): 1058–1070.

Gordon, M.H. and Roedig-Penman, A. (1998). Antioxidant activity of quercetin and myricetin in liposomes. *Chemistry and Physics of Lipids* 97 (1): 79–85.

Gośliński, M., Nowak, D., and Kłębukowska, L. (2020). Antioxidant properties and antimicrobial activity of manuka honey versus Polish honeys. *Journal of Food Science and Technology* 57 (4): 1269–1277.

Graf, B.A., Milbury, P.E., and Blumberg, J.B. (2005). Flavonols, flavones, flavanones, and human health: epidemiological evidence. *Journal of Medicinal Food* 8 (3): 281–290.

Gray, A.M. and Flatt, P.R. (1997). Nature's own pharmacy: the diabetes perspective. *Proceedings of the Nutrition Society* 56 (1B): 507–517.

Guardia, T., et al. (2001). Anti-inflammatory properties of plant flavonoids. Effects of rutin, quercetin and hesperidin on adjuvant arthritis in rat. *Il Farmaco* 56 (9): 683–687.

Guerrero, J., et al. (2005). Flavonoids inhibit platelet function through binding to the thromboxane A2 receptor. *Journal of Thrombosis and Haemostasis* 3 (2): 369–376.

Guerrero, M., et al. (2002). Quercetin 3, 7-dimethyl ether: a vasorelaxant flavonoid isolated from Croton schiedeanus Schlecht. *Journal of Pharmacy and Pharmacology* 54 (10): 1373–1378.

Gulati, N., et al. (2006). The antiproliferative effect of Quercetin in cancer cells is mediated via inhibition of the PI3K-Akt/PKB pathway. *Anticancer Research* 26 (2A): 1177–1181.

Guo, N., et al. (2020). Comparison of the chemical composition and biological activity of mature and immature honey: an HPLC/QTOF/MS-based metabolomic approach. *Journal of Agricultural and Food Chemistry* 68 (13): 4062–4071.

Halagarda, M., et al. (2020). Antioxidant activity and phenolic profile of selected organic and conventional honeys from Poland. *Antioxidants* 9 (1): 44.

Ham, J.R., et al. (2016). Anti-steatotic and anti-inflammatory roles of syringic acid in high-fat diet-induced obese mice. *Food and Function* 7 (2): 689–697.

Hanahan, D. and Weinberg, R.A. (2000). The hallmarks of cancer. *Cell* 100 (1): 57–70.

Hanasaki, Y., Ogawa, S., and Fukui, S. (1994). The correlation between active oxygens scavenging and antioxidative effects of flavonoids. *Free Radical Biology and Medicine* 16 (6): 845–850.

Harborne, J.B. and Williams, C.A. (2000). Advances in flavonoid research since 1992. *Phytochemistry* 55 (6): 481–504.

Heiss, C., et al (2003). Vascular effects of cocoa rich in flavan-3-ols. *JAMA* 290 (8): 1030–1031.

Heiss, C., et al. (2005). Acute consumption of flavanol-rich cocoa and the reversal of endothelial dysfunction in smokers. *Journal of the American College of Cardiology* 46(7): 1276–1283.

Hii, C. and Howell, S. (1985). Effects of flavonoids on insulin secretion and 45Ca2+ handling in rat islets of Langerhans. *Journal of Endocrinology* 107 (1): 1–8.

Hsu, Y.-L., et al. (2009). Tricetin, a dietary flavonoid, inhibits proliferation of human breast adenocarcinoma mcf-7 cells by blocking cell cycle progression and inducing apoptosis. *Journal of Agricultural and Food Chemistry* 57 (18): 8688–8695.

Hu, P., et al. (2008). Quercetin relieves chronic lead exposure-induced impairment of synaptic plasticity in rat dentate gyrus in vivo. *Naunyn-Schmiedeberg's Archives of Pharmacology* 378 (1): 43–51.

Huang, W.W., et al. (2010). Kaempferol induced apoptosis via endoplasmic reticulum stress and mitochondria-dependent pathway in human osteosarcoma U-2 OS cells. *Molecular Nutrition & Food Research* 54 (11): 1585–1595.

Hubbard, G., et al. (2004). Ingestion of quercetin inhibits platelet aggregation and essential components of the collagen-stimulated platelet activation pathway in humans. *Journal of Thrombosis and Haemostasis* 2 (12): 2138–2145.

Hung, J.Y., et al. (2015). Tricetin, a dietary flavonoid, suppresses benzo (a) pyrene-induced human non-small cell lung cancer bone metastasis. *International Journal of Oncology* 46 (5): 1985–1993.

Hussain, S., Mahwi, T., and Aziz, T. (2012). Quercetin dampens postprandial hyperglycemia in type 2 diabetic patients challenged with carbohydrates load. *International Journal of Diabetes Research* 1 (3): 32–35.

Hwang, S.J., et al. (2014). Anti-inflammatory effects of chlorogenic acid in lipopolysaccharide-stimulated RAW 264.7 cells. *Inflammation Research* 63 (1): 81–90.

Inanami, O., et al. (1998). Oral administration of (-) catechin protects against ischemia-reperfusion-induced neuronal death in the gerbil. *Free Radical Research* 29 (4): 359–365.

Ishige, K., Schubert, D., and Sagara, Y. (2001). Flavonoids protect neuronal cells from oxidative stress by three distinct mechanisms. *Free Radical Biology and Medicine* 30 (4): 433–446.

Ishikawa, Y. and Kitamura, M. (2000). Bioflavonoid quercetin inhibits mitosis and apoptosis of glomerular cells in vitro and in vivo. *Biochemical and Biophysical Research Communications* 279 (2): 629–634.

Jaganathan, S.K. and Mandal, M. (2009). Antiproliferative effects of honey and of its polyphenols: a review. *BioMed Research International* 2009.

Jakopic, J., et al. (2007). How much do cultivar and preparation time influence on phenolics content in walnut liqueur? *Food Chemistry* 104 (1): 100–105.

Jandera, P., et al. (2005). RP-HPLC analysis of phenolic compounds and flavonoids in beverages and plant extracts using a CoulArray detector. *Journal of Separation Science* 28 (9-10): 1005–1022.

Jang, S., Dilger, R.N., and Johnson, R.W. (2010). Luteolin inhibits microglia and alters hippocampal-dependent spatial working memory in aged mice. *The Journal of Nutrition* 140 (10): 1892–1898.

Jellinger, K. (2001). Cell death mechanisms in neurodegeneration. *Journal of Cellular and Molecular Medicine* 5 (1): 1–17.

Jin, B.-H., et al. (2009). Apigenin protects endothelium-dependent relaxation of rat aorta against oxidative stress. *European Journal of Pharmacology* 616(1): 200–205.

Jung, U.J., et al. (2004). The hypoglycemic effects of hesperidin and naringin are partly mediated by hepatic glucose-regulating enzymes in C57BL/KsJ-db/db mice. *The Journal of Nutrition* 134 (10): 2499–2503.

Kamalakkannan, N. and Prince, P.S.M. (2006). Antihyperglycaemic and antioxidant effect of rutin, a polyphenolic flavonoid, in streptozotocin-induced diabetic Wistar rats. *Basic & Clinical Pharmacology & Toxicology* 98 (1): 97–103.

Kamaruzzaman, M.A., et al. (2019). A review of potential beneficial effects of honey on bone health. *Evidence-Based Complementary and Alternative Medicine* 2019.

Katiyar, S.K., et al. (2001). Green tea polyphenol (–)-epigallocatechin-3-gallate treatment of human skin inhibits ultraviolet radiation-induced oxidative stress. *Carcinogenesis* 22 (2): 287–294.

Kelly, B.B. and Fuster, V. (2010). *Promoting Cardiovascular Health in the Developing World: A Critical Challenge to Achieve Global Health*. Washington, DC: National Academies Press.

Kenjerić, D., et al. (2007). Flavonoid profile of Robinia honeys produced in Croatia. *Food Chemistry* 102 (3): 683–690.

Kesh, S.B., et al. (2013). Promising role of ferulic acid, atorvastatin and their combination in ameliorating high fat diet-induced stress in mice. *Life Sciences* 92 (17): 938–949.

Khan, N., et al. (2006). Targeting multiple signaling pathways by green tea polyphenol (−)-epigallocatechin-3-gallate. *Cancer Research* 66 (5): 2500–2505.

Kim, -H.-H., Bae, Y., and Kim, S.-H. (2013). Galangin attenuates mast cell-mediated allergic inflammation. *Food and Chemical Toxicology* 57: 209–216.

Kim, J.-H., et al. (2005). Luteolin prevents PDGF-BB-induced proliferation of vascular smooth muscle cells by inhibition of PDGF β-receptor phosphorylation. *Biochemical Pharmacology* 69 (12): 1715–1721.

Kim, J.-S., Kwon, C.-S., and Son, K.H. (2000). Inhibition of alpha-glucosidase and amylase by luteolin, a flavonoid. *Bioscience, Biotechnology, and Biochemistry* 64 (11): 2458–2461.

Koh, P.-O. (2012). Ferulic acid prevents the cerebral ischemic injury-induced decrease of Akt and Bad phosphorylation. *Neuroscience Letters* 507 (2): 156–160.

Kong, L., et al. (2013). The anti-inflammatory effect of kaempferol on early atherosclerosis in high cholesterol fed rabbits. *Lipids in Health and Disease* 12 (1): 1.

Kostyuk, V.A., et al. (2011). Antioxidant and signal modulation properties of plant polyphenols in controlling vascular inflammation. *European Journal of Pharmacology* 658 (2): 248–256.

Kumar, A., et al. (2008). Protective effect of quercetin against ICV colchicine-induced cognitive dysfunctions and oxidative damage in rats. *Phytotherapy Research* 22 (12): 1563–1569.

Kumar, S., Prahalathan, P., and Raja, B. (2011). Antihypertensive and antioxidant potential of vanillic acid, a phenolic compound in L-NAME-induced hypertensive rats: a dose-dependence study. *Redox Report* 16 (5): 208–215.

Kuntz, S., Wenzel, U., and Daniel, H. (1999). Comparative analysis of the effects of flavonoids on proliferation, cytotoxicity, and apoptosis in human colon cancer cell lines. *European Journal of Nutrition* 38 (3): 133–142.

Kwon, S.-H., et al. (2010). Neuroprotective effects of chlorogenic acid on scopolamine-induced amnesia via anti-acetylcholinesterase and anti-oxidative activities in mice. *European Journal of Pharmacology* 649 (1): 210–217.

Lafay, et al. (2005). Caffeic acid inhibits oxidative stress and reduces hypercholesterolemia induced by iron overload in rats. *International Journal for Vitamin and Nutrition Research* 75 (2): 119–125.

Larocca, L.M., et al. (1995). Quercetin inhibits the growth of leukemic progenitors and induces the expression of transforming growth factor-beta 1 in these cells. *Blood* 85 (12): 3654–3661.

Lee, C.-C., et al. (2012). Fagopyrum tataricum (buckwheat) improved high-glucose-induced insulin resistance in mouse hepatocytes and diabetes in fructose-rich diet-induced mice. *Experimental Diabetes Research* 2012.

Lee, J.-H., et al. (2007). Anti-inflammatory mechanisms of apigenin: inhibition of cyclooxygenase-2 expression, adhesion of monocytes to human umbilical vein endothelial cells, and expression of cellular adhesion molecules. *Archives of Pharmacal Research* 30 (10): 1318–1327.

Li, S. and Pu, X.-P. (2011). Neuroprotective effect of kaempferol against a 1-methyl-4-phenyl-1, 2, 3, 6-tetrahydropyridine-induced mouse model of Parkinson's disease. *Biological and Pharmaceutical Bulletin* 34 (8): 1291–1296.

Li, Y., et al. (2008). Neuroprotective effects of chlorogenic acid against apoptosis of PC12 cells induced by methylmercury. *Environmental Toxicology and Pharmacology* 26 (1): 13–21.

Liu, H., et al. (2015). (2R, 3S)-Pinobanksin-3-cinnamate improves cognition and reduces oxidative stress in rats with vascular dementia. *Journal of Natural Medicines* 69 (3): 358–365.

Londesborough, J., et al. (2015). Analysis of beers from an 1840s' shipwreck. *Journal of Agricultural and Food Chemistry* 63 (9): 2525–2536.

Lungkaphin, A., et al. (2015). Pinocembrin reduces cardiac arrhythmia and infarct size in rats subjected to acute myocardial ischemia/reperfusion. *Applied Physiology, Nutrition, and Metabolism* 40 (10): 1031–1037.

Luo, Y., et al. (2002). Inhibition of amyloid-β aggregation and caspase-3 activation by the Ginkgo biloba extract EGb761. *Proceedings of the National Academy of Sciences* 99 (19): 12197–12202.

M Alvarez-Suarez, J., Giampieri, F., and Battino, M. (2013). Honey as a source of dietary antioxidants: structures, bioavailability and evidence of protective effects against human chronic diseases. *Current Medicinal Chemistry* 20 (5): 621–638.

Mahesh, T. and Menon, V.P. (2004). Quercetin alleviates oxidative stress in streptozotocin-induced diabetic rats. *Phytotherapy Research* 18 (2): 123–127.

Mantawy, E.M., et al. (2014). Chrysin alleviates acute doxorubicin cardiotoxicity in rats via suppression of oxidative stress, inflammation and apoptosis. *European Journal of Pharmacology* 728: 107–118.

Mantena, S.K., Baliga, M.S., and Katiyar, S.K. (2006). Grape seed proanthocyanidins induce apoptosis and inhibit metastasis of highly metastatic breast carcinoma cells. *Carcinogenesis* 27 (8): 1682–1691.

Mao, T., et al. (2002). Modulation of TNF-α secretion in peripheral blood mononuclear cells by cocoa flavanols and procyanidins. *Journal of Immunology Research* 9 (3): 135–141.

Mao, W., Schuler, M.A., and Berenbaum, M.R. (2013). Honey constituents up-regulate detoxification and immunity genes in the western honey bee Apis mellifera. *Proceedings of the National Academy of Sciences* 110 (22): 8842–8846.

Marles, R. and Farnsworth, N. (1995). Antidiabetic plants and their active constituents. *Phytomedicine* 2 (2): 137–189.

Mathur, S., et al. (2002). Cocoa products decrease low density lipoprotein oxidative susceptibility but do not affect biomarkers of inflammation in humans. *The Journal of Nutrition* 132 (12): 3663–3667.

McDougall, G.J., et al. (2005). Different polyphenolic components of soft fruits inhibit α-amylase and α-glucosidase. *Journal of Agricultural and Food Chemistry* 53 (7): 2760–2766.

McDougall, G.J. and Stewart, D. (2005). The inhibitory effects of berry polyphenols on digestive enzymes. *Biofactors* 23(4): 189–195.

Milner, J.A. (1994). *Reducing the Risk of Cancer, in Functional Foods*. Boston, MA: Springer. 39–70.

Min, S.-W., Ryu, S.-N., and Kim, D.-H. (2010). Anti-inflammatory effects of black rice, cyanidin-3-O-β-D-glycoside, and its metabolites, cyanidin and protocatechuic acid. *International Immunopharmacology* 10 (8): 959–966.

Molina, M.F., et al. (2003). Quercetin, a flavonoid antioxidant, prevents and protects against ethanol-induced oxidative stress in mouse liver. *Biological and Pharmaceutical Bulletin* 26 (10): 1398–1402.

Nabavi, S.F., et al. (2013). Hepatoprotective effect of gallic acid isolated from Peltiphyllum peltatum against sodium fluoride-induced oxidative stress. *Industrial Crops and Products* 44: 50–55.

Nabavi, S.F., et al. (2015). Role of quercetin as an alternative for obesity treatment: you are what you eat! *Food Chemistry* 179: 305–310.

Naczk, M. and Shahidi, F. (2004). Extraction and analysis of phenolics in food. *Journal of Chromatography A* 1054 (1): 95–111.

Nagai, T., et al. (2006). Characterization of honey from different floral sources. Its functional properties and effects of honey species on storage of meat. *Food Chemistry* 97 (2): 256–262.

Namsi, A., et al. (2018). Induction of neuronal differentiation of murine N2a cells by two polyphenols present in the Mediterranean diet mimicking neurotrophins activities: resveratrol and apigenin. *Diseases* 6 (3): 67.

Nanjundaiah, S.M., Annaiah, H.N.M., and Dharmesh, S.M. (2011). Gastroprotective effect of ginger rhizome (Zingiber officinale) extract: role of gallic acid and cinnamic acid in H. *Evidence-Based Complementary and Alternative Medicine* 2011.

Nishitani, Y., et al. (2013). Intestinal anti-inflammatory activity of luteolin: role of the aglycone in NF-κB inactivation in macrophages co-cultured with intestinal epithelial cells. *Biofactors* 39 (5): 522–533.

Nolan, V.C., Harrison, J., and Cox, J.A.G. (2019). Dissecting the antimicrobial composition of honey. *Antibiotics (Basel, Switzerland)* 8 (4): 251.

O'malley, T., et al. (1995). Platelet size in stroke patients. *Stroke* 26 (6): 995–999.

Ozguner, F., et al. (2005). Mobile phone-induced myocardial oxidative stress: protection by a novel antioxidant agent caffeic acid phenethyl ester. *Toxicology and Industrial Health* 21 (7–8): 223–230.

Panchal, S.K., Ward, L., and Brown, L. (2013). Ellagic acid attenuates high-carbohydrate, high-fat diet-induced metabolic syndrome in rats. *European Journal of Nutrition* 52 (2): 559–568.

Pandey, K.B., Mishra, N., and Rizvi, S.I. (2009). Protective role of myricetin on markers of oxidative stress in human erythrocytes subjected to oxidative stress. *Natural Product Communications* 4 (2): 221–226.

Panickar, K., Polansky, M., and Anderson, R. (2009). Myricetin and quercetin attenuate ischemic injury in glial cultures by different mechanisms. In: *Society for Neuroscience Abstracts and Proceedings*.

Park, C.H., et al. (2005). Quercetin, a potent inhibitor against β-catenin/Tcf signaling in SW480 colon cancer cells. *Biochemical and Biophysical Research Communications* 328 (1): 227–234.

Patel, D., et al. (2012). Diabetes mellitus: an overview on its pharmacological aspects and reported medicinal plants having antidiabetic activity. *Asian Pacific Journal of Tropical Biomedicine* 2 (5): 411–420.

Pauliuc, D., Dranca, F., and Oroian, M. (2020). Antioxidant activity, total phenolic content, individual phenolics and physicochemical parameters suitability for Romanian honey authentication. *Foods (Basel, Switzerland)* 9 (3): 306.

Prabhakar, P.K. and Doble, M. (2009). Synergistic effect of phytochemicals in combination with hypoglycemic drugs on glucose uptake in myotubes. *Phytomedicine* 16 (12): 1119–1126.

Promsan, S., et al. (2016). Pinocembrin attenuates gentamicin-induced nephrotoxicity in rats. *Canadian Journal of Physiology and Pharmacology* (ja) 94 (08) 808–818.

Pyrzynska, K. and Biesaga, M. (2009). Analysis of phenolic acids and flavonoids in honey. *TrAC Trends in Analytical Chemistry* 28 (7): 893–902.

Rein, D., et al. (2000). Epicatechin in human plasma: in vivo determination and effect of chocolate consumption on plasma oxidation status. *The Journal of Nutrition* 130 (8): 2109S–2114S.

Rein, D., et al. (2000). Cocoa inhibits platelet activation and function. *The American Journal of Clinical Nutrition* 72 (1): 30–35.

Rice-Evans, C.A., Miller, N.J., and Paganga, G. (1996). Structure-antioxidant activity relationships of flavonoids and phenolic acids. *Free Radical Biology and Medicine* 20 (7): 933–956.

Romero, M., et al. (2009). Quercetin inhibits vascular superoxide production induced by endothelin-1: role of NADPH oxidase, uncoupled eNOS and PKC. *Atherosclerosis* 202 (1): 58–67.

Rosado, T., et al. (2012). Methyl syringate: an efficient phenolic mediator for bacterial and fungal laccases. *Bioresource Technology* 124: 371–378.

Rouse, M., Younès, A., and Egan, J.M. (2014). Resveratrol and curcumin enhance pancreatic β-cell function by inhibiting phosphodiesterase activity. *Journal of Endocrinology* 223(2): 107–117.

Ruggeri, Z.M. (2002). Platelets in atherothrombosis. *Nature Medicine* 8 (11): 1227–1234.

Sabu, M., Smitha, K., and Kuttan, R. (2002). Anti-diabetic activity of green tea polyphenols and their role in reducing oxidative stress in experimental diabetes. *Journal of Ethnopharmacology* 83 (1): 109–116.

Saeed, S., et al. (2020). Antidotal or protective effects of honey and chrysin, its major polyphenols, against natural and chemical toxicities. *Acta Bio Medica: Atenei Parmensis* 90 (4): 533.

Samarghandian, S., Afshari, J.T., and Davoodi, S. (2011). Chrysin reduces proliferation and induces apoptosis in the human prostate cancer cell line pc-3. *Clinics* 66 (6): 1073–1079.

Samarghandian, S., Farkhondeh, T., and Samini, F. (2017). Honey and health: a review of recent clinical research. *Pharmacognosy Research* 9 (2): 121–127.

Samhan-Arias, A.K., Martín-Romero, F.J., and Gutiérrez-Merino, C. (2004). Kaempferol blocks oxidative stress in cerebellar granule cells and reveals a key role for reactive oxygen species production at the plasma membrane in the commitment to apoptosis. *Free Radical Biology and Medicine* 37 (1): 48–61.

Sanchez, M., et al. (2006). Quercetin downregulates NADPH oxidase, increases eNOS activity and prevents endothelial dysfunction in spontaneously hypertensive rats. *Journal of Hypertension* 24 (1): 75–84.

Scazzocchio, B., et al. (2011). Cyanidin-3-O-β-glucoside and protocatechuic acid exert insulin-like effects by upregulating PPARγ activity in human omental adipocytes. *Diabetes* 60 (9): 2234–2244.

Schramm, D.D., et al. (2003). Food effects on the absorption and pharmacokinetics of cocoa flavanols. *Life Sciences* 73 (7): 857–869.

Shakya, G., et al. (2014). Hepatoprotective role of kaempferol during alcohol- and DeltaPUFA-induced oxidative stress. *Journal of Basic and Clinical Physiology and Pharmacology* 25 (1): 73–79.

She, S., et al. (2019). Discrimination of geographical origins of Chinese acacia honey using complex 13C/12C, oligosaccharides and polyphenols. *Food Chemistry* 272: 580–585.

Siekmann, T., et al. (2013). The antiproliferative effect of pinostrobin on human umbilical vein endothelial cells (HUVEC). *European Review for Medical and Pharmacological Sciences* 17 (5): 668–672.

Solano, F. (2014). Melanins: skin pigments and much more—types, structural models, biological functions, and formation routes. *New Journal of Science* 2014.

Solayman, M., et al. (2015). Polyphenols: Potential Future Arsenals in the Treatment of Diabetes. *Current Pharmaceutical Design* 22 (5): 549–565.

Son, H.J., et al. (2012). Methyl syringate, a low-molecular-weight phenolic ester, as an activator of the chemosensory ion channel TRPA1. *Archives of Pharmacal Research* 35 (12): 2211–2218.

Spires, T.L. and Hannan, A.J. (2005). Nature, nurture and neurology: gene–environment interactions in neurodegenerative disease. *FEBS Journal* 272 (10): 2347–2361.

Stagos, D., et al. (2018). Antibacterial and antioxidant activity of different types of honey derived from Mount Olympus in Greece. *International Journal of Molecular Medicine* 42 (2): 726–734.

Stein, J.H., et al. (1999). Purple grape juice improves endothelial function and reduces the susceptibility of LDL cholesterol to oxidation in patients with coronary artery disease. *Circulation* 100 (10): 1050–1055.

Swartz, M.E. (2005). UPLC™: an introduction and review. *Journal of Liquid Chromatography & Related Technologies* 28 (7–8): 1253–1263.

Swellam, T., et al. (2003). Antineoplastic activity of honey in an experimental bladder cancer implantation model: in vivo and in vitro studies. *International Journal of Urology* 10 (4): 213–219.

Tadera, K., et al. (2006). Inhibition of. ALPHA.-Glucosidase and. ALPHA.-Amylase by Flavonoids. *Journal of Nutritional Science and Vitaminology* 52 (2): 149–153.

Taubert, D., et al. (2007). Effects of low habitual cocoa intake on blood pressure and bioactive nitric oxide: a randomized controlled trial. *Jama* 298 (1): 49–60.

Terzo, S., Mulè, F., and Amato, A. (2020). Honey and obesity-related dysfunctions: a summary on health benefits. *The Journal of Nutritional Biochemistry* 82 (2020): 108401.

Todd, F.E. and Vansell, G.H. (1942). Pollen grains in nectar and honey. *Journal of Economic Entomology* 35 (5): 728–731.

Tomás-Barberán, F.A., et al. (2001). HPLC flavonoid profiles as markers for the botanical origin of European unifloral honeys. *Journal of the Science of Food and Agriculture* 81 (5): 485–496.

Tundis, R., Loizzo, M., and Menichini, F. (2010). Natural products as α-amylase and α-glucosidase inhibitors and their hypoglycaemic potential in the treatment of diabetes: an update. *Mini Reviews in Medicinal Chemistry* 10 (4): 315–331.

Türk, G., et al. (2008). Improvement of cisplatin-induced injuries to sperm quality, the oxidant-antioxidant system, and the histologic structure of the rat testis by ellagic acid. *Fertility and Sterility* 89 (5): 1474–1481.

Ujiki, M.B., et al. (2006). Apigenin inhibits pancreatic cancer cell proliferation through G2/M cell cycle arrest. *Molecular Cancer* 5 (1): 1.

Ulbricht, T. and Southgate, D. (1991). Coronary heart disease: seven dietary factors. *The Lancet* 338 (8773): 985–992.

Umadevi, S., et al. (2012). Studies on the cardio protective role of gallic acid against age-induced cell proliferation and oxidative stress in H9C2 (2-1) cells. *Cardiovascular Toxicology* 12 (4): 304–311.

Urquiaga, I. and Leighton, F. (2000). Plant polyphenol antioxidants and oxidative stress. *Biological Research* 33 (2): 55–64.

Uzun, F.G. and Kalender, Y. (2013). Chlorpyrifos induced hepatotoxic and hematologic changes in rats: the role of quercetin and catechin. *Food and Chemical Toxicology* 55: 549–556.

Varma, S.D. and Kinoshita, J.H. (1976). Inhibition of lens aldose reductase by flavonoids—their possible role in the prevention of diabetic cataracts. *Biochemical Pharmacology* 25 (22): 2505–2513.

Vauzour, D., et al. (2007). Activation of pro-survival Akt and ERK1/2 signalling pathways underlie the anti-apoptotic effects of flavanones in cortical neurons. *Journal of Neurochemistry* 103 (4): 1355–1367.

Vessal, M., Hemmati, M., and Vasei, M. (2003). Antidiabetic effects of quercetin in streptozocin-induced diabetic rats. *Comparative Biochemistry and Physiology Part C: Toxicology & Pharmacology* 135 (3): 357–364.

Vijayababu, M., et al. (2005). Quercetin-induced growth inhibition and cell death in prostatic carcinoma cells (PC-3) are associated with increase in p21 and hypophosphorylated retinoblastoma proteins expression. *Journal of Cancer Research and Clinical Oncology* 131 (11): 765–771.

Wan, Y., et al. (2001). Effects of cocoa powder and dark chocolate on LDL oxidative susceptibility and prostaglandin concentrations in humans. *The American Journal of Clinical Nutrition* 74 (5): 596–602.

Wang, G., et al. (2012). Luteolin ameliorates cardiac failure in type I diabetic cardiomyopathy. *Journal of Diabetes and Its Complications* 26 (4): 259–265.

Wang, S., et al. (2004). Study of the electrophoretic behaviour of flavonoids. *Chromatographia* 59(7–8): 507–511.

Weng, M.-S., Ho, Y.-S., and Lin, J.-K. (2005). Chrysin induces G1 phase cell cycle arrest in C6 glioma cells through inducing p21 Waf1/Cip1 expression: involvement of p38 mitogen-activated protein kinase. *Biochemical Pharmacology* 69 (12): 1815–1827.

World Health Organization. (2007). Prevention of cardiovascular disease : guidelines for assessment and management of total cardiovascular risk. World Health Organization. https://apps.who.int/iris/handle/10665/43685

Xian, Y.-F., et al. (2012). Protective effects of pinostrobin on β-amyloid-induced neurotoxicity in PC12 Cells. *Cellular and Molecular Neurobiology* 32 (8): 1223–1230.

Xu, B., et al. (2010). Luteolin promotes long-term potentiation and improves cognitive functions in chronic cerebral hypoperfused rats. *European Journal of Pharmacology* 627 (1): 99–105.

Xu, D., et al. (2014). Tetrachlorobenzoquinone induces acute liver injury, up-regulates HO-1 and NQO1 expression in mice model: the protective role of chlorogenic acid. *Environmental Toxicology and Pharmacology* 37 (3): 1212–1220.

Yang, Z., et al. (2018). Chrysin attenuates carrageenan-induced pleurisy and lung injury via activation of SIRT1/NRF2 pathway in rats. *European Journal of Pharmacology* 836: 83–88.

Yao, L., et al. (2004). Quantitative high-performance liquid chromatography analyses of flavonoids in Australian Eucalyptus honeys. *Journal of Agricultural and Food Chemistry* 52 (2): 210–214.

Yordi, E.G., et al. (2012). *Antioxidant and Pro-oxidant Effects of Polyphenolic Compounds and Structure-activity Relationship Evidence*. 2 (2012): 23–48. INTECH Open Access Publisher.

Youdim, K.A. and Joseph, J.A. (2001). A possible emerging role of phytochemicals in improving age-related neurological dysfunctions: a multiplicity of effects. *Free Radical Biology and Medicine* 30 (6): 583–594.

Youl, E., et al. (2010). Quercetin potentiates insulin secretion and protects INS-1 pancreatic β-cells against oxidative damage via the ERK1/2 pathway. *British Journal of Pharmacology* 161 (4): 799–814.

Zarzuelo, A., et al. (1996). Effects of luteolin 5-O-β-rutinoside in streptozotocin-induced diabetic rats. *Life Sciences* 58 (25): 2311–2316.

Zhang, H., et al. (2013). Galangin inhibits proliferation of HepG2 cells by activating AMPK via increasing the AMP/TAN ratio in a LKB1-independent manner. *European Journal of Pharmacology* 718 (1): 235–244.

Zhao, H., et al. (2018). Hepatoprotective effects of the honey of apis cerana fabricius on bromobenzene-induced liver damage in mice. *Journal of Food Science* 83 (2): 509–516.

Zhou, L.-T., et al. (2015). Pinocembrin inhibits lipopolysaccharide-induced inflammatory mediators production in BV2 microglial cells through suppression of PI3K/Akt/NF-κB pathway. *European Journal of Pharmacology* 761: 211–216.

# 10

## Aroma Compounds

*Md. Mijanur Rahman, Nusrat Fatima, and Nur-E-Alam*

### Introduction

Honey is a natural sweetener produced from flower nectar by honeybees (*Apis mellifera*; family, Apidae). It is generally accepted that honey has been used since prehistoric time by human society. History tracing demonstrates that humans started using honey about 8000 years ago. Different communities used honey for distinct purposes (Pearce 2015). The ancient Assyrians and Chinese used honey for wounds, skin diseases, and cataracts. In Egypt, honey was offered to gods and was used as a balm for wounds. Honey has antique use as sweetener in unfermented grape juice beverages. Honey was also used as cosmetic ingredients in ancient Rome. Thus, we can conclude that honey is the only insect-derived natural product with therapeutic, religious, nutritional, and cosmetic value (Eteraf-Oskouei and Najafi 2013; Rahman et al. 2014).

Nutritionally, carbohydrates and proteins are the major components of honey (White 1980; White et al. 1962). Honey also contains every essential and nonessential amino acid except for glutamine and asparagine (National Nutrient Database for Standard Reference 2015). Most of the water-soluble vitamins are present in honey, and vitamin C is the most abundant among them. Honey contains about 31 different minerals, including the major minerals such as calcium, phosphorus, potassium, sulfur, sodium, chlorine, and magnesium. Although honey contains a substantial amount of several essential trace elements to partially meet the recommended dietary intake for children, honey is considered a good source of potassium for adults (Rahman et al. 2014).

Recently, honey has drawn significant attention from the scientific community because of its medicinal value. Besides the traditional use of honey in therapeutic treatment, honey has reported to have antiseptic, antimicrobial, antiviral, anti-inflammatory, antioxidant, wound-healing, antipyretic, antifungal, cardioprotective, diabetic-ameliorating, and memory-ameliorating effects (Eteraf-Oskouei and Najafi 2013; Jeffrey and Echazarreta 1996). Many of these effects have been attributed to the physiologically active secondary metabolites available in honey such as polyphenols, alkaloids, anthraquinone glycosides, cardiac glycosides, flavonoids, and reducing compounds (Bogdanov et al. 2008; Helmy and El-Soud 2012; Islam et al. 2012; Montenegro et al. 2009a, 2009b). Currently, a good number of research studies are going on to further address the medicinal impact of honey.

Approximately 300 varieties of honey have been identified (The National Honey Board 2015) so far. Different varieties of honey possess distinct tastes, smells, and physiochemical properties. The characteristic aroma of honey is attributed to the volatile organic compounds present in it. More than 600 volatile compounds of different chemical classes have been identified in honey (Castro-Várquez et al. 2003). At the beginning of this chapter, we will discuss the chemical classes of aroma compounds and their occurrence in honey. Not all volatile chemicals impart aroma. The aroma profile of the volatile organic compound present in honey holds special importance. Besides ensuring organoleptic characterization, aroma compounds in honey have been implicated in honey's identity, quality, bioactivity, and toxicity. This chapter elaborates on these attributes of honey's aroma compounds.

### Classification of the Aroma Compounds of Honey

Volatile and semi-volatile compounds are primarily responsible for the fragrance of honey. The color of honey depends on the chemicals present in the honey and reflects the aroma quality of honey. In general, bright-colored honey is mild in flavor, and darker honey has a stronger flavor (Honey 2003; Kaskoniene et al. 2008). Honey's aroma also varies largely

*Honey: Composition and Health Benefits,* First Edition. Edited by Md. Ibrahim Khalil, Gan Siew Hua, and Bey Hing Goh.

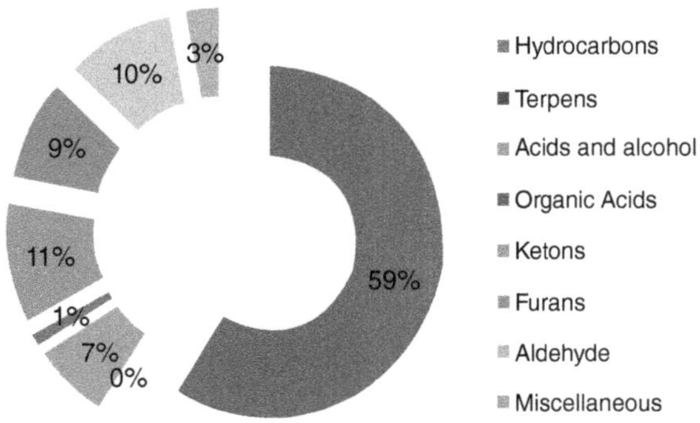


**Figure 10.1** Volatile compounds present in honey (Syazana et al. 2013).

depending on its nectar source. When nectar is collected from only a single floral source, the aroma of the honey remains nearly unique and highly distinct. On the other hand, multifloral honey is relatively less flavorful (Guyot et al. 1999; Jerkovic and Marijanovic 2010).

What is the source of the aroma compounds of honey? If we focus on the conversion process of nectar to honey, it becomes clear that the plant or nectar source is also the source of the volatile organic compounds present in honey. Some of these phytochemicals remain unchanged, escaping enzymatic digestion in the bees' stomach during honey ripening. Some others are changed by the metabolism of the bees, heating, or handling during honey processing and storage or by microbial and environmental contamination (Guyot-Declerck et al. 2002; Jerkovic et al. 2006; Jerkovic and Marijanovic 2009).

In general, honey's volatile organic compounds belong to one or more chemical classes, including hydrocarbon; aldehyde; alcohol; ketone; acid and ester; benzene and its derivatives, furan and pyran; terpenes and their derivatives; and sulfur as well as cyclic compounds (Wolski et al. 2006). The volatile aromatic compounds with their aroma profiles are summarized in Table 10.1. The characteristic honey aroma may result from a single prominent aroma compound present in honey. Sometimes honey's aroma result from multiple volatile organic compounds. Some volatile compounds are common to a group of honey. To date, a good number of studies have been carried out to detect the volatile compounds of honey. Not all volatile compounds have a significant impact on honey's aroma.

What type of volatile aromatic compound has major contribution to the aroma of honey? This question is very baffling. The extent of different classes of volatile compounds present in honey is relatively uncertain. Some other studies have estimated a rough and relative amount of hydrocarbons, terpenes, acids and alcohol, organic acids, furans, ketones, aldehyde, and miscellaneous volatile compounds in Malaysian Tualang honey. Although not comprehensive, the study provided an assumption of the relative amounts of distinct classes of volatile compound present in honey (Figure 10.1). According to this study, hydrocarbons are the major volatile compounds present in honey. Interestingly, aroma profiles of hydrocarbons match less with the aroma of honey. In most of the cases, other volatile compound classes play a significant role in the aroma of honey. Even some odorless compounds occur in honey with high frequency and in relatively large concentrations.

## Implications of the Presence of Aroma Compounds in Honey

### Tracing the Floral Origin of Honey

The floral origin of honey is an important characteristic in the evaluation of honey's quality. Indeed, the consumer value of a honey variety largely depends on its organoleptic characteristics, which are strongly dependent on the botanical origin of honey (Baroni et al. 2006). Generally, a unifloral honey contains highly characteristic aroma profiles compared with polyfloral honey. A number of studies have reported the same volatile compounds in honey, in its precursor nectar, and in the flowers from which honey is collected (Kaškonienė and Venskutonis 2010).

The microscopic examination of pollen present in honey is the first study used to determine the botanical source of honey (Alissandrakis et al. 2003). This method is specifically popular because of the simplicity of its operational procedure. However, melissopalinology is a relatively old time-consuming procedure with its more than 30-year-old specification limits. Precise melissopalinologycal method performance requires highly specialized personnel and a pollen library. Furthermore, pollen analysis may be misleading for the honey collected from sterile plants (Manyi-Loh et al. 2011). Thus, melissopalinology has failed to be a standard method for the botanic origin determination of honey in this molecular arena of science.

A good number of studies have demonstrated that honey from the different botanical sources usually has distinct aroma profiles (especially applicable for unifloral honeys). Some honey types may be distinguished by a single floral or

**Table 10.1** Classes of honey aromatic compounds.

| Chemical Class | | Aroma Profile |
|---|---|---|
| 1 | **Hydrocarbons** (Cajka et al. 2009; Silici 2011; Wolski et al. 2006) | |
| | Hexane | Gasoline-like odor |
| | Octane | Gasoline-like odor |
| | Cyclohexane | Mild sweet odor |
| | Isooctane | Gasoline-like odor |
| | n-Decanal | Citrus odor |
| | Nonanal | Orange-rose odor |
| | Cyclotetrasiloxane | – |
| | Benzene | Gasoline-like odor |
| | Alpha-pinene | Odor of pine |
| | Cyclohexasiloxane | – |
| | n-Undecane | – |
| | n-Tetracosane | – |
| | Hexanal | Strong green grass odor |
| | Heptanal | Fatty, pungent odor |
| | Octanal | Strong, fruity odor |
| 2 | **Acids and Esters** (Cajka et al. 2009; Silici 2011; Wolski et al. 2006) | |
| | Nonanoic acid | Coconut aroma |
| | Propanoic acid | Slightly pungent disagreeable, rancid odor |
| | Hexadecanoic acid | – |
| | Hexyl butyrate | – |
| | Butyl butyrate | – |
| | Butyric acid | Unpleasant rancid odor |
| | Benzoic acid | Odorless or with a slight benzaldehyde odor |
| | Methyl salicylate | Wintergreen-like odor |
| | Propanoic acid | Slightly pungent disagreeable, rancid odor |
| | Phosphoric acid | Odorless |
| | Fumaric acid | Odorless |
| | Pentanoic acid | Unpleasant odor |
| | Methyl dihydrojasmonate | – |
| | Benzeneacetic acid | Floral odor |
| | Hexanedioic acid | Odorless |
| | Thiocyanic acid | Odorless |
| | Ethyl octanoate | – |
| | Ethyl nonanoate | – |
| | Ethyl decanoate | – |
| | Ethyl heptanoate | – |
| | Butanoic acid | Unpleasant rancid odor |
| | Formic acid | Pungent, penetrating odor |
| | Acetic acid | Sour, vinegar-like odor |
| | 2-Methyl propanoic acid | Sharp, butter fat–like odor |
| | Propanoic acid | Slightly pungent, disagreeable, rancid odor |

*(Continued)*

**Table 10.1** (Continued)

| Chemical Class | | Aroma Profile |
|---|---|---|
| | 2-Ethyl-hexanoic acid | Mild odor |
| | 3-Methyl butanoic acid | Disagreeable rancid-cheese odor |
| | 2-Methyl butanoic acid | – |
| 3 | **Aldehydes** (Cajka et al. 2009; Silici 2011; Wolski et al. 2006) | |
| | Benzene acetaldehyde | – |
| | Butanal | Characteristic, pungent, aldehyde odor |
| | Succinaldehyde | – |
| | 3-Cyclohexene-1-acetaldehyde | – |
| | Benzaldehyde | Characteristic odor or volatile oil of almond |
| | Lilac aldehyde II | – |
| | Lilac aldehyde III | – |
| | 2-Phenylacetaldehyde | – |
| | Lilac aldehyde I | – |
| | Nonanal | Orange-rose odor |
| | Decanal | Citrus odor |
| | Benzene acetaldehyde | Gasoline-like odor |
| 4 | **Ketones** (Cajka et al. 2009; Silici 2011; Wolski et al. 2006) | |
| | 2-Cyclopentanone | Distinctive ethereal odor, somewhat like peppermint |
| | Acetone | Fruity odor |
| | 2,5-Furandione | Irritating, choking odor |
| | 1-Phenylethanone | Orange blossom or jasmine-like |
| | 2,3-Butanodione | Odorless |
| | 1-Hydroxy-2-propanone | – |
| | 3-Hydroxy-2-butanone | Bland, woody, yogurt odor |
| | 3,4-Dimethyl-3-hexen-2-one | – |
| | 6-Methyl-5-octen-2-one | – |
| | Benzyl methyl ketone | – |
| 5 | **Benzene Derivatives** (Barra et al. 2010; Cajka et al. 2009; Wolski et al. 2006) | |
| | Toluene | Sweet, pungent, benzene-like odor |
| | Vinylbenzene | Sweet, floral odor |
| | 1,2-Dimethyl benzene | Sweet |
| | 1,3-Dimethyl benzene | Sweet odor; like benzene; characteristically aromatic |
| | 1,3-Benzenediamine | – |
| | 1-Methyl-4-propan-2-yl-benzene | Sweetish aromatic odor |
| 6 | **Terpenes and Their Derivatives** (Montenegro et al. 2009a, 2009b) | |
| | Epoxylinalool | – |
| | Linalool | Floral, spicy, wood odor |
| | 1-Octen-7-methylocta-1,3(Z),5(E) triene | – |
| | α-Felandrene | – |
| | Iso terpinolene | – |
| | α-Pinene | Odor of turpentine |
| | β-Pinene | Piney, turpentine-like odor |

**Table 10.1** (Continued)

| Chemical Class | | Aroma Profile |
|---|---|---|
| | Limonene | Odor of orange peels (orange oil sweet) |
| | Sabinene | – |
| | Camphene | Camphor-like odor |
| | linalool oxide | – |
| | Linalool | Floral, spicy, wood odor |
| | Hotrienol | – |
| | Myrtenol | – |
| | Neryl acetone | – |
| | Eucarvone | – |
| | lilac alcohol | – |
| | Trans anethol | Anise oil odor |
| | 8-Hydroxy-6,7-dihydrolinalool | – |
| | p-Cymene | Sweetish aromatic odor |
| | Cis rose oxide | – |
| | Nerol | Sweet rose odor |
| | α-Calacorene | – |
| | Cumene | Gasoline-like odor |
| | Cadalene | – |
| 7 | **Furan Derivatives** (Barra et al. 2010; Cajka et al. 2009; Odeh et al. 2013; Silici 2011; Wolski et al. 2006) | |
| | Furfuryl alcohol | Faint burning odor |
| | 5-Hydroxymethylfurfural | Odor of chamomile flowers |
| | Furfural | Peculiar odor resembling the odor of benzaldehyde |
| | 2-Furancarboxaldehyde | Almond-like odor |
| | 2-Furanomethanol | – |
| | 2-Furanmethanol | Faint burning odor |
| | Butyrolactone | Pleasant odor |
| | Valerolactone | – |
| 8 | **Alcohols** (Barra et al. 2010; Cajka et al. 2009; Odeh et al. 2013; Silici 2011; Wolski et al. 2006) | |
| | Phenylethyl alcohol | Characteristic rose-like odor |
| | Ethanol | Mild, rather pleasant; like wine or whiskey |
| | 1-Butanol | Harsh fusel odor with banana |
| | Methanol | Characteristic pungent odor |
| | 1-Propanol | Similar to ethanol |
| | 1,3-Butanediol | Odorless when pure |
| | Oxiranemethanol | – |
| | 1-(1-Cyclopentenyl)-propanol | – |
| | 2-Methyl-1-propanol | Sweet, musty odor |
| | 1-Butanol | Harsh fusel odor with banana |
| | 3-Methyl-3-buten-1-ol | – |
| | 2-Methyl-1-butanol | – |
| | 1,3-Propanodiol | – |
| | 2,3-Butanodiol | Odorless |

*(Continued)*

Table 10.1 (Continued)

| Chemical Class | | Aroma Profile |
|---|---|---|
| | 2-Furanmethanol | Faint burning odor |
| | Benzyl alcohol | Faint aromatic odor |
| | Hotrienol | – |
| | Linalool | Floral, spicy, wood odor |
| | Isoborneol | Pungent camphor-like |
| | Nonanol | Floral odor |
| | 4-Hexen-3-ol | – |
| | 1-Penten-3-ol | – |
| | Thymol or carvacrol | Odor reminiscent of thyme |
| | Hexan-1-ol | Characteristic; sweet alcohol; pleasant |
| | Heptan-1-ol | Faint, aromatic, fatty |
| | 2-Phenylethanol | Characteristic rose-like odor |
| | Nonan-1-ol | Floral odor |
| | Decan-1-ol | Resembles orange flowers |
| | Octan-1-ol | Fresh, orange-rose odor |
| | p-Cresol | Phenolic odor |
| | 2,3-Butanediol | Odorless |
| | (Benzenol) or phenol | Somewhat sickening sweet and acrid odor |
| | (α-Phenethyl alcohol) | Mild floral odor |
| | Acetophenone | Sweet, pungent odor of acacia |
| | Lilac alcohol | – |
| 9 | **Pyran Derivatives** (Barra et al. 2010; Wolski et al. 2006) | |
| | 2,3-Dihydro-3,5-dihydroxyl-6-methyl-4H-pyran-4-one | – |
| | Valerolactone | – |
| 10 | **Others** (Silici 2011; Wolski et al. 2006) | |
| | Thiophene | Slight aromatic odor resembling benzene |
| | Methanamine hydrochloride | – |
| | N-Methyl-D3-arizidine | – |
| | Propanenitrile | Pleasant, sweetish, ethereal odor |
| | 2(3H)-Benzothiazolone | – |
| | 1,3-Benzenediamine | – |
| | 2-Gydroxymethylbenzimidazole | – |
| | Acetamide | Odorless when pure but frequently has a mousy odor |
| | Guanidine | – |
| | Dibuthyl phythalat | – |

phytochemical marker. For example, methyl anthranilate indicates citrus honey, and isophorone is the marker of strawberry honey (Kaškonienė and Venskutonis 2010). In the case of multifloral honey, a diverse and high number of volatile organic compounds indeed give a distinct profile that could represent the fingerprint of each honey type (Guyot et al. 1999). However, there is disagreement over which compounds could serve as floral markers for a given honey type because differences exist between plant varieties, geographical origins, and beekeeping practices (Castro-Várquez et al. 2006).

**Table 10.2** Main marker aroma compounds assumed to the possible indicator of respective honeys.

| Botanic Source | Floral Marker | Reference |
|---|---|---|
| Thyme | 1-Phenyl-2,3-butanedione, 3-hydroxy-4-phenyl-2-butanone, 3-hydroxy-1-phenyl-2-butanone, phenylacetonitrile, and carvacrol | Alissandrakis et al. 2007 |
| Oak honeydew | Trans-oak lactone, aminoacetophenone, propylanisol | Castro-Várquez et al. 2006 |
| *Tolpis virgata* | 3,5-Dihydroxytoluene and tridecane | Odeh et al. 2007 |
| *Thymus capitatus* | 1,3-Diphenyl-2-propanone, (3-methylbutyl) benzene, 3,4,5-trimethoxybenzaldehyde, and 3 4-dimethoxybenzaldehyde | |
| *Thymalae hirsuta* | Benzene propanol, benzylalcohol, nonanal, hexanol and 4-methoxyphenol | |
| Acacia | cis-Linalool oxide and heptanal | Wardencki et al. 2009 |
| Buckwheat | Pentanal, furfural, and 2-ethylhexanol | |
| Lime | *p*-Methyl acetophenone | |
| Honeydew | Methyl butanal and lilac aldehyde | |
| Rape | p-Cymene | |
| Citrus | Methyl anthranilate and limonene diol | Piasenzotto et al. 2003 |
| Lime tree | p-Methyl acetophenone, carvacrol, and 8-p-menthen-1,2-diol | |
| Thyme | Ethenyl phenyl acetate | |
| Dandelion | Phenyl acetonitrile | |
| Chestnut | Aminoacetophenone | |
| Eucalyptus | Nonanoic acid and acetoin | |
| *Salvia officinalis L.* | Benzoic acid and phenyl acetic acid | Jerkovic et al. 2006 |
| *Calluna vulgaris* | Phenyl acetic acid and dehydrovomifoliol | Montenegro et al. 2009a, 2009b |
| *Erica arborea* | 4-Methoxybenzaldehyde, 4-methoxybenzoic acid, and methyl vanillate | |
| *Quillaja saponaria* | Megastigmatrienone, 2-p-hydroxyphenyl alcohol, β-pineme, and linalool oxide | Guyot et al. 1999 |
| *Escallonia pulverulenta* | Safranal | |
| *Eucryphia cordifolia* | Isophorone and cetosiophorone | |
| Eucalyptus | 3-Caren-2-ol, p-cymene and its derivate alcohol | Castro-Várquez et al. 2009 |
| Citrus | Linalool oxide, lilac alcohol, and lilac aldehyde | |
| Lavender | Nerolidol oxide | |
| Rhododendron | n-Decane, lilac aldehyde, 2-aminoaceto phenone, benzenedicarboxylic, nonanal, isobutylphthalate, and damascenone | Senyuva et al. 2009 |
| Christ's thorn | Nonanal, four isomers of lilac aldehyde, decanal, methyl nonanoate, hexanoic and 2-ethylhexanoic acids | Jerković et al. 2009 |
| Blue gum | 8,9-Dehydrotheaspirone and 3-oxo-α-ionone | D'arcy et al. 1997 |
| Sunflower | α-Pinene or 3-methyl-2-butanol | Radovic et al. 2001 |
| Willow | Methyl salycilate | De La Fuente et al. 2007 |
| Rosemary | Lilac aldehyde isomers | Castro-Várquez et al. 2009 |
| Marmeleiro | Isovaleric acid, γ-decalactone, benzoic acid, and vanillin | Moreira et al. 2002 |
| Turkish pine | 3-Carene and unidentified compound | Tananaki et al. 2007 |

Additionally, methods of extracting the volatile fractions may display a degree of selectivity and effectiveness depending on the compounds involved (Castro-Várquez et al. 2003). Whatever the deviation, the aroma profile of honey is now considered one of the main factors for differentiating honeys from different botanical origin. The main marker aroma compounds assumed to be the possible indicators of respective honeys are listed in Table 10.2.

## Aroma in Tracing the Geographical Origin of Honey

The aroma profile of the honey can also be optimized in the determination of the geographic origin of honey. With geographic change, climatic changes occur. Several aromatic components of honey show seasonal variations. Honey of the same floral source can vary because of seasonal climatic changes or a different geographical origin (Castro-Várquez et al. 2010). Indeed, accumulation of phytochemicals is dependent on climatic conditions, soil characteristics, and other factors. Therefore, it seems reasonable that differences between honeys obtained from different countries results from different compositions of nectar as well as the aroma composition of honey (Kaškonienė and Venskutonis 2010).

A number of studies suggest a finer differentiation could be possible to trace the geographic origin of honey by aroma compound analysis. In a recent study by Giovanna Aronne et al. (2014), the researchers selected four samples of *Robinia* honey from different geographical locations (named samples from sites 1, 2, 3, and 4) of the Italian peninsula and analyzed the organic aroma components present. The study results reveal that a panel of volatile components was present with distinct geographic specificity. For example, 1-decanol, 2-heptanone, and dihydro-alpha-terpinyl acetate were only present in the sample from site 1; 2-methylbutanal, 1-decanol, furan-2,3-dihydro-4-methyl, and nonane were only present in the sample from site 2; hexanal-3-methyl, 1-heptanolwerw were present in the sample from site 3; and myrtenal and β-pinene were present in the sample from site 4 (Aronne et al. 2014). Another study by Kádár et al. (2010). also showed from aroma compound analysis of acacia honey (botanic source, *Robinia pseudoacacia*) that compounds such as ethyl acetate and 2-propanol were exclusively present in honey from Romania. Likewise, 2-methylpropanol and acetic acid buthyl ester were present only in acacia honey from the Czech Republic, but 2-methyl propanoic acid and 2-methyl-2-butenal were present in the Spanish acacia honey. It was suggested that English honey samples can be distinguished by the presence of 1-penten-3-ol and honey from Denmark by the absence of 3-methylbutanal. 2,2,6-Trimethylcyclohexanone, ethyl-2-hydroxypropanoate, and 3-hexenyl-formate have been suggested as possible marker compounds for honey from Portugal.

The aroma and aroma-active compounds of astragalus and citrus honeys were determined by gas chromatography–mass spectrometry (GC-MS) olfactometry. The amount of aroma compounds of citrus honey (14391.6 μg/kg) was higher than astragalus honey (4270.1 μg/kg). Terpenes and acids in citrus honey and esters and aldehydes in astragalus honey were the most dominant aroma compounds (Uckun and Selli 2017).

A study reported that based on GC-MS analysis, a total of 55 volatile compounds with different species and concentrations were identified in the five honey samples, indicating that the flavor qualities of these samples clearly differed according to their flower sources (Tian et al. 2018).

In a recent study, the botanical origins and volatile profiles of 10 off-odor honeys (H1-–10) and flowers of *Banksia verticillata* were evaluated. There was no quantitative correlation between the *B. verticillata* pollen content and the off-odor attributes of the honeys. Skatole was identified in all the honeys except H4, H5, and the control honeys, suggesting that skatole contributed to the off-odor attributes of the products (Carlos et al. 2021).

## Aroma Compounds in Honey Quality Issue

Honey is a natural sweetener with several quality factors that ensure its purity, safety, and taste. According to the Codex Alimentarius, there are three quality factors of honey. First, honey should be free from any objectionable matter, flavor, and aroma. Second, it should not begin to ferment or effervesce. Third, honey should not be processed to such an extent that it changes the essential composition or quality of honey. How can one confirm these quality factors in routine laboratory testing? Determination of carbohydrates, physiochemical parameters, and pollen analysis are some of the ways to confirm the quality of honey. Interestingly, honey's aroma compounds are now also being used to indicate the quality of types of honey.

The water content of honey harvested in countries with a moderate climate is below 18%. When nectar is harvested before proper ripening and maturation, the water content may easily exceed 25%. This "green" honey ferments easily, which results in production of an "off" taste and high levels of dead yeast, glycerol and butanediol (Huidobro et al. 1993), and ethanol (Huidobro et al. 1994), causing a significant loss of aroma compounds. Fresh honey has a very low hydroxymethylfurfural (HMF) level, not exceeding 25 mg/kg. When honey is heated, the fructose is dehydrated to produce ionic HMF, leading to a consequential increase in the HMF levels in the honey.

## Biological Significance: Physiological Activity and Toxicity

The honey aromatic compounds include both physiological active compounds as well as toxic compounds. Some data are present in support of their physiological or toxic effects. However, a number of supplementary studies clearly support the biological significance of honey's aromatic compounds.

### Hydrocarbons

Hydrocarbons are organic compounds that contain only carbon and hydrogen. Except for several compound, hydrocarbons are less significant to contribute the aroma property of honey. They are more important in terms of toxicologic considerations. Hexane is the highly toxic alkane member and is reported to induce liver tumors in laboratory mice upon inhalation (Daughtrey et al. 1999). N-hexane also causes peripheral polyneuropathy and testicular atrophy. The mechanism of hexane toxicity is due to the haxen metabolite 2,5-hexanedione, which produces gonadal toxicity by altering testicular tubulin. 2,5-Hexanedione collapses the cytoskeletal network by decreasing phosphorylation of neurofilaments. When neurofilament proteins accumulates and crosslinks in the axon, they produce the giant axonal swelling that characterizes this axonopathy (Lopachin et al. 2002). Inhalation of cyclohexane causes developmental and reproductive toxicities (Kreckmann et al. 2000). Long-term exposure of toluene has a significant neurotoxic impact along with adverse behavioral outcomes. Prolonged toluene exposure causes upregulation of γ-aminobutyric acid (GABA) receptor in the cerebellar cortex (Miyagi et al. 1999). Additionally, low-level toluene exposure may cause altered differentiation of astrocyte precursor cells (Yamaguchi et al. 2002). Benzene is an occupational toxicant and an environmental pollutant that increases the risk of cancer and other illnesses. Hematotoxicity and hematological malignancies that benzene is associated with include acute myeloid leukemia, aplastic anemia, myelodysplastic syndrome, acute lymphoblastic leukemia, and chronic myeloid leukemia. However, cyclotetrasiloxane is a beneficial hydrocarbon present in honey. It is a well-known adjuvant and is used as a skin- and hair-conditioning agent.

### Acids and Esters

Acids and esters are another group of volatile organic compounds that have a significant contribution in the aroma of honey. Benzoic acid is an antifungal agent that inhibits the growth of mold, yeast, and some bacteria (Warth 1991). The antifungal activity of benzoic acid relies on its acidifying capacity. Upon absorption into cells, benzoic acid lowers intracellular pH. When intracellular pH falls to 5 or lower, the anaerobic fermentation of glucose decreased drastically by approximately 90% (Pastrorova et al. 1997). Methyl salicylate is a well-known honey ester that is widely used to treat muscle pain, backache, and pain caused by sciatica or other rheumatic conditions. It has topical anti-inflammatory and cooling actions. Butyric acid is a volatile fatty acid that inhibits colonic tumors and inflammation-mediated ulcerative colitis (Vanhoutvin et al. 2009). Indeed, butyrate possesses both preventive and therapeutic potential to counteract inflammation-mediated ulcerative colitis and colorectal cancer by inhibition of interferon-γ–STAT1 (signal transducer and activator of transcription) signaling, a pathway responsible for chronic inflammation (Zimmerman et al. 2012). Propionic acid is another honey acid that functions as a fungicide and bactericide. Propionic acid inhibits the growth of mold and some bacteria (Haque et al. 2009). Besides propionic acid, formic acid was also found to act as a disinfectant. Fumaric acid is an approved food additive and has antimicrobial effects. Both of propionic acid and formic acid have antibacterial activity against *Salmonella typhimurium* and *Listeria monocytogenes* (Raftari et al. 2012).

### Aldehydes

Aldehydes are an important class of biomolecules present in different metabolic pathways as precursors of by-products. The involvement of honey aldehydes in physiological function has been studied little to date, but they are considered significant in honey's fragrance.

Phenylacetaldehyde is an aldehyde with a natural source of flowers and imparts the characteristic aroma of honey, which include sweet, rose, green, and grassy aromas. Recently, Zhu et al. (2011) have demonstrated that phenylacetaldehyde possesses antibacterial activities. Phenylacetaldehyde is also responsible for the antibiotic activity of maggot therapy.

### Ketones

Ketones are another class of volatile compounds responsible for honey's aroma. Besides their fragrance properties, ketones may be biologically active or toxic compounds. For example, cyclopentanone is an innocuous fragrance, but at a high concentration, cyclopentanone is a central nervous system (CNS) depressant. 1-Phenylethanone is the simplest aromatic ketone that naturally occurs in many foods, including apples, cheeses, apricots, bananas, beef, and cauliflower. Acetophenone and its derivatives are known to have hypnotic and anticonvulsant activities (Sameem et al. 2012). 3,5,5-Trimethylcyclohex-2-en-1-one (isophorone) is an important aroma compound in honey that imports a camphor-like odor. Unfortunately, isophorone has been graded as a possible human carcinogen. However, an animal study demonstrated limited evidence of carcinogenicity with isophorone, but it increased preputial gland carcinomas in male rats (Integrated Risk Information System (IRIS) 2000). Acetoin is another honey ketone with pronounced aroma quality. Acetoin is naturally found in apples, butter, yogurt, asparagus, black currants, blackberries, wheat, broccoli, brussels sprouts, and cantaloupe. Acetoin has a limited toxic effect, though it is not serious. A laboratory study using rats showed that acetoin exposure causes slight anemia (Gaunt et al. 1972).

#### Benzene Derivatives

Normally, benzene derivatives are sweet in smell. They add considerable aroma to honey. At low concentrations, toluene interferes with the dopaminergic mechanisms of the basal ganglia and causes consequential functional changes in sensory–motor integration (Fuxe et al. 1982). Biochemically, toluene exposure augments glutamine synthetase activity in the cerebellar hemisphere. Glutamine synthetase is a glial enzyme, and glutamine synthetase's upregulation indicates glial proliferation, which is frequently associated with CNS damage (Bjornaes and Naalsund 1988). Vinylbenzene is another benzene derivative that causes toxicity in both the lung and liver. However, the exact mechanism of vinylbenzene's toxicity is not completely clear. An experimental study suggests that vinylbenzene interacts with epithelial cells, which stimulates the expression of a variety of inflammatory mediators, including the chemotactic cytokine monocyte chemoattractant protein-1 (Röder-Stolinski et al. 2008). 1-Methyl-4-propan-2-yl-benzene (p-cymene) is a physiologically active benzene derivative present in honey. It halts lipopolysaccharide (LPS)-induced acute lung injury. p-Cymene treatment significantly reduces the pro-inflammatory cytokines (tumor necrosis factor-α, interleukin-1β, and interleukin-6), lung water gain, inflammatory cell infiltration, and lung tissue myeloperoxidase activity (Xie et al. 2012). In addition, p-cymene has a potential anti-inflammatory action on cytokine production by blocking nuclear factor-kβ and (MAPK) signaling pathways (Zhong et al. 2013).

#### Terpenes and Their Derivatives

Terpenes are a diverse class of organic compounds naturally synthesized by plants and are probably the most widespread group of natural products. In terms of physiological function, terpenes are important volatile compounds present in honey. Terpene's aroma compound is known to be active against a wide range of micro-organisms, including gram-negative and -positive bacteria, viruses, and fungi. Linalool is a naturally occurring monoterpene found in many flowers and spice plants predominately as (-)-linalool enantiomer. The racemate form of linalool has effect on the brain neurotransmitters glutamic acid, GABA, acetylcholine, and dopamine. Studies recently performed by our research group have shown that (-)-linalool possesses anti-inflammatory, antinociceptive, and antihyperalgesic activities in different animal models (Peana et al. 2002; Peana and Moretti 2008).

D-limonene possesses chemopreventive and chemotherapeutic effects against liver, kidney, and gastric tumors in animal models. It is also potential anti-inflammatory effects. An in vitro study showed that D-limonene suppresses LPS-induced production of nitric oxide, prostaglandin E2, and pro-inflammatory cytokines (Yoon et al. 2010).

Two naturally occurring pinene structural isomers, α-pinene and β-pinene, are found in honey. Pinene has been reported to have significant antimicrobial and cytotoxic activities Pinenes have isomer- and enantiomer-selective anti-inflammatory effects, with (+)-α-pinene being a promising antiosteoarthritic drug (Rufino et al. 2014). Recently, a study by Chen et al. (2015) demonstrated α-pinene's inhibitory effect on hepatoma carcinoma BEL-7402 cells. Myrtenol has other interesting properties such as antifungal, antimicrobial, anticancer, antihyperglycemic, and hypotensive effects (Moreira et al. 2014).

#### Furan Derivatives

Furans are a class of heterocyclic aromatic compounds characterized by a ring structure composed of one oxygen atom and four carbon atoms. Furan derivatives are a panel of furan constituents. They are the only class of volatile compounds used as an indicator of purity or maturity.

Furfural is a toxic heterocyclic aldehyde that occurs in many foods and flavorants. A study demonstrated the carcinogenicity of furfural in an animal model. Interestingly, the carcinogenicity of furfural in humans is controversial because of a lack of proper evidence. Besides, inhalation of furfural caused pulmonary irritation, parenchyma injury, and the regenerative proliferation of type II pneumocytes (Gupta et al. 1991).

HMF is an organic compound derived from dehydration of certain sugars. It is a purity indicator of honey. High concentrations of HMF in honey reflect overheating, poor storage conditions, and old honey (Khalil et al. 2010). Recent clinical studies strongly suggest that HMF has the potency to inhibit tumor growth (Michail et al. 2007). These results suggest that HMF does not have any serious health risk (Janzowski et al. 2000).

Furfuryl alcohol is a furan compound with a dual role of benignity and thickness. Furfuryl alcohol is a colorless to yellow liquid with a faint burning odor. Furfuryl alcohol has been reported to increase the incidences of nasal and renal tubule neoplasms (National Toxicology Program 1999). Furfuryl alcohol and furfural have antimicrobial activity and can inhibit the proliferation of *Salmonella* bacteria and *Bacillus subtilis* (Chai et al. 2013).

#### Alcohols

Alcohols are a group of organic compounds characterized by the presence of hydroxyl group (-OH). Honey alcohol constitute approximately 5% of its volatile organic compound content. In general, alcohol is a perfect food preservative that kills contaminating germs.

1,3-Butanediol has been reported to depresses CNS activity and induce physical dependence. It is also used as a blood glucose–lowering agent. 1,3-Butanediol can be converted into ketone bodies and β-hydroxybutyrate and salvaged in brain metabolism as an energy source other than glucose (Christine et al. 1987).

2-Phenylethanol is naturally found in rose, carnation, hyacinth, Aleppo pine, orange blossom, ylang-ylang, geranium, neroli, and champaca. It is therapeutically used as an anti-infective agent, preservative, and local disinfectant. Interestingly, the fungus Candida albicans produces 2-phenylethanol as an autoantibiotic (Lingappa et al. 1969).

Thymol is therapeutically used in the treatment of patients with oral infections because of its potent fungicidal, bactericidal, and antioxidant properties (Ogaard et al. 1997). Thymol is a strong anti-inflammatory agent that works by reducing the release of inflammatory metabolites such as prostanoids, interleukins, and leukotrienes (Yucel-Lindberg et al. 1999). A recent study has suggested that thymol possess cognitive-enhancing activity in dementia rat models (Azizi et al. 2012).

#### Pyran Derivative

A pyran is a six-membered heterocyclic, non-aromatic ring consisting of five carbon atoms and one oxygen atom. Pyrans are relatively less significant in fundamental chemistry. However, they hold important considerations in biochemistry, being important biological molecules.

Commonly, there are two types of butyrolactone, γ-butyrolactone and β-butyrolactone. γ-Butyrolactone is an endogenous brain metabolite. γ-Butyrolactone has a euphoric effect at low doses and a sedative effect at a high doses (Nieuwenhuijzen and Mcgregor 2009). It acts as a prodrug for γ-hydroxybutyrate. γ-Butyrolactone is rapidly converted to γ-hydroxybutyrate in blood and the liver. This may account for its subsequent CNS depressant effects.

Valerolactone can exist in β, δ, or γ form. γ-Valerolactone (GBV) is a prodrug to γ-hydroxyvaleric acid. γ-Valerolactone is converted to 4-methyl-gamma hydroxybutyrate, which in turn exerts neurologic effects. The effects of both γ-butyrolactone and γ-valerolactone do not appear to be mediated by activation of GABA-B receptors in and around the lateral ventricle of the rat brain (Marinetti 2003).

## Effects of Gastrointestinal Digestion on Aromatic Compounds

Through proton nuclear magnetic resonance analysis, the signal intensity of the aromatic compounds phenyllactic acid and 4-methoxyphenyllactic acid remained stable during digestion (Seraglio et al. 2021). These important findings were associated with the partial antibacterial activity of manuka honey during the digestive process, mainly after duodenal digestion, and they bring new perspectives about the role of other compounds beyond methylglyoxal and other α-dicarbonyls on the antibacterial activity of honeys during digestion. Although fecal fermentation provides different results according to the unique gut microbial composition of each person, it is clear that this phase of digestion can vastly change the profile and amount of many compounds, and consequently, it affects the bioaccessible compounds, considering the whole digestion process (Seraglio et al. 2021).

## Conclusion

Honey is a nutritionally important food with traditional and cultural value. Elaborate research is now being conducted throughout the world to address its medicinal benefits; there is no way to avoid research into honey in the medical sciences. Honey containing active compounds such as volatile aromatic compounds contributes to the aroma profile of honey, and the composition and quality of honey depend on their botanical and geographical derivation. It is evident that differential properties are found in honey from different origins with varying aromatic compound compositions. As a result, priority can be given to making inquiries into the aromatic compounds of honey in the context of contemporary medicinal research. In addition, aromatic honey possesses both health benefits and unwanted toxicologic effects. Although it is reported that as a novel food, honey exerts some stunning biological effects, its constituents may undergo various notorious metabolic conversions that may be life threatening. A future research prospect could include the collection of honey from different sources to evaluate changing its properties, new aromatic compound screening, construction of a database with existing and upcoming aromatic compounds, and experimental investigation of notorious aromatic compounds to discover honey's obscure toxicologic effects. Further exclusive investigation of honey could create a new horizon in the remediation aspect of several clinical conditions.

## References


Alissandrakis, E., Daferera, D., Tarantilis, P.A., et al. (2003). Ultrasound-assisted extraction of volatile compounds from citrus flowers and citrus honey. *Food Chemistry* 82: 575–582.

Alissandrakis, E., Tarantilis, P.A., Harizans, P.C., and Polissons, M. (2007). Comparison of the volatile composition of thyme honeys from several origins in Greece. *Journal of Agricultural and Food Chemistry* 55: 8152–8157.

Aronne, G., Giovanetti, M., Sacchi, R., and De Micco, V. (2014). From flower to honey bouquet: possible markers for the botanical origin of robinia honey. *Scientific World Journal* 2004: 1–7.

Azizi, Z., Ebrahimi, S., Saadatfar, E., et al. (2012). Cognitive-enhancing activity of thymol and carvacrol in two rat models of dementia. *Behavioural Pharmacology* 23 (3): 241–249.

Baroni, M.V., Nores, M.L., Díaz, M.D.P., et al. (2006). Determination of volatile organic compound patterns characteristics of five unifloral honeys by solid- phase microextraction—Gas chromatography-mass spectrometry coupled to chemimetrics. *Journal of Agricultural and Food Chemistry* 54: 7235–7241.

Barra, M.P.G., Ponce-Díaz, M.C., and Venegas-Gallegos, C. (2010). Volatile compounds in honey produced in the central valley of ñuble province, chile. *Chilean Journal of Agricultural Research* 70 (1): 75–84.

Bjornaes, S. and Naalsund, L.U. (1988). Biochemical changes in different brain areas after toluene inhalation. *Toxicology* 49 (2–3): 367–374.

Bogdanov, S., Jurendic, T., Sieber, R., and Gallmann, P. (2008). Honey for nutrition and health: a review. *American Journal of the College of Nutrition* 27: 677–689.

Cajka, T., Hajslova, J., Pudil, F., and Riddellova, K. (2009). Traceability of honey origin based on volatiles pattern processing by artificial neural networks. *Journal of Chromatography A* 1216 (9): 1458–1462.

Carlos, B.R.S., Carvalho, D., Larissa, C.B., et al. (2021). Volatile compounds in off-odor honey. *Ciência Rural* 51 (7): e20200387.

Castro-Várquez, L., Díaz-Maroto, M.C., González-Viñas, M.A., and Pérez-Coello, M.S. (2009). Differentiation of monofloral citrus, rosemary, eucalyptus, lavender, thyme and heather honeys based on volatile composition and sensory descriptive analysis. *Food Chemistry* 112: 1022–1030.

Castro-Várquez, L., Pérez-Coello, M.S., and Cabezudo, M.D. (2003). Analysis of volatile compounds of rosemary honey. Comparison of different extraction techniques. *Chromatographia* 57: 227–233.

Castro-Várquez, L.M., Díaz-Maroto, M.C., De Tores, C., and Pérez-Coello, M.S. (2010). Effects of geographical origins on the chemical and sensory characteristics of chestnut honeys. *Food Research Internationa* 43: 2335–2340.

Castro-Várquez, L.M., Díaz-Maroto, M.C., and Pérez-Coello, M.S. (2006). Volatile composition and contribution to the aroma of Spanish honeydew honeys. Identification of a new chemical marker. *Journal of Agricultural and Food Chemistry* 54: 4809–4813.

Chai, W.M., Liu, X., Hu, Y.H., et al. (2013). Antityrosinase and antimicrobial activities of furfuryl alcohol, furfural and furoic acid. *International Journal of Biological Macromolecules* 57: 151–155.

Chen, W., Liu, Y., Li, M., et al. (2015). Anti-tumor effect of α-pinene on human hepatoma cell lines through inducing G2/M cell cycle arrest. *Journal of Pharmacological Sciences* 127 (3): 332–338.

Christine, M., Anne-Marie, B., and Jean, B. (1987). Protective action of 1,3-butanediol in cerebral ischemia. A neurologic, histologic, and metabolic study. *Journal of Cerebral Blood Flow & Metabolism* 7 (6): 794.

D'arcy, B.R., Rintoul, G.B., Rowland, C.Y., and Blackman, A.J. (1997). Composition of Australian honey extractives. 1. Norisoprenoids, monoterpenes, and other natural volatiles from blue gum (Eucalyptus leucoxylon) and yellow box (Eucalyptus melliodora) honeys. *Journal of Agricultural and Food Chemistry* 45: 1834–1843.

Daughtrey, W., Newton, P., Rhoden, R., et al. (1999). Chronic inhalation carcinogenicity study of commercial hexane solvent in F-344 rats and B6C3F1 mice. *Toxicological Sciences* 48 (1): 21–29.

De La Fuente, E., Sanz, M.L., Martínez-Castro, I., et al. (2007). Volatile and carbohydrate composition of rare unifloral honeys from Spain. *Food Chemistry* 105: 84–93.

Eteraf-Oskouei, T. and Najafi, M. (2013). Traditional and modern uses of natural honey in human diseases: a review. *Iranian Journal of Basic Medical Sciences* 16: 731–742.

Fuxe, K., Andersson, K., Nilsen, O.G., et al. (1982). *Toxicology Letters* 12 (2–3): 115–123.

Gaunt, I.F., Brantom, P.G., Kiss, I.S., et al. (1972). Short-term toxicity of acetoin (acetylmethylcarbinol) in rats. *Food and Cosmetics Toxicology* 10 (2): 131–141.

Gupta, G.D., Misra, A., and Agarwal, D.K. (1991). Inhalation toxicity of furfural vapours: an assessment of biochemical response in rat lungs. *Journal of Applied Toxicology* 11 (5): 343–347.

Guyot, C., Scheirman, V., and Collin, S. (1999). Floral origin markers of Heather honeys: Calluna vulgaris and Erica arborea. *Food Chemistry* 64: 3–11.

Guyot-Declerck, C., Renson, S., Bouseta, A., and Collins, S. (2002). Floral quality and discrimination of Lavandula angustifoliaxlatifolia honeys. *Food Chemistry* 79: 453–459.

Haque, M.N., Chowdhury, R., Islamand, K.M.S., et al. (2009). Propionic acid is an alternative to antibiotics in poultry diet. *Journal of Animal Science* 38 (1&2): 115–122.

Helmy, N. and El-Soud, A. (2012). Honey between traditional uses and recent medicine. *Macedonian Journal of Medical Sciences* 5 (2): 205–214.

Honey, R. (2003). Comparison of different extraction techniques. *Chromatographia* 57: 227–233. http://dx.doi.org/10.1155/2014/547275 (accessed 30/016/2015).

Huidobro, J.F., Rea, M.E., Branquinho De Andrade, P.C., et al. (1993). Enzymatic determination of glycerol in honey. *Journal of Agricultural and Food Chemistry* 41: 557–559.

Huidobro, J.F., Rea, M.E., De Andrade, P.C.B., et al. (1994). Enzymatic determination of primary normal alcohols as apparent ethanol content in honey. *Journal of Agricultural and Food Chemistry* 42: 1975–1978.

Integrated Risk Information System (IRIS). (2000) Summary on Isophorone (78-59-1). http://www.epa.gov/iris (accessed July 2015).

Islam, A., Khalil, I., Islam, N., et al. (2012). Physicochemical and antioxidant properties of Bangladeshi honeys stored for more than one year. *BMC Complementary and Alternative Medicine* 12 (177): 1–10. http://www.biomedcentral.com/1472-6882/12/177 [Accessed 11/07/2015].

Janzowski, C., Glaab, V., Samimi, E., et al. (2000). 5-Hydroxymethylfurfural: assessment of mutagenicity, DNA-damaging potential and reactivity towards cellular glutathione. *Food and Chemical Toxicology* 38 (9): 801–809.

Jeffrey, A.E. and Echazarreta, C.M. (1996). Medical uses of honey. *Revista Biomédica* 7: 43–49.

Jerkovic, I. and Marijanovic, Z. (2009). Screening of volatile composition of Lavandula hybrida REVERCHON II honey using headspace solid-phase micro extraction and ultrasonic solvent extraction. *Chemistry & Biodiversity* 6: 421–430.

Jerkovic, I. and Marijanovic, Z. (2010). Volatile composition screening of Salix spp. Nectar honey. Benzenecarboxylic acids, norisoprenoids, terpenes and others. *Chemistry & Biodiversity* 7: 2309–2325.

Jerković, I., Tuberoso, C.I.G., Marijanović, Z., et al. (2009). Headspace, volatile and semi-volatile patterns of Paliurus spina-Christi unifloral honey as markers of botanical origin. *Food Chemistry* 112: 239–245.

Jerkovic, J., Mastelic, J., and Marijanovic, Z. (2006). A variety of volatile compounds as markers in unifloral honey from Dalmatian sage (Salvia officinalis L.). *Chemistry & Biodiversity* 3: 1307–1316.

Kádár, M., Juan-Borrás, M., Hellebrandova, M., et al. (2010). Volatile fraction composition of Acacia (Robinia pseudoacacia) honey from Romania, Spain and Czech Republic. *Bulletin of the University of Agricultural Sciences and Veterinary Medicine Cluj-Napoca Agriculture* 67 (2): 259–265.

Kaškonienė, V. and Venskutonis, P.R. (2010). Floral markers in honey of various botanical and geographic origins: a review. *Comprehensive Reviews in Food Science and Food Safety* 9 (6): 620–634.

Kaskoniene, V., Venskutonis, P.R., and Ceksteryte, V. (2008). Composition of volatile compounds of honey of various floral origin and beebread collected in Lithuania. *Food Chemistry* 111: 988–997.

Khalil, M.I., Sulaiman, S.A., and Gan, S.H. (2010). High 5-hydroxymethylfurfural concentrations are found in Malaysian honey samples stored for more than one year. *Food and Chemical Toxicology* 48 (8): 2388–2392.

Kreckmann, K.H., Baldwin, J.K., Roberts, L.G., et al. (2000). Inhalation developmental toxicity and reproduction studies with cyclohexane. *Drug and Chemical Toxicology* 23 (4): 555–573.

Lingappa, B.T., Prasad, M., Lingappa, Y., et al. (1969). Phenethyl alcohol and tryptophol: autoantibiotics produced by the fungus Candida albicans. *Science* 163 (3863): 192–194.

Lopachin, R.M., Ross, J.F., Reid, M.L., et al. (2002). Neurological evaluation of toxic axonopathies in rats: acrylamide and 2,5-hexanedione. *Neurotoxicology* 23 (1): 95–110.

Manyi-Loh, C.E., Ndip, R.N., and Clarke, A.M. (2011). Volatile compounds in honey: a review on their involvement in aroma, botanical origin determination and potential biomedical activities. *International Journal of Molecular Sciences* 12: 9514–9532.

Marinetti, L. (2003) The pharmacology of gamma valerolactone (GVL) as compared to gamma hydroxybutyrate (GHB), gamma butyrolactone (GBL), 1,4 butanediol (1,4 BD), ethanol (EtOH) and baclofen (BAC) in the rat. A Thesis Submitted in partial fulfillment of the Requirements of Wayne State University for the Degree of Doctor of Philosophy. Wayne State University. Detroit: Wayne State University.

Michail, K., Matzi, V., Maier, A., et al. (2007). Hydroxymethylfurfural: an enemy or a friendly xenobiotic? A bioanalytical approach. *Analytical and Bioanalytical Chemistry* 387 (8): 2801–2814.

Miyagi, Y., Shima, F., Ishido, K., et al. (1999). Tremor induced by toluene misuse successfully treated by a Vim thalamotomy. *Journal of Neurology Neurosurgery and Psychiatry* 66 (6): 794–796.

Montenegro, G., Gómez, M., Casaubon, G., et al. (2009a). Analysis of volatile compounds in three unifloral native Chilean honeys. *International Journal of Experimental Botany* 78: 61–65.

Montenegro, G., Gómez, M., Casaubon, G., et al. (2009b). Analysis of volatile compounds in three unifloral native Chilean honeys. *Fyton* 78: 61–65.

Moreira, M.R., Salvadori, M.G., De Almeida, A.A., et al. (2014). Anxiolytic-like effects and mechanism of (-)-myrtenol: a monoterpene alcohol. *Neuroscience Letters* 579: 119–124.

Moreira, R.F.A., Trugo, L.C., Pietroluongo, M., and De Maria, C.A.B. (2002). Flavor composition of cashew (Anacardium occidentale) and marmeleiro (Croton species) honeys. *Journal of Agricultural and Food Chemistry* 50: 7616–7621.

The National Honey Board. (2015) Honey Varietals. https://honey.com/about-honey/honey-varietals.

National Nutrient Database for Standard Reference. (2015) Basic Report: 19296, Honey. https://fdc.nal.usda.gov/fdc-app.html#/food-details/169640/nutrients.

National Toxicology Program. (1999) Toxicology and carcinogenesis studies of furfuryl alcohol (CAS No. 98-00-0) in F344/N rats and B6C3F1 mice (inhalation studies). *National Toxicology Program Technical Report Series* 482: 1–248.

Nieuwenhuijzen, P.S. and Mcgregor, I.S. (2009). Sedative and hypothermic effects of gamma-hydroxybutyrate (GHB) in rats alone and in combination with other drugs: assessment using biotelemetry. *Drug and Alcohol Dependence* 103 (3): 137–147.

Odeh, I., Abu-Lafi, S., and Al-Najjar, I. (2013). Determination of potential volatiles markers from citrus, eucalyptus, cotton and wildflower Palestinian honeys using SPME followed by GCMS analysis. *International Food Research Journal* 20 (3): 1243–1247.

Odeh, I., Abu-Lafi, S., Dewik, H., et al. (2007). A variety of volatile compounds as markers in Palestinian honey from Thymus capitatus, Thymelaea hirsute and Tolpis virgata. *Food Chemistry* 101: 1393–1397.

Ogaard, B., Larsson, E., Glans, R., et al. (1997). Antimicrobial effect of a chlorhexidine-thymol varnish (Cervitec) in orthodontic patients. A prospective, randomized clinical trial. *Journal of Orofacial Orthopedics* 58: 206–213.

Pastrorova, I., De Koster, C.G., and Boom, J.J. (1997). Analytic study of free and ester bound benzoic and cinnamic acids of gum benzoin resins by GC-MS HPLC-frit FAB-MS. *Phytochemical Analysis* 8 (2): 63–73.

Peana, A.T., D'aquila, P.S., Panin, F., et al. (2002). Anti-inflammatory activity of linalool and linalyl acetate constituents of essential oils. *Phytomedicine* 9 (8): 721–726.

Peana, A.T. and Moretti, M.D.L. (2008). Linalool in essential plant oils: pharmacological effects. In: *Botanical Medicine in Clinical Practice* (ed. R.R. Watson and V.R. Preedy). Oxfordshire, UK: CABInternational.

Pearce, C.J. (2015). Review of honey in traditional and modern medicine. *Journal of Natural Products* 78 (4): 967–967.

Piasenzotto, L., Gracco, L., and Conte, L. (2003). Solid phase micro extraction (SPME) applied to honey quality control. *Journal of the Science of Food and Agriculture* 83: 1037–1044.

Radovic, B.S., Careri, M., Mangia, A., et al. (2001). Contribution of dynamic headspace GC–MS analysis of aroma compounds to authenticity testing of honey. *Food Chemistry* 72: 511–520.

Raftari, M., Jalilian, F.A., Abdulamir, A.S., et al. (2012). Antibacterial activity of organic acids on Salmonella and Listeria. *The Asian International Journal of Life Sciences* 21 (1): 13–30.

Rahman, M.M., Gan, S.H., and Khalil, M.I. (2014). Neurological effects of honey: current and future prospects. *Evidence-Based Complementary and Alternative Medicine* 2014 (958721): 13. http://dx.doi.org/10.1155/2014/958721. (accessed 20 June 2015).

Röder-Stolinski, C., Fischäder, G., Oostingh, G.J., et al. (2008). Styrene induces an inflammatory response in human lung epithelial cells via oxidative stress and NF-kappaB activation. *Toxicology and Applied Pharmacology* 231 (2): 241–247.

Rufino, A.T., Ribeiro, M., Judas, F., et al. (2014). Anti-inflammatory and chondroprotective activity of (+)-α-pinene: structural and enantiomeric selectivity. *Journal of Natural Products* 77 (2): 264–269.

Sameem, S., Kumar, N., and Pathak, D. (2012). Synthesis and anticonvulsant activity of some newer semicarbazone derivatives. *International Journal of Pharmaceutical Sciences and Drug Research* 4 (3): 195–198.

Senyuva, H.Z., Gilbert, J., Silici, S., et al. (2009). Profiling Turkish honeys to determine authenticity using physical and chemical characteristics. *Journal of Agricultural and Food Chemistry* 57: 3911–3919.

Seraglio, S.K.T., Schulz, M., Gonzaga, L.V., et al. (2021). Current status of the gastrointestinal digestion effects on honey: a comprehensive review. *Food Chem* 19 (357): 129807.

Silici, S. (2011). Determination of volatile compounds of pine honeys. *Turkish Journal of Biology* 35 (5): 641–645.

Tananaki, C.H., Thrasyvoulou, A., Giraudel, J.L., and Montury, M. (2007). Determination of volatile characteristics of Greek and Turkish pine honey samples and their classification by using Kohonen self organising maps. *Food Chemistry* 10: 1687–1693.

Tian, H., Shen, Y., Yu, H., and Chen, C. (2018). Aroma features of honey measured by sensory evaluation, gas chromatography-mass spectrometry, and electronic nose. *International Journal of Food Properties* 21 (1): 1755–1768.

Uckun, O. and Selli, S. (2017). Characterization of key aroma compounds in a representative aromatic extracts from citrus and astragalus honeys based on aroma extract dilution analyses. *Food Measure* 11: 512–522.

Vanhoutvin, S.A., Troost, F.J., Hamer, H.M., et al. (2009). Butyrate-induced transcriptional changes in human colonic mucosa. *PLoS ONE* 4 (8): 1–7. https://doi.org/10.1371/journal.pone.0006759.

Wardencki, W., Chmiel, T., Dymerski, T., et al. (2009). Application of gas chromatography, mass spectrometry and olfactometry for quality assessment of selected food products. *Ecological Chemistry and Engineering* 16: 287–300.

Warth, A.D. (1991). Mechanism of action of benzoic acid on Zygosaccharomyces bailii: effects on glycolytic metabolite levels, energy production, and intracellular pH. *Applied and Environmental Microbiology* 57 (12): 3410–3414.

White, J.W., Riethof, M.J., Subers, M.H., et al. (1962). *Composition of American Honeys*. Washington, DC: Department of Agriculture.

White, J.W., Jr. (1980). Detection of honey adulteration by carbohydrate analysis. *Journal of the Association of Official Analytical Chemists* 63 (1): 11–18.

Wolski, T., Tambor, K., Rybak-Chmielewska, H.B., and Keddzia, B. (2006). Identification of honey volatile components by solid phase microextraction (SPME) and Gas Chromatography/ Mass Spectrometry (GC/MS). *Journal of Apicultural Science* 50 (2): 115–126.

Xie, G., Chen, N., Soromou, L.W., et al. (2012). p-cymene protects mice against lipopolysaccharide-induced acute lung injury by inhibiting inflammatory cell activation. *Molecules* 17: 8159–8173.

Yamaguchi, H., Kidachi, Y., and Ryoyama, K. (2002). Toluene at environmentally relevant low levels disrupts differentiation of astrocyte precursor cells. *Archives of Environmental Health* 57 (3): 232–238.

Yoon, W.J., Lee, N.H., Hyun, C.G., et al. (2010). Limonene suppresses lipopolysaccharide-induced production of nitric oxide, prostaglandin E2, and pro-inflammatory cytokines in RAW 264.7 macrophages. *Journal of Oleo Science* 59 (8): 415–421.

Yucel-Lindberg, T., Twetman, S., Skold-Larsson, K., and Modeer, T. (1999). Effect of an antibacterial dental varnish on the levels of prostanoids, leukotriene B4, and interleukin-1 beta in gingival crevicular fluid. *Acta Odontologica Scandinavica* 57: 23–27.

Zhong, W., Chi, G., Jiang, L., et al. (2013). p-Cymene modulates in vitro and in vivo cytokine production by inhibiting MAPK and NF-κB activation. *Inflammation* 36 (3): 529–537.

Zhu, Y., Zhou, H., Hu, Y., et al. (2011). Antityrosinase and antimicrobial activities of 2-phenylethanol, 2-phenylacetaldehyde and 2-phenylacetic acid. *Food Chemistry* 124 (1): 298–302.

Zimmerman, M.A., Singh, N., Martin, P.M., et al. (2012). Butyrate suppresses colonic inflammation through HDAC1-dependent Fas upregulation and Fas-mediated apoptosis of T cells. *The American Journal of Physiology-Gastrointestinal and Liver Physiology* 302 (12): G1405–15.

# 11

# Furfural and Hydroxymethylfurfural

*Md. Solayman, Ummay Mahfuza Shapla, and Md. Ibrahim Khalil*

## Introduction

Honey is widely considered a nutritional and medicinal food, but some constituents, for example, heavy metals (in trace amounts), some alkaloids, and hydroxymethylfurfural (HMF) and its derivatives are, unfortunately, toxic (Islam et al. 2014; Solayman et al. 2016). HMF is a cyclic aldehyde produced from sugar degradation occurring via a Maillard reaction during food processing or as the result of a long storage period (Markowicz Bastos et al. 2012). Because the presence of simple sugars (glucose and fructose) and many acids and minerals in honey can stimulate the production of the substance (Choudhary et al. 2020), HMF is a widely recognized parameter related to honey's freshness. It is typically absent or present in very low levels in fresh honeys, but its concentration tends to rise during processing and because of aging (Shapla et al. 2018). Previous studies have reported that honey stored at lower temperatures or fresh conditions has lower or zero HMF content. On the contrary, aged or stored honey at comparatively higher or medium temperatures has a higher HMF content. Metallic containers, storage conditions, and floral sources of honey are also the critical factors relating the level of HMF content (Table 11.1). Hence, a higher HMF concentration offers messages of poor storage conditions or excess heating of honey (Fallico et al. 2004b; Khalil et al. 2010).

## Formation of Hydroxymethylfurfural in Honey

HMF is a six-carbon heterocyclic organic compound containing both aldehyde and alcohol (hydroxymethyl) functional groups. The ring of the structure is centered on furan moieties, and the two functional groups, formyl and hydroxy-methyl groups, are linked with it at positions 2 and 5, respectively (Figure 11.1). It is a yellow low-melting solid that is highly soluble in water (Rosatella et al. 2011).

HMF is the most important intermediate product that is formed though two reactions: (1) the acid-catalyzed degradation of hexose and (2) the decomposition of 3-deoxyosone in a Maillard reaction (Figure 11.2) (Belitz et al. 2009). The formation of HMF has been correlated with chemical characteristics such as pH, free acids, total acidity, lactones, and mineral content, which are related to the floral source of the different honeys. The simple sugars such as glucose and fructose as well as many acids present in honey are favorable for the production of this substance (Fallico et al. 2004a).

Although the level is very low or even absent in fresh honey, heat-treated or stored honeys have higher quantities of HMF (Shapla et al. 2018). Moniruzzaman et al. (2013) showed that the mean HMF content in Malaysian honeys (stored for 2 months at 4°–5°C) was 35.98 mg/kg. On the contrary, Khalil et al. (2010) reported that HMF concentrations were high (118.47–1139.95 mg/kg) in Malaysian honey when stored for more than 1 year at 25°–30°C. Another study (Islam et al. 2012) showed that the HMF level in honey samples from Bangladesh honey sored for more than 1 year at room temperature (20°–25°C) was 3.18–703.10 mg/kg. Therefore, the value of HMF depends on the freshness, storage period, and conditions of the honey (see Table 11.1).

*Honey: Composition and Health Benefits*, First Edition. Edited by Md. Ibrahim Khalil, Gan Siew Hua, and Bey Hing Goh.

**Table 11.1** The concentration of hydroxymethylfurfural in honey samples at various storage times and geographical sources.

| Country | Storage Time | Storage Temperature (°C) | Concentration of Hydroxymethylfurfural (mg/kg) | Reference |
|---|---|---|---|---|
| **Asia** | | | | |
| Bangladesh | >1.5 yr | 20–25 | 3.18–703.10 | Islam et al. 2012 |
| India | Fresh | – | 0.15–1.70 | Singh and Bath 1997 |
| | – | – | 4.99–136.70 | Saxena et al. 2010 |
| Malaysia | <1 yr | 4–5 | 0.26–68.99 | Moniruzzaman et al. 2013 |
| | >1 yr | 25–30 | 206.06–383.39 | Khalil et al. 2010 |
| | >2 yr | 25–30 | 986.57–1131.76 | |
| Nepal | – | 25–29 | 30.36 –56.10 | Qamer et al. 2008 |
| Oman | <6 mo | 25 | 0.00–3.97 | Al-Farsi et al. 2018 |
| Pakistan | – | 4–5 | ND–6.00 | Akhtar et al. 2014 |
| | – | – | 23.18–27.37 | Iftikhar et al. 2011 |
| Saudi Arabia | 45 days | 25 | 31.7–37.80 | Al-Ghamdi et al. 2019 |
| Iran | – | – | 0.04–17.20 | Mehryar et al. 2013 |
| Turkey | Fresh | – | 0.00–11.50 | Yilmaz and Küfrevioğlu 2001 |
| | <6 mo | – | 19.20–28.60 | Küçük et al. 2007 |
| | 1 yr | 20 ± 5 | 8.60–39.00 | Yilmaz and Küfrevioğlu 2001 |
| **Europe** | | | | |
| Italy | Fresh | – | 1.23–5.95 | Fallico et al. 2004a |
| Croatia | – | – | 0.00–23.69 | Tucak et al. 2007 |
| Czech Republic | 4 yr | 20 ± 2 | 10.30–44.20 | Kalabova et al. 2003 |
| Poland | – | – | 0.70–3.50 | Krauze and Zalewski 1991 |
| Portugal | – | 25 | 18.00–94.00 | Gomes et al. 2010 |
| Spain | <3 yr | – | 0.00–21.39 | Nalda et al. 2005 |
| Northwest Spain | <2 yr | −30 | 0.00–1.60 | Olga et al. 2012 |
| Switzerland | <7 yr | 4 | 0.00–112.00 | Ruoff et al. 2007 |
| **Africa** | | | | |
| Algeria | <1 yr | 4–6 | 1.73–480.00 | Chefrour et al. 2009 |
| | 9 mo | 4–35 | 11.04–82.00 | Mouhoubi-Tafinine et al. 2018 |
| Burkina Faso | <1 yr | 0–4 | 3.00–27.50 | Meda et al. 2005 |
| Ethiopia | <6 mo | – | 0.68–6.56 | Belay et al. 2013; Kinati et al. 2011 |
| Kenya | <1 yr | 25 ± 2 | 3.70–389.36 | Muli et al. 2007 |
| Morocco | – | 4 | 3.20–52.60 | Terrab et al. 2003 |
| Nigeria | <6 mo | – | 0.66–1.43 | Ayansola and Banjo 2011 |
| Tanzania | – | – | 5.00–26.40 | Muruke 2014 |
| **South America** | | | | |
| Argentina | – | 23 ± 2 | 0.05–2.94 | Baroni et al. 2009 |
| | Within 2 mo | – | 1.48–34.08 | Cantarelli et al. 2008 |
| Argentinean Patagonia | >3 yr | – | 0.00–14.70 | Aloisi 2010 |
| Brazil | – | – | 1.50–115.20 | Alfredo et al. 2004; Sodré et al. 2011 |
| **North America** | | | | |
| Cuba | <1 year | 4 | 3.30–15.90 | Alvarez-Suarez et al. 2010 |
| Mexico | – | 25 | 9.01–21.96 | Mondragón-Cortez et al. 2013 |
| **Australia** | | | | |
| Australia | 2 yr | −18 | 1.30–12.40 | D'Arcy et al. 1997 |

ND, not detected.

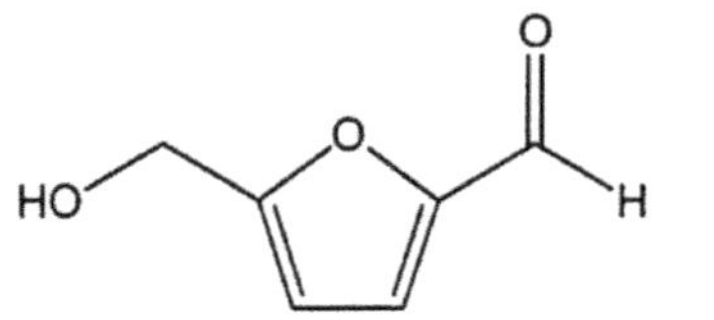


**Figure 11.1** The chemical structure of hydroxymethylfurfural.

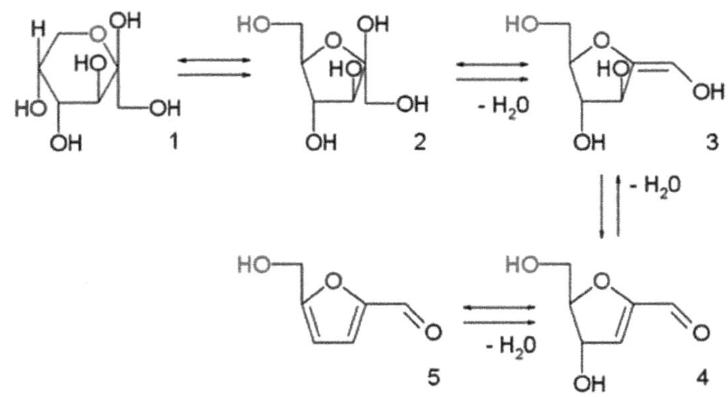


**Figure 11.2** Formation of hydroxymethylfurfural in honey: (1) fructopyranose, (2) fructofuranose, (3 and 4) two intermediate stages of dehydration (not isolated), and (5) hydroxymethylfurfural.

## Factors Affecting the Formation of Hydroxymethylfurfural in Honey

HMF can be produced from sugars termed *hexoses*. It is an intermediate compound formed by the degradation of hexoses in acidic condition at high temperatures or during a Maillard reaction (a non-enzymatic browning reaction) (Arribas-Lorenzo and Morales 2010). It is also created from oligo- and polysaccharides that can yield hexoses on hydrolysis. However, it seems to be more selectively produced from keto-hexoses such as fructose (Román-Leshkov et al. 2006). Interestingly, there are two reasons behind the higher yields of HMF from fructose (ketose) than that from glucose (aldose). First, glucose reactivity is lower than that of fructose, and the enolization rate is low (Kuster 1990). The enolization is believed to be the rate-determining step of HMF production. Furthermore, fructose forms an equilibrium mixture of di-fructose as well as dianhydrides and thus internally blocks the most reactive groups, leading to the formation of some by-products. On the contrary, glucose forms true oligosaccharides, which still contain reactive reducing groups, leaving more risk for cross-polymerization with reactive intermediates and HMF (Kuster 1990).

A great deal of different types of sugars is found in honey. A study (Solayman et al. 2016) has shown that honey contains fructose (39.44%), glucose (28.15%), and sucrose (3.19%). Cyclic aldehyde HMF is also produced in honey by the degradation of these sugars (Khalil et al. 2010). Although it is not produced in fresh and untreated honey, its concentration increases because of processing and long-term storage (Fallico et al. 2004b). Actually, heating during honey processing requires decreasing viscosity to prevent crystallization and fermentation (Singh and Singh 2018). There are also several factors that influence the formation of HMF in honey such as the physicochemical properties (the pH, free acids, total acidity, lactones, and mineral content) of honey itself, water activity ($a_w$), the use of metallic containers, and thermal and photochemical stress (Spano et al. 2006; Yang et al. 2019).

In low-pH or acidic conditions, HMF can be formed at low temperature (Lee and Nagy 1990), but a high temperature and storage time extremely increase its concentration. In case of dry and pyrolytic conditions, a different pathway has been proposed that leads to HMF formation from fructose and sucrose. In addition to temperature and pH, the rate of HMF formation in honey is also reported to be dependent on the water content (Gökmen et al. 2008; Gökmen and Şenyuva 2007). Therefore, a low moisture content in honey samples should be maintained to minimize HMF formation.

The rate of HMF formation depends on the fructose-to-glucose ratio and type of sugars. At a pH of 4.6, fructose has five times more reactivity than glucose because a high fructose-to-glucose ratio accelerates the reaction (Lee and Nagy 1990). Turhan et al. (2009) has shown that temperature as well as the duration of heat treatment can influence the formation of

HMF in honey samples. They reported that heating honey samples that have been collected from the center of Anatolia, Turkey, at 135ºC for 100 seconds produces similar amounts of HMF as that done at 150ºC for 40 seconds.

Sancho et al. (1992) reported that a logarithmic relationship between storage time and HMF levels present in honey exists. Khalil et al. (2010), in their study on nine Malaysian honey samples, showed that the samples stored for only 3–6 months had a HMF value, falling in the range of internationally acceptable value (80 g/kg). However, honey samples stored for 12–24 months have higher HMF concentrations. To investigate the correlation between the physicochemical properties of honey and HMF formation, Khalil et al. (2010) have shown that free acids and total acidity exhibit strong correlation, but pH and lactones exhibit moderate correlation. In another study with four honey samples (eucalyptus, Sulla, orange, chestnut) conducted by Fallico et al. (2004), the relationship between HMF content and the range of heating time, acidity, and pH has also been found. During the storage period, a proportional relation exists between the amount of HMF formation and the degree of temperature, moisture content, and water activity (Bulut and Kilic 2009). The concentrations of the metallic ions such as manganese, zinc, magnesium, and iron (II) also have a positive effect on the formation of HMF during storage (Wen et al. 2016).

## Hydroxymethylfurfural in Foods Correlates with Urine Metabolites

After oral or intravenous administration, HMF is metabolized into mainly three major metabolites – 5-hydroxymethylfuroic acid (HMFA), 2,5-furandicarboxylic acid (FDCA), and 5-(hydroxymethyl)-2-furoyl glycine (HMFG) – and possibly a forth metabolic substance, termed 5-sulphoxymethylfurfural (SMF) (Hardt-Stremayr et al. 2013). Studies designed to trace the distribution of radiolabeled [U-$^{14}$C]-HMF and its metabolites by whole-body autoradiography have revealed that neither HMF nor its metabolites accumulate. It is estimated that 80%–100% of the total amount of radioactivity is excreted within the first 24 hours after administration (Germond et al. 1987; Godfrey et al. 1999). HMF is first oxidized to carboxylic acid and then conjugated with glycine, leading to the formation of HMFG. The concentration of free glycine is rate limiting in this pathway. In case of its unavailability, HMF undergoes further oxidation yielding FDCA. In an in vivo study, Godfrey et al. (1999) have detected three metabolites in the urine after oral administration of labeled HMF at a dose of 10–500 mg/kg. The relative amounts of FDCA, HMFA, and HMFG excreted were up to 2%–6%, 78%–85%, and 5%–8%, respectively, depending on the species and doses. In humans, HMF is completely cleared out upon oral administration through the juice of dried plums (Prior et al. 2006). Similarly, another clinical study by Hardt-Stremayr et al. (2013) showed that HMF has (orally 240 mg/day) completely excreted through the urine within 48 hours; SMF was not detected in urine as in previous studies.

## Detection Methods of Hydroxymethylfurfural

Accurate quantitative analysis of HMF is of great importance because HMF is used as a marker of quality deterioration, thermal processing, and other abuse practices. HMF is also important in clinical research and therapeutics. The International Honey Commission has recommended three main methods for the determination of HMF: two routinely used spectrophotometric methods and a reverse-phase high-performance liquid chromatography (HPLC) method (Zappala et al. 2005). Prior to the spectrophotometric method, optical and chemical methods were used. A new spectrophotometric method used 0.1% sodium bicolpate containing a clarified honey sample as reference and a honey solution without sodium bisulfate as the sample (White 1979). On the other hand, Winkler (1955) used honey solutions with p-toluidine and barbituric acid. These methods are fast but lack sensitivity and specificity. Besides, the method described by Winkler uses p-toluidine, which is also carcinogenic. Because of being devised with sophisticated equipment, the HPLC method is expensive and slow. But it is advantageous with respect to labor and time. It is an automated and sensitive method excluding all interferences by other related compounds (Wootton and Ryall 1985). The methods are not satisfactory enough, so they require further development and modification (Anklam 1998). An electrochemical approach has been described by Reyes-Salas et al. (2006). In this method, a single and sharp reduction signal was created at −1100 mV versus Ag/AgCl, and borate was used as a supporting electrolyte. Another method has been described as ion-exchange liquid chromatography photodiode array detection approached by Yuan and Chen (1998), the result of which was consistent with the method by Winkler (1955). Automated flow injection is a method devised by Iglesia et al. (1997) principled on Winkler's method and having a detection range of 5–40 ppm. Micellar electrokinetic capillary chromatography is another rapid method that uses caffeine as the standard. It is suitable for rapid quantification of HMF, especially in honey samples without any pretreatment

(Rizelio et al. 2012). A unique and efficient rapid screening technique is direct analysis in real time coupled with high-resolution time-of-flight mass spectrometry (Rajchl et al. 2013). This method can quantitatively analyze HMF content. This method requires no or very little extent pretreatment of the sample. In 2020, Salhi et al. (2020) first used a cyclic and square-wave voltammetry method to determine HMF in honey, which has very low detection and quantification limits.

## Hydroxymethylfurfural Hazard to Honeybees

The productivity of apiculture depends on the health, survival, and quality of honeybees. To have better health for bees, quality food is one of the priorities. Normally, the carbohydrate sources of honeybees are nectar, honeydew collected from plants and processed by foragers. But in the case of commercial honey production, the food and nutrients of cultured bees become depleted upon the collection of the autumn yield from the hive. Besides, natural resources for honeybees in the environment decrease during winter. Therefore, beekeepers have to use different alternative sources to replenish the carbohydrate demand of bees. The sources include sucrose, high fructose corn syrup (HFCS), various fruit sugars, and invert sugars (Neupane and Thapa 2005). Among the sources, HFCS is considered as a flexible source to be used compared with sucrose because of its storage, transportation facility, and liquidity as opposed to the crystallization method. HFCS is produced commercially using the enzymes alpha amylase, amyloglucosidase, and invertase from corn starch (Blanchard and Geiger 1984; Buck 2001), and sucrose is produced by acid hydrolysis (Bailey 1966). However, these are not safe sources. HMF and its daughter products, formic and levulinic acids, are formed in these foods and pose potential threats to bees (Figure 11.3). In fact, beet sugar, which is also used for bee feeding, contains an HMF content of up to 47 5mg/kg depending on adverse reactions and storage conditions (van der Zee and Pisa 2010).

In another study, a sugar solution containing 30–150 mg/kg of HMF caused a 15%–58.7% mortality rate to the caged bees within 20 days (Jachimowicz and El Sherbiny 1975). Krainer et al. (2016) reported that median lethal dose ($LD_{50}$) values of HMF for bee larvae fed for more than 6 days were 4280 and 2424 mg/kg on days 7 and 22, respectively. Interestingly, they found that 7-day larvae emerging after being fed HMF for 6 days were more sensitive than were adult bees, but the emerging adult bees at day 22 were less sensitive. Moreover, the authors concluded that HMF used as a supplement in food does compromise colony fitness but may not cause great losses of broods. It is hypothesized that HMF causes bees to experience dysentery-like symptoms and ulcers in their gastrointestinal tracts (LeBlanc et al. 2009), leading to their death.

## Effects of Hydroxymethylfurfural on Human Health

HMF and other congeners exert both detrimental and positive effects on human health (see Figure 11.3).

### Adverse Effects on Human Health

HMF and its derivatives have been proved to confer genotoxic, mutagenic, carcinogenic, DNA-damaging, organotoxic, and enzyme-inhibitory effects (Shapla et al. 2018).

#### Hydroxymethylfurfural as an Indirect Mutagen

Florin et al. (1980) aimed to investigate the mutagenic activity of some compounds, including HMF, to four mutant strains of *Salmonella typhimurium*. Liver extract from methylcholanthrene-induced rats were used for metabolic activation of the test compounds. The study showed that HMF is not a mutagen. On the contrary, Lee et al. (1995) demonstrated that HMF is an indirect mutagen because it gets converted to an active metabolite, sulfuric acid ester, SMF, and it shows mutagenicity toward *S. typhimurium* TA104. HMF becomes enzymatically activated to SMF by sulphotransferases (SULTs) contained in rat liver extract enriched with the sulpho-group donor 3′-phosphoadenosine-5′-phosphosulfate. This finding is further supported by another study in which HMF, 2,5-bishydroxymethylfuran (a metabolite of HMF), FFA (furfuryl alcohol), and 5-methyl-FFA showed mutagenicity to *S. typhimurium* TA100 expressing human SULT1C2 but not to parental strains (Glatt et al. 2011). Another study designed with FVB/N (FVB) mice expressing hSULT1A1/1A2 revealed that HMF has a genotoxic effect. In the mice in which a single dose of 900 or 1300 mg/kg body weight was orally administered, DNA damage of renal cells has been induced at a statistical significant level detected by alkaline single-cell gel electrophoresis assay. 2,5-Dimethylfuran (DMF), another furan derivative, also poses DNA damage in the kidney and colon (Høie et al.

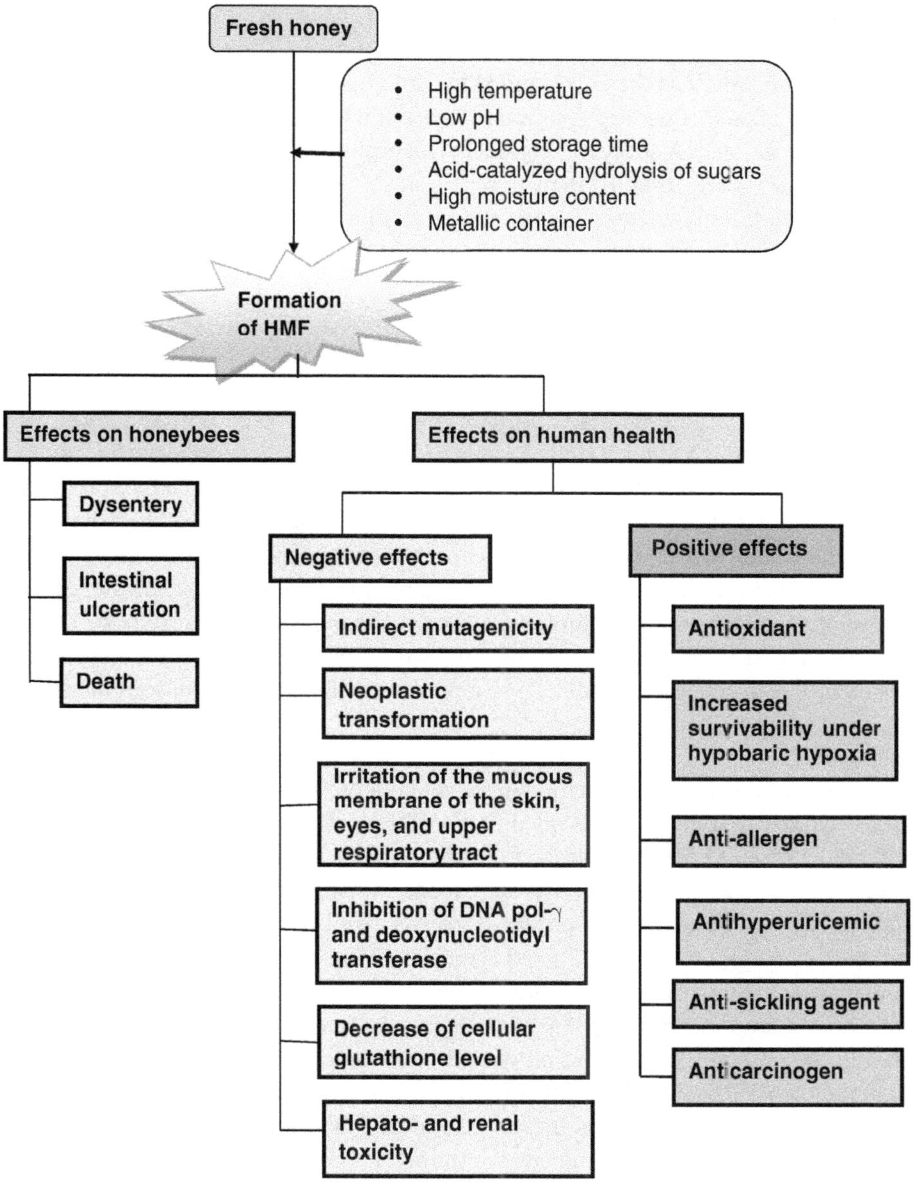


**Figure 11.3** The effects of hydroxymethylfurfural (HMF) on honeybee and human health.

2015). In an in vitro erythropoietic micronucleus assay, rat bone marrow cells were exposed to DMF at a concentration of 0.1 mm for 1 hour. This study showed that DMF has genotoxicity on hematopoietic cells (Fromowitz et al. 2012).

In another study conducted on five cell lines having SULT1A1 activity at different levels (mouse L5178Y, no activity; Chinese hamster [V79-Hp-PST], high activity; V79, negligible activity; human [HEK293], higher activity; and Caco-2, low activity), HMF exerted a DNA-damaging effect exposed at a concentration of 100 mm for 3 hours (Durling et al. 2009). However, it was also determined that HMF poses a DNA-damaging effect irrespective of SULT1A1 activities of the cell lines. 5-HMF also causes chromosomal aberrations of the Chinese hamster V79–derived cell line constitutively expressing human sulphotransferase SULT1A1 and CYP2E1. 5-HMF is a potent inducer of sister–chromatid exchange at a concentration of 19.8–3808 μM in these cells exposed for 32 hours (Glatt et al. 2005). Nishi et al. (1989) have also confirmed that 5-HMF causes chromosomal aberrations of V79 cells at a concentration of 15.8 mM. Cueva et al. (2017), in their preclinical and clinical study, have shown that orally administered HMF, upon being absorbed through gastrointestinal tract, gets converted into reactive SMF. The SMF forms DNA SMF adduct in the kidneys, leukocytes, and liver cells in mice and in leukocytes in the pre-adolescent population. In addition, SMF is not properly excreted through urine because of renal reabsorption, thus allowing SMF to accumulate in the plasma and making it available to react with cellular proteins and DNA.

#### Hydroxymethylfurfural as a Dual Player in Carcinogenesis

HMF and its derivative SMF have been confirmed to be potent carcinogens in several studies at the preclinical level. They cause neoplastic transformations in several organs such as the colon and skin. Colon cancer includes multiple steps to develop. Micro-adenomas and aberrant crypt foci (ACF) are morphological markers of it (Archer et al. 1992). HMF acts as an initiator as well as promoter of ACF (Bruce et al. 1993). A study was conducted on multiple intestinal neoplasia (min/ +) mice that were administered with a single dose of 500 mg/kg of HMF or 25 mg/kg of SMF subcutaneously within 3–6 days after birth. At 12 weeks, it was seen that HMF and SMF increased the number of small intestinal adenomas and flat dysplastic lesions (flat ACF) in the large intestine (Svendsen et al. 2009). Min/ + mice have a mutant copy of the tumor suppressor gene adenomatous polyposis coli (*APC*). Mutation in the *APC* gene leads to the development of adenomas in the small as well as large intestines, analogous to human familial adenomatous polyposis syndrome (Paulsen et al. 2005; Preston et al. 2008). In a study by Zhang et al. (1993), 45% of the F344 female rats administered with HMF at a dosage of 250 mg/kg twice a day by oral gavage developed large intestinal ACF by day 30. HMF increases both the number and size of ACF. But Florian et al. (2012), in their extensive study (with a two-mouse model, wild-type FVB/N and genetically engineered mice having several copies of human SULT1A1 and SULT1A2 at chromosome 9), had the opposite finding that neither HMF nor its metabolite SMF causes the induction of ACF or colonic tumors. Instead, HMF also induces skin papilloma. Upon topical application of sulfoxymethyl and chloromethyl derivatives of HMF, mice developed papillomas on their skin. Another derivative of HMF, 5-chloromethyfurfural, induces hepatocarcinoma in male B6C3F1 rats at a very early age (Surh et al. 1994). Schoental et al. (1971) have shown that the rats treated with HMF at a dose of 200 mg/kg body weight subcutaneously consequently developed renal lipomatous tumors. In contrast to these studies, a study by Zhao et al. (2014) with the A375 cell line concluded that HMF can induce apoptosis and G0/G1 arrest in the DNA-damaged cells through the ROS-mediated signal transduction pathway. Thus, HMF can be a potent therapeutic as an anticarcinogen.

#### Hydroxymethylfurfural as an Organotoxic Agent

HMF also has cytotoxic effects at high concentrations. It causes irritation in the mucous membranes, skin, eyes, and upper respiratory tract (Morales 2008). In an in vivo study, it has been revealed that SMF is a strong nephrotoxic agent. In this study, the male FVB/N mice intraperitoneally received SMF at a dose of 250 mg/kg body weight. After 5–11 days of the treatment, the mice either died or became morbid condition because of liver damage and severe renal damage, particularly at proximal tubules. Histopathological study revealed extensive proteinaceous casts and necrosis at that area (Bauer-Marinovic et al. 2012). The organic anion transporters type 1 and 2 (OAT1 and OAT2) are highly conserved in different species, including rats and humans, and usually get expressed on the cellular basolateral membrane of proximal tubules. They mediate the uptake of different organic anions, including SMF, as their substrates from the bloodstream and concentrate into tubular cells. That is why the proximal tubules are mostly affected (Burckhardt and Burckhardt 2003). Another study with human embryonic kidney cells (HEK293) that constitutively express OAT1 and OAT2 transporters also supported this phenomenon (Bakhiya et al. 2009). Glutathione is an important endogenous antioxidant in the body. In an ex vivo study with two mammalian cell lines V79 and Caco-2, HMF decreased the level of cellular glutathione in a concentration-dependent manner at doses of 50 and 120 mM, respectively (Janzowski et al. 2000).

#### Hydroxymethylfurfural as Enzyme Inhibitor

The human genome encodes 16 DNA-dependent DNA polymerases. Among them, three are involved in nuclear DNA replication, and others are involved in the repair system (Bebenek and Kunkel 2004; Friedberg et al. 2000). Human polymerase $\gamma$ is a multifunctional enzyme having DNA polymerase, terminal transferase, 5′-deoxyribose phosphate (dRP) lyase, and polynucleotide synthetase activities. Its primary structure includes a nuclear localization signal, BRCA1 carboxy terminus domain, proline-rich region, pol $\beta$-like region, and pol X region (Bork et al. 1997; Ramadan et al. 2003, 2004). DNA pol $\gamma$ shares a sequence homology with terminal deoxynucleotidyltransferase (TdT), which catalyzes the addition of deoxyribonucleotide at the 3′ end of double-stranded DNA or single-stranded DNA in a template-independent manner (Pandey and Modak 1987; Roychoudhury et al. 1976). HMF competitively inhibits DNA pol $\gamma$ and TdT with respect to deoxynucleotide substrate and DNA template primer, showing 50% minimal inhibitory concentration ($IC_{50}$) values of 26.1 and 5.5 $\mu$M, respectively (Mizushina et al. 2006).

### Positive Effects of Hydroxymethylfurfural on Human Health

Reactive oxygen species (ROS) are produced as toxic by-products of aerobic metabolism in our bodies. They oxidize cellular macromolecules such as proteins, membrane lipids, and DNA and cause cellular damage. The consequences range from stress to metabolic defects, neurodegenerative diseases, and even neoplastic transformations (Sies 1985). HMF has been proved to have a dose-dependent free radical–scavenging capacity in a study by Zhao et al. (2013). HMF also has significant protective effect on erythrocytes against ROS-induced damage. To investigate this protective effect, the oxidative stress induced by 2,2′-azobis (2-amidinopropane) dihydrochloride (AAPH); the level of ROS and MDA (malondialdehyde, an indirect determinant of lipid peroxidation) production; and activity of antioxidant enzymes glutathione peroxidase, superoxide dismutase, and catalase in erythrocytes pretreated with HMF were determined. A study revealed that the contents of ROS and MDA are reduced in HMF-treated cells, and the activities of these enzymes get elevated in comparison with negative control cells (Zhao et al. 2013). HMF has also protective effect at the morphological as well as biochemical levels on hepatocytes damaged by $H_2O_2$-induced oxidative stress. Hepatocytes oxidatively stressed by $H_2O_2$ undergo some morphological changes such as wrinkling, condensation of chromatin, and splitting of nuclei, which are hallmarks of apoptotic cells and necrotic cells. However, in $H_2O_2$-stressed human liver cell line L02, the cells that have been treated with HMF (0.79 μM) can preserve better morphology than that of nontreated cells. HMF also reduces the level of caspases 3 and 9 (executioner of apoptosis) in these cells (Wang et al. 2010). Ding et al. (2010) have explained the underlying protective biochemical mechanism. The protective effect is caused by inhibition of apoptosis via accelerating the transition of S phase to G2/M phase and a decreased level of nitric oxide and caspase 3.

Oxygen is essential for survival. Deficiency of oxygen (hypoxic condition) has numerous detrimental and even life-changing effects on health. A hypoxic condition may be created by many factors. There are environmental factors such as altitude and self-related factors such as ischemia, atherosclerosis, and cancer (Pattinson et al. 2005). Several cellular mechanisms are triggered to manage hypoxic conditions. Extracellular signal-regulated kinase (ERK)–mediated transactivation of the transcription factor and hypoxia-inducible factors participate in this process (Sang et al. 2003). Mitochondrial membrane potential also declines and negatively affects hypoxic cells (Iijima 2006). Li et al. (2011), in their in vitro study with the cell line ECV304 (human umbilical cord vain endothelial cell), showed that the cells pretreated with HMF (200 μg/mL for 1 hour) before exposed to hypoxic conditions (0.3% oxygen for 24 hours) have an increased mitochondrial membrane potential and a decreased phosphorylated ERK level. The number of apoptotic and necrotic cells also significantly declines. In their further study with Kunming mice model, it was revealed that pre-exposure to HMF (100 μg/mL for 1 hour) significantly attenuates the extent of hypobaric hypoxia–induced permeability of the blood–brain barrier. It also decreases the extent of neuronal damage at the CA1 region of the hippocampus. Thus, HMF increases survivability under hypobaric hypoxic condition. It may also be a potent therapeutic against acute mountain sickness, high-altitude cerebral edema, and high-altitude pulmonary edema (Li et al. 2011).

Basophiles and mast cells participate in the pathogenesis and manifestations of allergic reactions such as asthma, atopic dermatitis, and allergic rhinitis. RBL-2H3 cells are mast cells located in the mucosal layer. They express immunoglobulin Fc epsilon receptor type I (FcεRI) on their surfaces. The crosslinking of IgE with specific protein antigens and then the binding of crosslinked IgE to FcεRI trigger molecular switch to launch an intracellular signal transduction cascades. These events lead to $Ca^{2+}$influx and mediator release by means of degranulation, mitogen-activated protein kinase phosphorylation, upregulation of cytokine gene expression, and increased ROS generation (Beaven and Metzger 1993; Kim et al. 2008; Schroeder et al. 1995). Yamada et al. (2011) have shown that HMF acts on different stages to inhibit degranulation. Additionally, HMF interferes with antigen–antibody crosslinking and antibody–receptor binding. HMF also blocks $Ca^{2+}$influx into immunoglobulin E (IgE)–sensitized bovine serum albumin–stimulated RBL-2H3 cells. In another study with ovalbumin (OVA)-immunized BALB/c mice, it was found that HMF decreases the level of total IgE and OVA-specific IgE. The study also showed that HMF-treated immunized mice have lower levels of interferon-γ and IL-4 than untreated mice. So, HMF may be a potent anti-allergic compound (Alizadeh et al. 2017).

Uric acid is the end product of purine catabolism. The last two steps of the purine catabolic pathway are catalyzed by a critical enzyme, xanthine oxidase (XO). Uric acid is mainly excreted through urine. A high level of uric acid in the blood leads to the development of hyperuricemia (Pacher et al. 2006). Hyperuricemia is the main cause of gout. Many pathological states are associated with hyperuricemia, including metabolic syndrome, heart failure, pulmonary disorder, and type 2 diabetes mellitus (Hayden and Tyagi 2004). Increased XO activity downregulates the anti-inflammatory transcription factor peroxisome proliferator–activated receptor γ and accelerates inflammatory effects (Gibbings et al. 2011). XO is also an endogenous producer of superoxide, a potent activator of nuclear factor-κB (NF-κB) (Lorne et al. 2008). NF-κB acts as a transcription factor and upregulates the expression of nitric oxide synthase 2 and IL-8. HMF exerts anti-inflammatory effect by downregulating NF-κB (Kitts et al. 2012) and inhibits the activity of XO (Lin et al. 2012).

Hemoglobinopathies such as sickle cell disease are life threatening. The underlying mechanism beyond its vicious effect is the polymerization of abnormal hemoglobin (sickle hemoglobin, HbS) under hypoxic conditions. The pathophysiology is manifested by the deformation of red blood cells, loss of resilience, and occlusion of tiny blood capillaries, leading to morbidity (Goldberg et al. 1992; Magdoff-Fairchild et al. 1972; Stetson 1966). There are many antisickling agents, but they are not free of inconvenience, ranging from different side effects, lack of bioavailability in minimal doses, and reaction to nontarget proteins. HMF can act as an effective antisickling agent. Abdul Malik et al. (2005) worked on transgenic (Tg) sickle mice and showed that orally administered HMF at a lower concentration can get absorbed into the bloodstream through the gastrointestinal tract. Then it can penetrate the erythrocytes and form stable Schiff-base adduct with HbS. Upon formation of such adduct, HMF allosterically shifts the $O_2$ equilibrium curve to the left and prevents the sickling of erythrocytes. It is found that the Tg sickle mice pretreated with HMF can survive longer under hypoxic conditions than untreated mice.

## Mitigation Strategies

There is no particular strategy to mitigate the formation of HMF in honey. This inability is attributable to the numerous precursors of HMF and types of reaction orders (Capuano et al. 2008). HMF is formed following a zero-order kinetic and exponential pattern (Gentry and Roberts 2004). Various mechanistic strategies may be adapted to mitigate its formation. There is positive relation between a high temperature–time profile and acidic pH with the formation of HMF, which can be modified (Gökmen et al. 2007). Reduction of thermal input can significantly decrease the formation of HMF. The reduction can be achieved by various means, for example, by adjusting the temperature of the oven when the moisture content is high or decreasing the temperature when the moisture content is low. Reduced thermal input can also be obtained by making sure food is heated at a lower temperature for a prolonged time by vacuum frying or dielectric heating (microwave, radiofrequency). In the case of the dielectric heating technique, an external AC electric field is applied. In response to this, the dielectric molecules within the food undergo rotations and frictions. These ensure the generation of uniform and rapid heat. In addition, water molecules are the main target of this mechanism, so heat is generated in the portion of food containing water (Zhao et al. 2000). The longer the period of storage, the higher the amount of HMF formed in honey (Khalil et al. 2010), so shortening the storage period can cause its mitigation.

## Strategy of Removing Hydroxymethylfurfural and Other Furans

To remove already formed HMF, postprocess vacuum technology can be exploited. In this technology, depending on the physical and chemical properties of the compounds to be removed, the temperature, pressure, and time profiles are adjusted. The efficacy of this technology depends on the composition of the food, the moisture content, and the nature of the compound (Anese et al. 2010; Quarta and Anese 2012). The viscosity constraint minimizes the diffusion of molecules through the matrix and thus compromises the effect of this process on dry foods (Flink and Karel 1970). Because honey is liquid or semiliquid, this process can effectively remove HMF from honey. This technology can be applied to any finished product without any modification of the process parameters (Anese et al. 2013).

## Conclusion

The generation process of HMF is multifactorial. So, it is not easy to reduce its formation. Because of adoption of different processes to determine its concentration and their inefficiency, it is not so easy to establish a complete database depicting the concentration of HMF in different mono- and multifloral honeys. As a constituent of processed honey, HMF has profound effects on human health. HMF has not only detrimental but also has therapeutic effects. Some effects of HMF on health such as its carcinogenic versus anticarcinogenic effects are also inconclusive. Besides, most of the studies have been conducted at a preclinical level. Therefore, it demands extensive study at a clinical level to understand the well-defined action and therapeutic application of HMF on human health.

## References

Abdulmalik, O., Safo, M.K., Chen, Q., et al. (2005). 5-hydroxymethyl-2-furfural modifies intracellular sickle haemoglobin and inhibits sickling of red blood cells†. *British Journal of Haematology* 128 (4): 552–561.

Akhtar, S., Ali, J., Javed, B., et al. (2014). Comparative physiochemical analysis of imported and locally produced Khyber Pakhtunkhwa honey. *Global Journal of Biotechnology & Biochemistry* 9 (3): 55–59.

Al-Farsi, M., Al-Belushi, S., Al-Amri, A., et al. (2018). Quality evaluation of Omani honey. *Food Chemistry* 262: 162–167.

Alfredo, B.D.A.S.C., da Silva, L.D.C.G., and Marchini, S.L.C. (2004). Características físico-químicas de amostras de mel de Melipona asilvai (Hymenoptera: apidae). *Ciência Rural* 34 (5): 644–650.

Al-Ghamdi, A., Mohammed, S.E.A., Ansari, M.J., and Adgaba, N. (2019). Comparison of physicochemical properties and effects of heating regimes on stored Apis mellifera and Apis florea honey. *Saudi Journal of Biological Sciences* 26 (4): 845–848.

Alizadeh, M., Khodaei, H., Mesgari Abbasi, M., and Saleh-Ghadimi, S. (2017). Assessing the effect of 5-hydroxymethylfurfural on selected components of immune responses in mice immunised with ovalbumin. *Journal of the Science of Food and Agriculture* 97: 3979–3984.

Aloisi, P.V. (2010). Determination of quality chemical parameters of honey from Chubut (Argentinean Patagonia). *Chilean Journal of Agricultural Research* 70 (4): 640–645.

Alvarez-Suarez, J.M., Tulipani, S., Díaz, D., et al. (2010). Antioxidant and antimicrobial capacity of several monofloral Cuban honeys and their correlation with color, polyphenol content and other chemical compounds. *Food and Chemical Toxicology* 48 (8): 2490–2499.

Anese, M., Manzocco, L., Calligaris, S., and Nicoli, M.C. (2013). Industrially applicable strategies for mitigating acrylamide, furan and 5-hydroxymethylfurfural in food. *Journal of Agricultural and Food Chemistry* 61 (43): 10209–10214.

Anese, M., Suman, M., and Nicoli, M.C. (2010). Acrylamide removal from heated foods. *Food Chemistry* 119 (2): 791–794.

Anklam, E. (1998). A review of the analytical methods to determine the geographical and botanical origin of honey. *Food Chemistry* 63 (4): 549–562.

Archer, M.C., Bruce, W.R., Chan, C.C., et al. (1992). Aberrant crypt foci and microadenoma as markers for colon cancer. *Environmental Health Perspectives* 98: 195.

Arribas-Lorenzo, G. and Morales, F.J. (2010). Estimation of dietary intake of 5-hydroxymethylfurfural and related substances from coffee to Spanish population. *Food and Chemical Toxicology* 48 (2): 644–649.

Ayansola, A.A. and Banjo, A.D. (2011). Physico-chemical evaluation of the authenticity of honey marketed in Southwestern Nigeria. *Journal of Basic and Applied Scientific Research* 1 (12): 3339–3344.

Bailey, L. (1966). The effect of acid-hydrolysed sucrose on honeybees. *Journal of Apicultural Research* 5 (3): 127–136.

Bakhiya, N., Monien, B., Frank, H., et al. (2009). Renal organic anion transporters OAT1 and OAT3 mediate the cellular accumulation of 5-sulfooxymethylfurfural, a reactive, nephrotoxic metabolite of the Maillard product 5-hydroxymethylfurfural. *Biochemical Pharmacology* 78 (4): 414–419.

Baroni, M.V., Arrua, C., Nores, M.L., et al. (2009). Composition of honey from Córdoba (Argentina): assessment of North/South provenance by chemometrics. *Food Chemistry* 114 (2): 727–733.

Bauer-Marinovic, M., Taugner, F., Florian, S., and Glatt, H. (2012). Toxicity studies with 5-hydroxymethylfurfural and its metabolite 5-sulphooxymethylfurfural in wild-type mice and transgenic mice expressing human sulphotransferases 1A1 and 1A2. *Archives of Toxicology* 86 (5): 701–711.

Beaven, M.A. and Metzger, H. (1993). Signal transduction by Fc receptors: the Fc&RI case. *Immunology Today* 14 (5): 222–226.

Bebenek, K. and Kunkel, T. (2004). DNA repair and replication. *Advances in Protein Chemistry* 69: 137–165.

Belay, A., Solomon, W., Bultossa, G., et al. (2013). Physicochemical properties of the Harenna forest honey, Bale, Ethiopia. *Food Chemistry* 141 (4): 3386–3392.

Belitz, H.-D., Grosch, W., and Schieberle, P. (2009). Milk and dairy products. In: *Food Chemistry*, 498–545. Berlin: Springer.

Blanchard, P. and Geiger, E. (1984). Production of high fructose corn syrup in the USA. A review article. *Sugar Technology Reviews* 11: 1–94.

Bork, P., Hofmann, K., Bucher, P., et al. (1997). A superfamily of conserved domains in DNA damage-responsive cell cycle checkpoint proteins. *The FASEB Journal* 11 (1): 68–76.

Bruce, W.R., Archer, M.C., Corpet, D.E., et al. (1993). Diet, aberrant crypt foci and colorectal cancer. *Mutation Research/Fundamental and Molecular Mechanisms of Mutagenesis* 290 (1): 111–118.

Buck, A.W. (2001). High fructose corn syrup. In: *Alternative sweeteners*, 391–412. New York: CRC Press.

Bulut, L. and Kilic, M. (2009). Kinetics of hydroxymethylfurfural accumulation and color change in honey during storage in relation to moisture content. *Journal of Food Processing and Preservation* 33 (1): 22–32.

Burckhardt, B. and Burckhardt, G. (2003). Transport of organic anions across the basolateral membrane of proximal tubule cells. *Reviews of Physiology, Biochemistry and Pharmacology* 146: 95–158. Springer.

Cantarelli, M., Pellerano, R., Marchevsky, E., and Camiña, J. (2008). Quality of honey from Argentina: study of chemical composition and trace elements. *The Journal of Argentine Chemical Society* 96 (1–2): 33–41.

Capuano, E., Ferrigno, A., Acampa, I., et al. (2008). Characterization of the Maillard reaction in bread crisps. *European Food Research and Technology* 228 (2): 311–319.

Chefrour, C., Draiaia, R., Tahar, A., et al. (2009). Physicochemical characteristics and pollen spectrum of some north-east Algerian honeys. *African Journal of Food, Agriculture, Nutrition and Development* 9 (5): 1276–1293.

Choudhary, A., Kumar, V., Kumar, S., et al. (2021). 5-Hydroxymethylfurfural (HMF) formation, occurrence and potential health concerns: recent developments. *Toxin Reviews* 40 (4): 545–561.

D'Arcy, B.R., Rintoul, G.B., Rowland, C.Y., and Blackman, A.J. (1997). Composition of Australian honey extractives. 1. Norisoprenoids, monoterpenes and other natural volatiles from blue gum (Eucalyptus leucoxylon) and yellow box (Eucalyptus melliodora) honeys. *Journal of Agricultural and Food Chemistry* 45 (5): 1834–1843.

de la Iglesia, F., Lázaro, F., Puchades, R., and Maquieira, A. (1997). Automatic determination of 5-hydroxymethylfurfural (5-HMF) by a flow injection method. *Food Chemistry* 60 (2): 245–250.

Ding, X., Wang, M.-Y., Yao, Y.-X., et al. (2010). Protective effect of 5-hydroxymethylfurfural derived from processed Fructus Corni on human hepatocyte LO2 injured by hydrogen peroxide and its mechanism. *Journal of Ethnopharmacology* 128 (2): 373–376.

Durling, L.J., Busk, L., and Hellman, B.E. (2009). Evaluation of the DNA damaging effect of the heat-induced food toxicant 5-hydroxymethylfurfural (HMF) in various cell lines with different activities of sulfotransferases. *Food and Chemical Toxicology* 47 (4): 880–884.

Fallico, B., Zappala, M., Arena, E., and Verzera, A. (2004a). Effects of conditioning on HMF content in unifloral honeys. *Food Chemistry* 85 (2): 305–313.

Fallico, B., Zappalà, M., Arena, E., and Verzera, A. (2004b). Effects of heating process on chemical composition and HMF levels in Sicilian monofloral honeys. *Food Chemistry* 85 (2): 305–313.

Flink, J. and Karel, M. (1970). Retention of organic volatiles in freeze-dried solutions of carbohydrates. *Journal of Agricultural and Food Chemistry* 18 (2): 295–297.

Florian, S., Bauer-Marinovic, M., Taugner, F., et al. (2012). Study of 5-hydroxymethylfurfural and its metabolite 5-sulfooxymethylfurfural on induction of colonic aberrant crypt foci in wild-type mice and transgenic mice expressing human sulfotransferases 1A1 and 1A2. *Molecular Nutrition & Food Research* 56 (4): 593–600.

Florin, I., Rutberg, L., Curvall, M., and Enzell, C.R. (1980). Screening of tobacco smoke constituents for mutagenicity using the Ames' test. *Toxicology* 15 (3): 219–232.

Friedberg, E.C., Feaver, W.J., and Gerlach, V.L. (2000). The many faces of DNA polymerases: strategies for mutagenesis and for mutational avoidance. *Proceedings of the National Academy of Sciences* 97 (11): 5681–5683.

Fromowitz, M., Shuga, J., Wlassowsky, A.Y., et al. (2012). Bone marrow genotoxicity of 2, 5-dimethylfuran, a green biofuel candidate. *Environmental and Molecular Mutagenesis* 53 (6): 488–491.

Gentry, T.S. and Roberts, J.S. (2004). Formation kinetics and application of 5-hydroxymethylfurfural as a time–temperature indicator of lethality for continuous pasteurization of apple cider. *Innovative Food Science & Emerging Technologies* 5 (3): 327–333.

Germond, J.E., Philippossian, G., Richli, U., et al. (1987). Rapid and complete urinary elimination of [14C]-5-hydroxymethyl-2-furaldehyde administered orally or intravenously to rats. *Journal of Toxicology and Environmental Health, Part A Current Issues* 22 (1): 79–89.

Gibbings, S., Elkins, N.D., Fitzgerald, H., et al. (2011). Xanthine oxidoreductase promotes the inflammatory state of mononuclear phagocytes through effects on chemokine expression, peroxisome proliferator-activated receptor-γ sumoylation and HIF-1α. *Journal of Biological Chemistry* 286 (2): 961–975.

Glatt, H., Schneider, H., and Liu, Y. (2005). V79-hCYP2E1-hSULT1A1, a cell line for the sensitive detection of genotoxic effects induced by carbohydrate pyrolysis products and other food-borne chemicals. *Mutation Research/Genetic Toxicology and Environmental Mutagenesis* 580 (1): 41–52.

Glatt, H., Schneider, H., Murkovic, M., et al. (2011). Hydroxymethyl-substituted furans: mutagenicity in Salmonella typhimurium strains engineered for expression of various human and rodent sulphotransferases. *Mutagenesis* 27 (1): 41–48.

Godfrey, V.B., Chen, L.-J., Griffin, R.J., et al. (1999). Distribution and metabolism of (5-hydroxymethyl) furfural in male F344 rats and B6C3F1 mice after oral administration. *Journal of Toxicology and Environmental Health Part A* 57 (3): 199–210.

Gökmen, V., Açar, Ö.Ç., Köksel, H., and Acar, J. (2007). Effects of dough formula and baking conditions on acrylamide and hydroxymethylfurfural formation in cookies. *Food Chemistry* 104 (3): 1136–1142.

Gökmen, V., Açar, Ö.Ç., Serpen, A., and Morales, F.J. (2008). Effect of leavening agents and sugars on the formation of hydroxymethylfurfural in cookies during baking. *European Food Research and Technology* 226 (5): 1031–1037.

Gökmen, V. and Şenyuva, H.Z. (2007). Effects of some cations on the formation of acrylamide and furfurals in glucose–asparagine model system. *European Food Research and Technology* 225 (5–6): 815–820.

Goldberg, M.A., Brugnara, C., Dover, G.J., et al. (1992). Hydroxyurea and erythropoietin therapy in sickle cell anemia. *Seminars in Oncology* 19 (3): 74–81.

Gomes, S., Dias, L.G., Moreira, L.L., et al. (2010). Physicochemical, microbiological and antimicrobial properties of commercial honeys from Portugal. *Food and Chemical Toxicology* 48 (2): 544–548.

Hardt-Stremayr, M., Mattioli, S., Greilberger, J., et al. (2013). Determination of metabolites of 5-hydroxymethylfurfural in human urine after oral application. *Journal of Separation Science* 36 (4): 670–676.

Hayden, M.R. and Tyagi, S.C. (2004). Uric acid: a new look at an old risk marker for cardiovascular disease, metabolic syndrome and type 2 diabetes mellitus: the urate redox shuttle. *Nutrition & Metabolism* 1 (1): 10.

Høie, A.H., Svendsen, C., Brunborg, G., et al. (2015). Genotoxicity of three food processing contaminants in transgenic mice expressing human sulfotransferases 1A1 and 1A2 as assessed by the in vivo alkaline single cell gel electrophoresis assay. *Environmental and Molecular Mutagenesis* 56 (8): 709–714.

Iftikhar, F., Masood, M.A., and Waghchoure, E.S. (2011). Comparison of Apis cerana, Apis dorsata, Apis florea and Apis mellifera honey from different areas of Pakistan. *Asian Journal of Experimental Biological Sciences* 2 (3): 399–403.

Iijima, T. (2006). Mitochondrial membrane potential and ischemic neuronal death. *Neuroscience Research* 55 (3): 234–243.

Islam, A., Khalil, I., Islam, N., et al. (2012). Physicochemical and antioxidant properties of Bangladeshi honeys stored for more than one year. *BMC Complementary and Alternative Medicine* 12 (1): 177.

Islam, M., Khalil, M., and Gan, S.H. (2014). Toxic compounds in honey. *Journal of Applied Toxicology* 34 (7): 733–742.

Jachimowicz, T. and El Sherbiny, G. (1975). Zur Problematik der Verwendung von Invertzucker für die Bienenfütterung. *Apidologie* 6 (2): 121–143.

Janzowski, C., Glaab, V., Samimi, E., et al. (2000). 5-Hydroxymethylfurfural: assessment of mutagenicity, DNA-damaging potential and reactivity towards cellular glutathione. *Food and Chemical Toxicology* 38 (9): 801–809.

Kalabova, K., Vorlova, L., Borkovcova, I., et al. (2003). Hydroxymethylfurfural in Czech honeys. *Czech Journal of Animal Science* 48 (12): 551–557.

Khalil, M., Sulaiman, S., and Gan, S. (2010). High 5-hydroxymethylfurfural concentrations are found in Malaysian honey samples stored for more than one year. *Food and Chemical Toxicology* 48 (8): 2388–2392.

Kim, K., Kim, Y., Kim, H.Y., et al. (2008). Inhibitory mechanism of anti-allergic peptides in RBL2H3 cells. *European Journal of Pharmacology* 581 (1): 191–203.

Kinati, C., Tolemariam, T., and Debele, K. (2011). Quality evaluation of honey produced in Gomma Woreda of South Western Ethiopia. *Livestock Res. Rural Dev* 23 (9): 190.

Kitts, D.D., Chen, X.-M., and Jing, H. (2012). Demonstration of antioxidant and anti-inflammatory bioactivities from sugar–amino acid Maillard reaction products. *Journal of Agricultural and Food Chemistry* 60 (27): 6718–6727.

Krainer, S., Brodschneider, R., Vollmann, J., et al. (2016). Effect of hydroxymethylfurfural (HMF) on mortality of artificially reared honey bee larvae (Apis mellifera carnica). *Ecotoxicology* 25 (2): 320–328.

Krauze, A. and Zalewski, R.I. (1991). Classification of honeys by principal component analysis on the basis of chemical and physical parameters. *Zeitschrift fuer Lebensmittel-Untersuchung und Forschung* 192 (1): 19–23.

Küçük, M., Kolaylı, S., Karaoğlu, Ş., et al. (2007). Biological activities and chemical composition of three honeys of different types from Anatolia. *Food Chemistry* 100 (2): 526–534.

Kuster, B. (1990). 5-Hydroxymethylfurfural (HMF). A review focussing on its manufacture. *Starch-Stärke* 42 (8): 314–321.

LeBlanc, B.W., Eggleston, G., Sammataro, D., et al. (2009). Formation of hydroxymethylfurfural in domestic high-fructose corn syrup and its toxicity to the honey bee (Apis mellifera). *Journal of Agricultural and Food Chemistry* 57 (16): 7369–7376.

Lee, H.S. and Nagy, S. (1990). Relative reactivities of sugars in the formation of 5-hydroxymethylfurfural in sugar-catalyst model systems. *Journal of Food Processing and Preservation* 14 (3): 171–178.

Lee, Y.-C., Shlyankevich, M., Jeong, H.-K., et al. (1995). Bioactivation of 5-hydroxymethyl-2-furaldehyde to an electrophilic and mutagenic allylic sulfuric acid ester. *Biochemical and Biophysical Research Communications* 209 (3): 996–1002.

Li, M.-M., Wu, L.-Y., Zhao, T., et al. (2011). The protective role of 5-hydroxymethyl-2-furfural (5-HMF) against acute hypobaric hypoxia. *Cell Stress and Chaperones* 16 (5): 529.

Lin, S.-M., Wu, J.-Y., Su, C., et al. (2012). Identification and mode of action of 5-hydroxymethyl-2-furfural (5-HMF) and 1-methyl-1, 2, 3, 4-tetrahydro-β-carboline-3-carboxylic Acid (MTCA) as potent xanthine oxidase inhibitors in vinegars. *Journal of Agricultural and Food Chemistry* 60 (39): 9856–9862.

Lorne, E., Zmijewski, J.W., Zhao, X., et al. (2008). Role of extracellular superoxide in neutrophil activation: interactions between xanthine oxidase and TLR4 induce proinflammatory cytokine production. *American Journal of Physiology-Cell Physiology* 294 (4): C985–C993.

Magdoff-Fairchild, B., Swerdlow, P.H., and Bertles, J.F. (1972). Intermolecular organization of deoxygenated sickle haemoglobin determined by x-ray diffraction. *Nature* 239 (5369): 217–219.

Markowicz Bastos, D., Monaro, E., Siguemoto, E., et al. (2012). Maillard reaction products in processed food: pros and cons.

Meda, A., Lamien, C., Millogo, J., et al. (2005). Physiochemical analyses of Burkina fasan honey. *Acta Veterinaria Brno* 74 (1): 147–152.

Mehryar, L., Esmaiili, M., and Hassanzadeh, A. (2013). Evaluation of some physicochemical and rheological properties of Iranian honeys and the effect of temperature on its viscosity. *American-Eurasian Journal of Agricultural & Environmental Sciences* 13 (6): 807–819.

Mizushina, Y., Yagita, E., Kuramochi, K., et al. (2006). 5-(Hydroxymethyl)-2-furfural: a selective inhibitor of DNA polymerase λ and terminal deoxynucleotidyltransferase. *Archives of Biochemistry and Biophysics* 446 (1): 69–76.

Mondragón-Cortez, P., Ulloa, J., Rosas-Ulloa, P., et al. (2013). Physicochemical characterization of honey from the West region of México. *CyTA-Journal of Food* 11 (1): 7–13.

Moniruzzaman, M., Khalil, M.I., Sulaiman, S.A., and Gan, S.H. (2013). Physicochemical and antioxidant properties of Malaysian honeys produced by Apis cerana, Apis dorsata and Apis mellifera. *BMC Complementary and Alternative Medicine* 13 (1): 43.

Morales, F.J. (2008). Hydroxymethylfurfural (HMF) and related compounds. In: *Process-induced Food Toxicants: Occurrence, Formation, Mitigation and Health Risks*, 135–174. Chichester, UK: Wiley-Blackwell.

Mouhoubi-Tafinine, Z., Ouchemoukh, S., Louaileche, H., and Tamendjari, A. (2018). Effect of storage on hydroxymethylfurfural (HMF) and color of some Algerian honey. *International Food Research Journal* 25 (3).

Muli, E., Munguti, A., and Raina, S. (2007). Quality of honey harvested and processed using traditional methods in rural areas of Kenya. *Acta Veterinaria Brno* 76 (2): 315–320.

Muruke, M.H. (2014). Assessment of quality of Tanzanian honey based on physicochemical properties. *Food Science and Quality Management* 33: 61–72.

Nalda, M.N., Yagüe, J.B., Calva, J.D., and Gómez, M.M. (2005). Classifying honeys from the Soria Province of Spain via multivariate analysis. *Analytical and Bioanalytical Chemistry* 382 (2): 311–319.

Neupane, K. and Thapa, R. (2005). Alternative to off-season sugar supplement feeding of honeybees. *Journal of the Institute of Agriculture and Animal Science* 26: 77–81.

Nishi, Y., Miyakawa, Y., and Kato, K. (1989). Chromosome aberrations induced by pyrolysates of carbohydrates in Chinese hamster V79 cells. *Mutation Research Letters* 227 (2): 117–123.

Olga, E., María, F.-G., and Carmen, S.M. (2012). Differentiation of blossom honey and honeydew honey from Northwest Spain. *Agriculture* 2 (1): 25–37.

Pacher, P., Nivorozhkin, A., and Szabó, C. (2006). Therapeutic effects of xanthine oxidase inhibitors: renaissance half a century after the discovery of allopurinol. *Pharmacological Reviews* 58 (1): 87–114.

Pandey, V. and Modak, M.J. (1987). Biochemistry of terminal deoxynucleotidyltransferase (TdT): characterization and mechanism of inhibition of TdT by P1, P5-bis (5′-adenosyl) pentaphosphate. *Biochemistry* 26 (7): 2033–2038.

Pastoriza de la Cueva, S., Álvarez, J., Végvári, Á., et al. (2017). Relationship between HMF intake and SMF formation in vivo: an animal and human study. *Molecular Nutrition & Food Research* 61 (3).

Pattinson, K.T., Sutherland, A.I., Smith, T.G., et al. (2005). Acute mountain sickness, Vitamin C, free radicals and HIF-1α. *Wilderness & Environmental Medicine* 16 (3): 172–175.

Paulsen, J.E., Løberg, E.M., Ølstørn, H.B., et al. (2005). Flat dysplastic aberrant crypt foci are related to tumorigenesis in the colon of azoxymethane-treated rat. *Cancer Research* 65 (1): 121–129.

Preston, S., Leedham, S., Oukrif, D., et al. (2008). The development of duodenal microadenomas in FAP patients: the human correlate of the Min mouse. *The Journal of Pathology* 214 (3): 294–301.

Prior, R.L., Wu, X., and Gu, L. (2006). Identification and urinary excretion of metabolites of 5-(hydroxymethyl)-2-furfural in human subjects following consumption of dried plums or dried plum juice. *Journal of Agricultural and Food Chemistry* 54 (10): 3744–3749.

Qamer, S., Ahmad, F., Latif, F., et al. (2008). Physicochemical analysis of Apis dorsata honey from terai forests, Nepal. *Pakistan Journal of Zoology* 40 (1): 53–58.

Quarta, B. and Anese, M. (2012). Furfurals removal from roasted coffee powder by vacuum treatment. *Food Chemistry* 130 (3): 610–614.

Rajchl, A., Drgová, L., Grégrová, A., et al. (2013). Rapid determination of 5-hydroxymethylfurfural by DART ionization with time-of-flight mass spectrometry. *Analytical and Bioanalytical Chemistry* 405 (14): 4737–4745.

Ramadan, K., Maga, G., Shevelev, I.V., et al. (2003). Human DNA polymerase λ possesses terminal deoxyribonucleotidyl transferase activity and can elongate RNA primers: implications for novel functions. *Journal of Molecular Biology* 328 (1): 63–72.

Ramadan, K., Shevelev, I.V., Maga, G., and Hübscher, U. (2004). De novo DNA synthesis by human DNA polymerase λ, DNA polymerase μ and terminal deoxyribonucleotidyl transferase. *Journal of Molecular Biology* 339 (2): 395–404.

Reyes-Salas, E.O., Manzanilla-Cano, J.A., Barceló-Quintal, M.H., et al. (2006). Direct electrochemical determination of hydroxymethylfurfural (HMF) and its application to honey samples. *Analytical Letters* 39 (1): 161–171.

Rizelio, V.M., Gonzaga, L.V., Borges, G.D.S.C., et al. (2012). Development of a fast MECK method for determination of 5-HMF in honey samples. *Food Chemistry* 133 (4): 1640–1645.

Román-Leshkov, Y., Chheda, J.N., and Dumesic, J.A. (2006). Phase modifiers promote efficient production of hydroxymethylfurfural from fructose. *Science* 312 (5782): 1933–1937.

Rosatella, A.A., Simeonov, S.P., Frade, R.F., and Afonso, C.A. (2011). 5-Hydroxymethylfurfural (HMF) as a building block platform: biological properties, synthesis and synthetic applications. *Green Chemistry* 13 (4): 754–793.

Roychoudhury, R., Jay, E., and Wu, R. (1976). Terminal labeling and addition of homopolymer tracts to duplex DNA fragments by terminal deoxynudeotidyl transferase. *Nucleic Acids Research* 3 (1): 101–116.

Ruoff, K., Luginbühl, W., Bogdanov, S., et al. (2007). Quantitative determination of physical and chemical measurands in honey by near-infrared spectrometry. *European Food Research and Technology* 225 (3–4): 415–423.

Salhi, I., Samet, Y., and Trabelsi, M. (2020). Direct electrochemical determination of very low levels of 5-hydroxymethyl furfural in natural honey by cyclic and square wave voltametric techniques. *Journal of Electroanalytical Chemistry* 873: 114326.

Sancho, M.T., Muniategui, S., Huidobro, J.F., and Simal Lozano, J. (1992). Aging of honey. *Journal of Agricultural and Food Chemistry* 40 (1): 134–138.

Sang, N., Stiehl, D.P., Bohensky, J., et al. (2003). MAPK signaling up-regulates the activity of hypoxia-inducible factors by its effects on p300. *J Biol Chem* 278 (16): 14013–14019.

Saxena, S., Gautam, S., and Sharma, A. (2010). Physical, biochemical and antioxidant properties of some Indian honeys. *Food Chemistry* 118 (2): 391–397.

Schoental, R., Hard, G., and Gibbard, S. (1971). Histopathology of renal lipomatous tumors in rats treated with the "natural" products, pyrrolizidine alkaloids and α, β-unsaturated aldehydes. *Journal of the National Cancer Institute* 47 (5): 1037–1044.

Schroeder, J., Kagey-Sobotka, A., and Lichtenstein, L. (1995). The role of the basophil in allergic inflammation. *Allergy* 50 (6): 463–472.

Shapla, U.M., Solayman, M., Alam, N., et al. (2018). 5-Hydroxymethylfurfural (HMF) levels in honey and other food products: effects on bees and human health. *Chemistry Central Journal* 12 (1): 1–18.

Sies, H. (1985). Oxidative stress: introductory remarks.In: *Oxidative Stress*, 1–8. London: Academic Press.

Singh, I. and Singh, S. (2018). Honey moisture reduction and its quality. *Journal of Food Science and Technology* 55 (10): 3861–3871.

Singh, N. and Bath, P.K. (1997). Quality evaluation of different types of Indian honey. *Food Chemistry* 58 (1): 129–133.

Sodré, G.D.S., Marchini, L.C., Moreti, A.C.D.C.C., et al. (2011). Physico-chemical characteristics of honey produced by Apis mellifera in the Picos region, state of Piauí, Brazil. *Revista Brasileira de Zootecnia* 40 (8): 1837–1843.

Solayman, M., Islam, M.A., Paul, S., et al. (2016). Physicochemical properties, minerals, trace elements and heavy metals in honey of different origins: a comprehensive review. *Comprehensive Reviews in Food Science and Food Safety* 15 (1): 219–233.

Spano, N., Casula, L., Panzanelli, A., et al. (2006). An RP-HPLC determination of 5-hydroxymethylfurfural in honey: the case of strawberry tree honey. *Talanta* 68 (4): 1390–1395.

Stetson, C.A., Jr (1966). The state of hemoglobin in sickled erythrocytes. *The Journal of Experimental Medicine* 123 (2): 341.

Surh, Y.-J., Liem, A., Miller, J.A., and Tannenbaum, S.R. (1994). 5-Sulfooxymethylfurfural as a possible ultimate mutagenic and carcinogenic metabolite of the Maillard reaction product, 5-hydroxymethylfurfural. *Carcinogenesis* 15 (10): 2375–2377.

Svendsen, C., Husøy, T., Glatt, H., et al. (2009). 5-Hydroxymethylfurfural and 5-sulfooxymethylfurfural increase adenoma and flat ACF number in the intestine of Min/+ mice. *Anticancer Research* 29 (6): 1921–1926.

Terrab, A., González, A.G., Díez, M.J., and Heredia, F.J. (2003). Characterisation of Moroccan unifloral honeys using multivariate analysis. *European Food Research and Technology* 218 (1): 88–95.

Tucak, Z., Periškić, M., Škrivanko, M., and Konjarević, A. (2007). The influence of the botanic origin of honey plants on the quality of honey. *Agriculture* 13 (1): 234–236.

Turhan, K. (2009). Effects of thermal treatment and storage on hydroxymethylfurfural (HMF) content and diastase activity of honeys collected from middle Anatolia in Turkey. In: *Innovations in Chemical Biology*, 233–239. Dordrecht, Netherlands: Springer.

van der Zee, R. and Pisa, L. (2010). Bijensterfte 2009-10 en toxische invertsuikersiroop. *NCB Rapp* 2 (2010): 1–15.

Wang, M.-Y., Zhao, F.-M., Peng, H.-Y., et al. (2010). Investigation on the morphological protective effect of 5-hydroxymethylfurfural extracted from wine-processed Fructus corni on human L02 hepatocytes. *Journal of Ethnopharmacology* 130 (2): 424–428.

Wen, C., Shi, X., Wang, Z., et al. (2016). Effects of metal ions on formation of acrylamide and 5-hydroxymethylfurfural in asparagine–glucose model system. *International Journal of Food Science & Technology* 51 (2): 279–285.

White, J., Jr (1979). Spectrophotometric method for hydroxymethylfurfural in honey. *Journal-Association of Official Analytical Chemists* 62 (3): 509–514.

Winkler, O. (1955). Beitrag zum Nachweis und zur Bestimmung von Oxymethylfurfurol in Honig und Kunsthonig. *Zeitschrift für Lebensmitteluntersuchung und-Forschung A* 102 (3): 161–167.

Wootton, M. and Ryall, L. (1985). A comparison of Codex Alimentarius Commission and HPLC methods for 5-hydroxymethyl-2-furaldehyde determination in honey. *Journal of Apicultural Research* 24 (2): 120–124.

Yamada, P., Nemoto, M., Shigemori, H., et al. (2011). Isolation of 5-(hydroxymethyl) furfural from Lycium chinense and its inhibitory effect on the chemical mediator release by basophilic cells. *Planta medica* 77 (05): 434–440.

Yang, W., Zhang, C., Li, C., et al. (2019). Pathway of 5-hydroxymethyl-2-furaldehyde formation in honey. *Journal of Food Science and Technology* 56 (5): 2417–2425.

Yilmaz, H. and Küfrevioğlu, Ö.İ. (2001). Composition of honeys collected from Eastern and South-Eastern Anatolia and effect of storage on hydroxymethylfurfural content and diastase activity. *Turkish Journal of Agriculture and Forestry* 25 (5): 347–349.

Yuan, J.-P. and Chen, F. (1998). Separation and identification of furanic compounds in fruit juices and drinks by high-performance liquid chromatography photodiode array detection. *Journal of Agricultural and Food Chemistry* 46 (4): 1286–1291.

Zappala, M., Fallico, B., Arena, E., and Verzera, A. (2005). Methods for the determination of HMF in honey: a comparison. *Food Control* 16 (3): 273–277.

Zhang, X.-M., Chan, C.C., Stamp, D., et al. (1993). Initiation and promotion of colonic aberrant crypt foci in rats by 5-hydroxymethy1-2-furaldehyde in thermolyzed sucrose. *Carcinogenesis* 14 (4): 773–775.

Zhao, L., Chen, J., Su, J., et al. (2013). In vitro antioxidant and antiproliferative activities of 5-hydroxymethylfurfural. *Journal of Agricultural and Food Chemistry* 61 (44): 10604–10611.

Zhao, L., Su, J., Li, L., et al. (2014). Mechanistic elucidation of apoptosis and cell cycle arrest induced by 5-hydroxymethylfurfural, the important role of ROS-mediated signaling pathways. *Food Research International* 66: 186–196.

Zhao, Y., Flugstad, B., Kolbe, E., et al. (2000). Using capacitive (radio frequency) dielectric heating in food processing and preservation–a review. *Journal of Food Process Engineering* 23 (1): 25–55.

# 12

## Other Possible Contaminants, Toxic Compounds, and Microbial Growth

*Fahmida Alam, Kashif Maroof, Ng Choon Ming, Md. Ibrahim Khalil, and Siew Hua Gan*

### Introduction

Honey has innumerable health benefits and possesses the image of being natural, healthy, and clean. It is consumed as a wholesome product by many people around the world because it fulfils necessary nutritional and safety standards in addition to the specific health-promoting qualities. Consumer acceptance and satisfaction have led to continuous high demand in the market. Therefore, it is important to supply honey enriched with nutritional aspects while meeting quality standards. To maintain such good attributes, the safety of honey for human consumption is also critical. However, in the past and at present, honey and other bee products are being polluted by different sources of contamination from the environment and from beekeeping practices that include pesticides, antibiotics, various microorganisms, and radioactive materials.

### Pesticides and Honey Contamination

Globally, pesticides are used to control bee diseases and pests. However, pesticide use is often uncontrolled and applied without authorized protocols (Al-Waili 2012). Although pesticides play a vital role in agricultural productivity by protecting crops from pests, their usage in agriculture is generally unwelcome (Sharma 2019). Uncontrolled pesticide administration possesses serious contamination and increases environmental, animal, and human health risks (Aktar et al. 2009). Undoubtedly, farm workers and their family members are particularly vulnerable. Children are also susceptible and sensitive to pesticides because of growth, development, physiology, and diet factors (Roberts et al. 2012). For instance, unintended consumption among children can potentially be at a much higher dose than adults because of increased food and fluids per kilogram of body weight. The negative consequences of pesticide exposure include both acute and chronic effects, such as mild skin inflammation, food poisoning, birth defects, tumor formation, mutagenic effects, blood and nerve disorders, endocrine disruption, and even coma or death (Singh 2018). Moreover, chronic exposure to pesticides can lead to various conditions and illnesses, including cancer, neurobehavioral disorders, and infertility (Nicolopoulou-Stamati 2016).

A variety of pesticides residues exist, such as the acaricides, organic acids, insecticides, fungicides, herbicides, and bactericides. These chemical substances, including acaricides amitraz, celazole, bromopropylate, coumaphos, flumethrin, and taufluvalinate, are usually administered to manage varroatosis and ascospheriosis. However, the usage of these toxic chemicals inside beehives may lead to high direct contamination risk to honey and other bee products (Korta 2002). For instance, it was reported that 79 and 56 distinct pesticides and metabolites were found in 168 pollen and 142 wax comb samples, respectively (Ostiguy 2019).

In addition, organic pollutants such as polychlorinated biphenyls (PCBs) used in past for industrial and commercial purpose could potentially contaminate bee products. Despite the ban on PCBs relating to the harmful effects on humans, the stability and lipophilic nature of PCBs have led to accumulation and residues in the environment (Thomas 2008). For example, PCBs were detected highest in honeybees followed by honey and pollen, all of which was greater in urban areas than semi-urban areas (Sari et al. 2020). Meanwhile, a review elucidated that organochlorine pesticides found in honey varied substantially across countries (Bogdanov 2006).

The substances approved for use in apiculture are listed in Table 12.1 along with the acceptable daily intakes established by the Joint Food and Agriculture Organization/World Health Organization (WHO) Expert Committee on Food Additives

*Honey: Composition and Health Benefits*, First Edition. Edited by Md. Ibrahim Khalil, Gan Siew Hua, and Bey Hing Goh.

**Table 12.1** List of products approved in apiculture.

| Name of Product | Major Application | Acceptable Daily Intake (mg/kg body weight per day) |
|---|---|---|
| Acrinathrine | Pesticide and acaricide | – |
| Amitraz (Apivar) | Pesticide and acaricide | 0–0.01 |
| Bromopropylate | Pesticide and acaricide | 0–0.03 |
| Chlorobenzilate | Pesticide and acaricide | 0–0.02 |
| Coumaphos (Perizin)[a] | Pesticide | – |
| Cymiazole hydrochloride (Apitol) | Pesticide and acaricide | – |
| Enilconazol (imazalil) | Pesticide and acaricide | 0–0.03 |
| Fenproximate | Pesticide | – |
| Fipronil | Pesticide | 0–0.0002 |
| Flumethrin (Bayvarol) | Pesticide | 0–0.004 |
| Fumagillin | Pesticide | – |
| Malathion | Pesticide | 0–0.30 |
| Methyl bromide | Pesticide | – |
| Oxalic acid | Pesticide | – |
| Paradichlorobenzene | Pesticide | – |
| Permethrin | Pesticide | 0–0.05 |
| Propargite | Pesticide | 0–0.01 |
| Spinosad | Pesticide | 0–0.02 |
| Tau-fluvalinate (Apistan) | Pesticide | – |
| Thymol[b] | Pesticide | – |

[a] Acceptable daily intake (ADI) withdrawn in 1980; no allocated ADI in 1990.
[b] Substances generally regarded as safe by numerous national authorities.

(WHO 2009, 2018). Generally, the maximum residue limits (MRLs) for pesticides in honey are set at levels as low as parts per billion. Yet challenges remain in international trade because of the lack of standardization for pesticide residues permitted across varying national regulations (Al-Waili 2012). For instance, countries such as Germany, Italy, and Switzerland have differing MRLs levels for amitraz, bromopropylate, coumaphos, cyamizole, flumetrine, and fluvalinate (Bogdanov 2006). On the other hand, the Codex Alimentarius did not set MRLs for pesticide residues in honey (Codex Alimentarius 1998). As for the European Commission, technical guidelines are drawn up to assess the extent of pesticide residues found in honey. European Union (EU) legislation regulates the MRLs of pesticide residues in honey, including amitraz, coumaphos, and cyanamide set at MRLs of 0.2, 0.1, and 0.01 mg/kg, respectively (European Commission 2022). In 2012, the European Commission further restricted the use of several insecticides (neonicotinoids and fipronil) owing to the acute risks reported (European Commission 2022). More recently in 2021, the EU published a report to modify the existing MRL for ametoctradin in honey from 0.05 to 5 mg/kg with compelling evidence of low risks to humans (EFSA 2021). On the other hand, the Food and Drug Administration (FDA) of the United States regulate the MRLs for amitraz, coumaphos, and fluvalinate, set at 1.0, 0.10, and 0.05 mg/kg, respectively (FDA).

### Asian Countries

In India, an exploratory study comprising 51 honey samples collected from various parts of Himachal Pradesh was conducted to find out the levels of pesticide residue contamination. A total of 18 samples (35.29%) were contaminated with organochlorine and cyclodiene pesticide residues in which hexachlorocyclohexane (HCH) and its isomers were present predominantly (17.64%), and dichloro-diphenyl-trichloroethane (DDT) and its isomers made up 13.72%. Contamination of commercial apiary honeys was higher (31.37%) compared with the honeys from natural vegetation (9.80%) produced by *Apis cerana*. Most of the contaminated samples were of mustard and rapeseed floral origins (90.90%), and 41.12% of multifloral honeys were contaminated (Choudhary and Sharma 2008). Although the use of DDT and HCH in agriculture has been banned in India for decades with restricted use, their presence in honey samples implies their long-term stability in soil and water resources (Eto and Eto 1979).

In another study, a total of 27 market honey samples collected from Haryana, India, from 1993 to 1995 were examined for the presence of organochlorine, synthetic pyrethroids, organophosphates, and carbamate insecticides. The findings demonstrated that all honey samples tested were contaminated with at least one type of pesticide. Residues of 10 distinct organochlorines pesticides found ranged between 0.01–6.0 mg/kg, whereas residues of 8 distinct organophosphates and 3 carbamate pesticides were found in relatively higher levels (between 0.1 and 9.0 mg/kg) in the honey samples studied (Anju 1997). In southern Karnataka, India, multifloral honey samples from wild colonies of *Apis dorsata* were analyzed for organochlorine, organophosphorus, pyrethroids, and herbicides. Interestingly, all the honeys tested were not contaminated with pesticide within detectable limits, suggesting that honey originated from the combs of *A. dorsata* in the wild is safe for consumption (Raghunandan and Basavarajappa 2013).

In Malaysia, 14 varying honey samples collected from various regions from 2010 to 2011 were analyzed for seven organophosphorus pesticide residues (diazinon, chlorpyrifos, acephate, dimethoate, malathion, phenthoate, and fenitrothion) and three carbamate pesticides (carbaryl, carbofuran, and carbosulfan) to explore their contamination levels. After investigation, no pesticide residues were detected in any of the honey samples, suggesting the purity of honey in uncontaminated regions (Moniruzzaman 2014).

In Turkey, analysis of a total 109 different market honey samples collected from Konya reported the presence of organochlorine pesticide residues in which aldrin (0.0032 μg/g), cis-chlordane (0.0043 μg/g), trans-chlordane (0.0146 μg/g), oxy-chlordane (0.0540 μg/g), 2,4′-dichloro-diphenyl-dichloroethylene (0.0072 μg/g), and 4,4′-DDE (dichlorodiphenyldichloroethylene) (0.0077 μg/g) were found in all honey samples (Arnon 1979). The pesticide residues of oxy-chlordane in 55 samples exceeded the limits of Turkish Alimentarius Codex MRLs. Furthermore, all the honey samples tested were contaminated, and the majority of them exceeded the MRLs (Codex Alimentarius 1997, 2005; Yavuz 2010). In the same country, analysis of nine honey samples from the area Kahramanmaraş showed the presence of γ-HCH (0.71–2.47 ng/g) in all the samples. However, the levels of pesticides were low, and all samples were within the limits set by Turkish Codex Alimentarius (Erdoğrul 2007). Similarly, a study assessed 15 organophosphorus insecticides in 275 honey samples from 33 cities in Turkey; which no insecticide residues were detected with limit of determination values ranging from 0.25 to 9.55 ng/g (Das and Kaya 2009). However, another study determined the presence of naphthalene in 100 samples of honey (53 from markets and 47 from street bazaars) in Turkey, revealing that 80.4% of the samples tested were contaminated with 1.0 μg/g of naphthalene (Beyoglu and Omurtag 2007).

In Jordan, analysis of 26 honey samples in 1995 reported the presence of organochlorine pesticides, including α-HCH, β-HCH, and lindane (0.01–0.026 μg/g), in the majority of the locally produced honey samples (86%), and others were contaminated with organophosphorus pesticides (Al-Rifai and Akeel 1997).

A recent study in Saudi Arabia identified the pesticide residue content in natural honeys produced in various regions (Osman et al. 2020). The study findings indicated the presence of chlorothalonil (4.56 mg/kg), carbaryl (4.12 mg/kg), and dicofol (0.24 mg/kg) in honey samples produced in Gizan, Belgarshy, and Al-Taaf, respectively. The highest amount of monocrotophos found was in samples from Belgarshy (49.13 mg/kg), and chlorothalonil was detected in honey from Riyadh (2.38 mg/kg) and Unayzah (0.253 mg/kg). The control of pesticides usage and contamination is imperative given the numerous groups of pesticides found in honey samples from different locations; if this is not monitored, it can pose serious risks to human health and the environment.

## European Countries

In a Spanish study, 111 honey samples of varying botanical origins from the Aragon area were analyzed. The analysis revealed the presence of α-HCH, PCB-28, and PCB-138 in more than half of the samples, of which PCB-138 was found predominantly (83.75%). The average concentrations of the organochlorine pesticide residues ranged from 0.17–7.90 μg/kg, whereas PBSs alone ranged from 0.28–25.40 μg/kg. Overall, these concentrations of persistent organic pollutants represent a very low level of contamination in honey samples (Herrera 2005). In Portugal and Spain, 50 honey samples obtained from local markets were examined for 42 organochlorine, carbamate, and organophosphorus pesticide residues (Blasco 2003). The majority of the pesticides found were organochlorines, of which γ-HCH was commonly detected (in 50% of the samples) followed by hexachlorobenzene (HCB) (32%), α-HCH (28%), β-HCH (26%), DDT, and their metabolites (20%). In general, Portuguese honeys contained greater amounts of pesticides than Spanish honeys. However, based on the comparison between estimated honey consumed and the respective ADI levels of pesticides present, the toxicologic effect was found to be minimal with relatively low health risks.

In Italy, 21 honey samples botanically originated from carob (*Ceratonia siliqua* L.), chestnut (*Castanea sativa*), and eucalyptus (*Eucalyptus globulus* Labill.) obtained in various areas of Sicily in 2012 were analyzed. Concentrations of all analyzed pesticides, including organochlorines, organophosphates, pyrethrins, and pyrethroids, in honey samples were below the limit of detection (<0.01 mg/kg) (Naccari 2014). The results indicated that honey produced in Sicily poses a minimal risk of pesticide exposure to consumers. In another Italian study, 72 honey samples collected from various geographic regions were analyzed, and 94% of the samples contained no less than one pesticide with concentrations lower than the MRLs. The presence of organochlorine pesticides – DDT, dichloro-diphenyl-dichloro-ethane (DDD), and DDE were found more often in honey produced in the industrialized area (>80% of the samples). Chlorpyrifos (100%) and quinoxyfen (90%) residues were found in samples that originated from the apple orchard area, where numerous insecticides are applied for crop protection, followed by captan, trifloxystrobin, iprodione, and boscalid (75%). The study results showed that pesticide contamination in honey could be an indicator of environmental pollution in the specific area (Panseri 2014).

In Greece, 50 honey samples were randomly collected from apiaries representing six agricultural areas where insecticides were used largely for crop protection. The most contaminated honeys were produced from citrus groves containing chlorpyrifos (52.63%) followed by chlorfenvinphos (21.05%) and phorate (10.53%). Samples originating from cotton fields contained mainly phorate (35.29%) followed by chlorfenvinphos (5.88%) and chlorpyrifos (5.88%), whereas samples from sunflower fields contained mostly phorate residues (44.44%). The study also reported that although 10 different insecticides are used in the representing areas, only 3 of the 50 analyzed honey samples contained residues of phorate (24%) followed by chlorpyrifos (22%) and chlorfenvinphos (10%). Fortunately, these three insecticides were present in low levels ranging from 0.70 to 0.89 μg/kg (Balayiannis and Balayiannis 2008). The detected chlorfenvinphos, chlorpyrifos, and phorate residues, despite low levels, could be due to their stability to hydrolytic, oxidative, and metabolic processes and low volatility as well as excessive usage by the farmers.

In Brazil, locally produced honey samples collected from 2003–2004 from Bauru (State of São Paulo) were analyzed for 48 pesticides, including organohalogen, organophosphorous, pyrethroids, and organonitrogen (Rissato 2007). The findings reported the presence of pesticides that belong to the organohalogen (endosulfan sulfate, HCB, and tetradifon) and organophosphorous (atrazine, simazine, and tebuconazole) groups. Malathion residues were found to be of higher concentrations in all the samples tested because of its usage to in controlling dengue mosquitoes in the studied location.

In Poland, fungicides typically applied to fruit trees for pest control can be contaminated by the surrounding environment, implicating bees and their products. For instance, studies detected the residues of vinclozolin, iprodione, methyl thiophanate, captan, and difenoconazole in honey samples. Vinclozolin was found in the highest level (0.11 mg/kg), and the other fungicides detected ranged between 0.001 and 0.06 mg/kg (Fléché 1997; Rexilius 1986). In another Polish study conducted in 2008, 13 different pesticides were tested in 40 honey samples obtained from apiaries of five areas. Of the 13 pesticides, 8 were detected in the samples analyzed, and residue of clothianidin was found most commonly (65%) followed by carfentrazone ethyl (32%), fluroxypyr-meptyl (32%), metalaxyl (28%), and methidathion (25%). Meanwhile, imazalil (12%), chlorpyrifos (8%), and bitertanol (2%) were detected in some of the samples. The levels of clothianidin (2.36 ng/g) and imazalil (4.98 ng/g) were the highest (Kujawski and Namieśnik 2011). Furthermore, another study conducted in the northern part of Poland in 2010 analyzed 45 honey samples of different botanical origins. In the study, 29% of the samples contained some of the target compounds. Concentrations of bifenthrin (14.5 ng/g), fenpyroximate (16.3 ng/g), methidathion (25.7 ng/g), spinosad (20.6 ng/g), thiamethoxam (20.2 ng/g), and triazophos (20.3 ng/g) exceeded the MRLs in five samples (11%), correlating to the agriculture practices in that region (Bargańska et al. 2013).

In Switzerland, a study detected the residues of dithianon, pyrifenox, penconazole, and cyproconazole fungicides in honey samples ranging from 0.02 to 0.09 mg/kg (Rüegg 1995). These contaminants were commonly administered to fruit trees situated in the surrounding areas.

In France, an analysis of 141 honey samples collected from 2008–2009 from a western area reported that 135 samples were contaminated with a minimum of one pesticide contaminant. Of 80 analyzed pesticides, 28 were detected in honey samples. The most frequently detected residues were coumaphos (78.0%), amitraz II (68.8%), carbendazim (64.5%), phosmet (12.8%), and cyproconazole (11.3%). The highest amounts detected were the acaricide amitraz II (116.10 ng/g) and the fungicide carbendazim (87.90 ng/g). Honey was mostly contaminated by triazole fungicides such as cyproconazole, which was detected in some of the samples tested (11.3%). Coumaphos (78.0%), triphenylphosphate (2.1%), and phosmet (12.8%) were the three most dominant cholinesterase inhibitor insecticides found (Lambert 2013).

In Romania, 18 honey samples of varying botanical origins and geographical counties were analyzed for the presence and quantification of fungicides using a high-resolution mass spectrometry technique. The results showed the presence of 11 contaminants from agriculture practices, namely, carbendazim, enilconazole, hexaconazole, penconazole, tebuconazole, flusilazole, thiabendazole, terconazole, cyproconazole, propiconazole, and metalaxyl. Furthermore, the most abundant antifungal was enilconazole (imazalil), identified in 14 honey samples. Nevertheless, all the detected contaminants were below the MRLs set by legislative body, ranging between 1.7 and 7.2 μg kg − 1. Hence, it was concluded that the honeys did not have adverse health effects to consumers (Blaga 2020).

### South American Countries

In Colombia, 61 multifloral honey samples collected from beekeepers of four geographical locations in 2011 were analyzed. The analysis detected pesticide residues in 32 samples, mainly from the organochlorine and organophosphorus groups. Primarily, chlorpyrifos was detected (36.1%) followed by profenofos (16.4%), DDT (6.6%), HCB, γ-HCH (4.9%), and fenitrothion (1.6%). Despite the ban on the use of organochlorine pesticides in Colombia for the past decades, the concentration of DDT residues (0.038 mg/kg) detected was the highest among other pesticides (López 2014). Overall, 4.9% of the honey samples tested surpassed the EU MRLs.

### North American Countries

In the United States, the use of parasitic mites (*Acarapis woodi* and *Varroa destructor*) demonstrated a considerable negative effect on beekeeping production (Sammatara et al. 2000). Mite infestations have profoundly decreased the number of beekeeper colonies and contributed to the extensive application of chemical miticides in beehives. However, a study was carried out in the United States in 2007 in which a total of 50 honey samples from Virginia were collected for the detection of fluvalinate and coumaphos residues. Samples were retrieved from both hives (22) and bottled honey (28) supplied by beekeepers. Fortunately, coumaphos or fluvalinate residues did not exceed the limit of quantification (0.05 mg/kg) in any of the tested samples. Additionally, no residues were found in any of the bottled honey (Fell and Cobb 2009).

### African Countries

In Ghana, honey samples collected in 2014 from wild forest and beekeepers were analyzed. Pesticides, mainly synthetic pyrethroids, organophosphorus, and organochlorines, were detected. Although the sampling sites were situated in agricultural farmlands that were constantly exposed to various pesticides for the control of insects or weeds, organophosphorus pesticides, particularly chlorpyrifos, dimethoate, methoxychlor, and malathion, identified were below their respective MRLs set by the EU (Darko 2017). However, cyfluthrin (concentration was equal to the MRL, 0.02 mg/kg, in the forest honey samples) and permethrin (concentration close to the MRL) were the most detected pyrethroids. The concentration of ΣDDT from the organochlorines was equal to the MRL and thus warrants attention as suggested by the researchers.

In Egypt, 46 pesticides, including organochlorine, organophosphorous, pyrethroid, and organonitrogen, were examined in honey samples from 18 apiaries in the year 2011. A total of 55.6% of the samples contained residues of the tested pesticides, mostly belonging to the organophosphorous group followed by the organochlorine group. Residues of dicofol was found in 38.9% of the samples in addition to other acaricides, including bromopropylate, tetradifon, and malathion, relating to their application in controlling *V. destructor*. Other pesticides such as β-HCH, γ-HCH, and carbamate pesticide pirimicarb were also found. Unfortunately, the majority of the pesticides found (81.8%) exceeded the EU MRLs.

To protect consumer health, researchers thus suggested the regular monitoring of pesticide residues in honey at a widespread country level (Eissa et al. 2014). Likewise, an Egyptian pilot study comprising 100 honey samples collected in 2013 from different botanical origins detected the presence of organochlorine HCB (0.0085 mg/kg) most frequently (70%) followed by permethrin and heptachlor epoxide. Surprisingly, only one sample was detected with a higher level of γ-HCH exceeding the MRL (0.01 mg/kg) (Malhat 2015). Although the use of organochlorine pesticides has been banned in Egypt since 1986, its extensive usage in the past and its highly persistent nature have led to their lasting presence in the environment.

## Antibiotics and Honey Contamination

Antibiotics are medicines that are applied therapeutically for human and animal health. Antibiotics function by inhibiting the growth of microorganisms, which include bacteria, fungi, and protozoa. Generally, beekeeping practices that use antibiotics are less labor intensive and more cost effective. Typically, antibiotic residues found in honey mainly originate from beekeeping practices (Johnson 2010). As a result, these antibiotics can affect consumers' health by either (1) directly producing toxic reactions or

(2) generating allergies or hypersensitivity reactions (Velicer 2004). For instance, cutaneous eruptions, dermatitis, gastrointestinal symptoms, and anaphylaxis can occur even with a minimal dose of lactam antibiotics (Paige et al. 1997). Moreover, the resistance capability of the bacterial population can increase if antibiotics are consumed along with food and honey.

Antibiotics that have been approved worldwide in apiculture for controlling bee diseases are listed in Table 12.2 (WHO 2009). Antibacterial substances such as sulfonamides, tetracyclines, erythromycin, lincomycin, tylosin, and streptomycin are used to treat bacterial brood diseases (Sereia 2010). Yet the MRLs of sulfonamides and TCs have not been determined for honey (Mutinelli 2003).

The EU regulation on honey's requirements falls under the Council Directive 2001/110/EC (European Commission 2021). According to EU regulation 37/2010, the MRLs for all antibiotics must be established before the usage in any food-producing species. However, there is no MRL set for antibiotics use in honey based on the EU regulations (Mutinelli 2003). Therefore, this implies that the EU does not permit the usage of antibiotics in honeybees, thereby restricting the sale and trade of any honey that contains antibiotic residues. Likewise, the usage of antibiotics in bee products is regarded as illegal in certain EU countries. For the import of honey, EU has established the Reference Points for Action (RPAs), which reflect the lowest antibiotics concentration permitted without harm to humans and detectable by official laboratories (EFSA 2018). When the RPA limit is exceeded, the shipment of the imported honey will be rejected by being either destroyed or sent back to the origin country. Thus far, RPAs are set for the presence of chloramphenicol and nitrofurans. Additionally, the EU has established MRL of 25 μg/kg for oxytetracycline residues in honey. Additionally, some countries, including Switzerland, the United Kingdom, and Belgium, have set Action Limits for antibiotics in honey corresponding to the maximum amount allowed, which generally ranges from 0.01 to 0.05 mg/kg for every antibiotic group (Johnson 2010).

In United States, the FDA authorized 11 antimicrobials, including oxytetracycline, tylosin, and lincomycin, for the use in beekeeping practices, strictly with licensed veterinarian and professional supervision (FDA 2022). The use of veterinary drugs must also adhere to the indications, oral dosage limit, and time set. The FDA and authoritative bodies in India and Australia have allowed for oxytetracycline use with MRL set at 300 ppb. In Canada, both MRL and working residue level (WRL) have been established for various antimicrobial residues in honey (Canada Government 2005). Basically, the WRLs are values obtained by extrapolating the lowest permitted MRL in other food-producing animals. Examples of antimicrobial veterinary drugs include lincomycin (WRL, 0.03 ppm), tetracycline (MRL, 0.075 ppm), and erythromycin (MRL, 0.03 ppm). Meanwhile, chloramphenicol antibiotics and 5-nitrofuran compounds are strictly prohibited in Canada. Furthermore, the Ministry of Commerce and Industries India has set up a Residue Monitoring Plan to regulate antibiotic residues in honey for export to the EU and United States (Johnson 2010). Yet there remains lack of monitoring of antibiotics level in the domestic market and imported honey in India itself.

**Table 12.2** List of veterinary drugs approved in apiculture.

| Drug | Acceptable Daily Intake (mg/kg body weight per day) |
|---|---|
| Chlortetracycline | 0–0.003 |
| Erythromycin | 0–0.0007 |
| Formic acid[a] | 0–3 |
| Lactic acid[a] | Not limited |
| Lincomycin hydrochloride | 0–0.03 |
| Menthol[a] | 0–4 |
| Monensin | 0–0.01 |
| Oxytetracycline | 0–0.003 |
| Rifampicin | – |
| Streptomycin/dihydrostreptomycin | 0–0.05 |
| Sulfathiazole | No ADI allocated |
| Tylosin tartrate | 0–0.03 |
| Thymol[a] | Acceptable |

[a] Substances generally regarded as safe by authoritative bodies in various countries.

## Asian Countries

In China, antibiotics, namely TC (20 μg/L), oxytetracycline (20 μg/L), chloramphenicol (40 μg/L), and doxycycline (20 μg/L), were identified using high-performance capillary electrophoresis in honey samples with respective detection limits in the year 2001 (Chen 2001). Later in 2008, five tetracyclines – including TC, oxytetracycline, CTC, doxycycline, and metacycline – were determined in eight honey samples of varying origins using high-performance liquid chromatography analysis in which the limits of detection ranged from 5 to 12 ng/g. Researchers concluded that the tested honey samples were not contaminated with TC residues at the minimum detectable level (Li 2008). Similarly, in 2009, six honey samples collected from the local markets and beekeepers located in Changchun and Beishan areas of China were analyzed with a polymer-based method to which no sulfonamide residues were found at detectable levels (Chen 2009).

In Japan, a study in 2004 reported an analysis of 40 commercial honey samples (28 imported, 12 locally produced). Streptomycin at a residue level of 0.01–0.17 μg/g was detected in 17 samples of the imported honeys and 2 of the domestically produced honeys. Nevertheless, these residue levels of streptomycin contamination were considered a low risk factor for causing significant health hazards (Horie 2004).

In India, nectar and honey produced during the flowering periods of rubber (March–April) and banana crops (December–January) south of Tamil Nadu were examined for antibiotic contamination. Findings revealed that the tested samples contained 4.0–17.0 and 11.0–29.0 ng/kg of streptomycin, 2.0–29.0 and 3.0–44.0 ng/kg of ampicillin, and 17.0–34.0 and 26.0–48.0 ng/kg of kanamycin, respectively (Solomon et al. 2006). According to a report by authoritative body in India, whereas approximately 14% of collected honey samples were contaminated with TC in 2006, approximately 28% were contaminated with the same antibiotic in 2007 to 2008 (Johnson 2010). Subsequently, in 2009–2010, 29.2% of 362 honey samples exceeded the antibiotic levels allowed.

On the other hand, only a small number of honey samples produced locally in Belgium from 2000 to 2001 were found to be contaminated with streptomycin in 4 of 248 samples, TC in 2 of 72 samples, and sulfonamides in 4 of 72 samples (Reybroeck 2002). Nonetheless, the study reported that imported honey was highly contaminated with streptomycin (51 of 108 samples), tetracyclines (29 of 98 samples), sulfonamides (31 of 98 samples), and chloramphenicol (40 of 85 samples).

In Iran, 135 honey samples collected from the northern part were analyzed using an enzyme-linked immunosorbent assay (ELISA) method to detect antibiotics, including enrofloxacin, penicillin, chloramphenicol, gentamicin, tylosin, TC, and sulfonamide (Mahmoudi et al. 2014). Among all these antibiotics, enrofloxacin was present in the highest percentage (20.7%) in 28 honey samples with increased concentration (0.6–72.1 ng/g) followed by gentamicin (18.15%), sulfonamide (14.81%), TC (14.07%), and chloramphenicol (8.9%). The greatest and smallest mean contamination value (ng/g) was enrofloxacin (10.8 ± 1.6) followed by PN (4.4 ± 2.9) and chloramphenicol (0.1 ± 0.1). PN was found in all samples ranging from 0.2–25.8 ng/g. Similarly, a study conducted in 2014 analyzed 145 honey samples obtained from the northwest region of Iran, in which oxytetracycline residue were detected in 34 samples using ELISA at minimum and maximum levels of 5.32 and 369.1 ng/g, respectively (Mahmoudi 2014).

In Turkey, the use of all antibiotics, including oxytetracycline and erythromycin, has been prohibited because of increased resistance, antibiotic residue-derived health problems, and requirements set by EU. In 2008, 50 honey samples, which included chestnut, pine, linden, and multiflower honeys obtained directly from the hives of Southern Marmara in Turkey, were analyzed for erythromycin antibiotics. Results showed that only 8% of the honey samples contained erythromycin residues at concentrations of 50.0–1776.0 ng/g (Gunes 2008). In 2009, another similar Turkish study from the same area analyzed 50 honey samples for the presence of oxytetracycline and sulfonamide residues. However, none of the tested antibiotic residues were detected, indicating their presence in concentrations below the detection limit or absence in the honey samples (Gunes et al. 2009).

## European Countries

In Switzerland, 75 commercial samples (34 produced in Asian countries) were analyzed in which 13 samples were contaminated with chloramphenicol residues (Darko 2017). Contamination of chloramphenicol residues were identified in 38% of Asian honey samples, indicative of the use of chloramphenicol in Asian countries. The chloramphenicol concentrations in honey samples ranged from 0.4 to 6.0 μg/kg, in which six samples contained about 0.8–0.9 μg/kg (almost exceeding the Swiss MRL of 1.0 μg/kg) and two samples contained roughly 5.0 μg/kg of chloramphenicol.

In Greece, 251 honey samples from pine tree or spring wildflower analyzed using liquid chromatography detected TC residues in 29% samples. Among the TCs, oxytetracycline residue was found in 14% and doxycycline in 10% of the samples. Most of the samples had residues in the amount of 0.018–0.055 mg/kg of honey, and certain samples contained residues in excess of 0.100 mg/kg (Saridaki-Papakonstadinou 2006), indicating the presence of TC drug residues in Greek honey. Later in 2012, a total of

30 honey samples that were either from Greece or imported from other countries were analyzed using liquid chromatography. Only two samples were found to contain sulfathiazole (5.3 and 5.9 mg/kg) and sulfamethoxazole (1.8 and 3.4 mg/kg), and trimethoprim (0.4 and 0.6 mg/kg) was detected in both cases, but its concentration was below the detection capability (Economou 2012).

In Poland, 241 commercial honey samples were examined using liquid chromatography in from 2007 to 2008. Findings revealed the presence of sulfonamides in 41 samples (17%) that were of higher levels than the quantification limit (50 μg/kg) set by the Polish National Monitoring Program for honey. The total content of sulfonamides ranged from 5.0 to 891.0 μg/kg (Szczêsna 2009). In essence, the study indicated the use of sulfonamides in beekeeping practices in Poland.

In Brazil, 22 honey samples of distinct floral origins and from regions of Brazil (south, southeast, northeast) were analyzed in 2010 for TC detection, but no TC residues were found (Peres et al. 2010). In 2016, 43 monofloral and multifloral honey samples originated from fruit plants (e.g. orange trees, apple trees, peach trees, and grape vines), eucalyptus, sunflower blossom, and canola were collected from beekeepers of various areas in the state of Rio Grande do Sul. Of the analyzed samples, 44% were detected with residues of at least one analyte (pesticides, antibiotics), ranging between 0.12 and 10.0 μg/kg in each sample (Orso 2016).

In Germany in 2012, a total of eight honey samples produced from forest plants (heath, forest blossom, Robinia, lime) and field plants (spring blossom, rape, buckwheat, sunflower) were analyzed and detected with the presence of streptomycin and dihydro-streptomycin at concentrations of 18.9 and 19.9 mg/kg, respectively (Bohm et al. 2012).

In Italy, 112 honey samples were tested for residues of TC and oxytetracycline from 2001 to 2002. Results showed the presence of TC in 2.7% and oxytetracycline in 4.5% of samples (Sabatini 2003). In another Italian study, 5303 of honey samples (imported, local, and honey from the Italian market) acquired from 2001 to 2007 were analyzed for TCs, sulfonamides, streptomycin, chloramphenicol, and tylosin. In total, 6.3% of all samples contained all the antibiotics analyzed; specifically, 6.8% of imported samples and 6.1% of samples from Italian market. Only 1.7% of local honey was found to be contaminated with antibiotic residues. Sulfonamides were found to be the most used antibiotic followed by TCs, streptomycin, tylosin, and chloramphenicol (Baggio 2009). In 2015, 74 honey samples from various botanical origins and geographical regions obtained from the Italian market were analyzed for 27 basic drugs. Trace levels of sulfonamides were confirmed in nine honey samples (12%) with amounts lower than 2 mg/kg, indicating no serious concern for consumer health (Galarini 2015).

In Spain, 567 Basque honey samples of distinct floral origins were tested using liquid chromatography in which 68 samples were probable positive for sulfonamides, 24 samples positive for TCs, and all negative samples for chloramphenicol (<0.3 μg/kg) residues. The residues were confirmed based on EU requirement for veterinary drug residues (2002/657/EC). Among 68 sulfonamide presumptive positive samples, 19 (3.4%) were presumptive to be contaminated with sulfathiazole residues at values ranging from 20.0 to 210.0 μg/kg. Among the 24 TC presumptive positive samples, TC (4.22% of samples) were present at levels from 15.0 to 920.0 μg/kg (Bonvehí and Gutierrez 2009). Meanwhile, another Spanish study in 2007 reported an analysis of seven Spanish honey samples originating from different flowers, in which all samples were not contaminated with the eight TCs tested (Casado-Terrones 2007). Similarly, in 2008, a study reported the detection analysis of TCs in seven Spanish honey samples in which all samples showed no presence of TCs residues at detectable levels (Carrasco-Pancorbo 2008). Furthermore, another Spanish study conducted in 2009 analyzed commercial honey samples for sulfonamides, with reported results of sulfonamide absence in all honey samples (Bernal 2009).

### North American Countries

In Canada, a study conducted in 2005 reported sulfonamides' presence in 23 different honey samples produced within the province of Alberta, ranging from the 0.5–2.0 μg/kg detection limit (Thompson and Noot 2005). Another study from the similar researcher group reported the presence of lincomycin and tylosin at the 5.0 and 2.0 mg/kg detection limit in Alberta honey samples (Thompson 2005).

## Microorganisms and Honey Contamination

Generally, microorganisms present in honey are those that are able to endure the antimicrobial properties of honey, alongside the low water activity, high sugar content, and acidity (Olaitan et al. 2007). The survival of these microorganisms can negatively affect honey's quality and safety for consumption. Some of the common sources of microorganism contamination include pollen, nectar, digestive tracts of honeybees, dirt, dust, and air, which are typically beyond control. Other possible sources relate to after harvest processes, such as air (during packing), food handlers (skin infections, sneezing or fecal contamination), equipment, containers, buildings, and cross-contamination that take place because of a lack of good manufacturing practice (GMP).

Primarily, the microorganisms of concern particularly in terms of safety aspects are yeasts and spore-forming bacteria. Typically, sugar-tolerant yeast originates from flowers and soil. Because of the antimicrobial properties of honey, it is it difficult for most bacteria and various microbes to survive, grow, or multiply in honey (Snowdon and Cliver 1996). For instance, it was elucidated that the inoculation of several bacteria into aseptically collected honey lost their viability at 20°C within 8–24 days (Olaitan et al. 2007). Despite this, the spore-forming microbes can survive in honey at low temperatures. For example, when *Bacillus cereus*, *Clostridium perfringens*, and *Clostridium botulinum* spores were inoculated into honey, the *C. botulinum* population remained unchanged at 4°C over the a 1-year period (Al-Waili 2012). As such, it is unsurprising that cases of honey contamination with spores of *Clostridium* spp. have been reported worldwide. Of concern is infant botulism, which commonly occurs after the ingestion of *C. botulinum* spores; multiplication of spores leads to accumulation of botulinum toxin in the gastrointestinal tracts of newborns and infants (Cox and Hinkle 2002).

In addition, it was reported that microbes readily found in honeycomb and adult bees were *Bacillus*, *Micrococcus*, and *Saccharomyces* spp. (Sackett 1919). A number of microbial species – most prevalently *Bacillus* spp. followed by gram-variable pleomorphic bacteria, mold, actinomycetes, gram-negative rods (likely the Enterobacteriaceae), and yeast – are isolated from the feces of bee larvae (White 1996). Furthermore, the microbial load of adult honeybees' intestines mainly originates from pollen, predominantly the gram-variable pleomorphic bacteria of ambiguous taxonomy, *Bacillus* spp., *Enterobacteriaceae*, *Penicillium* spp., *Aspergillus* spp., and *Torulopsis* spp (Gilliam and Prest 1987). Additionally, various microorganisms were detected in the intestines of bees, particularly, 1% yeast, 27% gram-positive bacteria (*Bacillus*, *Bacteridium*, *Streptococcus*, and *Clostridium* spp), 70% gram-negative or gram-variable bacteria (*Escherichia coli* and *Achromobacter*, *Citrobacter*, *Enterobacter*, *Erwinia*, *Flavobacterium*, *Klebsiella*, *Proteus*, and *Pseudomonas* spp.) (Tysset and Roy 1991).

## Asian Countries

In Japan, a study detected the survival of *C. botulinum* type F spores (1–60/g) in different containers of honey purchased from local retail shops in Hiroshima (Nakano and Sakagucki 1991). Despite similar pH, hydroxymethyl furfural contents, or diastase activity between contaminated and uncontaminated honey, the aerobe count was greater in the contaminated honey. Aerobic microbes predominantly detected were the motile colony-forming *Bacillus alvei*, which increased the toxin produced by *C. botulinum* type F in the culture medium incubated in the presence of oxygen. Therefore, the survival of *C. botulinum* in the contaminated samples could be attributed to the stimulation by *B. alvei* or other microorganisms during the honey-ripening processes (Nakano and Sakagucki 1991).

In Iran, an analysis of 100 honey samples collected from retail markets in Shiraz in 2004 showed contamination with *C. botulinum*. The findings confirmed that 2% of the samples were positive for type B *C. botulinum* carrying the neurotoxin B gene (Poormontaseri et al. 2014).

In Pakistan, a study in 2013 reported the analysis of six commercially available honey samples in which mesophilic bacteria ($3.2 \times 10^3 - 4.0 \times 10^3$ CFU/g) was present in four samples, total coliforms ($2 \times 10$ and $4 \times 10$ CFU/g) in two samples, and *Bacillus* spore ($1.2 \times 10^3 - 9.0 \times 10^3$ CFU/g) in five samples, indicating insanitary conditions and unhygienic standards of commercial honey processing (Sherwani, 2013).

In Malaysia, meliponini honey samples were obtained from hives in the Besut area of Terengganu state. Microbiological analysis detected the growth of aerobic bacteria and yeasts between $10^2$–$10^5$ and $10^2$–$10^6$ CFU/g, respectively, and negative results for *Clostridium* (Lani 2017).

In Kazakhstan, 110 honey samples collected from different areas reported the presence of *C. botulinum* type A strain in one sample, and *Clostridium* spp. was detected in 43 (39%) of the tested samples (Mustafina 2015).

In Saudi Arabia, analysis of 45 packed honey samples collected from various retail markets was studied in 2004 in which a total of 14 mold species (e.g. *Aspergillus flavus, Aspergillus niger, Aspergillus fumigatus* and *Aspergillus versicolor*) were identified, representing a low level of microbial contamination (Nasser 2004). In 2012, a total of 91 bee honey samples collected from different plant sources and regions of Saudi Arabia were examined. Aerobic mesophilic bacteria, molds, and yeasts present were less than $10^5$ CFU/g with no detection of total coliforms, *Salmonella* spp., *Shigella* spp, *E. coli*, or *Clostridium* spp. sulfite reducers (Sadik and Ali 2012).

In Turkey, 48 honey samples consisting of 6 comb honeys, 34 flower honeys, and 8 honeydew honeys collected from the retail markets in Ankara were analyzed for *C. botulinum* spores in 2006. Six (12.5%) of the 48 analyzed samples contained *C. botulinum* (Küplülü 2006). In 2007, 26 marketed and raw honey samples randomly chosen from 15 superstores and 11 apiaries located across the Marmara and Black Sea regions were analyzed where mostly the bacteria *Bacillus* spp. were isolated in 23 (88.5%) samples (Özakin 2007). In 2013, analysis of 20 artisanal (n = 10) and retail (n = 10) honey samples produced in Turkey detected total aerobic mesophilic and spore-forming bacteria in minimal amounts; meanwhile, yeasts and molds were present in 15% and 20% of the tested samples, respectively (Tornuk 2013). Subsequently in 2013, 250 honey samples collected from several markets in Ankara showed the presence of 19 *C. botulinum* strains, including 12 type A strains, 3 type B strains, 2 type F strains, and 2 unconfirmed strains type (Koluman 2013).

## European Countries

In Argentina, a study in 1997 detected *C. botulinum* type A in 2 of 177 honey samples produced in the rural area (Satorres 1997). In 2005, a study reported the analysis of 70 polyfloral unpasteurized honeys, including commercial honeys, apiary honeys, and honeys purchased in bulk for industry uses (Iurlina and Fritz 2005). The aerobic mesophilic bacteria, mold, and yeast counts in tested samples were not more than $10^3$ CFU/g. Additionally, fecal coliforms, *E. coli*, *Salmonella* spp., *Shigella* spp., and *Clostridium* sulfite reducers were absent, but *Paenibacillus larvae*, *B. cereus*, *Bacillus pumilus*, and *Bacillus laterosporus* were detected. In 2007, 23 honey samples collected from local producers of unidentified botanical origins were analyzed. The yeast and fungi count in the examined samples was not more than $1 \times 10^2$ CFU/g. Moreover, vegetative cells and spores of *Clostridia* were present in low quantities (Finola et al. 2007). In 2014, 28 beehive samples of yateí honey were retrieved from various regions of the Province of Misiones. After the analysis, *Clostridium* spp. (64%) and *Bacillus* spp. (39%) were the most common genera found in the tested samples (Pucciarelli 2014).

In Brazil, 85 honey samples from different origins were analyzed in 1999, among which 6 were contaminated with *C. botulinum* (7.06%) and ascertained as the producers of type A, B, and D toxins (Schocken-Iturrino 1999). In 2003, 100 honey samples collected from the São Paulo state tested negative for *Salmonella* spp., *Shigella* spp., and total coliforms. However, the spores of *C. botulinum* were detected in 3% of the samples, but molds and yeasts were identified in 64% of the samples and small sporogenic gram-positive rods in 42% of the samples (Rall 2003). In 2010, analysis of 11 organic and 6 non-organic Brazilian honey samples from different regions detected yeast and mold growth at the highest count of $1.1 \times 10^3$ CFU/g in organic honey and $2.5 \times 10^2$ CFU/g in non-organic honey samples (Sereia 2010). In 2011, 33 organic honey samples of *Apis mellifera* produced in the Parana River islands were analyzed between 2005 and 2006. Among 33 samples, 11 were from beekeepers, and 22 samples were obtained and processed in accord with ideal personal hygiene practices and GMP. Microbiological analysis showed a maximum level of $3.8 \times 10^1$ CFU/g of molds and yeasts in honeys harvested with GMP and $1.1 \times 10^3$ CFU/g for honeys harvested without the GMP regulations (Sereia 2011).

In Spain, a study involving 86 honey samples collected from the northwestern area of Galicia from 2008 to 2009 reported the highest presence of 325/g *Cladosporium*, 135/g *Myxomycetes*, and 151/g *Penicillium* spores in more than 40% of the samples. In addition, *Metschnikowia* yeast was present commonly in 79.1% of the samples with a maximum number of 46,217 cells/g (Olga et al. 2012).

In Portugal, 73 organic honey samples from the northeast part of the country that were of monofloral origin were analyzed in 2011. Findings showed minimal counts of yeast, molds, and aerobic mesophiles with no detection of fecal coliforms, *Salmonella* spp., or sulphite-reducing *Clostridium* spp. (Gomes 2011).

In Poland, 245 unifloral (lime, buckwheat, acacia, rape) and multifloral honeys acquired from apiaries were microbiologically tested in which the total aerobic bacteria count ranged between $1.0 \times 10^1$ and $7.5 \times 10^4$ CFU/g. Although *Salmonella* spp. was undetected in the tested honey, spore-forming anaerobic bacteria were present in 14.3% of acacia honey samples and up to 36.4% of rape honey samples. Meanwhile, yeast and mold counts were minimal and sporadically exceeded $1.0 \times 10^2$ CFU/g (Różańska 2011). In another study conducted in 2018, 240 honey samples obtained from apiaries located in Polish provinces were examined for the presence of *Clostridium* spp., including *C. perfringens*. *Clostridium* spp. was noted in 56% (136 of 240) of the samples, and the titers varied from 0.1 g to 0.001 g. *C. perfringens* was identified in 27.5% (66 of 240) of the tested honey. All isolates were categorized as toxinotype A. Overall, the considerable number of samples containing *Clostridium* spp. reflects a possible food safety risk to human health (Grenda 2018).

In France, 116 randomly collected honey samples from retail outlets in the eastern part of the country were analyzed in 1994. *C. sporogenes* was detected in one sample from Romania, and *C. perfringens* was present in three samples from France, two samples from China, and one sample from Spain (Delmas et al. 1994).

In Slovakia, 50 samples of honey collected from various areas were studied in 2009. All the samples were detected with the presence of microscopic fungi. In 15 of 50 honey samples, *P. larvae* was found most frequently (Kačániová 2009).

In Finland, 190 honey samples were analyzed in 2002. Spores of *C. botulinum* were present in 8 (7%) of the 114 Finnish samples and 12 (16%) of the 76 imported honey samples. In addition, strains of *C. botulinum* type A were detected from 14 samples, and type B were present in 2 of the 20 polymerase chain reaction (PCR)–positive samples. Researchers also reported the detection of neurotoxin gene sequences via PCR analysis, which corresponded to *C. botulinum* type A in 17 samples, proteolytic type B in 12 samples, and both types A and B in 9 samples (Nevas 2002). According to another study conducted in 2006, 1168 honey samples from 100 apiaries were analyzed for the occurrence of *C. botulinum*, and 216 samples were found positive. The spore count in the beehive and soil samples ranged between 60 and 1200/kg and 50 and 1050/kg respectively. Notably, the greatest number of spores were identified in honey that was contaminated with *C. botulinum* type B (Nevas 2006).

### North American Countries

In Mexico, 1920 honey samples sourced from eight states were examined, and 45% of the samples (777 of 1920) did not conform to the mesophilic aerobic microorganism specifications established. Similarly, 17% (327 of 1920) of the samples did not adhere to the specifications set for molds, and 18.1% (348 of 1920) of the samples were out of the specifications for yeasts. Moreover, *C. perfringens* was detected in concentration exceeding 100 CFU/g (Vázquez-Quiñones 2018). In California, 9 of 90 honey samples contained *C. botulinum*, in which 6 of the positive samples were fed to infants, possibly causing infant botulism (Midura 1979). Likewise in another study, *C. botulinum* organisms were detected in 6 honey samples that were consumed by three patients who developed infant botulism (Arnon 1979).

### African Countries

In Algeria, 19 honey samples from different parts (mainly Kabylia) were analyzed in 2010. Of the 19 samples, only 2 samples showed yeast growth, indicating nectar contamination of the hive, and 4 samples showed mold growth, suggesting a lack of hygiene practices during harvesting, packaging, or storage (Amir 2010). In Ghana, a study was carried out in 2017 with honey collected from different areas within the Tamale metropolis. The highest bacterial count was detected at $1.1 \times 10^5$ CFU/mL, which included *E. coli*, *Staphylococcus* spp., *Shigella* spp., *Streptococcus* spp., and *Bacillus* spp., indicative of improper hygienic conditions (Adadi and Obeng 2017).

In Morocco, a total of 10 Moroccan honey samples of unifloral (Eucalyptus, Thymus, Euphorbia) and multifloral origins were analyzed in 2005. Microbiology analysis showed that the microbial profile were very low (0–200 CFU/g) for all the microorganisms, there were no total coliforms in any sample, and fungi (yeasts and molds) growth was at low levels (at times below 100 CFU/g) (Malika et al. 2005).

In Nigeria, honey samples collected from six states in the southwestern region in 2012 contained seven heterotrophic fungi species (*Cladosporium*, *Cladosporium herbarum*, *Cephalosporium* spp., *Mucor mucedo*, *Rhizopus rubrum*, *Trichophyton rubrum*, *Scopulariopsis brevicaulis*) and four bacterial species (*Klebsiella edwardsii*, *K. pneumoniae*, *Pseudomonas aeruginosa*, *Staphylococcus aureus*) (Ayansola and Banjo 2012). In 2013, another Nigerian study tested 18 honey samples collected from various areas in the northeast subregion. Results showed that the *Bacillus* spp. was most prevalent in nearly all tested samples, except honey obtained from Tongo (contaminated with *Klebsiella* spp.) and honey from Alkaleri (contaminated with *S. aureus*). Moreover, molds were also present in all tested samples, suggesting the predominance of *A. niger* in most of the analyzed honey samples (Buba et al. 2013). A 2014 study examined 15 natural honey samples obtained directly from beekeepers, retail outlets, and supermarkets of Edo state. The bacterial count was reported in 4 samples from markets and roadsides with a maximum count of $2.0 \times 10^4$ CFU/mL, whereas the coliforms were detected only in one sample with a count of $2.0 \times 10^4$ CFU/mL (Oshomah 2014).

## Radionuclides

Radionuclides are radioactive isotopes that occur naturally in soils, rock, groundwater, and surface water either as normal constituents or because of global atmospheric contamination resulting from dry and wet depositional processes. Radioactivity is measured in becquerels (Bq) per kilogram. Typically, the sources of artificial radionuclides include atomic weapon tests, accident or release from nuclear plants, and other sources such as the emissions of radioactive substances from atomic or medical laboratories as well as nuclear processing facilities (Herrero-Latorre 2017). The study of radioactivity in honey is of significance as a bioindicator for tracking radionuclide contamination. With regards to this, radionuclides can enter the nectar of flowers and secretions of plants through migration from soils and water by plant uptake processes (Barišić 1994; Bunzl et al. 1988; Horn 1996; Jones 1987). For instance, floral and honeydew honeys are widely used as bioindicators of environment pollution because of the capacity to detect immediate environment conditions and changes (Bromenshenk 1985; Kezic 1997; Leita 1996; Tonelli 1990).

In today's world, many countries are actively involved in developing nuclear-related materials and producing nuclear energy, which consequently aids the production of radioisotopes. The greatest industrial disaster of all time was the explosion of one of the nuclear reactor core at Chernobyl, Ukraine, in 1986, resulting in the accidental release of various fission products (e.g. $^{133}$Xe, $^{131}$I, $^{134}$Cs, $^{137}$Cs, $^{132}$Te, $^{89}$Sr, $^{90}$Sr, $^{140}$Ba, $^{95}$Zr, $^{99}$Mo, $^{103}$Ru, $^{106}$Ru, $^{141\ CE}$, $^{144\ CE}$, $^{239}$Np, $^{238}$Pu, $^{239}$Pu, $^{240}$Pu, $^{241}$Pu, and $^{242}$Cm) into the atmosphere (Eisler 1994). Radionuclide releases were estimated roughly at 1760, 54, 85, 115, and 10 petabecquerels for $^{131}$I, $^{134}$Cs, $^{137}$Cs, $^{89}$Sr, and $^{90}$Sr, respectively (Pokarzhevskii 1998). Of note, the iodine, cesium, and strontium were particularly harmful (Pokarzhevskii 1998; Steinhauser et al. 2014). First, $^{131}$I adversely affect the thyroid gland, thereby increasing thyroid cancer risks. On the other hand, cesium and strontium are easily absorbed into biological tissues as the homologues of potassium and calcium correspondingly. The Chernobyl accident has driven numerous research projects on honey samples globally, in which the concentration of cesium has been of particular interest (Sereia 2010).

In a Ukrainian study, the average concentration of radioactive cesium detected in locally produced honey in the years 1986 to 1989 was 4430 Bq/kg (Alexenitser and Bodnarchuk 1999).

In Germany, approximately 6000 floral and honeydew honeys were analyzed in the period between 1986 and 1991 (Horn and Vorwohl 1986a, 1986b). Despite the relatively increased concentration of cesium detected in honey during the first year (51.0 Bq/kg), the contamination level was reduced by approximately half in the following year and remained at an average concentration of roughly10.0 Bq/kg in the subsequent years. German heather honey was highly contaminated, with an average of 532.0 Bq/kg Cs in 1987 (Dustmann and von der Ohe 1988) and up to about 650.0 Bq/kg Cs in 1988 (Assmann-Werthmüller et al. 1991). The differences in cesium values might be due to the variable distribution of the radioactive contamination on the earth surface and local rain, varying affinity of cesium to the plants, or different isotopic concentrations in the soil (Kubik 1991).

In Croatia, the $^{137}$Cs values in honey samples mainly originated from acacia, meadow, lime, heather, chestnut, and rosemary varied from 0.8 to 15.9 Bq/kg in 1990 (Barišić 1992). In 1999, the $^{137}$Cs concentrations were found to be 4.8–36.2 Bq/kg in honeydew honey and 0.9–6.8 Bq/kg in mixed meadow and honeydew honey in Croatia. Additionally, meadow honey obtained from similar areas contained i0.0–1.0 Bq/kg of $^{137}$Cs (Barišić 1999). In Scotland, 11 samples of heather honey harvested in 1993 were analyzed for radioactive isotopes of cesium. All samples contained $^{137}$Cs (52.0–690.0 Bq/kg), and 10 samples contained $^{134}$Cs (1.3–21.0 Bq/kg). Overall, the majority of the honey samples tested were close to the detection limit (Fisk 1995).

In Italy, honey samples collected from different regions in 1986 (May–October) were reported to contain 131I (1051 Bq/kg), $^{103}$Ru (575 Bq/kg), $^{137}$Cs (363 Bq/kg), and $^{134}$Cs (171 Bq/kg). The melissopalynology study of pollen revealed that these honey samples were originated from honeydew, chestnut, and acacia (Tonelli 1990). In a recent study, 27 honey samples (25 wildflower, 1 honeydew, and 1 acacia) collected in 2013 from central-eastern Italy were found to contain $^{137}$Cs concentrations of less than 0.10 Bq/kg in all samples, which proves that residual contamination from anthropogenic radioactivity is not present in these areas (Meli 2016). In another recent study published in 2020, a radioactivity determination of 28 honey obtained from retail outlets in Italy was conducted based on the presence of natural radionuclides and $^{137}$Cs. Study findings reported that the concentrations of $^{40}$ K, $^{238}$U, $^{234}$U, and $^{210}$Po were 50.6 ± 46.3, 0.023 ± 0.010, 0.019 ± 0.017, and 0.27 ± 0.40 Bq/kg, respectively. The levels of $^{235}$U, $^{228}$Th, and $^{232}$Th detected were consistently below 0.007 Bq/kg, and $^{226}$Ra was below 0.200 Bq/kg. Also, the level of $^{137}$Cs was less than 2.1 Bq/kg in the majority of the honey samples (93%). Moreover, only 2 tested samples exceeded the minimum detectable level for radiocesium, in which the detected samples were in accordance with the allowed concentration for food products. Similarly, the effective doses of $^{210}$Po merely accounted for 0.0026%–5.31% of worldwide exposure to natural radiation. In summary, the tested samples adhered to the quality analysis of honey in terms of radionuclides level, thereby concurring with honey's reputation as health food (Meli 2020).

Furthermore, radioactivity analysis of 13 honeys produced in various countries, including Japan, China, Hungary, Italy, and Australia, demonstrated the presence of $^{226}$Ra, $^{214}$Pb, $^{214}$Bi, and $^{40}$K. The radionuclide $^{137}$Cs was detected in 4 samples (acacia honey from Hungary, chestnut honey from Italy, and lime honey from Japan), and $^{134}$Cs was found in 1 Italian chestnut honey. The minimum concentration of $^{40}$K and $^{226}$Ra was found in acacia honey (15.0 and 24.0

Bq/kg, respectively), and the maximum concentration was present in chestnut honey (211.0 and 45.0 Bq/kg, respectively) (Handa 1997). The $^{40}K$ radioactivity of various types of honey originating from Poland ranged between 39.0 and 123.0 Bq/kg (Borawska 2000). Meanwhile Croatian honeydew, Croatian meadow honey, and Italian honey contained 43.8–143.3 Bq/kg, 8.4–63.6 Bq/kg, and 7.28–101.0 Bq/kg of $^{40}K$ radionuclide (Barišić 1999; Meli 2016). Being a natural radioisotope, $^{40}K$ is readily taken up by the plant roots via ion channels from soil. Thus, it is unsurprising that $^{40}K$ is found abundantly in honey.

Furthermore, another nuclear disaster that occurred at Fukushima Daiichi, Japan, in 2011 led to the release of various radioactive materials to the environment. After the nuclear power plant accident, a study explored the radioactivity of 106 honey samples (e.g. lime, multifloral, buckwheat, honeydew, and other types of honey), obtained from the northeastern region of Poland before (2010) and after the Fukushima accident (2011–2012) (Borawska 2013). The radioactivity of $^{137}Cs$ and $^{40}K$ was found at mean concentrations of 1.19 and 32.92 Bq/kg in 2010, 0.90 and 31.13 Bq/kg in 2011, and 1.31 and 36.06 Bq/kg in 2012, respectively. The analysis showed no significant differences in honey samples throughout the years. Thus, it can be concluded that the tested samples obtained after the Fukushima nuclear leak are safe for human consumption, with minimal concentration of $^{137}Cs$ and $^{40}K$, which is insufficient to pose serious health risks (Borawska 2013).

Although the level of radioactive isotopes in honey could be a useful indicator of environment radiation, especially after nuclear accidents, the radioactivity concentration is not an issue in honey at present, given the low levels tested. Nevertheless, after all thermonuclear accidents, honey and other bee products should be tested and monitored closely before consumption.

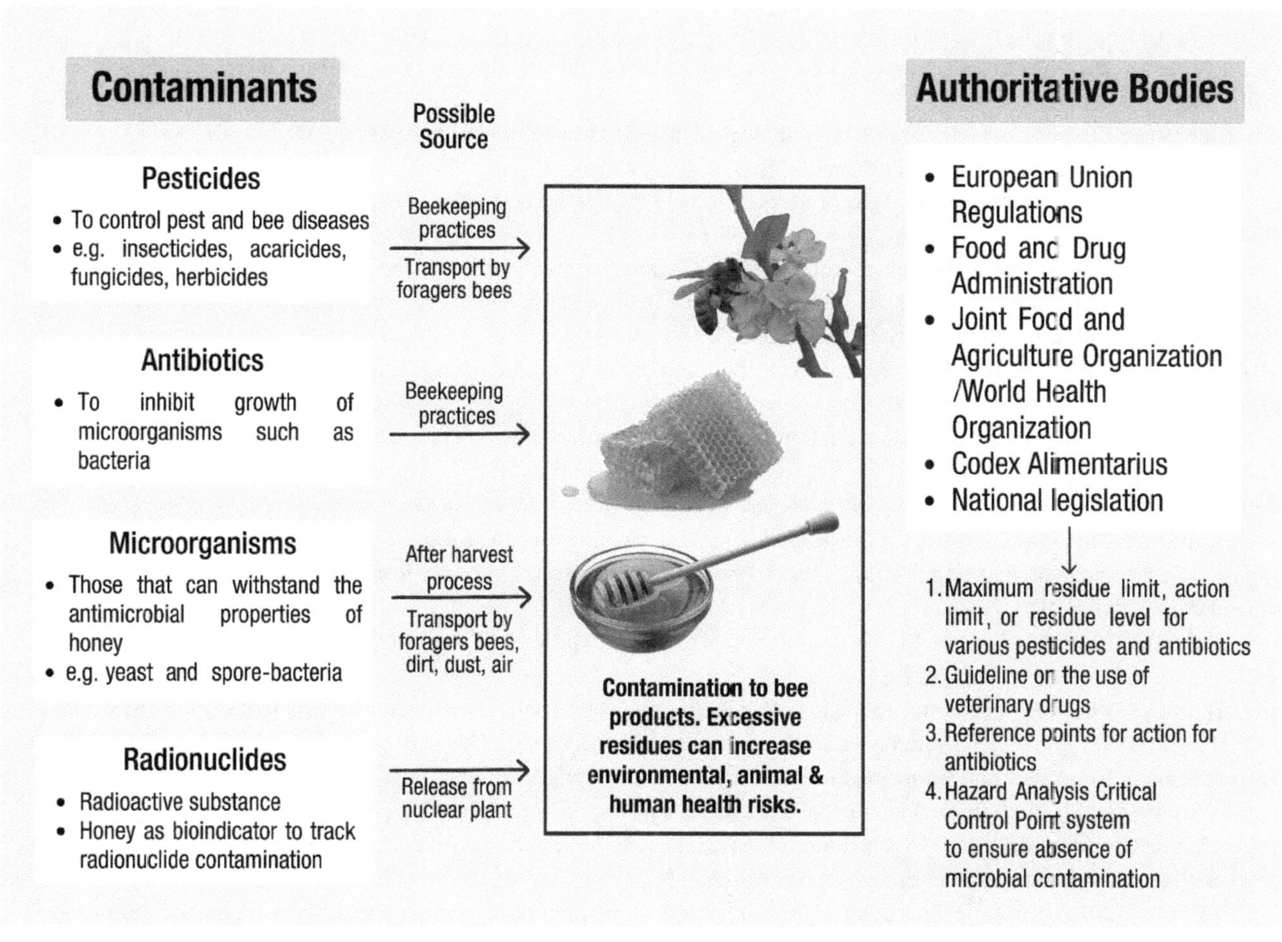


**Figure 12.1** Possible sources of honey contamination and the regulatory bodies involved.

## Conclusion

In global trade, honey has achieved an esteemed place because of its wide range of beneficial properties. Therefore, a definite international standard for marketing and consumption of honey is required for maintaining its exponential spread of trading worldwide. Setting up a universal regulation for controlling honey production, handling, and analysis will limit the exposure to toxic chemicals caused by their uncontrolled administration in apiarian practices and will ensure honey's safety for consumption (Figure 12.1). Ingestion of honey without confirming the safety aspects and sources may have serious implications. Thus, proper labelling of honey ought to be mandatory to confirm its origin, composition, and safety in terms of contaminant and residue levels. Raw honey without proper analysis or sterilization must not be used for consumption and other indirect applications in humans. GMP should be monitored in every step of honey production until it is marketed to avoid high residual levels of contaminants.

## References

Adadi, P. and Obeng, A.K. (2017). Assessment of bacterial quality of honey produced in Tamale metropolis (Ghana). *Journal of Food and Drug Analysis* 25 (2): 369–373.

Aktar, M.W., Sengupta, D., and Chowdhury, A. (2009). Impact of pesticides use in agriculture: their benefits and hazards. *Interdisciplinary Toxicology* 2 (1): 1–12.

Alexenitser, M. and Bodnarchuk, L. (1999). Beekeeping products as bioindicators of radioactive contamination in a certain zone. *Apiacta* 34: 93–96.

Al-Rifai, J. and Akeel, N. (1997). Determination of pesticide residues in imported and locally produced honey in Jordan. *Journal of Apicultural Research* 36 (3–4): 155–161.

Al-Waili, N., et al. (2012). Antibiotic, pesticide and microbial contaminants of honey: human health hazards. *The Scientific World Journal* 930849-930849.

Amir, Y., et al. (2010). Physico-chemical and microbiological assessment of honey from Algeria. *Electronic Journal of Environmental, Agricultural & Food Chemistry* 9 (9).

Anju, R., et al. (1997). Multiresidue analysis of market honey samples for pesticidal contamination. *Pesticide Research Journal* 9: 226–230.

Arnon, S.S., et al. (1979). Honey and other environmental risk factors for infant botulism. *The Journal of Pediatrics* 94 (2): 331–336.

Assmann-Werthmüller, U., Werthmüller, K., and Molzahn, D. (1991). Cesium contamination of heather honey. *Journal of Radioanalytical and Nuclear Chemistry* 149 (1): 123–129.

Ayansola, A. and Banjo, A. (2012). Microbiological examination of honey marketed in Southwestern Nigeria. *Journal of Basic and Applied Scientific Research* 2: 1701–1705.

Baggio, A., et al. (2009). Residues of antibacterial drugs in honey from the Italian market. *Food Additives and Contaminants: Part B* 2 (1): 52–58.

Balayiannis, G. and Balayiannis, P. (2008). Bee honey as an environmental bioindicator of pesticides' occurrence in six agricultural areas of Greece. *Archives of Environmental Contamination and Toxicology* 55 (3): 462.

Bargańska, Ż., Ślebioda, M., and Namieśnik, J. (2013). Pesticide residues levels in honey from apiaries located of Northern Poland. *Food Control* 31 (1): 196–201.

Barišić, D., et al. (1992). 137Cs in flowers, pollen and honey from the Republic of Croatia four years after the Chernobyl accident. *Apidologie* 23 (1): 71–78.

Barišić, D., et al. (1994). 40K, 134Cs and 137Cs in pollen, honey and soil surface layer in Croatia. *Apidologie* 25 (6): 585–595.

Barišić, D., et al. (1999). Radionuclides and selected elements in soil and honey from Gorski Kotar, Croatia. *Apidologie* 30 (4): 277–287.

Bernal, J., et al. (2009). A new and simple method to determine trace levels of sulfonamides in honey by high performance liquid chromatography with fluorescence detection. *Journal of Chromatography A* 1216 (43): 7275–7280.

Beyoglu, D. and Omurtag, G.Z. (2007). Occurrence of naphthalene in honey consumed in Turkey as determined by high-pressure liquid chromatography. *Journal of Food Protection* 70 (7): 1735–1738.

Blaga, G.-V., et al. (2020). Antifungal residues analysis in various Romanian honey samples analysis by high resolution mass spectrometry. *Journal of Environmental Science and Health, Part B*: 1–11.

Blasco, C., et al. (2003). Assessment of pesticide residues in honey samples from Portugal and Spain. *Journal of Agricultural and Food Chemistry* 51 (27): 8132–8138.

Bogdanov, S. (2006). Contaminants of bee products. *Apidologie* 37 (1): 1–18.

Bohm, D.A., Stachel, C.S., and Gowik, P. (2012). Confirmatory method for the determination of streptomycin and dihydrostreptomycin in honey by LC–MS/MS. *Food Additives & Contaminants: Part A* 29 (2): 189–196.

Bonvehí, J.S. and Gutierrez, A.L. (2009). Residues of antibiotics and sulfonamides in honeys from Basque Country (NE Spain). *Journal of the Science of Food and Agriculture* 89 (1): 63–72.

Borawska, M., et al. (2000). Radioactivity of honeybee honey. *Bulletin of Environmental Contamination and Toxicology* 64 (5): 617–621.

Borawska, M.H., et al. (2013). Radioactivity of honeys from Poland after the Fukushima accident. *Bulletin of Environmental Contamination and Toxicology* 91 (5): 489–492.

Bromenshenk, J.J., et al. (1985). Pollution monitoring of Puget Sound with honey bees. *Science* 227: 632–635.

Buba, F., Gidado, A., and Shugaba, A. (2013). Physicochemical and microbiological properties of honey from North East Nigeria. *Biochemistry and Analytical Biochemistry* 2 (142): 61–67.

Bunzl, K., Kracke, W., and Vorwohl, G. (1988). Transfer of Chernobyl-derived 134Cs, 137Cs, 131I and 103Ru from flowers to honey and pollen. *Limits of pesticide residue* 6 (3): 261–269.

Canada Government. (2005). *Working Residue Levels (WRLs) For Antimicrobial Residues In Honey*. https://www.canada.ca/en/health-canada/services/drugs-health-products/veterinary-drugs/legislation-guidelines/policies/working-residue-levels-wrls-antimicrobial-residues-honey-policy-administrative-maximum-residue-limits-working-residue-levels-veterinary-drugs-food.html.

Carrasco-Pancorbo, A., et al. (2008). Reversed-phase high-performance liquid chromatography coupled to ultraviolet and electrospray time-of-flight mass spectrometry on-line detection for the separation of eight tetracyclines in honey samples. *Journal of Chromatography A* 1195 (1–2): 107–116.

Casado-Terrones, S., et al. (2007). Determination of tetracycline residues in honey by CZE with ultraviolet absorbance detection. *Electrophoresis* 28 (16): 2882–2887.

Chen, L., et al. (2009). Fast and selective extraction of sulfonamides from honey based on magnetic molecularly imprinted polymer. *Journal of Agricultural and Food Chemistry* 57 (21): 10073–10080.

Chen, T.B., et al. (2001). Detection of residual antibiotics in honey with capillary electrophoresis. *Se Pu= Chinese Journal of Chromatography* 19 (1): 91–93.

Choudhary, A. and Sharma, D.C. (2008). Pesticide residues in honey samples from Himachal Pradesh (India). *Bulletin of Environmental Contamination and Toxicology* 80 (5): 417–422.

Codex Alimentarius (1997). Limits of pesticide residue [Turkish]. *The Official Gazette of Republic of Turkey* Alimentarius (1998). Draft Revised for Honey at Step 6 of the Codex Procedure. CX 5/10.2. CL 1998/12-S. Supplement 17 (16.11): 1997–23172.

Codex Alimentarius (1998). *Draft Revised for Honey at Step 6 of the Codex Procedure. CX 5/10.2.* CL 1998/12-S.

Codex Alimentarius (2005). Honey rescript. *The Official Gazette of Republic of Turkey* 17: 2005–26026.

Cox, N. and Hinkle, R. (2002). Infant botulism. *American Family Physician* 65 (7): 1388.

Darko, G., et al. (2017). Pesticide residues in honey from the major honey producing forest belts in Ghana. *Journal of Environmental and Public Health* 2017.

Das, Y.K. and Kaya, S. (2009). Organophosphorus insecticide residues in honey produced in Turkey. *Bulletin of Environmental Contamination and Toxicology* 83 (3): 378–383.

Delmas, C., Vidon, D.-M., and Sebald, M. (1994). Survey of honey for Clostridium botulinum spores in eastern France. *Food Microbiology* 11 (6): 515–518.

Dustmann, J. and von der Ohe, W. (1988). Radioaktivität in Honig aus niedersächsischen Trachtgebieten, Nord-Westdtsch. *Imkerztg* 40: 129–131.

Economou, A., et al. (2012). Determination of a liquid chromatography–tandem mass spectrometry method for the determination of sulfonamides, trimethoprim and dapsone in honey and validation according to Commission Decision 2002/657/EC for banned compounds. *Talanta* 97: 32–41.

EFSA, et al. (2018). Update: methodological principles and scientific methods to be taken into account when establishing Reference Points for Action (RPAs) for non-allowed pharmacologically active substances present in food of animal origin. *EFSA Journal* 16 (7): e05332.

EFSA, et al. (2021). Modification of the existing maximum residue level for ametoctradin in honey. *EFSA Journal* 19 (11): e06943.

Eisler, R. (1994). *Radiation Hazards to Fish, Wildlife and Invertebrates: A Synoptic Review*. Report 29: US Department of the Interior, National Biological Service.

Eissa, F., El-Sawi, S., and Zidan, N.E.-H. (2014). Determining pesticide residues in honey and their potential risk to consumers. *Polish Journal of Environmental Studies* 23 (5).

Erdoğrul, Ö. (2007). Levels of selected pesticides in honey samples from Kahramanmaraş, Turkey. *Food Control* 18 (7): 866–871.

Eto, M. and Eto, M. (1979). *Organophosphorus Pesticides: Organic and Biological Chemistry*. pp. 387 (57–78). Cleveland, OH: CRC Press. CRC.

European Commission, E. (2022). *EU Pesticides Database*. 8 June 2022. https://ec.europa.eu/food/plants/pesticides/eu-pesticides-database_en.

European Commission, E. (2021). *Guidance on EU requirements for imports of products of animal origin: monitoring for residues of veterinary medicines, pesticides and contaminants*.

European Commission, E. (2022). *Pesticides and Bees*. 08 June 2022. https://ec.europa.eu/food/animals/live-animal-movements/honey-bees/pesticides-and-bees_en.

FDA. (2022). *Using Medically Important Antimicrobials in Bees*. https://www.fda.gov/animal-veterinary/development-approval-process/using-medically-important-antimicrobials-bees-questions-and-answers.

Fell, R.D. and Cobb, J.M. (2009). Miticide residues in Virginia honeys. *Bulletin of Environmental Contamination and Toxicology* 83 (6): 822.

Finola, M.S., Lasagno, M.C., and Marioli, J.M. (2007). Microbiological and chemical characterization of honeys from central Argentina. *Food Chemistry* 100 (4): 1649–1653.

Fisk, S. (1995). Is Scottish honey radioactive? *Scottish Beekeeper* 72 (2): 26–29.

Fléché, C., et al. (1997). Contamination des produits de la ruche et risques pour la santé humaine: situation en France. *Revue scientifique et technique-Office international des épizooties* 16 (2): 609–619.

Food and Drug Administration of the United States. (2003). *Pesticide tolerances*. http://www.cfsan.fda.gov.

Galarini, R., et al. (2015). Multiclass determination of 27 antibiotics in honey. *Food Control* 48: 12–24.

Gilliam, M. and Prest, D. (1987). Microbiology of Lanial honey bee (Apis melifera). *Journal of Invertebrate Pathology* 49: 70–75.

Gomes, T., et al. (2011). Study of organic honey from the northeast of Portugal. *Molecules* 16 (7): 5374–5386.

Grenda, T., et al. (2018). Clostridium perfringens spores in Polish honey samples. *Journal of Veterinary Research* 62 (3): 281–284.

Gunes, M., Gunes, N., and Cibik, R. (2009). Oxytetracycline and sulfonamide residues analysis of honey samples from Southern Marmara Region in Turkey. *Bulgarian Journal of Agricultural Science* 15 (2): 163–167.

Gunes, N., et al. (2008). Erythromycin residue in honey from the Southern Marmara region of Turkey. *Food Additives and Contaminants* 25 (11): 1313–1317.

Handa, Y., et al. (1997). Radioactivity due to several radionuclides detected in honey of different geographical origins. *American Bee Journal* 137 (4): 307–309.

Herrera, A., et al. (2005). Determination of pesticides and PCBs in honey by solid-phase extraction cleanup followed by gas chromatography with electron-capture and nitrogen–phosphorus detection. *Analytical and Bioanalytical Chemistry* 381 (3): 695–701.

Herrero-Latorre, C., et al. (2017). The use of honeybees and honey as environmental bioindicators for metals and radionuclides: a review. *Environmental Reviews* 25 (4): 463–480.

Horie, M., et al. (2004). Determination of streptomycin and dihydrostreptomycin in honey by liquid chromatography–electrospray mass spectrometry. *Journal of Liquid Chromatography & Related Technologies* 27 (5): 863–874.

Horn, H. and Vorwohl, G. (1986a). Radioaktive Rückstände im Honig–eine Zwischenbilanz. *Allg. dtsch. lmkerztg* 20 (11): 345–348.

Horn, H. and Vorwohl, G. (1986b). *Radioaktive Rückstände im Honig*. Belastung mit Caesium-137 im Verlauf der Honigsaison, 213–216.

Horn, U., et al. (1996). The transfer of 226 Ra to honey and the possible use of the honey bee as a bioindicator in the uranium mining area of the Wismut region (Germany)[Natural radionuclide, soil-plant-honey Transfer]. *Apidologie (France)* 27 (4): 211–217.

Iurlina, M.O. and Fritz, R. (2005). Characterization of microorganisms in Argentinean honeys from different sources. *International Journal of Food Microbiology* 105 (3): 297–304.

Johnson, S., et al. (2010). *Antibiotic residues in honey*. Report September.

Jones, K. (1987). Honey as an indicator of heavy metal contamination. *Water, Air, & Soil Pollution* 33 (1): 179–189.

Kačániová, M., et al. (2009). Microbial communities in bees, pollen and honey from Slovakia. *Acta Microbiologica Et Immunologica Hungarica* 56 (3): 285–295.

Kezic, N., et al. (1997). *Honey-dew Honey as a Long Term Indicator of 137 Cs Pollution*.

Koluman, A., et al. (2013). Clostridium botulinum in honey: prevalence and antibiotic susceptibility of isolated strains. *Turkish Journal of Veterinary and Animal Sciences* 37 (6): 706–711.

Korta, E., et al. (2002). Study of an accelerated solvent extraction procedure for the determination of acaricide residues in honey by high-performance liquid chromatography–diode array detector. *Journal of Food Protection* 65 (1): 161–166.

Kubik, M., et al. (1991). Caesium-137 and potassium-40 levels in selected fruits in Poland in 1989. *Fruit Science Reports* 18 (4): 147–154.

Kujawski, M.W. and Namieśnik, J. (2011). Levels of 13 multi-class pesticide residues in Polish honeys determined by LC-ESI-MS/MS. *Food Control* 22 (6): 914–919.

Küplülü, Ö., et al. (2006). Incidence of Clostridium botulinum spores in honey in Turkey. *Food Control* 17 (3): 222–224.

Lambert, O., et al. (2013). Widespread occurrence of chemical residues in beehive matrices from apiaries located in different landscapes of western France. *PLoS One* 8 (6): e67007.

Lani, M.N., et al. (2017). Microbiological quality and pH changes of honey produced by stingless bees, Heterotrigona itama and Geniotrigona thoracica stored at ambient temperature. *Malaysian Applied Biology* 46 (3): 89–96.

Leita, L., et al. (1996). Investigation of the use of honey bees and honey bee products to assess heavy metals contamination. *Environmental Monitoring and Assessment* 43 (1): 1–9.

Li, J., et al. (2008). Determination of tetracyclines residues in honey by on-line solid-phase extraction high-performance liquid chromatography. *Talanta* 75 (5): 1245–1252.

López, D.R., et al. (2014). Evaluation of pesticide residues in honey from different geographic regions of Colombia. *Food Control* 37: 33–40.

Mahmoudi, R., et al. (2014). Detection of oxytetracycline residues in honey samples using ELISA and HPLC methods. *Pharmaceutical Sciences* 19 (4): 145.

Mahmoudi, R., Norian, R., and Pajohi-Alamoti, M. (2014). Antibiotic residues in Iranian honey by ELISA. *International Journal of Food Properties* 17 (10): 2367–2373.

Malhat, F.M., et al. (2015). Residues of organochlorine and synthetic pyrethroid pesticides in honey, an indicator of ambient environment, a pilot study. *Chemosphere* 120: 457–461.

Malika, N., Mohamed, F., and Chakib, E. (2005). Microbiological and physicochemical properties of Moroccan honey. *International Journal of Agriculture and Biology* 7 (5): 773–776.

Meli, M.A., et al. (2016). Radioactivity in honey of the central Italy. *Food Chemistry* 202: 349–355.

Meli, M.A., et al. (2020). Assessment of radioactivity in commercially available honey in Italy. *Food Control* 110: 107001.

Midura, T., et al. (1979). Isolation of Clostridium botulinum from Honey. *Journal of Clinical Microbiology* 9 (2): 282–283.

Moniruzzaman, M., et al. (2014). Determination of mineral, trace element and pesticide levels in honey samples originating from different regions of Malaysia compared to Manuka honey. *BioMed Research International* 2014.

Mustafina, R., et al. (2015). Contamination of honey produced in the Republic of Kazakhstan with Clostridium botulinum. *Bulletin of the Veterinary Institute in Pulawy* 59 (2): 241–246.

Mutinelli, F. (2003). Practical application of antibacterial drugs for the control of honey bee diseases. *Apiacta* 38: 149–155.

Naccari, C., et al. (2014). Risk assessment of heavy metals and pesticides in honey from Sicily (Italy). *Journal of Food Research* 3 (2): 107.

Nakano, H. and Sakagucki, G. (1991). An unusually heavy contamination of honey products by Clostridium botulinum type F and Bacillus alvei. *FEMS Microbiology Letters* 79 (2–3): 171–177.

Nasser, L.A. (2004). Isolation and characterization of fungi contaminating packaged honey commonly consumed in Saudi Arabia. *Assiut University Bulletin for Environmental Researches* 7 (1): 1–7.

Nevas, M., et al. (2002). High prevalence of Clostridium botulinum types A and B in honey samples detected by polymerase chain reaction. *International Journal of Food Microbiology* 72 (1–2): 45–52.

Nevas, M., et al. (2006). Contamination routes of Clostridium botulinum in the honey production environment. *Environmental Microbiology* 8 (6): 1085–1094.

Nicolopoulou-Stamati, P., et al. (2016). Chemical pesticides and human health: the urgent need for a new concept in agriculture. *Frontiers in Public Health* 4: 148-148.

Olaitan, P.B., Adeleke, O.E., and Iyabo, O. (2007). Honey: a reservoir for microorganisms and an inhibitory agent for microbes. *African Health Sciences* 7 (3).

Olga, E., María, F.-G., and Carmen, S.M. (2012). Differentiation of blossom honey and honeydew honey from Northwest Spain. *Agriculture* 2 (1): 25–37.

Orso, D., et al. (2016). Simultaneous determination of multiclass pesticides and antibiotics in honey samples based on ultra-high performance liquid chromatography-tandem mass spectrometry. *Food Analytical Methods* 9 (6): 1638–1653.

Oshomah, M.U. (2014). Microbiological evaluation of commercial honey from Edo State, Nigeria. *International Journal of Scientific & Engineering Research* 5 (12): 796–799.

Osman, K.A., Al-Doghairi, M.A., and Al-Otaibi, N.D. (2020). Spatial distribution of environmental pollutants in natural honeys collected from some regions of Saudi Arabia. *Journal of Apicultural Research* 1–10.

Ostiguy, N., et al. (2019). Honey bee exposure to pesticides: a four-year nationwide study. *Insects* 10 (1): 13.

Özakin, C., et al. (2007). Bacterial analysis of marketed and raw honey in Turkey. *Uludağ Arıcılık Dergisi* 7 (1): 30–34.

Paige, J., Tollefson, L., and Miller, M. (1997). Public health impact on drug residues in animal tissues. *Veterinary and Human Toxicology* 39 (3): 162–169.

Panseri, S., et al. (2014). Occurrence of pesticide residues in Italian honey from different areas in relation to its potential contamination sources. *Food Control* 38: 150–156.

Peres, G.T., Rath, S., and Reyes, F.G.R. (2010). A HPLC with fluorescence detection method for the determination of tetracyclines residues and evaluation of their stability in honey. *Food Control* 21 (5): 620–625.

Pokarzhevskii, A. (1998). J. Roembke and JF Moltmann, Applied Ecotoxicology, Boca Raton, FL: Lewis Publishers, 1996. *Russian Journal of Ecology C/C of Ekologiia* 29: 293–294.

Poormontaseri, M., Hosseinzadeh, S., and Shekarforoush, S. (2014). *Characterization of clostridium botulinum spores and its toxin in honey*.

Pucciarelli, A.B., et al. (2014). Microbiological and physicochemical analysis of yateí (Tetragonisca angustula) honey for assessing quality standards and commercialization. *Revista Argentina de Microbiologia* 46 (4): 325–332.

Raghunandan, K. and Basavarajappa, S. (2013). Analysis of multifloral honey of the giant honeybee, Apis Dorsata F., for pesticide residues in Southern Karnataka, India. *European Journal of Zoological Research* 2 (3): 22–28.

Rall, V., et al. (2003). Honey consumption in the state of São Paulo: a risk to human health? *Anaerobe* 9 (6): 299–303.

Rexilius, L. (1986). Rückstände von Pflanzenbehandlungsmitteln in Rapshonig der Ernte 1984 aus Schleswig-Holstein—Eine-Statusuntersuchung. *Nachrichtenbl Dtsch Pflanzenschutzd* 38: 49–56.

Reybroeck, W. (2002). Residues of antibiotics and sulfonamides in honey on the Belgian market. *Group* 200: 50.

Rissato, S.R., et al. (2007). Multiresidue determination of pesticides in honey samples by gas chromatography–mass spectrometry and application in environmental contamination. *Food Chemistry* 101 (4): 1719–1726.

Roberts, J.R., Karr, C.J., and Council On Environmental Health. (2012). Pesticide exposure in children. *Pediatrics* 130 (6): e1765–e1788.

Różańska, H. (2011). Microbiological quality of Polish honey. *Bulletin- Veterinary Institute in Pulawy* 55: 443–445.

Rüegg, J. (1995). MoniliaimObstbau—Prüfung von Fungiziden in der biologi schenind Integrierten Produktion. *Obst-und Weinbau* 131: 228–230.

Sabatini, A.G., et al. (2003). Presence of acaricides and antibiotics in samples of Italian honey. *Apiacta* 38: 46–49.

Sackett, W.G. (1919). *Honey as a Carrier of Intestinal Diseases*. CO: The Experiment Station.

Sadik, M. and Ali, M. (2012). *Survey and identification of microorganisms in bee honey samples collected from different plant sources and regions in Saudi Arabia*.

Sari, M.F., Gurkan Ayyildiz, E., and Esen, F. (2020). Determination of polychlorinated biphenyls in honeybee, pollen and honey samples from urban and semi-urban areas in Turkey. *Environmental Science and Pollution Research International* 27 (4): 4414–4422.

Saridaki-Papakonstadinou, M., et al. (2006). Determination of tetracycline residues in Greek honey. *Trakia Journal of Sciences* 4 (1): 33–36.

Satorres, S., et al. (1997). Detection of Clostridium botulinum spores in honey. *Revista Argentina de Microbiología* 29 (3): 147–151.

Schocken-Iturrino, R.P., et al. (1999). Study of the presence of the spores of Clostridium botulinum in honey in Brazil. *FEMS Immunology & Medical Microbiology* 24 (3): 379–382.

Sereia, M.J., et al. (2010). Microorganisms in organic and non organic honey samples of Africanized honeybees. *Journal of Apicultural Science* 54 (1): 49–54.

Sereia, M.J., et al. (2011). Microbial flora in organic honey samples of Africanized honeybees from Parana River islands. *Food Science and Technology* 31 (2): 462–466.

Sharma, A., et al. (2019). Worldwide pesticide usage and its impacts on ecosystem. *SN Applied Sciences* 1: 1446.

Sherwani, S.K., et al. (2013). Microbiological quality assessment of different commercially available honey products in Karachi, Pakistan. *International Journal of Advances in Pharmaceutical Research* 4 (3): 1531–1535.

Singh, N.S., et al. (2018). *Pesticide Contamination and Human Health Risk Factor, in Modern Age Environmental Problems and Their Remediation* (Editors M. Oves, M. Zain Khan, and I.M.I. Ismail), 49–68. Cham, Switzerland: Springer International Publishing.

Snowdon, J.A. and Cliver, D.O. (1996). Microorganisms in honey. *International Journal of Food Microbiology* 31 (1–3): 1–26.

Solomon, R.J., Santhi, V.S., and Jayaraj, V. (2006). Prevalence of antibiotics in nectar and honey in South Tamilnadu, India. *Integrative Biosciences* 10 (3): 163–167.

Steinhauser, G., Brandl, A., and Johnson, T.E. (2014). Comparison of the Chernobyl and Fukushima nuclear accidents: a review of the environmental impacts. *Science of the Total Environment* 470: 800–817.

Szczêsna, T., et al. (2009). Study on sulfonamide residues in honey. *Journal of Apicultural Science* 53 (2).

Thomas, G.O. (2008). Polychlorinated Biphenyls. In: *Encyclopedia of Ecology* (ed. S.E. Jørgensen and B.D. Fath), 2872–2881. Oxford, UK: Academic Press.

Thompson, T.S., et al. (2005). Determination of lincomycin and tylosin residues in honey by liquid chromatography/tandem mass spectrometry. *Rapid Communications in Mass Spectrometry* 19 (3): 309–316.

Thompson, T.S. and Noot, D.K. (2005). Determination of sulfonamides in honey by liquid chromatography–tandem mass spectrometry. *Analytica Chimica Acta* 551 (1–2): 168–176.

Tonelli, D., et al. (1990). Honey bees and their products as indicators of environmental radioactive pollution. *Journal of Radioanalytical and Nuclear Chemistry* 141 (2): 427–436.

Tornuk, F., et al. (2013). Quality characterization of artisanal and retail Turkish blossom honeys: determination of physicochemical, microbiological, bioactive properties and aroma profile. *Industrial Crops and Products* 46: 124–131.

Tysset, C.D.C. and Roy, D.R.D.L. (1991). *Contribution to the Study of Intestinal Microbial Infection of Healthy Honeybees: Inventory of Bacterial Population by Negative Organisms*, 21–55. Philadelphia: Department of Agriculture, SEA-AR, Eastern Region Research Centers.

Vázquez-Quiñones, C.R., et al. (2018). Microbiological assessment of honey in México. *Revista Argentina de microbiologia* 50 (1): 75–80.

Velicer, C.M., et al. (2004). Antibiotic use in relation to the risk of breast cancer. *JAMA* 291 (7): 827–835.

White, P. (1996). The normal flora of the bee. In: *Agricultural Research Service*, 301–309. US Department of Agriculture Washington AC.

WHO. (2009). *Evaluation of certain veterinary drug residues in food: seventieth report of the Joint FAO/WHO Expert Committee on Food additives (WHO technical report series; no. 954).*

World Health Organization. et al. (2018). *Evaluation of Certain Veterinary Drug Residues in Food: Eighty-fifth Report of the Joint FAO/WHO Expert Committee on Food Additives.* WHO technical report series;1008. Geneva: World Health Organization.

Yavuz, H., et al. (2010). Determination of some organochlorine pesticide residues in honeys from Konya, Turkey. *Environmental Monitoring and Assessment* 168 (1): 277–283.

# 13

## Antimicrobial Properties of Honey

*Mahendran Sekar[1,2,*], Nur Zulaikha Azwa Zuraini[3], Nur Najihah Izzati Mat Rani[3], Pei Teng Lum[2], and Siew Hua Gan[1]*

[1] *School of Pharmacy, Monash University Malaysia, Bandar Sunway, Subang Jaya 47500, Selangor, Malaysia*
[2] *Department of Pharmaceutical Chemistry, Faculty of Pharmacy and Health Sciences, Royal College of Medicine Perak, Universiti Kuala Lumpur, Ipoh 30450, Perak, Malaysia*
[3] *Faculty of Pharmacy and Health Sciences, Royal College of Medicine Perak, Universiti Kuala Lumpur, Ipoh 30450, Perak, Malaysia*
* *Corresponding author*

## Introduction

Since ancient times, bees' honey has been one of the most valued and appreciated natural substances known to humankind. It is used in many civilizations because of its wide range of therapeutic benefits. Honey's medicinal value has been documented in the world's oldest medical literature because it has a long history of human consumption (Eteraf-Oskouei and Najafi 2013). It is usually used as a sweetener and flavoring in a variety of foods and beverages (Bagde et al. 2013). Carbohydrates, such as monosaccharides, fructose, glucose, disaccharides, maltose, isomaltose, maltulose, sucrose, and turanose, are the most essential constituents in honey, and they are responsible for its sweetness. The flavor of a certain variety of honey is determined by the liquid sources found in the flowers and plants that the honeybees visit (Meo et al. 2017). It is also one of the foods that has religious significance. Honey was mentioned in the Holy Quran more than 1400 years ago as a source of healing. Almighty Allah mentioned honey's unique potential to heal and cure sickness (Purbafrani et al. 2014). Allah said,

> And your Lord inspired to the bee: Make hives in the mountains, houses and among the trees in what they build. Then eat from all the fruits and walk in the ways of your lord laid down [for you]. There emerges from their bellies a drink, varying in colours, in which there is healing for people. Verily in this, sign for people who give thought.
> (Surah Al-Nahl; verses 68 and 69)

Honey is also one of the foods of Paradise mentioned in the Quran – there will be a sea of water, a sea of wine, a sea of milk, and a sea of honey in Paradise, and the rivers of Paradise will flow out of these seas. There will be also many springs in Paradise, and its people will drink from these seas, rivers, and springs (surah Muhammad; verses 15). The prophet Muhammad also lauded honey's healing power (Aurongzeb and Azim 2011). The Divine Claim of the Almighty that honey contains healing and medication for humankind has been confirmed to be truthful. Scientific statements and notions abound in the Noble Quran. What's even more astounding is that all of these scientific assertions and hypotheses have been demonstrated to be in perfect harmony with science and modern scientific discoveries. Also, in Hinduism, one of the five elixirs of immortality is Madhu (honey). At a ritual known as Madhu abhisheka, honey is poured over deities in temples. According to the Vedas and other ancient texts, honey is a powerful therapeutic and health food (Bagde et al. 2013).

In traditional Chinese medicine (TCM), honey, also known as Feng Mi, may nourish yin energy while also strengthening the spleen (Zhang et al. 2015). Honey is a Chinese favorite that is regarded as a "neutral" meal with healing properties in addition to its well-known nutritional benefits. Honey has been used in TCM for more than 2000 years to cure a number of disorders. Furthermore, honey is used in medical therapy with bee venom, propolis, royal jelly, pollen, and other natural medicines. Bacterial infections, gastrointestinal diseases, allergic and immunologic disorders, and traumatic wounds can all be prevented when using it (Horn 2013). Honey was also used to cure wounds topically in pre-ancient Egypt. It has been used in Egyptian therapeutic formulations for more than five millennia. It was also believed by the ancient Greeks to help humans live longer lives (Purbafrani et al. 2014).

*Honey: Composition and Health Benefits*, First Edition. Edited by Md. Ibrahim Khalil, Gan Siew Hua, and Bey Hing Goh.

Honey has a broad range of antimicrobial properties that protect against infections from bacteria, fungi, and viruses (Aggad and Guemour 2014; Cooper et al. 1999; Nasir et al. 2010). Tualang and Manuka honeys are the most effective and well-studied honeys for antimicrobial properties (Al Somal et al. 1994; Willix et al. 1992). A brief overview of the research carried out on the antimicrobial properties of honey is presented in this chapter.

## Chemistry of Honey

According to a study by Ediriweera and Premarathna (2012), the flowers in the forest also provide food for bees. Many types of bees, including *Apis koschevnikov*, *Apis cerana*, *Apis mellifera*, *Apis floria*, *Apis dorsata*, *Apis andreniformis*, and *Apis laborisa*, produce honey. They collect nectar from flowers or sweet juices from specific plant species (honeydew or forest honey) and process it into honey. When bees collect nectar, they store it in their honey stomachs, which are separate from their regular stomachs. The nectar is blended with enzymes that break down the nectar's major sugar, particularly sucrose, into smaller sugars such as glucose and fructose (Figure 13.1). Sucrose is the predominant or exclusive sugar of many nectars. But sucrose, glucose, galactose, raffinose, and fructose are present in almost the same amounts in some nectars (Ball 2007).

The nectar is then passed on to a house bee, who regurgitates and redrinks it for 20 minutes, thereby breaking down the bigger sugars. The nectar is placed in the honeycomb, and the bees fan it to speed up the evaporation of the water until the concentration of honey is around 17%. Honey has a low water content because it removes excess from its surroundings, which allows it to dehydrate microorganisms and prevent rotting (Olaitan et al. 2007). According to a study by Purbafrani et al. (2014), honey is high in minerals, vitamins, amino acids, and antioxidants. Honey contains vitamins such as riboflavin, niacin, and pantothenic acid. As for minerals, honey contains copper, calcium, magnesium, manganese, iron, phosphorus, zinc, and potassium. Furthermore, the flavonoids and phenolic acids that are present in honey were found to act as antioxidants, scavenging and eliminating free radicals.

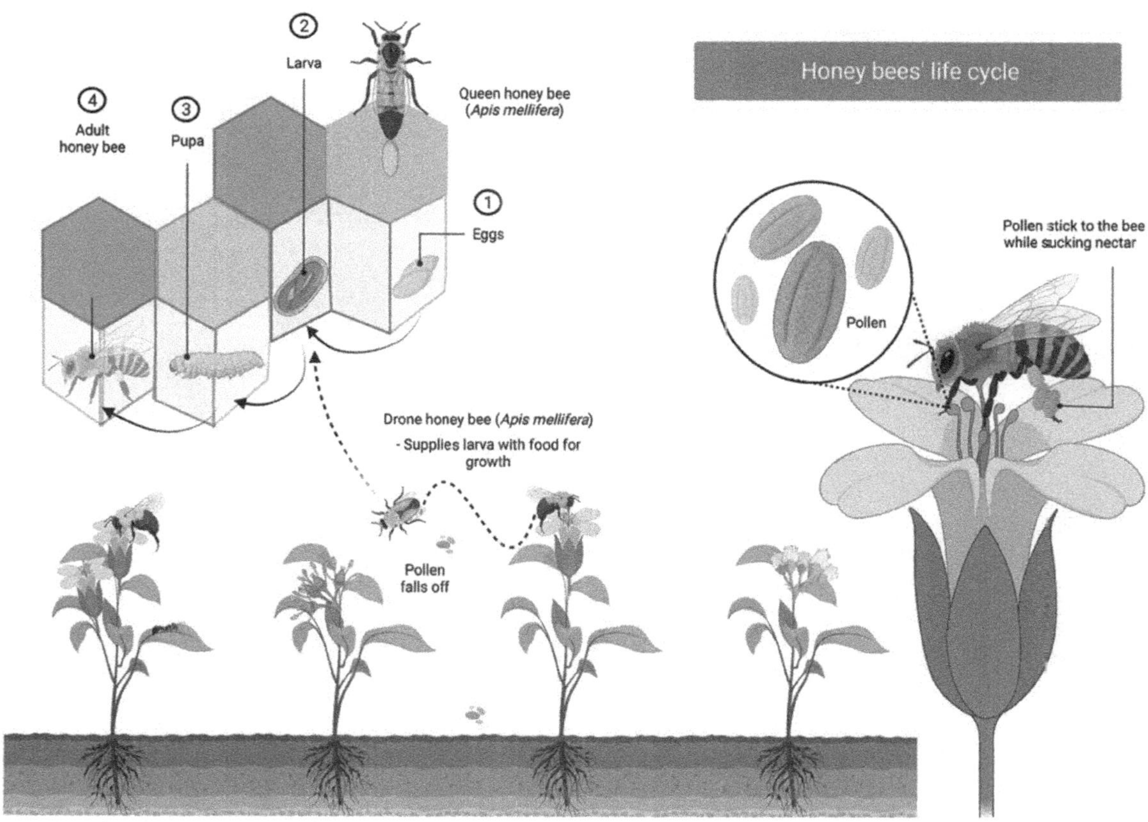


**Figure 13.1** The life cycle of honeybees.

## Antimicrobial Properties

Honey has been reported to have potential antimicrobial properties (Figure 13.2), according to research findings. Honey inhibits a wide range of bacterial species (aerobic, anaerobic, gram-positive, and gram-negative bacteria). Furthermore, honey possesses antimicrobial characteristics that are effective against both pathogenic and non-pathogenic microorganisms (yeasts and fungi). Depending on the concentration, the antimicrobial effects of honey could be bacteriostatic or bactericidal (Nair et al. 2004). Principles from honey have been shown to have antibacterial, antifungal, and antiviral activities.

## Antibacterial Properties

Honey has been shown to have antibacterial action (in vitro) in clinical case studies in a number of studies. These studies claim that applying honey to infected cutaneous wounds can help to decrease infection and speed up healing. Honey's antibacterial activities are believed to be helped by a number of factors, including (1) the osmotic effect of the high sugar content in honey can reduce the amount of water required for microorganism growth; (2) its high acidity (low pH) and composition can deprive bacteria of nitrogen, which is required for reproduction; (3) bacterial growth can be inhibited by the presence of hydrogen peroxide ($H_2O_2$) and antioxidants in honey (Purbafrani et al. 2014); and (4) the presence of various phytochemicals, including alkaloids and flavonoids (caffeic acid, quercetin, pinobanksin, chrysin, and galangin), as well as bee-added peptides, such as apidaecin and bee-defensin-1 (Figure 13.3) (Nolan et al. 2019).

According to Visavadia et al. (2008), honeys that comes from specific floral sources –specifically, *Leptospermum* spp., which is found mostly in Australia and New Zealand – have also been shown to have higher antibacterial activity. These honeys have been certified for use as therapeutic dressings, and it has been proven by Tafere (2021) that honey is safe to use because it shows no symptoms of allergic reaction or adverse effects after applying it onto wounds. The risk of wound infection can be prevented because it reduces bacterial invasion and growth. Cooper et al. (2002) investigated the

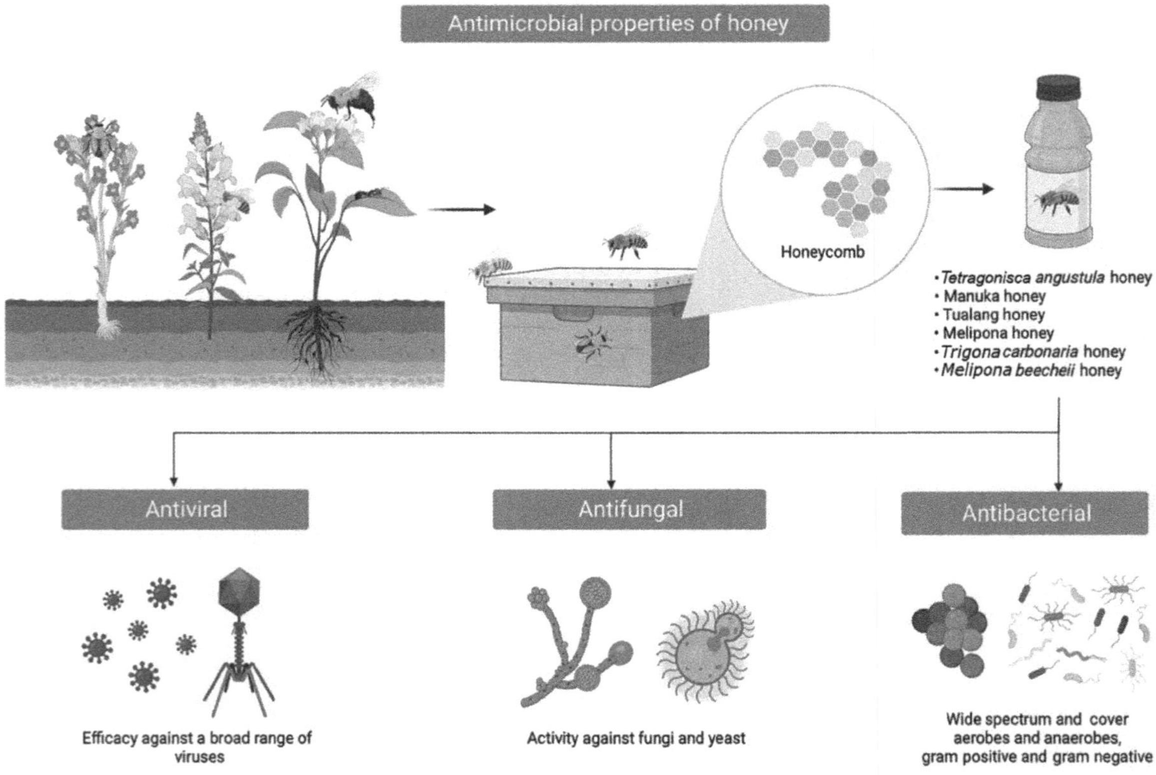


**Figure 13.2** The antimicrobial properties of honey. Honey has a wide range of biological effects, including suppressing bacteria, yeasts, and fungi and even having antibiotic-resistant strains. Depending on the concentration, the outcome may differ.

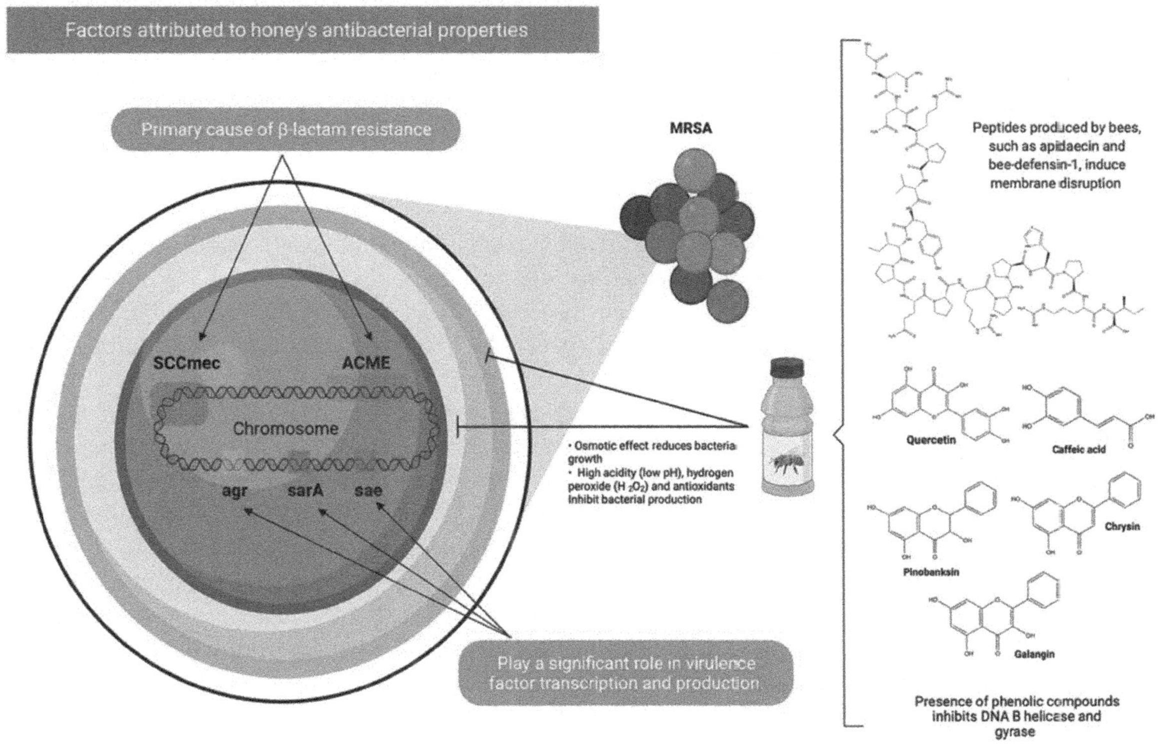


**Figure 13.3** Antimicrobial properties of honey. Methicillin-resistant *Staphylococcus aureus* (MRSA) is a type of bacteria that causes a wide range of local and systemic illnesses in humans, and global regulators such as agr, sar, and sae regulate the expression of most virulence factors. Honey is said to be antibacterial because of its high sugar content, acidity (low pH), and composition, which deprive bacteria of nitrogen, which is required for reproduction. Besides, $H_2O_2$ and antioxidants contained in honey can also aid in bacterial growth prevention. ACME, arginine catabolic mobile element; SSmec, staphylococcal cassette chromosome mec.

antibacterial activity of pasture honey and manuka honey against 17 *Pseudomonas aeruginosa* strains isolated from infected burns. Artificial honey had a mean minimum inhibitory concentration (MIC) of 19.2% (v/v). Manuka and pasture honey had MICs of 7.5% and 6.8% (v/v), respectively. In vitro experiments revealed that natural honeys were 2.5 and 2.8 times more effective than manufactured honey at inhibiting pseudomonas. In tube dilution tests, mean minimum bactericidal concentration (MBC) values for manuka and pasture honey were only marginally higher than MIC values, indicating a bactericidal rather than bacteriostatic effect. Furthermore, in a study by Cooper et al. (2002), the antibacterial effect was shown to most likely penetrate beneath the wound's surface. Although topical honey is commonly bactericidal rather than bacteriostatic in action, it has been demonstrated in vitro that honey can give quick clearance of deep-seated infection and boils with unbroken skin. This implies that honey's antibacterial activity may diffuse through the skin to deeper tissues. Honey is high in phenolic acids and flavonoids, which may be responsible for its antibacterial and anti-inflammatory properties (Samarghandian et al. 2017). Honey is very effective against a variety of human infections, including boils, abscesses, carbuncles, wound infection, and impetigo, and even may include diarrhea, septicemia, urinary infections, and typhoid (Tafere 2021). These conditions occur because of common bacterial species such as *Escherichia coli, Enterobacter aerogenes, Salmonella typhimurium, and Staphylococcus aureus.*

The stingless bee *Tetragonisca angustula* is abundantly available in Brazil and Mexico. The antibacterial activity of *T. angustula* honey was reported against numerous bacterial strains, including *S. aureus* (Miorin et al. 2003), *Bacillus cereus*, and *P. aeruginosa*. A study found that *T. angustula* honey had antibacterial action against gram-positive bacteria, including *S. aureus* and *Enterococcus faecalis*, as well as coagulase-negative methicillin-sensitive bacteria such as *P. aeruginosa* and *E. coli* (Sgariglia et al. 2010). Based on a study by George and Cutting (2007), the honey has also been shown to be effective against methicillin-resistant *S. aureus* (MRSA), hemolytic streptococci, vancomycin-resistant enterococci, and other multidrug-resistant (MDR) gram-negative organisms, including *P. aeruginosa*, in vitro studies. Honey generated by *Trigona laeviceps*, a stingless bee native to Thailand, showed antibacterial properties against *E. coli* and *S. aureus* bacteria (Chanchao 2009).

The antibacterial action of honey against bacterial pathogens can also be found in honey other than commercially produced antibacterial honeys such as manuka honey. Manuka honey is a monofloral honey produced from the manuka tree (*Leptospermum scoparium*), which grows as a shrub or small tree mostly in New Zealand and Eastern Australia and belongs to the Myrtaceae family (Brophy et al. 2000). Sugars, free amino acids, proteins, enzymes, vital minerals, vitamins, and a variety of secondary metabolites are all abundant in Manuka honey (flavonoids, phenolic acids, and 1,2-dicarbonyl compounds), which are responsible for its biological actions (Alvarez-Suarez et al. 2014). Manuka honey has been shown to exhibit antimicrobial activity against wide range of bacteria, including *Streptococcus pyogenes*, *Streptococcus mutans*, *Proteus mirabilis*, *P. aeruginosa*, *Enterobacter cloacae*, and *S. aureus* (Alandejani et al. 2009; Maddocks et al. 2012, 2013; Majtan et al. 2014). Manuka honey has been found to eradicate MRSA from colonized wounds and to inhibit MRSA in vitro by interrupting cell division. Furthermore, manuka honey restores MRSA's susceptibility to oxacillin; molecular investigation revealed that it also changes the regulation of the *mecR1* gene, which could explain the recovered susceptibility (Jenkins and Cooper 2012)

A synergistic effect of rifampicin and Food and Drug Administration–approved manuka honey (Medihoney, Medihoney Ltd, Slough, United Kingdom) on clinical *S. aureus* isolates, including MRSA strains, was demonstrated in another investigation. Unlike rifampicin alone, which caused resistance after an overnight incubation on plates, the combination of Medihoney and rifampicin maintained *S. aureus* susceptible to rifampicin (Müller et al. 2013). Thus, manuka honey appears to have considerable potential in offering novel synergistic antibiotic combinations for treating wound infections caused by MDR bacteria. It is worth noting that the antibiotics that have exhibited synergy with manuka honey come from a variety of antibiotic classes and target different targets, including the 30S ribosome, RNA polymerase, membranes, and penicillin-binding proteins. This finding supports with the theory that honey is a complex drug with various active ingredients that activate multiple cellular target sites (Jenkins and Cooper 2012).

Tualang honey can also inhibit the growth of bacteria that cause stomach disorders, such as *S. typhi*, *Shigella flexneri*, and *E. coli*, at low concentrations, according to the study reported by Lusby et al. (2005). Tualang honey has been shown to be effective against *E. coli*, *S. typhi*, and *S. pyogenes*, and when consumed orally in its natural, concentrated form, this honey may aid in the recovery of people with these diseases. Honey also works well as a glucose substitute in oral rehydration, and its antibacterial properties help to decrease the length of bacterial diarrhea (Tumin et al. 2005). Overall, assessing different types of honey against a wide range of bacteria revealed that it might be used as an antibacterial agent to treat bacterial infections.

## Antifungal Properties

In most countries around the world, the rise of opportunistic fungal diseases is a major problem. This sort of infection is usually caused by the *Candida* genus. Invasive candidiasis is becoming more common, and there have been numerous reports of strains resistant to conventional therapy based on the usage of azoles and echinocandins (Akins and Sobel 2017). Honey (30%–50%) suppresses the growth of numerous pathogenic microorganisms, including *Candida albicans* (Al-Waili 2001). Irish et al. (2006) found that several honeys were antifungal against clinical isolates of *C. albicans*, *Candida glabrata*, and *Candida dubliniensis*. Honey exhibits antifungal activity against *Candida* species such as *C. albicans*, *Candida parapsilosis*, *Candida tropicalis*, *Candida kefyr*, *C. glabrata*, and *C. dubliniensis* (Khosravi et al. 2008).

Aside from the studies of honey's bactericidal and bacteriostatic activities, there have been few findings on its antifungal activity as well (Theunissen et al. 2001). Because there has been an increase interest in the research for novel antifungals among natural resources in recent years, the honey from bee species is unexceptional in the current studies to treat a wide range of fungal illnesses. Fungal infections are becoming more widespread in both community and hospital settings, with several causative agents, including yeasts such as *Candida* spp., which is an opportunistic polymorphic fungus that is thought to be the most important yeast pathogen (Arisawa et al. 2007). Honey has been shown to have antifungal properties against the yeast *C. albicans* as well as most species of *Aspergillus baumannii*, *Penicillium chrysogenum*, *Cryptococcus neoformans, and Candida krusei* (Ansari et al. 2013). Honey's antifungal properties are hypothesized to be aided by the same reasons as mentioned in the discussion of its antibacterial properties. It is due to its osmotic effect, high sugar content (low pH), and presence of $H_2O_2$ and phytochemicals such as flavonoids. However, not all of these reasons apply to all kinds of honey. This is confirmed by the findings of Hau-Yama et al. (2020), who revealed that osmotic pressure had no antifungal influence on *C. albicans* by honey from stingless bee *Melipona beecheii*. On the other hand, an increase in fungal growth was seen, most likely the result of increased sugar availability. According to Feás and Estevinho (2011), honey's mechanism for inhibiting fungal growth is unrelated to the osmotic shock caused by the presence of sugar in the culture media. Wahdan (1998) further suggested that honey's high sugar content causes a high osmolarity, which results in antibacterial action. They further discovered that the sugar solutions have no inhibitory efficacy against *Trichophyton mentagrophytes* and *C. albicans* and that fungi are generally more tolerant to strong osmotic effects than bacteria.

Furthermore, the da Silva et al. (2014) study found that flavonoids, including quercetin, epigallocatechin gallate, and catechin, when combined with fluconazole, can actually increase apoptosis of cells by exposing phosphatidylserine in the plasma membrane and morphological alterations in *C. tropicalis* resistant to fluconazole (Figure 13.4). The flavonoids can be seen mostly in Melipona honey. The natural cure for *Candida* infections is more effective and able to last the long term because it focuses on curing the infection before causing irritation. Candiracci et al. (2011) researched how to suppress yeast proliferation in body by using honey.

All honey production and honey categories are dependent on the natural vegetative flowers that bloom at different times of the year. Thus, the flowers from which bees collect nectar to make honey may contribute to the difference in antifungal activity. This is because not all honeys being produced have similar actions. Boorn et al. (2010) used Meliponinae, a family of plants that includes species such as *M. beecheii*, in their study. These species of honey have been discovered to treat ocular and gastrointestinal infections (fungal infection) in different cultures in Latin America and Australia, respectively. However, *Trigona carbonaria* honey from Australia is proven to have a minimal antifungal activity. The honey of the stingless bee *T. carbonaria* exhibits weak antifungal activity against *C. albicans* and *C. glabrata* (Boorn et al. 2010). Alvarez-Suarez et al. (2018) studied the antifungal activity of honey generated by *A. mellifera* and *M. beecheii* in Cuba, finding that honey from both species inhibits *C. albicans* growth.

The antifungal action by honey depends on its dosage as well. Higher concentrations of honey with varied entomological sources have the potential to inhibit the growth of many fungal infections. This has been stated because Hau-Yama et al. (2020)'s study shows that there has been a minor decrease of colonies at 25% concentration of honey, while 30% of colonies have undergone major colony losses. The same studies also show that the major antifungal effect can only be detected by applying concentrations of 40% honey and above.

In addition, an investigation by Hermanns et al. (2020) was made on the safety, quality, and efficacy of medical-grade honey subjected to stringent testing. It has also been reported that it is effective in suppressing *Candida* species such as

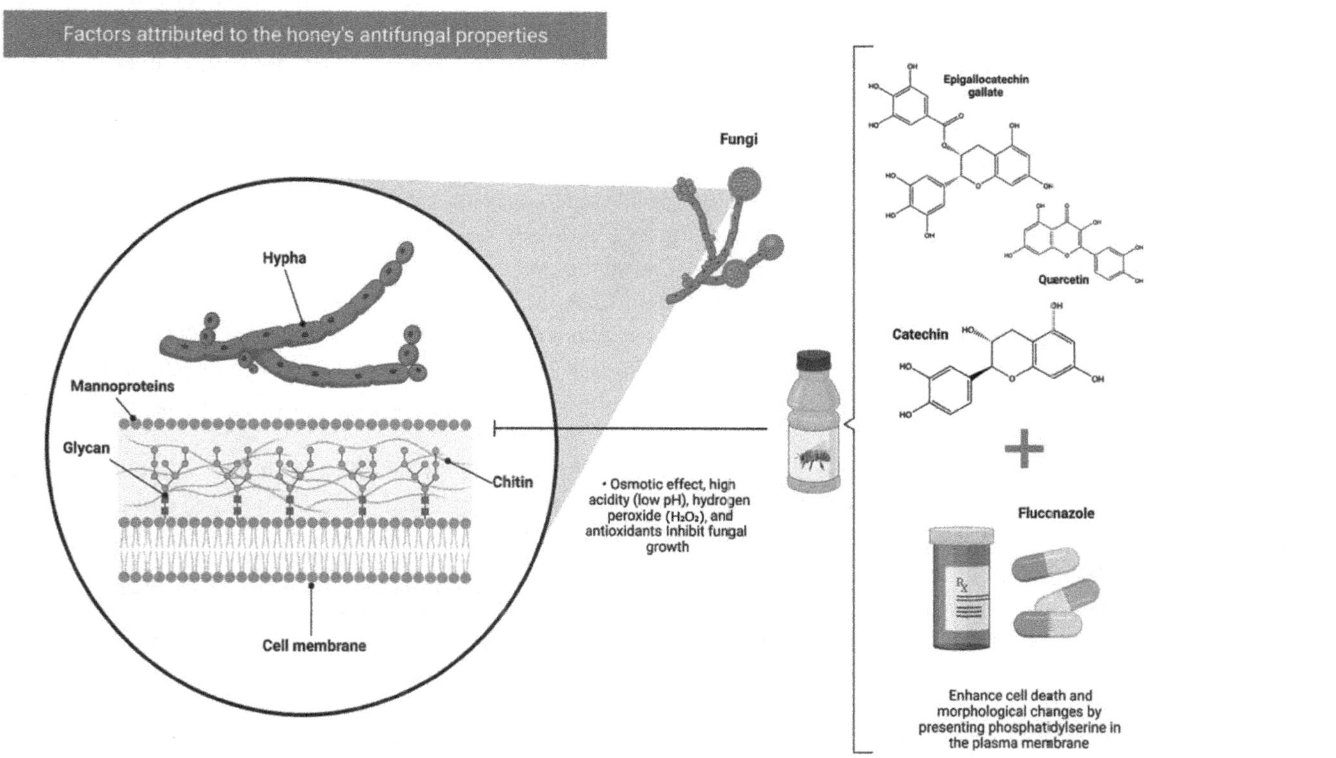


**Figure 13.4** Factors linked to honey's antifungal effects. Different studies have shown that the same parameters, such as osmotic pressure, play a key role in its antifungal capabilities, and some contradict. When flavonoids such as quercetin, epigallocatechin gallate, and catechin are combined with the antibiotic fluconazole, it can actually accelerate cell apoptosis.

*C. albicans* and *C. tropicalis* and is proven to be very effective in reducing fungal infections. Similarly, other researchers have done a study regarding whether a higher concentration of medical-grade honey (40%) affects the growth of *Candida auris* and the other *Candida* species (de Groot et al. 2021). A comparison has been made in terms of its efficacy between unprocessed local honey and medical-grade honey. It turns out that the medical-grade honey is more effective in reducing all *Candida* species except *C. krusei*, in which there is no large difference found (de Groot et al. 2021). Morroni et al. (2018) examined four types of honey from three different countries, including honey from the stingless bee *M. beecheii*. They found that *M. beecheii* honey exhibits antifungal action against *C. albicans* even at low concentrations. Thus, investigations based on the effects in *Candida* planktonic cells are essential in evaluating the antifungal potential of stingless bee honey.

Honey samples have different antifungal efficacies depending on their botanical sources and concentrations. This variance could be caused by the floral source or the location of the flowers. The composition of a certain honey sample is highly dependent on the nectar's content as well as its source. Honey can only be used for topical therapies and cannot be used to treat candidemia, the most serious form of candidiasis (Irish et al. 2006). However, the usage of topical treatments containing antifungal components such as flavonoids or antioxidants was influenced by skin fungal infections. Honey could be used to prevent more serious infections and in the treatment of oral and vaginal candidiasis.

## Antiviral Properties

Honey has antiviral efficacy against a variety of viruses, including herpes simplex virus (HSV), human immunodeficiency virus, respiratory syncytial virus, varicella-zoster virus (VZV), adenovirus, and influenza viruses, according to several studies (Esawy et al. 2011; Hermanns et al. 2020; Shahzad and Cohrs 2012; Watanabe et al. 2014; Zareie 2011). Honey also has anti-inflammatory properties and is known to be a potent immune booster, making it an excellent therapy for reducing the severity of viral illnesses (Abuharfeil et al. 1999; Tonks et al. 2003).

Herpes labia and genitalia are common viral conditions nowadays. Many medicines, including famciclovir, foscarnet, acyclovir, and valacyclovir, are available to treat HSV lesions. A number of topical medicines – 1% viroptic, 1% penciclovir, and 5% acyclovir – are used in the treatment of herpes. Although acyclovir is a safe medicine, nephrotoxicity and neurotoxicity can be the side effects for people who use it. The usage of acyclovir has also been found in cases of allergic dermatitis (Bourezane et al. 1996; Ernst and Franey 1998).

Thus, acceptable natural healing agents can be used to substitute the usage acyclovir. The previous study investigated the probable therapeutic effect on the recurrence lesions of labial and vaginal herpes lesions of the topical application of honey compared with treatment by topical application of acyclovir cream (Al-Waili 2004). Honey contains a variety of nutrients, including vitamin B complex, vitamin E, flavonoids, vitamin C, zinc, amino acids, and copper (Godfrey et al. 2001). Appropriate management of viral infections with the use of all these elements has been confirmed. The treatment with honey was found to exhibit a shorter healing period compared with acyclovir treatment (Al-Waili 2004). However, the findings acquired in this study require further validation because the experimental design is deemed incomplete.

Honey is not only efficient for antibacterial and antifungal uses, but it is also effective in the treatment of herpetic lesions (antiviral), as discovered by Godfrey et al. (2001). The presence of ascorbic acid and $H_2O_2$ in honey can enhance the inactivation of HSV, with the help of copper as well (Figure 13.5). It was shown that cupric ions (Cu II) inactivate numerous viruses, including herpesvirus and arenavirus members, several in vitro bacteriophages, and intracellular immunodeficiency virus.

Besides, it is found that both manuka and clover honey displayed strong antiviral efficacy with an approximate $EC_{50}$ (concentration effective in producing 50% of the maximal response) of 4.5% wt/vol against VZV isolated from patients in an in vitro investigation (Shahzad and Cohrs 2012). Honeys from manuka and clover are used in quantities ranging from 0% to 6% wt/vol. VZV-infected human malignant melanoma cells had their tissue culture media supplemented with varying amounts of manuka and clover honey. This indicates that honey can be administered topically to treat zoster rash. Therefore, honey may be great therapy for treating zoster rash in impoverished nations where antiviral medications are expensive or unavailable. It is also deemed convenient to use because it could be easily applied on the skin, is generally well accessible, and is inexpensive.

In the context of COVID-19, there are both direct and indirect lines of evidence showing that honey possesses great potential to be used as a supplementary treatment or preventive natural substance against this infection. According to Al-Hatamleh et al. (2020), honey consumption may assist to reduce the severity of COVID-19 infection, either directly through antiviral actions against SARS-CoV-2 or indirectly through immune response enhancement. Also, according to

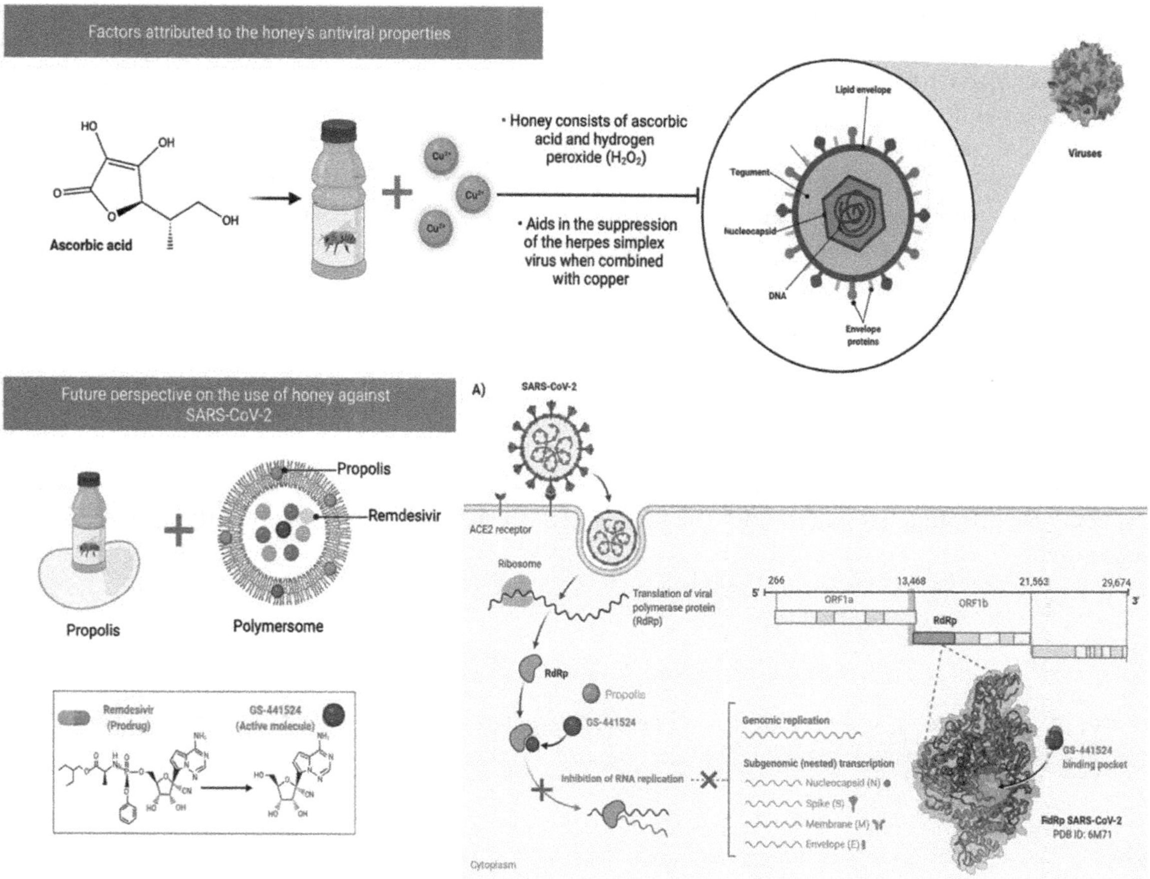


**Figure 13.5** Factors attributed to the antiviral properties of honey. Honey, which includes a lot of phytochemical substances, combined with hydrogen peroxide and copper ions, helps to block viruses, which eventually speeds up viral eradication. Future perspectives on the use of propolis and antiviral medicines, such as remdesivir loaded in a polymersome for a dual drug delivery strategy to halt viral replication, may help in combating severe acute respiratory syndrome coronavirus 2 (SARS-CoV-2 or COVID-19). ACE, angiotensin-converting enzyme.

research by Kalediene et al. (2021), Camelyn, which is a honey extract, has also been proven to have antiviral activity against SARS-CoV-2. It is found to be safe and to have no cytotoxic effects and is available as a natural supplement. Honey's direct and indirect anti–COVID-19 effects are mostly due to its high concentration of antioxidant, phenolic components and other major attributions already discussed. Despite the many advantages of honey, it does not supersede requesting medical advice and taking medicines. Note that honey is not as effective as other well-established medicines in treating COVID-19 because of its limited therapeutic capability.

To fully understand the mechanisms of action of honey against COVID-19, more preclinical and clinical research are needed. In the future, more detailed molecular investigation studies of honey's impact on viral multiplication toward the immune system as well as a thorough and comprehensive examination of the pharmacokinetics of honey-derived phenolic chemicals are essential to unravel its biological functions in a much more detailed manner.

## Conclusion

Honey's broad-spectrum antimicrobial properties, including its antibacterial, antifungal, and antiviral properties, have been showcased in numerous in vitro and limited clinical studies. Generally, its functionalities may be attributed to its acidity (low pH); osmotic impact; high sugar concentration; existence of $H_2O_2$ and antioxidants; and presence of phytochemicals, including alkaloids, flavonoids, lysozyme, and bee peptides. This indicates that honey contains antimicrobial properties that are mediated through a range of mechanisms. Because of the attribution differences described in this chapter, not all honeys have the same antibacterial effectiveness. Other variables, such as storage conditions and the susceptibility of different bacteria strains, might also be some of the challenges. It is also postulated that certain types of honey might possess no antibacterial action. Therefore, there is an urgent need to establish an acceptable standard quantitative testing method for honey's antibacterial properties because it is rather common to face challenge in aspects such as the dose and formulation. It is mainly characterized by the wide variations in honey composition, strength, quality, and processing methods such as filtering and heating, as well as the standardization of storage conditions and adulteration. Because there are no rules governing the standardization of honey for use in a therapeutic environment, more research is needed to determine whether its antimicrobial activity has any clinical utility. Therefore, identifying the bioactive compounds in honey and their clinical assessment and pharmacological standardization is deemed essential and significant in paving the way for any future standardization initiatives.

## Acknowledgment

The authors thank Universiti Kuala Lumpur Royal College of Medicine Perak, Malaysia, for providing necessary facilities and resources to complete this study for publication. The figures were created with the support of https://biorender.com under a paid subscription.

## Conflict of Interest

We wish to confirm that there are no known conflicts of interest associated with this publication, and there has been no significant financial support for this work that could have influenced its outcome.

## References

Abuharfeil, N., Al-Oran, R., and Abo-Shehada, M. (1999). The effect of bee honey on the proliferative activity of human B-and T-lymphocytes and the activity of phagocytes. *Food and Agricultural Immunology* 11 (2): 169–177.

Aggad, H. and Guemour, D. (2014). Honey antibacterial activity. *Medicinal & Aromatic Plants* 3: 152.

Akins, R.A. and Sobel, J.D. (2017). Antifungal targets, mechanisms of action, and resistance in *Candida albicans*. In: Mayers, D.L. (ed). *Antimicrobial Drug Resistance. Infectious Disease*, 347-407. New York: Humana Press.

Al-Hatamleh, M.A., Hatmal, M.M., Sattar, K., et al. (2020). Antiviral and immunomodulatory effects of phytochemicals from honey against COVID-19: potential mechanisms of action and future directions. *Molecules* 25 (21): 5017.

Al-Waili, N. (2001). Therapeutic and prophylactic effects of crude honey on chronic seborrheic dermatitis and dandruff. *European Journal of Medical Research* 6 (7): 306–308.

Al-Waili, N.S. (2004). Topical honey application vs. acyclovir for the treatment of recurrent herpes simplex lesions. *Medical Science Monitor* 10 (8): MT94–MT98.

Al Somal, N., Coley, K., Molan, P., and Hancock, B. (1994). Susceptibility of Helicobacter pylori to the antibacterial activity of manuka honey. *Journal of the Royal Society of Medicine* 87 (1): 9.

Alandejani, T., Marsan, J., Ferris, W., et al. (2009). Effectiveness of honey on Staphylococcus aureus and Pseudomonas aeruginosa biofilms. *Otolaryngology—Head and Neck Surgery* 141 (1): 114–118.

Alvarez-Suarez, J.M., Gasparrini, M., Forbes-Hernández, T.Y., et al. (2014). The composition and biological activity of honey: a focus on Manuka honey. *Foods* 3 (3): 420–432.

Alvarez-Suarez, J.M., Giampieri, F., Brenciani, A., et al. (2018). Apis mellifera vs Melipona beecheii Cuban polifloral honeys: a comparison based on their physicochemical parameters, chemical composition and biological properties. *LWT* 87: 272–279.

Ansari, M.J., Al-Ghamdi, A., Usmani, S., et al. (2013). Effect of jujube honey on Candida albicans growth and biofilm formation. *Archives of Medical Research* 44 (5): 352–360.

Arisawa, T., Arisawa, S., Yokoi, T., et al. (2007). Endoscopic and histological features of the large intestine in patients with atopic dermatitis. *Journal of Clinical Biochemistry and Nutrition* 40 (1): 24–30.

Aurongzeb, M. and Azim, M.K. (2011). Antimicrobial properties of natural honey: a review of literature. *Pakistan Journal of Biochemistry and Molecular Biology* 44 (3): 118–124.

Bagde, A., Sawant, R., Bingare, S., et al. (2013). Therapeutic and nutritional values of honey (madhu). *International Research Journal of Pharmacy* 4 (3): 19–22.

Ball, D.W. (2007). The chemical composition of honey. *Journal of Chemical Education* 84 (10): 1643.

Boorn, K., Khor, Y.Y., Sweetman, E., et al. (2010). Antimicrobial activity of honey from the stingless bee Trigona carbonaria determined by agar diffusion, agar dilution, broth microdilution and time-kill methodology. *Journal of Applied Microbiology* 108 (5): 1534–1543.

Bourezane, Y., Girardin, P., Aubin, F., et al. (1996). Allergic contact dermatitis to Zovirax® cream. *Allergy (Copenhagen)* 51 (10): 755–756.

Brophy, J.J., Goldsack, R.J., Bean, A.R., et al. (2000). Leaf essential oils of the genus Leptospermum (Myrtaceae) in eastern Australia, Part 6. Leptospermum polygalifolium and allies. *Flavour and Fragrance Journal* 15 (4): 271–277.

Candiracci, M., Citterio, B., Diamantini, G., et al. (2011). Honey flavonoids, natural antifungal agents against Candida albicans. *International Journal of Food Properties* 14 (4): 799–808.

Chanchao, C. (2009). Antimicrobial activity by Trigona laeviceps (stingless bee) honey from Thailand. *Pakistan Journal of Medical Sciences* 25 (3): 364–369.

Cooper, R.A., Halas, E., and Molan, P.C. (2002). The efficacy of honey in inhibiting strains of Pseudomonas aeruginosa from infected burns. *The Journal of Burn Care & Rehabilitation* 23 (6): 366–370.

Cooper, R.A., Molan, P.C., and Harding, K. (1999). Antibacterial activity of honey against strains of Staphylococcus aureus from infected wounds. *Journal of the Royal Society of Medicine* 92 (6): 283–285.

da Silva, C.R., de Andrade Neto, J.B., de Sousa Campos, R., et al. (2014). Synergistic effect of the flavonoid catechin, quercetin, or epigallocatechin gallate with fluconazole induces apoptosis in Candida tropicalis resistant to fluconazole. *Antimicrobial Agents and Chemotherapy* 58 (3): 1468–1478.

de Groot, T., Janssen, T., Faro, D., et al. (2021). Antifungal activity of a medical-grade honey formulation against Candida auris. *Journal of Fungi* 7 (1): 50.

Ediriweera, E. and Premarathna, N. (2012). Medicinal and cosmetic uses of bee's honey–a review. *Ayu* 33 (2): 178.

Ernst, M.E. and Franey, R.J. (1998). Acyclovir-and ganciclovir-induced neurotoxicity. *Annals of Pharmacotherapy* 32 (1): 111–113.

Esawy, M.A., Ahmed, E.F., Helmy, W.A., et al. (2011). Production of levansucrase from novel honey Bacillus subtilis isolates capable of producing antiviral levans. *Carbohydrate Polymers* 86 (2): 823–830.

Eteraf-Oskouei, T. and Najafi, M. (2013). Traditional and modern uses of natural honey in human diseases: a review. *Iranian Journal of Basic Medical Sciences* 16 (6): 731.

Feás, X. and Estevinho, M.L. (2011). A survey of the in vitro antifungal activity of heather (Erica sp.) organic honey. *Journal of Medicinal Food* 14 (10): 1284–1288.

George, N.M. and Cutting, K.F. (2007). Antibacterial honey (Medihoney™): in-vitro activity against clinical isolates of MRSA, VRE, and other multiresistant gram-negative organisms including Pseudomonas aeruginosa. *Wounds* 19 (9): 231.

Godfrey, H.R., Godfrey, N.J., Godfrey, J.C., and Riley, D. (2001). A randomized clinical trial on the treatment of oral herpes with topical zinc oxide/glycine. *Alternative Therapies in Health and Medicine* 7 (3): 49.

Hau-Yama, N.E., Magaña-Ortiz, D., Oliva, A., and Ortiz-Vázquez, E. (2020). Antifungal activity of honey from stingless bee Melipona beecheii against Candida albicans. *Journal of Apicultural Research* 59 (1): 12–18.

Hermanns, R., Mateescu, C., Thrasyvoulou, A., et al. (2020). Defining the standards for medical grade honey. *Journal of Apicultural Research* 59 (2): 125–135.

Horn, H. (2013). Honey in medicine. *Deutsche medizinische Wochenschrift (1946)* 138 (51–52): 2647–2652.

Irish, J., Carter, D.A., Shokohi, T., and Blair, S.E. (2006). Honey has an antifungal effect against Candida species. *Medical Mycology* 44 (3): 289–291.

Jenkins, R. and Cooper, R. (2012). Improving antibiotic activity against wound pathogens with manuka honey in vitro. *PloS One* 7 (9): e45600.

Kalediene, L., Baz, M., Liubaviciute, A., et al. (2021). Antiviral effect of honey extract Camelyn against SARS-CoV-2. *Journal of Advanced Biotechnology and Experimental Therapeutics* 4 (3): 290–297.

Khosravi, A.R., Shokri, H., Katiraee, F., et al. (2008). Fungicidal potential of different Iranian honeys against some pathogenic Candida species. *Journal of Apicultural Research* 47 (4): 256–260.

Lusby, P.E., Coombes, A.L., and Wilkinson, J.M. (2005). Bactericidal activity of different honeys against pathogenic bacteria. *Archives of Medical Research* 36 (5): 464–467.

Maddocks, S.E., Jenkins, R.E., Rowlands, R.S., et al. (2013). Manuka honey inhibits adhesion and invasion of medically important wound bacteria in vitro. *Future Microbiology* 8 (12): 1523–1536.

Maddocks, S.E., Lopez, M.S., Rowlands, R.S., and Cooper, R.A. (2012). Manuka honey inhibits the development of Streptococcus pyogenes biofilms and causes reduced expression of two fibronectin binding proteins. *Microbiology* 158 (3): 781–790.

Majtan, J., Bohova, J., Horniackova, M., et al. (2014). Anti-biofilm effects of honey against wound pathogens Proteus mirabilis and Enterobacter cloacae. *Phytotherapy Research* 28 (1): 69–75.

Meo, S.A., Al-Asiri, S.A., Mahesar, A.L., and Ansari, M.J. (2017). Role of honey in modern medicine. *Saudi Journal of Biological Sciences* 24 (5): 975–978.

Miorin, P.L., Levy Junior, N., Custodio, A., et al. (2003). Antibacterial activity of honey and propolis from Apis mellifera and Tetragonisca angustula against Staphylococcus aureus. *Journal of Applied Microbiology* 95 (5): 913–920.

Morroni, G., Alvarez-Suarez, J.M., Brenciani, A., et al. (2018). Comparison of the antimicrobial activities of four honeys from three countries (New Zealand, Cuba, and Kenya). *Frontiers in Microbiology* 9: 1378.

Müller, P., Alber, D.G., Turnbull, L., et al. (2013). Synergism between Medihoney and rifampicin against methicillin-resistant Staphylococcus aureus (MRSA). *PloS One* 8 (2): e57679.

Nair, H.K., Rao, K.V., Aalinkeel, R., et al. (2004). Inhibition of prostate cancer cell colony formation by the flavonoid quercetin correlates with modulation of specific regulatory genes. *Clinical and Vaccine Immunology* 11 (1): 63–69.

Nasir, N.-A.M., Halim, A.S., Singh, K.-K.B., et al. (2010). Antibacterial properties of Tualang honey and its effect in burn wound management: a comparative study. *BMC Complementary and Alternative Medicine* 10 (1): 1–7.

Nolan, V.C., Harrison, J., and Cox, J.A. (2019). Dissecting the antimicrobial composition of honey. *Antibiotics* 8 (4): 251.

Olaitan, P.B., Adeleke, O.E., and Iyabo, O. (2007). Honey: a reservoir for microorganisms and an inhibitory agent for microbes. *African Health Sciences* 7 (3).

Purbafrani, A., Ghazizade Hashemi, S.A., Bayyenat, S., et al. (2014). The benefits of honey in Holy Quran. *International Journal of Pediatrics* 2 (3): 67–73.

Samarghandian, S., Farkhondeh, T., and Samini, F. (2017). Honey and health: a review of recent clinical research. *Pharmacognosy Research* 9 (2): 121.

Sgariglia, M.A., Vattuone, M.A., Vattuone, M.M.S., et al. (2010). Properties of honey from Tetragonisca angustula fiebrigi and Plebeia wittmanni of Argentina. *Apidologie* 41 (6): 667–675.

Shahzad, A. and Cohrs, R.J. (2012). In vitro antiviral activity of honey against varicella zoster virus (VZV): a translational medicine study for potential remedy for shingles. *Translational Biomedicine* 3 (2): 2–7.

Tafere, D.A. (2021). Chemical composition and uses of Honey: a review. *Journal of Food Science and Nutrition Research* 4 (3): 194–201.

Theunissen, F., Grobler, S., and Gedalia, I. (2001). The antifungal action of three South African honeys on Candida albicans. *Apidologie* 32 (4): 371–379.

Tonks, A.J., Cooper, R., Jones, K., et al. (2003). Honey stimulates inflammatory cytokine production from monocytes. *Cytokine* 21 (5): 242–247.

Tumin, N., Halim, N., Shahjahan, M., et al. (2005). Antibacterial activity of local Malaysian honey. *Malaysian Journal of Pharmaceutical Sciences* 3 (2): 1–10.

Visavadia, B.G., Honeysett, J., and Danford, M.H. (2008). Manuka honey dressing: an effective treatment for chronic wound infections. *British Journal of Oral and Maxillofacial Surgery* 46 (1): 55–56.

Wahdan, H. (1998). Causes of the antimicrobial activity of honey. *Infection* 26 (1): 26–31.

Watanabe, K., Rahmasari, R., Matsunaga, A., et al. (2014). Anti-influenza viral effects of honey in vitro: potent high activity of manuka honey. *Archives of Medical Research* 45 (5): 359–365.

Willix, D., Molan, P.C., and Harfoot, C. (1992). A comparison of the sensitivity of wound-infecting species of bacteria to the antibacterial activity of manuka honey and other honey. *Journal of Applied Bacteriology* 73 (5): 388–394.

Zareie, P.P. 2011. Honey as an antiviral agent against respiratory syncytial virus. *Doctoral dissertation*, University of Waikato, Hamilton, New Zealand.

Zhang, Y., Wang, M., Yang, J., and Li, X. (2015). Research process of chemical constituents and pharmacological effects of baked licorice. *Acta Universitatis Traditionis MedicalisSinensis Pharmacologiaeque Shanghai* 3: 99–102.

# 14

# Use of Honey in Cardiovascular Diseases

*Shridhar C. Ghagane and Aimen A. Akbar*

## Introduction

Globally, one of the major causes of death is cardiovascular disease (CVD), with an estimated of about 17.9 million deaths per year. A total of 7.2 million deaths were caused by heart attack, and about 5.7 million deaths were caused by stroke. The disease includes a group of the heart and blood vessel diseases that include rheumatic heart disease, coronary heart disease (CHD), cerebrovascular disease, and other conditions. About 80% of these deaths occur in low-income countries. By 2030, an estimated 23.6 million deaths will occur from CVD if there is no change in the existing trend. Numerous medications are presently used for the treatment of patients with these diseases and their comorbidities. The side effects are inevitable, and eventually, the ideal goal is barely attained. Since time immemorial, honey, along with other bee products, has traditionally been used as medicine for disease management. Honey in modern medicine is used in management of wound and burn healing. It is being used as a food and medicinal product because of its high antioxidant, anti-inflammatory, and antimicrobial activities. Honey is a natural product produced naturally by honeybees scientifically known as *Apis* from the nectar of flowers.

Honey's high acidic nature and high sugar content have led to prodigious use of it for medicinal purposes because its nature does not allow the growth of microorganisms in sugar. Also, honey's high sugar content leads to an osmotic effect in some microorganisms. Moreover, hydrogen peroxide, which is known to produce strong antibacterial activity, is found in honey. Certain honeys are also known to have high levels of anti-inflammatory and antioxidant properties, which primarily only depend on the blooms or the flora from which the bees collect the nectar. Recently, several studies have proven the effectiveness of honey for various medical purposes because of its components and its antibacterial, anti-inflammatory, antioxidant, antiviral, antifungal, and anticancer properties. This chapter focuses on the antioxidant properties of honey and its ability to protect against CVDs.

CVD is an assembly of numerous disorders and severe heart- and blood vessel–related conditions. The increase in mortality rates worldwide is caused by CVDs, estimating about one-third of the deaths globally until 2019 and continuing to rise further (Sjogren et al. 2015). The main cause of CVD is build-up of cholesterol and fats, resulting in plaque formation in and on the artery walls, leading to restriction of blood flow (Mensah et al 2020). CVD, sometimes also called heart disease, includes heart failure, acute myocardial infraction, coronary heart and artery disease, atherosclerosis, ischemic heart disease, cardiomegaly, cardiomyopathy, arrhythmia, peripheral arterial disease, rheumatic heart disease, and other sever conditions, among which coronary artery disease is considered the most common type (Virani et al. 2020). Most CVDs occur because of interference with the heart's normal function. Increased oxidative stress, inflammatory factors, and diabetes along with insulin resistance play distinct roles in the development of CVD (Roth et al. 2020). Accordingly, lipids, high blood pressure, glucose, blood clots, and overweight and obesity can contribute to CVD. However, many of these conditions are preventable by physical activity (exercising), diet, and maintaining a healthy lifestyle (Prabhakaran et al. 2018). Each condition has different symptoms, but the most common alarming signs include shortness of breath, discomfort in the chest or chest pain (angina), numb or cold extremities, lightheadedness, and fatigue (Abdul-Aziz et al. 2019). Some people may have or feel these symptoms, and some do not; hence, one of the best ways to protect oneself from any CVD is to get screened regularly.

*Honey: Composition and Health Benefits, First Edition. Edited by Md. Ibrahim Khalil, Gan Siew Hua, and Bey Hing Goh.*

## Burden of Cardiovascular Disease

Being one of the leading primary causes of death in both men and women in developed as well as developing nations, CVD accounts for nearly 17.8 million or 31% of deaths globally as per the World Health Report in 2017 (Roth et al. 2020). China tops the list with the highest number of deaths from CVD followed by India, Russia, the United States, and Indonesia up to 2019. France, Japan, and Peru had the lowest death rates from CVD (Preidt 2021). According to a data review, more than 75% of global deaths occur from CVD, especially in low- and middle-income countries (LMICs) according to the Global Burden of Diseases Study (Prabhakaran et al. 2018). This is mostly caused by less equitable health care services and unawareness in LMIC countries. Additionally, the population in these countries lacks awareness; hence, they do not benefit from integrated primary health care programs and treatment when risk factors and symptoms are encountered. The burden of CVDs has been increasing in India as well with the prevalence estimated at about 54.5 million. One in four deaths in India is caused by CVD with ischemic heart disease and stroke responsible for more than approximately 80% of this burden (Abdul-Aziz et al. 2019; Roth et al. 2015). Between 2005 and 2015, the mortality rate of CVD decreased substantially, particularly in young children, although recent data suggest this rate of decline has been slowing (Andersson and Vasan 2018; Norheim et al. 2015; Sanchis-Gomar et al. 2016). Nearly 50% of all noncommunicable disease deaths are caused by CVD. The systematic cost analysis for CVD reveals its economic burden to be increasing. The estimated global cost of CVD is set to rise from approximately US$863 billion in 2010 to a staggering cost of US$1044 billion by 2030 (Carnethon et al. 2017).

## Risk Factors of Cardiovascular Disease

The risk factors leading to the onset of CVDs are well recognized and can be categorized into modifiable and nonmodifiable risk factors (Figure 14.1). Nonmodifiable risk factors are those that cannot be altered. These include ethnicity, age, and genetics (family history). Modifiable risk factors include adiposity, raised blood sugar levels, alcohol consumption, smoking, high blood pressure, high body mass index (BMI), high cholesterol, dietary habits, and physical inactivity (Rodriguez et al. 2014; Volgman et al. 2018). The prevalence of these modifiable CVD risk factors has been observed to be steady or increasing in the young population of the developed countries along with the increased usage of substance abuse (anabolic steroids, cocaine, electronic cigarettes, and opioids) (Lawes et al. 2008). The World Health Organization (WHO) states that 61% of deaths are caused by CVDS, but the modifiable CVD risk factors and CVD can be largely preventable by addressing these well-established modifiable risk factors (Lewington 2002).

### Nonmodifiable Risk Factors

Risk factors are certain characteristic that give an increase likelihood of developing a certain medical condition. CVD is associated with several risk factors; some of them are nonmodifiable, and some are modifiable. Nonmodifiable risk factors are age, gender, ethnicity, and family history; these are beyond the control of one individual.

#### Age

The American Heart Association's computations indicated that around 80% of deaths caused by CVD are in people 65 years of age and older. The prevalence of CVD is observed to increase in both men and women after the age of 35 years. The lifetime risk of developing CVD in women is 32% and in men is 49% (Garrison et al. 1987).

#### Gender

Men are at higher risk for onset of CVD at an earlier age than women (Poirier and Eckel 2002).

#### Ethnicity

Ethnic groups including Latinos, Southeast Asians, African Americans, and Hispanics are at a higher risk of cardiovascular mortality and morbidity. The higher risk of CVD in these groups could be due their backgrounds because they are more likely to have other risk factors for CVD, such as high blood pressure, high cholesterol, and obesity (Carson et al. 2020; Prabhakaran et al. 2018; Prasad and Das 2009).

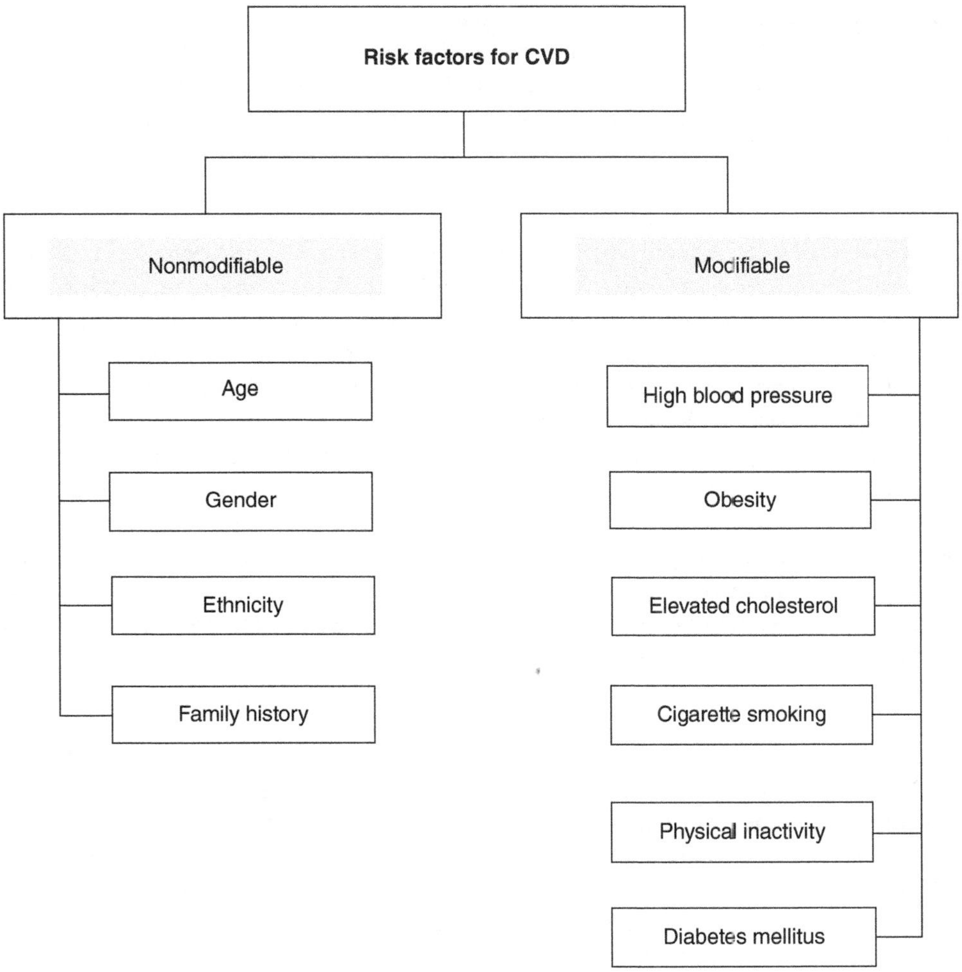


**Figure 14.1** Risk factors of cardiovascular diseases (CVDs).

### Family History

Developing a risk of CVD is higher for an individual if the family has a history of CVD. A study suggested that having a mother or sister diagnosed before the age of 65 years along with father or brother diagnosed with CVD before 55 years of age are considered risk factors (Hossen et al. 2017). Patients younger than 50 years of age with premature cardiac disease along with a family history of CVD have an increased CVD mortality risk.

## Modifiable Risk Factors

Strong evidence from several studies suggests that a decrease in CVD risk factors is possible through adherence to healthy lifestyle choices. Abstaining from factors such as smoking, drinking alcohol, and using tobacco may help in improving health. Additionally, performing regular physical activity and consuming a healthy diet lower the risk of CVD. Modifiable risk factors are high blood pressure, obesity, elevated cholesterol, smoking, and physical activity.

### High Blood Pressure

Among the established CVD risk factors, high blood pressure (also known as hypertension) is associated with the strongest evidence for causation. It is estimated that approximately 47% of CHDs and 54% of strokes worldwide can be attributed to raised blood pressure (Estevinho et al. 2008). High blood pressure gradually increases the pressure of blood flow and can damage the cells lining the arteries, resulting in narrowed arteries and consequently limiting blood flow throughout the body. There is a high positive correlation between diastolic blood pressure (DBP), systolic blood pressure (SBP), and risk of death from CVD above the usual DBP of 75 mm Hg and SBP of 115 mm Hg (Machado De-Melo et al. 2018). When the blood supply to the heart is affected, troubles such as irregular heart rhythms (arrhythmias), heart attack, and chest pain (angina) can occur. This also forces the heart to work harder so it can pump blood to other organs. Over time, strain on the heart weakens it and lessens its efficiency, thereby increasing the risk of heart attack, heart failure, and sudden cardiac death (Uthurry et al. 2011). The increasing risk of CVD mortality correlated with high blood pressure persists but is more predominant in older age people.

#### Obesity

Obesity, one of the most common risk factors of CVD, is found closely related to hypertension. It is also estimated that 65%–78% of cases of hypertension could be attributed to obesity (Farooqui and Farooqui 2011). The increasing urbanization and mechanization of the world fostered by a modern lifestyle along with a carbohydrate-dense diet have led to obesity, which has undoubtedly contributed to creating an atherosclerosis-prone environment characterized by macrophage infiltration, plaque instability, and thereby development of CVD. Obesity also causes ventricular enlargement, atrial enlargement, thromboembolic disease, promotion of sleep apnea, and the onset of metabolic diseases such as type 2 diabetes and dyslipidemia. Studies suggest that BMI (a measure of whole-body adiposity) is considered as a less strong predictor of CVD risk relative to the measure of central adiposity (waist circumference and waist-to-hip ratio). Adipose tissue is not only the storehouse of fat but is also endocrine in nature and has the capability of synthesizing and releasing many different molecules in the bloodstream. Concentrations of circulating plasminogen activator inhibitor 1, fibrinogen, angiotensin II, C-reactive protein (CRP), and tumor necrosis factor-α (TNF-α) are all linked to adiposity and have a clinical impact, particularly in relation to cardiovascular disease. Interleukin-6 (IL-6) modulates CRP production in the liver, which is important because studies have shown that about 30% of the total concentrations of IL-6 circulating originates in adipose tissue (Khalil et al. 2011).

#### Elevated Cholesterol

There is a positive correlation between high cholesterol levels and worsening of CVD. Globally, 16% of CVD deaths are caused by increased cholesterol levels (Volgman et al. 2018). High-density lipoprotein (HDL), also called "good cholesterol," and low-density lipoprotein (LDL) components of total cholesterol have opposing effects on CVD mortality and body homeostasis. HDL cholesterol is transported away from the bloodstream back to the liver for disposal, which eventually leads to lowering the possibility of CVD. Premenopausal women have a lower risk of CVDs because the female hormone estrogen increases HDL cholesterol levels. On the contrary, increases in concentration of LDL cholesterol lead to atherosclerosis, which eventually increases the risk of ischemic stroke and heart attack. Elevated levels of triglyceride concentration combined with higher levels of LDL cholesterol speeds up the formation of plaques (Yoshizumi et al. 2021). The diet is a major aspect of blood cholesterol levels, and a cohort study showed nonsignificant positive relationships between dietary cholesterol and risk of CVD. Whereas intake of saturated fatty acids increases blood cholesterol, polyunsaturated fatty acids reduce blood cholesterol levels (Edwards et al. 2007; Sanchez et al. 2006).

#### Cigarette Smoking

The use of tobacco is a strong independent predictor of cardiovascular death. It is estimated to cause almost 12% of CVDs globally (Crane 1983). Passive smoking is considered more noxious than smoke inhaled and exhaled by the smoker. Toxic components present in cigarette such as nicotine, heavy metals and their oxides, polycyclic aromatic hydrocarbons, and carbon monoxide have myriad effects on the vascular endothelium (White and Crane 1975). These chemicals induce and promote smoking-activated inflammatory cells to adhere to the inner walls of the vessels (endothelium) that have been damaged by smoking and accumulate under the surface (subendothelium), leading to chronic inflammation and endothelial dysfunction. Increased exposure to smoke can cause adverse rupture of the aorta, myocardial infarction (heart attack), and stroke (Alvarez-Suarez et al. 2010). Reduced availability of nitric oxide (NO), inflammation, endothelial dysfunction, insulin resistance hyperinsulinemia, and altered lipid metabolism are some of the major pathophysiological mechanisms of tobacco in the development of CVD. Other forms of tobacco smoking include cigar, bidi, pipe, waterpipe tobacco (hookah, shisha, or narghile), smokeless tobacco, and electronic cigarettes.

#### Physical Inactivity

Studies regarding physical activity and its health benefits are overwhelming, and a sedentary lifestyle is considered an important contributor to cardiovascular morbidity and mortality for all age groups across all continents. The WHO estimated that about 60% of the global population is not sufficiently physically active (BtHjIdrus et al. 2020). Regular physical activity is directly associated with a decline in the risk of developing and dying prematurely from CVD. Regardless of inherited factors, physical activity can help in regulating body weight, improving the body's use of insulin, lowering myocardial oxygen demand, improving endothelial function, enhancing fibrinolysis, and decreasing sympathetic tone, thus leading to fewer symptoms of CVD and an increased life span. Experts recommend vigorous physical activity for an hour every day or at least moderate exercise per week for about 150 minutes (Wang et al. 2019).

## Honey and Health: Historical significance

For many years, honey has been consumed as a food and as a medical product. It is a naturally occurring substance that is produced by honeybees (*Apis mellifera*), with the nectar of blossoms or exudates of trees and plants giving nectar honeys or honeydews, respectively (Khalil et al. 2015). In ancient times, honey was the only available sweetener ingredient. Indeed, the interrelationship between humankind and bees started during the prehistoric period. The first literature reference to honey, on a Sumerian tablet dating back to 2100–2000 BC, discussed the properties and use of honey as a drug and an ointment (Crane 1975). For decades, honey was known and used for medicinal and nutritional purpose. In human history, the belief that honeys is a nutrient that has medicinal properties has continued for a long period. Honey was an only preferred source of carbohydrates until the eighteenth century, when industrial sugar began to replace it (Eteraf-Oskouei and Najafi 2013). Along with its antimicrobial properties, traditionally honey has been used as a remedy for wound healing, burns, cataracts, and ulcers (Pekkarinen and Varis 2000). Owing to its high osmolarity, honey provides a moist wound-healing environment in the form of a solution that does not stick to wounded tissues and provides a protective barrier (Hossen et al. 2017). Bacterial colonization is prevented because of this moist environment, so honey decreases inflammation and reduces exudate formation more promptly than standard treatments. Today there is wide information available on using honey to treat many human diseases (Atkins 2016). At present, annual world honey production is about 1.2 million tons, which is less than 1% of the total sugar production. The consumption of honey differs greatly from country to country. The major honey-exporting countries, China and Argentina, have small annual consumption rates of 0.1–0.2 kg per capita. Honey consumption is higher in developed countries, where domestic production does not always meet market demand (Al-Waili et al. 2013). Thus, the growing literature on honey and its importance in curing aliments in humans has in recent years given birth to an alternative medicine branch called apitherapy.

## Composition of Honey

Primarily, the composition of honey depends on several factors such as the floral source, seasonal and environmental factors, and processing. Honey has high levels of monosaccharides, fructose, and glucose, and it contains about 70%–80% sugar, which provides its sweetness. Honey also has antiseptic and antibacterial properties. Modern medical science has managed to find uses for honey in chronic wound management and combating infection. There are at least 181 elements present in honey, and it is a supersaturated solution of sugars, mainly composed of glucose (31%) and fructose (38%) (Hossen et al. 2017; Pekkarinen and Varis 2000). It also contains (2%) minerals, (0.5%) proteins, (1%) free amino acids, enzymes, and vitamins (<1%). Honey also contains a significant quantity of antioxidant properties because of the presence of phenolic acids, flavonoids, several enzymes, glucose oxidase, catalase, and amino acids. The composition of honey is summarized in Table 14.1.

## Medicinal Properties of Honey

The importance of honey as a medicine has been known for a long time in many cultures. Because of the limited use in medicine, only a few scientific reports are available (Nega and Eshete 2018). However, in recent years, honey is becoming accepted as a reputable and effective therapeutic agent. Its beneficial role has been endorsed because of its antimicrobial, anti-inflammatory, and antioxidant activities as well as the way it boosts the immune system (Estevinho et al. 2008). In addition to just commonly being used as a sweetener, honey offers many medicinal uses described in traditional medicine. Honey also contains a natural resinous substance called propolis, which is collected by honeybees, mainly from poplar bud exudates. Propolis exhibits a broad spectrum of biological and pharmacological properties such as antimicrobial, antioxidant, anti-inflammatory, immunomodulatory, antitumor, anticancer, antiulcer, hepatoprotective, neuroprotective, and

**Table 14.1** General composition of honey.

| Component | Value/100 g |
|---|---|
| Total carbohydrates | 82.4 g |
| Fructose | 38.5 g |
| Glucose | 31.28 g |
| Sucrose | 1.31 g |
| Maltose | 7.31 g |
| Total acid as gluconic | 0.57 g |
| Moisture content | 17.1 g |
| Ash | 0.169 g |
| Fibre | 0.2 g |
| Amino acids and proteins | 0.3 g |
| Nitrogen | 0.041 g |
| Iron | 0.42 mg |
| Calcium | 52 mg |
| Potassium | 6.00 mg |
| Prosperous | 4.00 mg |
| Magnesium | 4.00 mg |
| Copper | 2.00 mg |
| Zinc | 1–100 μg/g |
| Vitamin $B_2$ | 0.22 mg |
| Vitamin $B_3$ | 0.038 mg |
| Vitamin $B_5$ | 0.21 mg |
| Vitamin $B_6$ | 0.068 mg |
| Vitamin $B_9$ | 0.024 mg |
| Vitamin C | 2 μg |
| Miscellaneous groups | – |

cardioprotective actions (Uthurry et al. 2011). The consumption of natural honey has been shown to inhibit oxidation of low-density protein, vasodilate blood vessels caused by NO production, decrease platelet aggregation, and exert analgesic and antiatherogenic effects, each of which may decrease cardiovascular risk.

## Honey Polyphenols for Cardiovascular Diseases

Phenolic compounds of honey are divided into phenolic acids and flavonoids such as quercetin, caffeic acid phenethyl ester (CAPE), kaempferol, acacetin, and galangin. These compounds exhibit protective effects such as antioxidant, vasorelaxant, anti-ischemic, and antithrombotic effects against CVD (Hossen et al. 2017). Recent evidence indicates that phenolic compounds constitute about 30% of the content of honey; however, the contents and profiles of different types of honeys vary greatly depending on the plant source, sensitivity or reliability of the detection method, and geographical origin. For instance, quercetin, luteolin, kaempferol, galangin, and isorhamnetin are present in all types of honey, but naringenin and hesperetin are found only in specific varieties (Atkins 2016).

### Quercetin

Studies have reported that daily intake of bioflavonoids may affect angiotensin II–induced mitogen-activated protein kinase activation in cultured rat aortic smooth muscle cells (RASMCs). Angiotensin II–induced rapid and significant activation of extracellular signal-regulated kinase (ERK) 1/2, c-Jun N-terminal kinase (JNK), and p38 in RASMC. Quercetin

(3,3′,4′,5,7-pentahydroxyflavone) inhibited angiotensin II–induced JNK but ERK1/2, and p38 activation were not affected. Quercetin also inhibited Src homology and collagen (Shc) caused by tyrosine phosphorylation of angiotensin II, Shc.p85 association, and consequent activation of the phosphatidylinositol 3-kinase (PI3-K)/Akt pathway in RASMCs (Yoshizumi et al. 2021). In another study, it was suggested that quercetin also prevents the oxidative damage of LDL. This is essential because oxidative modifications of LDL are considered critical in the onset of atherosclerosis. Furthermore, a quercetin derivative along with a PI3-K inhibitor (LY294002) inhibited angiotensin II–induced JNK activation as well as Akt phosphorylation. These findings suggest that the quercetin is useful for prevention and treatment of CVDs (Sanchez et al., 2006).

Several studies have proposed that chronic treatment with the flavonoid quercetin showed antihypertensive effects and restored endothelial dysfunction in spontaneously hypertensive rats by upregulating endothelial nitric oxide synthase (eNOS) activity and decreasing nicotinamide adenine dinucleotide phosphate (NADPH) oxidase–mediated superoxide anion generation associated with reduced p47 protein expression (Farooqui and Farooqui 2011). Population studies have consistently reported that daily intake of quercetin, an antioxidant flavanol, demonstrated a reduced risk of ischemic heart diseases (known as the "French paradox"), CHD, and stroke. Contradicting animal-based studies, there was no quercetin-induced reduction in systemic markers of oxidative stress (Khalil et al. 2011).

## Acacetin

A study investigated whether the natural compound acacetin is an atrium-selective anti-atrial fibrillation agent. The effects of acacetin suppressed $I_{Kur}$ (also referred as ultrarapid delayed rectifier K + current) and extended the action potential duration in human atrial myocytes. Not only did it block acetylcholine-activated K + current but it also did not affect the L-type Ca2 + current, Na + current, or inward-rectifier K + current in guinea pig cardiac myocytes (Edwards et al. 2007; Sanchez et al. 2006). Also, acacetin caused a weak reduction in the hERG and hKCNQ1/hKCNE1 channels stably expressed in HEK 293 cells, but it did not prolong the corrected QT interval in the hearts of rabbits. In another experiment, acacetin (5 mg/kg) prolonged the atrial effective refractory period in the left and right and left atria of anesthetized dogs 1–4 hours after intraduodenal administration without prolongation of the corrected QT interval (Crane 1983).

## Caffeic Acid

Propolis of honeybee hives has a phenolic active compound called CAPE, which reduced blood pressure and heart rate in rats. Hypotensive effects, the role of vagal activity, and atropine blockage on the bradycardic caused by CAPE were researched in 2005. The results of the study revealed that CAPE caused hypotension and intense and transient bradycardia. Also, vagotomy completely abolished bradycardia, just as the effects of bradycardia were attenuated by atropine (Crane 1983). However, the hypotensive effect of CAPE was affected neither by bilateral vagotomy nor by atropine treatment. It was thought that CAPE may exert its effects on heart rate via a central parasympathetic control mechanism but not on the central parasympathetic blood pressure control system (Crane 1983).

## Kaempferol

Investigations on the vascular effects of kaempferol in isolated porcine coronary artery rings were done, and they demonstrated no significant effect at a lower concentration (1nM - 100$\mu M$). However, it was noticed that kaempferol (it is microM (10$\mu M$)) enhanced relaxation produced by bradykinin, isoproterenol, the calcium ionophore A23187, sodium nitroprusside, levcromakalim, and nifedipine in endothelium-intact porcine coronary arteries at a higher concentration. However, antioxidant agents did not affect the enhancement effect of kaempferol or bradykinin-evoked relaxation. This study concluded that the action of kaempferol is unrelated to its antioxidant properties, and it has the capability to enhance relaxation in endothelium-dependent and endothelium-independent arteries (White and Crane 1975).

Another study examined if Bcl-2 and endoplasmic reticulum (ER) stress proteins are linked to the protective effect of kaempferol on ischemia-reperfusion–induced cardiac damage. This study demonstrated that kaempferol significantly decreased the level of BAX (Bcl-2 Associated X-protein), a pro-apoptotic protein; increased the expression level of Bcl-2, an anti-apoptotic protein; and mitigated the expressions of the ER stress proteins such as CHOP (C/EBP Homologous Protein), GRP78 (Glucose-Regulating Protein 78), ATF-6$\alpha$ (Activating Transcription Factor 6- $\alpha$), IRE1-$\alpha$ (Inositol-Requiring Enzyme-1$\alpha$), phosphor-eIF-2$\alpha$ (Eukaryotic Initiation factor-2$\alpha$), and XBP-2 (X-Box-Binding Protein-2) Thus, this study of kaempferol advocates a protective effect against ischemia-reperfusion-induced cardiac damage (Alvarez-Suarez et al. 2010).

### Galangin

This compound exerts an antioxidative effect on endothelial tissues and affects lipid peroxidation, consequently preventing heart disease. It helps to preserve other protective antioxidants as well, including vitamin C, vitamin E, and other flavonoids (BtHjIdrus et al. 2020). Evidence indicates the presence of about 30 types of phenolic compounds in honey; however, the profiles of these phenolic compounds can depend on various factors, including the floral source and climatic and geographical conditions. For example, quercetin, galangin, kaempferol, isorhamnetin, and luteolin are present in all types of honey, but hesperetin and naringenin are found only in specific varieties. Recently, Wang (et al. 2019) have suggested that galangin may also protect against cardiac remodeling by decreasing inflammatory responses and apoptosis. Its action was associated with the inhibition of PI3K-Akt-GSK3β signaling pathways.

## Overview of Honey in Treating Various Diseases

### Honey and Wound Healing

Honey is the oldest wound-healing agent known to humankind when some modern chemicals have failed in this regard. Experimental research illustrated more documents supporting its usage in wound healing because of its bioactivities, including antibacterial, antiviral, anti-inflammatory, and antioxidant activities (BtHjIdrus et al. 2020). Honey induces leukocytes to release cytokines, which is what begins the tissue repair cascades. Furthermore, it activates an immune response to infection. A great deal of evidence suggests the use of honey in the control and treatment of acute wounds and for mild to moderate superficial and partial-thickness burns (Al-Waili et al. 2013). Although some studies have indicated the efficacy of honey concerning wound treatment and leg ulcers, more studies are needed to strengthen the current evidence.

### Honey and Diabetes

There is strong evidence indicating the beneficial effects of honey in the treatment of patients with diabetes. These results point out the therapeutic prospects of using honey or other potent antioxidants as adjuncts to standard antidiabetic drugs in the control of diabetes (Samarghandian et al. 2017). Regarding the restrictions associated with using antioxidants, other interventions targeted at decreasing reactive oxygen species (ROS) generation may also be used as adjuncts to conventional diabetes therapy (Nega and Eshete 2018). In one of the clinical trials of type 1 and type 2 diabetes mellitus, the application of honey showed was associated with a dramatically lower glycemic index when compared to sucrose. Honey compared to dextrose induced a significant reduction in plasma glucose levels in diabetic subjects. In normal and hyperlipidemic patients, it also reduces blood lipids, homocysteine, and CRP contents (Eteraf-Oskouei and Najafi 2013). However, several questions remain, especially relating to the prospect of controlling diabetes by interventions that target both oxidative stress and hyperglycemia. Furthermore, the therapeutic effects of honey in the management of diabetes may not only be restricted to controlling glycemia but also may be extended to ameliorating the associated metabolic complication diseases.

### Honey and Cancer

Studies show that honey may exert anticancer effects through several mechanisms. Investigations have indicated that honey has anticancer properties through its interference with multiple cell-signaling pathways, including inducing apoptosis, antimutagenic, antiproliferative, and anti-inflammatory pathways. Honey modifies immune responses (Al-Waili 2004; Eteraf-Oskouei and Najafi 2013; Nega and Eshete 2018). Honey has been indicated to prevent cell proliferation, induce apoptosis, modify cell cycle progression, and cause mitochondrial membrane depolarization in several types of cancer such as skin cancer cells (melanoma), adenocarcinoma epithelial cells, cervical cancer cells, endometrial cancer cells, liver cancer cells, colorectal cancer cells, prostate cancer cells, renal cell carcinoma, bladder cancer cells, human non–small cell lung cancer, bone cancer cells (osteosarcoma), and leukemia and mouth cancer cells (oral squamous cell carcinoma) (Al-Waili 2004). In addition, honey might be able to inhibit several forms of tumors in animal modeling, including breast cancer, carcinoma, melanoma, colon carcinoma, hepatic cancer, and bladder cancer. However, more studies are needed to improve our understanding of the positive effect of honey on cancer.

## Honey and Cardiovascular Diseases

Antioxidants present in honey such as flavonoids, polyphenolics, vitamin C, and monophenolics may be associated with a reduced risk of cardiovascular failure (Hadi et al. 2021). In CHD, the protective effects of flavonoids such as antioxidant, antithrombotic, antiischemic, and vasorelaxant and flavonoids reduce the risk of coronary heart disorders through three mechanisms: (1) improving coronary vasodilatation, (2) reducing the ability of platelets in the blood to clot, and (3) inhibiting LDL from oxidizing. Although there is a wide spectrum of antioxidant types, caffeic acid, quercetin, phenethyl ester, kaempferol, galangin, and acacetin predominate in different types of honeys. Several investigations have showed that certain honey polyphenols have promising pharmacological functions in reducing cardiovascular disorders (Samarghandian et al. 2017). However, in vitro and in vivo research and clinical trials should be initiated to further validate these compounds in medical applications.

Polyphenols extracted from propolis show scavenging activity, protecting against the peroxidative damage induced by doxorubicin toxicity, implicating that the cardioprotective effects of propolis are attributable to its polyphenols such as flavonoids. The beneficial properties of honey depend on a wide range of flavonoids; therefore, honey is considered a promising potential therapeutic agent. The consumption of natural raw honey has been reported to reduce cardiovascular risk factors such as total cholesterol, LDL, triacylglycerol, body weight, fasting blood glucose, and CRP, particularly in participants with elevated risk factors, supporting the view that honey exerts therapeutic activity against CVDs (Erejuwa 2014). Endothelium-derived NO is a paracrine factor that controls the vascular tone (constant level of nervous stimulation to the muscles in the blood vessel walls, giving the muscles a resting level of contraction). An enhanced inactivation and/or reduced synthesis of NO and excess production of ROS in the vascular wall seem to occur in pathological conditions such as hypertension, hypercholesterolemia, diabetes, and cigarette smoking that are risk factors for CVDs. The cardioprotective effects of flavonoids from honey are shown in Table 14.2. This suggests that oxidative stress and endothelial dysfunction in the coronary and peripheral circulations have important prognostic implications for subsequent cardiovascular events. ROS are produced by various oxidase enzymes, including NADPH oxidase, xanthine oxidase, uncoupled eNOS, cyclooxygenase, glucose oxidase, and lipoxygenase, and mitochondrial electron transport. Among these enzymatic systems, NADPH oxidases are major sources of superoxide in vascular cells and myocytes, showing a link with hypertension and other pathological states associated with uncontrolled growth and inflammation that occur in atherosclerosis. Endothelial dysfunction promotes vasospasm, thrombosis, vascular inflammation, and proliferation of vascular smooth muscle cells. Vascular oxidative stress caused by increased production of ROS contributes to vascular dysfunction (Al-Waili 2004; Samarghandian et al. 2017). Therefore, oxidative stress and endothelial dysfunction have been suggested to promote atherogenesis. Therapeutically, drugs in clinical use (e.g. angiotensin-converting enzyme inhibitors, angiotensin II type 1 receptor antagonists, and statins) have pleiotropic actions that can improve endothelial function. Also, dietary polyphenolic antioxidants can reduce oxidative stress. However, larger interventional clinical trials by giving exogenous antioxidants (vitamins C and E) have failed to improve cardiovascular outcomes (Al-Waili et al. 2013). These clinical data do not justify the use of antioxidant vitamin supplements for cardiovascular risk reduction.

**Table 14.2** Cardioprotective effects of flavonoids from honey.

| Flavonoid | Cardioprotective Effects |
|---|---|
| Quercetin | Chain-breaking antioxidant; reduces blood pressure in patients with hypertension, cardiomyocyte cytoprotective effect, chelation of intracellular iron and suppression of OH radical production, |
| Acacetin | Vas-relaxing, inhibition of atrial fibrillation |
| Caffeic acid phenethyl ester | Inhibition of lipid peroxidation and induction in the expression of antioxidant enzymes in diabetic hearts; antiarrhythmic activity |
| Kaempferol | Cardiomyocyte cytoprotective effect; endothelium-independent and -dependent relaxation of coronary arteries |
| Galangin | Inhibition of lipid peroxidation |
| Chrysin and luteolin | Vasorelaxing; chelation of intracellular iron and suppression of OH radical production; reduction of blood pressure in patients with hypertension |
| Hesperetin | Vasorelaxing; antiplatelet activity |
| Apigenin | Protection of endothelium-dependent relaxation of aorta against oxidative stress; vasorelaxing; cardiomyocyte cytoprotective effect |
| Syringetin | Antiplatelet activity |

The human body is equipped with defence systems made up of endogenous antioxidants (glutathione, lipoic acid) and antioxidant enzymes (glutathione reductase, glutathione synthetase, superoxide dismutase, and catalase) to protect cells and tissues from ROS-mediated oxidative damage. However, there is also a need to take supplements that can provide minerals and phytochemicals (flavonoids) to fight and neutralize free radical–mediated damage, especially during aging and under pathological conditions (Olas 2020). Lately, plant polyphenols are beginning to acquire importance for the treatment of patients with acute and chronic free radical–mediated diseases.

The effects of ingestion of 75 g of natural honey compared with the same amount of artificial honey (fructose plus glucose) or glucose on plasma glucose, plasma insulin, cholesterol, triglycerides, blood lipids, CRP, and homocysteine, most of them being risk factors for CVDs, were studied in humans. Elevations of insulin and CRP were significantly higher after glucose intake than after honey consumption (Al-Waili 2004). Glucose reduced cholesterol and LDL cholesterol (Al-Waili 2004; Eteraf-Oskouei and Najafi 2013; Nega and Eshete 2018). Artificial honey slightly decreased cholesterol and LDL cholesterol and elevated triglycerides. Honey reduced cholesterol, LDL cholesterol, and triglycerides and slightly elevated HDL cholesterol. In patients with hypertriglyceridemia, artificial honey increased triglycerides, but honey decreased triglycerides. In patients with hyperlipidemia, artificial honey increased LDL cholesterol, but honey decreased LDL cholesterol. In patients with diabetes, honey compared with dextrose caused a significantly lower increase of plasma glucose (Al-Waili 2004). Honey can contain NO metabolites, which are known indicators of CVD risk. Increased levels of NO in honey might have a protective function in CVDs (Samarghandian et al. 2017). Total nitrite concentrations of different biological fluids from humans, including saliva, plasma, and urine, were measured after ingestion of 80 g of honey (Erejuwa 2014). Salivary, plasma, and urinary NO metabolite concentrations showed a tendency to increase. Different honey types contained various concentrations of NO metabolites, with darker or fresh honeys containing more NO metabolites than light or stored honey (Olas 2020). After heating, NO metabolites decreased in all honey types. Compared with fructose-fed rats, honey-fed rats had higher plasma tocopherol levels and higher tocopherol-to-triacylglycerol ratios, as well as lower plasma nitrate levels and a lower susceptibility of the heart to lipid peroxidation (Ahmed et al. 2018).

## Health Benefits of Honey: Implications for Treating Cardiovascular Diseases

Honey, a natural sweetener, is a remarkably complex viscous syrup that contains about 200 different substances (Alvarez-Suarez et al. 2010). The chemical composition of this nutritional product varies depending on the floral source and the geographical and climatic conditions (Figure 14.2). The health benefits of honey as an important substance have been very well known (Farooqui and Farooqui 2011). There is sufficient evidence from the past few decades recommending the use of honey for its medicinal benefits because of its antibacterial, apoptotic, anti-inflammatory, and antioxidant properties (Ranneh et al. 2021). Its beneficial effects on human health were described in early Greek, Vedic, Roman, and Islamic texts as well (Jibril et al. 2019). Laboratory and clinical investigation have revealed that regular intake of honey is associated with a reduced risk of CVDs. Although honey comprises various components, it mostly contains carbohydrates, which account for 95%–99% of its dry weight. Monosaccharides such as fructose (38%) and glucose (31%) are the most important contributors to this supersaturated solution of sugars (Ab Wahab et al. 2018). Additionally, there is a wide range of minor bioactive constituents such as flavonoids, polyphenolics, enzymes (e.g. glucose oxidase, catalase) carotenoid derivatives, organic acids, Maillard reaction products, amino acids, and proteins that

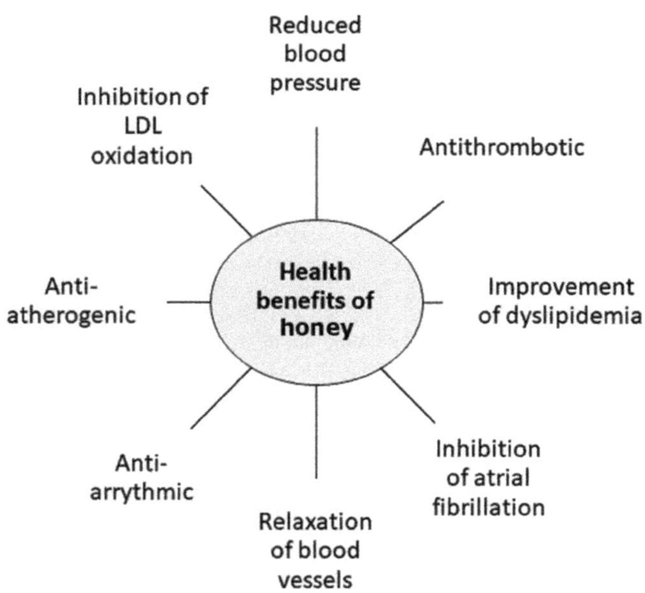


**Figure 14.2** Health benefits effects of honey for cardiovascular disease. LDL, low-density lipoprotein.

act as antioxidants (Ab Wahab et al. 2018). These antioxidant components have been reported to ameliorate CVD through mechanisms such as improving coronary vasodilatation, inhibiting LDL from oxidizing, reducing inflammatory responses, decreasing oxidative stress, improving endothelial function, providing antioxidant protection, and inhibiting platelet aggregation.

## Conclusion

Honey, a natural sweetener with potential medicinal benefits, has been in use since ancient times because of its biological properties such as its anti-inflammatory, antimicrobial, antioxidant, and immunomodulatory properties, which are majorly attributed to flavonoids. The cardioprotective effects of honey are owing to increased bioavailability of NO, inhibition of lipoxygenase and cyclooxygenase, and inhibition of lipid peroxidation, causing improvement in endothelial function and reduction in platelet aggregation and exerting beneficial effects on insulin resistance, blood pressure, and blood lipid profiles, leading to vasodilation. Because of the variation in botanical origins, honey differs in appearance, sensory perception, and composition. It contains mainly carbohydrates. The glycemic index of honey varies from 32 to 87, depending on botanical origin and on fructose content. The main nutrition- and health-relevant components are the carbohydrates, which make it an excellent energy source, especially for children and athletes. Besides its main components, the carbohydrates fructose and glucose, honey also contains a great number of other constituents in small and trace amounts, producing numerous nutritional and biological effects, including antimicrobial, antioxidant, antiviral, antiparasitic, anti-inflammatory, antimutagenic, anticancer, and immunosuppressive activities. Different nutritional studies have confirmed various effects after honey ingestion (e.g. enhanced gastroenterologic and cardiovascular health). Also, honey shows physiological effects on blood health indicators as well as effects on patients with hepatitis A and radiation mucositis. However, it should be pointed out that most of the studies in humans were based on relatively high honey intakes of 50–80 g. Honey compositions and their different biological effects depend to a great extent on the botanical origin of the honey. This fact was often not considered in the reviewed studies. However, this chapter has demonstrated that certain polyphenols have a promising role in the prevention and treatment CVDs. Although there are few negative results of some studies, collective epidemiologic and population studies suggest that honey is linked to a reduced risk of CVD. However, clinical trials should be initiated to further validate all the medicinal aspects of honey.

## Key Highlights

About 95% of honey's dry matter is composed of carbohydrates, mainly fructose and glucose. About 5%–10% of the total carbohydrates are oligosaccharides, in total about 25 different di- and trisaccharides.

- The glycemic index of honey varies from 32 to 85, depending on the botanical source. Fructose-rich honeys such as acacia honey have low glycemic indexes.
- Honey contains small amounts of proteins, enzymes, amino acids, minerals, trace elements, vitamins, aromatic compounds, and polyphenols.
- Honey has been shown to possess antimicrobial, antiviral, antiparasite, anti-inflammatory, antioxidant, antimutagenic, and antitumor properties.
- Because of its high carbohydrate content and functional properties, honey is an excellent source of energy for athletes.
- Most of the health-promoting properties of honey are only achieved by the application of rather high doses of honey such as 50–80 g per intake.

## Conflict of Interest

The authors declare they have no conflicts of interest.

## References


Ab Wahab, S.Z., Hussain, N.H., Zakaria, R., et al. (2018). Long-term effects of honey on cardiovascular parameters and anthropometric measurements of postmenopausal women. *Complementary Therapies in Medicine* 41: 154–160.

Abdul-Aziz, A.A., Desikan, P., Prabhakaran, D., and Schroeder, L.F. (2019). Tackling the burden of cardiovascular diseases in India: the essential diagnostics list. *Circulation Cardiovascular quality and outcomes* 12 (4): 1–4.

Ahmed, S., Sulaiman, S.A., Baig, A.A., et al. (2018). Honey as a potential natural antioxidant medicine: an insight into its molecular mechanisms of action. *Oxidative Medicine and Cellular Longevity* 2018: 1–19.

Alvarez-Suarez, J.M., Tulipani, S., Romandini, S., et al. (2010). Contribution of honey in nutrition and human health: a review. *Mediterranean Journal of Nutrition and Metabolism* 3 (1): 15–23.

Al-Waili, N., Salom, K., Al-Ghamdi, A., et al. (2013). Honey and cardiovascular risk factors, in normal individuals and in patients with diabetes mellitus or dyslipidemia. *Journal of Medicinal Food* 16 (12): 1063–1078.

Al-Waili, N.S. (2004 Spring). Natural honey lowers plasma glucose, C-reactive protein, homocysteine, and blood lipids in healthy, diabetic, and hyperlipidemic subjects: comparison with dextrose and sucrose. *Journal of Medicinal Food* 7 (1): 100–107.

Andersson, C. and Vasan, R.S. (2018). Epidemiology of cardiovascular disease in young individuals. *Nature Reviews Cardiology* 15 (4): 230–240.

Atkins, J.L. (2016). Body composition, dietary patterns, cardiovascular disease and mortality in older age. PhD diss. UCL (University College London).

BtHjIdrus, R., Sainik, N.Q., Nordin, A., et al. (2020). Cardioprotective effects of honey and its constituent: an evidence-based review of laboratory studies and clinical trials. *International Journal of Environmental Research and Public Health* 17 (10): 3613.

Carnethon, M.R., Pu, J., Howard, G., et al. (2017). Cardiovascular health in African Americans: a scientific statement from the American Heart Association. *Circulation* 136 (21): e393–423.

Carson, J.A., Lichtenstein, A.H., Anderson, C.A., et al. (2020). Dietary cholesterol and cardiovascular risk: a science advisory from the American Heart Association. *Circulation* 141 (3): e39–53.

Crane, E. (1975). History of honey. In: *Honey, a Comprehensive Survey* (ed., E. Crane). London: William Heinemann.

Crane, E. (1983). *The Archaeology of Beekeeping*. London: Gerald Duckworth & Co. Ltd. 1.

Edwards, R.L., Lyon, T., Litwin, S.E., et al. (2007). Quercetin reduces blood pressure in hypertensive subjects. *The Journal of Nutrition* 137 (11): 2405–2411.

Erejuwa, O.O. (2014 January 29). Effect of honey in diabetes mellitus: matters arising. *Journal of Diabetes and Metabolic Disorders* 13 (1): 23.

Estevinho, L., Pereira, A.P., Moreira, L., et al. (2008). Antioxidant and antimicrobial effects of phenolic compounds extracts of Northeast Portugal honey. *Food and Chemical Toxicology* 46 (12): 3774–3779.

Eteraf-Oskouei, T. and Najafi, M. (2013 June). Traditional and modern uses of natural honey in human diseases: a review. *Iranian Journal of Basic Medical Sciences* 16 (6): 731.

Farooqui, T. and Farooqui, A.A. (2011). *Health benefits of honey: implications for treating cardiovascular diseases*. (614): 232–252.

Farooqui, T. and Farooqui, A.A. (2011). Health benefits of honey: implications for treating cardiovascular diseases. *Current Nutrition & Food Science* 7 (4): 232–252.

Garrison, R.J., Kannel, W.B., Stokes III, J., and Castelli, W.P. (1987). Incidence and precursors of hypertension in young adults: the Framingham Offspring Study. *Preventive Medicine* 16 (2): 235–251.

Hadi, A., Rafie, N., and Arab, A. (2021 January 1). Bee products consumption and cardiovascular diseases risk factors: a systematic review of interventional studies. *International Journal of Food Properties* 24 (1): 115–128.

Hossen, M.S., Ali, M.Y., Jahurul, M.H., et al. (2017). Beneficial roles of honey polyphenols against some human degenerative diseases: a review. *Pharmacological Reports* 69 (6): 1194–1205.

Jibril, F.I., Hilmi, A.B., and Manivannan, L. (2019). Isolation and characterization of polyphenols in natural honey for the treatment of human diseases. *Bulletin of the National Research Centre* 43 (1): 1–9.

Khalil, M., Tanvir, E.M., Afroz, R., et al. (2015). Cardioprotective effects of Tualang honey: amelioration of cholesterol and cardiac enzymes levels. *BioMed Research International* 2015.

Khalil, M.I., Alam, N., Moniruzzaman, M., et al. (2011). Phenolic acid composition and antioxidant properties of Malaysian honeys. *Journal of Food Science* 76 (6): C921–8.

Lawes, C.M., Vander Hoorn, S., and Rodgers, A. (2008). Global burden of blood-pressure-related disease, 2001. *The Lancet* 371 (9623): 1513–1518.

Lewington, S. (2002). Prospective studies collaboration. Age-specific relevance of usual blood pressure to vascular mortality: a meta-analysis of individual data for one million adults in 61 prospective studies. *Lancet* 360: 1903–1913.

Machado De-Melo, A.A., de Almeida-Muradian, L.B., Sancho, M.T., and Pascual-Maté, A. (2018). Composición y propiedades de la miel de Apis mellifera: una revisión. *Journal of Apicultural Research* 57 (1): 5–37.

Nega, T. and Eshete, Y. (2018). Review of Ethiopia's global position in honey and other bee products production and marketing-analysis of sectoral opportunities and limitations. *Biomedical Journal of Scientific & Technical Research* 10 (3): 7879–7883.

Norheim, O.F., Jha, P., Admasu, K., et al. (2015). Avoiding 40% of the premature deaths in each country, 2010-30: review of national mortality trends to help quantify the UN Sustainable Development Goal for health. *Lancet* 385 (9964): 239–252.

Olas, B. (2020). Honey and its phenolic compounds as an effective natural medicine for cardiovascular diseases in humans? *Nutrients* 12 (2): 283.

Pekkarinen, A. and Varis, A.L. (2000 September 1). The world history of beekeeping and honey hunting. *Entomologica Fennica* 11 (3): 194.

Poirier, P. and Eckel, R.H. (2002). Obesity and cardiovascular disease. *Current Atherosclerosis Reports* 4 (6): 448–453.

Prabhakaran, D., Anand, S., Watkins, D., et al. (2018). Cardiovascular, respiratory, and related disorders: key messages from Disease Control Priorities. *The Lancet* 391 (10126): 1224–1236.

Prabhakaran, D., Jeemon, P., Sharma, M., et al. (2018). The changing patterns of cardiovascular diseases and their risk factors in the states of India: the global burden of disease study 1990–2016. *The Lancet Global Health* 6 (12): e1339–e1351.

Prasad, D. and Das, B. (2009). Physical inactivity: a cardiovascular risk factor. *Indian Journal of Medical Sciences* 63 (1): 33–42.

Mensah GA, Roth GA, Fuster V. The global burden of cardiovascular diseases and risk factors: 2020 and beyond. *Journal of the American College of Cardiology*. 2019 Nov 19;74(20):2529–32.

Ranneh, Y., Akim, A.M., Hamid, H.A., et al. (2021). Honey and its nutritional and anti-inflammatory value. *BMC Complementary Medicine and Therapies* 21 (1): 1–7.

Rodriguez, C.J., Allison, M., Daviglus, M.L., et al. (2014). Status of cardiovascular disease and stroke in Hispanics/Latinos in the United States: a science advisory from the American Heart Association. *Circulation* 130 (7): 593–625.

Roth, G.A., Forouzanfar, M.H., Moran, A.E., et al. (2015). Demographic and epidemiologic drivers of global cardiovascular mortality. *New England Journal of Medicine* 372 (14): 1333–1341.

Roth, G.A., Mensah, G.A., Johnson, C.O., et al. (2020 December 22). Global burden of cardiovascular diseases and risk factors, 1990–2019: update from the GBD 2019 study. *Journal of the American College of Cardiology* 76 (25): 2982–3021.

Samarghandian, S., Farkhondeh, T., and Samini, F. (2017). Honey and health: a review of recent clinical research. *Pharmacognosy Research* 9 (2): 121.

Sanchez, M., Galisteo, M., Vera, R., et al. (2006). Quercetin downregulates NADPH oxidase, increases eNOS activity and prevents endothelial dysfunction in spontaneously hypertensive rats. *Journal of Hypertension* 24 (1): 75–84.

Sanchis-Gomar, F., Perez-Quilis, C., Leischik, R., and Lucia, A. (2016). Epidemiology of coronary heart disease and acute coronary syndrome. *Annals of Translational Medicine* 4 (13): 1–12.

Sjögren, B., et al. (2015). Chapter 16 - Cardiovascular Disease. Handbook on the Toxicology of Metals (Fourth Edition). G. F. Nordberg, B. A. Fowler and M. Nordberg. San Diego, Academic Press: 313–331.

Uthurry, C.A., Hevia, D., and Gomez-Cordoves, C. (2011). Role of honey polyphenols in health. *Journal of ApiProduct and ApiMedical Science* 3 (4):141–159.

Virani, S.S., Alonso, A., Benjamin, E.J., et al. (2020). Heart disease and stroke statistics-2020 update: a report from the American Heart Association. *Circulation* 141 (9): e139–596.

Volgman, A.S., Palaniappan, L.S., Aggarwal, N.T., et al. (2018). Atherosclerotic cardiovascular disease in South Asians in the United States: epidemiology, risk factors, and treatments: a scientific statement from the American Heart Association. *Circulation* 138 (1): e1–34.

Wang, H.B., Huang, S.H., Xu, M., et al. (2019). Galangin ameliorates cardiac remodeling via the MEK1/2-ERK1/2 and PI3K-AKT pathways. *Journal of Cellular Physiology* 11: 1–9.

White, J.W. and Crane, E. (1975). *Honey a Comprehensive Survey*. ed., E. Crane. London: Morrison and Gibb Ltd. 194–206.

Yoshizumi, M., Tsuchiya, K., Kirima, K., et al. (2021). Quercetin inhibits Shc- and phosphatidylinositol 3- kinase-mediated c-Jun N-terminal kinase activation by angiotensin II in cultured rat aortic smooth muscle cells. (Cdc). 2–3.

# 15

# Use of Honey in Diabetes

*Mahendran Sekar[1,2,*], Nurul Amirah Mohd Zaid[3], Nur Najihah Izzati Mat Rani[3], and Siew Hua Gan[1]*

[1] *School of Pharmacy, Monash University Malaysia, Bandar Sunway, Subang Jaya 47500, Selangor, Malaysia*
[2] *Department of Pharmaceutical Chemistry, Faculty of Pharmacy and Health Sciences, Royal College of Medicine Perak, Universiti Kuala Lumpur, Ipoh 30450, Perak, Malaysia*
[3] *Faculty of Pharmacy and Health Sciences, Royal College of Medicine Perak, Universiti Kuala Lumpur, Ipoh 30450, Perak, Malaysia*
* *Corresponding author*

## Introduction

Diabetes is one of the common diseases that usually occur in people, especially adults, that lead to other severe complications. According to Kharroubi and Darwish (2015), the region that has the highest number of adults with diabetes is the Western Pacific region, which also has the countries with the highest frequency of diabetes (37.5%). The regions with the highest frequency of diabetes in adults are North Africa and the Middle East region (10.9%).

Glucose is transported across cell membranes by a distinct glucose transport mechanism in a normal or healthy state, which is triggered when insulin binds to cells. When glucose is internalized, glucokinase converts it to adenosine triphosphate (ATP), and energy is released through phosphorylation and several cascade mechanisms (Figure 15.1). Generally, diabetes mellitus is a disease that occurs because of abnormal action of insulin, secretion of insulin, or both in which insulin plays the crucial part as an anabolic hormone that can cause metabolic abnormalities (Kharroubi and Darwish 2015). Chronic hyperglycemia is a characteristic of diabetes that can cause deteriorated, long-term damage and failure of the organs such as the kidneys, heart, eyes, blood vessels, and nerves. The major causes of hyperglycemia are damage of insulin action and abnormalities in insulin secretion (Kharroubi and Darwish 2015). According to research evidence, insulin defection in target tissue can cause problems in fat, carbohydrate, and protein metabolism. This is because the level of insulin is too low to achieve adequate responses or because of insulin resistance of target tissue, especially in skeletal muscle, adipose tissue, and the liver, to the insulin receptor level, signal transduction system, and effector enzyme (American Diabetes Association 2010; Kharroubi and Darwish 2015).

The term *prediabetes* is associated with abnormalities in fasting glucose, glucose tolerance, or glycated hemoglobin (HbA1c), which leads to a high risk of developing diabetes and its complications (Kharroubi and Darwish 2015; Punthakee et al. 2018). The complications of diabetes include microvascular and macrovascular categories in which microvascular is more prevalent (Papatheodorou et al. 2018). Nephropathy, retinopathy, and neuropathy are among the microvascular complications, and cardiovascular disease (CVD), stroke, and peripheral artery disease are examples of macrovascular complications (Papatheodorou et al. 2018). Furthermore, diabetic foot ulcer is also associated with these complications. Other complications not included in these two categories are low resistance to infection, dental illness, and birth problems in women who have gestational diabetes (Papatheodorou et al. 2018).

In general, according to American Diabetes Association, the classification of diabetes includes type 1, type 2, and gestational diabetes mellitus (Kharroubi and Darwish 2015). According to Punthakee et al. (2018), type 1 diabetes results from an autoimmune process as the pancreatic β-cells are destroyed followed by insulin deficiency, which has a tendency toward ketoacidosis. Type 2 diabetes is caused by insulin resistance, insulin deficiency, and defects in secretion of insulin that cause insulin demand that cannot be met as the function of pancreatic β-cells are damaged (Figure 15.2). Gestational diabetes mellitus is associated with glucose intolerance during pregnancy (Punthakee et al. 2018).

In the past, honey was generally not suggested for patients with diabetes to consume because of its high carbohydrate content (Bobiş et al. 2018). However, honey has been studied to determine whether it has any antidiabetic properties. One of the latest studies highlighted that honey might have a potential hypoglycemic effect because of its potential antioxidant properties, and it might protect the pancreas against oxidative stress and damage (Bobiş et al.

*Honey: Composition and Health Benefits*, First Edition. Edited by Md. Ibrahim Khalil, Gan Siew Hua, and Bey Hing Goh.

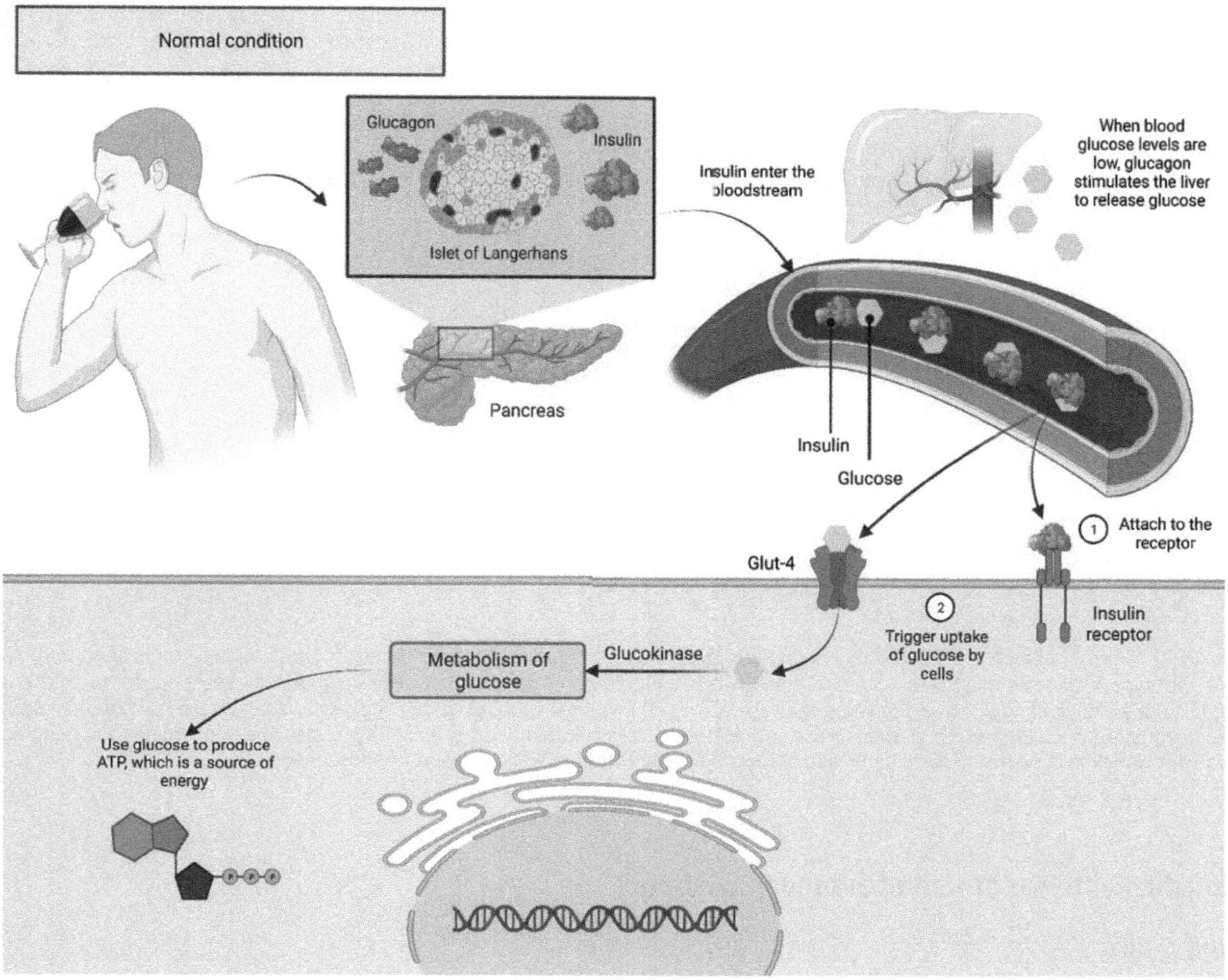


**Figure 15.1** Normal blood glucose control mechanism. To maintain the amount of glucose in the body within control, the pancreas secretes digestive enzymes as well as the hormones insulin and glucagon. Insulin allows glucose to enter the cells, where it is further processed for energy by glucokinase. ATP, adenosine triphosphate; Glut-4, glucose transporter type 4.

2018). Therefore, there are few past studies investigating the effect of honey in diabetes, and the results of those past studies varied. Omotayo et al. (2010) stated that honey had a few effects that focus on diabetes such as fructose, which is the main composition of honey. Fructose was thought to increase hepatic glucose phosphorylation through glucokinase activation and inhibition of glycogenolysis through phosphorylase suppression. Then supplementation of honey will also increase hepatic glucose intake and glycogen synthesis and storage, which will lead to improved glycemic control in those with diabetes (Figure 15.3).

Supplementation of honey also could contribute to hepatoprotective activities; honey might protect the liver against oxidative stress and damage, which is beneficial in diabetes mellitus (Eraslan et al. 2010; Kılıcoglu et al. 2008). Thus, these hepatoprotection properties will delay deterioration and enhance hepatic function, which lead to better glucose homeostasis (Klip and Vranic 2006; Omotayo et al. 2010). Some reports also show that the oligosaccharides that exist in honey might lead to antidiabetic properties of honey through the modulation of gut microbiota (Erejuwa et al. 2012b; Kajiwara et al. 2002; Omotayo et al. 2010; Shamala et al. 2000; Shin and Ustunol 2005) or via a systemic effect of oligosaccharides (Cani et al. 2006; Erejuwa et al. 2012b; Omotayo et al. 2010). This chapter discusses the scientific evidence from research that has been conducted to prove the potential benefits of honey on individuals with diabetes.

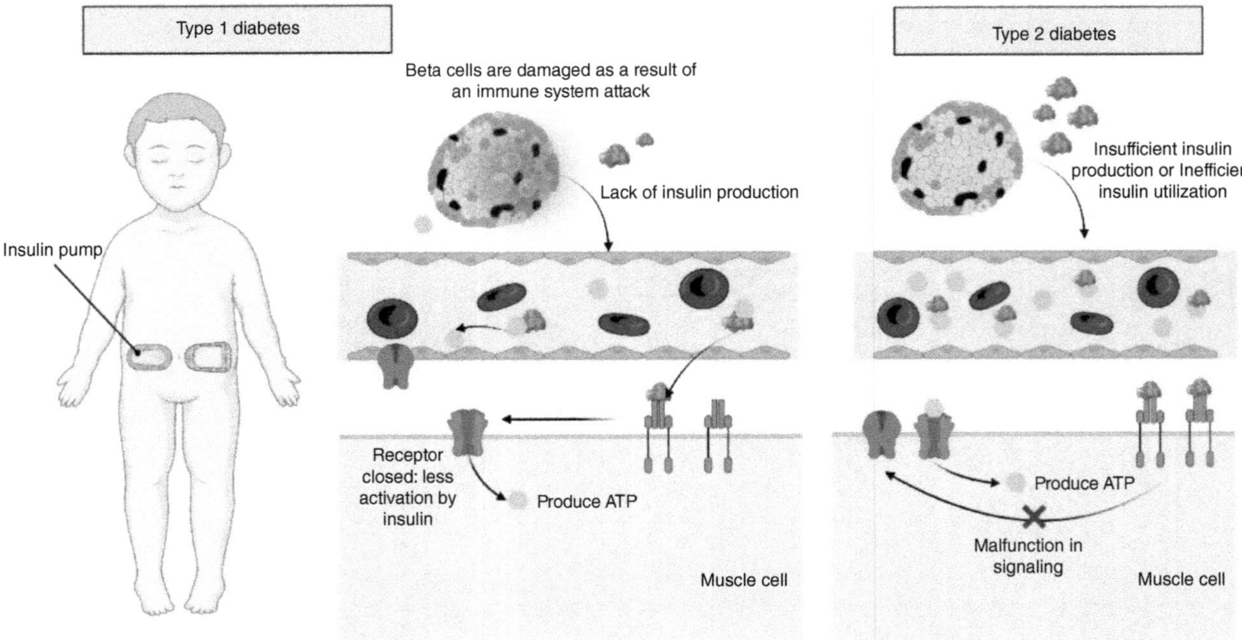


**Figure 15.2** Types of diabetes. When the body's immune system destroys pancreatic β-cells, insulin production is diminished, and people with type 1 diabetes must take insulin to manage blood glucose on a daily basis to survive. Type 2 diabetes occurs when the body does not produce enough insulin or when the cells disregard insulin. Insulin resistance, a disease in which the cells do not use insulin adequately, is usually served as the starting point. The pancreas gradually loses its ability to make insulin when blood glucose is consistently elevated, resulting in diabetes and associated complications. ATP, adenosine triphosphate.

## Scientific Evidence of Use of Honey in Diabetes

### In Vitro Studies

The two primary enzymes involved in blood glucose elevation are α-amylase and α-glucosidase. Inhibition of these two enzymes has an antidiabetic effect, which has been determined by in vitro methods (Alqahtani et al. 2020). Ali et al. (2020) found that stingless bee honey (SBH) from mangroves had the highest phenolic content, and honey from coconut had the highest flavonoid content. Tualang honey showed the highest α-amylase inhibition abilities (74.83% at 100 μg/mL), and SBH samples from coconut, acacia, and mangrove honey also gave high α-amylase inhibition with more than 70% at 100 μg/mL. Meanwhile, honey samples from coconut produced the highest α-glucosidase inhibition abilities (68.33% at 100 μg/mL). The findings showed that the honey type had a potential in decreasing blood glucose level in the body and acting as antidiabetic through α-amylase and α-glucosidase inhibition.

Similarly, in another report, raw and processed *Trigona* honey samples were analyzed to have the highest α-amylase and α-glucosidase enzyme inhibition abilities; they were about 77.61% and 80.46% at 500 μg/mL and 64.84% and 78.29% at 500 μg/mL, respectively (Krishnasree and Ukkuru 2017). In addition, the honeys were found to have a moderate glycemic index (GI) and glycemic load value, which showed the beneficial effect of honey in the treatment of patients with diabetes (Krishnasree and Ukkuru 2017).

### In Vivo Studies

Many studies have involved animal model or in vivo experiments. The majority of the experiments used streptozotocin and alloxan to induce diabetes in animal models (Bobiş et al. 2018). In one of the experiments, 40 6-week-old Sprague-Dawley rats were fed a powdered diet that consisted of 8% mixed sugars, 8% sucrose as in honey or sugar free, or 10% honey freely

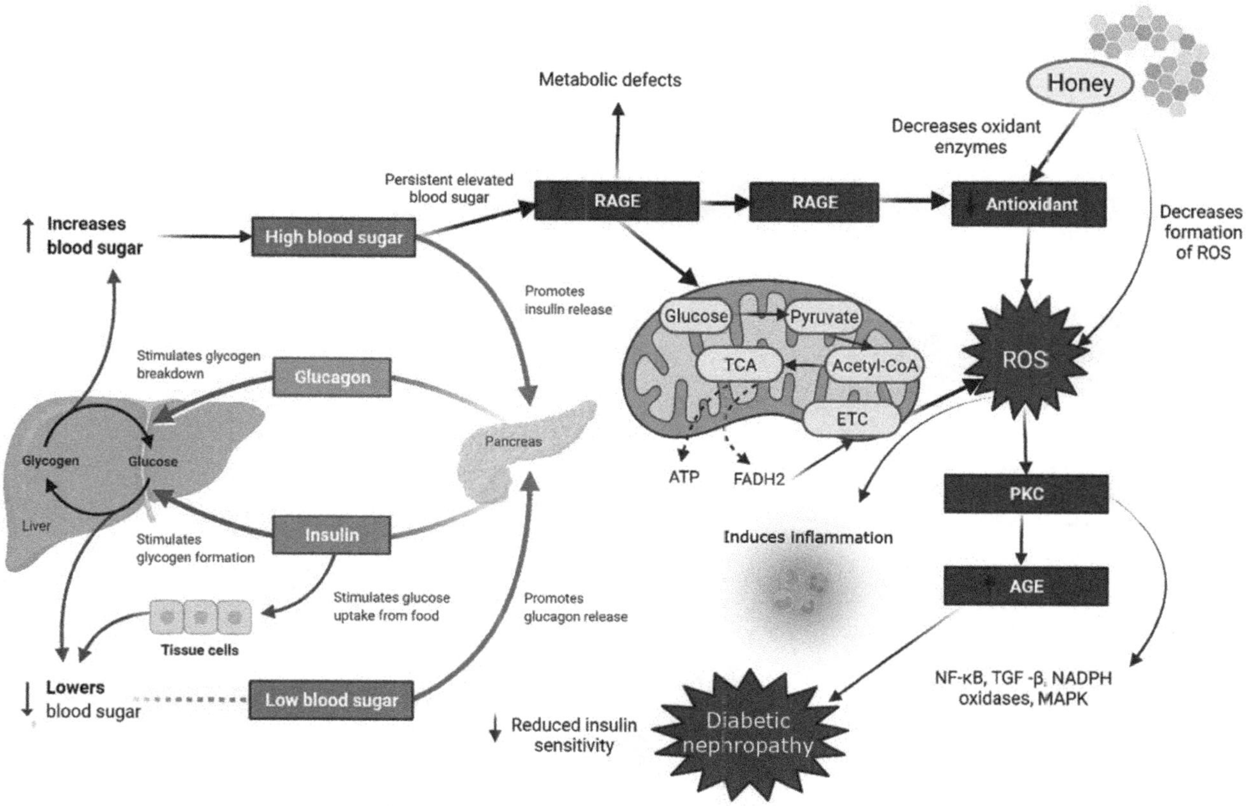


**Figure 15.3** Efficacy of honey in the management of diabetes. The hormones glucagon and insulin are involved in normal control of blood glucose. However, glucose utilization is lower in both insulin-susceptible and non–insulin-susceptible tissues in people with diabetes. Persistently high blood glucose levels might overcome the antioxidant defense mechanism, allowing reactive species to damage cells and induce inflammation. Activation of the signaling cascade leads to additional tissue damage and potentially causes diabetic nephropathy, leading to a reduction in insulin sensitivity. Honey's dietary polyphenols have been shown to increase peripheral glucose absorption into tissues and to reduce oxidant enzymes and reactive oxygen species (ROS), both of which are important for maintaining balance and vitality. acetyl-CoA, acetyl coenzyme A; AGE, advanced glycation end products; ATP, adenosine triphosphate; ETC, electron transport chain; FADH2, flavin adenine dinucleotide; MAPK, mitogen-activated protein kinase; NADPH oxidase, nicotinamide adenine dinucleotide phosphate oxidase; NF-κB, nuclear factor κ light chain enhancer of activated B cells; PKC, protein kinase C; RAGE, receptor for advanced glycation end products; TCA, tricarboxylic acid cycle; TGF-β, transforming growth factor β.

for 6 weeks. The results showed that the amount of weight gain of honey-fed rats was reduced, but the HbA1c and triglyceride levels were higher in all-sugar-treatment rats than sugar-free fed rats (Chepulis 2007). This showed that honey gives a good result, but no reduction of HbA1c was observed. Furthermore, long-term honey feeding in a 2-month-old rat model showed a reduction in HbA1c level, but high-density lipoprotein (HDL) cholesterol levels were increase compared with sucrose-fed rats and rats fed a sugar-free diet (Chepulis and Starkey 2008). However, the weight gains for honey-fed rats and sugar-free diet-fed rats were the same but lower than for the sucrose-fed rats. A previous study also demonstrated that streptozotocin-induced diabetic male rats have upregulated catalase activity, downregulated glutathione peroxidase (GPx), and decreased lipid peroxidation (thiobarbituric acid-reactive substances) when honey was combined with glibenclamide and metformin (Erejuwa et al. 2010b).

Furthermore, blood glucose also decreased when honey was combined with these two antidiabetic drugs compared with the drugs alone. This result showed that an additional antioxidant effect by honey increased the protection effect on the pancreas by glibenclamide and metformin against oxidative stress, which also exhibits the hypoglycemic effect of honey. Interestingly, Erejuwa et al. (2012a) found that supplementing Malaysian Tualang honey in streptozotocin-induced diabetic rats reduced serum levels of alanine transaminase (ALT), aspartate transaminase (AST), and alkaline phosphatase (ALP), implying that Tualang honey has a hepatoprotective effect in diabetic rats. Generally, honey has potential as a cure for diabetes as well as CVD because it contains beneficial effects such as hypoglycemic, antioxidant, and hepatoprotective effects. In addition, Asaduzzaman et al. (2015) found that honey lowers blood glucose levels in diabetic rats and reduces

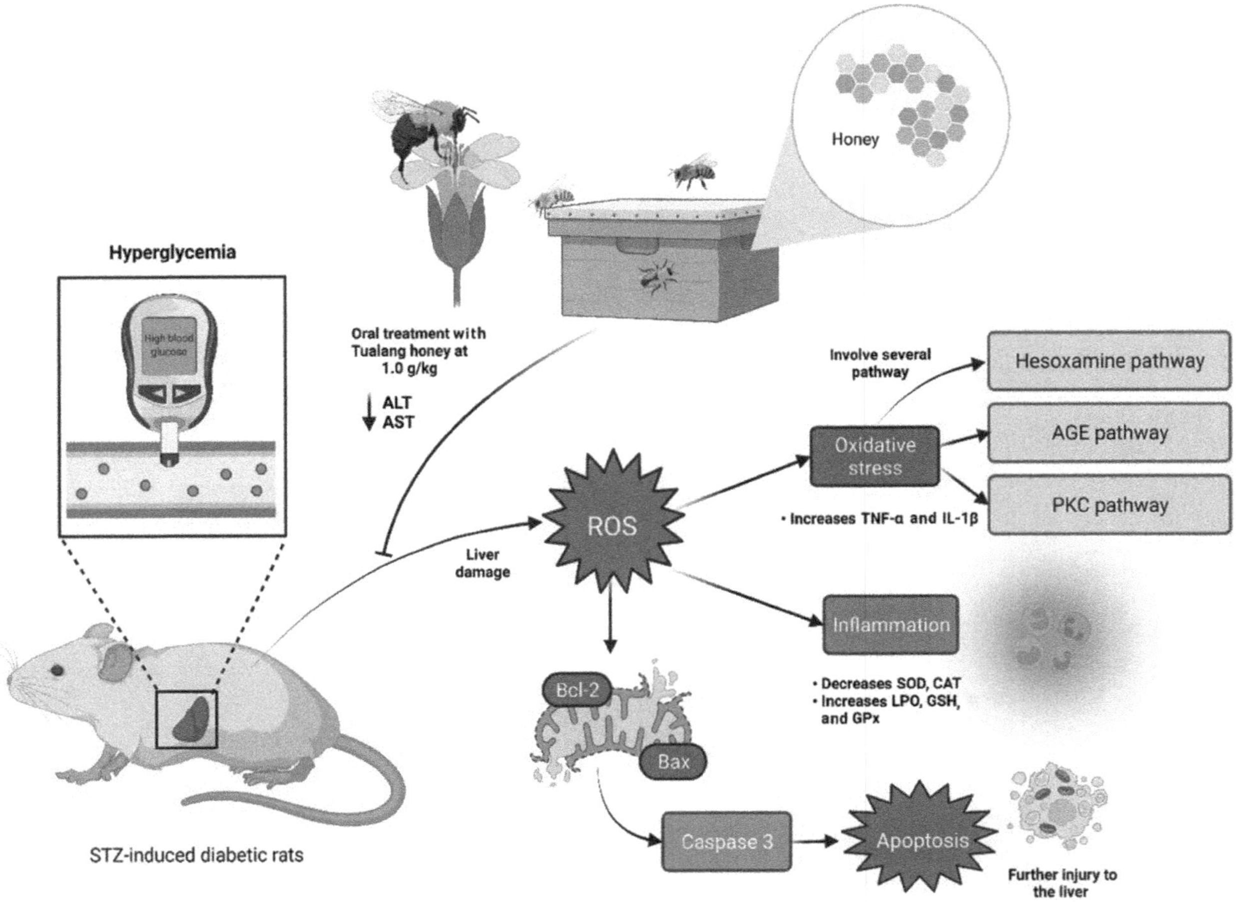


**Figure 15.4** Hepatoprotective effects of honey in diabetic animals. Increased oxidative stress is attributed to the increase in the production of ROS and/or a reduction in reactive oxygen species (ROS) scavenging efficiency, which leads to tissue damage. At certain concentrations, honey may help to lower the alanine aminotransferase (ALT) and aspartate aminotransferase (AST) enzymes released by the liver by reducing the synthesis of ROS and pro-inflammatory cytokines. Bax, Bcl-2-associated X protein; Bcl-2, B-cell lymphoma 2; CAT, catalase; Gp, glutathione peroxidase; GSH, glutathione; IL-1β, interleukin-1 β; LPO, lactoperoxidase; SOD, superoxide dismutase; STZ, streptozotocin; TNF-α, tumor necrosis factor-α.

total cholesterol (TC), triglycerides, low-density lipoprotein (LDL), and very-low-density lipoprotein (VLDL) levels while increasing HDL levels. Other than that, AST, ALT, and C-reactive protein (CRP) were also reduced. These findings suggest that honey is able to control blood glucose level as well as other lipid profiles in diabetic rats.

In streptozotocin-induced diabetic rats, honey (1.0 g/kg) increased glutathione, total antioxidant status, catalase, and glutathione reductase (GR) levels and decreased fasting plasma glucose (FPG) and malondialdehyde levels, and superoxide dismutase (SOD) activity (Erejuwa et al. 2009). Similarly, honey at 0.2, 1.2, and 2.4 g/kg/day for 4 weeks significantly enhanced the levels of catalase, GPx, GR, and GST in streptozotocin-induced diabetic rats (Erejuwa et al. 2010a). Oral treatment with Tualang honey at 1.0 g/kg decreases the levels of AST and ALT, indicating potential hepatoprotection (Figure 15.4). (Erejuwa et al. 2011). All of these findings suggest that honey has a hypoglycemic effect and reduces oxidative stress in a diabetic animals whose diabetes is caused by streptozotocin.

The feeding of healthy rats with a honey-based diet prepared by substitute honey for refined carbohydrate for 2 weeks resulted in an increase in higher plasma α-tocopherol level, a higher α-tocopherol–to–triacylglycerol ratio, a lower plasma nitrate concentrations, and less susceptibility of the heart to lipid peroxidation, showing that substituting honey for fructose gives beneficial effect (Busserolles et al. 2002). Tocopherol is a substance with vitamin E activity. In one another study, after 33 days of feeding the rats a diet containing 20% honey, Nemoseck et al. (2011) noticed that body weight, epididymal fat weight, serum triglyceride concentrations, and leptin levels were reduced, and HDL cholesterol, total cholesterol, CRP, glucose, monocyte chemoattractant protein-1, and adiponectin levels were unchanged. The result suggested that when honey was used compared with sucrose, there was the potential for beneficial effects such as a decrease in triglyceride level and improved regulation of weight (lower weight gain) because of the high antioxidant content of honey, which has a lower GI than sucrose. A previous study included an experiment that involved diabetic rats and nondiabetic rats that were administered distilled water and Tualang honey. Results demonstrated that the blood glucose level decrease (8.8 [5.8] mmol/L; median [interquartile range]) compared with the diabetic control group. Tualang honey was also able to decrease the elevated malondialdehyde level and restored SOD and catalase activities because of its antioxidative effect on the pancreas (Erejuwa et al. 2010a). Furthermore, Adesoji and Oluwakemi (2008) conducted an experiment in which alloxan-induced diabetic rats were fed honey and found that after 3 weeks, honey significantly lowered blood glucose and hyperglycemia. Honey was also able to reduce hyperglycemia in the rats that were given long-term fructose ingestion.

Another study reported the effect of three types of honey – (1) honey from *Apis florea* (small bee) bees, (2) honey from *Apis dorsata* (large bee) bees, and (3) honey adulterated with saturated sucrose solution – on an alloxan-induced diabetic rabbit group and a normal rabbit group (Akhtar and Khan 1989; Bobiş et al. 2018). At low doses of pure *A. florea* and *A. dorsata* honey, the blood glucose levels did not increase in both rabbit model groups, but it did increase in higher doses of these types of honey. Meanwhile, the adulterated honey showed an increase in blood glucose level at both low and high doses. Therefore, it was assumed that pure natural honey in low doses can be suggested as a sweetening agent and carbohydrate sources in place of sucrose to treat diabetes (Akhtar and Khan 1989; Bobiş et al. 2018).

## Clinical Studies

The effects of honey have also been demonstrated and evaluated in clinical trials. In one report, healthy participants, participants with hypertriglyceridemia, and participants with diabetes were compared regarding the effect of honey in their diets. In healthy participants, honey was able to reduce cholesterol, triglycerides, and LDL cholesterol while increasing the HDL cholesterol level. Also, after 15 days of consumption, honey decreased cholesterol (7%), CRP (7%), triglycerides (2%), LDL cholesterol (1%), homocysteine (6%), and plasma glucose level (PGL) (6%) while increasing HDL cholesterol (2%) (Al-Waili 2004). In patients with hypertriglyceridemia, honey reduced triglycerides, and artificial honey increased triglycerides.

Meanwhile in patients with hyperlipidemia, honey decrease LDL cholesterol, cholesterol, and CRP after 15 days. For participants with diabetes, honey was able to lower the increase of PGL compared with dextrose, and after 60, 120, and 180 minutes, elevation of PGL in honey was lower than that of sucrose (Al-Waili 2003). Honey also contributed to a bigger elevation in insulin than sucrose. Honey caused a lower elevation of PGL in the diabetic group compared with dextrose and sucrose, and honey decreased the blood lipid, CRP, and homocysteine levels in the normal and hyperlipidemia groups (Al-Waili 2003). These clinical studies show that honey has a beneficial effect in healthy, diabetic, and hypertriglyceridemic patient.

In addition, clinical trials with a group of type 2 adult patients with diabetes who use honey as their sole treatment and stopped using their modern medications have shown promising results (Abdulrhman 2016). The findings showed that consumption of honey caused more hyperglycemia in these patients, but the diabetic ketoacidosis and hyperglycemic hyperosmolar states were absent. Longer term consumption of honey caused weight reductions in all patients and controlled blood pressure in patients who had hypertension before intervention with honey. Moreover, patients with macrovascular complications, especially coronary heart disease (CHD), before the intervention showed improved CHD status. The researchers suggested that the result of this clinical trial might help patients with diabetes mellitus because the honey intervention showed its potency. In one clinical trial, the researchers found that honey consumption in the experimental group was able to slightly reduce body fat (1.1%) and body weight (1.3%); decrease LDL cholesterol (5.8%), fasting blood glucose by 4.2%, total cholesterol (3%), triacylglycerol by 11%, and CRP (3.2%); and increase HDL cholesterol (3.3%) in normal participants. Meanwhile, in patients with elevated variables, honey decreased LDL cholesterol by 4.3%, CRP by 3.2%, total cholesterol by 3.3%, and triacylglycerol by 19% (Yaghoobi et al. 2008). This shows that natural honey decreases CVD risk factors, which is also beneficial in treating patients with diabetes.

A past report also showed that patients with diabetic type 2 who ate natural honey for 8 weeks showed a reduction in total cholesterol, LDL cholesterol, body weight, and triglycerides; their HDL cholesterol increased (Figure 15.5) (Bahrami et al. 2009). However, the HbA1c level increased significantly. It can be concluded that honey consumption can improve the body weight and body lipids of patients with diabetes; however, they need to be cautious regarding consumption of honey because of HbA1C increases.

According to Abdulrhman et al. (2011), honey had a lower GI and peak incremental index (PII) than sucrose in 20 children and adolescents with type 1 diabetes and 10 healthy nondiabetic children and adolescents as control participants. Compared with glucose or sucrose, the increase in C-peptide level after taking honey was not significant. Hence, honey could be used as a sugar alternative in patients with type 1 diabetes because it has a lower PII and GI level than sucrose

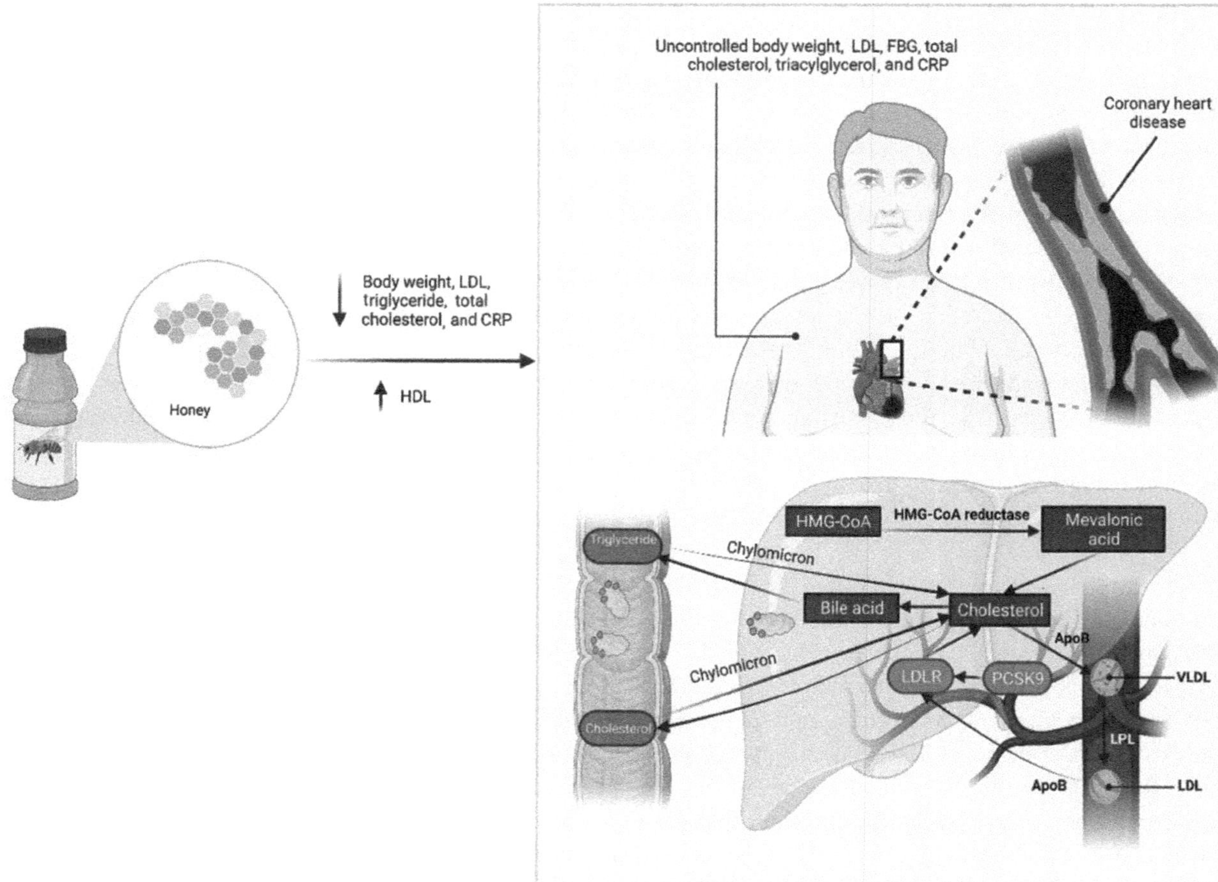


**Figure 15.5** Efficacy of honey in the management of hyperlipidemia. Coronary heart disease is the most common type of heart disease caused by lipid accumulation in the inner lining of artery walls, which can obstruct as plaque formation progresses. In normal circumstances, the liver is where cholesterol is synthesized, excreted, and converted to bile acids and very-low-density lipoprotein (VLDL). Excess lipids and/or lipoproteins in the blood may be a highly modifiable risk factor for cardiovascular disease, and honey, which contains a number of phenolic compounds, can provide an alternative by influencing cholesterol modulation. ApoB, apolipoprotein B; CRP, C-reactive protein; FBG, fasting blood glucose; HDL, high-density lipoprotein; HMG-CoA, β-hydroxy β-methylglutaryl-CoA; LDL, low-density lipoprotein; LDLR, low-density lipoprotein receptor; LPL, lipoprotein lipase; PCSK9, proprotein convertase subtilisin/kexin type 9.

(Abdulrhman et al. 2011). Furthermore, the consumption of honey decreases the postprandial glycemic response in diabetic and nondiabetic participants and reduces the serum concentration of glucose in patients with type 1 and type 2 diabetes compared with other sweeteners such as fructose and sucrose; hence, honey is recommended over sucrose because it has a gentler effect on level on blood sugar on a per-gram basis (Cianciosi et al. 2018; Shambaugh et al. 1990).

Caffeic acid phenethyl ester, a flavonoid compound that has several such as antioxidant and anti-inflammatory effects, is one of the major components of honeybee propolis. A clinical study has been conducted with three groups of patients with type 2 diabetes. Group 1 was the 75-g honey group, group 2 was 30-g honey group, and group 3 was the 75-g glucose group. The results showed that after 2 hours, the mean increases in blood glucose were 85 mg/dL for group 1, 30 mg/dL for group 2, and 170 mg/dL for group 3 (Nazir et al. 2014). Furthermore, the PGL in response to honey peaked at 60 minutes and rapidly declined in response to glucose, which showed that honey has a lower GI.

In another note, some participants even exhibit a glucose-lowering effect after being given a low dose of honey. The authors concluded that a low dose of honey was a good substitute for sugar in patients with diabetes (Nazir et al. 2014). Another study found that after 60 minutes, those who took natural honey had a lower increase in PGL (~20%) compared with those who consumed simulated honey and D-glucose, who had increases of 47% and 52%, respectively (Ahmad et al. 2008). However, after 180 minutes, the D-glucose group showed a 20% reduction in PGL compared with the natural honey group, which was about a 9.75% decrease. Post-hoc testing also showed that in the natural honey group, the response of glucose was significantly lower at all time points compared with the other two groups, showing that natural honey preserves the physiological glycemic response with rebound retrieval of PGL (Ahmad et al. 2008).

In patients with type 2 diabetes, intrapulmonary administration of honey (60% w/v) significantly reduced blood glucose levels (Al-Waili 2003). Another study on patients with type 2 diabetes found that 30 g of honey in the diet elevated the blood glucose level by 32.4% (Ionescu-Tîrgovişte et al. 1983). This finding was also in line with another study on patients with type 2 diabetes who were given honey as part of their diet (Agrawal et al. 2007). In patients with type 1 diabetes, oral administration of honey at 1.75 g/kg resulted in a considerably lower GI than sucrose (Abdulrhman et al. 2013).

## Conclusion

Honey appears to provide potential benefits in the control of diabetes, according to scientific evidence from in vitro, in vivo, and clinical investigations. Honey has been shown to have a hypoglycemic impact in many investigations. This may be due to the presence of polyphenols in honey, including quercetin, apigenin, luteolin, catechin, rutin, and kaempferol, which have been discovered to have antidiabetic activities. Hence, honey could be used as a potential antidiabetic agent. However, the mechanism behind this action is unknown, and more long-term clinical studies on humans with type 2 diabetes mellitus are needed, with a larger number of patients. Furthermore, randomized clinical trials to be conducted for patients with different doses of honey and in patients with different levels of diabetes.

## Acknowledgment

The authors thank Universiti Kuala Lumpur Royal College of Medicine Perak, Malaysia, for providing necessary facilities and resources to complete this study for publication. The figures were created with the support of https://biorender.com under a paid subscription.

## Conflict of Interest

We wish to confirm that there are no known conflicts of interest associated with this publication and there has been no significant financial support for this work that could have influenced its outcome.

## References

Abdulrhman, M., El Hefnawy, M., Ali, R., et al. (2013). Effects of honey, sucrose and glucose on blood glucose and C-peptide in patients with type 1 diabetes mellitus. *Complementary Therapies in Clinical Practice* 19 (1): 15–19.

Abdulrhman, M., El-Hefnawy, M., Hussein, R., and Abou El-Goud, A. (2011). The glycemic and peak incremental indices of honey, sucrose and glucose in patients with type 1 diabetes mellitus: effects on C-peptide level—a pilot study. *Acta Diabetologica* 48 (2): 89–94.

Abdulrhman, M.A. (2016). Honey as a sole treatment of type 2 diabetes mellitus. *Endocrinology & Metabolic Syndrome* 5 (232): 2161–1017. 1000232.

Adesoji, F. and Oluwakemi, A. (2008). Differential effect of honey on selected variables in alloxan-induced and fructose-induced diabetic rats. *African Journal of Biomedical Research* 11 (2).

Agrawal, O., Pachauri, A., Yadav, H., et al. (2007). Subjects with impaired glucose tolerance exhibit a high degree of tolerance to honey. *Journal of Medicinal Food* 10 (3): 473–478.

Ahmad, A., Azim, M., Mesaik, M., et al. (2008). Natural honey modulates physiological glycemic response compared to simulated honey and D-glucose. *Journal of Food Science* 73 (7): H165–H167.

Akhtar, M.S. and Khan, M.S. (1989). Glycaemic responses to three different honeys given to normal and alloxan-diabetic rabbits. *Journal of the Pakistan Medical Association* 39 (4): 107–113.

Al-Waili, N. (2003). Intrapulmonary administration of natural honey solution, hyperosmolar dextrose or hypoosmolar distill water to normal individuals and to patients with type-2 diabetes mellitus or hypertension: their effects on blood glucose level, plasma insulin and C-peptide, blood pressure and peaked expiratory flow rate. *European Journal of Medical Research* 8 (7): 295–303.

Al-Waili, N.S. (2004). Natural honey lowers plasma glucose, C-reactive protein, homocysteine, and blood lipids in healthy, diabetic, and hyperlipidemic subjects: comparison with dextrose and sucrose. *Journal of Medicinal Food* 7 (1): 100–107.

Ali, H., Abu Bakar, M., Majid, M., et al. (2020). In vitro anti-diabetic activity of stingless bee honey from different botanical origins. *Food Research* 4 (5): 1421–1426.

Alqahtani, A.S., Hidayathulla, S., Rehman, M.T., et al. (2020). Alpha-amylase and alpha-glucosidase enzyme inhibition and antioxidant potential of 3-oxolupenal and katononic acid isolated from Nuxia oppositifolia. *Biomolecules* 10 (1): 61.

American Diabetes Association. (2010). Diagnosis and classification of diabetes mellitus. *Diabetes Care* 33 (Supplement 1): S62–S69.

Asaduzzaman, M., Sohanur Rahman, M., Munira, S., et al. (2015). Effects of honey supplementation on hepatic and cardiovascular disease (CVD) marker in streptozotocin-induced diabetic rats. *Journal of Diabetes and Metabolism* 6 (592): 2.

Bahrami, M., Ataie-Jafari, A., Hosseini, S., et al. (2009). Effects of natural honey consumption in diabetic patients: an 8-week randomized clinical trial. *International Journal of Food Sciences and Nutrition* 60 (7): 618–626.

Bobiş, O., Dezmirean, D.S., and Moise, A.R. (2018). Honey and diabetes: the importance of natural simple sugars in diet for preventing and treating different type of diabetes. *Oxidative Medicine and Cellular Longevity* 2018.

Busserolles, J., Gueux, E., Rock, E., et al. (2002). Substituting honey for refined carbohydrates protects rats from hypertriglyceridemic and prooxidative effects of fructose. *The Journal of Nutrition* 132 (11): 3379–3382.

Cani, P.D., Knauf, C., Iglesias, M.A., et al. (2006). Improvement of glucose tolerance and hepatic insulin sensitivity by oligofructose requires a functional glucagon-like peptide 1 receptor. *Diabetes* 55 (5): 1484–1490.

Chepulis, L. (2007). The effect of honey compared to sucrose, mixed sugars, and a sugar-free diet on weight gain in young rats. *Journal of Food Science* 72 (3): S224–S229.

Chepulis, L. and Starkey, N. (2008). The long-term effects of feeding honey compared with sucrose and a sugar-free diet on weight gain, lipid profiles, and DEXA measurements in rats. *Journal of Food Science* 73 (1): H1–H7.

Cianciosi, D., Forbes-Hernández, T.Y., Afrin, S., et al. (2018). Phenolic compounds in honey and their associated health benefits: a review. *Molecules* 23 (9): 2322.

Eraslan, G., Kanbur, M., Silici, S., and Karabacak, M. (2010). Beneficial effect of pine honey on trichlorfon induced some biochemical alterations in mice. *Ecotoxicology and Environmental Safety* 73 (5): 1084–1091.

Erejuwa, O., Sulaiman, S., Ab Wahab, M., et al. (2009). Effects of Malaysian Tualang honey supplementation on glycemia, free radical scavenging enzymes and markers of oxidative stress in kidneys of normal and streptozotocin-induced diabetic rats. *International Journal of Cardiology* 137: S45.

Erejuwa, O., Sulaiman, S., Wahab, M., et al. (2010a). Antioxidant protection of Malaysian Tualang honey in pancreas of normal and streptozotocin-induced diabetic rats. *Annales d'endocrinologie* 71: 291–296.

Erejuwa, O.O., Sulaiman, S.A., Wahab, M., et al. (2012a). Hepatoprotective effect of Tualang honey supplementation in streptozotocin-induced diabetic rats. *International Journal of Applied Research in Natural Products* 4 (4): 37–41.

Erejuwa, O.O., Sulaiman, S.A., and Wahab, M.S.A. (2012b). Oligosaccharides might contribute to the antidiabetic effect of honey: a review of the literature. *Molecules* 17 (1): 248–266.

Erejuwa, O.O., Sulaiman, S.A., Wahab, M.S.A., et al. (2011). Comparison of antioxidant effects of honey, glibenclamide, metformin, and their combinations in the kidneys of streptozotocin-induced diabetic rats. *International Journal of Molecular Sciences* 12 (1): 829–843.

Erejuwa, O.O., Sulaiman, S.A., Wahab, M.S.A., et al. (2010b). Antioxidant protective effect of glibenclamide and metformin in combination with honey in pancreas of streptozotocin-induced diabetic rats. *International Journal of Molecular Sciences* 11 (5): 2056–2066.

Ionescu-Tîrgovişte, C., Popa, E., Sintu, E., et al. (1983). Blood glucose and plasma insulin responses to various carbohydrates in type 2 (non-insulin-dependent) diabetes. *Diabetologia* 24 (2): 80–84.

Kajiwara, S., Gandhi, H., and Ustunol, Z. (2002). Effect of honey on the growth of and acid production by human intestinal Bifidobacterium spp.: an in vitro comparison with commercial oligosaccharides and inulin. *Journal of Food Protection* 65 (1): 214–218.

Kharroubi, A.T. and Darwish, H.M. (2015). Diabetes mellitus: the epidemic of the century. *World Journal of Diabetes* 6 (6): 850.

Kılıcoglu, B., Gencay, C., Kısmet, K., et al. (2008). The ultrastructural research of liver in experimental obstructive jaundice and effect of honey. *The American Journal of Surgery* 195 (2): 249–256.

Klip, A. and Vranic, M. (2006). Muscle, liver, and pancreas: three musketeers fighting to control glycemia. *American Journal of Physiology-Endocrinology and Metabolism* 291 (6): E1141–E1143.

Krishnasree, V. and Ukkuru, M.P. (2017). In vitro antidiabetic activity and glycemic index of bee honeys.

Nazir, L., Samad, F., Haroon, W., et al. (2014). Comparison of glycaemic response to honey and glucose in type 2 diabetes. *Journal of the Pakistan Medical Association* 64 (1): 69–71.

Nemoseck, T.M., Carmody, E.G., Furchner-Evanson, A., et al. (2011). Honey promotes lower weight gain, adiposity, and triglycerides than sucrose in rats. *Nutrition Research* 31 (1): 55–60.

Omotayo, E.O., Gurtu, S., Sulaiman, S.A., et al. (2010). Hypoglycemic and antioxidant effects of honey supplementation in streptozotocin-induced diabetic rats. *International Journal for Vitamin and Nutrition Research* 80 (1): 74.

Papatheodorou, K., Banach, M., Bekiari, E., et al. (2018). Complications of diabetes 2017. *Journal of Diabetes Research* 2018: 3086167.

Punthakee, Z., Goldenberg, R., and Katz, P. (2018). Definition, classification and diagnosis of diabetes, prediabetes and metabolic syndrome. *Canadian Journal of Diabetes* 42: S10–S15.

Shamala, T., Shri Jyothi, Y., and Saibaba, P. (2000). Stimulatory effect of honey on multiplication of lactic acid bacteria under in vitro and in vivo conditions. *Letters in Applied Microbiology* 30 (6): 453–455.

Shambaugh, P., Worthington, V., and Herbert, J. (1990). Differential effects of honey, sucrose, and fructose on blood sugar levels. *Journal of Manipulative and Physiological Therapeutics* 13 (6): 322–325.

Shin, H.-S. and Ustunol, Z. (2005). Carbohydrate composition of honey from different floral sources and their influence on growth of selected intestinal bacteria: an in vitro comparison. *Food Research International* 38 (6): 721–728.

Yaghoobi, N., Al-Waili, N., Ghayour-Mobarhan, M., et al. (2008). Natural honey and cardiovascular risk factors; effects on blood glucose, cholesterol, triacylglycerole, CRP, and body weight compared with sucrose. *The Scientific World Journal* 8: 463–469.

# 16

# Use of Honey in Kidney Disease

*R. B. Nerli, Saziya R. Bidi, and Shridhar C. Ghagane*

## Introduction

The kidneys' functions are to filter excess water and wastes from the blood and make urine. They also regulate blood pH, blood pressure, synthesis of hemoglobin, and vitamin D. The structure consists of millions of filtration units called nephrons. When the kidneys get damaged and cannot filter blood, the condition is known as kidney disease (Coresh et al. 2007). Different types of kidney diseases include acute kidney failure, chronic kidney diseases, diabetic kidney failure, polycystic kidney disease and pyelonephritis (kidney infection), diabetes, and hypertension. A family history of kidney disease and heart disease are the common risk factors. The treatment strategy for chronic kidney disease (CKD) depends on its cause (Hsu et al. 2004). Although CKD has no cure, its treatment usually consists of measures to reduce complications, relieve symptoms, and slow progression. The prevalence of CKD is increasing throughout the world (Plantinga et al. 2008). Because the risk for cardiovascular events increases as CKD progresses, prevention is essential to improve the disease's prognosis. The common pathway in CKD is kidney fibrosis, characterized by the replacement of functional renal tissue with the extracellular matrix, and provides a greater contribution to the risk for end-stage renal disease (ESRD) (Jha et al. 2013).

Several factors, including aging, impact negatively on renal function, which may, in turn, lead to CKD (Rule et al. 2010). With the increase in life expectancy, especially in developed countries, there has been a rise in the incidence of impaired renal function and CKD in the aging population (Coresh et al. 2003). Lifestyle-related diseases such as diabetes mellitus, obesity, hypertension, and metabolic syndrome are on the rise, and these diseases are associated with a decline in renal function, which accelerates CKD's progression (Wahba and Mak 2007). Another aspect of this disease is infection, which plays a major cause of morbidity at all stages of CKD and can directly contribute patient death (McDonald et al. 2014). The causes of infection may be low immunity and use of catheters for hemodialysis or peritoneal dialysis (PD). Therefore, the need for holistic care arises in infection prevention and management of patients with CKD. The incidence of acute kidney disease is on the rise in both developed and developing countries (Lameire et al. 2013). It is associated with severe complications, and it increases the risk of short- and long-term mortality, development of CKD, and rapid progression to ESRD (Rewa and Bagshaw 2014). Also, although there is growing insight into the causes and mechanisms of kidney diseases, preventive and therapeutic measures are still few (Mohd Effendy et al. 2012). Natural products were used for medicinal purposes long before recorded history, and presently their use coupled with other alternative treatment options has accelerated because of the rise in prevalence of diseases, increasing the costs of medical care and public concerns about the rising adverse effects of conventional medicine (Silver 2011).

In this era of modern medicine, the treatment of infection relies on the use of antibiotics. However, the emergence of antimicrobial resistance has become a new problem for the treatment and prevention of infections, limiting the use of antibiotics. In the past few decades, multidrug-resistant organisms with a simultaneous decrease in new antibiotics have emerged in the market (Silver 2011). In patients on dialysis, the rate of antibiotic-resistant bacteria is highest, which has triggered interest in research in alternative approaches for prevention and management of infections. One such promising alternative approach is the use of topical honey (D'Agata 2002; Hadley et al. 2007).

Honey is a saturated complex containing fructose (38%), glucose (31%), and sucrose (7%) as major constituents, and other minor constituents include phenolic acids, flavonoids, ascorbic acid, certain antioxidant enzymes such as glucose oxidase and catalase, carotenoid-like substances, organic acids, and Maillard reaction products (El Denshary et al. 2012; White 1975).

*Honey: Composition and Health Benefits*, First Edition. Edited by Md. Ibrahim Khalil, Gan Siew Hua, and Bey Hing Goh.

These minor substances are reported to exhibit the antioxidant properties of honey. The phenolic and flavonoids act as antitumor or anti-inflammatory agents as reported by epidemiological studies (Facino 2004; Galal et al. 2012).

## Outcome of Honey on Handling Infection in Chronic Kidney Disease

A number of studies have been assessing the safety and efficacy of honey in the remedy of skin and wound infections in patients vulnerable to chronic kidney ailments. However, there have been no medical or preclinical studies for CKD that have used honey as potent healer. A study on rats showed bacterial peritonitis next to cecal ligation and puncture. Yuzbasioglu et al. (2009) confirmed that intraperitoneal control of honey resulted in lesser peritoneal adhesion rankings and tissue oxidative strain ranges at day 14 related to rats receiving intraperitoneal 5% dextrose or no treatment at all. This extended a first-rate perspective that honey may additionally have a capacity beneficial adjunctive role inside the management of PD-associated peritonitis, although the study is limited via its small pattern size, uncertain generalizability of the version, and shortage of blinding of outcome assessors to the medical situation of PD-related peritonitis. Clearly, further studies of the healing efficiency of honey in patients with CKD and infections are vital.

## Nephroprotective Characteristics of Honey

In a study by Ajibola et al. (2012), stingless bee honey (SBH) was shown to be defensive against oxidative stress and lipopolysaccharide (LPS)-induced chronic subclinical systemic inflammation (CSSI) in rats. Meta inflammation, mini-inflammation, and low-grade chronic inflammation are characterized by increased levels of inflammatory mediators such as nuclear factor-κB (NF-κB) and pro-inflammatory cytokines, regardless of the lack of clinical signs. CSSI plays an imperative part in homeostasis and is triggered by oxidative stress and metabolic malfunction, which is known to have a positive feedback mechanism at the systematic or local level (Ajibola et al. 2012). Hence, the pathogenesis of various diseases such as cancer, type 2 diabetes, cardiovascular disease, and CKD has been thought to involve CSSI and oxidative stress. (Molan 1998). With treatment with SBH, modulations of NF-κB, nuclear factor erythroid 2–related factor (Nrf2), and bvp38 mitogen-activated protein kinase (MAPK) were seen in the kidney, hearts, and lungs of rats. The kidneys showed a significant attenuation in inflammatory cell infiltration and inflammation score (Al-Quassemi 2003; Molan 1999).

LPS-induced rats displayed higher serum levels of C-reactive protein, tumor necrosis factor-α, interleukins (IL-1β, IL-6, and IL-8), monocyte chemoattractant protein-1, malondialdehyde, and 8-hydroxy-2′-deoxyguanosine (8-OHdG), as well as significant leucocytosis. Contrarily, SBH decreased NF-κB, p65, and p38 MAPK and increased Nrf2 expression in the liver, kidneys, hearts, and lungs of LPS-induced rats (Molan 2001). Histopathology reports of LPS-induced rats suggest that the hematoxylin and eosin–stained sections of the kidneys showed degenerative changes in the tubular cells, glomerular atrophy, granuloma, and blood cell exudation into the tubulointerstitial spaces and aggregation of inflammatory cells. Compared with control rats, the LPS-induced rats exhibited significantly increased inflammation scores in the kidney (McDonald et al. 2014). SBH has been claimed to prevent LPS-induced histologic and functional alterations in the liver, kidneys, hearts, and lungs of rats and is known to have a substantial protective role against LPS-induced CSSI in rats mediated through oxidative stress; amelioration of inflammation; and p38 MAPK, NF-κB, and Nrf2 signaling (Silver 2011). A total of 13 compounds, consisting of 8 phenolic acids and 5 flavonoids, were investigated in SBH, which was found to be rich in polyphenols compared with other types of honey (D'Agata 2002; Hadley et al. 2007).

## Role of Honey in Suppressing Cisplatin-Induced Acute Kidney Injury

Inflammation and oxidative stress are the major contributors of drug-induced nephrotoxicity such as cisplatin-induced acute kidney injury (AKI). Currently, no drugs are available to reduce cisplatin-induced AKI. Moreover, available anti-inflammatory and antioxidant therapies may reduce cisplatin's efficacy against tumors. Therefore, it is important to develop new methods to treat patients with cisplatin nephrotoxicity. Natural compounds such as honey are rich in antioxidants and may have anti-inflammatory effects that have been used to treat a wide variety of illnesses in Ayurvedic medicine in Asia. However, honey's usefulness in reducing cisplatin nephrotoxicity is unknown. A study conducted by Hamad et al. (2015) demonstrates for the first time that oral administration of crude honey is effective in preventing cisplatin-induced kidney injury by reducing inflammation and oxidative stress through inhibition of NF-κB activation.

Feeding honey to animals suppressed cisplatin-induced inflammation in the kidneys, including leukocyte infiltration, cytokine and chemokine production, and cytokine excretion in urine. Honey is known to have phenolic compounds (Khalil 2012; White 1975) that may be responsible for the observed effects on inflammation. NF-κB activation by cisplatin is a major mediator of inflammatory response in the kidneys in response to cisplatin administration (Miller et al. 2010; Ramesh and Reeves 2002, 2004). The study showed that cisplatin-induced a large increase in phospho-P65 (NF-κB) levels in the kidneys, which was largely reduced with honey feeding, suggesting that honey may reduce inflammation through suppression of NF-κB activation. The mechanism of NF-κB inhibition is unknown. Oxidative stress is a hallmark of cisplatin-induced nephrotoxicity. Hydroxyl radicals and oxidants such as superoxide and hydrogen peroxide are known activators of NF-κB (Baliga et al. 1999). Reverse transcription polymerase chain reaction analysis of oxidative stress markers such as heme oxygenases (HO-1) is highly elevated after cisplatin nephrotoxicity, which was significantly downregulated with honey treatment, suggesting a reduction of oxidative stress with honey. Therefore, it is possible that reduction of oxidative stress may reduce NF-κB activation and thereby reduce inflammation in the kidneys.

In summary, oral administration of honey is effective in preventing cisplatin nephrotoxicity through the suppression of inflammation. Suppression of inflammation may be related to the reduction of oxidative stress. Moreover, honey intake could benefit patients who receive chemotherapy for cancer (Morgan and Liu 2011).

## References

Ajibola, A., Chamunorwa, J.P., and Erlwanger, K.H. (2012 December). Nutraceutical values of natural honey and its contribution to human health and wealth. *Nutrition & Metabolism* 9 (1): 1–2.

Al-Quassemi, R. (2003). Robinson RK Some special nutritional properties of honey - a brief review. *Nutrition & Food Science* 33: 254–260.

Baliga, R., Ueda, N., Walker, P.D., and Shah, S.V. (1999). Oxidant mechanisms in toxic acute renal failure. *Drug Metabolism Reviews* 31: 971–997. [PubMed: 10575556].

Coresh, J., Astor, B.C., Greene, T., et al. (2003). Prevalence of chronic kidney disease and decreased kidney function in the adult US population: Third National Health and Nutrition Examination Survey. *American Journal of Kidney Diseases* 41: 1–12.

Coresh, J., Selvin, E., Stevens, L.A., et al. (2007). Prevalence of chronic kidney disease in the United States. *JAMA* 298 (17): 2038–2047. doi: 10.1001/jama.298.17.2038 [PubMed: 17986697].

D'Agata, E.M.C. (2002). Antimicrobial-resistant, gram-positive bacteria among patients undergoing chronic hemodialysis. *Clinical Infectious Diseases* 35 (10): 1212–1218.

El Denshary, E.S., Al Gahazali, M.A., Mannaa, F.A., et al. (2012). Dietary honey and ginseng protect against carbon tetrachloride-induced hepatonephrotoxicity in rats. *Experimental and Toxicologic Pathology* 64: 753–760.

Facino, R.M. (2004). Honey in tumor surgery. *Archives of Surgery* 139: 802.

Galal, R.M., Zaki, H.F., Seif El-Nasr, M.M., and Agha, A.M. (2012). Potential protective effect of honey against paracetamol-induced hepatotoxicity. *Archives of Iranian Medicine* 15: 674–680. [PubMed: 23102243].

Hadley, C., Karchmer, T.B., Russell, G.B., et al. (2007). The prevalence of resistant bacterial colonization in chronic hemodialysis patients. *American Journal of Nephrology* 27 (4): 352–359.

Hamad, R., Jayakumar, C., Ranganathan, P., et al. (2015). Honey feeding protects kidney against cisplatin nephrotoxicity through suppression of inflammation. *Clinical and Experimental Pharmacology and Physiology* 42 (8): 843–848. doi: 10.1111/1440-1681.12433.

Hsu, C.Y., Vittinghoff, E., Lin, F., and Shlipak, M.G. (2004). The incidence of end-stage renal disease is increasing faster than the prevalence of chronic renal insufficiency. *Annals of Internal Medicine* 141 (2): 95–101. doi: 10.7326/0003-4819-141-2-200407200-00007 [PubMed: 15262664].

Jha, V., Garcia-Garcia, G., Iseki, K., et al. (2013). Chronic kidney disease: global dimension and perspectives. *Lancet* 382 (9888): 260–272. doi: 10.1016/S0140-6736(13)60687-X [PubMed: 23727169].

Khalil, M.I., Sulaiman, S.A., and Boukraa, L. (2012) Antioxidant properties of honey and its role in preventing health disorder. *The Open Nutraceuticals Journal* 3: 6–16.15.

Lameire, N.H., Bagga, A., Cruz, D., et al. (2013). Acute kidney injury: an increasing global concern. *Lancet (London, England)* 382 (9887): 170–179.

McDonald, H.I., Thomas, S.L., and Nitsch, D. (2014). Chronic kidney disease as a risk factor for acute community-acquired infections in high-income countries: a systematic review. *BMJ Open* 4 (4): Article ID e004100.

Miller, R.P., Tadagavadi, R.K., Ramesh, G., and Reeves, W.B. (2010). Mechanisms of Cisplatin nephrotoxicity. *Toxins (Basel)* 2: 2490–2518. [PubMed: 22069563].

Mohd Effendy, N., Mohamed, N., Muhammad, N., et al. (2012). The effects of Tualang honey on bone metabolism of postmenopausal women. In: *Evidence-Based Complementary and Alternative Medicine* 2012:938574.

Molan, P. (1998). Brief review of honey as a clinical dressing. *Australian Journal of Wound Management* 6: 148–158.

Molan, P. (1999). Why honey is effective as a medicine. 1. Its use in modern medicine. *Bee World* 80: 79–92.

Molan, P. (2001). Why honey is effective as a medicine. 2. The scientific explanation of its effects. *Bee World* 82: 22–40.

Morgan, M.J. and Liu, Z. (2011). Crosstalk of reactive oxygen species and NF-[kappa]B signaling. *Cell Research* 21: 103–115. [PubMed: 21187859].

Plantinga, L.C., Boulware, L.E., Coresh, J., et al. (2008). Patient awareness of chronic kidney disease: trends and predictors. *Archives of Internal Medicine* 168 (20): 2268–2275. doi: 10.1001/archinte.168.20.2268 [PubMed: 19001205].

Ramesh, G. and Reeves, W.B. (2002). TNF-alpha mediates chemokine and cytokine expression and renal injury in cisplatin nephrotoxicity. *Journal of Clinical Investigation* 110: 835–842. [PubMed: 12235115].

Ramesh, G. and Reeves, W.B. (2004). Salicylate reduces cisplatin nephrotoxicity by inhibition of tumor necrosis factor-alpha. *Kidney International* 65: 490–499. [PubMed: 14717919].

Rewa, O. and Bagshaw, S.M. (2014). Acute kidney injury-epidemiology, outcomes and economics. *Nature Reviews Nephrology* 10 (4): 193–207.

Rule, A.D., Amer, H., Cornell, L.D., et al. (2010). The association between age and nephrosclerosis on renal biopsy among healthy adults. *Annals of Internal Medicine* 152: 561–567.

Silver, L.L. (2011). Challenges of antibacterial discovery. *Clinical Microbiology Reviews* 24 (1): 71–109.

Wahba, I.M. and Mak, R.H. (2007). Obesity and obesity-initiated metabolic syndrome: mechanistic links to chronic kidney disease. *Clinical Journal of the American Society of Nephrology* 2: 550–562.

White, J.W. (1975). Composition of honey. In: *Honey: A Comprehensive Survey* (ed. E. Crane), 157-206. London: Bee Research Association and Chalfont.

Yuzbasioglu, M.F., Kurutas, E.B., Bulbuloglu, E., et al. (2009). Administration of honey to prevent peritoneal adhesions in a rat peritonitis model. *International Journal of Surgery* 7 (1): 54–57.

# 17

# Use of Honey in Liver Disease

*Mahendran Sekar*[1,2,*], *Pei Teng Lum*[2], *Srinivasa Reddy Bonam*[3], *and Siew Hua Gan*[1]

[1] *School of Pharmacy, Monash University Malaysia, Bandar Sunway, Subang Jaya 47500, Selangor, Malaysia*
[2] *Department of Pharmaceutical Chemistry, Faculty of Pharmacy and Health Sciences, Royal College of Medicine Perak, Universiti Kuala Lumpur, Ipoh 30450, Perak, Malaysia*
[3] *Institut National de la Santé et de la Recherche Médicale; Centre de Recherche des Cordeliers, Equipe-Immunopathologie et Immunointervention Thérapeutique, Sorbonne Université, Université De Paris, Paris, France*
* *Corresponding author*

## Introduction

The liver is an organ mainly involved in body metabolic functions and a series of other actions, such as storage of glycogen, decomposition of blood, synthesis of plasma protein, endocrine function, detoxification, and others. Accordingly, despite significant advances in disease treatment, more than 10% of the world's population still has liver disease, either acute or chronic liver disease followed by cirrhosis and even liver cancer (Muriel 2017). The liver, which serves as a detoxifier, is impaired by reactive oxygen species (ROS), which depletes the body's antioxidant stocks. The occurrence of ROS in all chronic liver disorders contributes to oxidative stress (OS). Consumption of foods rich in antioxidants provides the body with necessary antioxidants to break down ROS, reducing inflammatory responses and avoiding liver injury (Abdelrahman et al. 2018). Chronic liver disorders are also caused by elevated OS, irrespective of the cause of liver dysfunction (Cichoż-Lach and Michalak 2014).

Honey has a long history of use in traditional medicine. It mainly contains water and carbohydrates (sugars), a naturally sweet and thick substance made by *Apis mellifera* bees from plant secretions or floral nectar. It also contains trace amounts of amino acids, organic acids, vitamins, minerals, enzymes, pollen, fragrance, and flavor compounds (Ball 2007). Honey components have been reported to have antioxidant, antimicrobial, anti-inflammatory, hepatoprotective, antiproliferative, anticancer, and antimetastatic properties (Samarghandian et al. 2017). There are several honey types that are primarily categorized based on the type of flower (unifloral, bifloral, or polyfloral) from which the nectar was obtained. The Fabaceae, Lamiaceae, and Rhamnaceae families, which make up the majority of flowering plants, are well-known as good sources of nectar for honeybees. Mad honey, however, differs from standard natural or commercially available honey because it is composed of rhododendron nectar in which grayanotoxins are contaminants, leading to poisoning (Ullah et al. 2018). Although fatalities are uncommon, intake of "mad honey" can cause arrhythmias, which can be life threatening and complicated. Several symptoms can develop after ingestion of mad honey, from mild gastrointestinal discomfort to life-threatening bradyarrhythmias and hypotension.

Honey is well known for its antioxidant activity, which can theoretically allow it to quench the formation of resonance-stabilized phenoxyl radicals from free radicals (Shahidi and Ambigaipalan 2015). It has recently been used as a dietary supplement and treatment for a number of hepatic diseases, such as hepatitis, cirrhosis, acute liver failure, non-alcoholic liver fatty disease, alcoholic liver disease, fibrosis, steatosis, hepatoma, and inflammation of the hepatic portal. The key emphasis of this chapter is to highlight the use of honey in liver diseases caused by various hepatotoxicants (Figure 17.1).

## Effect of Honey on Hepatotoxicant-Induced Liver Toxicity

### Effects of Honey on $CCl_4$-Induced Liver Toxicity

Carbon tetrachloride ($CCl_4$) is a toxic substance commonly used to trigger liver damage in animals (Xie et al. 2014; Zangar et al. 2000). After exposure to the body, $CCl_4$ is metabolized by cytochrome P450 2E1 to create free radical metabolites (trichloromethyl and trichloromethyl peroxyl radicals) that cause toxicity. Finally, $CCl_4$ causes OS and severely damages the liver (Al-Yahya et al. 2013; Botsoglou et al. 2008; Moneim and El-Deib 2012). Several studies have therefore shown that

*Honey: Composition and Health Benefits*, First Edition. Edited by Md. Ibrahim Khalil, Gan Siew Hua, and Bey Hing Goh.

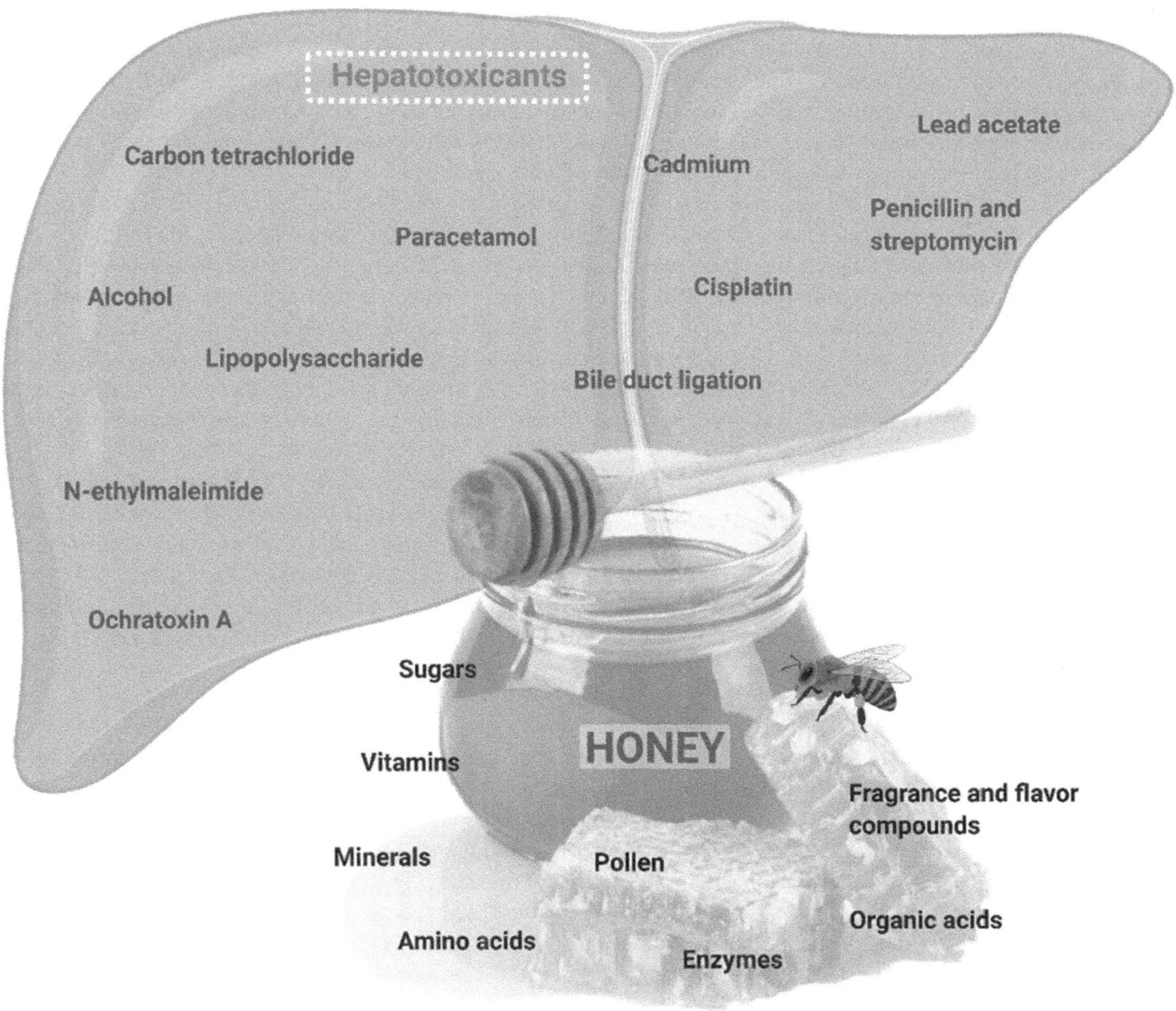


**Figure 17.1** The effect of honey on hepatotoxicants.

natural antioxidants can protect organs against toxic substances, including $CCl_4$ (Hermenean et al. 2017; Said et al. 2018). Indeed, $CCl_4$ increased the profibrotic cytokines, tumor necrosis factor-α (TNF-α) and transforming growth factor-β (TGF-β), production (Sadek et al. 2016). Nuclear factor erythroid 2–related factor 2 (Nrf2) activation could play a role in preventing and treating liver damage and cirrhosis (Osburn et al. 2008). In addition, acute liver injury caused by $CCl_4$ has been shown to be followed by inflammatory process activation (Hermenean et al. 2017).

Carob honey is made by honeybees from the nectar of *Ceratonia silique* flowers. It is a prominent tree that also known as the carob tree in Morocco (El-Haskoury et al. 2018b). Carob honey possesses a strong antioxidant potential and has shown to have a protective role by preserving the antioxidant protection mechanism against $CCl_4$-induced liver toxicity (El-Haskoury et al. 2018a). Carob honey administration (2 g/kg orally [PO] for 12 days) ameliorates the $CCl_4$ (1 mL/kg intraperitoneally [IP]) induced elevation of liver enzymes (El-Haskoury et al. 2018a), including alkaline phosphatase (ALP), alanine transaminase (ALT), aspartate transaminase (AST), lactate dehydrogenase, and oxidative markers, including catalase, advanced oxidation protein products, lipid peroxidation (LPO), and protein carbonyl (PCO).

In addition, El Denshary et al. (2012) stated that honey administration (5 g/kg PO for 4 weeks) had a protective effect against $CCl_4$-induced (100 mg/kg PO) liver toxicity and improved the levels of ALT, AST, and ALP. Honey therapy, on the other hand, has shown improvement in the oxidative biomarkers malondialdehyde, nitric oxide (NO), and total antioxidant ability (TAC). Another research (Al-Waili et al. 2006) found that feeding honey (50% concentration) significantly alters the enhanced ALT, AST, and hematological improvements acquired after $CCl_4$ (2.4 mL/kg intramuscularly). Treatment with honey (50 mg/kg PO for 4 weeks) in $CCl_4$-induced rats (1 mL/kg of 10% v/v $CCl_4$ in olive oil) showed improvement in antioxidant status and histopathological findings and decreases in fibrotic areas (Khadr et al. 2007). Al-Waili (2003) examined the protection of honey administration (intravenous or intrapulmonary) and its effects on liver damage caused by $CCl_4$ in healthy sheep aged 6–8 months. The research indicated that slow intravenous (IV) infusion or rapid IV injection was safe to inject honey at various concentrations and could enhance liver function.

Moreover, with two apitherapy diet formulations containing honey, pollen, propolis, and Apilarnil with or without royal jelly has been studied for hepatoprotection. Biochemical results indicate that the two formulations of the apitherapy diet have a beneficial impact on enhancing the lipids and proteins profiles of the enzymes. The histopathological findings showed the advantages of the two formulations of the apitherapy diet to minimize liver toxicity in laboratory animals (Andritoiu et al. 2014). Very recently, Abdelrahman et al. (2018) recorded that the pretreatment with honey extract (200 mg/kg PO for 5 weeks three times a week) developed antioxidant activity with the potential to reverse $CCl_4$-induced liver fibrosis (0.4 mL of 40% $CCl_4$). It further significantly reduced TGF-β, TNF-α, epidermal growth factor, ALT, AST, and γ-glutamyl transferase (γ-GT). The study indicated that repeated intake of honey may minimize liver damage and prevent the development of liver disease.

A recent study conducted on laboratory animals demonstrated that diethylnitrosamine (50 mg/kg IP) and $CCl_4$ (2 mL/kg IP) caused OS and elevated levels of both α-fetoprotein (AFP) and α-L-fucosidase (AFU) in the liver (Mohamed et al. 2019). Honey therapy (2 g/day) in combination with chemotherapeutic drugs showed an increase in AFP, AFU, and liver function enzymes. In addition, the highest percentages of improvement in catalase, superoxide dismutase (SOD), glutathione peroxidase (GPx), glutathione-s-transferase (GST), and OS biomarkers were recorded by this combination administered to carcinogenic rats; malondialdehyde correlated well with its antioxidant content (Mohamed et al. 2019).

### Effects of Honey on Alcohol-Induced Liver Toxicity

Globally, alcoholic liver disease is considered the most common root cause that contributes to the preventable mortality and morbidity of liver diseases (Addolorato et al. 2016; Mathurin and Bataller 2015). In the liver, ethanol is oxidized by alcohol dehydrogenases (ADHs) into acetaldehyde followed by the conversion of acetaldehyde into acetic acid by acetaldehyde dehydrogenases (ALDH) (Liu 2014; Mello et al. 2008). This oxidation process produces ROS and P4502E1 (CYP2E1) hepatic cytochrome, consequently leading to an increased activity of the microsomal ethanol oxidation system (MEOS) as well as worsening liver toxicity (Cederbaum 2012; Lu and Cederbaum 2016; Neuman et al. 2015). Liver damage could also be caused by the combination of OS and elevated levels of CYP2E1. Therefore, avoidance of OS is considered a key hallmark mechanism to protect the liver from alcoholic injury (Yang et al. 2014). In fact, acute alcoholic liver injury is typically caused by acute alcohol consumption, which is the common way of consuming alcohol. People increasingly suffer from acute liver alcoholic damage (Stahre et al. 2014). Another type is chronic consumption of ethanol, which leading to smooth endoplasmic reticulum proliferation of the liver. The findings indicated that the metabolism was mediated by cytochrome P450 2E1 (CYP2E1) along with the increased activity of MEOS (Lieber 1997, 1999). The produced ROS, such as superoxide (O2•), hydrogen peroxide ($H_2O_2$), and hydroxyl radical (OH•), react with the lipid components of membrane, induce lipid peroxidation subsequent by increased permeability of membrane and even hepatic apoptosis (Adachi and Ishii 2002; Albano 2008; Bailey and Cunningham 2002; García-Ruiz et al. 2013). Hence, the main mechanisms pathologically involved in the mediation of chronic alcoholic liver injury are OS and free radical damage.

Pretreatment with *Apis cerana* honey (5, 10, and 20 g/kg intragastrically twice daily for 12 weeks) inhibited acute alcohol-induced (50% ethanol 5 g/kg intragastrically) liver toxicity in experimental animals. Treatment with honey increases serum ALT and AST, decreases hepatic malondialdehyde production, and promotes glutathione peroxidase (Px) and SOD function. More significantly, the amount of TGF-β in the serum and liver was also strikingly inhibited. These findings indicate that the pretreatment of *A. cerana* honey avoids acute liver harm caused by alcohol owing to honey's antioxidant properties and ability to resist OS (Zhao et al. 2017). Another study reported that treatment with jujube honey (*Shanxi linxian*) (13.5 and 27 g/kg twice daily for 12 weeks) remarkably inhibited serum LPO, reduced the effect of alcoholism (2–4.8 g/kg for 12 weeks) on AST and ALT. It also inhibited 8-hydroxy-2-deoxyguanosine (8-OHdG) production, increased hepatic glutathione–Px function, and reduced malondialdehyde levels. This study shows that jujube honey has strong antioxidant potential and remarkable protection against chronic alcoholism-related liver disorders (Cheng et al. 2014).

Edrees and EL-Said (2007) reported that experimental rats administered ethanol (20%, exceeds 2.5% weekly PO for 8 weeks) showed increasing serum glucose, total lipids, cholesterol, low-density lipoprotein, triglycerides, urea, liver glucose-6-phosphatase (G6P), pancreas and liver malondialdehyde, and PCO. At the same time, there was a decrease in serum insulin, high-density lipoprotein, total protein, Na, K, Ca, Mg, Cu, liver glycogen, pancreas and liver glucose-6-phosphate dehydrogenase (G6PD), GST, glutathione, catalase, and SOD. Administration of honey (2.5 g/kg PO for 8 weeks) prevented to a great extent the lesions caused by chronic alcohol administration (Edrees and EL-Said 2007).

### Effects of Honey on Paracetamol-Induced Liver Toxicity

High or toxic doses of paracetamol, a mild analgesic and antipyretic drug, can cause life-threatening liver dysfunction (Pingili et al. 2020). For paracetamol, the mechanisms of cell death are the formation of the presumed reactive metabolite N-acetyl-p-benzoquinone imine (NAPQI), developed primarily in mice and humans by the cytochrome P450 enzymes such as CYP2E1 and 1A2 (Raucy et al. 1989; Zaher et al. 1998). Paracetamol-induced liver injury includes mitochondrial disruption, OS, c-Jun N-terminal kinase (JNK) activation, and nuclear DNA fragmentation in mouse models and in humans. Oncotic necrosis is the mode of cell death in these models (Jaeschke et al. 2014).

Vitex honey (*Vitex negundo* Linn) is one of the most precious unifloral honeys on the Chinese market because it is highly valued for its flavor and taste. Several researchers have been paying attention to the therapeutic properties of Vitex honey in recent years. The high antioxidant potential of Vitex honey, conceived in association with its high phenolic content, was stated by Cao et al. (2005). The pretreatment of vitex honey (5 and 20 mg/kg PI twice daily for 10 weeks) to paracetamol-induced (0.4 g/kg PO) mice showed protection against liver damage in mice (Wang et al. 2015). Similarly, pretreatment with Indian honey (1 g/kg PO for 10 days) a showed protective effect against paracetamol-induced (3 g/kg PO) OS and liver damage in experimental rats (Mahesh et al. 2009). Another study also reflected the same results that the pretreatment of honey (5, 10, and 20 mg/kg PO for 7 days) showed a protective effect against paracetamol-induced (2 g/kg PO) hepatotoxicity in experimental rats (Galal et al. 2012). All of these experiments with honey alter the changes of AST, ALT, ALP, SOD, malondialdehyde, catalase, GPx, and glutathione levels caused by paracetamol toward a normal level (Galal et al. 2012; Mahesh et al. 2009). A synergistic liver protective effect was also observed when *Picrorhiza kurroa* was used along with honey in the paracetamol-induced Balb/c mice (Gupta et al. 2016).

### Effects of Honey on Other Hepatotoxicants-Induced Liver Toxicities

Treatment of stingless bee honey (4.6 and 9.3 g/kg PO for 30 days) on lipopolysaccharide (LPS)-induced (1 mg/kg IP three times a week for 4 weeks) chronic subclinical systemic inflammation was significantly reduced inflammatory markers, malondialdehyde, and 8-OHdG and enhanced the antioxidants. In addition, stingless bee honey treatment reduced nuclear factor κB (NF-κB), p65, and p38 mitogen-activated protein kinase (MAPK) and increased Nrf2 expression in the livers of LPS-induced rats (Ranneh et al. 2019). The protective effect of Anzer honey (0.275 g/kg PO for 30 days) against N-ethylmaleimide (NEM)–induced (0.075 mg/kg subcutaneously for 30 days) liver injury in rats was reported by Korkmaz and Kolankaya (2009). Treatment with Anzer honey reversed all the changes caused by NEM in the liver enzymes. Hepatoprotection was observed when ochratoxin A (1 mg/kg once weekly for 4 weeks) was given with bee honey (1 mg/kg PO for 4 weeks) in rats (El-Khayat et al. 2009). In the experimental obstructive jaundice model, Kılıcoglu et al. (2008), studied the effect of honey on OS and apoptosis. All changes in liver enzymes and antioxidant parameters caused by common bile duct ligation were reversed by oral supplementation of honey (10 g/kg PO for 7 days) in experimental rats.

Administration of honey with royal jelly (500 mg/kg and 100 mg/kg, respectively, every day for 15 days) enhanced cisplatin-induced liver toxicity (7 mg/kg IP for 15 days) in rats (Bhalchandra and Alqadhi 2018). This combination treatment reversed the liver function biomarkers and antioxidant parameters caused by cisplatin. The possible effect of natural honey (0.05 mL/rat PO for 4 weeks) against cadmium-induced hepatotoxicity (0.5 mg/kg IP for 4 weeks) in rats was shown by Abdel-Moneim and Ghafeer (2007). Cadmium-induced alterations in the liver enzymes and antioxidant parameters were reversed in honey-treated rats. Pretreatment with Sumbawa forest honey (25, 50, and 75 mg/kg PO for 28 days) showed important liver safety by enhancing liver enzymes caused by lead acetate-induced hepatotoxicity (10 mg/kg PO for 14 days) in rats (Widiarko and Permata 2020). Pretreatment with Nabk honey (600 mg/kg PO for 30 days) ameliorated the level of hepatic enzymes caused by penicillin and streptomycin (50000 IU/kg and 50 mg/kg, respectively, PO for 30 days) in guinea pigs (SAAl-Awar et al. 2013).

### Effects of Honey on Non-alcoholic Fatty Liver Disease

The most common chronic liver condition worldwide with no curative treatment is non-alcoholic fatty liver disease (NAFLD). Treatment with trihoney (0.6 g/kg PO for 12 weeks) in hypercholesterolemic rabbits retained normal liver function and exhibited hepatoprotection by hypocholesterolemic and antioxidant functions against NAFLD progression to non-alcoholic steatohepatitis (NASH). Trihoney improved the parameters of the liver enzymes, including ALT, AST, ALP, and γ-GT and the parameters of OS, including SOD, GPx, and malondialdehyde (Alfarisi et al. 2020). This study witnessed the establishment of trihoney as an adjuvant therapy for hypercholesterolemic NAFLD patients with atorvastatin.

## Inhibitory Effect of Honey on Liver Cancer Cells

The most common type of liver cancer is hepatocellular carcinoma (HCC). HCC is associated to a number of factors, including infection with the hepatitis B or C virus, diabetes, obesity, genetic risk factors, and social risk factors such as excessive alcohol intake (Venook et al. 2010). HepG2 cells have been shown to express various phase I enzymes at a basal level that may be increased by induction reagents such 3-methylcholanthrene for CYP1A2 or phenobarbital for CYP2B6 and CYP3A4, and they are also used for metabolism and toxicologic research (Steinbrecht et al. 2019). Treatment of HepG2 cells with bee honey (5%–20%) led to a significant decrease in both the number of viable HepG2 cells and NO levels but also to an improvement in overall antioxidant status and caspase-3 activity, especially in HepG2 cells treated with higher doses of bee honey (20%) and for longer durations (72 hours). Bee honey is effective in reducing HepG2 cell viability, enhancing its antioxidant status, and inducing apoptotic death (Hassan et al. 2012). Different laboratory studies have examined the anticancer effects of honey on liver cancer cells. Honey treatment with HepG2 cells decreased the amount of NO levels in the cells and significantly reduced the number of HepG2 cells. This enhanced the whole antioxidant profile of the cells. ROS stimulated the survival of HepG2 cells, and adequate levels of ROS activated the proliferation and differentiation of cells. Reduced ROS and improved antioxidant efficacy inhibit the proliferation of cancer cells and reduced the amount of HepG2 cells (Erejuwa et al. 2014). The effects of honey on HepG2 cell lines were also explored by Abdel Aziz et al. (2009). The study showed that honey had cytotoxic, antimetastatic, and anti-angiogenic effects on HepG2 cells (Abdel Aziz et al. 2009).

## Clinical Studies on the Effect of Honey in Liver Disease

The effects of honey ingestion on Egyptian children with hepatitis A have been studied by Abdulrhman et al. (2016). Overall hepatitis A recovery is characterized as the absence of all the symptoms and signs of hepatitis A and the return to normal levels of serum bilirubin and liver enzymes (ALT, AST, and γ-GT). Fifty patients were treated with honey (2 mL/kg PO twice weekly for 4 weeks) and reported a 92% recovery rate at the end of the fourth week of the trial. The recommended doses of honey relieved the symptoms and accelerated the hepatitis A recovery. The study suggested that honey can also be prescribed to children during the course of hepatitis A treatment as a dietary supplement. Baltuškevičius and Čeksterytė (1998) examined patients with hepatitis A treated with monofloral clover honey and water solution, which showed a significant improvement in their hepatic function (liver size, ALT, and serum bilirubin level). Another study conducted by Ismaeil et al. (1996) reported that patients with hepatitis A with a healthy diet inclusion of honey had increased liver function. The effect of honey on alcohol dependence in patients with chronic hepatitis was examined by Čeksterytė et al. (2012). After treatment with Livosan (83.33 mg of milk thistle extract) and bee bread mixed with honey (15 g/day, 1:2), there was a significant reduction in CRP, AST, and ALT levels, suggesting a hepatoprotective effect and improved liver function (Čeksterytė et al. 2012).

## Discussion

Because the liver plays a key role in the transformation and clearance of chemicals, the organ may be injured by some medicinal agents when taken in excess. Hepatotoxicity may also be caused by other chemical agents or industrial agents. Such chemicals also cause liver subclinical damage, increasing liver enzymes but not causing pathological complications such as changes in the status of histologic and hepatosteatosis. A significant clinical problem is drug-induced acute liver failure (Ostapowicz et al. 2002), and it is therefore crucial to establish therapeutic and preventive strategies against acute liver toxicity. Through OS, inflammation, fibrogenesis, and liver necrosis, many hepatotoxicants (Malhi et al. 2010; Yang et al. 2014) could induce liver injury. OS prevention must therefore be the main mechanism for protecting the liver from damage caused by hepatotoxicants. Furthermore, hepatotoxicity or liver injury has been identified by the Food and Drug Administration working group for drug-induced hepatotoxicity, in which serum ALT level increases threefold and serum ALP level increases twofold (Lee 2003; Navarro and Senior 2006).

Honey has been used in the conventional treatment of people diverse diseases since ancient times. The use of honey for these indications has re-emerged in recent years, primarily because of clinical observations of hepatoprotective activity, gastroprotective effects, and improved wound healing (French et al. 2005; Molan 2001). Several preclinical studies have been conducted and documented to have possible liver defense against hepatotoxidants. Overall, AST, ALT, ALP, ROS, malondialdehyde, γ-GT, OS, TNF-α, and TGF-β were greatly reduced with pretreatment with honey. At the same time, all the studies reported that dramatic gains in SOD, GST, GPx, catalase, LPO, NO, glutathione, GST, and GPx were observed. It is hypothesized that honey's antioxidant properties are the basis behind the benefits provided by honey by eliminating free radicals from the cellular environment and thereby providing protection against ROS-mediated damage to membrane lipids and macromolecules. The possible mechanisms involved in the hepatoprotective effect of honey are summarized and shown in Table 17.1 and Figure 17.2. Further studies are, however, necessary investigate the specific mechanism of honey in liver protection.

**Table 17.1** Possible mechanism involved in the hepatoprotective effect of honey.

| Hepatotoxicant | Metabolites | Hepatoprotective Mechanism of Honey | References |
|---|---|---|---|
| Carbon tetrachloride ($CCl_4$) | Trichloromethyl radical and trichloromethyl peroxyl radical | ↓ AST, ALT, and ALP<br>↓ ROS, malondialdehyde, and OS<br>↑ SOD, GST, GPx, catalase, LPO, PCO, AOPP, NO, and TAC<br>↑ AFP and AFU<br>↓ TNF-α, TGF-β, and γ-GT | Abdelrahman et al. 2018; Al-Waili 2003; Al-Waili et al. 2006; Andritoiu et al. 2014; El Denshary et al. 2012; El-Haskoury et al. 2018a); Mohamed et al. 2019 |
| Alcohol | Acetaldehyde and acetic acid or acetate | ↓ ROS, malondialdehyde, and OS<br>↑ CYP2E1<br>↓ AST, ALT, and ALP<br>↓ CRP<br>↓ 8-OHdG<br>↓ LDL, G6P, malondialdehyde, and PCO<br>↑ G6PD, GST, glutathione, catalase, and SOD | Čeksterytė et al. 2012; Cheng et al. 2014; Edrees and EL-Said 2007; Zhao et al. 2017 |
| Paracetamol | N-acetyl-p-benzoquinone imine | ↓ AST, ALT, and ALP<br>↓ Malondialdehyde<br>↑ SOD, GST, and GPx | Galal et al. 2012; Gupta et al. 2016; Mahesh et al. 2009; Wang et al. 2015 |
| Lipopolysaccharide (LPS) | NA | ↓ Malondialdehyde, 8-OHdG, and OS<br>↓ NF-κB, p65, and p38 MAPK<br>↑ Nrf2 expression | Ranneh et al. 2019 |
| N-ethylmaleimide (NEM) | NEM-glutathione conjugate and N-ethylmaleamic acid | ↓ ROS<br>↑ Glutathione | Korkmaz and Kolankaya 2009; Mojica et al. 2008 |
| Ochratoxin A | Ochratoxin A (OTA)–glutathione conjugate, OTA–quinone, OTA–hydroquinone and Ochratoxin B | ↓ ROS, malondialdehyde, and oxidized glutathione (GSSG)<br>↑Glutathione<br>↑ SOD and GPx<br>↓ γ-GT<br>↓ Serum total cholesterol | El-Khayat et al. 2009; Kőszegi and Poór 2016 |
| Bile duct ligation | NA | ↓ Malondialdehyde<br>↑Glutathione | Kılıcoglu et al. 2008 |
| Cisplatin | Platinum (II) triamine complexcis-diamminechloro-[2-(diethylamino) ethyl 4-amino-benzoate and N4]-chlorideplatinum(II) monohydrochloride monohydrate | ↓ AST, ALT, and ALP | Bhalchandra and Alqadhi 2018; Zicca et al. 2002 |

(Continued)

**Table 17.1** (Continued)

| Hepatotoxicant | Metabolites | Hepatoprotective Mechanism of Honey | References |
|---|---|---|---|
| Cadmium | Propanoic acid, aminomalonic acid, glutamine, acetyl glutamine, and gluconic acid | ↑ GPx and glutathione<br>↓ AST and ALT | Abdel-Moneim and Ghafeer 2007; Xu et al. 2020 |
| Lead acetate | Reactive oxygen metabolites | ↓ AST and ALT | Widiarko and Permata 2020 |
| Penicillin and streptomycin | Penicillin; penillic acid, penilloic acid, and penicilloic acid; there is no metabolite product for streptomycin | ↓ AST, ALT, and ALP | Canzani and Aldeek 2017; SAAl-Awar et al. 2013 |

8-OHdG, 8-hydroxy-2-deoxyguanosine; AFP, α-fetoprotein; AFU, α-L-fucosidase; ALP, alkaline phosphatase; ALT, alanine transaminase; AOPP, advanced oxidation protein product; AST, aspartate transaminase; CRP, C-reactive protein; G6P, glucose-6-phosphatase; G6PD, glucose-6-phosphate dehydrogenase; GPx, glutathione peroxidase; GST, glutathione-s-transferase; LDL, low-density lipoprotein; LPO, lipid peroxidation; LPS, lipopolysaccharide; MAPK; -activated protein kinase; NA; not applicable; NEM, N-ethylmaleimide; NF-κB, nuclear factor κB; NO, nitric oxide; Nrf2, nuclear factor erythroid 2–related factor 2; OS, oxidative stress; PCO, protein carbonyl; ROS, reactive oxygen species; SOD, superoxide dismutase; TAC, total antioxidant ability; TGF-β, transforming growth factor β; TNF-α, tumor necrosis factor-α; γ-GT, γ-glutamyl transferase.

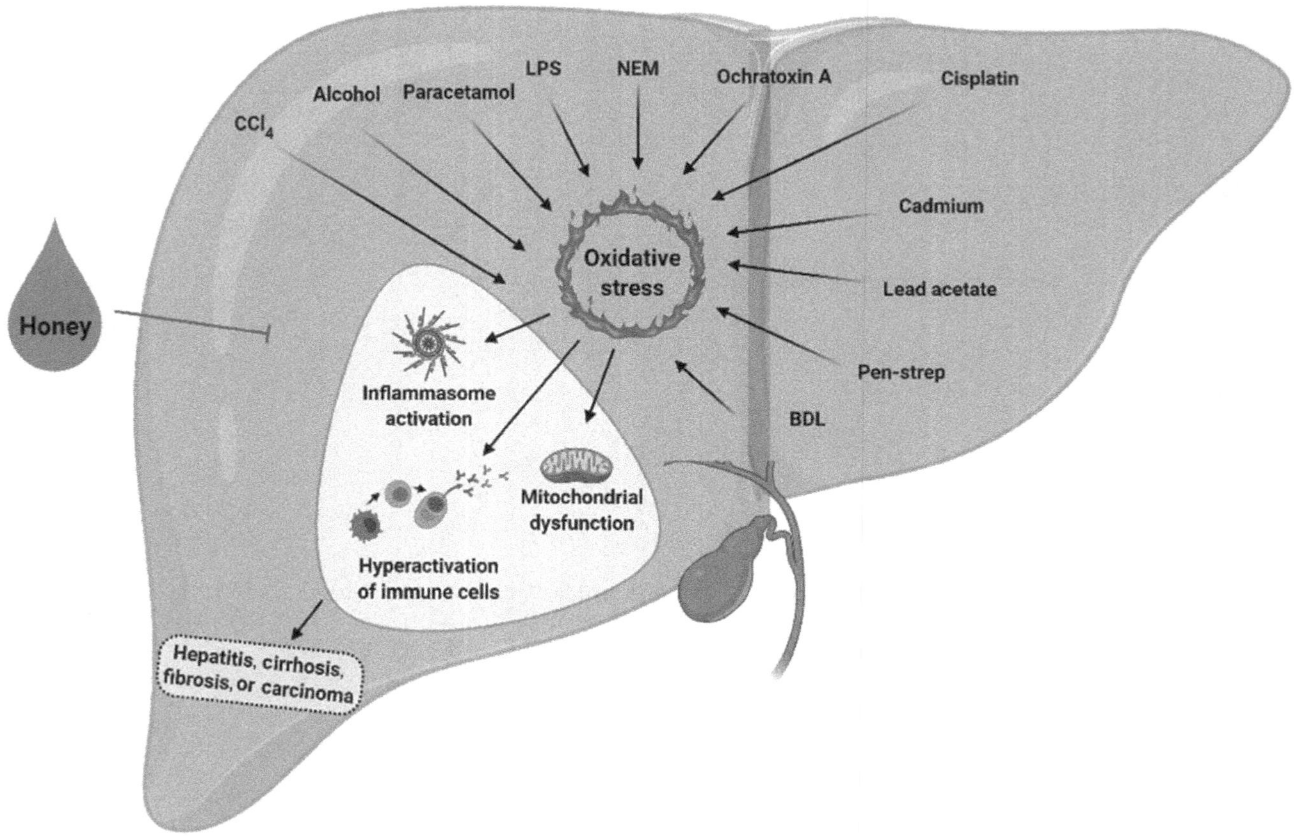


**Figure 17.2** Possible mechanism involved in the hepatoprotective effect of honey. Drugs, chemicals, and experimental conditions that induce liver diseases are depicted. The majority of the signaling is mediated via induction of oxidative stress followed by activation of inflammasome, mitochondrial dysfunction, hyper activation of immune cells, and others, leading to the hepatitis, cirrhosis, fibrosis, and carcinoma. The details are provided in the main text. BDL, bile duct ligation, $CCl_4$, carbon tetrachloride; LPS, lipopolysaccharide; NEM, N-ethylmaleimide; Pen-strep, penicillin–streptomycin. Figure created in biorender.com under a paid subscription.

## Conclusion and Future Perspectives

Diverse preclinical studies have confirmed the effectiveness of different varieties of honey in the treatment of patients with liver diseases. This chapter has presented an overview of the possible mechanisms of how honey decreases the hepatotoxicity of various hepatotoxicants because of its antioxidant, free radical scavenging, anti-inflammatory, and CYP2E1 inhibitory activities. In conclusion, honey may possess medicinal benefits in various types of liver disorders, primarily those associated to oxidative disorders. Honey's effects on NASH, liver disease, hepatotoxicity, cirrhosis, liver fibrosis, and liver injury are attributable to this capacity. Honey is also effective against liver cancer cells. The daily intake of honey or as dietary supplementation along with limited treatment options may provide ideal prevention and therapy for patients with liver disease. To establish the exact bioefficacy, bioavailability, and signaling pathways of honey for cell transduction in the management of oxidative-associated liver disorders, further preclinical and clinical studies are encouraged. This will help to develop novel honey-containing dietary products or pharmaceutical products for liver and other diseases.

## Acknowledgment

The authors thank the Universiti Kuala Lumpur Royal College of Medicine Perak, Malaysia, for providing the necessary facilities and resources to complete this study for publication. The figures were created with the support of https://biorender.com under a paid subscription.

## Conflict of Interest

We wish to confirm that there are no known conflicts of interest associated with this publication, and there has been no significant financial support for this work that could have influenced its outcome.

## References

Abdel Aziz, A., Rady, H., Amer, M., and Kiwan, H. (2009). Effect of some honey bee extracts on the proliferation, proteolytic and gelatinolytic activities of the hepatocellular carcinoma Hepg2 cell line. *Australian Journal of Basic and Applied Sciences* 3 (3): 2754–2769.

Abdel-Moneim, W.M. and Ghafeer, H.H. (2007). The potential protective effect of natural honey against cadmium-induced hepatotoxicity and nephrotoxicity. *Mansoura Journal of Forensic Medicine and Clinical Toxicology* 15 (2): 75–98.

Abdelrahman, M.T., Maina, E.N., and Elshemy, H.A. (2018). Clove (Syzygium aromaticum) and honey extracts significantly reduce inflammatory cytokines and liver function enzymes in experimental rats fed on carbon tetrachloride (CCl4). *Journal of Radiation Research and Applied Sciences* 11 (4): 416–422.

Abdulrhman, M., Shatla, R., and Mohamed, S. (2016). The effects of honey supplementation on Egyptian children with hepatitis A: a randomized double blinded placebo-controlled pilot study. *Journal of Apitherapy* 1: 23.

Adachi, M. and Ishii, H. (2002). Role of mitochondria in alcoholic liver injury. *Free Radical Biology and Medicine* 32 (6): 487–491.

Addolorato, G., Mirijello, A., Barrio, P., and Gual, A. (2016). Treatment of alcohol use disorders in patients with alcoholic liver disease. *Journal of Hepatology* 65 (3): 618–630.

Albano, E. (2008). Oxidative mechanisms in the pathogenesis of alcoholic liver disease. *Molecular Aspects of Medicine* 29 (1–2): 9–16.

Alfarisi, H.A.H., Ibrahim, M.B., Mohamed, Z.B.H., et al. (2020). Hepatoprotective effects of a novel trihoney against nonalcoholic fatty liver disease: a comparative study with Atorvastatin. *The Scientific World Journal* 2020: 1–14.

Al-Waili, N.S. (2003). Intravenous and intrapulmonary administration of honey solution to healthy sheep: effects on blood sugar, renal and liver function tests, bone marrow function, lipid profile, and carbon tetrachloride-induced liver injury. *Journal of Medicinal Food* 6 (3): 231–247.

Al-Waili, N.S., Saloom, K.Y., Al-Waili, T.N., et al. (2006). Influence of various diet regimens on deterioration of hepatic function and hematological parameters following carbon tetrachloride: a potential protective role of natural honey. *Natural Product Research* 20 (13): 1258–1264.

Al-Yahya, M., Mothana, R., Al-Said, M., et al. (2013). Attenuation of CCl4-induced oxidative stress and hepatonephrotoxicity by Saudi Sidr honey in rats. *Evidence-Based Complementary and Alternative Medicine* 2013: 569037.

Andritoiu, C.V., Ochiuz, L., Andritoiu, V., and Popa, M. (2014). Effect of apitherapy formulations against carbon tetrachloride-induced toxicity in Wistar rats after three weeks of treatment. *Molecules* 19 (9): 13374–13391.

Bailey, S.M. and Cunningham, C.C. (2002). Contribution of mitochondria to oxidative stress associated with alcoholic liver disease. *Free Radical Biology and Medicine* 32 (1): 11–16.

Ball, D.W. (2007). The chemical composition of honey. *Journal of Chemical Education* 84 (10): 1643–1646.

Baltuškevičius, A. and Čeksterytė, V. (1998). Influence of monofloral honey on human gastric and hepatic functions. *Acta Zoologica Lituanica* 8 (3): 89–91.

Bhalchandra, W. and Alqadhi, Y.A. (2018). Administration of honey and royal jelly ameliorate cisplatin induced changes in liver and kidney function in rat. *Biomedical and Pharmacology Journal* 11 (4): 2191–2199.

Botsoglou, N., Taitzoglou, I., Botsoglou, E., et al. (2008). Effect of long-term dietary administration of Oregano on the alleviation of carbon tetrachloride-induced oxidative stress in rats. *Journal of Agricultural and Food Chemistry* 56 (15): 6287–6293.

Canzani, D. and Aldeek, F. (2017). Penicillin G's function, metabolites, allergy, and resistance. *Journal of Nutrition and Human Health* 1: 28–40.

Cao, W., Lu, K., Chen, W.-J., et al. (2005). Study on antioxidation effects of different honeys. *Food Science* 8: 118.

Cederbaum, A.I. (2012). Alcohol metabolism. *Clinics in Liver Disease* 16 (4): 667–685.

Čeksterytė, V., Balžekas, J., Baltuškevičius, A., and Jurgevičius, E. (2012). The use of beebread-honey mixture in the treatment of liver diseases in alcohol-dependent patients. *Chemical Technology* 60 (2): 62–66.

Cheng, N., Du, B., Wang, Y., et al. (2014). Antioxidant properties of jujube honey and its protective effects against chronic alcohol-induced liver damage in mice. *Food & Function* 5 (5): 900–908.

Cichoż-Lach, H. and Michalak, A. (2014). Oxidative stress as a crucial factor in liver diseases. *World Journal of Gastroenterology* 20 (25): 8082.

Edrees, G. and EL-Said, F. (2007). Protective effect of natural honey, urtica diocia and their mixture against oxidative stress caused by chronic ethanol consumption. *The Egyptian Journal of Hospital Medicine* 27 (1): 223–233.

El Denshary, E.S., Al-Gahazali, M.A., Mannaa, F.A., et al. (2012). Dietary honey and ginseng protect against carbon tetrachloride-induced hepatonephrotoxicity in rats. *Experimental and Toxicologic Pathology* 64 (7–8): 753–760.

El-Haskoury, R., Al-Waili, N., Kamoun, Z., et al. (2018a). Antioxidant activity and protective effect of carob honey in CCl4-induced kidney and liver injury. *Archives of Medical Research* 49 (5): 306–313.

El-Haskoury, R., Kriaa, W., Lyoussi, B., and Makni, M. (2018b). Ceratonia siliqua honeys from Morocco: physicochemical properties, mineral contents, and antioxidant activities. *Journal of Food and Drug Analysis* 26 (1): 67–73.

El-Khayat, Z., Ezzat, A., Arbid, M., et al. (2009). Potential effects of bee honey and propolis against the toxicity of ochratoxin A in rats. *Macedonian Journal of Medical Sciences* 2 (4): 311–318.

Erejuwa, O.O., Sulaiman, S.A., and Wahab, M.S.A. (2014). Effects of honey and its mechanisms of action on the development and progression of cancer. *Molecules* 19 (2): 2497–2522.

French, V.M., Cooper, R.A., and Molan, P.C. (2005). The antibacterial activity of honey against coagulase-negative staphylococci. *Journal of Antimicrobial Chemotherapy* 56 (1): 228–231.

Galal, R.M., Zaki, H.F., Seif, E.N.M.M., and Agha, A.M. (2012). Potential protective effect of honey against paracetamol-induced hepatotoxicity. *Archives of Iranian Medicine* 15 (11): 674–680.

García-Ruiz, C., Kaplowitz, N., and Fernandez-Checa, J.C. (2013). Role of mitochondria in alcoholic liver disease. *Current Pathobiology Reports* 1 (3): 159–168.

Gupta, P., Tripathi, A., Agrawal, T., et al. (2016). Synergistic protective effect of picrorhiza with honey in acetaminophen induced hepatic injury. *Indian Journal of Experimental Biology* 54 (8): 530–536.

Hassan, M.I., Mabrouk, G.M., Shehata, H.H., and Aboelhussein, M.M. (2012). Antineoplastic effects of bee honey and Nigella sativa on hepatocellular carcinoma cells. *Integrative Cancer Therapies* 11 (4): 354–363.

Hermenean, A., Mariasiu, T., Navarro-González, I., et al. (2017). Hepatoprotective activity of chrysin is mediated through TNF-α in chemically-induced acute liver damage: an in vivo study and molecular modeling. *Experimental and Therapeutic Medicine* 13 (5): 1671–1680.

Ismaeil, N., Mekawy, A.A., and Ragab, S.H. (1996). Dietary modification and its relation to anorexia in Egyptian school children suffering from uncomplicated hepatitis A virus. *The Medical Journal of Cairo University* 64: 183–196.

Jaeschke, H., Xie, Y., and McGill, M.R. (2014). Acetaminophen-induced liver injury: from animal models to humans. *Journal of Clinical and Translational Hepatology* 2 (3): 153.

Khadr, M.E., Mahdy, K.A., El-Shamy, K.A., et al. (2007). Antioxidant activity and hepatoprotective potential of black seed, honey and silymarin on experimental liver injuries induced by CCl 4 in rats. *Journal of Applied Sciences* 7 (24): 3909–3917.

Kılıcoglu, B., Gencay, C., Kısmet, K., et al. (2008). The ultrastructural research of liver in experimental obstructive jaundice and effect of honey. *The American Journal of Surgery* 195 (2): 249–256.

Korkmaz, A. and Kolankaya, D. (2009). Anzer honey prevents N-ethylmaleimide-induced liver damage in rats. *Experimental and Toxicologic Pathology* 61 (4): 333–337.

Kőszegi, T. and Poór, M. (2016). Ochratoxin A: molecular interactions, mechanisms of toxicity and prevention at the molecular level. *Toxins* 8 (4): 111.

Lee, W.M. (2003). Drug-induced hepatotoxicity. *New England Journal of Medicine* 349 (5): 474–485.

Lieber, C.S. (1997). Cytochrome P-4502E1: its physiological and pathological role. *Physiological Reviews* 77 (2): 517–544.

Lieber, C.S. (1999). Microsomal ethanol-oxidizing system (MEOS): the first 30 years (1968–1998)–a review. *Alcoholism: Clinical and Experimental Research* 23 (6): 991–1007.

Liu, J. (2014). Ethanol and liver: recent insights into the mechanisms of ethanol-induced fatty liver. *World Journal of Gastroenterology* 20 (40): 14672.

Lu, Y. and Cederbaum, A.I. (2016). Alcohol upregulation of CYP2A5: role of reactive oxygen species. *Reactive Oxygen Species (Apex, NC)* 1 (2): 117.

Mahesh, A., Shaheetha, J., Thangadurai, D., and Rao, D.M. (2009). Protective effect of Indian honey on acetaminophen induced oxidative stress and liver toxicity in rat. *Biologia* 64 (6): 1225–1231.

Malhi, H., Guicciardi, M.E., and Gores, G.J. (2010). Hepatocyte death: a clear and present danger. *Physiological Reviews* 90 (3): 1165–1194.

Mathurin, P. and Bataller, R. (2015). Trends in the management and burden of alcoholic liver disease. *Journal of Hepatology* 62 (1): S38–S46.

Mello, T., Ceni, E., Surrenti, C., and Galli, A. (2008). Alcohol induced hepatic fibrosis: role of acetaldehyde. *Molecular Aspects of Medicine* 29 (1–2): 17–21.

Mohamed, N.Z., Aly, H.F., Moneim El-mezayen, H.A., and El-Salamony, H.E. (2019). Effect of co-administration of Bee honey and some chemotherapeutic drugs on dissemination of hepatocellular carcinoma in rats. *Toxicology Reports* 6: 875–888.

Mojica, E.-R.E., Kim, S., and Aga, D.S. (2008). Formation of N-ethylmaleimide (NEM)-glutathione conjugate and N-ethylmaleamic acid revealed by mass spectral characterization of intracellular and extracellular microbial metabolites of NEM. *Applied and Environmental Microbiology* 74 (1): 323–326.

Molan, P.C. (2001). Potential of honey in the treatment of wounds and burns. *American Journal of Clinical Dermatology* 2 (1): 13–19.

Moneim, A.A. and El-Deib, K.M. (2012). The possible protective effects of Physalis peruviana on carbon tetrachloride-induced nephrotoxicity in male albino rats. *Life Science Journal* 9 (3): 38–52.

Muriel, P. (2017). The liver: general aspects and epidemiology. In: *Liver Pathophysiology*, 3–22. St. Louis: Elsevier.

Navarro, V.J. and Senior, J.R. (2006). Drug-related hepatotoxicity. *New England Journal of Medicine* 354 (7): 731–739.

Neuman, M.G., Malnick, S., Maor, Y., et al. (2015). Alcoholic liver disease: clinical and translational research. *Experimental and Molecular Pathology* 99 (3): 596–610.

Osburn, W.O., Yates, M.S., Dolan, P.D., et al. (2008). Genetic or pharmacologic amplification of nrf2 signaling inhibits acute inflammatory liver injury in mice. *Toxicological Sciences* 104 (1): 218–227.

Ostapowicz, G., Fontana, R.J., Schiødt, F.V., et al. (2002). Results of a prospective study of acute liver failure at 17 tertiary care centers in the United States. *Annals of Internal Medicine* 137 (12): 947–954.

Pingili, R.B., Challa, S.R., Pawar, A.K., et al. (2020). A systematic review on hepatoprotective activity of quercetin against various drugs and toxic agents: evidence from preclinical studies. *Phytotherapy Research* 34 (1): 5–32.

Ranneh, Y., Akim, A.M., Ab Hamid, H., et al. (2019). Stingless bee honey protects against lipopolysaccharide induced-chronic subclinical systemic inflammation and oxidative stress by modulating Nrf2, NF-κB and p38 MAPK. *Nutrition & Metabolism* 16 (1): 15.

Raucy, J.L., Lasker, J.M., Lieber, C.S., and Black, M. (1989). Acetaminophen activation by human liver cytochromes P450IIE1 and P450IA2. *Archives of Biochemistry and Biophysics* 271 (1): 270–283.

SAAl-Awar, M., MASalih, E., and AYAl-Eryani, M. (2013). The protective effect of nabk honey against pathological effects of penicillin and streptomycin: histological structure and functions of Guinea pigs liver. *Journal of Applied Pharmaceutical Science* 3 (4): S1.

Sadek, K., Beltagy, D., Saleh, E., and Abouelkhair, R. (2016). Camel milk and bee honey regulate profibrotic cytokine gene transcripts in liver cirrhosis induced by carbon tetrachloride. *Canadian Journal of Physiology and Pharmacology* 94 (11): 1141–1150.

Said, M.M., Azab, S.S., Saeed, N.M., and El-Demerdash, E. (2018). Antifibrotic mechanism of pinocembrin: impact on oxidative stress, inflammation and TGF-β/Smad inhibition in rats. *Annals of Hepatology* 17 (2): 307–317.

Samarghandian, S., Farkhondeh, T., and Samini, F. (2017). Honey and health: a review of recent clinical research. *Pharmacognosy Research* 9 (2): 121–127.

Shahidi, F. and Ambigaipalan, P. (2015). Phenolics and polyphenolics in foods, beverages and spices: antioxidant activity and health effects – a review. *Journal of Functional Foods* 18: 820–897.

Stahre, M., Roeber, J., Kanny, D., et al. (2014). Peer reviewed: contribution of excessive alcohol consumption to deaths and years of potential life lost in the United States. *Preventing Chronic Disease* 11: 1–12.

Steinbrecht, S., Kammerer, S., and Küpper, J.H. (2019). HepG2 cells with recombinant cytochrome P450 enzyme overexpression: their use and limitation as *in vitro* liver model. *Journal of Cellular Biotechnology* 5: 55–64.

Ullah, S., Khan, S.U., Saleh, T.A., and Fahad, S. (2018). Mad honey: uses, intoxicating/poisoning effects, diagnosis, and treatment. *RSC Advances* 8 (33): 18635–18646.

Venook, A.P., Papandreou, C., Furuse, J., and de Guevara, L.L. (2010). The incidence and epidemiology of hepatocellular carcinoma: a global and regional perspective. *Oncologist* 15: 5–13.

Wang, Y., Li, D., Cheng, N., et al. (2015). Antioxidant and hepatoprotective activity of Vitex honey against paracetamol induced liver damage in mice. *Food & Function* 6 (7): 2339–2349.

Widiarko, O.P. and Permata, F.S. (ed.) (2020). Preventive potency of Sumbawa forest honey on rats exposed by lead acetate based on liver histopathology and AST-ALT level. *Journal of Physics: Conference Series*. 1430: 012030. United Kingdom: IOP Publishing https://iopscience.iop.org/article/10.1088/1742-6596/1430/1/012030/meta

Xie, Y., Hao, H., Wang, H., et al. (2014). Reversing effects of lignans on CCl4-induced hepatic CYP450 down regulation by attenuating oxidative stress. *Journal of Ethnopharmacology* 155 (1): 213–221.

Xu, M.-Y., Wang, P., Sun, Y.-J., et al. (2020). Identification of metabolite biomarkers in serum of rats exposed to chlorpyrifos and cadmium. *Scientific Reports* 10 (1): 1–10.

Yang, L., Wu, D., and Cederbaum, A. (2014). CYP2E1, oxidative stress and MAPK signaling pathways in alcohol-induced hepatotoxicity. *Journal of Biochemical and Pharmacological Research* 2: 16.

Zaher, H., Buters, J.T., Ward, J.M., et al. (1998). Protection against acetaminophen toxicity in CYP1A2 and CYP2E1 double-null mice. *Toxicology and Applied Pharmacology* 152 (1): 193–199.

Zangar, R.C., Benson, J.M., Burnett, V.L., and Springer, D.L. (2000). Cytochrome P450 2E1 is the primary enzyme responsible for low-dose carbon tetrachloride metabolism in human liver microsomes. *Chemico-biological Interactions* 125 (3): 233–243.

Zhao, H., Cheng, N., He, L., et al. (2017). Antioxidant and hepatoprotective effects of A. cerana honey against acute alcohol-induced liver damage in mice. *Food Research International* 101: 35–44.

Zicca, A., Cafaggi, S., Mariggiò, M.A., et al. (2002). Reduction of cisplatin hepatotoxicity by procainamide hydrochloride in rats. *European Journal of Pharmacology* 442 (3): 265–272.

# 18

# Use of Honey in Immune Disorders and Human Immunodeficiency Virus

*Wan Nazirah Wan Yusuf, Suk Peng Tang, Noor Suryani Mohd Ashari, and Che Badariah Abd Aziz*

## Overview of the Immune System

*Immunity* refers to the ability of the body to protect against disease, specifically infectious diseases. The cells and molecules involved in immunity constitute the immune system, and the coordinated response of these system following introduction of foreign substances is known as the immune response (Abbas et al. 2017). The immune system responsible for defense against infectious microorganisms such as viruses, bacteria, and fungi is mediated by an early response of innate immunity (nonspecific) and later responses of adaptive immunity (specific).

Innate immunity, which is also known as natural or native immunity, is the first line of defense that protects the body against pathogens in which it blocks the entry of microbes or limits the growth of many microbes that colonize tissues. Its response does not change with repeated exposure to infection (Pal and Chakravarty 2020). The mechanisms of innate immunity are not specific and cannot distinguish fine differences between microbes. The components of innate immunity include the physical barriers (epithelium of the skin, respiratory tract, gastrointestinal [GI] tract), phagocytic cells (neutrophils, macrophages), natural killer (NK) cells, dendritic cells, innate lymphoid cells, and blood proteins such as the complements and cytokines (Marshall et al. 2018). Innate immunity immediately kills or removes pathogens and provides protection to the host without the need for pre-exposure to the pathogen.

On the other hand, adaptive or acquired immunity is stimulated by exposure to infectious agents and increases in magnitude and capabilities with repeated exposure to a particular microorganism. It is more specific, and the system can distinguish between different organisms and respond more vigorously with repeated exposure to the same microorganism because of the presence of memory cells. The major components of adaptive immunity include B lymphocytes and T lymphocytes. The B lymphocytes mediate the humoral immune response by secreting antibodies, which attach to pathogens, marking the pathogens for destruction by other immune cells. The T lymphocytes mediate the cellular responses via the action of helper T lymphocytes (CD4 T cells), which produces cytokines that activates the macrophages and cytotoxic T lymphocytes (CD8 T cell). Activated macrophages kill the pathogens via phagocytosis, and the activated cytotoxic T lymphocytes directly kill the pathogens via cell destructions. Figure 18.1 illustrates the innate and adaptive immunity after invasion of pathogens.

## Immune System Disorders

### Immunodeficiency

The integrity of immune system is essential for the survival of all individuals. It is important for the body's defense toward a wide variety of pathogens, from bacteria and viruses to parasitic worms, as well as cancer cells. Impairment of one or more components of the immune system can lead to immunodeficiency diseases, manifested by increased frequency and severity of infections. These impairments can also lead to autoimmune disorders and malignancies.

Immunodeficiency diseases are broadly classified into two groups, primary immunodeficiencies and secondary immunodeficiencies (Abbas et al. 2017). Primary immunodeficiency diseases are inherited groups of disorders resulting from genetic mutations that impair the immune response. Most cases are diagnosed in infants younger than 1 year of age, although milder

*Honey: Composition and Health Benefits*, First Edition. Edited by Md. Ibrahim Khalil, Gan Siew Hua, and Bey Hing Goh.

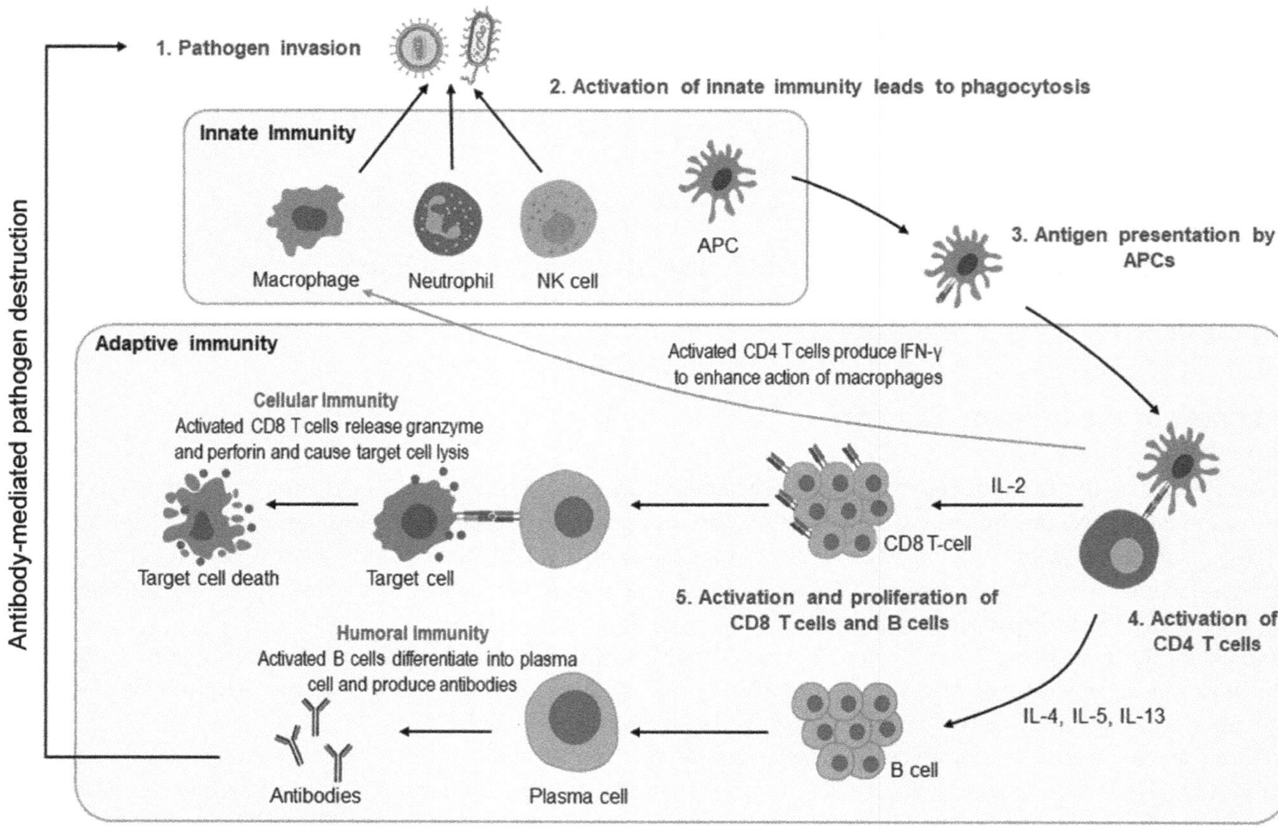


**Figure 18.1** Immune response after invasion of pathogens. APC, antigen-presenting cell; IFN, interferon; IL, interleukin; NK, natural killer.

forms may not be identified until adulthood. It was estimated that six million people live with primary immunodeficiency worldwide (Bousfiha et al. 2013).

Acquired or secondary immunodeficiencies are more common than primary immunodeficiency and can be defined as the occurrence of immunodeficiency in an individual who had an intact immune system before the onset of the condition (Chinen and Shearer 2010; Sánchez-Ramón et al. 2019). It develops as the result of environmental factors such as malnutrition, chronic diseases, disseminated cancer, treatment with immunosuppressive drugs, or infection of the immune system. Acquired immunodeficiency syndrome (AIDS) is one of the most notable secondary immunodeficiency and is caused by infection with human immunodeficiency virus (HIV) (Abbas et al. 2017).

### Hypersensitivity

Hypersensitivity refers to inappropriate or overreaction of the immune system, causing the normally protective immune system become undesirable and occasionally fatal. The Gell and Coombs classification of hypersensitivity is the most widely used and differentiate four types (types I–IV) of immune responses that lead to bystander tissue damage (Table 18.1).

## Human Immunodeficiency Virus Infection

Human immunodeficiency virus attacks the body's immune system. Globally, HIV infection is reported to be among the top 10 leading causes of death in low-income countries (World Health Organization 2020). There were an estimated 38.4 million (33.9–43.8 million) people living with HIV in 2021, including 1.7 million (1.3–2.1 million) children (aged 0–14 years). Meanwhile, approximately 1.5 million (1.1–2.0 million) people were newly infected with HIV in 2021. Approximately 5.9 million people were unaware that they were living with HIV, hence posing risks to others (UNAIDS 2022).

**Table 18.1** Gell and Coombs classification of hypersensitivity.

| Type | Mediator | Onset of time | Mechanisms | Example |
|---|---|---|---|---|
| I | IgE | Immediate or within minutes to hours | Allergen-specific IgE antibody binds to mast cells via Fc receptor. Subsequent exposure leads to cross-linking of IgE and induces degranulation of mast cells. | • Allergic rhinitis<br>• Atopic dermatitis<br>• Atopic asthma<br>• Anaphylaxis |
| II | IgG<br>IgM | 5–8 hr | IgG or IgM antibody binds to cell-bound antigen, leading to<br>• Complement activation and cell lysis<br>• antibody-dependent cell-mediated cytotoxicity | • Autoimmune hemolytic anemia<br>• Graves' disease |
| III | IgG<br>IgM | 2–8 hr | Excess antigen–antibody complexes that cannot be cleared by the mononuclear-phagocyte system trigger activation of the complement system and cell-mediated local reaction. | • Systemic lupus erythematosus<br>• Rheumatoid arthritis |
| IV | T cells | After 24 hr, usually 48–72 hr after exposure | Activated T helper cells secrete cytokines, which activate macrophages and T cytotoxic cells. | • Contact dermatitis<br>• Chronic graft rejection |

Ig, immunoglobulin.

Two major types of HIV known to cause AIDs include HIV-1 and HIV-2. Whereas HIV-1 is prevalent worldwide, HIV-2 is more confined to West Africa (Bbosa et al. 2019; Sharp and Hahn 2011). HIV-2 infection results in a much slower decline in CD4 T cells with lower levels of viral load compared with HIV-1 infection (Vidya Vijayan et al. 2017). HIV is known to be present in the blood, semen, and other body fluids of HIV-infected individuals and can be introduced into another individual via sexual contact, sharing HIV-contaminated needles, or transplacental passage (Chen et al. 2019). Upon entering the bloodstream, HIV infects vital cells in the human immune system. If left untreated, HIV infection can progress to AIDS, one of the common examples of secondary immunodeficiency.

## HIV Infection Mechanism

The main target of HIV is the CD4 T cells. Upon entering the blood circulation, HIV binds to the target cells. Then the viral envelope fuses to the host cell membrane, releasing its nucleic material (HIV virion) into the host cytoplasm. Subsequently, the viral RNA is reverse transcribed to produce viral DNA, which migrates to the nucleus and integrates with the host's DNA. The product of integration, also known as proviral DNA, may remain latent or transcribed to proviral RNA. Translation of proviral RNA produces viral protein, which undergoes cleavage for functional maturation and assembles with the viral RNA. The assembled virus buds out from the host and begin to infect other target cells (Balakrishna and Kondapi 2016). The life cycle of HIV is illustrated in Figure 18.2. HIV infection causes damage of infected cells either directly by the virus itself or by CD8-mediated cytotoxicity and leads to progressive loss of CD4 T cells. When the numbers of CD4 T cells severely decline, cellular-mediated immunity is compromised.

## Natural History of HIV Infection

The natural history of untreated HIV infection generally undergoes three distinct phases: the primary HIV infection, the second phase of chronic HIV infection (asymptomatic), and the third phase of advanced HIV infection (AIDS) (Figure 18.3). Upon transmission, HIV replicates extensively, resulting in high plasma levels of HIV along with transient marked reduction in the circulating CD4 T-cell count and expansion of CD8 T cells. The antibodies to HIV can be detectable at about 2–6 weeks after infection (Melhuish and Lewthwaite 2018). The CD4 T-cell count later increases, but the count is lower than in the pre-infective stage. Then the CD4 T-cell count and viral load usually stabilize for several years before the CD4 T-cell count begins to decline while the viral load increases. As the CD4 T-cell count continues to decline to less than 200 cells/mm$^3$, the patient

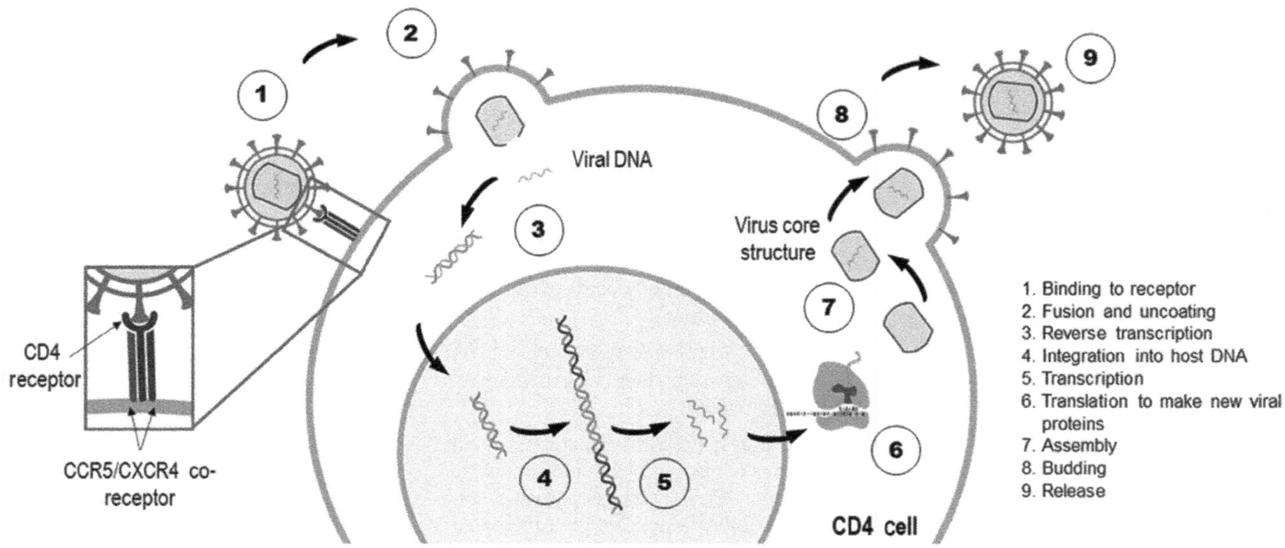


**Figure 18.2** Life cycle of human immunodeficiency virus (HIV).

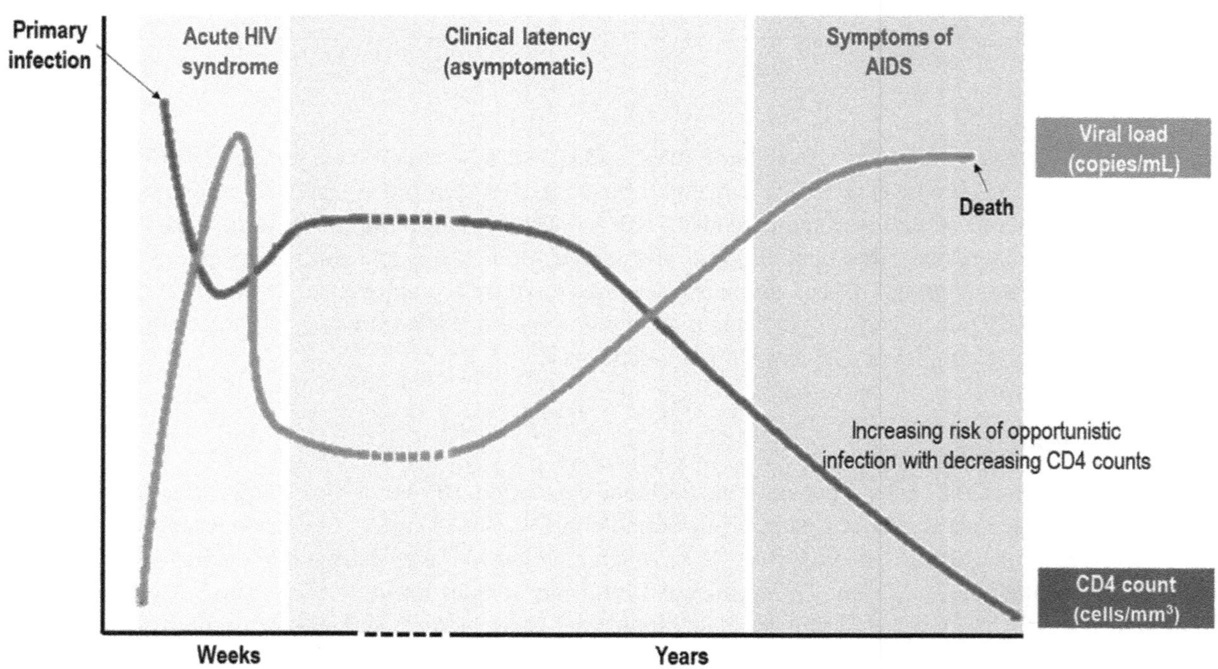


**Figure 18.3** The natural history of untreated human immunodeficiency syndrome (HIV) infection. AIDS, acquired immunodeficiency syndrome.

eventually develops opportunistic infections and AIDS. Some examples of AIDS-defining conditions include cervical cancer, HIV encephalopathy, histoplasmosis (disseminated or extra pulmonary), Kaposi's sarcoma, and Burkitt's lymphoma.

## Challenges in HIV Treatment

To date, there is no cure for HIV. The standard HIV treatment involves the use of highly active antiretroviral therapy (HAART), a drug regimen consisting of three or more antiretroviral drugs that act on different targets in the virus. Since its introduction, HAART has dramatically changed the once fatal diagnosis to a manageable chronic disease with a significant reduction in morbidity and mortality rates along with improvement in a quality of life (Ramana 2014). However, the extensive of HAART use has also led to the emergence of viral resistance and subsequent suboptimal responses. The situation is

**Table 18.2** Antiretroviral drugs.

| Drug Group | Examples | Common Adverse Effects |
|---|---|---|
| Nucleoside reverse transcriptase inhibitors (NRTIs) | Zidovudine, didanosine, stavudine, lamivudine, abacavir, tenofovir, emtricitabine | Anemia, neutropenia, myopathy, sensory neuropathy, pancreatitis, lipoatrophy, hypersensitivity syndrome (fever, abdominal pain, rash), nephrotoxicity, decreased bone mineral density with chronic use |
| Non-nucleoside reverse transcriptase inhibitors (NNRTIs) | Nevirapine, efavirenz, rilpivirine, delavirdine, etravirine | Rash, hepatitis, Stevens-Johnson syndrome, central nervous system toxicity, risk of prolonged QT interval |
| Protease inhibitors | Saquinavir, ritonavir, nelfinavir, lopinavir, atazanavir, fosamprenavir, tipranavir | Dyslipidemia, diarrhea, nausea, vomiting, hyperbilirubinemia, nephrolithiasis, cholelithiasis, skin rash, hepatotoxicity |
| Entry inhibitors | Enfuvirtide, maraviroc | Injection site reaction, subcutaneous nodules, concentration-dependent postural hypotension |
| Integrase inhibitors | Raltegravir, elvitegravir, dolutegravir | Diarrhea, nausea, fatigue, headache, muscle ache, increase creatinine kinase |

compounded when most patients did not adhere to the antiretroviral drugs because of various adverse effects and complex regimens (Iacob et al. 2017; Oliveira et al. 2018). Table 18.2 shows examples of antiretroviral drugs and their common adverse effects. Other factors that hinder successful treatment include social stigma, discrimination, the patient's willingness to receive treatment, and challenges associated with long-term follow-up (Iacob et al. 2017; Mahajan et al. 2008).

Most HIV treatment guidelines highlight the importance of ensuring good compliance to prevent the development of resistant strains of HIV. Because of limited resources, not all HIV-infected patients have access to free antiretroviral therapy depending on the criteria set by each country's policies. If left untreated, patients eventually develop AIDS and can transmit HIV to others. Various research studies to slow the progression of HIV based on natural products or complementary medicine have been conducted as an alternative to current treatment strategy (Lorenc and Robinson 2013; Pan et al. 2019). Natural products are considered more easily accessible, less expensive, and associated with fewer adverse effects, which also improve patient's compliance and quality of life.

## Honey in HIV

Honey is a natural food rich in vitamins, minerals, trace elements, proteins, enzymes, amino acids, and carbohydrates (Ahmed et al. 2018a; Ranneh et al. 2021). For instance, honey contains water-soluble vitamins, such as ascorbic acid, nicotinic acid, riboflavin, and thiamine, that have been found to be highly correlated with free radical–scavenging activity (Chua et al. 2013). In addition, honey contains phenolic and flavonoid compounds (Ahmed et al. 2018a; Miguel et al. 2017; Ranneh et al. 2021), which are important antioxidants that can alleviate chronic inflammation and stimulate immune cells (Tonks et al. 2007). Honey also exerts anti-inflammatory properties (Ranneh et al. 2021; Vallianou et al. 2014), which can help quench the inflammatory effects of residual immune activation and inflammation. Besides being rich in nutrients and antioxidants, honey provides energy whereby 1 tablespoon of honey can supply 64 calories of energy (Khan et al. 2007). Considering increased resting energy expenditure as common abnormalities in patients with HIV, honey supplementation may be useful by providing additional nutrient and calories to these patients to match their increased energy demand (Chang et al. 2007; World Health Organization 2003).

### Viral Load and CD4 T-Cell Count

A study by Heidari et al. (2012) reported that 30 days of honey supplementation (80 g/day) causes a 1% increase in CD4 and CD8 T-cell count in an untreated HIV patient. Another study by Onifade et al. (2013) reported that six patients with HIV consuming *Nigella sativa* (black cumin) and honey concoction for 4 months showed reductions of viral load to undetectable levels. A similar seronegative conversion result was seen in one woman with HIV who started the concoction during her third trimester of pregnancy for 1 year (Onifade et al. 2015). In both studies, CD4 counts were increased after consumption of the concoction. However, it was not ascertained whether the effect was due to *N. sativa* or honey.

So far, only one study reported on the direct effects of honey supplementation on a group of untreated patients with HIV (Wan Yusuf et al. 2019). In this study, the authors reported that Tualang honey supplementation at 20 and 60 g daily for 6 months resulted in insignificant viral load reduction. In comparison, the control group that did not receive honey showed a significant increase (130%) after 6 months. As for CD4 T-cell counts, a significant reduction was seen in the control and honey low-dose group (20 g/day), whereas patients with HIV supplemented with 40 and 60 g/day Tualang honey for 6 months showed an insignificant reduction in the CD4 T-cell levels (Wan Yusuf et al. 2019). The natural history of untreated HIV infection resulted in an increase of the viral load and reduction in the CD4 T-cell counts over time. The relatively maintained viral load and CD4 in the honey-treated group suggested that honey retards HIV proliferation and slows down CD4 T-cell death, thus potentially delaying HIV disease progression.

### Hematologic and Immunologic Parameters

Hematologic changes (bone marrow defects and immune cytopenia) and immunologic changes (mainly CD4 T-cell depletion, cytokine dysregulation, immune dysfunction) associated with HIV may progressively lead to profound immunodeficiency. The cause of the changes may be infection of the bone marrow by HIV or other opportunistic infections, malignancies such as lymphoma, and adverse effects of drugs used. For instance, increased incidences of anemia, leukocytopenia, thrombocytopenia, and pancytopenia were commonly reported among adults and children with HIV who had low CD4 T-cell counts and were not on antiretroviral treatment (Bhardwaj et al. 2020; Bhowmik and Banerjee 2015; Kathuria et al. 2016; Thulasi et al. 2016).

The effects of honey on hematologic and immunologic parameters have been reported in various animal and human models. For instance, both Tualang honey and Manuka honey supplementation in female rats showed significant increases in hemoglobin, red blood cells, total white blood cells, lymphocytes, and platelets counts after 4 months compared with the control group that did not receive any honey (Ahmed et al. 2018b). The study also reported elevated serum level of apoptotic protease-activating factor-1, interferon-γ, and tumor necrosis factor-α (TNF-α) and reduced estradiol, suggesting their potential role in modulating immune responses. Beneficial effects in hematologic and immunologic profiles were also reported in healthy human participants after 2 weeks of honey consumption (Al-Waili 2003). There were also studies that suggest that honey could boost the immune system by stimulating production of inflammatory cytokines (TNF-α, interleukin [IL]-1β, and IL-6) and enhance action of macrophages (Nikaein et al. 2014; Tonks et al. 2003). Altogether, these studies suggest the potential use of honey to improve hematologic and immunologic outcomes in patients with HIV.

A report has demonstrated that consumption of multifloral honey for 21 days was associated with a slight reduction in white blood cell count with an increased percentage of lymphocytes and reduced percentage of neutrophils. There were also increased platelet level and nitric oxide production, reduced prostaglandin level, and improved liver and renal function (Al-Waili et al. 2006). Another recent study reported that the hematologic and immunologic functions of patients with HIV supplemented with 20 g of Tualang honey twice and thrice daily for 6 months were relatively maintained. In contrast, the control group (no honey given) and those receiving a single daily dose showed significant deterioration after a 6-month period (Tang et al. 2021).

### Physiological and Biochemical Profiles

Tualang honey supplementation for 6 months did not significantly affect body weight, body mass index, blood pressure, lipid profiles, or renal and liver function of untreated patients with HIV; this honey consumption is safe in patients with HIV. However, there was a significant increase in the glucose level in patients with HIV supplemented with 60 g/day of honey. Although the level was still within the normal range, precaution is needed for long-term consumption of high-dose Tualang honey in patients with HIV (Tang et al. 2020). It is noteworthy in this study that the total cholesterol levels were relatively maintained in honey-treated groups compared with a 32% increase in control group after a 6-month period. Patients with HIV/AIDS frequently presented with altered lipid metabolism caused by HIV infection (Funderburg and Mehta 2016). For instance, hypertriglyceridemia, hypercholesterolemia, and low high-density lipoprotein cholesterol are common among untreated patients with HIV (Salami et al. 2009). Therefore, honey supplementation may delay the progression of dyslipidemia, which is one of the major risk factors for cardiovascular diseases, a common non–AIDS-related cause of death among patients with HIV.

## Quality of Life

Patients with HIV who are asymptomatic may seem to be in good health. However, they may experience social stigma, difficulty in relationships, or depression, which may negatively affect their quality of life. In a study using the World Health Organization's Quality of Life – HIV BREF questionnaire to assess quality of life in patients with HIV, Tualang honey supplementation showed to significantly improve the physical and psychological scores in all patients with HIV when compared with control participants (not taking honey) at 3 and 6 months after honey treatment (Wan Yusuf et al. 2019). Honey, which is rich in carbohydrates, may provide extra energy to patients and result in improved physical well-being. In addition, honey contains amino acids such as tryptophan and phenylalanine, which are the precursors for neurotransmitters such as serotonin and dopamine, respectively (Cools et al. 2011; Kıvrak 2017; Soto et al. 2011). These neurotransmitters plays important role in regulating mood, social behavior, appetite, digestion, sleep, memory, and sexual desire and function (Jenkins et al. 2016; Young 2013), hence leading to improved psychological function of patients with HIV.

## Other HIV-Related Symptoms and Diseases

Diarrhea is a common GI symptom among patients with HIV. It can occur because of bacterial, viral, fungal or parasitic infections. Other non-infectious causes include adverse effects from antiretroviral treatment and idiopathic HIV enteropathy (Gómez Venegas et al. 2018). Chronic diarrhea is common in patients with advanced HIV, which can lead to electrolyte imbalance and nutrients deficiencies, thus increasing the risk of morbidity and poor quality of life.

Although there was no report on the use of honey to treat diarrhea in patients with HIV, the beneficial effects of honey in improving various GI symptoms were previously reported. For instance, honey was reported to reduce the incidence of diarrhea among critically ill patients on enteral nutrition feeding (Vahdat Shariatpanahi et al. 2018). Honey also decreased the duration of diarrhea and improved stool consistency in those with acute infantile diarrhea (Elnady et al. 2011). Honey may improve GI function via antimicrobial properties against pathogenic microorganisms (Badawy et al. 2004; Omafuvbe and Akanbi 2009; Tan et al. 2009) as well as its prebiotic potential in improving intestinal microbial balance (Jiang et al. 2020; Mohan et al. 2017; Zhao et al. 2019).

Another common disease in the advanced stage of HIV is cancer. Kaposi's sarcoma, B-cell lymphomas, and invasive cervical cancer are among the common AIDS-defining cancers. Other non–AIDS-defining cancers (presence of cancer but not necessarily indicating AIDS) include lung cancer, breast cancer, liver cancer, renal cancer, Hodgkin's lymphoma, and oropharyngeal cancers (Chiu et al. 2017; Facciolà et al. 2018; Yarchoan and Uldrick 2018). The risk of malignancies decreases with improved CD4 T-cell count in patients on antiretroviral treatment (Puronen et al. 2019).

Although no study has been conducted in cancers among patients with HIV, there were many studies that addressed the anticancer effects of different types of honey using in vitro and in vivo cancer models. As an example, honey was found to have dose-dependent cytotoxicity against breast cancer cells and alleviate breast carcinogenesis by modulating the hematologic, hormonal, and apoptotic activities of the cells (Ahmed and Othman 2017; Portokalakis et al. 2016). Honey was reported to induce apoptosis and demonstrate antiproliferative, anti-inflammatory, and antimetastatic properties, hence modulating host immune function (Afrin et al. 2020; Waheed et al. 2019).

## The Plausible Protective Mechanisms by Honey in HIV

As mentioned earlier, HIV infection causes a progressive failure of the host's immune system, which leads to an increased risk of opportunistic infections and cancers. One possible mechanism is that HIV induces oxidative stress, which disrupts the cellular redox status and alters various cellular responses. Such responses implicated in HIV infection include enhanced viral replication and transcription, CD4 T-cell apoptosis, and a persistent inflammatory state, which have been linked to the development of a wide array of HIV-associated diseases (Conway and Tossonian 2014; Gil et al. 2003; Ivanov et al. 2016). A nested case-control study conducted in Spain also suggested that oxidative stress is a predictor for all causes of death in patients with HIV, including both AIDS and non-AIDS conditions (Masiá et al. 2016). Therefore, supplementation with exogenous antioxidants appears to be a potential alternative in the management of HIV (Mandal et al. 2020).

Honey is regarded as a good source of antioxidants, which contributes to its valuable medicinal properties. It is a natural food rich in macronutrients and micronutrients. The primary constituent of honey are carbohydrates, including fructose, glucose, and a mixture of other oligosaccharides. It also contains organic acids, amino acids, minerals, vitamins, polyphenol aromatic compounds, and various bioactive compounds. The compositions vary depending on their geographical and botanical origins (Santos-Buelga and González-Paramás 2017). Nonetheless, some of these bioactive compounds were shown to inhibit HIV activities via different mechanisms of action.

Vitamins B and C are examples of water-soluble vitamins found in honey. Vitamin C at noncytotoxic concentrations can suppress HIV production by interfering with the viral transcription in HIV-infected T cells (Harakeh et al. 1990, 1994; Hayash 2016). Meanwhile, vitamin $B_3$ intake was associated with decreased serum C-reactive protein in patients with HIV, suggesting its potential anti-inflammatory roles (Poudel-Tandukar and Chandyo 2016). Increased vitamin $B_3$ intake was also associated with increased plasma tryptophan and CD4 T-cell levels (Abrams et al. 1993; Murray et al. 2001). It was postulated that vitamin $B_3$ reduced tryptophan oxidation and prevented depletion of circulating tryptophan, thereby reduced systemic inflammation and improved CD4 T-cell recovery (Lebouché et al. 2014; Zhang et al. 2017).

Phenolic compounds are another important composition of honey that exhibit various biological activities such as antioxidant, anti-inflammatory, antibacterial, antifungal, antiviral, and anticancer properties. Examples of phenolic acids found in honey include caffeic, chlorogenic, coumaric, ellagic, ferulic, gallic, syringic, and vanillic acids. Some of the flavonoids that have been identified in honey include catechin, kaempferol, luteolin, naringenin, pinobanksin, and quercetin (Chew et al. 2018; Chua et al. 2013; Jasicka-Misiak et al. 2017; Khalil et al. 2011). Many of these bioactive compounds were reported to elicit anti-HIV activities by targeting different steps in the viral life cycle.

For instance, polyphenol extract and gallic acid were shown to prevent HIV entry through interaction with viral envelope protein gp120, thus inhibiting the binding of HIV to target cells (Zhang et al. 2017). Flavonoids such as quercetin may also interfere with the binding of HIV to target cells via interaction with viral envelope proteins or downregulation of co-receptors, thus preventing viral entry (Li et al. 2000; Nair et al. 2002, 2009). On the other hand, gallic acid and ellagic acid demonstrated antiviral effect against HIV by inhibiting the activity of reverse transcriptase and protease, respectively (Nutan et al. 2013; Zhang et al. 2017). Other phenolic compounds that demonstrated inhibitory action against reverse transcriptase include catechin, apigenin, caffeic acid, and naringenin (Chinsembu 2019).

Caffeic acid phenethyl ester and quercetin O-glycoside are examples of flavonoids that were shown to inhibit HIV-1 integrase activity, thereby preventing integration of viral DNA into the host DNA (Fesen et al. 1994; Kim et al. 1998). Meanwhile, flavonoids such as quercetin, kaempferol, and luteolin were also reported to inhibit HIV-1 protease, which in turn prevents proteolytic maturation necessary for HIV infectivity (Mahmood et al. 1996; Xu et al. 2000). Ko et al. (2009) suggested that the flavonols (kaempferol, myricetin, quercetin) or flavones (apigenin, luteolin) were more likely to inhibit reverse transcriptase and proteases simultaneously than the other flavonoid subgroups.

Methylglyoxal is another important antibacterial component found in honey that demonstrated potent anti–HIV-1 activity. An in vitro study by Behbahani (2014) suggested that methylglyoxal acts at the later stage of HIV infection by blocking the new virion assembly and maturation of HIV. Figure 18.4 illustrated the potential anti-HIV mechanisms of honey in various stages of HIV replicative cycle.

In addition to antiviral mechanisms, honey may confer beneficial effects via modulation of immune functions. Honey was previously showed to regulate pro-inflammatory cytokines such as IL-6, IL-8, and TNF-α (Hussein et al. 2012; Kalantari et al. 2016). Furthermore, flavonoids in honey may maintain the immune cell function; apigenin was shown to suppress antiapoptotic pathways in chronically activated human T cells. Phenolic compounds such as catechin and caffeic acids also demonstrated protective effects against oxidative stress–induced lymphocytes damage and increased lymphocyte proliferation in HIV-positive patients (Greenrod and Fenech 2003; Winkler et al. 2004). Nevertheless, most of these studies were conducted in vitro, and little is known about their actual contribution in the management of patients with HIV.

## Honey in Other Immune Disorders

Previous studies demonstrated that high-dose Tualang honey ingestion improved the overall and individual symptoms of allergic rhinitis such as sneezing, nasal pruritus, nasal blockage, and rhinorrhea (Asha'ari et al. 2013). Allergic rhinitis is a common disease affecting 20% to 30% of adults and up to 40% of children (Meltzer 2016). It is an example of type I hypersensitivity reactions caused by infiltration of allergens, characterized by an immunoglobulin (IgE)-mediated inflammatory disorder of the nasal mucosa. The common clinical manifestations of allergic rhinitis include sneezing, nasal obstruction, nasal pruritus, and rhinorrhea, and the symptoms are similar to those of an upper respiratory infection (Siti Sarah et al.

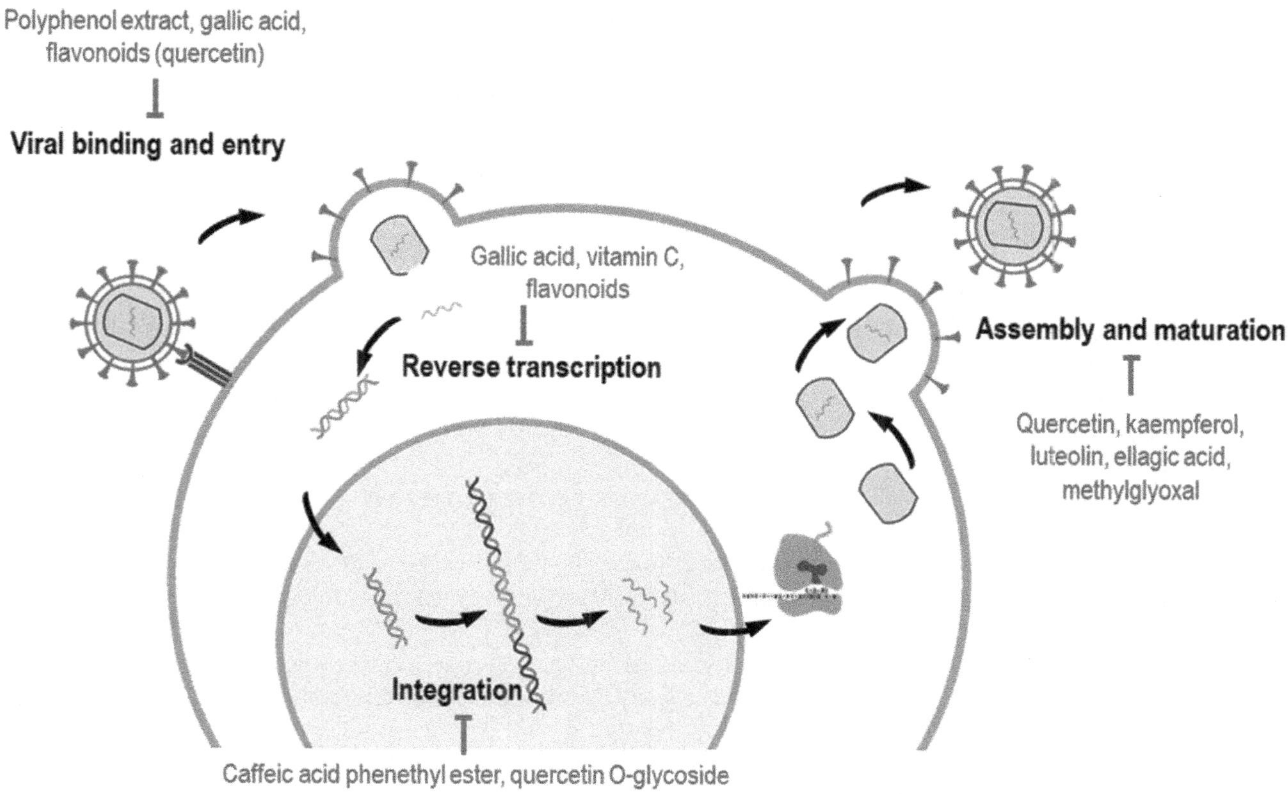


**Figure 18.4** The postulated anti–human immunodeficiency virus mechanisms of honey at different stages of the viral life cycle.

2020; Small et al. 2018). Indeed, honey is considered as a cheap, popular, and safe demulcent to be used in the treatment of young children (World Health Organization 2001). This is supported by a study in which honey was the preferred choice among parents to improve nocturnal cough and sleep difficulty among children compared with dextromethorphan or no treatment (Paul et al. 2007). Honey may improve the symptoms of allergic rhinitis by suppressing the ovalbumin-specific IgE antibody responses, the mediators of type 1 hypersensitivity reactions (Duddukuri et al. 1997).

## Honey Hypersensitivity

Honey is considered safe in general population, but hypersensitivity to honey is possible. Although rare, cases of anaphylactic reaction after honey consumption have been previously reported. Some patients developed mild symptoms such as urticaria, throat itchiness, and cough (Aguiar et al. 2017; Vezir et al. 2014). The reasons for the hypersensitivity reactions may be the pollen from floral sources or the presence of proteins residuals produced by honeybees (Bauer et al. 1996). As honey is being recognized as a superfood, allergic incidence may increase in conjunction with increased honey consumption in the populations. Allergen specific-IgE test and skin-prick test may be helpful to avoid hypersensitivity to honey in high-risk groups.

## Summary

There were relatively few reports on the effects of honey in immune-related disorder. So far, honey was shown to be beneficial in HIV and allergic rhinitis. However, most of the research was carried out in vitro, and relatively few studies were conducted in real patients. Indeed, it is difficult to conduct large clinical trial in vulnerable groups. Research ethics involving vulnerable group are generally challenging because they require close and careful attention in the study design, making participant recruitments difficult. In addition, long-term follow-up increases risk of dropouts and requires large research funding. Honey standardization is another challenging issue because its composition is affected by its geographical and botanical origins. However, if the efficacy of honey can be proven in vivo, it will provide an affordable and accessible alternative or as an adjunct for management of chronic immune disorders.

## References


Abbas, A. K., Lichtman, A. H., & Pillai, S. (2017). *Cellular and Molecular Immunology E-Book*. Elsevier Health Sciences.

Abrams, B., Duncan, D., & Hertz-Picciotto, I. (1993). A prospective study of dietary intake and acquired immune deficiency syndrome in HIV-seropositive homosexual men. *Journal of Acquired Immune Deficiency Syndromes*, 6(8): 949–958.

Afrin, S., Haneefa, S. M., Fernandez-Cabezudo, M. J., et al. (2020). Therapeutic and preventive properties of honey and its bioactive compounds in cancer: an evidence-based review. *Nutrition Research Reviews*, 33(1): 50–76. https://doi.org/DOI:10.1017/S0954422419000192

Aguiar, R., Duarte, F. C., Mendes, A., et al. (2017). Anaphylaxis caused by honey: a case report. *Asia Pacific Allergy*, 7(1): 48–50. https://doi.org/10.5415/apallergy.2017.7.1.48

Ahmed, S., & Othman, N. H. (2017). The anti-cancer effects of Tualang honey in modulating breast carcinogenesis: an experimental animal study. *BMC Complementary and Alternative Medicine*, 17(1): 208. https://doi.org/10.1186/s12906-017-1721-4

Ahmed, S., Sulaiman, S. A., Baig, A. A., et al. (2018a). Honey as a potential natural antioxidant medicine: an insight into its molecular mechanisms of action. *Oxidative Medicine and Cellular Longevity*, 2018: 8367846. https://doi.org/10.1155/2018/8367846

Ahmed, S., Sulaiman, S. A., Ibrahim, M., et al. (2018b). Effect of daily supplementation of Malaysian Jungle Tualang Honey and Australian/New Zealand Manuka honey on hematological and some biochemical variables in female rats. *Annals of Life Sciences*, 2(5): 10–22.

Al-Waili, N. S. (2003). Effects of daily consumption of honey solution on hematological indices and blood levels of minerals and enzymes in normal individuals. *Journal of Medicinal Food*, 6(2): 135–140. https://doi.org/10.1089/109662003322233549

Al-Waili, N. S., Al-Waili, T. N., Al-Waili, A. N., et al. (2006). Influence of natural honey on biochemical and hematological variables in AIDS: a case study. *The Scientific World Journal*, 6: 1985–1989. https://doi.org/10.1100/tsw.2006.331

Asha'ari, Z. A., Ahmad, M. Z., Jihan, W. S., et al. (2013). Ingestion of honey improves the symptoms of allergic rhinitis: evidence from a randomized placebo-controlled trial in the East Coast of Peninsular Malaysia. *Annals of Saudi Medicine*, 33(5): 469–475. https://doi.org/10.5144/0256-4947.2013.469

Badawy, O. F. H., Shafii, S. S. A., Tharwat, E. E., et al. (2004). Antibacterial activity of bee honey and its therapeutic usefulness against *Escherichia coli* O157: H7 and *Salmonella typhimurium* infection. *OIE Revue Scientifique et Technique (International Office of Epizootics)*, 23(3): 1011–1022. https://doi.org/10.20506/rst.23.3.1543

Balakrishna, L. S., & Kondapi, A. K. (2016). Role of host proteins in HIV-1 early replication. In: Saxena, S.K. (ed.), *Advances in Molecular Retrovirology*. IntechOpen. https://doi.org/10.5772/62108

Bauer, L., Kohlich, A., Hirschwehr, R., et al. (1996). Food allergy to honey: pollen or bee products? Characterization of allergenic proteins in honey by means of immunoblotting. *Journal of Allergy and Clinical Immunology*, 97(1): 65–73. https://doi.org/10.1016/s0091-6749(96)70284-1

Bbosa, N., Kaleebu, P., & Ssemwanga, D. (2019). HIV subtype diversity worldwide. *Current Opinion in HIV and AIDS*, 14(3): 153–160. https://doi.org/10.1097/COH.0000000000000534

Behbahani, M. (2014). Anti-HIV-1 activity of eight monofloral Iranian honey types. *PLoS One*, 9(10): e108195. https://doi.org/10.1371/journal.pone.0108195

Bhardwaj, S., Almaeen, A., Ahmed Wani, F., et al. (2020). Hematologic derangements in HIV/AIDS patients and their relationship with the CD4 counts: a cross-sectional study. *International Journal of Clinical and Experimental Pathology*, 13(4): 756–763.

Bhowmik, A., & Banerjee, P. (2015). Hematological manifestation in HIV infected children. *Journal of the College of Physicians and Surgeons--Pakistan*, 25(2): 119–123.

Bousfiha, A. A., Jeddane, L., Ailal, F., et al. (2013). Primary immunodeficiency diseases worldwide: more common than generally thought. *Journal of Clinical Immunology*, 33(1): 1–7. https://doi.org/10.1007/s10875-012-9751-7

Chang, E., Sekhar, R., Patel, S., et al. (2007). Dysregulated energy expenditure in HIV-infected patients: a mechanistic review. *Clinical Infectious Diseases*, 44(11): 1509–1517. https://doi.org/10.1086/517501

Chen, Q., Zeng, D., She, Y., et al. (2019). Different transmission routes and the risk of advanced HIV disease: a systematic review and network meta-analysis of observational studies. *EClinicalMedicine*, 16: 121–128. https://doi.org/10.1016/j.eclinm.2019.10.003

Chew, C. Y., Chua, L. S., Soontorngun, N., et al. (2018). Discovering potential bioactive compounds from Tualang honey. *Agriculture and Natural Resources*, 52(4): 361–365. https://doi.org/10.1016/j.anres.2018.10.011

Chinen, J., & Shearer, W. T. (2010). Secondary immunodeficiencies, including HIV infection. *The Journal of Allergy and Clinical Immunology*, 125(2 Suppl 2), S195–S203. https://doi.org/10.1016/j.jaci.2009.08.040

Chinsembu, K. C. (2019). Chemical diversity and activity profiles of HIV-1 reverse transcriptase inhibitors from plants. *Revista Brasileira de Farmacognosia*, 29(4): 504–528. https://doi.org/10.1016/j.bjp.2018.10.006

Chiu, C. G., Smith, D., Salters, K. A., et al. (2017). Overview of cancer incidence and mortality among people living with HIV/AIDS in British Columbia, Canada: implications for HAART use and NADM development. *BMC Cancer*, 17: 270. https://doi.org/10.1186/s12885-017-3229-1

Chua, L. S., Rahaman, N. L. A., Adnan, N. A., et al. (2013). Antioxidant activity of three honey samples in relation with their biochemical components. *Journal of Analytical Methods in Chemistry*, 2013: 313798. https://doi.org/10.1155/2013/313798

Conway, B., & Tossonian, H. (2014). HIV/AIDS – A model of chronic oxidative stress and immune activation. In: Laher, I. (ed.), *Systems Biology of Free Radicals and Antioxidants* (pp. 3217–3238). Springer, Berlin, Heidelberg. https://doi.org/10.1007/978-3-642-30018-9_149

Cools, R., Nakamura, K., & Daw, N. D. (2011). Serotonin and dopamine: unifying affective, activational, and decision functions. *Neuropsychopharmacology*, 36(1): 98–113. https://doi.org/10.1038/npp.2010.121

Duddukuri, G. R., Kumar, P. S., Kumar, V. B., et al. (1997). Immunosuppressive effect of honey on the induction of allergen-specific humoral antibody response in mice. *International Archives of Allergy and Immunology*, 114(4): 385–388. https://doi.org/10.1159/000237699

Elnady, H., Aly, N., El Hussieny, N., et al. (2011). Honey, an adjuvant therapy in acute infantile diarrhea. *Pediatric Research*, 70(5): 95. https://doi.org/10.1038/pr.2011.320

Facciolà, A., Ceccarelli, M., Venanzi Rullo, E., et al. (2018). Prostate cancer in HIV-positive patients: a review of the literature. *World Cancer Research Journal*, 5(3): e1136. https://www.wcrj.net/article/1136

Fesen, M. R., Pommier, Y., Leteurtre, F., et al. (1994). Inhibition of HIV-1 integrase by flavones, caffeic acid phenethyl ester (CAPE) and related compounds. *Biochemical Pharmacology*, 48(3): 595–608. https://doi.org/10.1016/0006-2952(94)90291-7

Funderburg, N. T., & Mehta, N. N. (2016). Lipid abnormalities and inflammation in HIV inflection. *Current HIV/AIDS Reports*, 13(4): 218–225. https://doi.org/10.1007/s11904-016-0321-0

Gil, L., Martínez, G., González, I., et al. (2003). Contribution to characterization of oxidative stress in HIV/AIDS patients. *Pharmacological Research*, 47(3): 217–224. https://doi.org/10.1016/s1043-6618(02)00320-1

Gómez Venegas, Á. A., Moreno Castaño, L. A., & Roa Chaparro, J. A. (2018). Approach to diarrhea in HIV patients. *Revista Colombiana de Gastroenterologia*, 33(2): 150–160. https://doi.org/10.22516/25007440.192

Greenrod, W., & Fenech, M. (2003). The principal phenolic and alcoholic components of wine protect human lymphocytes against hydrogen peroxide- and ionizing radiation-induced DNA damage in vitro. *Mutagenesis*, 18(2): 119–126. https://doi.org/10.1093/mutage/18.2.119

Harakeh, S., Jariwalla, R. J., & Pauling, L. (1990). Suppression of human immunodeficiency virus replication by ascorbate in chronically and acutely infected cells. *Proceedings of the National Academy of Sciences of the United States of America*, 87(18): 7245–7249. https://doi.org/10.1073/pnas.87.18.7245

Harakeh, S., Niedzwiecki, A., & Jariwalla, R. J. (1994). Mechanistic aspects of ascorbate inhibition of human immunodeficiency virus. *Chemico-Biological Interactions*, 91(2–3): 207–215. https://doi.org/10.1016/0009-2797(94)90041-8

Hayash, T. (2016). Preventive effect of ascorbic acid against biological function of human immunodeficiency virus trans-activator of transcription. *Journal of Intercultural Ethnopharmacology*, 5(2): 205–209. https://doi.org/10.5455/jice.20160316010322

Heidari, A., Zia, S. N., Amiri, G. H., et al. (2012). Has the natural raw honey any effect on HIV infection? *International Journal of Pharmaceutical Research and Bio-Science*, 1(5): 205–210.

Hussein, S. Z., Mohd Yusoff, K., Makpol, S., et al. (2012). Gelam honey inhibits the production of proinflammatory, mediators NO, PGE(2), TNF-α, and IL-6 in carrageenan-induced acute paw edema in rats. *Evidence-Based Complementary and Alternative Medicine : eCAM*, 2012: 109636. https://doi.org/10.1155/2012/109636

Iacob, S. A., Iacob, D. G., & Jugulete, G. (2017). Improving the adherence to antiretroviral therapy, a difficult but essential task for a successful HIV treatment-clinical points of view and practical considerations. *Frontiers in Pharmacology*, 8: 831. https://doi.org/10.3389/fphar.2017.00831

Ivanov, A. V., Valuev-Elliston, V. T., Ivanova, O. N., et al. (2016). Oxidative stress during HIV infection: mechanisms and consequences. *Oxidative Medicine and Cellular Longevity*, 2016: 8910396. https://doi.org/10.1155/2016/8910396

Jasicka-Misiak, I., Gruyaert, S., Poliwoda, A., et al. (2017). Chemical profiling of polyfloral Belgian honey: ellagic acid and pinocembrin as antioxidants and chemical markers. *Journal of Chemistry*, 2017: 5393158. https://doi.org/10.1155/2017/5393158

Jenkins, T. A., Nguyen, J. C., Polglaze, K. E., et al. (2016). Influence of tryptophan and serotonin on mood and cognition with a possible role of the gut-brain axis. *Nutrients*, 8(1): 56. https://doi.org/10.3390/nu8010056

Jiang, L., Xie, M., Chen, G., et al. (2020). Phenolics and carbohydrates in Buckwheat honey regulate the human intestinal microbiota. *Evidence-Based Complementary and Alternative Medicine*, 2020: 6432942. https://doi.org/10.1155/2020/6432942

Kalantari, N., Ghasemi, M., Bayani, M., et al. (2016). Effect of honey on mRNA expression of TNF-α, IL-1β and IL-6 following acute toxoplasmosis in mice. *Cytokine*, 88: 85–90. https://doi.org/10.1016/j.cyto.2016.08.029

Kathuria, S., Bagga, P. K., & Malhotra, S. (2016). Hematological manifestations in HIV infected patients and correlation with CD4 counts and anti retroviral therapy. *International Journal of Contemporary Medical Research*, 3(12): 3495–3498.

Khalil, M. I., Alam, N., Moniruzzaman, M., et al. (2011). Phenolic acid composition and antioxidant properties of Malaysian honeys. *Journal of Food Science*, 76(6): C921–C928. https://doi.org/10.1111/j.1750-3841.2011.02282.x

Khan, F. R., Ul Abadin, Z., & Rauf, N. (2007). Honey: nutritional and medicinal value. *International Journal of Clinical Practice*, 61(10): 1705–1707. https://doi.org/10.1111/j.1742-1241.2007.01417.x

Kim, H. J., Woo, E. R., Shin, C. G., et al. (1998). A new flavonol glycoside gallate ester from *Acer okamotoanum* and its inhibitory activity against human immunodeficiency virus-1 (HIV-1) integrase. *Journal of Natural Products*, 61(1): 145–148. https://doi.org/10.1021/np970171q

Kıvrak, Ş. (2017). Analysis of amino acid and phenolic content in honey by UPLC-ESI-MS/MS. In: de Alencar Arnaut de Toledo, V. (ed.), *Honey Analysis*. InTechOpen. https://doi.org/10.5772/67317

Ko, Y. J., Oh, H. J., Ahn, H. M., et al. (2009). Flavonoids as potential inhibitors of retroviral enzymes. *Journal of the Korean Society for Applied Biological Chemistry*, 52(4): 321–326. https://doi.org/10.3839/jksabc.2009.057

Lebouché, B., Jenabian, M. A., Singer, J., et al. (2014). The role of extended-release niacin on immune activation and neurocognition in HIV-infected patients treated with antiretroviral therapy - CTN PT006: study protocol for a randomized controlled trial. *Trials*, 15: 390. https://doi.org/10.1186/1745-6215-15-390

Li, B. Q., Fu, T., Dongyan, Y., et al. (2000). Flavonoid baicalin inhibits HIV-1 infection at the level of viral entry. *Biochemical and Biophysical Research Communications*, 276(2): 534–538. https://doi.org/10.1006/bbrc.2000.3485

Lorenc, A., & Robinson, N. (2013). A review of the use of complementary and alternative medicine and HIV: issues for patient care. *AIDS Patient Care and STDs*, 27(9): 503–510. https://doi.org/10.1089/apc.2013.0175

Mahajan, A. P., Sayles, J. N., Patel, V. A., et al. (2008). Stigma in the HIV/AIDS epidemic: a review of the literature and recommendations for the way forward. *AIDS*, 22(Suppl 2): S67–S79. https://doi.org/10.1097/01.aids.0000327438.13291.62

Mahmood, N., Piacente, S., Pizza, C., et al. (1996). The anti-HIV activity and mechanisms of action of pure compounds isolated from *Rosa damascena*. *Biochemical and Biophysical Research Communications*, 229(1): 73–79. https://doi.org/https://doi.org/10.1006/bbrc.1996.1759

Mandal, A., Biswas, D., & Hazra, B. (2020). Chapter 9 - Natural products from plants with prospective anti-HIV activity and relevant mechanisms of action. In: Atta-ur-Rahman (ed.), *Bioactive Natural Products*, In: *Studies in Natural Products Chemistry* (Vol. 66, pp. 225–271). Elsevier. https://doi.org/10.1016/B978-0-12-817907-9.00009-X

Marshall, J. S., Warrington, R., Watson, W., et al. (2018). An introduction to immunology and immunopathology. *Allergy, Asthma & Clinical Immunology*, 14(Suppl 2): 49. https://doi.org/10.1186/s13223-018-0278-1

Masiá, M., Padilla, S., Fernández, M., et al. (2016). Oxidative stress predicts all-cause mortality in HIV-infected patients. *PLoS ONE*, 11(4): e0153456. https://doi.org/10.1371/journal.pone.0153456

Melhuish, A., & Lewthwaite, P. (2018). Natural history of HIV and AIDS. *Medicine*, 46(6): 356–361. https://doi.org/10.1016/j.mpmed.2018.03.010

Meltzer, E. O. (2016). Allergic rhinitis: burden of illness, quality of life, comorbidities, and control. *Immunology and Allergy Clinics of North America*, 36(2): 235–248. https://doi.org/10.1016/j.iac.2015.12.002

Miguel, M. G., Antunes, M. D., & Faleiro, M. L. (2017). Honey as a complementary medicine. *Integrative Medicine Insights*, 12: 1178633717702869. https://doi.org/10.1177/1178633717702869

Mohan, A., Quek, S.-Y., Gutierrez-Maddox, N., et al. (2017). Effect of honey in improving the gut microbial balance. *Food Quality and Safety*, 1(2): 107–115. https://doi.org/10.1093/fqsafe/fyx015

Murray, M. F., Langan, M., & MacGregor, R. R. (2001). Increased plasma tryptophan in HIV-infected patients treated with pharmacologic doses of nicotinamide. *Nutrition*, 17(7–8): 654–656. https://doi.org/10.1016/s0899-9007(01)00568-8

Nair, M. P., Kandaswami, C., Mahajan, S., et al. (2002). Grape seed extract proanthocyanidins downregulate HIV-1 entry coreceptors, CCR2b, CCR3 and CCR5 gene expression by normal peripheral blood mononuclear cells. *Biological Research*, 35(3–4): 421–431. https://doi.org/10.4067/s0716-97602002000300016

Nair, M. P., Saiyed, Z. M., Gandhi, N. H., et al. (2009). The flavonoid, quercetin, inhibits HIV-1 infection in normal peripheral blood mononuclear cells. *American Journal of Infectious Diseases*, 5(2): 135–141. https://doi.org/10.3844/ajidsp.2009.135.141

Nikaein, D., Khosravi, A. R., Moosavi, Z., et al. (2014). Effect of honey as an immunomodulator against invasive aspergillosis in BALB/c mice. *Journal of Apicultural Research*, 53(1): 84–90. https://doi.org/10.3896/IBRA.1.53.1.08

Nutan, Modi, M., Goel, T., et al. (2013). Ellagic acid & gallic acid from *Lagerstroemia speciosa* L. inhibit HIV-1 infection through inhibition of HIV-1 protease & reverse transcriptase activity. *The Indian Journal of Medical Research*, 137(3): 540–548.

Oliveira, L., Caixeta, L. M., Martins, J., et al. (2018). Adherence to antiretroviral therapy and correlation with adverse effects and coinfections in people living with HIV/AIDS in the municipality of Goiás State. *Revista da Sociedade Brasileira de Medicina Tropical*, 51(4): 436–444. https://doi.org/10.1590/0037-8682-0467-2017

Omafuvbe, B. O., & Akanbi, O. O. (2009). Microbiological and physico-chemical properties of some commercial Nigerian honey. *African Journal of Microbiology Research*, 3(12): 891–896.

Onifade, A. A., Jewell, A. P., Ajadi, T. A., et al. (2013). Effectiveness of a herbal remedy in six HIV patients in Nigeria. *Journal of Herbal Medicine*, 3(3): 99–103. https://doi.org/10.1016/j.hermed.2013.04.006

Onifade, A. A., Jewell, A. P., & Okesina, A. B. (2015). Seronegative conversion of an HIV positive subject treated with *Nigella sativa* and honey. *African Journal of Infectious Diseases*, 9(2): 47–50. https://doi.org/10.4314/ajid.v9i2.6

Pal, A., & Chakravarty, A. K. (2020). Chapter 5 - Genetics for innate immunity. In: Pal, A., & Chakravarty, A. K. (eds.) *Genetics and Breeding for Disease Resistance of Livestock* (pp. 103–117). Academic Press. https://doi.org/10.1016/B978-0-12-816406-8.00005-X

Pan, X., Zhang, A., Henderson, G. E., et al. (2019). Traditional, complementary, and alternative medical cures for HIV: rationale and implications for HIV cure research. *Global Public Health*, 14(1): 152–160. https://doi.org/10.1080/17441692.2017.1413122

Paul, I. M., Beiler, J., McMonagle, A., et al. (2007). Effect of honey, dextromethorphan, and no treatment on nocturnal cough and sleep quality for coughing children and their parents. *Archives of Pediatrics & Adolescent Medicine*, 161(12): 1140–1146. https://doi.org/10.1001/archpedi.161.12.1140

Portokalakis, I., Mohd Yusof, H. I., Ghanotakis, D. F., et al. (2016). Manuka honey-induced cytotoxicity against MCF7 breast cancer cells is correlated to total phenol content and antioxidant power. *Journal of Advances in Biology & Biotechnology*, 8(2): 1–10. https://doi.org/10.9734/jabb/2016/27899

Poudel-Tandukar, K., & Chandyo, R. K. (2016). Dietary B vitamins and serum C-reactive protein in persons with human immunodeficiency virus infection: the positive living with HIV (POLH) study. *Food and Nutrition Bulletin*, 37(4): 517–528. https://doi.org/10.1177/0379572116657268

Puronen, C. E., Ford, E. S., & Uldrick, T. S. (2019). Immunotherapy in people with HIV and cancer. *Frontiers in Immunology*, 10: 2060. https://doi.org/10.3389/fimmu.2019.02060

Ramana, K. V. (2014). Effect of highly active antiretroviral therapy (HAART) on human immunodeficiency virus disease pathogenesis and progression. *American Journal of Public Health Research*, 2(3): 68–74. https://doi.org/10.12691/ajphr-2-3-1

Ranneh, Y., Akim, A. M., Hamid, H. A., et al. (2021). Honey and its nutritional and anti-inflammatory value. *BMC Complementary Medicine and Therapies*, 21(1): 30. https://doi.org/10.1186/s12906-020-03170-5

Salami, A. K., Akande, A. A., & Olokoba, A. B. (2009). Serum lipids and glucose abnormalities in HIV/AIDS patients on antiretroviral therapies. *West African Journal of Medicine*, 28(1): 10-15. https://doi.org/10.4314/wajm.v28i1.48417

Sánchez-Ramón, S., Bermúdez, A., González-Granado, L. I., et al. (2019). Primary and secondary immunodeficiency diseases in oncohaematology: warning signs, diagnosis, and management. *Frontiers in Immunology*, 10:586. https://doi.org/10.3389/fimmu.2019.00586

Santos-Buelga, C., & González-Paramás, A. M. (2017). Chemical composition of honey. In: Alvarez-Suarez, J. M. (ed.), *Bee Products - Chemical and Biological Properties* (pp. 43–82). Springer, Cham. https://doi.org/10.1007/978-3-319-59689-1_3

Sharp, P. M., & Hahn, B. H. (2011). Origins of HIV and the AIDS pandemic. *Cold Spring Harbor Perspectives in Medicine*, 1(1): a006841. https://doi.org/10.1101/cshperspect.a006841

Siti Sarah, C. O., Md Shukri, N., Mohd Ashari, N. S., et al. (2020). Zonula occludens and nasal epithelial barrier integrity in allergic rhinitis. *PeerJ*, 8: e9834. https://doi.org/10.7717/peerj.9834

Small, P., Keith, P. K., & Kim, H. (2018). Allergic rhinitis. *Allergy, Asthma & Clinical Immunology*, 14(Suppl 2): 51. https://doi.org/10.1186/s13223-018-0280-7

Soto, M. E., Ares, A. M., Bernal, J., et al. (2011). Simultaneous determination of tryptophan, kynurenine, kynurenic and xanthurenic acids in honey by liquid chromatography with diode array, fluorescence and tandem mass spectrometry detection. *Journal of Chromatography A*, 1218(42): 7592–7600. https://doi.org/10.1016/j.chroma.2011.06.105

Tan, H. T., Rahman, R. A., Gan, S. H., et al. (2009). The antibacterial properties of Malaysian tualang honey against wound and enteric microorganisms in comparison to manuka honey. *BMC Complementary and Alternative Medicine*, 9: 34. https://doi.org/10.1186/1472-6882-9-34

Tang, S. P., Wan Yusuf, W. N., Abd Aziz, C. B., et al. (2020). Effects of six-month Tualang honey supplementation on physiological and biochemical profiles in asymptomatic, treatment-naïve HIV-infected patients. *Tropical Journal of Natural Product Research*, 4(12): 1116–1123. https://www.tjnpr.org/index.php/home/article/view/889

Tang, S. P., Abdul Aziz, C. B., Mustafa, M., et al. (2021). Tualang honey potentially delay deterioration in haematological and immunological parameters in asymptomatic, treatment-naïve HIV-infected patients. *Malaysian Journal of Medicine and Health Sciences*, 17(4): 166–173.

Thulasi, R. R., Manimaran, D., Hemanathan, G., et al. (2016). Hematological abnormalities in HIV infected individuals in correlation to CD4 counts and ART status. *Asian Journal of Medical Sciences*, 7(4): 14–18. https://doi.org/10.3126/ajms.v7i4.14033

Tonks, A. J., Cooper, R. A., Jones, K. P., et al. (2003). Honey stimulates inflammatory cytokine production from monocytes. *Cytokine*, 21(5): 242–247. https://doi.org/10.1016/s1043-4666(03)00092-9

Tonks, A. J., Dudley, E., Porter, N. G., et al. (2007). A 5.8-kDa component of manuka honey stimulates immune cells via TLR4. *Journal of Leukocyte Biology*, 82(5): 1147–1155. https://doi.org/10.1189/jlb.1106683

UNAIDS. (2022). Global HIV & AIDS statistics — Fact sheet. Available at https://www.unaids.org/en/resources/fact-sheet [10 October 2022]

Vahdat Shariatpanahi, Z., Jamshidi, F., Nasrollahzadeh, J., et al. (2018). Effect of honey on diarrhea and fecal microbiotain in critically ill tube-fed patients: a single center randomized controlled study. *Anesthesiology and Pain Medicine*, 8(1): e62889. https://doi.org/10.5812/aapm.62889

Vallianou, N. G., Gounari, P., Skourtis, A., et al. (2014). Honey and its anti-inflammatory, anti-bacterial and anti-oxidant properties. *General Medicine: Open Access*, 2: 2. https://doi.org/10.4172/2327-5146.1000132

Vezir, E., Kaya, A., Toyran, M., et al. (2014). Anaphylaxis/angioedema caused by honey ingestion. *Allergy and Asthma Proceedings*, 35(1): 71–74. https://doi.org/10.2500/aap.2014.35.3718

Vidya Vijayan, K. K., Karthigeyan, K. P., Tripathi, S. P., et al. (2017). Pathophysiology of CD4+ T-Cell depletion in HIV-1 and HIV-2 infections. *Frontiers in Immunology*, 8: 580. https://doi.org/10.3389/fimmu.2017.00580

Waheed, M., Hussain, M. B., Javed, A., et al. (2019). Honey and cancer: A mechanistic review. *Clinical Nutrition*, 38(6): 2499–2503. https://doi.org/10.1016/j.clnu.2018.12.019

Wan Yusuf, W. N., Wan Mohammad, W. M. Z., Gan, S. H., et al. (2019). Tualang honey ameliorates viral load, CD4 counts and improves quality of life in asymptomatic human immunodeficiency virus infected patients. *Journal of Traditional and Complementary Medicine*, 9(4): 249–256. https://doi.org/10.1016/j.jtcme.2018.05.003

WHO. (2001). *Cough and cold remedies for the treatment of acute respiratory infections in young children* (p. 39). Available at https://apps.who.int/iris/handle/10665/66856/ [6 December 2021]

WHO. (2003). *Nutrient requirements for people living with HIV/AIDS : report of a technical consultation, 13-15 May 2003, Geneva*. World Health Organization. Available at https://apps.who.int/iris/handle/10665/42853 [6 December 2021]

WHO. (2020). *The top 10 causes of death, 9 December 2020*. World Health Organization. Available at https://www.who.int/en/news-room/fact-sheets/detail/the-top-10-causes-of-death [6 December 2021]

Winkler, P., Ellinger, S., Boetzer, A. M., et al. (2004). Lymphocyte proliferation and apoptosis in HIV-seropositive and healthy subjects during long-term ingestion of fruit juices or a fruit-vegetable-concentrate rich in polyphenols and antioxidant vitamins. *European Journal of Clinical Nutrition*, 58: 317–325. https://doi.org/10.1038/sj.ejcn.1601784

Xu, H. X., Wan, M., Dong, H., et al. (2000). Inhibitory activity of flavonoids and tannins against HIV-1 protease. *Biological & Pharmaceutical Bulletin*, 23(9): 1072–1076. https://doi.org/10.1248/bpb.23.1072

Yarchoan, R., & Uldrick, T. S. (2018). HIV-associated cancers and related diseases. *The New England Journal of Medicine*, 378(11): 1029–1041. https://doi.org/10.1056/NEJMra1615896

Young, S. N. (2013). The effect of raising and lowering tryptophan levels on human mood and social behaviour. *Philosophical Transactions of the Royal Society of London. Series B, Biological Sciences*, 368(1615): 20110375. https://doi.org/10.1098/rstb.2011.0375

Zhang, X., Xia, Q., Yang, G., et al. (2017). The anti-HIV-1 activity of polyphenols from *Phyllanthus urinaria* and the pharmacokinetics and tissue distribution of its marker compound, gallic acid. *Journal of Traditional Chinese Medical Sciences*, 4(2): 158–166. https://doi.org/10.1016/j.jtcms.2017.07.013

Zhao, H., Cheng, N., Zhou, W., et al. (2019). Honey polyphenols ameliorate DSS-induced ulcerative colitis via modulating gut microbiota in rats. *Molecular Nutrition & Food Research*, 63(23): e1900638. https://doi.org/10.1002/mnfr.201900638

# 19

## Use of Honey in Sports Medicine

*Foong Kiew Ooi and Chee Keong Chen*

### Introduction

Consuming honey has the potential to play a vital role in sports medicine, and this is attributed to its nutritional constituents, which consist of glucose, fructose, amino acids, multivitamins, and multiminerals (Bogdanov et al. 2008; Silva et al. 2016). In addition, honey has been shown to possess antioxidant, anti-inflammatory, antimicrobial, and antitumor properties (Abouda et al. 2011; Aljadi and Kamaruddin 2004; Erejuwa et al. 2012; Maruyama et al. 2010; Nagai et al. 2001; Schneider et al. 2013; Yang et al. 2007). The composition of honey can vary considerably because it is dependent on the variety of plants and flowers where the nectar is collected and the environmental conditions where the honey is obtained (Ahmed and Othman 2013; Batista et al. 2012; Bogdanov et al. 2008; Schneider et al. 2013). These diverse properties of honey have led to its usage and investigations on its effectiveness on various types of sports performance and health.

The main carbohydrates found in honey are fructose (35%–40%) and glucose (30%–35%) (Ahmed and Othman 2013; Bogdanov et al. 2008; Schneider et al. 2013). The glycemic index (GI) of honey is due to its proportionate content of fructose in which a greater fructose-to-glucose ratio represents a lower GI value (Deibert et al. 2010), and it has been reported that fructose-rich variety honey (e.g. acacia honey) has a low GI value (Atkinson et al. 2008).

Carbohydrate is a vital macronutrient to sustain muscular activity, and it is widely known that higher pre-exercise muscle glycogen storage improves sports performance, particularly during prolonged exercise lasting more than 1 hour (Close et al. 2016; Coyle et al. 1986; Hawley et al. 1997; Nicholas et al. 1995). Furthermore, carbohydrate intake before, during, and after exercise has been shown to augment performance in various types of sports that require endurance (Burke et al. 2011; Coyle et al. 1986; Jeukendrup 2008, 2014). Supplementation of carbohydrates has been postulated to enhance physical performance via a glycogen-sparing effect, maintaining or increasing blood glucose concentration and contributing directly to the central nervous system (CNS) as the major substrate (Khong et al. 2018).

The influx of oxygen to exercising skeletal muscle has been shown to increase the production of free radicals (Di Meo and Venditti 2001; Ji 1999; Leeuwenburgh and Heinecke 2001). Performing physical activities raises the oxygen uptake by 10- to 20-fold and initiates the creation of free radicals, which are also known as reactive oxygen species (ROS) (Sen 1995). ROS have been suggested as the main cause of disruptions in muscle homeostasis, which could lead to muscular exhaustion (Lecarpentier 2007; Powers and Howley 1997; Shindoh et al. 1990). The detrimental role of ROS in muscular fatigue has been reported in several studies (Barclay and Hansel 1991; Lecarpentier 2007; Novelli et al. 1990; Shindoh et al. 1990). For instance, Shindoh et al. (1990) demonstrated that application of *N*-acetylcysteine diminished diaphragmatic fatigue in rabbits. The precise mechanism of how ROS triggers muscle exhaustion remains unknown. Several researchers have hypothesized that free radical production may result in damage to the sarcoplasmic reticulum, which causes a reduction in calcium release during the depolarization phase in the muscle and subsequently the myofilament interaction and mitochondrial metabolism (Davies et al. 1982; Essig and Nosek 1997; Lovlin et al. 1987; Powers and Howley 1997; Powers and Jackson 2008; Reid et al. 1994). It was also postulated that the generation of ROS during strenuous physical activity may lead to muscular fatigue, and the supplementation of antioxidants might improve sports performance via the reduction of oxidative stress (Barclay and Hansel 1991; Novelli et al. 1990; O'Neill et al. 1996; Powers et al. 2004; Reid et al. 1992; Shindoh et al. 1990). Morillas-Ruiz et al. (2006) have demonstrated that the level of

*Honey: Composition and Health Benefits*, First Edition. Edited by Md. Ibrahim Khalil, Gan Siew Hua, and Bey Hing Goh.

protein carbonyl was attenuated during a prolonged cycling test when participants consumed polyphenolic antioxidants. Because honey has been shown to possess antioxidant properties, consuming adequate amount of honey may improve sports performance by ameliorating the deleterious effect of free radicals during exercise (Nagai et al. 2001; Aljadi and Kamaruddin 2004; Tartibian and Maleki 2012).

Studies have demonstrated that supplementing a combined carbohydrate source containing both fructose and glucose was able to improve prolonged sports performance compared with intake of glucose alone (Currell and Jeukendrup 2008; Triplett et al. 2010). Honey consists of various composition of different types of carbohydrates, including fructose, glucose, sucrose, and raffinose. Thus, it is highly possible that honey can be consumed as an ergogenic aid for enhancing sports performance. According to Stevenson et al. (2017), certain honey types such as Acacia honey have low GIs, which elicits a reduced insulinemic response and delays the distribution of glucose into the systemic circulation and hence could sustain plasma glucose levels for a longer period during exercise. Their study also demonstrated that blood glucose levels are better maintained during the second half of a soccer game when low-GI carbohydrate was ingested before the match and at halftime.

Postmenopausal women are predisposed to osteoporosis because of the decline in estrogen level, which results in increased oxidative stress that causes apoptosis of the osteoblasts (Effendy et al. 2012). In addition, pro-inflammatory cytokines that are increased as a result of estrogen deficiency are vital determinants of osteoclast differentiation and bone resorption activity. Even though estrogen replacement therapy is the primary intervention for postmenopausal osteoporosis (Boonen et al. 2005; Gambacciani and Levancini 2014), it has been known to cause several adverse effects (Lobo 1995; Santen et al. 2001). Thus, another type of intervention that uses antioxidants that can elicit an anti-inflammatory response has been suggested to replace this conventional therapy (Effendy et al. 2012). In this regard, honey seems to be an excellent option because it contains antioxidant properties that scavenge free radicals, subsequently ameliorating oxidative stress. Also, honey has anti-inflammatory effects that inhibit pro-inflammatory cytokines. Hence, a reduction in bone loss is expected because of the reduced osteoclastogenic activity. Thus, ingestion of honey can be used as an alternative modality for the management of postmenopausal osteoporosis. In addition, it is believed that honey consumption is beneficial in increasing and maintaining bone health in younger populations.

It has also been well documented that strenuous or prolonged physical activity may compromise immune functions and consequently escalate the risk of sustaining upper respiratory tract infections (Gleeson et al. 2004; Walsh et al. 2011; Weidner and Schurr 2003). Nieman et al. (1998) and Gleeson et al. (2004) have reported that training in a low-carbohydrate state could elicit increased production of blood stress hormones, disrupting certain markers of immune function. Moreover, ingesting carbohydrates at a rate of 30 to 60 $g.h^{-1}$ during prolonged physical activity could attenuate some negative responses via a larger supply of blood glucose and a concurrent decline in levels of stress hormone. Honey contains carbohydrates, so it is believed that honey consumption could elicit beneficial effects on immune function parameters in prolonged physical activities.

## Effects of Honey and Bee Bread Supplementation on Sports Performance and Antioxidant Status

Maintaining blood glucose concentration has been shown to improve sports performance among the athletic population. In addition, the role of blood glucose in the normal activity of the CNS is clear because inadequate circulating blood glucose has been postulated to result in central fatigue (Khong et al. 2018). The vital role of blood glucose concentrations on cognitive function was demonstrated by enhanced speed of fine-motor, psycho-motor, and visual discrimination following higher glucose availability during a soccer match (Bandelow et al. 2010). Thus, strategies used to sustain or raise blood glucose concentration could be beneficial for team sports athletes, particularly during the latter stages of a match (Hills and Russell 2017). Therefore, honey (particularly those with low GI carbohydrate) appears to be a logical and practical supplementation for sports activity.

A laboratory analysis carried out in the Department of Molecular Medicine, Universiti Malaya, Malaysia, has shown that Acacia honey consists of 22.9% glucose, 31.2% fructose, 3.3% maltose, and 9.9% sucrose. It also contains 13 mg of sodium and 75 g of carbohydrate per 100 g of serving (Ahmad et al. 2020). Ahmad et al. (2015) also reported that recreational runners who consumed a 6.8% Acacia honey solution to replenish approximately 150% of body weight loss during a 2-hour recovery period after a 60-minute run improved their time trial running performance compared with consuming a similar amount of plain water. Harper et al. (2017) also reported improvements in dribbling speed and self-paced sports performance among participants during a simulated football match after the participants ingested 12% carbohydrate-electrolyte solution before the start of each half compared with a placebo trial. These authors mentioned that future studies are warranted to examine the effects of consuming higher concentrations of honey (>10% carbohydrate) on cognitive, physical, and skilled performance among players participating in team sports.

The GI is the rate of blood glucose surge following the ingestion of a food compared with a reference (e.g. white bread) by determining the amount of available carbohydrate in the food that actually increases the blood glucose level. Ahmad et al. (2015) reported that a low-GI honey drink (1.9 g/kg body weight) significantly improved a 20-minute time trial performance compared with a placebo trial. It was postulated that delayed release of glucose affords a steady supply during prolonged physical activity, and muscle glycogen depletion only occurred at a later stage of exercise (Wu and Williams 2006). It was also speculated that consuming low-GI carbohydrates may reduce carbohydrate oxidation and favor fat oxidation compared with an intermediate- or high-GI meal (Bonen et al. 1989), which could optimally be beneficial during prolonged exercise (Klein et al. 1994; Lecoultre et al. 2010).

The observation of the better time trial performance with a comparatively low amount of carbohydrate in the honey (Ahmad et al. 2015) might be due to the composition of glucose and fructose in honey, which has been demonstrated to improve the total carbohydrate oxidation rate compared with glucose intake alone (Lecoultre et al. 2010). Jeukendrup et al. (2006) reported that participants who were prescribed a 90 $g.h^{-1}$ glucose-fructose drink during 5 hours of cycling had lower ratings of perceived exertion (RPEs), and cadence was better maintained at the end of the exercise session compared with those with glucose intake alone. Subsequently, other investigations have also corroborated the findings that improvements in prolonged performance lasting more than 2 hours are associated with ingesting exogenous supply of glucose-fructose solution (Rowlands et al. 2012; Triplett et al. 2010)

The effect of low- and high-GI carbohydrates during a 64-km cycling time trial was investigated by Earnest and colleagues (2004). Nine participants were given 15 g of low-GI (honey) or high-GI (dextrose) honey or placebo at 16-km intervals during a cycling timed trial. It was found that the times to finish the 64-km cycling timed trial were 128 ± 3.6 minutes, 128 ± 3.8 minutes, and 131 ± 3.9 minutes in the honey, dextrose, and placebo trials, respectively. This finding indicated that consumption of low-GI honey (15 g every 16 km) significantly reduced the time to complete the distance compared with the placebo trial, thus indicating the beneficial effects of low-GI honey consumption on cycling performance.

Aly et al. (2019) examined the effects of honey on a 1500-m run. Fifty participants were identified and randomly divided into two groups. The experimental group was given 50 g of honey, 0.5 g of royal jelly, and 0.5 g of bee pollen 30 minutes before running, and the control group ingested placebo. The research data indicated that the times to complete a 1500-m run in the honey group (5.01 ± 0.24 minutes) and placebo group (5.16 ± 0.11 minutes) were not different, although the blood glucose level was higher in the honey group compared with those in the placebo group. Nevertheless, the plasma lactic acid level was higher in the honey group than the placebo group (19.5 ± 3.8 vs 11.0 ± 3.8 mmol/L respectively), which was probably due to a faster pace in the honey group during the run. The authors concluded that despite an insignificant time difference to complete 1500 m, it seems that honey, when combined with bee pollen and royal jelly, elicited an improved 1500-m run.

Lancaster et al. (2001) reported that participants performed significantly better cycling performance in the dextrose and honey trials compared with the placebo trial. The total times required to perform the time trial were 131.3 ± 3.6, 128.3 ± 3.8, and 128.8 ± 3.5 minutes in the placebo, dextrose gel, and honey gel trials, respectively. In addition, mean power was significantly greater when participants consumed the carbohydrate gels during exercise. Average heart rate and blood glucose level were also higher in the carbohydrate-supplemented groups during the timed trials. These data demonstrated that consuming dextrose and honey gels during prolonged cycling improves performance, and it was probably attributed to the enhanced glucose availability.

Regarding the potential role of Malaysian honey on sports performance, a study by Ahmad et al. (2015) investigated the effects of an Acacia honey drink ingestion during a rehydration phase at post-exercise on selected physiological parameters and subsequent running performance in a hot environment. Ten participants were recruited, and they ran on a treadmill at 65% maximal oxygen consumption ($VO_2max$) for 60 minutes and then rested for 2 hours. During this recovery phase, the participants consumed either plain water or a honey drink that was similar to 150% of body mass loss. Subsequently, the participants were required to perform a 20-minute trial performance. The distance covered by the participants in the timed trial was significantly further in the honey trial compared with those in the plain water trial (3.42 ± 0.35 km vs 3.12 ± 0.34 km, respectively). There were also significant higher levels of blood glucose and insulin in the honey group during rehydration compared with the plain water group. It was concluded that ingesting honey drink during post-exercise recovery offered more favorable effects than plain water and thus could be recommended as an ergogenic aid while people exercised or trained in a hot and humid environment.

In a more recent study, Ahmad et al. (2020) investigated the effects of a sodium-enriched Acacia honey drink as an ergogenic aid in a hot and humid environment (31°C, 70% rh). Participants ingested plain water, an Acacia honey drink (H), or a sodium-enriched Acacia honey drink (HS) with a similar amount to 150% of body mass loss during the 2-hour rehydration phase. At the end of this phase, plasma glucose and plasma insulin levels in the H and HS trials were significantly higher than in the water trial. Thus, it was concluded that both Acacia honey and sodium-enriched Acacia honey drinks can be advocated as a recovery aid for physically active individuals training or competing in a hot environment.

Another recent study investigated the effect of Acacia honey intake on sports performance and selected biomarkers in the heat (Samsani et al. 2021). Ten recreational athletes ran at 70% $VO_2max$ for 1 hour (run 1) followed by a rehydration period for 2 hours and subsequently ran a 20-minute time trial (run 2). After run 1, participants ingested an Acacia honey, a sports drink, or water with an amount similar to 150% of body mass loss. Then the participants completed a 20-minute time trial. The results showed that Acacia honey ingestion improved running time trial performance compared with plain water trial, but it was not different from the sports drink trial. During the rehydration period, blood glucose, insulin, and free fatty acid concentrations were significantly higher in the honey and sports drink trials compared with the water trial. Thus, we conclude that Acacia honey could be advocated as a rehydration drink to enhance sports performance in the heat.

However, some contradictory findings were reported. According to Abbey and Rankin (2009), soccer players consuming a honey drink containing 1.0 g/kg carbohydrate at prematch and during halftime did not have any benefit in their physical or skilled performance (high-intensity run, agility and ball shooting, progressive shuttle run) compared with a placebo trial. Maleki et al. (2016) investigated the effects of 16-week consumption of honey or placebo on cycling performance. Twenty-four participants were randomly assigned into the experiment or control groups. Participants consumed either 70 g of honey or placebo before 90 minutes of each training session for 16 weeks. Following this supplementation regimen, the participants completed a 5- and 40-km cycling ergometer test, and peak power was measured at pre- and post-tests. These authors concluded that peak power and the time taken to complete both cycling tests between the two trials were not different. Hence, it was concluded that supplementation of honey did not improve cycling performance in road cyclists.

In our opinion, these equivocal findings are probably due to different types of exercises (intensity and duration of exercise), honey supplementation (timing, dosage, and duration of supplementation), research participants (individual responsiveness and fitness level of the individual), experimental designs, and environmental conditions (Table 19.1). Nevertheless, most of the data obtain from our studies have demonstrated the potential benefits of honey to improve sports performance. However, more studies are warranted to substantiate these findings so that honey can be routinely recommended for athletes as an ergogenic aid.

We also investigated the effects of bee bread on sports performance (Chen et al. 2018; Ping et al. 2018). Bee bread consists of glucose, essential amino acids, free fatty acids, multivitamins, and minerals and has antioxidant properties (Nagai et al. 2004, 2005; Stanciu et al. 2009). We speculated that these properties of bee bread may offer synergetic ergogenic effects on sports performance. In the first study, 12 recreational athletes ran on a motorized treadmill for 90 minutes immediately followed by a 4-hour recovery phase (Chen et al. 2018). During this recovery period, participants ingested either 30 g/h of bee bread or placebo. Then the participants completed a 20-minute time trial on a treadmill. Our results indicated that plasma glucose concentration in the bee bread trial was significantly higher toward the end of the recovery period compared with the placebo trial. The distance run during the time trial was significantly farther in the bee bread trial than the

**Table 19.1** Probable reasons for the equivocal findings from various studies on the effect of honey on sports performance.

| |
|---|
| 1) Different types of exercise test used to determine performance |
| 2) Different intensity and duration of the exercise tests |
| 3) Different types of honey supplements |
| 4) Different dosage, timing of ingestion, and duration of honey supplementation |
| 5) Different fitness levels of the study participants |
| 6) Different environmental conditions during the exercise tests |

placebo trial (3.45 ± 0.4 km vs 3.24 ± 0.4 km respectively). However, other physiological parameters such as heart rate, oxygen uptake, tympanic temperature, and plasma volume were not significantly different between trials. We concluded that bee bread consumption during a recovery phase improved subsequent running time trial performance.

We also investigated the effects of 8 weeks of bee bread supplementation on endurance performance and antioxidant status among recreational athletes (Ping et al. 2018). The participants in this study were supplemented with 20 g of bee bread or placebo daily for 8 weeks. Then the participants ran at 60% $VO_{2max}$ for 90 minutes immediately followed by a 20-minute time trial. We found that the distance ran in the bee bread trial was significantly better compared with the placebo trial (3.41 ± 0.2 km vs 3.28 ± 0.2 km, respectively). However, exercise heart rate, oxygen uptake, tympanic temperature, RPEs, and plasma glucose were not different between trials. In addition, total antioxidant status was also significantly higher after supplementation, immediate after exercise, and 24 hours after exercise in the bee bread trial. Hence, it was concluded that ingestion of bee bread at a daily dosage of 20 g for 8 weeks improved the running performance in recreational athletes, and this was presumably caused by the increased total antioxidants.

To date, available data on the effects of antioxidant supplementation on improved sports performance are equivocal. Previous studies have demonstrated beneficial effects of antioxidants, including *N*-acetylcysteine (Kelly et al. 2009; Slattery et al. 2014), quercetin (Nieman et al. 2010), beetroot beverage (Cermak et al. 2012; Hoon et al. 2014), citrus flavonoid (Overdevest et al. 2018), Montmorency powdered tart cherries (Levers et al. 2016), and cashew apple juice (Prasertsri et al. 2016).

## Effects of Honey Supplementation and Exercise on Bone Health

Because osteoporosis is a leading health issue among older adults around the world, numerous efforts have been proposed to delay or prevent it. One of the strategies is to enhance bone mass during the formative years in life and consequently to sustain the gain or reduce the rate of bone loss (Ooi 2018). This strategy could be accomplished via proper and sufficient nutritional intake and performing regular weight-bearing exercises. Weight-bearing exercises, which include walking, running, dancing, and jumping, are particularly crucial to develop and maintain strong bones. Honey consists of carbohydrates, flavonoids, antioxidants, minerals, organic acids, proteins, phenolic acids, and vitamins (Aljadi and Kamaruddin 2004), and these components are believed to be vital for enhancing bone health. It was previously reported that consuming honey enhances absorption of calcium in rats and plays a role in enhancing bone health (Ariefdjohan et al. 2008). In another study on animals, it was demonstrated that young Sprague Dawley rats fed with honey for 52 weeks had significantly higher bone mineral density than control rats (Chepulis and Starkey 2008). Shafin et al. (2014) have also shown that Tualang honey (TH) consumption for 16 weeks in postmenopausal women elicited lower oxidative stress as shown by an increase in endogenous enzymatic antioxidant activities and a concomitant decrease in 4-hydroxynonenal. Thus, this estrogenic property of honey could protect against postmenopausal osteoporosis. A couple of studies by the present research team also demonstrated that combined jumping exercise and honey ingestion resulted in increased bone mass, bone mineral density, geometry, mechanical properties, and bone metabolism in female rats (Mosavat et al. 2014; Ooi et al. 2014; Tavafzadeh et al. 2011).

Using a human model, the effect of 6 weeks of combined aerobic dance exercise and honey ingestion on markers of bone metabolism in young women was investigated (Ooi et al. 2011). The results of this study showed that 6 weeks of aerobic dance exercise combined with daily supplementation of 20 g of honey elicited increases in a blood bone formation marker (serum alkaline phosphatase [ALP]) in young women compared with honey ingestion alone or exercise alone trials. Another study was also carried out to investigate the effects of circuit training combined with honey supplementation in young men (Ooi and Aziz 2017). The circuit training involved two sets of 10 different activities using dumbbells or elastic bands as the resistance. Our data demonstrated that 6 weeks of combined circuit training and 20 g of daily honey supplementation also elicited increases in the blood bone formation marker in young men. It was speculated that bone response differs with age, so bone metabolism may not be similar in older adults compared with young women after an intervention of combined exercise training and honey consumption. Another study investigated the effects of combined aerobic dance training and honey consumption on markers of bone metabolism in women (25–40 years old) (Rahim et al. 2016). Results from this study indicated that honey consumption alone elevated plasma total calcium concentration, but aerobic dance sessions alone significantly elevated bone resorption marker. When aerobic dance exercise was combined with honey consumption, it elicited beneficial effects by reducing bone resorption markers in previously sedentary older women.

Another recent study was carried out to examine the combined effects of TH and countermovement jumping (CJ) exercise for 6 weeks on bone mineral density (via bone speed of sound), maximum jumping height, and bone turnover markers in young men (18–24 years old) (Mohd Zuhri et al. 2020). The participants recruited in this study were randomly divided into four groups, that is, control (C), supplementation of TH (H), CJ exercise (J), and combined TH supplementation and CJ exercise (JH) groups. Participants in the J and JH groups were instructed to jump for 6 weeks (two sessions per day, three times per week). Participants in the H and JH groups ingested 20 g of TH daily for 6 weeks. The participants' leg bone speed of sound (SOS) was determined at before and after testing. It was found that there were no significant differences in participants' bone SOS among the groups. Blood samples collected were analyzed for serum ALP (bone formation marker) and serum C-terminal telopeptide of type 1 collagen (bone resorption marker). The JH group had a 56.3% reduction in serum carboxy-terminal telopeptide of type 1 collagen (1CTP), which was significantly lower than the C group. ALP also declined by 7.7% in the JH group. Jumping height was significantly higher at post-test compared with pre-test values in the H and JH groups. Moreover, maximum jumping height was also significantly higher in the JH group compared with the C group at post-test. Thus, it was concluded that combined TH supplementation with countermovement jumping exercise could raise maximum jumping height and exert a positive effect on bone turnover markers in young men.

Supplementation with honey has also been shown to increase markers of bone structure in animal studies (Zaid et al. 2012), which could be attributed to certain components in honey such as vitamin K, calcium, phosphorus, iron, and magnesium, which could improve bone health status, particularly when combined with an exercise regimen. It was also postulated that the improvement in bone mass and its mechanical properties may be due to these nutrients, which promote bone formation, resulting in a net gain in bone strength (Reid et al. 2014).

Mechanical loading has been shown to play an effective role in enhancing bone health, whereby an appropriate exercise program can facilitate bone development as well as prevent age-related bone loss (Berard et al. 1997). There is also sufficient evidence that mechanical loading is dependent on the magnitude, type, and rate of the applied load (Maimoun et al. 2006). High and dynamic magnitude loading such as jumping that causes high ground reaction force could elicit favorable effects on bone health status (Ooi et al. 2009; Tavafzadeh et al. 2011; Umemura et al. 1997). One of the contributing factors of osteoporosis is oxidative stress (Mironczuk-Chodakowska et al. 2018; Rao and Rao 2013). Free radicals have been reported to be involved in the formation of osteoclasts (Li et al. 2013) and reduce the activities of osteoblasts (Zhang et al. 2015). Honey is a natural product that possesses antioxidant properties that are able to ameliorate these deleterious effects of ROS in the pathogenesis of osteoporosis and consequently prevent bone loss.

In summary, human studies with different gender and age groups demonstrated that ingestion of honey combined with exercise has the potential to provide favorable effects on bone metabolism compared with exercise or honey consumption alone. Similarly, animal studies carried out in our laboratory also indicated that jumping exercise with honey resulted in more beneficial effects on bone metabolism markers and properties.

## Effects of Honey Supplementation and Exercise on Immune Functions

The relationship between exercise participation and the immune function shows a J-shaped trend, whereby moderate intensity exercise seems to augment immune response compared with baseline levels, while prolonged strenuous exercise impairs the immune system (Nieman 1994). Previous studies have shown that the immune systems of active individuals are transiently affected by both short-term (Abbey and Rankin 2009) and prolonged exercise (Gmünder et al. 1990; Salehian et al. 2014). It was postulated that the higher incidence of infection after performing strenuous and prolonged physical activity may be associated with the interference of monocytes and lymphocytes in an immune response during and several hours after exercise (Lancaster et al. 2004). This response has been shown to predispose the exercising individuals to upper respiratory tract infections (Nieman 1994). In addition, some of the biomarkers of inflammation that increases following exhaustive prolonged exercise include interleukin (IL)-1 (IL-1), IL-6, tumor necrosis factor-α, and C-reactive protein (Helge et al. 2003; Hiscock et al. 2003).

Several previous studies have demonstrated that honey seems to exert a vital role in enhancement of immune functions. For instance, it was shown that ingesting 70 g of honey before training sessions for 8 weeks reduced the detrimental immune response to this intense cycling session compared with a placebo trial (Tartibian and Maleki 2012). This finding was supported by another study that demonstrated that consuming 70 g of honey 90 minutes before each training session over 16 weeks of moderate to intense cycling elicited a decline in plasma cytokine levels immediately, 12 and 24 hours after training in the eighth week and immediately, 12 hours, 24 hours, and 7 days after training in week 16 (Maleki et al. 2016).

Abbey and Rankin (2009) also reported that honey deferred the increase in IL-1 at post-test compared with the sports drink and placebo trials. IL-1 is an anti-inflammatory cytokine generated in response to a rise in IL-6 from the working muscles during exercise (Gleeson 2007). In general, honey consumption reduces inflammatory cytokines with a concomitant increase in anti-inflammatory cytokines (Gmünder et al. 1990; Salehian et al. 2014) probably caused by the presence of antioxidants in honey (Tartibian and Maleki 2012).

Two studies have been carried out to examine the synergistic effect of honey supplementation with exercise on immune function parameters in young men and women, respectively. In the first study, the effects of 6 weeks of combined circuit training and honey consumption on leukocytes and natural killer (NK) cells in young men was investigated (Mohamed and Ooi 2013). Forty men between 19 and 29 years old were randomly assigned into four groups – sedentary without honey supplementation control (C), sedentary with honey consumption (H), circuit training only (Ex), or circuit training with honey consumption (HEx). The circuit training program was carried out for 6 weeks (1 hour per session, three sessions per week). The training stations prescribed were triceps extension with dumbbells, rope skipping, biceps curls with dumbbells, back extension, sit-ups, push-ups, burpees, split squats, and hand and leg exercises with elastic bands. Participants in the HEx and H groups ingested 300 mL of honey drink, which contained 20 g of TH daily. Blood samples were collected before and after the intervention period to determine the levels of leukocytes and NK cells that indicate immune functions. The results of this study show that there were no significant changes in leukocytes in C, H, Ex, and HEx groups. However, the highest increment in NK cells was found in the HEx group. These findings imply that combined circuit training and honey consumption lead to enhanced level of NK cells compared with circuit training or honey consumption alone in young men.

In the second study, 44 healthy sedentary women with age ranging from 25 to 40 years old were divided into four groups – sedentary without supplementation control (Con), honey consumption (H), aerobic dance exercise (D), and combined aerobic dance exercise with honey consumption (HD) (Rahim et al. 2017). Aerobic dance exercise was performed for 8 weeks (1 hour per session, three sessions per week). The honey drink was ingested by participants in the H and HD groups at a dosage of 20 g of Gelam honey diluted in plain water and ingested daily for 8 weeks. In the HD group, the participants were instructed to ingest the honey drink 30 minutes before performing the exercise sessions. After 8 weeks of intervention, cytotoxic T cells (CD8) in the HD group increased significantly compared with the baseline level. Moreover, the percentage increases in total lymphocyte, T helper (CD4), and T cytotoxic (CD8) counts after 8 weeks were highest in the HD group compared with the other groups. Hence, it was concluded that aerobic dance and honey consumption seems to have a synergistic effect to improve immune functions in women.

The potential roles of honey supplementation in the field of sports medicine are summarized in Table 19.2.

**Table 19.2** The potential roles of honey supplementation in the field of sports medicine.

| | |
|---|---|
| **Honey Supplements as Ergogenic Aids to Improve Sports Performance** | |
| 1. Honey supplements improved cycling performance in trained cyclists. | Lancaster et al. 2001 |
| 2. Honey supplements improved time trial running performance in recreational runners. | Ahmad et al. 2015 |
| 3. Honey supplements elicited a shorter time to complete a 1500-m run in trained runners. | Aly et al. 2019 |
| 4. Sodium-enriched honey increased plasma glucose during a 2-hour recovery period. | Ahmad et al. 2020 |
| 5. Acacia honey supplementation during the rehydration phase improved subsequent running time trial performance. | Samsani et al. 2021 |
| 6. Supplementation of bee bread during a recovery period seemed to improve subsequent running time trial performance. | Chen et al. 2018 |
| 7. Supplementation of bee bread for 8 weeks improved running performance | Ping et al. 2018 |
| **Effects of Honey Supplementation and Exercise on Bone Health** | |
| 1. Honey supplements enhanced calcium absorption in rats and could therefore play a role in boosting bone health. | Ariefdjohan et al. 2008 |
| 2. Honey supplements enhanced markers of bone structure in rodent studies. | Zaid et al. 2012 |
| 3. Rats fed with honey for 52 weeks had significantly higher bone mineral density. | Chepulis and Starkey 2008 |
| 4. Tualang honey supplementation for 16 weeks could protect women against postmenopausal osteoporosis. | Shafin et al. 2014 |

**Table 19.2** (Continued)

| | |
|---|---|
| 5. Combined jumping exercise and honey supplementation increased bone mass, bone mineral density, geometry, mechanical properties, and bone metabolism markers in female rats. | Mosavat et al. 2014; Ooi et al. 2014; Tavafzadeh et al. 2011 |
| 6. Combined exercise training and honey supplementation elicited increases in a blood bone formation marker in young women and men. | Ooi et al. 2011; Ooi and Aziz 2017 |
| 7. Combined aerobic dance exercise and honey supplementation elicited reduction in a bone resorption marker in women. | Rahim et al. 2016 |
| 8. Combined honey supplementation with countermovement jumping exercise increased maximum jumping height and influenced bone turnover markers in young men. | Mohd Zuhri et al. 2020 |
| **Effects of Honey Supplementation and Exercise on Immune Functions** | |
| 1. Honey attenuated the negative immune responses to intense cycling. | Tartibian and Maleki 2012 |
| 2. Honey reduced plasma cytokine concentrations immediately, at 12 and 24 hours after cycling training. | Maleki et al. 2016 |
| 3. Honey reduced plasma inflammatory cytokine concentrations after short term and prolonged exercise. | Abbey and Rankin 2009; Salehian et al. 2014 |
| 4. Honey supplementation, when combined with circuit training, elicited increased natural killer cells in young men. | Mohamed and Ooi 2013 |
| 5. Combined aerobic dance and honey supplementation seemed to enhance immune function in women. | Rahim et al. 2017 |

## Conclusion

Honey is a natural product consisting of approximately 80% carbohydrates (primarily glucose and fructose), and it possesses antioxidant, antimicrobial, and anti-inflammatory properties. With these properties, honey has the potential to be used as a nutritional supplement in the field of sports medicine.

Despite the absence of conclusive evidence, available scientific data support the use of honey as a potential ergogenic aid when ingested before, during, and after exercise. However, it is recommended that future research should include approaches to determine if honey may provide benefits to other short-term physical performance, sports-specific skills, or cognitive performance with different supplementation modalities (types of honey and dosages) and durations compared with an equivalent amount of carbohydrate ingestion.

Overall, honey plays a vital role in sustaining and enhancing sports performance, and when honey is combined with an appropriate exercise training program, this combination could exert favorable effects on bone health and immune function. In addition, factors that need to be considered when honey supplementation is recommended for any sports performance enhancement or for health purposes include the types, purity, and nutritional constituents of the honey prescribed.

## References


Abbey, E.L. and Rankin, J.W. (2009). Effect of ingesting a honey-sweetened beverage on soccer performance and exercise-induced cytokine response. *International Journal of Sport Nutrition and Exercise Metabolism* 19: 659–672.

Abouda, Z., Zerdani, I., Kalalou, I., et al. (2011). The antibacterial activity of Moroccan bee bread and bee-pollen (fresh and dried) against pathogenic bacteria. *Research Journal of Microbiology* 6: 376–384.

Ahmad, N.S., Ismail, M.S., Mohamed, M., and Ooi, F.K. (2020). Blood glucose metabolism, serum, and urine osmolality in response to sodium-enriched acacia honey drink consumption during rehydration after exercise in hot and humid environment. *Journal of Sustainability Science and Management* 15 (5): 72–83.

Ahmad, N.S., Ooi, F.K., Ismail, M.S., and Mohamed, M. (2015). Effects of post-exercise honey drink ingestion on blood glucose and subsequent running performance in the heat. *Asian Journal of Sports Medicine* 6: e24044.

Ahmed, S. and Othman, N.H. (2013). Review of the medicinal effects of Tualang honey and a comparison with Manuka honey. *Malaysian Journal of Medical Sciences (MJMS)* 20: 6–13.

Aljadi, A.M. and Kamaruddin, M.Y. (2004). Evaluation of the phenolic contents and antioxidant capacities of two Malaysian floral honeys. *Food Chemistry* 85: 513–518.

Aly, M.O., Elgohary, R., and Tayel, D.I. (2019). The effect of honey supplementation formula on delaying some fatigue markers on 1500 meters runners with no impact on performance. *International Journal of Sports Science* 9: 47–53.

Ariefdjohan, M.W., Martin, B.R., Lachcik, P.J., and Weaver, C.M. (2008). Acute and chronic effects of honey and its carbohydrate constituents on calcium absorption in rats. *Journal of Agricultural and Food Chemistry* 56 (8): 2649–2654.

Atkinson, F.S., Foster-Powell, K., and Brand-Miller, J.C. (2008). International tables of glycemic index and glycemic load values. *Diabetes Care* 31: 2281–2283.

Bandelow, S., Maughan, R., Shirrefs, S., et al. (2010). The effects of exercise, heat, cooling and rehydration strategies on cognitive function in football players. *Scandinavian Journal of Medicine & Science in Sports* 20: 148–160.

Barclay, J.K. and Hansel, M. (1991). Free radicals may contribute to oxidative skeletal muscle fatigue. *Canadian Journal of Physiology and Pharmacology* 69 (2): 279–284.

Batista, B., Da Silva, L., Rocha, B., et al. (2012). Multi-element determination in Brazilian honey samples by inductively coupled plasma mass spectrometry and estimation of geographic origin with data mining techniques. *Food Research International* 49: 209–215.

Berard, A., Bravo, G., and Gauthier, P. (1997). Meta-analysis of the effectiveness of physical activity for the prevention of bone loss in postmenopausal women. *Osteoporosis International* 7 (4): 331–337.

Bogdanov, S., Jurendic, T., Sieber, R., and Gallmann, P. (2008). Honey for nutrition and health: a review. *Journal of the American College of Nutrition* 27: 677–689.

Bonen, A., Mcdermott, J., and Hutber, C. (1989). Carbohydrate metabolism in skeletal muscle: an update of current concepts. *International Journal of Sports Medicine* 10 (6): 385–401.

Boonen, S., Body, J.J., Boutsen, Y., et al. (2005). Evidence-based guidelines for the treatment of post-menopausal osteoporosis: a consensus document of the Belgian Bone Club. *Osteoporosis International* 16 (3): 239–254.

Burke, L.M., Hawley, J.A., Wong, S.H., and Jeukendrup, A.E. (2011). Carbohydrates for training and competition. *Journal of Sports Sciences* 29: S17–S27.

Cermak, N.M., Gibala, M.J., and van Loon, L.J. (2012). Nitrate supplementation's improvement of 10-km time-trial performance in trained cyclists. *International Journal of Sport Nutrition and Exercise Metabolism* 22 (1): 64–71.

Chen, C.K., Ping, F.W.C., Ooi, F.K., and Mohamed, M. (2018). Effects of bee bread supplementation during four hours recovery on subsequent running time trial performance and selected physiological parameters. *International Journal of Pharma and Bio Sciences* 9 (2): 281–288.

Chepulis, L. and Starkey, N. (2008). The long-term effects of feeding honey compared with sucrose and a sugar-free diet on weight gain, lipid profiles, and dexa measurements in rats. *Journal of Food Science* 73 (1): 1–7.

Close, G.L., Hamilton, D.L., Philip, A., et al. (2016). New strategies in sport nutrition to increase exercise performance. *Free Radical Biology and Medicine* 98: 144–158.

Coyle, E.F., Coggan, A.R., Hemmert, M., and Ivy, J.L. (1986). Muscle glycogen utilization during prolonged strenuous exercise when fed carbohydrate. *Journal of Applied Physiology* 61: 165–172.

Currell, K. and Jeukendrup, A.E. (2008). Superior endurance performance with ingestion of multiple transportable carbohydrates. *Medicine & Science in Sports & Exercise* 40: 275–281.

Davies, K.J.A., Quintanilha, A.T., Brooks, G.A., and Packer, L. (1982). Free radicals and tissue damage produced by exercise. *Biochemical and Biophysical Research Communications* 107 (4): 1198–1205.

Deibert, P., König, D., Kloock, B., et al. (2010). Glycaemic and insulinaemic properties of some German honey varieties. *European Journal of Clinical Nutrition* 64: 762–764.

Di Meo, S. and Venditti, P. (2001). Mitochondria in exercise-induced oxidative stress. *Biological Signals and Receptors* 10 (1–2): 125–140.

Earnest, C.P., Lancaster, S.L., Rasmussen, C.J., et al. (2004). Low vs. high glycemic index carbohydrate gel ingestion during simulated 64-km cycling time trial performance. *Journal of Strength and Conditioning Research* 18 (3): 466–472.

Effendy, N.M., Mohamed, N., Muhammad, N., et al. (2012). The effects of Tualang honey on bone metabolism of postmenopausal women. *Evidence-Based Complementary and Alternative Medicine* ID 938574. doi: 10.1155/2012/938574.

Erejuwa, O.O., Sulaiman, S.A., and AbWahab, M.S. (2012). Honey: a novel antioxidant. *Molecules* 17: 4400–4423.

Essig, D.A. and Nosek, T.M. (1997). Muscle fatigue and induction of stress protein genes: a dual function of reactive oxygen species? *Canadian Journal of Physiology and Pharmacology* 22 (5): 409–428.

Gambacciani, M. and Levancini, M. (2014). Hormone replacement therapy and the prevention of postmenopausal osteoporosis. *Przeglad Menopauzalny* 13 (4): 213–220.

Gleeson, M. (2007). Immune function in sport and exercise. *Journal of Applied Physiology* 103 (2): 693–699.

Gleeson, M., Nieman, D.C., and Pedersen, B.K. (2004). Exercise, nutrition and immune function. *Journal of Sports Sciences* 22: 115–125.

Gmünder, F., Joller, P., Joller-Jemelka, H., et al. (1990). Effect of a herbal yeast food supplement and long-distance running on immunological parameters. *British Journal of Sports Medicine* 24 (2): 103–112.

Harper, L.D., Stevenson, E.J., Rollo, I., and Russell, M. (2017). The influence of a 12% carbohydrate-electrolyte beverage on self-paced soccer-specific exercise performance. *Journal of Science and Medicine in Sport* 20: 1123–1129.

Hawley, J.A., Schabort, E.J., Noakes, T.D., and Dennis, S.C. (1997). Carbohydrate-loading and exercise performance. *Sports Medicine* 24: 73–81.

Helge, J.W., Stallknecht, B., Pedersen, B.K., et al. (2003). The effect of graded exercise on IL-6 release and glucose uptake in human skeletal muscle. *Journal of Physiology* 546 (1): 299–305.

Hills, S. and Russell, M. (2017). Carbohydrates for soccer: a focus on skilled actions and half-time practices. *Nutrients* 10: 22.

Hiscock, N., Fischer, C.P., Pilegaard, H., and Pedersen, B.K. (2003). Vascular endothelial growth factor mRNA expression and arteriovenous balance in response to prolonged, submaximal exercise in humans. *American Journal of Physiology-Heart and Circulatory Physiology* 285 (4): H1759–1763.

Hoon, M.W., Jones, A.M., Johnson, N.A., et al. (2014). The effect of variable doses of inorganic nitrate-rich beetroot juice on simulated 2,000-m rowing performance in trained athletes. *International Journal of Sports Physiology and Performance* 9 (4): 615–620.

Jeukendrup, A.E. (2008). Carbohydrate feeding during exercise. *European Journal of Sport Science* 8: 77–86.

Jeukendrup, A.E. (2014). A step towards personalized sports nutrition: carbohydrate intake during exercise. *Sports Medicine* 44: 25–33.

Jeukendrup, A.E., Moseley, L., Mainwaring, G.I., et al. (2006). Exogenous carbohydrate oxidation during ultraendurance exercise. *Journal of Applied Physiology* 100 (4): 1134–1141.

Ji, L.L. (1999). Antioxidants and oxidative stress in exercise. *Proceedings of the Society for Experimental Biology and Medicine* 222 (3): 283–292.

Kelly, M.K., Wicker, R.J., Barstow, T.J., and Harms, C.A. (2009). Effects of N-acetylcysteine on respiratory muscle fatigue during heavy exercise. *Respiratory Physiology & Neurobiology* 165 (1): 67–72.

Khong, T.K., Selvanayagam, V.S., Hamzah, S.H., and Yusof, A. (2018). Effect of quantity and quality of pre-exercise carbohydrate meals on central fatigue. *Journal of Applied Physiology* 125 1021–1029.

Klein, S., Coyle, E.F., and Wolfe, R.R. (1994). Fat metabolism during low-intensity exercise in endurance-trained and untrained men. *American Journal of Physiology - Endocrinology and Metabolism* 267 (6): E934–940.

Lancaster, G.I., Halson, S.L., Khan, Q., et al. (2004). Effects of acute exhaustive exercise and chronic exercise training on type 1 and type 2 T lymphocytes. *Exercise Immunology Review* 10: 91–106.

Lancaster, S., Kreider, R.B., Rasmussen, C., et al. (2001). Effects of honey supplementation on glucose, insulin and endurance cycling performance. *FASEB Journal* 15: LB315.2001.

Lecarpentier, Y. (2007). Physiological role of free radicals in skeletal muscles. *Journal of Applied Physiology* 103: 1917–1918.

Lecoultre, V., Benoit, R., Carrel, G., et al. (2010). Fructose and glucose co-ingestion during prolonged exercise increases lactate and glucose fluxes and oxidation compared with an equimolar intake of glucose. *American Journal of Clinical Nutrition* 92 (5): 1071–1079.

Leeuwenburgh, C. and Heinecke, J.W. (2001). Oxidative stress and antioxidants in exercise. *Current Medicinal Chemistry* 8 (7): 829–838.

Levers, K., Dalton, R., Galvan, E., et al. (2016). Effects of powdered Montmorency tart cherry supplementation on acute endurance exercise performance in aerobically trained individuals. *Journal of the International Society of Sports Nutrition* 13: 22.

Li, D.Z., Zhang, Q.X., Dong, X.X., et al. (2013). Treatment with hydrogen molecules prevents RANKL-induced osteoclast differentiation associated with inhibition of ROS formation and inactivation of MAPK, AKT and NFkappa B pathways in murine RAW264.7 cells. *Journal of Bone and Mineral Metabolism* 32 (5): 494–504.

Lobo, R.A. (1995). Benefits and risks of estrogen replacement therapy. *American Journal of Obstetrics and Gynecology* 173 (3 Pt 2): 982–989.

Lovlin, R., Cottle, W., Pyke, I., et al. (1987). Are indices of free radical damage related to exercise intensity? *European Journal of Applied Physiology* 56 (3): 313–316.

Maimoun, L., Manetta, J., Couret, I., et al. (2006). The intensity level of physical exercise and the bone metabolism response. *International Journal of Sports Medicine* 27 (2): 105–111.

Maleki, B.H., Tartibian, B., Mooren, F.C., et al. (2016). A randomized controlled trial examining the effects of 16 weeks of moderate-to-intensive cycling and honey supplementation on lymphocyte oxidative DNA damage and cytokine changes in male road cyclists. *Cytokine* 88: 222–231.

Maruyama, H., Sakarmoto, T., Araki, Y., and Hara, H. (2010). Anti-inflammatory effects of bee pollen ethanol extract from Cistus sp. of Spanish on carrageenan-induced rat hind paw edema. *BMC Complementary and Alternative Medicine*10 (30): 1–11.

Mironczuk-Chodakowska, I., Witkowska, A.M., and Zujko, M.E. (2018). Endogenous non-enzymatic antioxidants in the human body. *Advances in Medical Sciences* 63 (1): 68–78.

Mohamed, H.A. and Ooi, F.K. (2013). Combined effects of a circuit training programme and honey supplementation on immune functions in young males. *Journal of Physical Activity, Sports and Exercise* 1 (1): 38–42.

Mohd Zuhri, M.I., Dawlath Mohamed, M.M., Ooi, F.K., and Ahmad, N.S. (2020). Combination of honey supplementation and countermovement jumping exercise influence bone turnover markers in young males. *Sains Malaysiana* 49 (7): 1639–1649.

Morillas-Ruiz, J., García, J.V., López, F., et al. (2006). Effects of polyphenolic antioxidants on exercise-induced oxidative stress. *Clinical Nutrition* 25 (3): 444–453.

Mosavat, M., Ooi, F.K., and Mohamed, M. (2014). Effects of honey supplementation combined with different jumping exercise intensities on bone mass, serum bone metabolism markers and gonadotropins in female rats. *BMC Complementary and Alternative Medicine* 4 (14): 126.

Nagai, T., Inoue, R., Inoue, H., and Suzuki, N. (2001). Antioxidative activities of some commercially honeys, royal jelly and propolis. *Food Chemistry* 75: 237–240.

Nagai, T., Nagashima, T., Myoda, T., and Inoue, R. (2004). Preparation and functional properties of extracts from bee bread. *Nahrung/Food* 48 (3): 226–229.

Nagai, T., Nagashima, T., Suzuki, N., and Inoue, R. (2005). Antioxidant activity and angiotensin i-converting enzyme inhibition by enzymatic hydrolysates from bee bread. *Zeitschrift für Naturforsch C* 60: 1–2.

Nicholas, C.W., Williams, C., Lakomy, H.K., et al. (1995). Influence of ingesting a carbohydrate-electrolyte solution on endurance capacity during intermittent, high-intensity shuttle running. *Journal of Sports Sciences* 13: 283–290.

Nieman, D.C. (1994). Exercise, upper respiratory tract infection, and the immune system. *Medicine & Science in Sports & Exercise* 26 (2): 128–139.

Nieman, D.C., Nehlsen-Cannarella, S.L., Fagoaga, O.R., et al. (1998). Influence of mode and carbohydrate on the cytokine response to heavy exertion. *Medicine & Science in Sports & Exercise* 30: 671–678.

Nieman, D.C., Williams, A.S., Shanely, R.A., et al. (2010). Quercetin's influence on exercise performance and muscle mitochondrial biogenesis. *Medicine & Science in Sports & Exercise* 42 (2): 338–345.

Novelli, G.P., Bracciotti, G., and Falsini, S. (1990). Spin trappers and vitamin E prolong endurance to muscle fatigue in mice. *Free Radical Biology and Medicine* 8 (1): 9–13.

O'Neill, C.A., Stebbins, C.L., Bonigut, S., et al. (1996). Production of hydroxyl radicals in contracting muscle of cats. *Journal of Applied Physiology* 81 (3): 1197–1206.

Ooi, F.K. (2018). Bone properties and blood bone metabolism markers in response to exercises and honey supplementation: a contemporary review. *Malaysian Journal of Medical Research (MJMR)* 2 (2): 68–74.

Ooi, F.K. and Aziz, A. (2017). Bone metabolism markers in response to circuit training and honey supplementation in young males. *Malaysian Journal of Medical Research (MJMR)* 1 (4): 28–34.

Ooi, F.K., Ismail, N., and Abdullah, N.Y. (2011). Effects of combined aerobic dance exercise and honey supplementation on bone turnover markers in young females. *Asian Journal of Exercise and Sports Science* 8 (1): 1–11.

Ooi, F.K., Singh, R., Singh, H.J., and Umemura, Y. (2009). Minimum level of jumping exercise required to maintain exercise-induced bone gains in female rats. *Osteoporosis International* 20 (6): 963–972.

Ooi, F.K., Tavafzadeh, S.S., Hung, L., et al. (2014). Tibial bone mineral density, geometry, and mechanical properties in response to high-impact exercise and honey supplementation in rats. *Asian Journal of Exercise and Sports Science* 11 (2): 12–24.

Overdevest, E., Wouters, J.A., Wolfs, K.H.M., et al. (2018). Citrus flavonoid supplementation improves exercise performance in trained athletes. *Journal of Sports Science and Medicine* 17 (1): 24–30.

Ping, F.W.C., Ooi, F.K., Mohamed, M., and Chen, C.K. (2018). Effects of bee bread supplementation on endurance running performance and total antioxidant status in recreational athletes. *International Journal of Applied Research in Natural Products* 11 (1): 17–23.

Powers, S.K., DeRuisseau, K.C., Quindry, J., and Hamilton, K. (2004). Dietary antioxidants and exercise. *Journal of Sports Sciences* 22 (1): 81–94.

Powers, S.K. and Howley, E.T. (1997). *Exercise Physiology: Theory and Application to Fitness and Performance*, 3e. Madison, WI: Brown and Benchmark Publishers.

Powers, S.K. and Jackson, M.J. (2008). Exercise-induced oxidative stress: cellular mechanisms and impact on muscle force production. *Physiological Reviews* 88 (4): 1243–1276.

Prasertsri, P., Tong-un, T., Roengrit, T., et al. (2016). Cashew apple juice supplementation increases endurance and strength performance in cyclists. *Journal of Exercise Physiology* 19 (5): 59–70.

Rahim, M., Ooi, F.K., and Hamid, W.Z.W.A. (2016). Changes of bone metabolism markers and muscular performance with combined aerobic dance exercise and honey supplementation in adult women. *BMJ Open Sport & Exercise Medicine* 1 (6): 186–197.

Rahim, M., Ooi, F.K., and Hamid, W.Z.W.A. (2017). Blood immune function parameters in response to combined aerobic dance exercise and honey supplementation in adult women. *Journal of Traditional and Complementary Medicine* 7 (2): 165–171.

Rao, L.G. and Rao, A.V. (2013). Oxidative stress and antioxidants in the risk of osteoporosis -Role of the antioxidants lycopene and polyphenols. *Topics in Osteoporosis* 5: 117–144.

Reid, I.R., Bolland, M.J., and Grey, A. (2014). Effects of vitamin D supplements on bone mineral density: a systematic review and meta-analysis. *Lancet* 383 (9912): 146–155.

Reid, M.B., Hack, K.E., Franchek, K.M., et al. (1992). Reactive oxygen in skeletal muscle. I. Intracellular oxidant kinetics and fatigue in vitro. *Journal of Applied Physiology* 73 (5): 1797–1804.

Reid, M.B., Stokić, D.S., Koch, S.M., et al. (1994). N-acetylcysteine inhibits muscle fatigue in humans. *Journal of Clinical Investigation* 94 (6): 2468–2474.

Rowlands, D.S., Swift, M., Ros, M., and Green, J.G. (2012). Composite versus single transportable carbohydrate solution enhances race and laboratory cycling performance. *Applied Physiology, Nutrition, and Metabolism* 37 (3): 425–436.

Salehian, O., Rashidi, M., and Sedaghat, M. (2014). Oral supplementation of natural honey and levels of inflammatory and anti-inflammatory plasma cytokines during 10-week of intensive tread-mill training in endurance-trained athletes. *Biomedical Research* 25 (4): 459–462.

Samsani, A.-N.M., Ooi, O.F.K., Chen, C.K., and Ismail, M.S. (2021). Effects of Acacia honey and sports drink after exercise-induced dehydration on selected physiological parameters, urine osmolality and subsequent running performance in the heat. *Sains Malaysiana* 50 (8): 2293–2308.

Santen, R.J., Pinkerton, J., McCartney, C., and Petroni, G.R. (2001). Risk of breast cancer with progestins in combination with estrogen as hormone replacement therapy. *Journal of Clinical Endocrinology and Metabolism* 86 (1): 16–23.

Schneider, M., Coyle, S., Warnock, M., et al. (2013). Anti-microbial activity and composition of manuka and portobello honey. *Phytotherapy Research* 27: 1162–1168.

Sen, C.K. (1995). Oxidants and antioxidants in exercise. *Journal of Applied Physiology* 79 (3): 675–686.

Shafin, N., Othman, Z., Zakaria, R., and Hussain NHN, N. (2014). Tualang honey supplementation reduces blood oxidative stress levels/activities in postmenopausal women, *ISRN Oxidative Medicine*, Article ID 364836.

Shindoh, C., DiMarco, A., Thomas, A., et al. (1990). Effect of N-acetylcysteine on diaphragm fatigue. *Journal of Applied Physiology* 68 (5): 2107–2113.

Silva, P.M., Gauche, C., Gonzaga, L.V., et al. (2016). Honey: chemical composition, stability and authenticity. *Food Chemistry* 196: 309–323.

Slattery, K.M., Dascombe, B., Wallace, L.K., et al. (2014). Effect of N-acetylcysteine on cycling performance after intensified training. *Medicine & Science in Sports & Exercise* 46 (6): 1114–1123.

Stanciu, O.G., Marghitas, L.A., and Dezmirean, D.S. (2009). Macro- and oligomineral elements from honey bee-collected pollen and bee bread harvested from Transylvania (Romania). *Journal of Animal Science and Biotechnology* 66 (1–2): 276–281.

Stevenson, E.J., Watson, A., Theis, S., et al. (2017). A comparison of isomaltulose versus maltodextrin ingestion during soccer-specific exercise. *European Journal of Applied Physiology* 117: 2321–2333.

Tartibian, B. and Maleki, B.H. (2012). The effects of honey supplementation on seminal plasma cytokines, oxidative stress biomarkers and antioxidants during 8 weeks of intensive cycling training. *Journal of Andrology* 3: 449–461.

Tavafzadeh, S.S., Ooi, F.K., Oleksandr, K., and Sulaiman, S. (2011). Effect of a combination of jumping exercise and honey supplementation on the mass, strength and physical dimensions of bones in young female rats. *Journal of ApiProduct & ApiMedical Science* 3 (1): 26–32.

Triplett, D., Doyle, J.A., Rupp, J.C., and Benardot, D. (2010). An isocaloric glucose-fructose beverage's effect on simulated 100-km cycling performance compared with a glucose-only beverage. *International Journal of Sport Nutrition and Exercise Metabolism* 20: 122–131.

Umemura, Y., Ishiko, T., Yamauchi, T., et al. (1997). Five jumps per day increase bone mass and breaking force in rats. *Journal of Bone and Mineral Research* 12 (9): 1480–1485.

Walsh, N.P., Gleeson, M., Shephard, R.J., et al. (2011). Position statement Part one: immune function and exercise. *Exercise Immunology Review* 17: 6–63.

Weidner, T. and Schurr, T. (2003). Effect of exercise on upper respiratory tract infection in sedentary subjects. *British Journal of Sports Medicine* 37: 304–306.

Wu, C.-L. and Williams, C. (2006). A low glycemic index meal before exercise improves endurance running capacity in men. *Journal of Sport Nutrition and Exercise Metabolism* 16 (5): 510–527.

Yang, X.P., Guo, D., Zhang, J.M., and Wu, M.C. (2007). Characterization and antitumor activity of pollen polysaccharide. *International Immunopharmacology* 7: 401–408.

Zaid, S.S.M., Sulaiman, S.A., Othman, N.H., et al. (2012). Protective effects of Tualang honey on bone structure in experimental postmenopausal rats. *Clinics* 67: 779–784.

Zhang, B., Xie, Q.Y., Quan, Y., et al. (2015). Reactive oxygen species induce cell death via Akt signaling in rat osteoblast-like cell line ROS 17/2.8. *Toxicology and Industrial Health* 31 (12): 1236–1242.

# 20

# Medicinal Properties of Royal Jelly

*Wendy Wai Yeng Yeo, Usha Sundralingam and Sathiya Maran*

The concept of apiculture, or "functional food," has been hyped in recent years because of awareness of promoting better physiological and psychological well-being compared with conventional remediated and nutritional food (Pasupuleti et al. 2017). Apiculture, or beekeeping, is the science of prolonging, sustaining, and retaining well-being using products from beehives, including honey, bee bread, bee venom, bee pollen, propolis, and royal jelly (RJ). Among these products, RJ, which is chemically synthesized by bees from plant sources and engineered for their own use, has been reported for its medical benefits (Schmidt 1997).

RJ is a creamy substance secreted by the worker *Apis mellifera* (honeybees) from the mandibular and hypopharyngeal glands (Kunugi and Mohammed Ali 2019). The queen larvae consume RJ throughout their lifetimes to become queen bees. The nutritional value of RJ contributes to the queen bees' large size (being double that of worker bees), functioning sexual organs, and long lifespan (5–6 years) (Isidorov et al. 2012; Izuta et al. 2009; Khazaei et al. 2018). Some studies have also reported that RJ is formed in the worker bees' stomachs from incomplete digestion of honeydew (Melliou and Chinou 2014; Moselhy et al. 2013). Besides the queen larvae, bee larvae also feeds on RJ for the first 3 days of their lives. To date, many postulations have been made on the medicinal benefits of RJ, ranging from promoting healthy aging and longevity to its antimicrobial properties. However, studies on the effects of RJ remain scarce mainly because of technical and ethical reasons. In this chapter, the antimicrobial activities and health benefits of RJ are explored.

## Chemical Compositions and Bioactive Compounds of RJ

RJ, a non-homogenous jelly-like viscous substance, has a particularly sharp odor and a sweet and sour taste. Being partially soluble in water, RJ is an acid colloid (pH, 3.4–4.5) with a density of 1.1 g/mL (Takenaka and Echigo 1980). The water content and age of RJ contribute to its viscosity, which increases gradually when stored at room temperature or in a cold environment (5°C). The increase in viscosity could be due to the continued enzymatic activities and interaction between the lipid and protein fractions, resulting in an increase in water0insoluble nitrogenous compounds, coupled with the reduction in soluble nitrogen and free amino acids (Ramanathan et al. 2018).

The main bioactive compounds of RJ, which are water, sugar, proteins, lipids, vitamins, and some mineral salts, have been widely reported in the literature. Water, the main composition, ranges between 60% and 70% (Melliou and Chinou 2005) followed by carbohydrates at 11%–23% (Sabatini et al. 2009; Sesta 2006), proteins at 9%–18% (Malka et al. 2009; Melliou and Chinou 2005; Ramadan and Al-Ghamdi 2012), and lipids at 4%–8% (Malka et al. 2009; Nagai and Inoue 2004; Sabatini et al. 2009). Studies have also shown the presence of polyphenols, vitamins, and mineral salts in low amounts with other unknown substances present in traces ranging between 0.8% and 3.0% (Šimúth et al. 2004; Zhang et al. 2012). These values obtained from various studies are somewhat similar, but there is high variability displayed for some parameters (sugars and lipids). However, it should be taken into account that natural variations of bioactive compounds depend on the biodiversity of flora species present in the different ecosystems.

*Honey: Composition and Health Benefits*, First Edition. Edited by Md. Ibrahim Khalil, Gan Siew Hua, and Bey Hing Goh.

## Carbohydrates

Carbohydrates represent about 30% of the dry weight (Sabatini et al. 2009; Sesta 2006) and are important indicators of the authenticity of RJ (Daniele and Casabianca 2012; Lercker et al. 1981; Serra Bonvehi 1992). Monosaccharides fructose, glucose, and sucrose are sugars found abundantly in RJ, and the total content of fructose and glucose accounts for 90% of the total sugars in RJ (El-Guendouz et al. 2020). The composition of these monosaccharides in RJ is relatively constant with a distribution similar to that of honey.

The constitution of glucose (50%–70%) in RJ varies according to the age and beekeeping foods. Sucrose is always present but often in highly variable concentrations (Lercker et al. 1986). Studies have also indicated the presence of oligosaccharides such as maltose, trehalose, melibiose, ribose, and other sugars in small amounts (Finke 2005; Kheyri et al. 2012; Lercker et al. 1981, 1986; Šimúth 2001). It has been established that determination of sugar content could provide important information for identifying a characteristic pattern of comparison with honey and to determine its authenticity. Establishing the quality of RJ can be done by identifying possible adulteration with honey or sugars.

## Proteins

Proteins represent a major component (>50%) of dry matter of RJ and can be classified into minor and major proteins (Bılikovà et al. 2002; Buttstedt et al. 2014; Furusawa et al. 2008). RJ is basically abundant with soluble major RJ proteins (MRJPs) known as apalbumin (Šimúth 2001). Consisting of numerous essential amino acids, similar to ovalbumin and casein, these proteins are responsible for specific physiological roles in the development of queen honeybees (Nozaki et al. 2012; Schmitzova et al. 1998). These MRJPs are synthesized in the cephalic glands of honeybees and denote the main intake of essential amino acids needed for the development of the queen bee and larvae (Hanes and Šimuth 1992).

The major RJ proteins are MRJP1, MRJP2, MRJP3, MRJP4, MRJP5, MRJP6, MRJP7, MRJP8, and MRJP9, ranging with molecular masses of 49–87 kDa (Albert and Klaudiny 2004). The 10 most commonly found essential amino acids in MRJPs are arginine, histidine, isoleucine, leucine, lysine, methionine, phenylalanine, threonine, tryptophan, and valine. The amino acid content in each MRJPs can be classified to MRJP1 (48%), MRJP2 (47%), MRJP3 (39.3%), MRJP4 (44.5%), MRJP5 (51.4%), MRJP6 (42%), MRJP7 (48.3%), MRJP8 (49.5%), and MRJP9 (47.3%). The predominant amino acids in MRJP1, MRJP2, and MRJP4 are leucine and valine; in MRJP3, they are arginine and lysine (Schmitzova et al. 1998). Whereas the major amino acids in MRJP5 are arginine and methionine, leucine is the predominant one MRJP6, MRJP7, and MRJP8. Isoleucine is the predominant amino acid in MRJP9 (Ramanathan et al. 2018).

Of the nine MRJPs, MRJPs 1–5 exhibit nutritional functions, while MRJPs 6–9 represent non-nutritional functions. MRJ1 and MRJ2 have been known as major allergens of RJ, but another in vivo study showed these proteins were responsible for the production of mouse macrophages tumor necrosis factor-α (Rosmilah et al. 2008; Šimúth et al. 2004). Studies have reported that MRJP3 influences immune responses of T cells by suppressing the production of interleukin (IL)-4, IL-2 and interferon (IFN)-ɤ, thus reducing production of immunoglobulin (Ig)E and IgG1, causing an anti-allergic reaction (Kohno et al. 2004; Okamoto et al. 2003; Tamura et al. 2009). A study by Kamakura et al. (2001b) showed that MRJP1 stimulates cell proliferation and induces production of albumin in the absence of fetal bovine serum. The study postulated that MRJP1 acts as a biosimilar or substitute for growth factors. Another study by Nakaya and colleagues (2007) also reported that crude RJ protein inhibits bisphenol A–induced proliferation of human breast cancer cell lines. A study looking into the antioxidative properties of RJ peptides, specifically three peptides (Lys-Tyr, Arg-Tyr, and Tyr-Tyr) that contain Tyr residues at the C-terminal, exhibited strong hydroxyl radical and hydrogen peroxide scavenging activity. This was attributed to their ability to act as free radical scavengers (Guo et al. 2009).

In addition to the bioactive properties of RJ proteins and peptides, MRJPs contribute to the nutritional aspect in the diet of queen bees (Tamura et al. 2009). MRJP1, MRJP4, and MRJP5 are responsible for the main intake of essential amino acids, and MRJP2 and MRJP5 are essential nitrogen reserves for the growth of queen bees (Albert et al. 1999). MRJP3, a polymorphic protein, is responsible for its nitrogen supply. In addition, MRJPs are also responsible in the production of other bee biproducts such as formation of pollen pellets and pollen bread (Malka et al. 2009). Furthermore, MRJPs have an important role in distinguishing between queen bees and worker bees (Buttstedt et al. 2013).

## Lipid Content

The lipid composition of RJ is relatively low with varied concentrations (8%–18%) of dry matter. Nevertheless, the lipid and fatty acid content of RJ is unique and is a crucial feature of RJ (Terada et al. 2011). In terms of lipid fraction in RJ, it is predominant with organic acids (80%–90%), with free mono- and dihydroxy acids and dicarboxylic acids containing 8–12 atoms, unlike the tri-, di-, and monoglyceride forms with minor amounts of free fatty acids found in animals and plants (Sabatini et al. 2009). It is composed of 80%–85% fatty acids, 4%–10% phenols, 5%–6% waxes, 3–4% steroids, and 0.4%–0.8% phospholipids (Terada et al. 2011). Studies have identified the main fatty acid fractions in RJ to be trans-10-hydroxy-2-decenoic (10-HDA) (32%), glucinic acid (24%), and 10-hydroxydecanoic acids (10-HDAA) (22%), accounting for more than 60% of the total amount of organic acids found in RJ (Xue et al. 2017). The presence of major fatty acid, trans-10-HDA in RJ, is unique in that its presence in other bee products or even other natural products has not been reported to date. Hence, the content of 10-HDA in RJ has been used as a benchmark to determine the authenticity and quality of the RJ. RJs with 10-HDA content of more than 1.4% are considered fresh and authentic samples (Antinelli et al. 2003; Sabatini et al. 2009). It is crucial to determine the authenticity of RJ because of cases of adulteration with honey or sugars, which is known to reduce the protein and lipid component while increasing the concentration of minor sugars, resulting in RJ's being insoluble in alkaline medium (Boselli et al. 2003; Jianke and Shenglu 2003).

These unique features of RJ are known to contribute to its various pharmacological effects. The two main fatty acids of RJ (10-HDA and 10-HDAA) also contribute to the important biological role in the development of the colony strategies (Fratini et al. 2016). Previous studies have also exhibited fatty acids in RJ to be responsible for its anticancer properties (Miyata and Sakai 2018; Yang et al. 2018). In addition, studies have shown that 10-HDA of RJ promoted the collagen production of treated fibroblast cell lines by inducing transforming growth factor-β1 (TGF-β1), which is a potent immunosuppressive cytokine (Koya-Miyata et al. 2004).

### Vitamins

The minor components of RJ include different classes of chemicals such as vitamins, trace and mineral elements, phenolic compounds, and nucleotides. RJ is known to contain some vitamins, specifically water-soluble vitamins. Fat-soluble vitamins such as vitamins A, D, E, and K are absent (Morita et al. 2012). In regard to the water-soluble vitamins, RJ is rich in vitamin B, predominantly pantothenic acid ($B_5$), thiamine ($B_1$), riboflavin ($B_2$), niacin ($B_3$), pyridoxal ($B_6$), biotin ($B_8$), folic acid ($B_9$), and cobalamin ($B_{12}$), whereas niacin ($B_3$) and ascorbic acid (vitamin C) are present in trace amounts. Because the principal sources of vitamins in RJ are derived from the pollen of the flowers collected by worker bees, variations in the vitamin content of RJ are subjected to seasonal changes (Sabatini et al. 2009).

### Minerals and Other Elements

Minerals and other elements represent about 4%–8% of RJ dry matter. The most abundant elements in RJ are potassium, calcium, sodium, magnesium, zinc, iron, copper, and manganese. In addition, trace elements such as aluminium, barium, strontium, bismuth, cadmium, mercury, lead, tin, tellurium, thallium, wolfram, antimony, chromium, nickel, titanium, titanium, cobalt, and molybdenum in RJ were also detected (Miguel and El-Guendouz 2017). RJ trace and mineral elements are known to be unstable quantitatively because they do not only rely on the biological factors of the bees but also external factors such as the environment, source of feed, and production period (Bărnuţiu et al. 2011).

The polyphenolic compounds such as flavonoids and cinnamic acid derivatives present in RJ are another group of compounds that are crucial in exhibiting numerous biological and therapeutic properties. Some of the primary flavonoids present in RJ include flavonoles (e.g. quercetin, kaempferol, galangin, and fisetin), flavanones (e.g. pinocembrin, naringin and hesperidin), flavones (e.g. apigenin, acacetin, chrysin, and luteolin), and isoflavonoids (genistein and formononetin) (López-Gutiérrez et al. 2014). A previous study attributed the antioxidant activity of RJ to the tyrosyl residues of amino acids that contain a phenolic group. The presence of the phenolic hydrogens is known to act as hydrogen donating radical scavengers based on their chemical structures (Buratti et al. 2007).

Finally, low quantities of nucleotides (e.g. adenosine, guanosine, cytidine, iridine); phosphates (50-monophosphate, adenosine 50-diphosphate, and adenosine 50-triphosphate); acetylcholine; and gluconic, benzoic, malic, citric, and lactic acids can also be found in RJ. These are assumed to have originated from nurse bees (Fratini et al. 2016; Xue et al. 2017).

## Antimicrobial Properties and Activities of Royal Jelly

The antibacterial and antiviral effects of honey have been recommended by the World Health Organization for symptomatic treatment, especially for RJ (World Health Organization 2001; Wahdan 1998). It has been reported that RJ reduces bacterial motility and synergistically increases the antioxidant activities (Paul et al. 2007).

Promising antibacterial properties of RJ have been reported since ancient times, and recent reports have described their inhibitory effect against a numbers of gram-positive and gram-negative bacteria (Blum et al. 1959; Brady et al. 1996; Cooper et al. 2002). It is largely attributed to RJ's acidity, low pH, osmolarity, and hydrogen peroxide production (Bogdanov et al. 2002).

The quality of RJ is denoted based on the content of trans-10-HDA acid, which is a major fatty acid component and the most active ingredient of RJ (Duong et al. 2020). Studies have reported that 10-HDA acid is the component that exhibits the antibacterial effect in RJ (Blum et al. 1959; Melliou and Chinou 2005). The said component has also been reported to promote immunity and contribute toward antibacterial and anticancer activities (Ramadan and Al-Ghamdi 2012; Ramanathan et al. 2018).

The common antibacterial peptides found in RJ are apisimin; apalbumin α; jelleins I, II, III, and IV; and royalactin (Bărnuţiu et al. 2011). These peptides and proteins in RJ have also been reported to exhibit strong antibacterial properties (Fontana et al. 2004; Fujiwara et al. 1990). The short extension of the peptides, poor cytolytic and mast cell degranulating activities, and broad-spectrum antibiosis are its key antibacterial properties (Fontana et al. 2004).

To date, several studies have elucidated the different mechanisms of which RJ exert antibacterial effect (Bulet et al. 1999; Fratini et al. 2016). However, the exact mode of action is yet to be explained in detail. The cleavage products of MRJPs, jelleins, have been reported to show broad-spectrum activity against gram-positive organisms *Staphylococcus aureus*, *Staphylococcus saprophyticus*, and *Bacillus subtilis*; the gram-negative organisms *Escherichia coli*, *Enterobacter cloacae*, *Klebsiella pneumoniae*, *Pseudomonas aeruginosa*; and the yeast *Candida albicans* (Fontana et al. 2004).

Another component of RJ is royalisn, which exhibits activity against various gram-positive strains, including *Staphylococcus* spp., *Streptococcus* spp., *B. subtilis*, *Micrococcus luteus*, *Sarcina lutea*, *Clostridium* spp., *Corynebacterium* spp., *Lactobacillus helveticus*, *Paenibacillus larvae*, and *Leuconostoc* spp. No inhibition has been observed against the gram-negative organisms *E. coli* and *Serratia marcescens* (Cornara et al. 2017).

McCleskey and Melampy (1939) were the first to show the antimicrobial properties of RJ against gram-positive and gram-negative bacteria followed by subsequent demonstration against a number of bacteria reported over the years. It is notable that antibacterial effects have been observed against *S. aureus*, *P. aeruginosa*, *Streptococcus alactolyticus*, *Salmonella cholearasuis*, *Vibrio parahaemolyticus*, *B, subtilis*, and *E. coli*. Table 20.1 summarizes recent findings of the antibacterial effects of RJ.

Benefits of RJ toward health-related properties have also been extensively studied in vivo, in vitro, and in randomized clinical studies. Animal studies have examined and reported benefits of RJ in anti-aging (Qiu et al. 2020), osteoporosis (Hidaka et al. 2006), fatigue (Kamakura et al. 2001a), wound healing (Kamakura et al. 2001a; Lin et al. 2020; Sedaghat et al. 2010), hyperglycemia (Yoshida et al. 2016), gastrointestinal motility (Miyauchi-Wakuda et al. 2019), and longevity (Inoue et al. 2003). Also, a growing number of clinical trials have been performed to access RJ's effect toward different medical conditions (Table 20.2).

**Table 20.1** Antibacterial properties of royal jelly.

| Bee Native | Bacterial Strain | MBC | MIC | Findings | Reference |
|---|---|---|---|---|---|
| Chile | *Staphylococcus aureus* | 0.1:1 | 0.05:1 | RJ demonstrated increased inhibitory action against the growth of bacteria in solution and biofilm formation on surfaces. | Schuh et al. 2019 |
| Japan | *Pseudomonas aeruginosa* | NA | NA | RJ exhibited a low antimicrobial activity against *P. aeruginosa*. | Susilowati et al. 2017 |
| Taiwan | Gram-positive bacteria (*Staphylococcus aureus, Staphylococcus intermedius B, Staphylococcus Xylosus*, and *Streptococcus alactolyticus*) and gram-negative bacteria (*hemolytic Escherichia coli, P. aeruginosa, Salmonella cholearasuis*, and *Vibrio parahaemolyticus*) | 23–44 μM (gram-positive bacteria) and 40–43 μM (gram-negative bacteria) | 33–66 μM (gram-positive bacteria) and 74–78 μM (gram-negative bacteria) | 10-HDA showed inhibition against the production of pro-inflammatory cytokines TNF-α, IL-1β, and IL-8 in WiDr cells. | Yang et al. 2018 |

*(Continued)*

**Table 20.1** (Continued)

| Bee Native | Bacterial Strain | MBC | MIC | Findings | Reference |
|---|---|---|---|---|---|
| Canada | *Bacillus subtilis* (ATCC 6633) and *E. coli* (ATCC 14948) | 13.8µ g/mL (*E. coli*) and 13.8–1.24 µg/mL (*B.s ubstillis*) | | Glycoproteins as active principal molecules that cause agglutination and a rapid concentration-dependent bactericidal effect on both gram-negative *E. coli* and gram-positive *B. subtilis*. | Brudzynski and Sjaarda 2015 |
| Egypt | *S. aureus* (RCMBA 2004) and *B. subtilis* (RCMBA 6005) as gram-positive bacteria and *P. aeruginosa* (RCMBA 1002) and *E. coli* (RCMBA 5003) as gram-negative bacteria | NA | 15.63 µg/mL (*S. aureus*) and 7.81 µg/mL (*B. subtilis*) | RJ inhibited the growth of fungi and gram-positive and gram-negative bacterial strains. | Moselhy et al. 2013 |
| Argentina | *Enterococcus faecalis* 1 (ATCC 29212), *S. aureus* MS 1 (ATCC 25923), *Streptococcus agalactiae* 1 (ATCC 27956), *Streptococcus dysgalactiae* 2 (ATCC 27957), and *Micrococcus luteus* (ATCC 9341) | NA | 2.3 mg/mL | 10-HDA standard showed the highest antibacterial activity against *S. agalactiae*. | Garcia et al. 2013 |

10-HDA, trans-10-hydroxy-2-decenoic; IL, interleukin; NA, not available; RJ, royal jelly; TNF-α, tumor necrosis factor-α

**Table 20.2** Clinical studies on royal jelly against various conditions.

| NCT Number | Study Design | Intervention | Duration | Outcomes | Reference |
|---|---|---|---|---|---|
| RCT201009263278N2 | Randomized double-blind placebo-controlled trial | 32 patients with ulcers allocated to receive topical RJ | October 2010–June 2011 | 5% topical RJ showed superiority over placebo for the treatment of DFU | Siavash et al. 2015 |
| NA | Randomized, double-blind, placebo-controlled trial | Patients were divided into four groups: two bacterial groups receiving either antibiotics or antibiotics + bee products and two viral groups treated with either placebo or bee products | 1 October 2015–February 2016 | Patients with viral infection receiving a mixture product showed earlier improvement compared with the placebo group | Seçilmiş and Silici 2020 |
| NCT02379468 | Phase IIIb, prospective, open-label, randomized, controlled, parallel group trial | 3 months' duration comparing topical application of RJ and panthenol ointment with panthenol ointment only for diabetic foot ulcers | April 2013–October 2014 | Patients with diabetes showed successful healing of foot ulceration with panthenol ointment; time frame, 12 weeks | |
| NA | Double-blind randomized trial | 200 postmenopausal women (45–60 years) received either 1000 mg of RJ capsules or placebo daily for 8 weeks | June–November 2018 | Daily consumption of oral RJ (1000 mg) for 8 weeks alleviated menopausal symptoms | Sharif and Darsareh 2019 |
| NA | Randomized, placebo-controlled trial | 40 participants with mild hypercholesterolemia were divided into two groups (experimental and placebo) and were given nine capsules (350 mg/capsule) of RJ or placebo/day, for 3 months | April 2013–July 2014 | Lowered total cholesterol and LDL cholesterol levels through improving the levels of DHEA-S | Chiu et al. 2017 |

*(Continued)*

**Table 20.2** (Continued)

| NCT Number | Study Design | Intervention | Duration | Outcomes | Reference |
|---|---|---|---|---|---|
| – | Prospective, randomized trial | 103 patients received mouthwash therapy and RJ | July 2011–January 2012 | Combination of mouthwash and RJ demonstrated improved signs and symptoms of oral mucositis | Erdem and Güngörmüs 2014 |
| IRCT201011131197N5 | Parallel design randomized trial | 50 women with type 2 diabetes in supplemented ($n = 25$) and placebo ($n = 25$) groups were given a daily dose of 1000 mg of RJ soft gel or placebo for 8 weeks | December 2010–March 2011 | Increased erythrocyte superoxidase dismutase and glutathione peroxidase activities and decreased malondialdehyde levels | Pourmoradian et al. 2014 |
| UMIN000004057 | Randomized, double-blind, placebo-controlled study | Participants received daily either oral tablets containing 1.2 g/day (low-dose group, 3 pRJ and 9 placebo tablets per day), 4.8 g/d pRJ (high-dose group, 12 pRJ tablets per day) or starch placebo tablets (12 placebo tablets per day) | June 2012 and September 2012 | Protease-treated RJ treatment might not improve but rather attenuate the progression of decrease in muscle strength in older adults | Meng et al. 2017 |
| NA | Double-blind, placebo-controlled trial | 50 participants were given 3000 mg of RJ capsules per day (Natural Life, Frengrove, Australia), and the placebo group received three capsules of glycerine | – | RJ intake (3 g/day) reduced serum glucose, increased serum ApoA-I, and modified the ApoB-to-ApoA-I ratio after 8 weeks | Khoshpey et al. 2016 |
| NA | Randomized, triple-blind, placebo-controlled trial | 110 medical sciences students with PMS were given one RJ capsule orally per day, starting on the first day of menstruation and continued the same treatment daily throughout two consecutive menstrual cycles | December 2011 and October 2012 | 2 months' consumption of RJ was effective in reducing PMS | Taavoni et al. 2014 |
| NA | Randomized placebo-controlled, double-blind trial | 61 healthy volunteers received 3000 mg of RJ for 3 months | – | Improved erythropoiesis, glucose tolerance and mental health | Morita et al. 2012 |

Apo, apolipoprotein; DFU, diabetic foot ulcers; DHEA-S, dehydroepiandrosterone sulfate; LDL, low-density lipoprotein; NA, not available; NCT, National Clinical Trial; PMS, premenstrual syndrome; pRJ, protease-treated royal jelly; RCT, randomized controlled trial; RJ, royal jelly.

## Health Benefits of Royal Jelly

RJ has been used as a dietary supplement worldwide with a list of healing properties largely because of its beneficial effects, including antioxidant, antitumor, and anti-inflammatory effects. The individually isolated bioactive compounds found in RJ offer a range of health benefits as shown in Figure 20.1

### Cancer

An in vitro study reported that RJ lipophilic fraction has the potential to exert antiproliferative effects on SH-SY5Y human neuroblastoma cells (Gismondi et al. 2017). Another study showed that combination of RJ and human interferon-$\alpha$

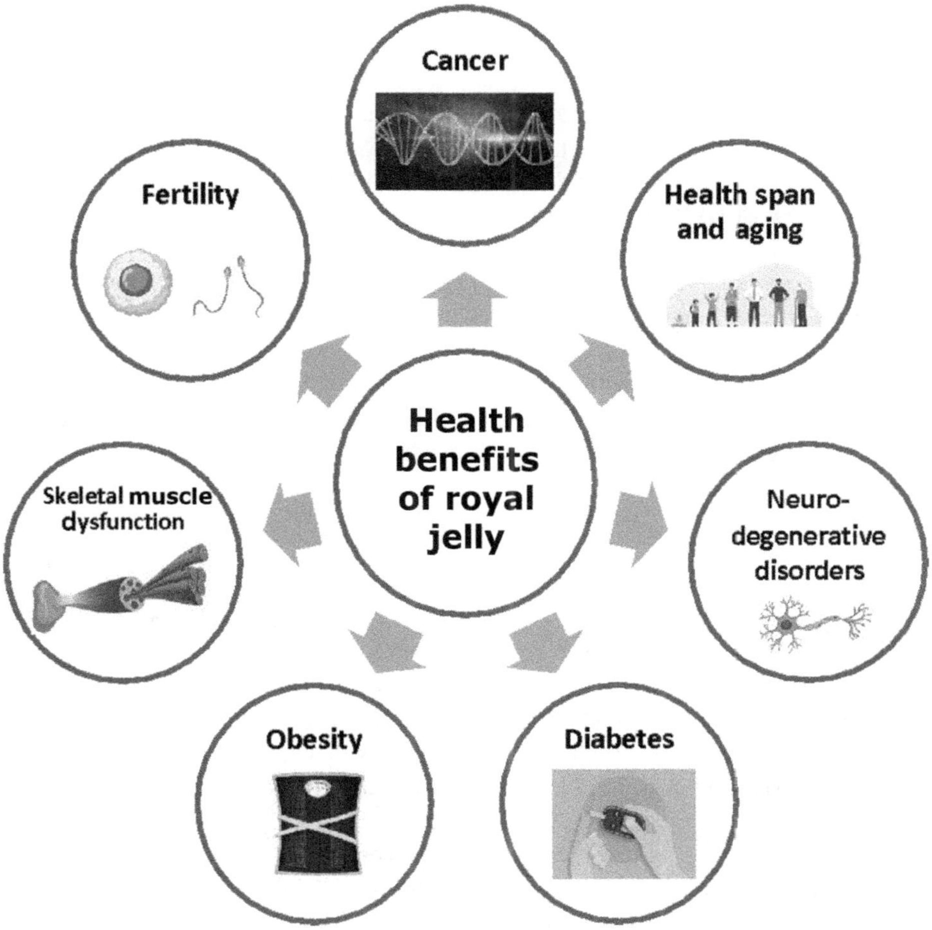


**Figure 20.1** Health benefits of royal jelly. Kirsty Pargeter / Adobe Stock.

(HuIFN-αN3) induces inhibition of CaCo-2 cell proliferation (Filipič et al. 2019). These results corroborated with the findings by Zhang and colleagues (2017), who suggested that the use of RJ reduces the weight of breast tumors in mice, further improving antioxidant activities of serum, the liver, and the kidneys. The antiproliferative and antioxidant effects of RJ suggest its antitumor potential in both the cell lines and animal studies.

On the other hand, clinical studies have reported that RJ has improved the signs and symptoms of oral mucositis in patients with various types of malignancies who underwent chemotherapy and radiotherapy (Erdem and Güngörmüs 2014; Yamauchi et al. 2014). A study related to cancer-related fatigue (CRF), which is common among 50%–90% of patients with cancer, demonstrated that RJ ameliorated symptoms of CRF without any adverse effects among patients who underwent hormone therapy, chemotherapy, chemoradiation, or radiotherapy (Mofid et al. 2016). However, this was a unicentric randomized clinical trial; therefore, a single source of participant may be insufficient to provide a viable size of clinical trial to further explicate the role of RJ in managing CRF.

## Health Span and Aging

The highly nutritious RJ fed by bee queens has lengthened its lifespan compared with other worker bees despite having the same genome. This may envision the effect of RJ on health maintenance and age-related disorders in humans. The nutritional content of RJ with its protein and lipid ingredients provides essential nutrients that extends the lifespan. It has been reported that RJ could improve the 50% survival rates of C3H/HeJ mice through a mechanism of reduced oxidative damage (Inoue et al. 2003). Also, the ability of RJ to prolong the lifespan of *Caenorhabditis elegans* suggests its longevity-promoting factors (Honda et al. 2011), which are likewise conserved in mice (Inoue et al. 2003). A recent review investigating the benefits of RJ in promoting healthy aging and longevity reported evidence of RJ and its components on lifespan and health span in different species ranging from *A. mellifera*, *Drosophila melanogaster*, and *C. elegans* to mice (Kunugi and Mohammed Ali 2019).

Skin, the protective layer of the body, is the most visible and widely exposed organ that generally represents the aging process in humans. It is believed that principal unsaturated fatty acid in RJ, which is the 10-HDA acid, may provide anti-aging effects and protect the skin from damaging effects of ultraviolet (UV) radiation. An in vitro study demonstrated the protective effects of 10-HDA acid on UVA-induced damage in human dermal fibroblasts (Zheng et al. 2013). Similarly, Park and colleagues (2011) found that RJ protects against UVB-induced photoaging in human skin fibroblasts by promoting collagen production via up-regulation of TGF-β1 expression.

Women past a certain age experience menopause as a natural part of aging caused by reduced production of progesterone and estrogen. RJ, which exerts estrogenic activities through interaction with estrogen receptors, is suggested to be consumed by postmenopausal women for improvement of quality of life and treatment of menopause-related complications (Bălan et al. 2020; Mishima et al. 2005). This is in concordance with some studies exhibiting the promising effects of RJ in relieving menopausal symptoms such as anxiety, backache, and low back pain (Asama et al. 2018) as well as hot flashes, sleep problems, depressive mood, anxiety, and physical and mental exhaustion in addition to muscle and joint pain (Sharif and Darsareh 2019).

### Neurodegenerative Disorders

Besides RJ's health benefits that fend off aging and its effects on longevity, studies have also examined the potential of RJ against neurodegenerative treatments. The unique component of RJ is the 10-HDA acid, which can pass through the blood–brain barrier, promoting neurogenesis (Hattori et al. 2007). The beneficial role of RJ on neural functions includes enhanced spatial learning and memory in a rat model of streptozotocin-induced sporadic Alzheimer's disease (AD) (Zamani et al. 2012). Another study demonstrated that RJ substantially improves cognitive deficits and alleviates β-amyloid accumulation in the APP/PS1 mouse model of AD (You et al. 2019). Similarly, RJ decreases β-amyloid in other animal models such as AD rabbits (Pan et al. 2018).

Further investigation on understanding the benefits of RJ in improving memory and exhibiting a neuroprotective action in the intracerebroventricular injection of streptozotocin rat model of sporadic AD confirmed the findings on the potential therapeutic value of RJ in treating patients with AD (Guardia de Souza et al. 2020). In addition to 10-HDA acid, Wan and colleagues (2018) found that royalactin, an active main compound of RJ that helps to maintain pluripotency by activating a ground-state pluripotency-like gene network, may play a crucial role in treating patients with disorders that are caused by cells dying off, such as AD. Taken together, several comprehensive reviews on RJ have supported the enormous therapeutic potential of RJ on neurologic disorders, including AD and Parkinson's disease, based on the different identified bioactive compounds (Ali and Kunugi 2020a, 2020b; El-Seedi et al. 2020).

### Diabetes

Investigations of the effect of RJ intake in reducing serum glucose levels in patients with type 2 diabetes have shown positive outcomes in different studies (Khoshpey et al. 2016; Mousavi et al. 2017). Interestingly, a study has also showed that RJ supplementation resulted in a significant reduction in serum glycosylated hemoglobin levels and an increase in insulin concentrations in addition to improvement of oxidative stress in a woman with type 2 diabetes (Pourmoradian et al. 2012). On top of this, previous studies reported that RJ reduces serum glucose levels in healthy participants and improves glucose tolerance (Morita et al. 2012; Münstedt et al. 2009).

Contrary to expectations, a study carried out by Mobasseri and colleagues (2015) reported that there is no significant difference of immediate effects on glycemic factors in patients with type 2 diabetes upon a single dose of fresh RJ as a complementary therapy. A possible explanation for these results may be the lack of an adequate sample size and requirement of optimization of different doses of RJ to achieve the therapeutic effects. The studies thus far have provided compelling evidence that RJ is more beneficial as a long-term dietary supplement rather than as an acute treatment for patients with diabetes (Omer et al. 2019).

### Obesity

Being overweight can be a challenge of obesity-related chronic diseases and its impact on increasing cardiovascular and diabetes risk factors. Obesity is typically associated with atherosclerosis, which is believed to involve lipid peroxidation

(Esterbauer et al. 1993; Yesilbursa et al. 2005). RJ peptides from water-soluble RJ protein hydrolysates that have strong antioxidant activity were reported to inhibit lipid peroxidation in vitro and in rats (Guo et al. 2008). Similarly, a study from Zahmatkesh and colleagues (2014) exhibited the protective effect of RJ on lipid peroxidation in mice treated with oxymetholone. On the other hand, RJ reduced cholesterol levels in rabbits (Pan et al. 2018). Interestingly, besides looking at RJ as a dietary food supplement, a study has showed the protective role of RJ on lipid peroxidation in *Allium cepa* L. root tips against the toxicity induced by petroleum wastewater (Türkmen et al. 2009).

Unfortunately, there are only a small number of clinical trials on humans regarding the effects of RJ in obesity. In a randomized clinical study, the body weights of overweight women with type 2 diabetes reduced significantly upon intake with RJ supplementation and decreased their mean daily total energy and fat intakes (Pourmoradian et al. 2012). Another study has also demonstrated that RJ supplementation significantly reduces the total cholesterol of asymptomatic overweight adults (Petelin et al. 2019). In this study, not limited to lipid profile, RJ administration was attributed to positive effects on satiety, inflammation, and antioxidant capacity in overweight adults (Petelin et al. 2019). Nevertheless, most of these studies have some limitations, including (1) the diets of participants were not controlled; (2) the small sample sizes; (3) the duration of study; (4) optimization of the amount of RJ; or (5) different sources of RJ used in the various studies, requiring further studies before concrete conclusions can be reached.

### Skeletal Muscle Dysfunction

RJ has been investigated for a variety of actions, including skeletal muscle dysfunction. Sarcopenia, which is known as age-related muscle loss, is characterized by generalized loss of skeletal muscle mass and strength that decrease life expectancy and quality of life (Santilli et al. 2014). A recent study has shown that daily oral administration of RJ prevents a decrease in myofiber size following denervation in the mice, which is useful for muscle atrophy disease such as sarcopenia (Shirakawa et al. 2020). Moreover, RJ-treated mice exhibited delays in age-related motor functional impairment compared with the control group in addition to affecting the muscle satellite cell markers and catabolic genes (Okumura et al. 2018).

Modification has been made to improve the absorption and digestion of RJ. A previous study showed that RJ together with protease-treated RJ diminished the decrease in muscle weight and grip strength in aged mice in addition to the regenerating capacity of injured muscles and serum insulin-like growth factor-1 levels (Niu et al. 2013). In the same study, the authors also demonstrated that modified protease-treated RJ alone increased the cell proliferation rate and promoted cell differentiation in isolated satellite cells from aged mice.

### Fertility

The increased popularity of RJ, which aids fertility, has been demonstrated in different animal models, including rats, rabbits, and hamsters. Ghanbari et al. (2015) reported that there were decreases in sperm deformity, DNA integrity, chromatin quality, and tissue malondialdehyde levels upon oral administration of RJ to diabetic adult male Wistar rats. RJ also increased testicular weight; sperm count, motility, and viability; and serum testosterone levels in diabetic rats in the same experiment. This is in concordance with the study by Zahmatkesh and colleagues (2014) showing the protective effect of RJ on sperm parameters and testosterone level in adult mice. On the other hand, RJ promotes folliculogenesis and increases ovarian hormones in immature female rats (Ghanbari et al. 2015).

In male hamsters, researchers have found that long-term feeding of RJ gives higher intratesticular free testosterone, which plays a vital role in spermatogenesis (Kohguchi et al. 2007). Meanwhile, supplementation of RJ in male rabbits resulted in an increase of sperm concentration, total sperm output, sperm motility, and live sperm and normal sperm compared with untreated control rabbits (El-Hanoun et al. 2014). With the exclusive ability to foster the fertility of queen bees, the potential estrogenic activities of RJ were investigated. A study has shown that RJ has estrogenic effects in vitro and in vivo, which indicates the potential of RJ in regulating sex hormones (Kunugi and Mohammed Ali 2019; Mishima et al. 2005).

## Conclusion and Future Prospects

This review has elucidated the antimicrobial properties, activities, and health benefits of RJ. The medicinal properties of RJ have been researched over the years, and these benefits are attracting increasing interest. Although clinical trials have provided evidence that RJ benefits patients with cancer, diabetes, neurodegenerative diseases, obesity, and fertility, the need for standardization of its compositions are crucial in ensuring its safety. The use of RJ compounds as curative drugs has not been instigated yet. Therefore, more clinical studies to determine its pharmacological mechanisms and safe dosages should be instigated.

## References

Albert, Š. and Klaudiny, J. (2004). The MRJP/YELLOW protein family of Apis mellifera: identification of new members in the EST library. *Journal of Insect Physiology* 50 (1): 51–59.

Albert, Š., Klaudiny, J., and Šimúth, J. (1999). Molecular characterization of MRJP3, highly polymorphic protein of honeybee (Apis mellifera) royal jelly. *Insect Biochemistry and Molecular Biology* 29 (5): 427–434.

Ali, A.M. and Kunugi, H. (2020a). Apitherapy for Parkinson's disease: a focus on the effects of propolis and royal jelly. *Oxidative Medicine and Cellular Longevity* 2020.

Ali, A.M. and Kunugi, H. (2020b). Royal jelly as an intelligent anti-aging agent—a focus on cognitive aging and Alzheimer's Disease: a review. *Antioxidants* 9 (10): 937.

Antinelli, J.-F., Zeggane, S., Davico, R., et al. (2003). Evaluation of (E)-10-hydroxydec-2-enoic acid as a freshness parameter for royal jelly. *Food Chemistry* 80 (1): 85–89.

Asama, T., Matsuzaki, H., Fukushima, S., et al. (2018). Royal jelly supplementation improves menopausal symptoms such as backache, low back pain, and anxiety in postmenopausal Japanese women. *Evidence-Based Complementary and Alternative Medicine* 2018.

Bălan, A., Moga, M.A., Dima, L., et al. (2020). Royal jelly—a traditional and natural remedy for postmenopausal symptoms and aging-related pathologies. *Molecules* 25 (14): 3291.

Bărnuţiu, L.I., Mărghitaş, L.A., Dezmirean, D.S., et al. (2011). Chemical composition and antimicrobial activity of royal jelly-review. *Scientific Papers Animal Science and Biotechnologies* 44 (2): 67–72.

Bılıková, K., Hanes, J., Nordhoff, E., et al. (2002). Apisimin, a new serine–valine-rich peptide from honeybee (Apis mellifera L.) royal jelly: purification and molecular characterization 1. *FEBS Letters* 528 (1–3): 125–129.

Blum, M.S., Novak, A.F., and Taber, S. (1959). 10-hydroxy-Δ2-decenoic acid, an antibiotic found in royal jelly. *Science* 130 (3373): 452–453.

Bogdanov, S., Martin, P., and Lullmann, C. (2002). Harmonised methods of the international honey commission. Swiss Bee Research Centre, FAM, Liebefeld.

Boselli, E., Caboni, M.F., Sabatini, A.G., et al. (2003). Determination and changes of free amino acids in royal jelly during storage. *Apidologie* 34 (2): 129–137.

Brady, N., Molan, P., and Harfoot, C. (1996). The sensitivity of dermatophytes to the antimicrobial activity of manuka honey and other honey. *Pharmacy and Pharmacology Communications* 2 (10): 471–473.

Brudzynski, K. and Sjaarda, C. (2015). Honey glycoproteins containing antimicrobial peptides, Jelleins of the Major Royal Jelly Protein 1, are responsible for the cell wall lytic and bactericidal activities of honey. *PloS one* 10 (4): e0120238–e0120238.

Bulet, P., Hetru, C., Dimarcq, J.-L., and Hoffmann, D. (1999). Antimicrobial peptides in insects; structure and function. *Developmental & Comparative Immunology* 23 (4–5): 329–344.

Buratti, S., Benedetti, S., and Cosio, M. (2007). Evaluation of the antioxidant power of honey, propolis and royal jelly by amperometric flow injection analysis. *Talanta* 71 (3): 1387–1392.

Buttstedt, A., Moritz, R.F., and Erler, S. (2013). More than royal food—major royal jelly protein genes in sexuals and workers of the honeybee Apis mellifera. *Frontiers in Zoology* 10 (1): 1–10.

Buttstedt, A., Moritz, R.F., and Erler, S. (2014). Origin and function of the major royal jelly proteins of the honeybee (Apis mellifera) as members of the yellow gene family. *Biological Reviews* 89 (2): 255–269.

Chiu, H.F., Chen, B.K., Lu, Y.Y., et al. (2017). Hypocholesterolemic efficacy of royal jelly in healthy mild hypercholesterolemic adults. *Pharmaceutical Biology* 55 (1): 497–502.

Cooper, R.A., Halas, E., and Molan, P.C. (2002). The efficacy of honey in inhibiting strains of Pseudomonas aeruginosa from infected burns. *The Journal of Burn Care & Rehabilitation* 23 (6): 366–370.

Cornara, L., Biagi, M., Xiao, J., and Burlando, B. (2017). Therapeutic properties of bioactive compounds from different honeybee products. *Frontiers in Pharmacology* 8 (412–432).

Daniele, G. and Casabianca, H. (2012). Sugar composition of French royal jelly for comparison with commercial and artificial sugar samples. *Food Chemistry* 134 (2): 1025–1029.

Duong, H.A., Vu, M.T., Nguyen, T.D., et al. (2020). Determination of 10-hydroxy-2-decenoic acid and free amino acids in royal jelly supplements with purpose-made capillary electrophoresis coupled with contactless conductivity detection. *Journal of Food Composition and Analysis* 87: 103422.

El-Guendouz, S., Machado, A.M., Aazza, S., et al. (2020). Chemical characterization and biological properties of royal jelly samples from the Mediterranean area. *Natural Product Communications* 15 (2): 1934578X20908080.

El-Hanoun, A., Elkomy, A., Fares, W., and Shahien, E. (2014). Impact of royal jelly to improve reproductive performance of male rabbits under hot summer conditions. *World Rabbit Science* 22 (3): 241–248.

El-Seedi, H.R., Khalifa, S.A., Abd El-Wahed, A., et al. (2020). Honeybee products: an updated review of neurological actions. *Trends in Food Science & Technology*. 101: 17–27.

Erdem, Ö. and Güngörmüs, Z. (2014). The effect of royal jelly on oral mucositis in patients undergoing radiotherapy and chemotherapy. *Holistic Nursing Practice* 28 (4): 242–246.

Esterbauer, H., Wäg, G., and Puhl, H. (1993). Lipid peroxidation and its role in atherosclerosis. *British Medical Bulletin* 49 (3): 566–576.

Filipič, B., Gradišnik, L., Rihar, K., et al. (2019). Royal jelly and human interferon-alpha (HuIFN-αN3) affect proliferation, glutathione level, and lipid peroxidation in human colorectal adenocarcinoma cells in vitro. In: M.A. Mansour (ed.), *Lipid Peroxidation Research*. London: IntechOpen.

Finke, M.D. (2005). Nutrient composition of bee brood and its potential as human food. *Ecology of Food and Nutrition* 44 (4): 257–270.

Fontana, R., Mendes, M.A., De Souza, B.M., et al. (2004). Jelleines: a family of antimicrobial peptides from the Royal Jelly of honeybees (Apis mellifera). *Peptides* 25 (6): 919–928.

Fratini, F., Cilia, G., Mancini, S., and Felicioli, A. (2016). Royal jelly: an ancient remedy with remarkable antibacterial properties. *Microbiological Research* 192: 130–141.

Fujiwara, S., Imai, J., Fujiwara, M., et al. (1990). A potent antibacterial protein in royal jelly. Purification and determination of the primary structure of royalisin. *Journal of Biological Chemistry* 265 (19): 11333–11337.

Furusawa, T., Rakwal, R., Nam, H.W., et al. (2008). Comprehensive royal jelly (RJ) proteomics using one-and two-dimensional proteomics platforms reveals novel RJ proteins and potential phospho/glycoproteins. *Journal of Proteome Research* 7 (8): 3194–3229.

Garcia, M.C., Finola, M.S., and Marioli, J.M. (2013). Bioassay directed identification of royal jelly's active compounds against the growth of bacteria capable of infecting cutaneous wounds.

Ghanbari, E., Nejati, V., Najafi, G., et al. (2015). Study on the effect of royal jelly on reproductive parameters in streptozotocin-induced diabetic rats. *International Journal of Fertility & Sterility* 9 (1): 113.

Guardia De Souza, E.S.T., Do Val de Paulo, M.E.F, Da Silva, J.R.M, et al. 2020. Oral treatment with royal jelly improves memory and presents neuroprotective effects on icv-STZ rat model of sporadic Alzheimer's disease. *Heliyon* 6: e03281.

Gismondi, A., Trionfera, E., Canuti, L., et al. (2017). Royal jelly lipophilic fraction induces antiproliferative effects on SH-SY5Y human neuroblastoma cells. *Oncology Reports* 38 (3): 1833–1844.

Guo, H., Ekusa, A., Iwai, K., et al. (2008). Royal jelly peptides inhibit lipid peroxidation in vitro and in vivo. *Journal of Nutritional Science and Vitaminology* 54 (3): 191–195.

Guo, H., Kouzuma, Y., and Yonekura, M. (2009). Structures and properties of antioxidative peptides derived from royal jelly protein. *Food Chemistry* 113 (1): 238–245.

Hanes, J. and Šimuth, J. (1992). Identification and partial characterization of the major royal jelly protein of the honey bee (Apis mellifera L.). *Journal of Apicultural Research* 31 (1): 22–26.

Hattori, N., Nomoto, H., Fukumitsu, H., et al. (2007). Royal jelly and its unique fatty acid, 10-hydroxy-trans-2-decenoic acid, promote neurogenesis by neural stem/progenitor cells in vitro. *Biomedical Research* 28 (5): 261–266.

Hidaka, S., Okamoto, Y., Uchiyama, S., et al. (2006). Royal jelly prevents osteoporosis in rats: beneficial effects in ovariectomy model and in bone tissue culture model. *Evidence-based Complementary and Alternative Medicine* 3.

Honda, Y., Fujita, Y., Maruyama, H., et al. (2011). Lifespan-extending effects of royal jelly and its related substances on the nematode Caenorhabditis elegans. *PLoS One* 6 (8): e23527.

Inoue, S.-I., Koya-Miyata, S., Ushio, S., et al. (2003). Royal Jelly prolongs the life span of C3H/HeJ mice: correlation with reduced DNA damage. *Experimental Gerontology* 38 (9): 965–969.

Isidorov, V., Bakier, S., and Grzech, I. (2012). Gas chromatographic–mass spectrometric investigation of volatile and extractable compounds of crude royal jelly. *Journal of Chromatography B* 885: 109–116.

Izuta, H., Chikaraishi, Y., Shimazawa, M., et al. (2009). 10-Hydroxy-2-decenoic acid, a major fatty acid from royal jelly, inhibits VEGF-induced angiogenesis in human umbilical vein endothelial cells. *Evidence-based Complementary and Alternative Medicine* 6.

Jianke, L. and Shenglu, C. (2003). Royal jelly and human health. *American Bee Journal* 143 (5): 398–402.

Kamakura, M., Mitani, N., Fukuda, T., and Fukushima, M. (2001a). Antifatigue effect of fresh royal jelly in mice. *Journal of Nutritional Science and Vitaminology* 47 (6): 394–401.

Kamakura, M., Suenobu, N., and Fukushima, M. (2001b). Fifty-seven-kDa protein in royal jelly enhances proliferation of primary cultured rat hepatocytes and increases albumin production in the absence of serum. *Biochemical and Biophysical Research Communications* 282 (4): 865–874.

Khazaei, M., Ansarian, A., and Ghanbari, E. (2018). New findings on biological actions and clinical applications of royal jelly: a review. *Journal of Dietary Supplements* 15 (5): 757–775.

Kheyri, H., Cribb, B.W., Reinhard, J., et al. (2012). Novel actin rings within the secretory cells of honeybee royal jelly glands. *Cytoskeleton* 69 (12): 1032–1039.

Khoshpey, B., Djazayeri, S., Amiri, F., et al. (2016). Effect of royal jelly intake on serum glucose, Apolipoprotein A-I (ApoA-I), Apolipoprotein B (ApoB) and ApoB/ApoA-I Ratios in Patients with type 2 diabetes: a randomized, double-blind clinical trial study. *Canadian Journal of Diabetes* 40 (4): 324–328.

Kohguchi, M., Inoue, S.-I., Ushio, S., et al. (2007). Effect of royal jelly diet on the testicular function of hamsters. *Food Science and Technology Research* 10 (4): 420–423.

Kohno, K., Okamoto, I., Sano, O., et al. (2004). Royal jelly inhibits the production of proinflammatory cytokines by activated macrophages. *Bioscience, Biotechnology, and Biochemistry* 68 (1): 138–145.

Koya-Miyata, S., Okamoto, I., Ushio, S., et al. (2004). Identification of a collagen production-promoting factor from an extract of royal jelly and its possible mechanism. *Bioscience, Biotechnology, and Biochemistry* 68 (4): 767–773.

Kunugi, H. and Mohammed Ali, A. (2019). Royal jelly and its components promote healthy aging and longevity: from animal models to humans. *International Journal of Molecular Sciences* 20 (19): 4662.

Lercker, G., Capella, P., Conte, L., et al. (1981). Components of royal jelly: i. Identification of the organic acids. *Lipids* 16 (12): 912–919.

Lercker, G., Savioli, S., Vecchi, M., et al. (1986). Carbohydrate determination of royal jelly by high resolution gas chromatography (HRGC). *Food Chemistry* 19 (4): 255–264.

Lin, Y., Zhang, M., Wang, L., et al. (2020). The in vitro and in vivo wound-healing effects of royal jelly derived from Apis mellifera L. during blossom seasons of Castanea mollissima Bl. and Brassica napus L. in South China exhibited distinct patterns. *BMC Complementary Medicine and Therapies* 20 (1): 357.

López-Gutiérrez, N., Del Mar Aguilera-luiz, M., Romero-González, R., et al. (2014). Fast analysis of polyphenols in royal jelly products using automated TurboFlow™-liquid chromatography–Orbitrap high resolution mass spectrometry. *Journal of Chromatography B* 973: 17–28.

Malka, O., Karunker, I., Yeheskel, A., et al. (2009). The gene road to royalty–differential expression of hydroxylating genes in the mandibular glands of the honeybee. *The FEBS Journal* 276 (19): 5481–5490.

McCleskey, C. and Melampy, R. (1939). Bactericidal properties of royal jelly of the honeybee. *Journal of Economic Entomology* 32 (4): 581–587.

Melliou, E. and Chinou, I. (2005). Chemistry and bioactivity of royal jelly from Greece. *Journal of Agricultural and Food Chemistry* 53 (23): 8987–8992.

Melliou, E. and Chinou, I. (2014). Chemistry and bioactivities of royal jelly. *Studies in Natural Products Chemistry* Elsevier. 43: 261–290.

Meng, G., Wang, H., Pei, Y., et al. (2017). Effects of protease-treated royal jelly on muscle strength in elderly nursing home residents: a randomized, double-blind, placebo-controlled, dose-response study. *Scientific Reports* 7 (1): 11416-11416.

Miguel, M.G. and El-Guendouz, S. (2017). Volatile compounds of royal jelly. In: J.M. Suarez (ed.), *Bee Products – Chemical and Biological Properties*, 191–197. New York: Springer.

Mishima, S., Suzuki, K.-M., Isohama, Y., et al. (2005). Royal jelly has estrogenic effects in vitro and in vivo. *Journal of Ethnopharmacology* 101 (1–3): 215–220.

Miyata, Y. and Sakai, H. (2018). Anti-cancer and protective effects of royal jelly for therapy-induced toxicities in malignancies. *International Journal of Molecular Sciences* 19 (10): 3270.

Miyauchi-Wakuda, S., Kagota, S., Maruyama-Fumoto, K., et al. (2019). Effect of royal jelly on mouse isolated ileum and gastrointestinal motility. *Journal of Medicinal Food* 22 (8): 789–796.

Mobasseri, M., Ghiyasvand, S., Ostadrahimi, A., et al. (2015). Effect of fresh royal jelly ingestion on glycemic response in patients with type 2 diabetes. *Iranian Red Crescent Medical Journal* 17 (9): e20074.

Mofid, B., Rezaeizadeh, H., Termos, A., et al. (2016). Effect of processed honey and royal jelly on cancer-related fatigue: a double-blind randomized clinical trial. *Electronic Physician* 8 (6): 2475.

Morita, H., Ikeda, T., Kajita, K., et al. (2012). Effect of royal jelly ingestion for six months on healthy volunteers. *Nutrition Journal* 11 (1): 77.

Moselhy, W., Fawzy, A., and Kamel, A. (2013). An evaluation of the potent antimicrobial effects and unsaponifiable matter analysis of the royal jelly. *Life Science Journal* 2 (10): 290–296.

Mousavi, S.N., Jazayeri, S., Khoshpay, B., et al. (2017). Royal jelly decreases blood pressure, serum glucose, and interleukin-6 in patients with type 2 diabetes on an iso-caloric diet. *Journal of Nutrition and Food Security* 2 (4): 300–307.

Münstedt, K., Bargello, M., and Hauenschild, A. (2009). Royal jelly reduces the serum glucose levels in healthy subjects. *Journal of Medicinal Food* 12 (5): 1170–1172.

Nagai, T. and Inoue, R. (2004). Preparation and the functional properties of water extract and alkaline extract of royal jelly. *Food Chemistry* 84 (2): 181–186.

Nakaya, M., Onda, H., Sasaki, K., et al. (2007). Effect of royal jelly on bisphenol A-induced proliferation of human breast cancer cells. *Bioscience, Biotechnology, and Biochemistry* 71 (1): 253–255.

Niu, K., Guo, H., Guo, Y., et al. (2013). Royal jelly prevents the progression of sarcopenia in aged mice in vivo and in vitro. *Journals of Gerontology Series A: Biomedical Sciences and Medical Sciences* 68 (12): 1482–1492.

Nozaki, R., Tamura, S., Ito, A., et al. (2012). A rapid method to isolate soluble royal jelly proteins. *Food Chemistry* 134 (4): 2332–2337.

Okamoto, I., Taniguchi, Y., Kunikata, T., et al. (2003). Major royal jelly protein 3 modulates immune responses in vitro and in vivo. *Life Sciences* 73 (16): 2029–2045.

Okumura, N., Toda, T., Ozawa, Y., et al. (2018). Royal Jelly delays motor functional impairment during aging in genetically heterogeneous male mice. *Nutrients* 10 (9): 1191.

Omer, K., Gelkopf, M.J., and Newton, G. (2019). Effectiveness of royal jelly supplementation in glycemic regulation: a systematic review. *World Journal of Diabetes* 10 (2): 96.

Pan, Y., Xu, J., Chen, C., et al. (2018). Royal jelly reduces cholesterol levels, ameliorates Aβ pathology and enhances neuronal metabolic activities in a rabbit model of Alzheimer's disease. *Frontiers in Aging Neuroscience* 10: 50.

Park, H.M., Hwang, E., Lee, K.G., et al. (2011). Royal jelly protects against ultraviolet B–induced photoaging in human skin fibroblasts via enhancing collagen production. *Journal of Medicinal Food* 14 (9): 899–906.

Pasupuleti, V.R., Sammugam, L., Ramesh, N., and Gan, S.H. (2017). Honey, propolis, and royal jelly: a comprehensive review of their biological actions and health benefits. *Oxidative Medicine and Cellular Longevity* 2017: 1259510.

Paul, I.M., Beiler, J., McMonagle, A., et al. (2007). Effect of honey, dextromethorphan, and no treatment on nocturnal cough and sleep quality for coughing children and their parents. *Archives of Pediatrics & Adolescent Medicine* 161 (12): 1140–1146.

Petelin, A., Kenig, S., Kopinč, R., et al. (2019). Effects of royal jelly administration on lipid profile, satiety, inflammation, and antioxidant capacity in asymptomatic overweight adults. *Evidence-Based Complementary and Alternative Medicine* 2019: 4969720.

Pourmoradian, S., Mahdavi, R., Mobasseri, M., et al. (2012). Effects of royal jelly supplementation on body weight and dietary intake in type 2 diabetic females. *Health Promotion Perspectives* 2 (2): 231.

Pourmoradian, S., Mahdavi, R., Mobasseri, M., et al. (2014). Effects of royal jelly supplementation on glycemic control and oxidative stress factors in type 2 diabetic female: a randomized clinical trial. *Chinese Journal of Integrative Medicine* 20 (5): 347–352.

Qiu, W., Chen, X., Tian, Y., et al. (2020). Protection against oxidative stress and anti-aging effect in Drosophila of royal jelly-collagen peptide. *Food and Chemical Toxicology* 135: 110881.

Ramadan, M.F. and Al-Ghamdi, A. (2012). Bioactive compounds and health-promoting properties of royal jelly: a review. *Journal of Functional Foods* 4 (1): 39–52.

Ramanathan, A.N.K.G., Nair, A.J., and Sugunan, V.S. (2018). A review on royal jelly proteins and peptides. *Journal of Functional Foods* 44: 255–264.

Rosmilah, M., Shahnaz, M., Patel, G., et al. (2008). Characterization of major allergens of royal jelly Apis mellifera. *Tropical Biomedicine* 25 (3): 243–251.

Sabatini, A.G., Marcazzan, G.L., Caboni, M.F., et al. (2009). Quality and standardisation of royal jelly. *Journal of ApiProduct and ApiMedical Science* 1 (1): 1–6.

Santilli, V., Bernetti, A., Mangone, M., and Paoloni, M. (2014). Clinical definition of sarcopenia. *Clinical Cases in Mineral and Bone Metabolism* 11 (3): 177.

Schmidt, J.O. (1997). Bee products. In: Y. Lensky (ed.), *Bee Products*. 15–26. New York: Springer.

Schmitzova, J., Klaudiny, J., Albert, Š., et al. (1998). A family of major royal jelly proteins of the honeybee Apis mellifera L. *Cellular and Molecular Life Sciences CMLS* 54 (9): 1020–1030.

Schuh, C.M.A.P., Aguayo, S., Zavala, G., and Khoury, M. (2019). Exosome-like vesicles in Apis mellifera bee pollen, honey and royal jelly contribute to their antibacterial and pro-regenerative activity. *The Journal of Experimental Biology* 222 (20): jeb208702.

Seçilmiş, Y. and Silici, S. (2020). Bee product efficacy in children with upper respiratory tract infections. *Turkish Journal of Pediatrics* 62 (4): 634–640.

Sedaghat, A., Ghasemi, S., and Shirzad, M. (2010). Effect of royal jelly on sterile wound healing in Balb/C mice. *Armaghane Danesh* 15 (1): 38–46.

Serra Bonvehi, J. (1992). Sugars, acidity and pH of royal jelly. *Anales de Bromatologia (Espana)* 44 (1): 65–69.

Sesta, G. (2006). Determination of sugars in royal jelly by HPLC. *Apidologie* 37 (1): 84–90.

Sharif, S.N. and Darsareh, F. (2019). Effect of royal jelly on menopausal symptoms: a randomized placebo-controlled clinical trial. *Complementary Therapies in Clinical Practice* 37: 47–50.

Shirakawa, T., Miyawaki, A., Matsubara, T., et al. (2020). Daily oral administration of protease-treated royal jelly protects against denervation-induced skeletal muscle atrophy. *Nutrients* 12 (10): 3089.

Siavash, M., Shokri, S., Haghighi, S., et al. (2015). The efficacy of topical royal jelly on healing of diabetic foot ulcers: a double-blind placebo-controlled clinical trial. *International Wound Journal* 12 (2): 137–142.

Šimúth, J. (2001). Some properties of the main protein of honeybee (Apis mellifera) royal jelly. *Apidologie* 32 (1): 69–80.

Šimúth, J., Bíliková, K., Kováčová, E., et al. (2004). Immunochemical approach to detection of adulteration in honey: physiologically active royal jelly protein stimulating TNF-α release is a regular component of honey. *Journal of Agricultural and Food Chemistry* 52 (8): 2154–2158.

Susilowati, H., Murakami, K., Yumoto, H., et al. (2017). Royal jelly inhibits Pseudomonas aeruginosa adherence and reduces excessive inflammatory responses in human epithelial cells. *BioMed Research International* 2017: 3191752–3191752.

Taavoni, S., Barkhordari, F., Goushegir, A., and Haghani, H. (2014). Effect of royal jelly on premenstrual syndrome among Iranian medical sciences students: a randomized, triple-blind, placebo-controlled study. *Complementary Therapies in Medicine* 22 (4): 601–606.

Takenaka, T. and Echigo, T. (1980). General chemical composition of the royal jelly [of honey bees]. In: J.M. Alvarez Suarez (ed.), *Bulletin of the Faculty of Agriculture Tamagawa University*. New York: Springer

Tamura, S., Kono, T., Harada, C., et al. (2009). Estimation and characterisation of major royal jelly proteins obtained from the honeybee Apis merifera. *Food Chemistry* 114 (4): 1491–1497.

Terada, Y., Narukawa, M., and Watanabe, T. (2011). Specific hydroxy fatty acids in royal jelly activate TRPA1. *Journal of Agricultural and Food Chemistry* 59 (6): 2627–2635.

Türkmen, Z., Çavuşoğlu, K., Çavuşoğlu, K., et al. (2009). Protective role of royal jelly (honeybee) on genotoxicity and lipid peroxidation, induced by petroleum wastewater, in Allium cepa L. root tips. *Environmental Technology* 30 (11): 1205–1214.

Wahdan, H. (1998). Causes of the antimicrobial activity of honey. *Infection* 26 (1): 26–31.

Wan, D.C., Morgan, S.L., Spencley, A.L., et al. (2018). Honey bee Royalactin unlocks conserved pluripotency pathway in mammals. *Nature Communications* 9 (1): 1–9.

World Health Organization. (2001). Cough and cold remedies for the treatment of acute respiratory infections in young children. World Health Organization. https://apps.who.int/iris/handle/10665/66856.

Xue, X., Wu, L., and Wang, K. (2017). Chemical composition of royal jelly. In: J.M. Alvarez-Suarez (ed.), *Bee Products-chemical and Biological Properties*. 181–190. New York: Springer.

Yamauchi, K., Kogashiwa, Y., Moro, Y., and Kohno, N. (2014). The effect of topical application of royal jelly on chemoradiotherapy-induced mucositis in head and neck cancer: a preliminary study. *International Journal of Otolaryngology* 2014: 974967.

Yang, Y.-C., Chou, W.-M., Widowati, D.A., et al. (2018). 10-hydroxy-2-decenoic acid of royal jelly exhibits bactericide and anti-inflammatory activity in human colon cancer cells. *BMC Complementary and Alternative Medicine* 18 (1): 202.

Yesilbursa, D., Serdar, Z., Serdar, A., et al. (2005). Lipid peroxides in obese patients and effects of weight loss with orlistat on lipid peroxides levels. *International Journal of Obesity* 29 (1): 142–145.

Yoshida, M., Hayashi, K., Watadani, R., et al. (2016). Royal jelly improves hyperglycemia in obese/diabetic KK-Ay mice. *Journal of Veterinary Medical Science* 79 (2): 299–307.

You, M., Pan, Y., Liu, Y., et al. (2019). Royal jelly alleviates cognitive deficits and β-amyloid accumulation in APP/PS1 mouse model via activation of the cAMP/PKA/CREB/BDNF pathway and inhibition of neuronal apoptosis. *Frontiers in Aging Neuroscience* 10: 428.

Zahmatkesh, E., Najafi, G., Nejati, V., and Heidari, R. (2014). Protective effect of royal jelly on the sperm parameters and testosterone level and lipid peroxidation in adult mice treated with oxymetholone. *Avicenna Journal of Phytomedicine* 4 (1): 43.

Zamani, Z., Reisi, P., Alaei, H., and Pilehvarian, A.A. (2012). Effect of royal jelly on spatial learning and memory in rat model of streptozotocin-induced sporadic Alzheimer's disease. *Advanced Biomedical Research* 1: 26.

Zhang, L., Fang, Y., Li, R., et al. (2012). Towards posttranslational modification proteome of royal jelly. *Journal of Proteomics* 75 (17): 5327–5341.

Zhang, S., Shao, Q., Geng, H., and Su, S. (2017). The effect of royal jelly on the growth of breast cancer in mice. *Oncology Letters* 14 (6): 7615–7621.

Zheng, J., Lai, W., Zhu, G., et al. (2013). 10-Hydroxy-2-decenoic acid prevents ultraviolet A-induced damage and matrix metalloproteinases expression in human dermal fibroblasts. *Journal of the European Academy of Dermatology and Venereology* 27 (10): 1269–1277.

# 21

## Medicinal Benefits of Propolis

*Kashif Maroof, Yim Yee Jin, Siew Liang Ching, and Siew Hua Gan*

## Introduction

Nature is abundant in providing us with a wide range of natural products, many of which remain unexplored and offer an immense possibility of new therapeutic agents. In fact, it is widely believed that natural products are healthier and pose better tolerability, attracting the interest of the scientific community, especially in plant products that are reformed by animals (Welz et al. 2018).

Honey is a delicious liquid that is produced by honeybees. Honey is well known over the world for its rich nutritional content, which is helpful to human health. Honey is also characterized as a sugar solution that is supersaturated (Durazzo et al. 2021). Natural honey is composed of 82.4% carbohydrates, 38.5% fructose, 31% glucose, 12.9% other sugars, 17.1% water, 0.5% protein, organic acids, multiminerals, amino acids, vitamins, phenols, and myriad other minor compounds (Pasupuleti et al. 2017). Honey also contains trace levels of beneficial compounds such as phenolic acid, flavonoid, and $\alpha$-tocopherol (Pasupuleti et al. 2017). Honey contains phenolic acids, flavonoids, ascorbic acid, proteins, carotenoids, and enzymes, including glucose oxidase and catalase, which have health advantages.

Propolis is one of the most popular natural products used to date. It is collected from the flowers, buds, and exudates as well as barks of trees and plants by honeybees and is composed of different types of material, including resins, beeswax, pollen, balsams, essential oils, and other organic compounds (Maroof and Gan 2021). Propolis is the third most important component of bee products. It is composed mainly of resin (50%), wax (30%), essential oils (10%), pollen (5%), and other organic compounds (5%) (Pasupuleti et al. 2017). Propolis contains phenolic compounds, esters, flavonoids, terpenoids, beta-steroids, aromatic aldehydes, and alcohols, among other compounds (Pasupuleti et al. 2017). By capillary zone electrophoresis, 12 different flavonoids, including pinocembrin, acacetin, chrysin, rutin, luteolin, kaempferol, apigenin, myricetin, catechin, naringenin, galangin, and quercetin; two phenolic acids, caffeic acid and cinnamic acid; and one stilbene derivative called resveratrol have been detected in propolis (Pasupuleti et al. 2017). Propolis also contains important vitamins, such as vitamins $B_1$, $B_2$, $B_6$, C, and E, and useful minerals, such as magnesium, calcium, potassium, sodium, copper, zinc, manganese, and iron. A few enzymes, such as succinic dehydrogenase, glucose-6-phosphatase, adenosine triphosphatase, and acid phosphatase, are also present in propolis (Anjum et al. 2019). A vast number of useful compounds in propolis have attracted the attention of cosmetic and dental products and pharmaceutical companies (Andrade et al. 2018; de Freitas et al. 2018; Maroof et al. 2020). Propolis is used to strengthen beehives. Propolis is confirmed to have promising sources of naturally derived antimicrobial compounds because it helps to keep the hive environment sterile and protects the colony from infection (Maroof and Gan 2021). Propolis has more phenolic components than honey and has proved to have significantly stronger antibacterial and antioxidant properties (Osés et al. 2016).

Surprisingly, distinct varieties of propolis may be distinguished even within the same climate region, with subgroups defined by color. In temperate zones (Europe, North America, some regions of Asia, and Australia), the most common propolis is the "poplar type" in addition to "birch type" (in Russia) and "Mediterranean type" (in Southern Europe). Similarly, in the tropical regions, a wide range of propolis is available in which two forms of Brazilian propolis are widely recognized (green from *Baccharis* spp. and red from *Dalbergia* spp.). Another form, known as the "clusia type," is common in Cuba and Venezuela, and the "macaranga type" is found in the Pacific zone and many other areas along the Pacific Ocean. Blue, green, and (most frequent) orange are the three primary botanically unique subtypes of propolis from Europe,

*Honey: Composition and Health Benefits*, First Edition. Edited by Md. Ibrahim Khalil, Gan Siew Hua, and Bey Hing Goh.

which is of the poplar type. Exudates from *Populus nigra* are the principal sources of propolis from the "orange poplar" variety, according to recent research based on phenolic profile analysis (Rojczyk et al. 2020).

Propolis' antiputrefactive effects have long been known, and Egyptians use it to embalm cadavers. Propolis' medicinal virtues were recognized by Greek and Roman doctors such as Aristoteles, Dioscorides, Pliny, and Galen (Castaldo and Capasso 2002). It was used as an antiseptic and cicatrizant in wound treatment and as a mouth disinfectant, and these usages carried on throughout the Middle Ages and among Arab physicians (Rojczyk et al. 2020). Propolis has also been known by people who are not tied to Old World traditions. Propolis, for example, was used by the Incas as an antipyretic. Propolis was first recognized as an official medicine in seventeenth century London pharmacopoeias. Because of its antibacterial properties, the medication became immensely popular in Europe during the seventeenth and twentieth centuries (Kuropatnicki et al. 2013). Today propolis is accessible in the form of capsules (either in pure form or in combination with aloe gel, rosacanina, or pollen), extracts (hydroalcoholic or glycolic), mouthwashes (mixed with melissa, sage, mallow, and/or rosemary), throat lozenges, lotions, and powder form (to be used in gargles or for internal use after being dissolved in water) (Pavlovic et al. 2020). It is also available commercially as a purified product that has had the wax removed. Propolis is also believed to be beneficial in cosmetics and as a food additive (Maroof and Gan 2021).

Propolis and its isolated compounds may be useful in a variety of pathological conditions such as tumors, infections, allergy, diabetes, and ulcers (Sforcin and Bankova 2011). However, before establishing a strategy using this bee product, it is necessary to understand the factors that cause variability in the presence of these compounds (Sforcin and Bankova 2011). Table 21.1 lists the compounds reported as being present in propolis. The presence of these bioactive compounds is strongly influenced by geographical location, plant source, and season (Bankova et al. 2021; Popova et al. 2021; Salatino et al. 2021).

Different solvents may be used for extraction of the active components of propolis. Among them, an ethanol/water mixture (70/30) is the most commonly used because it is non-toxic and very efficient at extraction, especially of polyphenols and flavonoids, commonly used as the quality criterion in this substance (Escriche and Juan-Borrás 2018). The final extraction of the bioactive compounds depends on the type and quantity of solvent, the temperature and time, and even the procedure used to interact with the crude propolis (Sawaya et al. 2011). Maceration is the traditional extraction procedure, although in recent years, sonication and microwaves have also been recommended because of their efficiency, time saving, and selectivity (Bankova et al. 2021; Sforcin 2016). Many researchers have looked at the effects of ethanol percentage and solid-to-liquid ratio. The ideal ethanol percentage in water was shown to be between 70% and 95%, most commonly 70%–80% (Bankova et al. 2021).

The final extraction of the bioactive compounds depends on the type and quantity of solvent, the temperature and time, and even the procedure used to interact with the crude propolis (Zhao et al. 2017). Many researchers have compared ultrasonic-assisted extraction (UAE) of propolis with maceration and discovered that extremely short extraction durations yield good results in terms of phenolic and flavonoid extraction percentages (Bankova et al. 2021). Numerous researchers have worked on UAE optimization, with the majority of them focusing on ethanol concentration, temperature, extraction time, solid-to-solvent ratio, and other factors (Table 21.2).

Teixeira et al. (2010) conducted a study to investigate the relationship between geographical location and seasonal variation. They reported the chemical composition of Brazilian propolis obtained from three different areas (Itapecerica, Paula Cândido, and Virginópolis). A total of 40 substances, including simple phenylpropanoids, prenylated phenylpropanoids, and sesqui- and diterpenoids, were detected. A single sample from Virginópolis displayed no antioxidant activity because of the minimal presence of phenolic constituents. In addition, there was considerable variation in the total phenolic contents of the samples from Itapecerica collected over different periods, confirming the influence of seasonal variation in propolis chemical composition.

Tomazzoli et al. (2020) studied the effect of the seasons on the chemical profile of propolis produced in southern Brazil. Field observation of the foraging behavior of *Apis mellifera* L. revealed its preference for the *Baccharis dracunculifolia* DC species *p*-Coumaric acid, 3, 4-dicaffeoylquinic acid, 3,5-dicaffeoylquinic acid, drupanin, and artepillin C, which were identified in both plant and propolis samples. Moreover, whereas higher artepillin C amounts have been detected in samples collected in the summer and autumn, the main compounds of *p*-coumaric acid and quercetin were available in spring and winter sampled propolis, respectively. The contents of balsam and total phenolic and flavonoid compounds significantly varied over the seasons. Compared with other seasons, propolis collected in the spring had the highest inhibitory effect on cancer cell development (Valencia et al. 2012), the best antioxidant activity (Sawaya et al. 2009), and the highest quantities of phenols (total phenols, flavones, flavonols, flavanones, and dihydroflavonols) (Miguel et al. 2010).

**Table 21.1** Compounds reported to be present in propolis.

| Compounds Present in Propolis | | |
|---|---|---|
| **Chemical Class** | **Compounds** | **References** |
| Benzenoic acid and derivatives | Benzoic acid, 4-aminohippurate, salicylic acid, gentisic acid, gallic acid, phenylmethyl ester of benzoic acid, phenylmethyl ester of salicylic acid, trans-coniferyl benzoate, trans-p-coumaroyl benzoate, protocatechuic acid | Anjum et al. 2019; Usman et al. 2018 |
| Fatty acids and their derivatives | Myristic acid, palmitic acid, linoleic acid, hexadecanoic acid, stearic acid, arachidic acid, 9,12-octadecenoic acid, ethyl oleate, 11-eicosenoic acid, eicosanoic acid, docosanoic acid, docosenamide, 1,15-hexadecadiene, 13-docosenamide | Ismail et al. 2018; Maroof and Gan 2021 |
| Phenols | Hydroginkol, 1,2-benzenediol, 3-pentadecylphenol, 1,3-benzenediol, vanillin, pyrogallol, torachrysone-O-hexose, torachrysone-O-(galloyl)-hexose, acetyleugenol | Bankova et al. 2021; Kuropatnicki et al. 2013; Popova et al. 2021 |
| Phenolic acids and their derivatives | Di-caffeic acid, cinnamic anhydride, ferulic acid derivative, isoferulic acid, ellagic acid vanillic acid, isomer of vanillic acid, syringic acid, benzoic acid, caffeic acid, prenyl caffeate, gallic acid, cinnamic acid | Castaldo and Capasso 2002; Sforcin and Bankova 2011 |
| Steroids | Cycloeucalenol, 19-cyclolanost-24-en-3-ol, 9,19-cyclolanostan-3-ol.24-methylene, 9,19-cyclolanost-24-en-3-ol, stigmasterol, lanosterol, ergosta-8,24(28)-dien-3-ol | Maroof and Gan 2021; Zhao et al. 2017 |
| Sugars and sugar alcohols | Erythritol, D-fructose, D-ribofuranose, D-galactose, arabinofuranose, galacturonic acid, glycerol, xylitol, erythritol, pentose, malonylhexose | Ismail et al. 2018; Usman et al. 2018 |
| Terpenoids | Delta-cadinene, nootkatone, caryophylleneoxide, squalene, lup-20(29)-ene-39,19-cyclolanostan-3-ol 24-methylene, aromadendrene α-eudesmol, caryophyllene oxide, squalene, α-cedrol, globulol, γ-gurjunene, δ-selinene, ledene, δ-cadinine, β-panasinsene, epiglobulol, 2,6,10,14,18,22-tetracosahexaene, lanosterol, β-amyrin, α-amyrin, bicyclo[3.1.1]hept-2-ene, eudesma-4(14), 11-diene, β-amyrin, α-amyrin, taraxasterol | Ismail et al. 2018; Maroof and Gan 2021; Zhao et al. 2017 |
| Flavonoids, flavanones, flavones, and flavonols | Pinobanksin-5-methyl ether acetate, kaempferol- dimethyl ether, pinocembrin, quercetin, luteolin, apigenin, vitexin-O-gallate, isomer of vitexin-O-gallate, Pinobanksin, kaempferol, islapinin, ermanin, pectolinarigenin, sakuranetin, isosakuranetin, quercetin-3,3′-dimethyl ether, 3-acetyl pinobanksin, betuletol, isorhamnetin, kaempferide, rhamnazin, rhamnetin, alnusin, alpinetin, alnusitol, pinostrobin, pinocembrin, chrysin, tectochrysin, acacetin, rhamnocitrin, quercetin, galangin, apigenin, pinobanksin | Anjum et al. 2019; Maroof and Gan 2021; Ong et al. 2017; Usman et al. 2018; Zhao et al. 2017 |

Bee species have a major impact on both chemical composition and propolis' quality. In a previous study, Ibrahim et al. (2016) research to compare the quality of propolis produced by different types of bee species – *Heterotrigona itama* and *Geniotrigona thoracica* – using a similar extraction method. They confirmed that significant differences in terms of the type of compounds extracted exists indicating that the bee species is important.

The antibacterial activity of propolis collected from the same apiary was found to be affected by bee variety; *A. mellifera carnica* colonies had poorer antibacterial activity than *A. mellifera anatolica* and *A. mellifera caucasica* hives. The three honeybee races did not use the same plant source or a single plant (Silici and Kutluca 2005). Benzophenones (but no flavonoids) have been identified as the major compounds in geopropolis produced by stingless bee species *Melipona scutellaris* (da Cunha et al. 2013); however, geopropolis produced by *Melipona fasciculate* contains high concentrations of polyphenols, flavonoids, triterpenoids, saponins, and even tannins (Dutra et al. 2014).

Although various honeybee species prefer different plants, propolis produced by the same species does not always have the same chemical composition. Both Brazilian green and red propolis are from Africanized *A. mellifera*; however, these types of propolis are higher in prenylated phenylpropanoids and isoflavonoids, respectively (Daugsch et al. 2008; Teixeira et al. 2005). The discrepancies are attributable to the plants that bees use as resin sources, specifically *B. dracunculifolia* and

**Table 21.2** Data on maceration, ultrasonic-assisted extraction (UAE), and microwave-assisted extraction (MAE) of propolis: optimal conditions.

| Method | Propolis Source | Solvent | Solid-to-Liquid Ratio (m/V) | Extraction Time | Temperature (°C) | Extraction Effectiveness | | | References |
|---|---|---|---|---|---|---|---|---|---|
| | | | | | | Yield | Total Flavonoid Content | Total Phenolic Content | |
| Maceration | Bulgarian | 70% ethanol | 1:10 | 72 hr | Room temperature | 55% | 28% | 43% | Trusheva et al. 2007 |
| | Chinese | 95% ethanol | 1:10 | 24 hr | | 38% | 199 mg/g | Not mentioned | YaZhou et al. 2012 |
| | Korean | 60–80% ethanol | 1:10 | 12 hr | | 20%–56% | 0–7 g QE/100 g | 12–24 g GAE/100 g | Bankova et al. 2021 |
| | Malaysian | 70% ethanol | 1:10 | 7 days | 28 | 10.6% | 7.25 mg QE/g | 26.1 mg GAE/g | Bankova et al. 2021 |
| UAE | Bulgaria | 70% ethanol | 1:10 | 30 min | 25 | 53% | 31.6% | 52% | Trusheva et al. 2007 |
| | Romanian | 70% ethanol | 1:50 | 15 min (two times) | 60 | 37.1–96.7 g/100 g | 107.1–276.8 mg/g | 193.6–504.2 mg GAE /g; 205.2–534.4 mg pinocembrin:galangin/g | Oroian et al. 2020 |
| MAE | Malaysian | 70% ethanol | 1:5 | 15 min | Maximum, 115 | 8–90 mg/mL | 140 mg QE/g extract | 130–140 mg GAE/g extract | Hamzah and Leo 2015 |

GAE, gallic acid equivalent; QE, quercetin equivalent.

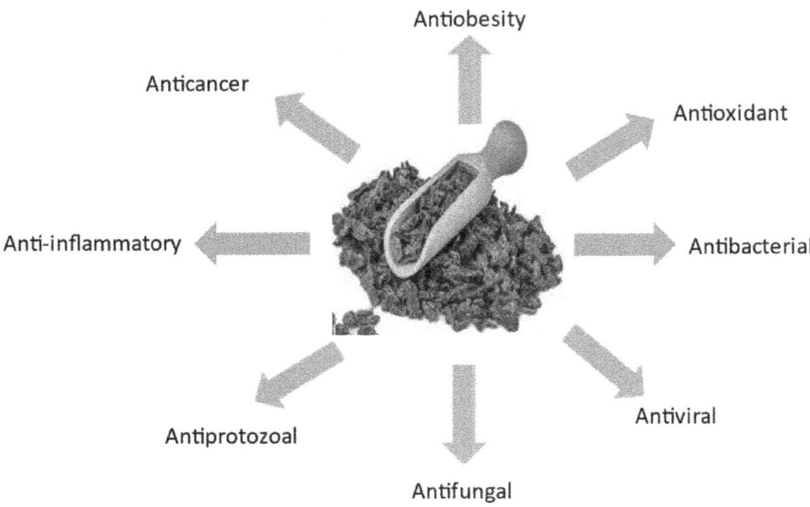


**Figure 21.1** Medicinal properties of propolis reported in the literature. Kosolovskyy / Getty Images.

*Dalbergia ecastophyllum*. C-methylated flavanones, terpenic acids, and phenolic acids, such as gallic acid, diterpenic acids of pimaric, and abietic type, are the most abundant chemicals in cerumen propolis from stingless bees (*Tetragonula carbonaria*), but this honey lacks the characteristic flavonoids and prenylated phenolics found in propolis from honeybee species in Australia (Massaro et al. 2011, 2014). As a result, the chemical content of propolis varies depending on the bees' botanical preferences (Huang et al. 2014) (Figure 21.1).

## Biomedical Applications of Propolis

### Antioxidant Activity

It is well established that cellular metabolism generates reactive oxygen species (ROS), such as hydrogen peroxide ($H_2O_2$), the superoxide anion ($O_2^{\cdot}$), and the highly reactive hydroxyl ion ($HO^{\cdot}$), as well as reactive nitrogen species (RNS), especially nitric oxide (NO). ROS and RNS are ideal signaling molecules because they are locally generated, are highly and rapidly diffusible, and can be neutralized by cellular antioxidants (Daleprane and Abdalla 2013). ROS are usually detoxified by intracellular enzymes, such as glutathione, superoxide dismutase (SOD), and catalase (Ighodaro and Akinloye 2019). However, unbalanced production and degradation of ROS and RNS can result in accumulation of these reactive species, commonly referred to as oxidative stress. Exposure of macromolecules (lipids, proteins, and DNA) to reactive species results in oxidative modifications with deleterious effects (Singh et al. 2019).

The flavonoids in propolis are powerful antioxidants that are capable of scavenging free radicals and thereby protecting cell membranes against lipid peroxidation (Bose et al. 2018; Jucá et al. 2018). Moreover, ROS and RNS, together with other factors, are involved in cellular aging and death in conditions such as cardiovascular disease, arthritis, cancer, diabetes, Parkinson's disease, and Alzheimer's disease (Bose et al. 2018; Munteanu and Apetrei 2021). Propolis can reduce cellular levels of $H_2O_2$ and NO, which may be involved in its anti-inflammatory effects (Daleprane and Abdalla 2013).

Diverse compounds from propolis have been described as potent inhibitors of oxidative stress. It is well known that the composition of propolis is variable; however, one of its major components, caffeic acid phenethyl ester (CAPE), blocks ROS production in several systems (Turan et al. 2020). In vivo, propolis can increase antioxidant capacity in animals and humans, leading to decreased lipid peroxidation (Darvishi et al. 2020; Oršolić et al. 2012).

Propolis is rich in flavonoids, polyphenols, vitamins C and E, tannins, and reducing sugars. These are good antioxidants, with 2,2-diphenyl-1-picrylhydrazyl (DPPH) and NO scavenging activities at half maximal inhibitory concentrations ($IC_{50}$) of 419.07 ± 0.98 μg/mL and 88.33 ± 0.15 μg/mL, respectively (Tanvir et al. 2018). Brazilian green propolis displays good antioxidant activities, which may be related to the presence of artepillin C in synergism with the other constituents of the extracts (Veiga et al. 2017). The antioxidant effect of Brazilian red propolis has been attributed to chalcones and isoflavonoids (including 7-O-methylvestitol, medicarpin, and 3,4,2′,3′-tetrahydrochalcone) that act as electron donors (Oldoni et al. 2011).

Teixeria et al. (2010) investigated the effect of seasonal variation on the total phenolic contents and antioxidant activity of Brazilian propolis from three different municipalities. They found that propolis from the municipalities Itapecerica and Paula Cândido had high phenolic contents with good antioxidant activity. In fact, some samples from Itapecerica collected over 1 year exhibit considerable variation in the total phenolic substances and antioxidant activity. However, propolis samples from the Virginópolis municipality, which had no detectable phenolic substances, exhibited no antioxidant activity, but the methanolic extracts from Itapecerica and Paula Cândido exhibited significant scavenging activity, indicating that the active constituent may again vary based on the geographical location. It has been reported that whereas compounds that contain phenolic hydroxyls in their structure tend to have higher antioxidant properties, compounds devoid of hydroxyl groups tend to have lower antioxidant activity (Derakhshan et al. 2018; Limmongkon et al. 2017).

Thus, the available data indicate that propolis of different origins and distinct compositions consistently exhibit antioxidant actions. In addition to this antioxidant effect, bioactive compounds in propolis influence a large number of biochemical signaling pathways and therefore physiological and pathological processes.

## Anticancer Activity

Propolis has largely been investigated for its anticancer effects. Propolis (200 μg/mL) from New Zealand, which is rich in chrysin, galangin, and CAPE, has antiproliferative effects against epithelial cells in colorectal adenocarcinoma (DLD-1), colon (HCT-116), gastric (NCI-N87), and esophageal squamous carcinomas (KYSE-30) (Catchpole et al. 2015).

Besides, propolis from Cuba containing the polycyclic poly-isoprenylated benzophenone nemorosone has been reported to confer anticancer effects on estrogen receptor–positive breast cancer cells (MCF-7) by stopping the cell cycle in G0/G1 phases, as well as by decreasing mitogen-activated protein kinase (MAPK) and serine/threonine kinase phosphorylations (Popolo et al. 2011).

Propolis from Bangladesh is a potential therapeutic agent against cancer because it exerts strong toxic effect against nauplii, possibly contributed by the active secondary metabolites, which include phenolic compounds, organic acids, tannins, and flavonoids (Tanvir et al. 2018). Propolis from Aydin, Turkey, which is abundant in flavonoids and CAPE, induces apoptosis in lymphoblastic leukemic cells by 1.5-fold compared with untreated cells at an $IC_{50}$ of 79.2 mg/mL by modulating the expressions of various miRNAs involved in the posttranscriptional regulation of mRNAs to control various cellular processes, including apoptosis (Yilmaz et al. 2016).

The combination of propolis from Botucatu, Brazil, and a lower concentration of an antitumor drug carboplatin has been shown to confer some cytotoxic effects against canine osteosarcoma cells after induction of apoptosis specific to tumor cells without affecting the mesenchymal stem cells. The activity indicates a promising combination candidate for canine osteosarcoma treatment (Bernardino et al. 2018). Another study on propolis from India also reported that administration of propolis to mice during and after chemotherapy treatment can mitigate mitomycin C–induced testicular toxicity, thereby ameliorating the loss of testicular function a chemotherapy (Zhang et al. n.d.).

Artepillin C present in propolis from Brazil has been shown to exert antiproliferative effect on prostate cancer cells by decreasing tumor necrosis factor (TNF)–related apoptosis-inducing ligand (TRAIL) resistance and inhibiting nuclear factor κB (NF-κB). On the other hand, galangin, a component present in propolis from China, Thailand, and Portugal, can induce MAPK phosphorylation to cause proapoptotic effects. Chrysin, a flavone found in propolis, has antiproliferative effects on human hepatoma cells (Hep-3B), transitional cell carcinoma, human lung carcinoma (A549), human cervix adenocarcinoma (HeLa), and colorectal cancer cells (Patel 2015).

Another interesting study on water-extract propolis from China reported chrysin, galangin, pinocembrin, pinobanksin, caffeic acid benzyl ester, CAPE, and apigenin (all at 80-μM concentrations) significantly inhibit the proliferation of four tumor cells, with cytotoxicity of pinocembrin > pinobanksin and CAPE > caffeic acid benzyl ester, indicating that the antitumor effects of propolis do not depend on a single component. These effective components induce apoptosis by activating caspase 3, inhibiting tumor cell migration, and promoting ROS production (Wang et al. 2016). Some compounds found in propolis that are known to promote cell proliferation or apoptosis are listed in Table 21.3.

**Table 21.3** Important compounds contributing to the cytotoxic activity of propolis.

| Compound | Chemical Class | Active Against | Assay Involved | Results | Reference |
|---|---|---|---|---|---|
| Caffeic acid phenethyl ester (CAPE) | Hydroxycinnamic acid | TNBC cell lines MDA-MB-231 and Hs578T | MTT assay | The morphological changes of breast carcinoma cells were observed after exposure to CAPE. The $IC_{50}$s for CAPE were 14.08 μM and 8.01 μM for the MDA-MB-231 and Hs578T cell line, respectively. | Rzepecka-Stojko et al. 2015 |
| Artepillin C | Prenylated p-coumaric acid | Gastric cancer cell line AGP-01 | Clonogenic survival assay | The compound was tested at 30, 15, and 7.5 μM. The $IC_{50}$ was found to be 15.39 μM. | Bastos et al. 2020 |
| 7-Hydroxy-6-methoxyflavanone | Isoflavanone | Murine B16-BL6 melanoma, murine Lewis lung carcinoma, human lung A549 adenocarcinoma, human HT-1080 fibrosarcoma | MTT assay | The compound showed the most potent activity against murine Lewis lung carcinoma cells with an $IC_{50}$ of 9.29 μM followed by human lung A549 adenocarcinoma ($IC_{50}$, 8.63 μM). There was also activity against human HT-1080 fibrosarcoma ($IC_{50}$, 7.94 μM) and murine B16-BL6 melanoma ($IC_{50}$, 6.66 μM). | Watanabe et al. 2011 |
| Mucronulatol | Isoflavonoid | Lewis lung carcinoma and human lung A549 adenocarcinoma | MTT assay | Activity against Lewis lung carcinoma was found to be more potent ($IC_{50}$, 8.38 μM) than activity against human lung A549 adenocarcinoma ($IC_{50}$,9.9 μM). | Watanabe et al. 2011 |
| Coniferyl aldehyde | Cinnamaldehyde | Human HT-1080 fibrosarcoma and murine colon 26-L5 carcinoma cells | MTT assay | $ED_{50}$ value for human HT-1080 fibrosarcoma was found to be less (4.05 μg/mL) than for murine colon 26-L5 carcinoma cells (10.44 μg/mL), showing that this compound has more potency toward Human HT-1080 fibrosarcoma cells. | Fokt et al. 2010 |
| Betuletol | Flavonoid | Human HT-1080 fibrosarcoma and murine colon 26-L5 carcinoma cells | MTT assay | $ED_{50}$ value for human HT-1080 fibrosarcoma was found to be more (5.83 μg/mL) than that for murine colon 26-L5 carcinoma cells (4.95 μg/mL), showing that this compound has more potency toward murine colon 26-L5 carcinoma cells. | Fokt et al. 2010 |
| Kaempferide | Flavonoid | Human HT-1080 fibrosarcoma and murine colon 26-L5 carcinoma cells | MTT assay | $ED_{50}$ value for human HT-1080 fibrosarcoma (2.91 μg/mL) was lower than that for murine colon 26-L5 carcinoma cells (5.95 μg/mL), showing that this compound has more potency toward human HT-1080 fibrosarcoma cells. | Fokt et al. 2010 |

**Table 21.3** (Continued)

| Compound | Chemical Class | Active Against | Assay Involved | Results | Reference |
|---|---|---|---|---|---|
| Ermanin | Flavonol | Human HT-1080 fibrosarcoma and murine colon 26-L5 carcinoma cells | MTT assay | $ED_{50}$ value for human HT-1080 fibrosarcoma (2.30 μg/mL) was lower than that for murine colon 26-L5 carcinoma cells (7.64 μg/mL), showing that this compound has more potency toward human HT-1080 fibrosarcoma cells. | Fokt et al. 2010 |
| Kaempferol | Flavonoid | HeLa cervical cancer cells | MTT assay | The compound stunted the growth of and impaired all the cells tested with an $ID_{50}$ of ~20 μg/mL. | Aminimoghadamfarouj and Nematollahi 2017 |

AGP, alpha-1-acid glycoprotein; $ED_{50}$, median effective dose; $IC_{50}$, half maximal inhibitory concentration; MTT, (3-(4,5-dimethylthiazol-2-yl)-2,5-diphenyltetrazolium bromide); TNBC, triple-negative breast cancer.

## Anti-inflammatory Effect

Inflammation is a physiological immune response that confers host protection from stimuli such as invasion of microbes and destruction of tissues. Inflammation is triggered by events occurring beyond the vasculature system, including hypoxia via hypoxia-inducible factor-1α (HIF-1α) (Checa and Aran 2020; Tejchman et al. 2021). Inflammation is a dynamic process, amplified by the generation of inflammatory cytokines, chemokines, eicosanoids, biogenic amines, and NO (Darvishi et al. 2020; Inui et al. 2021). The presence of inflammatory molecules as well as activated complement serve as a chemoattractant for immune cells – for instance, natural killer cells, monocytes, and neutrophils – to the site of inflammation. The migrated inflammatory cells, together with local endothelial cells, releases more pro-inflammatory signaling proteins, including TNF, macrophage inflammatory protein (MIP)-1α, interleukin (IL)-6, and IL-1β (Darvishi et al. 2020; Franchin et al. 2012).

Nuclear factor κB is the important transcription factor in response to different stimuli leading to inflammation. It increases the gene expression and production of the pro-inflammatory molecules. The innate response is initiated by inflammatory and adaptive immune responses, which is the ultimate process to clear the infection. However, failure of recovery and chronic inflammation can lead to many diseases, including hepatitis, arthritis, asthma, and inflammatory bowel disease (Taniguchi and Karin 2018).

Propolis from Turkey exhibits its anti-inflammatory effect by interfering with the NF-κB pathway and reducing HIF-1α levels, which further contribute to the low levels of TNF-α in the iris as well as ciliary body of Wistar male albino rats. The levels of lipopolysaccharide (LPS)-induced TNF-α and HIF-1α in rats that received intraperitoneal propolis (50 mg/kg) decreased significantly ($P < 0.01$) in comparison with the control group (Ertürküner et al. 2016). In human hepatic cell line HepG2, propolis from China (5–15 μg/mL) has a significant inhibitory effect on inflammation as induced by palmitic acid by suppressing the mRNA expression of TNF-α and IL-8. It also inhibits the activation of inflammatory transcription factor such as NF-κB and activator protein-1, thus confirming its anti-inflammatory activity on liver cells (Jin et al. 2017).

Funakoshi-Tago et al. (2015) identified four flavonoids in water and methanol-extracted propolis from Nepal that have an anti-inflammatory effect by interfering with the action of IL-33. Chrysin, 3′,4′-dihydroxy-4-methoxydalbergione, 4-methoxydalbergion, and cearoin inhibit the secretion of TNF-α, IL-6, and IL-13 caused by IL-33. IL-33 is the main regulator of several allergic diseases by stimulating the immune system. IL-33 stimulates the production of pro-inflammatory cytokines such as IL-1β, IL-6, monocyte chemoattractant protein-1 (MCP-1), C-X-C motif chemokine ligand 1 (CXCL1)/KC, C-C motif chemokine ligand 2 (CCL2)/MCP-1, TNF-α, and prostaglandin (PG) D2. The flavonoids interfere with the gene expression encoding for TNF-α, IL-6, and IL-13, thus further suppressing their production. Nevertheless, after treatment with up to four types of flavonoids (up to 30 μM), the cytotoxic effect was not seen in mouse bone marrow-derived mast cells, thus confirming that inhibition of the production of cytokines does not contribute to cytotoxicity. Additionally, 3′,4′-dihydroxy-4-methoxydalbergione shows the highest potency in inhibiting the expression of cytokines and NF-κB activation compared with other flavonoids possibly because of complete inhibition of IL-33–induced inhibitor of nuclear factor-kinase (IKK) activation, thus further diminishing NF-κB activation (Fanny et al. 2018; Funakoshi-Tago et al. 2015; Moulin et al. 2007).

Treatment for patients with cerebral ischemia with CAPE can extensively attenuate the increase in MCP-1 as induced by TNF-α in cortical ischemic core region. MCP-1 is a chemokine responsible for the migration and collection of monocytes and macrophages occurring via regulation of T helper-1 and -2 systems (Traynor et al. 2002). In the study, the amount of MCP-1–positive cells was successfully reduced by CAPE in C57 black 6 (C57BL/6) mice with permanent focal ischemia (142.3 ± 9.8 and 93.5 ± 14.6 positive cells/section at 2 and 5 mg/kg; $P < 0.05$ and $P < 0.001$, respectively). Treatment with CAPE twice daily (at 1 and 6 hours) after cerebral ischemia has also been reported to prevent further damage to the brain by reducing the production of IL-1α, a pro-inflammatory cytokine that causes inflammation in some neurodegenerative disease (Hwang et al. 2017; Rothwell and Luheshi 2000). CAPE has been demonstrated to be a potential anti-inflammatory agent by reducing inflammatory process in cerebral ischemia by inducing anti-inflammatory cytokine IL-10 from different types of inflammatory cells, especially macrophages. It significantly increases IL-10 levels when mice are treated with CAPE at 2 and 5 mg/kg ($P < 0.001$ and $P < 0.001$, respectively). Additionally, reduction of TNF-α expression in the cerebral cortex ipsilateral to the photothrombosis is also achieved after CAPE treatment (Hwang et al. 2017).

Kitamura et al. (2018) demonstrated that propolis from Brazil, particularly kaempferol, can cause induction of $CD11b^+$, $Gr\text{-}1^+$ myeloid-derived suppressor cells (MDSCs) from M1 macrophages, leading to the accumulation of MDSCs in the visceral adipose tissue of mice. MDSCs help to alleviate the inflammatory response by suppressing T-cell response in which an approximate 7.5-fold of increased in CD11b +, Gr-1 + MDSCs in the mesenteric adipose tissue occurs after intraperitoneal injection of propolis (100 mg/kg) twice weekly to mice with type 2 diabetes (Melero-Jerez et al. 2016). Additionally, kaempferol also reduces the number of M1 macrophage, the key factor involved in the production of pro-inflammatory cytokines, including TNF-α, especially in adipose tissue inflammation (Kitamura et al. 2018).

Pinocembrin, a compound reported to be present in propolis from many countries, including Mexico, China, Greece, and Thailand, has been reported to have good anti-inflammatory effects. According to a study by Hanieh et al. (2017), histamine decarboxylase activity is effectively inhibited by pinocembrin in 2,4-dinitrophenylated-bovine serum albumin (DNP-BSA)–stimulated rat basophilic leukemia (RBL) 2H3 cells after 24 hours of incubation with different doses of pinocembrin (25, 50, and 100 μM). Histamine decarboxylase works by converting L-histidine into histamine, a pro-inflammatory mediator that is part of the early phase of inflammatory cascade (Murata et al. 2013). In the study, docking analysis revealed that pinocembrin has a strong binding affinity to histamine decarboxylase as contributed by the multiple binding pockets on the enzyme where pinocembrin binds with a lower binding energy, thus confirming its better binding compared with many other natural substrates. Additionally, the amounts of cells that contain granules and β-hexosaminidase, a marker for cell degranulation, decrease when immunoglobulin E antigen–stimulated RBL-2H3 cells are treated with pinocembrin at 25 and 50 μM, purported to occur because of inhabitation of histamine decarboxylase. Thus, pinocembrin has the potential to be developed into an enzyme inhibitor to attenuate inflammation. In addition, pinocembrin also shows its anti-inflammatory effect by suppressing TNF-α, IL-6, and IκB-α phosphorylation, thus contributing to the production of low amounts of cyclooxygenase-2, inducible nitric oxide synthase (iNOS), and prostaglandin E2.

Another potential anti-inflammatory isoflavonoid, neovestitol, has been identified in red propolis from Northeastern Brazil. Franchin et al. (2016) confirmed that administration of neovestitiol reduces leucocyte rolling and adhesion by decreasing the expression of intercellular adhesion molecule type 1 (ICAM-1) in LPS-challenged mice after subcutaneous administration of neovestitol (10 mg/kg). Consequently, the migration of immune cells to inflammation site is ameliorated because of reduction in cell adhesion onto vascular endothelial cells (Vestweber 2015). Nevertheless, no similar effect on TNF-α, CXCL1/KC, and CXCL2/MIP-2 levels was seen in the peritoneal cavity of LPS-challenged mice, confirming that its anti-inflammatory effect does not depend on modulation of chemokines and cytokines. Therefore, it was hypothesized that neovestitol reduces leucocyte adherence and ICAM-1 expression via induction of iNOS pathway because NO levels tend to increase in the peritoneal cavity after 4 hours of pretreatment with neovestitol. Additionally, there is no direct inhibition of ICAM-1 expression in bEnd.3 endothelial cells (in vitro) that does not support iNOS expression, thus confirming that the action of neovestitol on peritonitis is NO dependent (rather than occurring via direct blockage of adhesion molecules expression). This finding was further confirmed in mice with collagen-induced arthritis in which it was observed that 12 days of treatment with neovestitol (10 mg/kg) also reduced the levels of IL-6 without affecting other mediators, including TNF-α, IL-17, and interferon-γ. All the aforementioned studies have shown that different propolis preparations or isolated components have diverse methods of inflammatory suppression.

## Antibacterial Activity

Generally, it is observed that the antimicrobial activity of propolis is higher in relation to gram-positive than gram-negative bacteria. This is explained by the species-specific structure of the outer membrane of the gram-negative bacteria and the production of hydrolytic enzymes that break down the active ingredients of propolis (Przybyłek and Karpiński 2019; Sforcin 2016). Research carried out in Brazil by Veiga et al. (2017) showed a higher concentration of artepillin C in ethanolic extracts of propolis compared with hexane extracts. These extracts also showed high antibacterial activity against methicillin-resistant *Staphylococcus aureus* (MRSA) (Darvishi et al. 2020). In studies against anaerobic bacterium *Porphyromonas* gingivalis, it was found that artepillin C has bacteriostatic activity with membrane blebbing (Yoshimasu et al. 2018).

The ethanolic extract of propolis containing high concentrations of kaempferide, artepillin-C, drupanin, and p-coumaric acid showed antioxidant activity and antibacterial against *S. aureus, Staphylococcus saprophyticus, Listeria monocytogenes, and Enterococcus faecalis* (Seibert et al. 2019).

Other flavonoids found in propolis are pinocembrin and apigenin. The study conducted by Veloz et al. (2019) on Chilean propolis found that antibacterial activity of both of these compounds against *Streptococcus mutans* is higher than the activity of polyphenols mixture or even chlorhexidine (minimum inhibitory concentration [MICs], 1.6 μg/mL), with MICs of 1.4 μg/mL and 1.3 μg/mL, respectively. Several studies have demonstrated the antibacterial activity of isolated pinocembrin *against S. mutans, Streptococcus sobrinus, S. aureus, E. faecalis, L. monocytogenes, Pseudomonas aeruginosa, and Klebsiella pneumoniae* (Przybyłek and Karpiński 2019). Isolated apigenin acts against gram-negative bacteria *P. aeruginosa, K. pneumoniae, Salmonella enterica* serotype *typhimurium, Proteus mirabilis,* and *Enterobacter aerogenes* (Nayaka et al. 2014). It was also observed the synergistic antibacterial effect of apigenin with β-lactam antibiotics against MRSA (Research et al. 2016) and apigenin with ceftazidime against ceftazidime-resistant *Enterobacter cloacae* (Eumkeb and Chukrathok 2013).

Recently, the evaluation of the antibacterial activity of propolis extracts has been based on the determination of total phenolic content and flavonoids. In their study, Bridi et al. (2015) found that total phenolic content and flavonoid tests do not always adequately reflect antimicrobial activity in vitro. The results of the total phenolic content in the samples with the highest and the lowest content were directly proportional to the content of flavonoids and antioxidant properties. However, they were not unambiguous in the case of antibacterial activity. It is suggested that other tests (e.g. oxygen radical absorbance capacity) and antimicrobial tests should be considered in setting international quality standards for propolis.

## Antifungal Action

Propolis extract shows excellent performance regarding in vitro tests against yeasts identified as onychomycosis agents. In low concentrations, propolis extract was not only found to be fungistatic but also fungicidal. *Trichosporon* spp. were the most vulnerable yeasts (Oliveira et al. 2006). The results reinforce the importance and the potential of propolis extract as a treatment for patients with onychomycosis. Similarly, the findings of a recent study suggest that the propolis extract is a promising natural compound for onychomycosis treatment. Results confirmed the in vitro antifungal activity against both the planktonic cells and the biofilm formed by species of *Trichophyton*, which are the most common agents of onychomycosis and are usually resistant to conventional antifungals. Patients treated with onychomycosis topically with propolis extract twice a day showed excellent clinical improvement within 6 months. Therefore, the extract of propolis is a potential new therapeutic agent for the topical treatment of *Trichophyton* onychomycosis (Veiga et al. 2018). Similarly, the antifungal activity of propolis extract on different strains of *Candida* yeast has also been tested, and it was found that the yeasts showed a clear antifungal activity against *Candida albicans, Candida tropicalis, Candida krusei,* and *Candida guilliermondii* (Khurshid et al. 2017).

Another study aiming to produce a vaginal dosage form for the treatment of vulvovaginal candidiasis (VVC) was conducted on Brazilian propolis. The patients' vaginal exudates were used for detection of 89 yeast strains. Most strains (n = 58) were *C. albicans*, and the remaining (n = 31) were non–*C. albicans* (17 *Candida glabrata*, 1 *Candida tropicalis*, 8 *Candida guilliermondii*, 5 *Candida parapsilosis*). Comparisons with the antifungal drugs (i.e. fluconazole, voriconazole, itraconazole, ketoconazole, miconazole, and amphotericin B) used in the treatment of patients with VVC were performed. The MICs were determined based on a standard broth microdilution method.

The results indicated that both *C. albicans* and non–*C. albicans* were inhibited by propolis extract up to a maximum concentration of 1100.63 μg/mL Some of the yeast isolates that showed resistance to itraconazole, ketoconazole, miconazole, and amphotericin B were inhibited by propolis, suggesting that propolis can be used to overcome antimicrobial resistance (Bruschi et al. 2011).

## Antiviral Activity

Propolis has been tested against various viral disease organisms; initial successes have prompted research to determine the most useful components, which may be modified to produce more active and specific pharmaceuticals (Zulhendri et al. 2021). Viruses that were controlled by propolis in animal models with suggestion for control in humans include influenza (Berretta et al. 2020; Kujumgiev et al. 1999), herpes simplex virus (HSV) types 1 and 2 (Mazia et al. 2016; Nolkemper et al. 2010), and human immunodeficiency virus (HIV) (Ito et al. 2001). Shimizu et al. (2011) evaluated three different types of propolis in ethanol extracts using a murine model of herpes simplex virus type 1. Despite the chemical differences caused by the different plant origins of the resins the bees used to produce the propolis (*Baccharis dracunculifolia*, *Baccharis eriodata*, and *Myrceugenia euosma*), all three propolis extracts not only had direct anti–HSV-1 effects but also stimulated immunologic activity against intradermal HSV-1 infection in mice.

The antiviral activity of propolis has been reported for DNA and RNA viruses (poliovirus, HSV, and adenovirus) in an in vitro model (cultured cells). The best results were obtained against poliovirus and HSV, with 99.9% inhibition of the latter, at a propolis concentration of 30 ug/mL (Amoros et al. 2004). The propolis components chrysine and kaempferol caused a concentration-dependent reduction of intracellular replication of HSV strains when host cell monolayers were infected and subsequently cultured in a drug-containing medium. Quercetin, another propolis component, had the same effect but only at the highest concentrations tested (60 μg/mL) against various HSV strains, with an intracellular replication reduction of approximately 65%, but it reduced the infectivity of bovine herpes virus, human adenovirus, human coronavirus, and bovine coronavirus about 50%. The reduction was 70% in the case of rotavirus (Tateo et al. 1990). All of the findings point to propolis as a potential antiviral agent; nonetheless, clinical research on people with viral disorders, in conjunction or not with antiviral medications or vaccines, should be conducted.

## Antidiabetic Effect

Given the relation between diabetes and inflammation and that flavonoids, major bioactive components of propolis, protect against free radicals and other pro-oxidative compounds, it is plausible that propolis consumption can reduce the risk of diabetes (Vinayagam and Xu 2015). Propolis has been shown to control oxidative stress, the build-up of advanced glycation end products (AGEs), and adipose tissue inflammation in preclinical studies, all of which lead to insulin resistance or insulin secretion abnormalities (Kitamura 2019). As a result, propolis therapy may help to reduce diabetes complications such nephropathy, retinopathy, foot ulcers, and non-alcoholic fatty liver disease.

Green propolis from Taiwan contains propolis C (a prenylflavanone compound, also known as nymphaeol-A), propolis D (nymphaeol-B), propolis F (a prenylated flavonoid compound; also known as isonymphaeol-B), and propolis G (a prenylflavanone), which have good anti–free radical activities. It can delay the progression of type 2 diabetes by scavenging ROS and increasing the amounts of antioxidants, such as SOD and glutathione peroxidase (GPx) (Chen et al. 2018). Another similar study also indicated that administration of Green propolis from Taiwan (200 and 400 mg/kg/day) can substantially boost the secretion of SOD and GPx in a dose-dependent manner in mice (Su et al. 2014). Green propolis from Taiwan has also been reported to regulate several hepatic genes in the liver, thereby improving lipid metabolism and delaying progression of type 2 diabetes (Chen et al. 2018).

Chrysin, a compound found in propolis, is useful for recovery from hepatotoxicity by ameliorating the elevation of malondialdehyde (an oxidative stress indicator) and reducing the antioxidant enzymes in the liver. Its protective effects against oxidative damage in the liver have also been confirmed in streptozotocin-induced diabetic rats (Samarghandian et al. 2015). Additionally, it can be used for prevention of neurotoxicity by increasing the antioxidant content and decreasing lipid peroxidation (Samarghandian et al. 2015). By modulating the oxygen radical production, chrysin is reported to be useful in the treatment of type 2 diabetes in which the underlying pathogenesis involves the overproduction of oxygen free radicals induced by hyperglycemia, auto-oxidation of glucose, lipid peroxide formation, and modification in antioxidant enzymes (Samarghandian et al. 2015). Green propolis from Taiwan was shown to extensively decrease TNF-α, IL-6, and IL-1β in streptozotocin/high-fat diet–induced rats with type 2 diabetes (Chen et al. 2018).

A variety of clinical research findings on the health effects of propolis on blood glucose levels and lipid profiles and have been published (Table 21.4). These clinical investigations have not provided extensive explanation of the processes behind propolis' therapeutic actions even though they imply potential applications of propolis in antidiabetic therapy.

**Table 21.4** Summary of findings on the effects of propolis on metabolic derangements in clinical studies.

| Location | Study Design | Gender | Mean Age | Sample Size at Baseline | Disease Background | Sample Size at End | Dose (Duration) | Findings | | Reference |
|---|---|---|---|---|---|---|---|---|---|---|
| | | | | | | | | Index | Effect | |
| Iran | RCT (double blind) | Both | Propolis: 55.4 yr<br>Placebo: 54.86 yr | Total: n = 100<br>Propolis: n = 50<br>Placebo: n = 50 | Type 2 diabetes | Total: n = 94<br>Propolis: n = 50<br>Control: n = 44 | 500 mg twice daily for 90 days | HbA1c, insulin, HOMA-IR, HOMA-β, TNF-α, BUN, ALT, AST, IL-1β | Decrease | Zakerkish et al. 2019 |
| | | | | | | | | 2hp, ALP, CRP, body weight, BMI, FBG, TGs, total cholesterol, LDL cholesterol, VLDL cholesterol, creatinine, uric acid, IL-6 | No significant change | |
| | | | | | | | | HDL cholesterol | Increase | |
| Iran | RCT (double blind) | Both | Propolis: 51.3 yr<br>Placebo: 56.07 yr | Total: n = 66<br>Propolis: n = 33<br>Placebo: n = 23 | Type 2 diabetes | Total: n = 57<br>Propolis: n = 30<br>Placebo: n = 27 | 300-mg propolis pills three times a day (12 wk) | FBG, HbA1c, BMI | Decrease | Samadi et al. 2017 |
| | | | | | | | | Total cholesterol, LDL cholesterol, body weight, TGs, HDL cholesterol, VLDL cholesterol | No significant change | |
| | | | | | | | | Insulin | Increase | |
| Iran | RCT (double blind) | Both | Propolis: 51.81 yr; placebo: 49.05 yr | Total: n = 62<br>Propolis: n = 31<br>Placebo: n = 31 | Type 2 diabetes | Total: n = 60<br>Propolis: n = 30<br>Placebo: n = 30 | 500-mg propolis capsules three times a day (8 wk) | FBG, 2hp, insulin, HOMA-IR, HbA1c, fructosamine, Ox-LDL | Decrease | Afsharpour et al. 2019, Hesami et al. 2019 |
| | | | | | | | | Body weight, BMI | No significant change | |
| | | | | | | | | TAC, SOD, GPx, CAT | Increase | |
| China | NRCT (double blind) | Both | Propolis: 57.7 yr<br>Placebo: 60.6 yr | Total: n = 6<br>Propolis: n = 31<br>Placebo: n = 30 | Type 2 diabetes | Total: n = 55<br>Propolis: *n* = 25<br>Placebo: *n* = 30 | 900 mg/day for 18 wk | BG, glycosylated hemoglobin, insulin | No significant change | Gao et al. 2018 |

*(Continued)*

**Table 21.4** (Continued)

| Location | Study Design | Gender | Mean Age | Sample Size at Baseline | Disease Background | Sample Size at End | Dose (Duration) | Findings | | Reference |
|---|---|---|---|---|---|---|---|---|---|---|
| | | | | | | | | Index | Effect | |
| Brazil | RCT (double blind) | Both | Propolis: 63.7 yr<br>Placebo: 62.9 yr | Total: n = 82<br>Propolis: n = 41<br>Placebo: n = 41 | Type 2 diabetes | Total: n = 80<br>Propolis: n = 41<br>Placebo: n = 39 | Propolis tablets 226.8 mg/day for 8 wk | HOMA-IR, FBG, HbA1c, insulin, uric acid, total cholesterol, LDL cholesterol, RLP cholesterol, TGs, TNF-α, IL-6, CRP | No significant change | Fukuda et al. 2015 |
| Brazil | RCT (double blind) | Both | Propolis: 59.5 yr<br>Placebo: 60.8 yr | Total: n = 70<br>Propolis: n = 35<br>Placebo: n = 35 | Type 2 diabetes | Total: n = 65<br>Propolis: n = 33<br>Placebo: n = 32 | Brazilian propolis in capsules at 900 mg/day for 18 wk | Carbonyls, LDH, TNF-α | Decreased | Zhao et al. 2016 |
| | | | | | | | | BG, glycosylated hemoglobin, insulin, FRAP, SOD, GPx, MDA, aldose reductase, adiponectin, Ox-LDL | No significant change | |
| | | | | | | | | GSH, polyphenols, IL-1β, IL-6 | Increased | |
| Egypt | RCT (double blind) | Both | Propolis: 48.9 yr<br>Placebo: 51.2 yr | Total: n = 50<br>Propolis: n = 26<br>Placebo: n = 24 | Type 2 diabetes | Total: n = 50<br>Propolis: n = 26<br>Placebo: n = 24 | 400 mg/day (3 and 6 months) | HbA1c, plasma glucose, CML | Decreased | El-Sharkawy et al. 2016 |

2hp, 2-hour postprandial blood glucose; ALP, alkaline phosphatase; ALT, alanine transaminase; AST, aspartate transaminase; BG, blood glucose; BMI, body mass index; BUN, blood urea nitrogen; CAT, catalase; CML, serum $N^{\epsilon}$-(carboxymethyl) lysine ; CRP, C-reactive protein; FBG, fasting blood glucose; FRAP, ferric-reducing antioxidant power; GSH, glutathione; GPx, glutathione peroxidase; HbA1c, glycated hemoglobin; HDL, high-density lipoprotein; HOMA-β, homeostatic model assessment for β-cell function; HOMA-IR, homeostatic model assessment for insulin resistance; IL, interleukin; iNOS, inducible isoform of nitric oxide synthase; LD low-density lipoprotein; MDA, malondialdehyde; NO, nitric oxide; NOS, nitric oxide synthase; Ox-LDL, oxidized low-density lipoprotein; NRCT, non-randomized controlled trial; Px, peroxidase; RCT, randomized control trial; RLP, remnant-like lipoprotein particles; SOD, superoxide dismutase; TAC, total antioxidant capacity: TG, triglyceride; TNF, tumor necrosis factor; VEGF, vascular endothelial growth factor; VLDL, very-low-density lipoprotein.

## Cardioprotective Activity

The anti-atherosclerotic effects of propolis are attributed to its ability to modulate both the serum lipoprotein profile and their handling by vascular tissues and infiltrating leucocytes (Silva et al. 2021). Propolis administered orally affected the lipoprotein profile in several animal models of atherosclerosis. In lead acetate–intoxicated mice, Lybian propolis administrated orally (200 mg/kg) for 30 days was shown to prevent an increase in the serum cholesterol, triglycerides, low-density lipoprotein, and very-low-density lipoprotein levels, but high-density lipoprotein levels increased (Silva et al. 2021). The same beneficial effect was found for Croatian propolis (50 mg/kg) in mice fed a high-fat diet (Elsayed Azab et al. 2015). Ethanolic extracts from southeastern Europe samples reduce the adenosine diphosphate–induced platelet aggregation when added to blood samples from young patients with thrombophilia, with higher activity being attributed to either low luteolin (2-(3,4-dihydroxyphenyl)-5,7-dihydroxychromen-4-one) content and/or higher pinocembrin-7-methyleter content (Bojić et al. 2018). Turkish propolis reduced blood pressure, which was attributed to the ROS scavenging activity of flavonoids, as well as to the presence of arginine, which stimulates nitric oxide synthase (NOS) activity and offsets the NOS-blocking activity of L-NG-nitroarginine methyl ester (L-NAME). Both of these actions increase the NO availability in the vascular smooth muscle, leading to vasorelaxation (Selamoglu Talas 2014).

## Antiprotozoal Activity

Propolis has been shown to have significant activity against a wide variety of protozoa as well (Anjum et al. 2019). For example, Brazilian red propolis has been reported to be effective against *Leishmania infantum* in vitro. In this study, nine compounds were identified: luteolin, apigenin, quercetin, and rutin, vitexin as well as caffeic, chlorogenic acid, ellagic, and p-coumaric acids (da Silveira Regueira-neto et al. 2018). In another study, the antiprotozoal effects of Brazilian green propolis were explored against *Leishmania braziliensis.* Caffeic acid, p-coumaric acid, aromadendrine-4′-methyl-ether, drupanin, and artepillin C were the compounds identified to have led to these effects (Pontin et al. 2008).Similarly, flavonoids such as quercetin, luteolin, and rutin have been reported to have high leishmanicidal activity (Do Nascimento et al. 2016). Other studies suggested that the probable mechanism of action is the presence of chalcones in the red propolis, which can hinder *Leishmania* growth by disturbing the cytoplasmic membrane and the mitochondrial complex of the parasite (Chen et al. 2001).

## Anti-obesity Activity

Propolis decreased inflammation and prevented obesity induced by hyperlipidemia and metabolic syndromes in highly caloric diets in mice. In propolis-fed mice, body weight gain, visceral adipose tissue, liver and serum triglycerides, cholesterol, and non-esterified fatty acids were all reduced (Koya-Miyata et al. 2009). After 10–20 days of treatment, oral administration of Croatian propolis (ethanol extract, 50 mg/kg/day) reduced body weight increase in C57BL/6N mice (Oršolić et al. 2019). As a result, the anti-obesity properties of propolis-derived compounds have been studied. In HFD-fed C57BL/6N mice, CAPE (0.02%–0.5%), a hydroxycinnamic acid found in propolis, dramatically reduced body weight gain in a dose-dependent manner (Shin et al. 2014). CAPE also reduced epididymal adipose tissue mass while having no effect on kidney or liver weights (Kitamura 2019). As a result, CAPE appears to be one of the components responsible for propolis' anti-obesity properties. Propolis extracts have therapeutic effects on obesity by modulating adipogenesis, adipokine production, food intake, and energy expenditure, according to growing evidence based on animal and cellular models (Kitamura 2019).

## Adverse Effects

Owing to the wide use of propolis for diversified purposes, there is increased incidence of allergic reactions to propolis. Among the reported cases, allergic erythematous contact dermatitis is the most common (Allergic Contact Dermatitis from Propolis: Dermatitis® n.d.). In one case a woman presented with an eczematous lesion on her extremities and abdomen after physical application of propolis on an insect bite (Lamoureux et al. 2017). In another case, a girl from Israel developed serious allergic contact dermatitis after propolis application for minor ailments (Lee et al. 2018). Another case in Korea reported a 36-year-old woman who experienced erythematous patches after ingestion of propolis solution for a few weeks (Cho et al. 2011). In yet another case, a Taiwanese man who used topical propolis to treat his acute pharyngitis was brought to the emergency department because of anaphylactic shock in which he experienced severe laryngeal edema (Hsu et al. 2004). All of these cases indicate that the occurrence of allergic reactions with propolis use is an important health issue.

Allergic contact dermatitis is the most common report among beekeepers (Münstedt and Kalder 2009). Because there are a higher number of cases of generalized allergy caused by topical or oral use of propolis, it is important to identify the actual compounds that contribute to the allergic response to develop a method to reduce the allergen present to enhance consumer safety.

To date, among the potential constituent to contribute to its allergic response, LB-1, which is reported to be present in propolis from China, Thailand, Korea, Malaysia, and Portugal, is believed to be the strongest sensitizer that causes eczematous contact dermatitis in consumers. LB-1 consists of caffeic acid, benzyl caffeate, CAPE, and three different types of pentenyl esters of caffeic acid, which are 2-methyl-2-butyl-caffeate (2M2B), 3-methyl-2-butyl-caffeate (3M2B), and 3-methyl-3-butyl-caffeate (3M3B) (Münstedt and Kalder 2009). In the study conducted by Hausen et al. (1987), 3M2B was identified as the major contact allergen in LB-1 as contributed by the presence of its aromatic rings. It is reported that the OH groups on the aromatic rings are readily oxidized to para-quinone or ortho-quinone. Para-quinone and ortho-quinone play important roles in redox reaction in cells by producing ROS. Excess ROS present in the body may affect the redox balance of cells and affect the antioxidant defense mechanism in the body, leading to oxidative stress (Bolton et al. 2018). Oxidative stress contributes to the mechanism of allergy by modulating the immune system such as by reducing the activation threshold for T cells (Van Rijt et al. 2017).

Aliboni et al. (2014) quantitatively revealed benzyl salicylate and benzyl cinnamate as two other minor compounds that contribute to many allergic reactions in propolis-sensitive consumers. However, two allergenic esters are present in minute amounts in raw propolis compared with that of balsam from Peru with the possibility of cross-sensitivity. Nevertheless, because many components in natural products including propolis tend to act synergistically, quantitative measurement of allergenic compounds is crucial for consumer safety, even if compounds are present in minute amounts (Olthof et al. 2001a, 2001b).

Gardana et al. (2012) designed a simple but selective strategy to reduce the amount of contact allergens found in propolis. They demonstrated that biotransformation using a type of dairy bacterium, *Lactobacillus helveticus MIMLh5*, can effectively eliminate the allergen without affecting other important bioactive components, including flavonoids. The allergenic compound, caffeic acid derivative, consists of one cinnamoyl ester bond, which can be easily hydrolyzed using cinnamoyl esterase to degrade the allergenic compounds into free caffeic acid and other products. Caffeic acid is claimed to be well absorbed, especially in the stomach and small intestine (Olthof et al. 2001a, 2001b). Free caffeic acid is a strong antioxidant that can effectively scavenge free radicals, thus preventing the imbalance of redox reaction within the cells and further reducing allergic reaction (Gülçin 2006). *L. helveticus* can produce high levels of cinnamoyl esterase in which for every gram of raw propolis, $10^{14}$ bacterial cells can successfully reduce the amount of caffeic acid esters by 98% with the addition of 50% polyethylene glycol additive as a solvent for propolis. A reduction of allergen in propolis via bacterial biotransformation does not affect propolis' health benefits (Guglielmetti et al. 2010a, 2010b). Moreover, *L. helveticus* confers some probiotic characteristics and has the ability to modulate the immune response by inhibiting the release of pro-inflammatory mediators.

## Discussion

Bees collect resinous and balsamic material from flowers, branches, pollen, and tree exudates to make propolis, a complex phytocompound. It is high in polyphenols, particularly flavonoids and phenolic acids, and possesses antiviral, antibacterial, antifungal, and antiparasitic activities that have been demonstrated in both in vitro and in vivo investigations. Propolis contains aromatic acids and esters, aldehydes and ketones, terpenoids and phenylpropanoids, steroids, amino acids, polysaccharides, and a variety of other organic and inorganic substances in addition to flavonoids. Propolis has a wide range of compositions, which are determined by the geographical region and the plants from which bees obtain nectar (Figure 21.2). Because of its reputation as a health-promoting drug with immunomodulatory, antiviral, antibacterial, antifungal, and antiparasitic qualities (see Figure 21.1), it is widely used around the world.

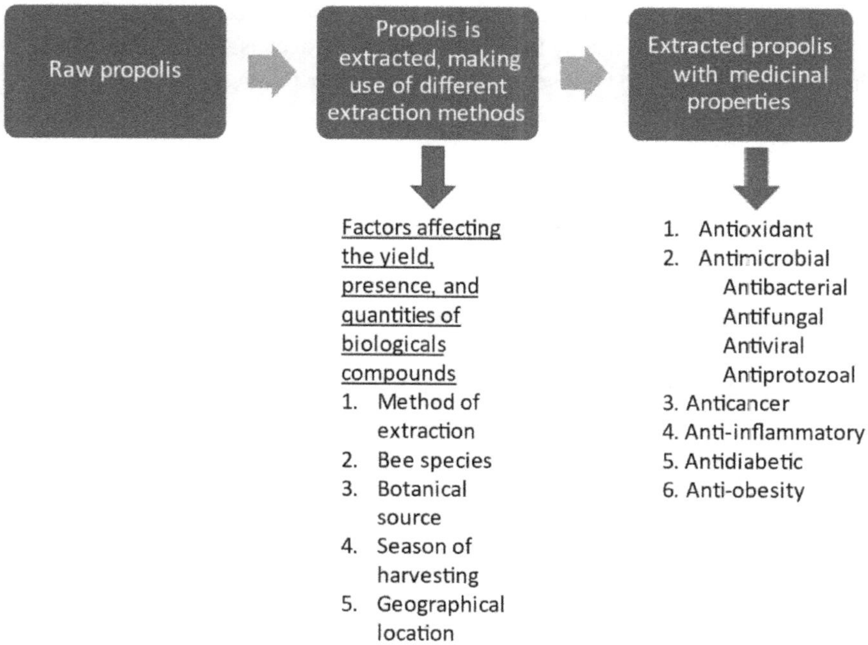


**Figure 21.2** Summary of factors affecting the yield, the presence, and the quantity of biological compounds and medicinal properties of propolis.

A variety of traditional and modern propolis extraction methods have been used and investigated. When it comes to solvents, water–ethanol mixes remain the most effective. When it comes to extraction methods, UAE appears to be the best option in terms of extraction time, yield, and cost-effectiveness. Of course, further research is needed to determine the best method for extracting bioactive chemicals from various types of propolis using chemometric optimization methodologies, especially for large-scale applications.

Propolis has antiproliferative effects against epithelial cells in colorectal adenocarcinoma (DLD-1), colon (HCT-116), gastric (NCI-N87), and esophageal squamous carcinomas (KYSE-30). Artepillin C present in propolis from Brazil has been shown to exert antiproliferative effect on prostate cancer cells by decreasing TRAIL resistance and inhibiting NF-κB. Other components of propolis induce apoptosis by activating caspase 3, inhibiting tumor cell migration, and promoting ROS production.

Propolis exhibits its anti-inflammatory effect by interfering with NF-κB pathway and reducing HIF-1α levels. Treatment of patients with cerebral ischaemia with CAPE can attenuate the increase in MCP-1 as induced by TNF-α in the cortical ischemic core region. CAPE has also been reported to prevent further damage to the brain by reducing the production of IL-1. Pinocembrin is present in propolis from many countries, including Mexico, China, and Greece. Study shows its anti-inflammatory effect by suppressing TNF-α, IL-6, and nuclear factor of kappa light polypeptide gene enhancer in B-cell inhibitor alpha (IκB-α) phosphorylation. Another potential anti-inflammatory isoflavonoid, neovestitol, has been identified in red propolis from northeastern Brazil. Administration of it reduces leucocyte adherence and ICAM-1 expression via induction of the iNOS pathway. This finding was further confirmed in mice with collagen-induced arthritis.

The antibacterial activity of propolis is higher in relation to gram-positive than gram-negative bacteria. Artepillin C has bacteriostatic activity with membrane blebbing against *S. mutans*. Other flavonoids with antibacterial activity found in propolis are pinocembrin and apigenin. Propolis extract shows excellent performance regarding in vitro tests against yeasts identified as onychomycosis agents. In low concentrations, propolis extract was found to be fungistatic and fungicidal. *Trichosporon* species were the most vulnerable yeasts. Propolis has been tested against various viral disease organisms. Viruses that were controlled by propolis in animal models with suggestion for control in humans include influenza, HSV-1 and HSV-2, and HIV. All of the findings point to propolis as a potential antimicrobial agent.

Propolis has been shown to control oxidative stress, the build-up of AGEs, and adipose tissue inflammation. Propolis therapy may help to reduce diabetes complications such nephropathy, retinopathy, foot ulcers, and non-alcoholic fatty liver disease. Propolis can modulate the lipoprotein profile and their handling by vascular tissues and infiltrating leucocytes. Ethanolic extracts reduce platelet aggregation in young patients with thrombophilia.

Because the composition of propolis varies from region to region and from the types of bees used for propolis collection, there is a need to standardize propolis extracts for future use by human beings. Also, because there is a myriad of compounds in propolis, there is an urgent need to develop tests to set some standards for the evaluation of propolis' biological activity.

## References


Afsharpour, F., Javadi, M., Hashemipour, S., et al. (2019). Propolis supplementation improves glycemic and antioxidant status in patients with type 2 diabetes: a randomized, double-blind, placebo-controlled study. *Complementary Therapies in Medicine* 43: 283–288. doi: 10.1016/J.CTIM.2019.03.001.

Allergic Contact Dermatitis from Propolis: Dermatitis® (n.d.). https://journals.lww.com/dermatitis/Abstract/2005/12000/Allergic_Contact_Dermatitis_from_Propolis.4.aspx (accessed December 27, 2021).

Aliboni, A. (2014). Propolis from northern California and Oregon: chemical composition, botanical origin, and content of allergens. *Zeitschrift für Naturforschung C* 69(1-2): 10–20.

Aminimoghadamfarouj, N. and Nematollahi, A. (2017). Propolis diterpenes as a remarkable bio-source for drug discovery development: a review. *International Journal of Molecular Sciences* 18: 1290. doi: 10.3390/IJMS18061290.

Amoros, M., Simões, C.M.O., Girre, L., et al. (2004). Synergistic effect of flavones and flavonols against herpes simplex virus type 1 in cell culture. Comparison with the antiviral activity of propolis. *Journal of Natural Products* 55: 1732–1740. doi: 10.1021/NP50090A003.

Andrade, J.K.S., Denadai, M., Andrade, G.R.S., et al. (2018). Development and characterization of microencapsules containing spray dried powder obtained from Brazilian brown, green and red propolis. *Food Research International* 109: 278–287. doi: 10.1016/J.FOODRES.2018.04.048.

Anjum, S.I., Ullah, A., Khan, K.A., et al. (2019). Composition and functional properties of propolis (bee glue): a review. *Saudi Journal of Biological Sciences* 26: 1695–1703. doi: 10.1016/J.SJBS.2018.08.013.

Bankova, V., Trusheva, B., and Popova, M. (2021). Propolis extraction methods: a review. *Journal of apicultural research* 60: 734–743. doi: 10.1080/00218839.2021.1901426.

Bastos, J.K., Arruda, C., Ribeiro, V.P., et al. (2020). Green propolis: cytotoxic and leishmanicidal activities of artepillin C, p-coumaric acid, and their degradation products. *Revista Brasileira de Farmacognosia* 30: 169. doi: 10.1007/S43450-020-00043-3.

Bernardino, P.N., Bersano, P.R.O., Lima Neto, J.F., and Sforcin, J.M. (2018). Positive effects of antitumor drugs in combination with propolis on canine osteosarcoma cells (spOS-2) and mesenchymal stem cells. *Biomedicine and Pharmacotherapy* 104: 268–274. doi: 10.1016/J.BIOPHA.2018.05.027.

Berretta, A.A., Silveira, M.A.D., Cóndor Capcha, J.M., and De Jong, D. (2020). Propolis and its potential against SARS-CoV-2 infection mechanisms and COVID-19 disease: running title: propolis against SARS-CoV-2 infection and COVID-19. *Biomedicine and Pharmacotherapy* 131: 110622. doi: 10.1016/J.BIOPHA.2020.110622.

Bojić, M., Antolić, A., Tomičić, M., et al. (2018). Propolis ethanolic extracts reduce adenosine diphosphate induced platelet aggregation determined on whole blood. *Nutrition Journal* 17: 1–8. doi: 10.1186/S12937-018-0361-Y/TABLES/3.

Bolton, J.L., Dunlap, T.L., and Dietz, B.M. (2018). Formation and biological targets of botanical o-quinones. *Food and Chemical Toxicology* 120: 700–707. doi: 10.1016/J.FCT.2018.07.050.

Bose, S., Sarkar, D., Bose, A., and Mandal, S.C. (2018). *Natural flavonoids and its pharmaceutical importance*. https://www.researchgate.net/publication/328577501 (accessed 30 July 2021).

Bridi, R., Montenegro, G., Nuñez-Quijada, G., et al. (2015). International regulations of propolis quality: required assays do not necessarily reflect their polyphenolic-related in vitro activities. *Journal of Food Science* 80: C1188–C1195. doi: 10.1111/1750-3841.12881.

Bruschi, M.L., Dota, K.F.D., Consolaro, M.E.L., and Svidzinski, T.I.E. (2011). Antifungal activity of Brazilian propolis microparticles against yeasts isolated from vulvovaginal candidiasis. *Evidence-Based Complementary and Alternative Medicine* 2011: 201953. https://doi.org/10.1093/ecam/neq029.

Castaldo, S. and Capasso, F. (2002). Propolis, an old remedy used in modern medicine. *Fitoterapia* 73: 1–6. doi: 10.1016/S0367-326X(02)00185-5.

Catchpole, O., Mitchell, K., Bloor, S., et al. (2015). Antiproliferative activity of New Zealand propolis and phenolic compounds vs human colorectal adenocarcinoma cells. *Fitoterapia* 106: 167–174. doi: 10.1016/J.FITOTE.2015.09.004.

Checa, J. and Aran, J.M. (2020). Reactive oxygen species: drivers of physiological and pathological processes. *Journal of Inflammation Research* 13: 1057. doi: 10.2147/JIR.S275595.

Chen, L.H., Chien, Y.W., Chang, M.L., et al. (2018). Taiwanese green propolis ethanol extract delays the progression of type 2 diabetes mellitus in rats treated with streptozotocin/high-fat diet. *Nutrition* 10: 503. doi: 10.3390/NU10040503.

Chen, M., Zhai, L., Christensen, S.B., et al. (2001). Inhibition of fumarate reductase in Leishmania major and l. donovani by chalcones. *Antimicrobial Agents and Chemotherapy* 45: 2023–2029. https://doi.org/10.1128/AAC.45.7.2023-2029.2001/ASSET/22498BC4-320F-4E8E-865A-E814B8E3EFAA/ASSETS/GRAPHIC/AC0710125007.JPEG.

Cho, E., Lee, J.D., and Cho, S.H. (2011). Systemic contact dermatitis from propolis ingestion. *Annals of Dermatology* 23: 85–88. doi: 10.5021/AD.2011.23.1.85.

da Cunha, M.G., Franchin, M., Galvão, L.C.C., et al. (2013). Antimicrobial and antiproliferative activities of stingless bee Melipona scutellaris geopropolis. *BMC Complementary and Alternative Medicine* 13: 1–9. doi: 10.1186/1472-6882-13-23/TABLES/3.

da Silveira Regueira-neto, M., Tintino, S.R., Rolón, M., et al. (2018). Antitrypanosomal, antileishmanial and cytotoxic activities of Brazilian red propolis and plant resin of Dalbergia ecastaphyllum (L) Taub. *Food and Chemical Toxicology* 119: 215–221. doi: 10.1016/J.FCT.2018.04.029.

Daleprane, J.B. and Abdalla, D.S. (2013). Emerging roles of propolis: antioxidant, cardioprotective, and antiangiogenic actions. *Evidence-Based Complementary and Alternative Medicine* 2013. doi: 10.1155/2013/175135.

Darvishi, N., Yousefinejad, V., Akbari, M.E., et al. (2020). Antioxidant and anti-inflammatory effects of oral propolis in patients with breast cancer treated with chemotherapy: a randomized controlled trial. *Journal of Herbal Medicine* 23: 100385. doi: 10.1016/J.HERMED.2020.100385.

Daugsch, A., Moraes, C.S., Fort, P., and Park, Y.K. (2008). Brazilian red propolis—chemical composition and botanical origin, evidence-based complement. *Alternative Medicine* 5: 435–441. doi: 10.1093/ECAM/NEM057.

de Freitas, M.C.D., de Miranda, M.B., de Oliveira, D.T., et al. (2018). Biological activities of red propolis: a review. *Recent Patents on Endocrine, Metabolic and Immune Drug Discovery* 11: 3–12. doi: 10.2174/1872214812666180223120316.

Derakhshan, Z., Ferrante, M., Tadi, M., et al. (2018). Antioxidant activity and total phenolic content of ethanolic extract of pomegranate peels, juice and seeds. *Food and Chemical Toxicology* 114: 108–111. doi: 10.1016/J.FCT.2018.02.023.

Do Nascimento, T.G., da Silva, P.F., Azevedo, L.F., et al. (2016). Polymeric nanoparticles of Brazilian red propolis extract: preparation, characterization, antioxidant and leishmanicidal activity. *Nanoscale Research Letters* 11 (1): 1–16. doi: 10.1186/S11671-016-1517-3.

Durazzo, A., Lucarini, M., Plutino, M., et al. (2021). Bee products: a representation of biodiversity, sustainability, and health. *Life* 11: 970. doi: 10.3390/LIFE11090970.

Dutra, R.P., De Barros Abreu, B.V., Cunha, M.S., et al. (2014). Phenolic acids, hydrolyzable tannins, and antioxidant activity of geopropolis from the stingless bee Melipona fasciculata smith. *Journal of Agricultural and Food Chemistry* 62: 2549–2557. doi: 10.1021/JF404875V/SUPPL_FILE/JF404875V_SI_001.PDF.

Elsayed Azab, A., El-Sayed Azab, A., Ammar Algridi, M., and Mohammed Lashkham, N. (2015). Hypolipidemic and antiatherogenic effects of aqueous extract of Libyan propolis in lead acetate intoxicated male albino mice hepato-renal protective effects of natural products view project chronic kidney diseases view project hypolipidemic and antiatherogenic effects of aqueous extract of Libyan propolis in lead acetate intoxicated male albino mice. *International Journal of Science and Research* 4. www.ijsr.net (accessed 27 December 2021).

El-Sharkawy, H.M., Anees, M.M., and Van Dyke, T.E. (2016). Propolis improves periodontal status and glycemic control in patients with type 2 diabetes mellitus and chronic periodontitis: a randomized clinical trial. *Journal of Periodontology* 87: 1418–1426. doi: 10.1902/JOP.2016.150694.

Ertürküner, S.P., Saraç, E.Y., Göçmez, S.S., et al. (2016). Anti-inflammatory and ultrastructural effects of Turkish propolis in a rat model of endotoxin-induced uveitis, Folia Histochem. *Cytobiologica* 54: 49–57. doi: 10.5603/FHC.A2016.0004.

Escriche, I. and Juan-Borrás, M. (2018). Standardizing the analysis of phenolic profile in propolis. *Food Research International* 106: 834–841. doi: 10.1016/J.FOODRES.2018.01.055.

Eumkeb, G. and Chukrathok, S. (2013). Synergistic activity and mechanism of action of ceftazidime and apigenin combination against ceftazidime-resistant Enterobacter cloacae. *Phytomedicine* 20: 262–269. doi: 10.1016/J.PHYMED.2012.10.008.

Fanny, M., Nascimento, M., Baron, L., et al. (2018). The IL-33 receptor ST2 regulates pulmonary inflammation and fibrosis to bleomycin. *Frontiers in Immunology* 9: 1476. doi: 10.3389/FIMMU.2018.01476/BIBTEX.

Fokt, H., Ferreira, A.M., Cunha, A., et al. (2010). How do bees prevent hive infections? The antimicrobial properties of propolis Bioactivities of Portuguese Propolis View project How do bees prevent hive infections? *The antimicrobial properties of propolis* https://www.researchgate.net/publication/265978408 (accessed December 20, 2021).

Franchin, M., Colón, D.F., Da Cunha, M.G., et al. (2016). Neovestitol, an isoflavonoid isolated from Brazilian red propolis, reduces acute and chronic inflammation: involvement of nitric oxide and IL-6. *Scientific Reports* 6 (1): 1–12. doi: 10.1038/srep36401.

Franchin, M., Da Cunha, M.G., Denny, C., et al. (2012). Geopropolis from Melipona scutellaris decreases the mechanical inflammatory hypernociception by inhibiting the production of IL-1β and TNF-α. *Journal of Ethnopharmacology* 143: 709–715. doi: 10.1016/J.JEP.2012.07.040.

Fukuda, T., Fukui, M., Tanaka, M., et al. (2015). Effect of Brazilian green propolis in patients with type 2 diabetes: a double-blind randomized placebo-controlled study. *Biomedical Reports* 3: 355–360. doi: 10.3892/BR.2015.436/HTML.

Funakoshi-Tago, M., Okamoto, K., Izumi, R., et al. (2015). Anti-inflammatory activity of flavonoids in Nepalese propolis is attributed to inhibition of the IL-33 signaling pathway. *International Immunopharmacology* 25: 189–198. doi: 10.1016/J.INTIMP.2015.01.012.

Gao, W., Pu, L., Wei, J., et al. (2018). Serum antioxidant parameters are significantly increased in patients with type 2 diabetes mellitus after consumption of chinese propolis: a randomized controlled trial based on fasting serum glucose level. *Diabetes Therapy* 9: 101–111. doi: 10.1007/S13300-017-0341-9/TABLES/5.

Gardana, C., Barbieri, A., Simonetti, P., et al. (2012). Biotransformation strategy to reduce allergens in propolis. *Applied and Environmental Microbiology* 78: 4654–4658. doi:10.1128/AEM.00811-12.

Guglielmetti, S., Taverniti, V., Minuzzo, M., et al. (2010a). A dairy bacterium displays in vitro probiotic properties for the pharyngeal mucosa by antagonizing group A streptococci and modulating the immune response. *Infection and Immunity* 78: 4734–4743. https://doi.org/10.1128/IAI.00559-10.

Guglielmetti, S., Taverniti, V., Minuzzo, M., et al. (2010b). Oral bacteria as potential probiotics for the pharyngeal mucosa. *Applied and Environmental Microbiology* 76: 3948–3958. https://doi.org/10.1128/AEM.00109-10.

Gülçin, I. (2006). Antioxidant activity of caffeic acid (3,4-dihydroxycinnamic acid). *Toxicology* 217: 213–220. doi: 10.1016/J.TOX.2005.09.011.

Hamzah, N. and Leo, C.P. (2015). Microwave-assisted extraction of trigona propolis: the effects of processing parameters. *International Journal of Food Engineering* 11: 861–870. doi: 10.1515/IJFE-2015-0106/PDF.

Hanieh, H., Islam, V.I.H., Saravanan, S., et al. (2017). Pinocembrin, a novel histidine decarboxylase inhibitor with anti-allergic potential in in vitro. *European Journal of Pharmacology* 814: 178–186. doi: 10.1016/J.EJPHAR.2017.08.012.

Hausen, B.M., Wollenweber, E., Senff, H., and Post, B. (1987). Propolis allergy. *Contact Dermatitis* 17: 163–170. doi: 10.1111/J.1600-0536.1987.TB02699.X.

Hesami, S., Hashemipour, S., Shiri-Shahsavar, M.R., et al. (2019). Administration of Iranian Propolis attenuates oxidative stress and blood glucose in type II diabetic patients: a randomized, double-blind, placebo-controlled, clinical trial. *Caspian Journal of Internal Medicine* 10: 48. doi: 10.22088/CJIM.10.1.48.

Hsu, C.Y., Chiang, W.C., Weng, T.I., et al. (2004). Laryngeal edema and anaphylactic shock after topical propolis use for acute pharyngitis. *American Journal of Emergency Medicine* 22: 432–433. doi: 10.1016/J.AJEM.2004.06.007.

Huang, S., Zhang, C.P., Wang, K., et al. (2014). Recent advances in the chemical composition of propolis. *Molecules* 19: 19610–19632. doi: 10.3390/MOLECULES191219610.

Hwang, S.A., Kim, C.D., and Lee, W.S. (2017). Caffeic acid phenethyl ester protects against photothrombotic cortical ischemic injury in mice. *Korean Journal of Physiology and Pharmacology* 22: 101–110. doi: 10.4196/KJPP.2018.22.1.101.

Ibrahim, N., Niza, N.F.S.M., Rodi, M M., et al. (2016). Chemical and biological analyses of Malaysian stingless bee propolis extracts. *Malaysian Journal of Analytical Sciences* 20(2): 413–422.

Ighodaro, O.M. and Akinloye, O.A. (2019). First line defence antioxidants-superoxide dismutase (SOD), catalase (CAT) and glutathione peroxidase (GPX): their fundamental role in the entire antioxidant defence grid. *Alexandria Journal of Medicine* 54: 287–293. doi: 10.4314/bafm.v54i4.

Inui, S., Hosoya, T., Yoshizumi, K., et al. (2021). Phytochemical and anti-inflammatory properties of Senegalese propolis and isolated compounds. *Fitoterapia* 151: 104861. doi: 10.1016/J.FITOTE.2021.104861.

Ismail, T.N.N.T., Sulaiman, S.A., Ponnuraj, K.T., et al. (2018). Chemical constituents of Malaysian apis mellifera propolis. *Sains Malaysiana* 47: 117–122. doi: 10.17576/jsm-2018-4701-14.

Ito, J., Chang, F.R., Wang, H.K., et al. (2001). Anti-AIDS agents. 48.1 anti-HIV activity of moronic acid derivatives and the new melliferone-related triterpenoid isolated from brazilian propolis. *Journal of Natural Products* 64: 1278–1281. doi: 10.1021/NP010211X.

Jin, X.L., Wang, K., Li, Q.Q., et al. (2017). Antioxidant and anti-inflammatory effects of Chinese propolis during palmitic acid-induced lipotoxicity in cultured hepatocytes. *Journal of Functional Foods* 34: 216–223. doi: 10.1016/J.JFF.2017.04.039.

Jucá, M.M., Filho, F.M.S.C., de Almeida, J.C., et al. (2018). Flavonoids: biological activities and therapeutic potential. *Natural Product Research* 34: 692–705. doi: 10.1080/14786419.2018.1493588.

Khurshid, Z., Naseem, M., Zafar, M.S., et al. (2017). Propolis: a natural biomaterial for dental and oral healthcare. *Journal of Dental Research, Dental Clinics, Dental Prospects* 11: 265. doi: 10.15171/JODDD.2017.046.

Kitamura, H. (2019). Effects of propolis extract and propolis-derived compounds on obesity and diabetes: knowledge from cellular and animal models. *Molecules* 24: 4394. doi: 10.3390/MOLECULES24234394.

Kitamura, H., Saito, N., Fujimoto, J., et al. (2018). Brazilian propolis ethanol extract and its component kaempferol induce myeloid-derived suppressor cells from macrophages of mice in vivo and in vitro. *BMC Complementary and Alternative Medicine* 18: 1–11. doi: 10.1186/S12906-018-2198-5/FIGURES/7.

Koya-Miyata, S., Arai, N., Mizote, A., et al. (2009). Propolis prevents diet-induced hyperlipidemia and mitigates weight gain in diet-induced obesity in mice. *Biological and Pharmaceutical Bulletin* 32: 2022–2028. doi: 10.1248/BPB.32.2022.

Kujumgiev, A., Tsvetkova, I., Serkedjieva, Y., et al. (1999). Antibacterial, antifungal and antiviral activity of propolis of different geographic origin. *Journal of Ethnopharmacology* 64: 235–240. doi: 10.1016/S0378-8741(98)00131-7.

Kurokawa, M., Shimizu, T., Takeshita, Y., et al. (2011). Efficacy of Brazilian propolis against herpes simplex virus type 1 infection in mice and their modes of antiherpetic efficacies. *Evidence-Based Complementary and Alternative Medicine* 2011: 976196. doi: 10.1155/2011/976196.

Kuropatnicki, A.K., Szliszka, E., and Krol, W. (2013). Historical aspects of propolis research in modern times. *Evidence-Based Complementary and Alternative Medicine* 2013: 964149. doi: 10.1155/2013/964149.

Lamoureux, A., Meharon, M., Durand, A.L., et al. (2017). A first case of erythema multiforme-like contact dermatitis caused by propolis. *Contact Dermatitis* 77: 263–264. doi: 10.1111/COD.12814.

Lee, S.X., Boontaveeyuwat, E., Thaiwat, S., et al. (2018). Severe allergic contact dermatitis mimicking angioedema caused by propolis used as a traditional remedy. *Contact Dermatitis* 79: 185–186. https://doi.org/10.1111/COD.13021.

Limmongkon, A., Janhom, P., Amthong, A., et al. (2017). Antioxidant activity, total phenolic, and resveratrol content in five cultivars of peanut sprouts. *Asian Pacific Journal of Tropical Biomedicine* 7: 332–338. doi: 10.1016/J.APJTB.2017.01.002.

Maroof, K. and Gan, S.H. (2021). A review on chemical compositions, biological activity and formulation techniques of malaysian honey bee and meliponine propolis. *Journal of Biologically Active Products from Nature* 10: 507–523. doi: 10.1080/22311866.2020.1856716.

Maroof, K., Lee, R.F.S., Siow, L.F., and Gan, S.H. (2020). Microencapsulation of propolis by spray drying: a review. *Drying Technology*. doi: 10.1080/07373937.2020.1850470.

Massaro, C.F., Katouli, M., Grkovic, T., et al. (2014). Anti-staphylococcal activity of C-methyl flavanones from propolis of Australian stingless bees (Tetragonula carbonaria) and fruit resins of Corymbia torelliana (Myrtaceae). *Fitoterapia* 95: 247–257. doi: 10.1016/J.FITOTE.2014.03.024.

Massaro, F.C., Brooks, P.R., Wallace, H.M., and Russell, F.D. (2011). Cerumen of Australian stingless bees (Tetragonula carbonaria): gas chromatography-mass spectrometry fingerprints and potential anti-inflammatory properties. *Naturwissenschaften* 98: 329–337. doi: 10.1007/S00114-011-0770-7/FIGURES/2.

Mazia, R.S., De Araújo Pereira, R.R., De Francisco, L.M.B., et al. (2016). Formulation and evaluation of a mucoadhesive thermoresponsive system containing Brazilian green propolis for the treatment of lesions caused by herpes simplex type I. *Journal of Pharmaceutical Sciences* 105: 113–121. doi: 10.1016/J.XPHS.2015.11.016.

Melero-Jerez, C., Ortega, M.C., Moliné-Velázquez, V., and Clemente, D. (2016). Myeloid derived suppressor cells in inflammatory conditions of the central nervous system. *Biochimica et Biophysica Acta - Molecular Basis of Disease* 1862: 368–380. doi: 10.1016/J.BBADIS.2015.10.015.

Miguel, M.G., Nunes, S., Dandlen, S.A., et al. (2010). Phenols and antioxidant activity of hydro-alcoholic extracts of propolis from Algarve, South of Portugal. *Food and Chemical Toxicology* 48: 3418–3423. doi: 10.1016/J.FCT.2010.09.014.

Moulin, D., Donzé, O., Talabot-Ayer, D., et al. (2007). Interleukin (IL)-33 induces the release of pro-inflammatory mediators by mast cells. *Cytokine* 40: 216–225. doi: 10.1016/J.CYTO.2007.09.013.

Münstedt, K. and Kalder, M. (2009). Contact allergy to propolis in beekeepers. *Allergologia et immunopathologia (Madr)* 37: 298–301. doi: 10.1016/J.ALLER.2009.04.005.

Munteanu, I.G. and Apetrei, C. (2021). Analytical methods used in determining antioxidant activity: a review. *International Journal of Molecular Sciences* 22: 3380. doi: 10.3390/IJMS22073380.

Murata, K., Takano, S., Masuda, M., et al. (2013). Anti-degranulating activity in rat basophil leukemia RBL-2H3 cells of flavanone glycosides and their aglycones in citrus fruits. *Journal of Natural Medicines* 67: 643–646. doi: 10.1007/S11418-012-0699-Y/TABLES/2.

Nayaka, H.B., Londonkar, R.L., Umesh, M.K., and Tukappa, A. (2014). Antibacterial attributes of apigenin, isolated from portulaca oleracea L. *International Journal of Bacteriology* 2014: 175851. doi: 10.1155/2014/175851.

Nolkemper, S., Reichling, J., Sensch, K.H., and Schnitzler, P. (2010). Mechanism of herpes simplex virus type 2 suppression by propolis extracts. *Phytomedicine* 17: 132–138. doi: 10.1016/J.PHYMED.2009.07.006.

Oldoni, T.L.C., Cabral, I.S.R., D'Arce, M.A.B.R., et al. (2011). Isolation and analysis of bioactive isoflavonoids and chalcone from a new type of Brazilian propolis. *Separation and Purification Technology* 77: 208–213. doi: 10.1016/J.SEPPUR.2010.12.007.

Oliveira, A.C.P., Shinobu, C.S., Longhini, R., et al. (2006). Antifungal activity of propolis extract against yeasts isolated from onychomycosis lesions. *Memorias do Instituto Oswaldo Cruz* 101: 493–497. doi: 10.1590/S0074-02762006000500002.

Olthof, M.R., Hollman, P.C., Zock, P.L., and Katan, M.B. (2001b). Consumption of high doses of chlorogenic acid, present in coffee, or of black tea increases plasma total homocysteine concentrations in humans. *American Journal of Clinical Nutrition* 73: 532–538. doi: 10.1093/AJCN/73.3.532.

Olthof, M.R., Hollman, P.C.H., and Katan, M.B. (2001a). Chlorogenic acid and caffeic acid are absorbed in humans. *Journal of Nutrition* 131: 66–71. doi: 10.1093/JN/131.1.66.

Ong, T.H., Chitra, E., Ramamurthy, S., et al. (2017). Chitosan-propolis nanoparticle formulation demonstrates anti-bacterial activity against Enterococcus faecalis biofilms. *PLoS One* 12: e0174888. doi: 10.1371/JOURNAL.PONE.0174888.

Oroian, M., Dranca, F., and Ursachi, F. (2020). Comparative evaluation of maceration, microwave and ultrasonic-assisted extraction of phenolic compounds from propolis. *Journal of Food Science and Technology* 57: 70–78. doi: 10.1007/S13197-019-04031-X/FIGURES/2.

Oršolić, N., Jurčević, I.L., Đikić, D., et al. (2019). Effect of propolis on diet-induced hyperlipidemia and atherogenic indices in mice. *Antioxidants* 8: 156. doi: 10.3390/ANTIOX8060156.

Oršolić, N., Sirovina, D., Končić, M.Z., et al. (2012). Effect of Croatian propolis on diabetic nephropathy and liver toxicity in mice, BMC Complement. *Alternative Medicine* 12 (1): 1–16. doi: 10.1186/1472-6882-12-117.

Osés, S.M., Pascual-Maté, A., Fernández-Muiño, M.A., et al. (2016). Bioactive properties of honey with propolis. *Food Chemistry* 196: 1215–1223. doi: 10.1016/J.FOODCHEM.2015.10.050.

Pasupuleti, V.R., Sammugam, L., Ramesh, N., and Gan, S.H. (2017). Honey, propolis, and royal jelly: a comprehensive review of their biological actions and health benefits. *Oxidative Medicine and Cellular Longevity* 2017. doi: 10.1155/2017/1259510.

Patel, S. (2015). Emerging adjuvant therapy for cancer: propolis and its constituents. *Journal of Dietary Supplements* 13: 245–268. doi: 10.3109/19390211.2015.1008614.

Pavlovic, R., Borgonovo, G., Leoni, V., et al. (2020). Effectiveness of different analytical methods for the characterization of propolis: a case of study in Northern Italy. *Molecules* 25: 504. doi: 10.3390/MOLECULES25030504.

Pontin, K., Da Silva Filho, A.A., Santos, F.F., et al. (2008). In vitro and in vivo antileishmanial activities of a Brazilian green propolis extract. *Parasitology Research* 103: 487–492. doi: 10.1007/S00436-008-0970-Z/TABLES/1.

Popolo, A., Piccinelli, A.L., Morello, S., et al. (2011). Cytotoxic activity of nemorosone in human MCF-7 breast cancer cells. *Canadian Journal of Physiology and Pharmacology* 89: 50–57. doi: 10.1139/Y10-100/ASSET/IMAGES/LARGE/Y10-100F5.JPEG.

Popova, M., Trusheva, B., and Bankova, V. (2021). Propolis of stingless bees: a phytochemist's guide through the jungle of tropical biodiversity. *Phytomedicine* 86: 153098. doi: 10.1016/J.PHYMED.2019.153098.

Przybyłek, I. and Karpiński, T.M. (2019). Antibacterial properties of propolis. *Molecules* 24: 2047. doi: 10.3390/MOLECULES24112047.

Research, O., Akilandeswari, K., and Ruckmani, K. (2016). Synergistic antibacterial effect of apigenin with Î$^2$-lactam antibiotics and modulation of bacterial resistance by a possible membrane effect against methicillin resistant Staphylococcus aureus. *Cellular and Molecular Biology* 62: 74–82. doi: 10.14715/CMB/.

Rojczyk, E., Klama-Baryła, A., Łabuś, W., et al. (2020). Historical and modern research on propolis and its application in wound healing and other fields of medicine and contributions by Polish studies. *Journal of Ethnopharmacology* 262: 113159. doi: 10.1016/J.JEP.2020.113159.

Rothwell, N.J. and Luheshi, G.N. (2000). Interleukin 1 in the brain: biology, pathology and therapeutic target. *Trends in Neurosciences* 23: 618–625. doi: 10.1016/S0166-2236(00)01661-1.

Rzepecka-Stojko, A., Kabala-Dzik, A., Mozdzierz, A., et al. (2015). Caffeic acid phenethyl ester and ethanol extract of propolis induce the complementary cytotoxic effect on triple-negative breast cancer cell lines. *Molecules* 20: 9242–9262. doi: 10.3390/MOLECULES20059242.

Salatino, A., Salatino, M.L.F., and Negri, G. (2021). How diverse is the chemistry and plant origin of Brazilian propolis? *Apidologie* 1: 1–23. doi: 10.1007/S13592-021-00889-Z.

Samadi, N., Mozaffari-Khosravi, H., Rahmanian, M., and Askarishahi, M. (2017). Effects of bee propolis supplementation on glycemic control, lipid profile and insulin resistance indices in patients with type 2 diabetes: a randomized, double-blind clinical trial. *Journal of Integrative Medicine* 15: 124–134. doi: 10.1016/S2095-4964(17)60315-7.

Samarghandian, S., Azimi-Nezhad, M., Samini, F., and Farkhondeh, T. (2015). Chrysin treatment improves diabetes and its complications in liver, brain, and pancreas in streptozotocin-induced diabetic rats. *Canadian Journal of Physiology and Pharmacology* 94: 388–393. https://doi.org/10.1139/CJPP-2014-0412.

Sawaya, A.C.H.F., Barbosa da Silva Cunha, I., and Marcucci, M.C. (2011). Analytical methods applied to diverse types of Brazilian propolis. *Chemistry Central Journal* 5. 1–10. doi: 10.1186/1752-153X-5-27.

Sawaya, A.C.H.F., Calado, J.C.P., Dos Santos, L.C., et al. (2009). Composition and antioxidant activity of propolis from three species of Scaptotrigona stingless bees. *Journal of ApiProduct and ApiMedical Science* 1: 37–42. doi: 10.3896/IBRA.4.01.2.03.

Seibert, J.B., Bautista-Silva, J.P., Amparo, T.R., et al. (2019). Development of propolis nanoemulsion with antioxidant and antimicrobial activity for use as a potential natural preservative. *Food Chemistry* 287: 61–67. doi: 10.1016/J.FOODCHEM.2019.02.078.

Selamoglu Talas, Z. (2014). Propolis reduces oxidative stress in l-NAME-induced hypertension rats. *Cell Biochemistry and Function* 32: 150–154. doi: 10.1002/CBF.2986.

Sforcin, J.M. (2016). Biological properties and therapeutic applications of propolis. *Phytotherapy Research* 30: 894–905. doi: 10.1002/PTR.5605.

Sforcin, J.M. and Bankova, V. (2011). Propolis: is there a potential for the development of new drugs? *Journal of Ethnopharmacology* 133: 253–260. doi: 10.1016/J.JEP.2010.10.032.

Shimizu, T., Takeshita, Y., Takamori, Y., et al. (2011). Efficacy of Brazilian propolis against herpes simplex virus type 1 infection in mice and their modes of antiherpetic efficacies. *Evidence-Based Complementary and Alternative Medicine* 2011: 976196.

Shin, S.H., Seo, S.G., Min, S., et al. (2014). Caffeic acid phenethyl ester, a major component of propolis, suppresses high fat diet-induced obesity through inhibiting adipogenesis at the mitotic clonal expansion stage, *Journal of Agricultural and Food Chemistry* 62: 4306–4312. 10.1021/JF405088F/SUPPL_FILE/JF405088F_SI_001.PDF.

Silici, S. and Kutluca, S. (2005). Chemical composition and antibacterial activity of propolis collected by three different races of honeybees in the same region. *Journal of Ethnopharmacology* 99: 69–73. doi: 10.1016/J.JEP.2005.01.046.

Silva, H., Francisco, R., Saraiva, A., et al. (2021). The cardiovascular therapeutic potential of propolis—a comprehensive review. *Biology* 10: 27. doi: 10.3390/BIOLOGY10010027.

Singh, A., Kukreti, R., Saso, L., and Kukreti, S. (2019). Oxidative stress: a key modulator in neurodegenerative diseases. *Molecules* 24: 1583. doi: 10.3390/MOLECULES24081583.

Su, K.Y., Hsieh, C.Y., Chen, Y.W., et al. (2014). Taiwanese green propolis and propolin G protect the liver from the pathogenesis of fibrosis via eliminating TGF-β-induced smad2/3 phosphorylation. *Journal of Agricultural and Food Chemistry* 62: 3192–3201. doi: 10.1021/JF500096C.

Taniguchi, K. and Karin, M. (2018). NF-κB, inflammation, immunity and cancer: coming of age. *Nature Reviews Immunology* 18 (5): 309–324. doi: 10.1038/nri.2017.142.

Tanvir, E.M., Sakib Hossen, M., Mahfuza Shapla, U., et al. (2018). Antioxidant, brine shrimp lethality and analgesic properties of propolis from Bangladesh. *Journal of Food Biochemistry* 42: e12596. doi: 10.1111/JFBC.12596.

Tateo, F., Pagani, L., Luini, M., and Romero, E. (1990). *Effects of propolis flavonoids on virus infectivity and replication.* https://www.researchgate.net/publication/21029715 (accessed 23 December 2021).

Teixeira, É.W., Message, D., Negri, G., et al. (2010). Seasonal variation, chemical composition and antioxidant activity of brazilian propolis samples. *Evidence-Based Complementary and Alternative Medicine* 7: 307–315. doi: 10.1093/ECAM/NEM177.

Teixeira, É.W., Negri, G., Meira, R.M.S.A., et al. (2005). Plant origin of green propolis: bee behavior, plant anatomy and chemistry. *Evidence-Based Complementary and Alternative Medicine* 2: 85–92. doi: 10.1093/ECAM/NEH055.

Tejchman, K., Kotfis, K., and Sieńko, J. (2021). Biomarkers and mechanisms of oxidative stress—last 20 years of research with an emphasis on kidney damage and renal transplantation. *International Journal of Molecular Sciences* 22: 8010. doi: 10.3390/IJMS22158010.

Traynor, T.R., Herring, A.C., Dorf, M.E., et al. (2002). Differential roles of CC chemokine ligand 2/monocyte chemotactic protein-1 and CCR2 in the development of T1 immunity. *Journal of Immunology* 168: 4659–4666. doi: 10.4049/JIMMUNOL.168.9.4659.

Trusheva, B., Trunkova, D., and Bankova, V. (2007). Different extraction methods of biologically active components from propolis; a preliminary study. *Chemistry Central Journal* 1: 1–4. doi: 10.1186/1752-153X-1-13/TABLES/1.

Tomazzoli, M.M., Zeggio, A.R.S., Pai Neto, R.D., et al. (2020). Botanical source investigation and evaluation of the effect of seasonality on Brazilian propolis from Apis mellifera L. *Scientia Agricola* 77.

Turan, D., Abdik, H., Sahin, F., and Avşar Abdik, E. (2020). Evaluation of the neuroprotective potential of caffeic acid phenethyl ester in a cellular model of Parkinson's disease. *European Journal of Pharmacology* 883: 173342. doi: 10.1016/J.EJPHAR.2020.173342.

Usman, U.Z., Bakar, A.B.A., Zin, A.A.M., and Mohamed, M. (2018). LC-MS analysis and effects of Malaysian propolis on insulin, glucagon, pancreas and oxidative stress status in streptozotocin-induced diabetic rats. *Journal of Medicine and Biomedical Research* 16: 15–27. doi: 10.4314/jmbr.v16i1.

Valencia, D., Alday, E., Robles-Zepeda, R., et al. (2012). Seasonal effect on chemical composition and biological activities of Sonoran propolis. *Food Chemistry* 131: 645–651. doi: 10.1016/J.FOODCHEM.2011.08.086.

Van Rijt, L.S., Utsch, L., Lutter, R., and Van Ree, R. (2017). Oxidative stress: promoter of allergic sensitization to protease allergens? *International Journal of Molecular Sciences* 18: 1112. doi: 10.3390/IJMS18061112.

Veiga, F.F., Gadelha, M.C., da Silva, M.R.T., et al. (2018). Propolis extract for onychomycosis topical treatment: from bench to clinic. *Frontiers in Microbiology* 9: 779. doi: 10.3389/FMICB.2018.00779/BIBTEX.

Veiga, R.S., De Mendonça, S., Mendes, P.B., et al. (2017). Artepillin C and phenolic compounds responsible for antimicrobial and antioxidant activity of green propolis and Baccharis dracunculifolia DC. *Journal of Applied Microbiology* 122: 911–920. doi: 10.1111/JAM.13400.

Veloz, J.J., Alvear, M., and Salazar, L.A. (2019). Antimicrobial and antibiofilm activity against streptococcus mutans of individual and mixtures of the main polyphenolic compounds found in Chilean propolis. *BioMed Research International* 2019. doi: 10.1155/2019/7602343.

Vestweber, D. (2015). How leukocytes cross the vascular endothelium. *Nature Reviews Immunology* 15 (11): 692–704. doi: 10.1038/nri3908.

Vinayagam, R. and Xu, B. (2015). Antidiabetic properties of dietary flavonoids: a cellular mechanism review. *Nutrition and Metabolism* 12 (1): 1–20. doi: 10.1186/S12986-015-0057-7.

Wang, W., Wang, Y.X., and Yang, H.B. (2016). Supramolecular transformations within discrete coordination-driven supramolecular architectures. *Chemical Society Reviews* 45: 2656–2693. doi: 10.1039/C5CS00301F.

Watanabe, M.A.E., Amarante, M.K., Conti, B.J., and Sforcin, J.M. (2011). Cytotoxic constituents of propolis inducing anticancer effects: a review. *Journal of Pharmacy and Pharmacology* 63: 1378–1386. https://doi.org/10.1111/J.2042-7158.2011.01331.X.

Welz, A.N., Emberger-Klein, A., and Menrad, K. (2018). Why people use herbal medicine: insights from a focus-group study in Germany. *BMC Complementary and Alternative Medicine* 18: 1–9. doi: 10.1186/S12906-018-2160-6/TABLES/4.

YaZhou, Z., WenLi, T., ZhanBao, G., et al. (2012). Optimization of ethanol extraction technology from propolis by response surface methodology. *Journal of Agricultural Science and Technology* 14: 85–93.

Yilmaz, U.C., Bagca, B.G., Karaca, E., et al. (2016). Evaluation of the miRNA profiling and effectiveness of the propolis on B-cell acute lymphoblastic leukemia cell line. *Biomedicine & Pharmacotherapy* 84: 1266–1273. doi: 10.1016/J.BIOPHA.2016.10.056.

Yoshimasu, Y., Ikeda, T., Sakai, N., et al. (2018). Rapid bactericidal action of propolis against porphyromonas gingivalis. *Journal of Dental Research* 97: 928–936. doi: 10.1177/0022034518758034.

Zakerkish, M., Jenabi, M., Zaeemzadeh, N., et al. (2019). The effect of Iranian propolis on glucose metabolism, lipid profile, insulin resistance, renal function and inflammatory biomarkers in patients with type 2 diabetes mellitus: a randomized double-blind clinical trial. *Scientific Reports* 9 (1): 1–11. doi: 10.1038/s41598-019-43838-8.

Zhang, W., Lan, Y., Huang, Q., and Hua, Z. (n.d.). *Galangin induces B16F10 melanoma cell apoptosis via mitochondrial pathway and sustained activation of p38 MAPK.* doi: 10.1007/s10616-012-9499-1.

Zhao, L., Pu, L., Wei, J., et al. (2016). Brazilian green propolis improves antioxidant function in patients with type 2 diabetes mellitus. *International Journal of Environmental Research and Public Health* 13: 498. https://doi.org/10.3390/IJERPH13050498.

Zhao, L., Yu, M., Sun, M., et al. (2017). Rapid determination of major compounds in the ethanol extract of geopropolis from Malaysian stingless bees, heterotrigona itama, by UHPLC-Q-TOF/MS and NMR. *Molecules* 22: 1935. doi: 10.3390/MOLECULES22111935.

Zulhendri, F., Chandrasekaran, K., Kowacz, M., et al. (2021). Antiviral, antibacterial, antifungal, and antiparasitic properties of propolis: a review. *Foods* 10: 1360. doi: 10.3390/FOODS10061360.

# 22

# Medicinal Benefits of Bee Venom

*Mahendran Sekar[1,2,*], Pei Teng Lum,[2] Srinivasa Reddy Bonam[3], and Siew Hua Gan[1]*

[1] *School of Pharmacy, Monash University Malaysia, Bandar Sunway, Subang Jaya 47500, Selangor, Malaysia*
[2] *Department of Pharmaceutical Chemistry, Faculty of Pharmacy and Health Sciences, Royal College of Medicine Perak, Universiti Kuala Lumpur, Ipoh 30450, Perak, Malaysia*
[3] *Institut National de la Santé et de la Recherche Médicale; Centre de Recherche des Cordeliers, Equipe-Immunopathologie et Immunointervention Thérapeutique, Sorbonne Université, Université De Paris, Paris, France*
* *Corresponding author*

## Introduction

Bee venom is a transparent and odorless liquid that possesses a variety of pharmacologically active components, including polypeptides (melittin, apamin, mast cell degranulating [MCD] peptide, and adolapin), enzymes (phospholipase $A_2$ and hyaluronidase), and some low-molecular-weight compounds such as sugars, amino acids, minerals and catecholamines (Wehbe et al. 2019). Owing to the anti-inflammatory, antioxidant, antibacterial, anticancer, analgesic, and anti-atherogenic properties of bee venom, it has been extensively studied in the management of different kinds of diseases, particularly neurological diseases, skin diseases, cancer, and arthritis (Figure 22.1). In recent years, bee venom has also been investigated in atherosclerosis, acute kidney injury, and acute gastric ulceration (Table 22.1). A brief overview of the research carried out on the potential medicinal benefits of bee venom is presented in this chapter.

## Composition of Bee Venom

Bee venom contains merely 0.1 µg of dry venom and 88% of water (Bellik 2015). It mainly consists of a complex mixture of polypeptides, including melittin, apamin, MCD peptide, and adolapin; melittin is the major component of bee venom. Melittin is a peptide with 26 residues; it represents 40%–60% of bee venom (Chen et al. 2016). It can be soluble in tetrameric and monomeric forms owing to its amphipathic properties. Therefore, melittin can be easily passed through membranes via the disruption of either synthetic or natural phospholipid bilayers. Apamin (18 amino acids) and MCD peptide (22 amino acids) correspond to 2%–3% of bee venom's composition (Wehbe et al. 2019), respectively. Both polypeptides have a similar structure with two disulfide bonds. Apamin is the smallest neurotoxin that possesses the potential to cross the blood–brain barrier and exerts its effects on the central nervous system (CNS) via various mechanisms of action (Son et al. 2007). On the other hand, MCD peptide is an epileptogenic neurotoxin that acts as a $K^+$ channels inhibitor to induce a decrease of blood pressure in in vivo models (Hanson et al. 1974; Wehbe et al. 2019). Adolapin accounts for about 1% of the composition of bee venom, containing 103 amino acids. It has been reported to have antipyretic, antinociceptive, and anti-inflammatory potentials via inhibition of cyclooxygenase (COX) activity and prostaglandin synthesis (Cherniack and Govorushko 2018).

Enzymes such as phospholipase $A_2$ and hyaluronidase are also present in bee venom. Phospholipase $A_2$, which accounts for 12%–15% of bee venom's dry weight, is the most lethal hydrolytic enzyme in bee venom (Wehbe et al. 2019). It consists of 128 amino acids along with four disulfide bonds on a single polypeptide chain. Accordingly, phospholipase $A_2$ in bee venom plays a crucial role in neuroprotection against various neurodegenerative diseases (Chung et al. 2015). Hyaluronidase corresponds to 1.5%–2% of bee venom's composition. Likewise, hyaluronidase is important for venom's spread through the patient's tissue by upregulating blood flow and affecting the structural integrity of the specific area (Hossen et al. 2017).

*Honey: Composition and Health Benefits*, First Edition. Edited by Md. Ibrahim Khalil, Gan Siew Hua, and Bey Hing Goh.

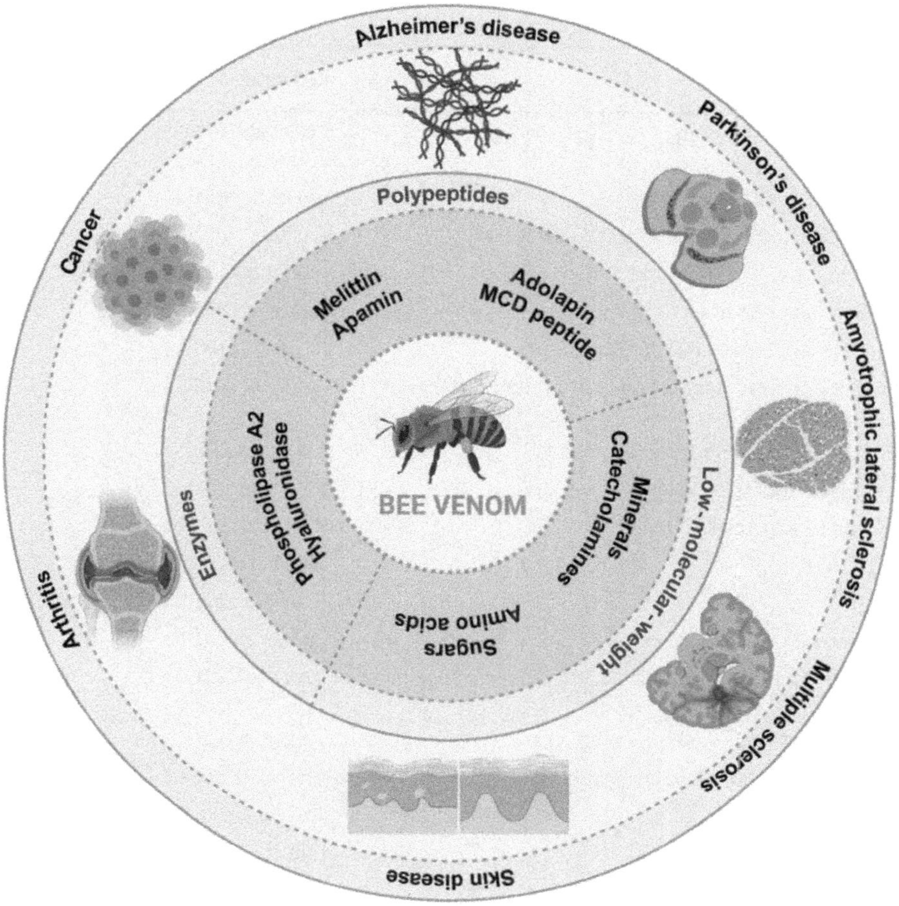


**Figure 22.1** Medicinal benefits of bee venom. MCD, mast cell degranulating.

**Table 22.1** Medicinal benefits of bee venom in a variety of diseases.

| Disease | Mechanism of Action | Reference |
|---|---|---|
| Alzheimer's disease | Mechanism of action; reduction of Aβ; anti-inflammatory | Ye et al. 2016 |
| | Anti-inflammatory | Baek et al. 2018 |
| Parkinson's disease | Protection of dopaminergic neurons | Doo et al. 2010 |
| | Anti-inflammatory | Kim et al. 2011 |
| | Anti-inflammatory | Tanner et al. 2011 |
| | Restoration of neuroinflammatory and apoptotic markers | Khalil et al. 2015 |
| | Increase of CD4 T-cell infiltration | Chung et al. 2012 |
| | Protection of dopaminergic neurons | Alvarez-Fischer et al. 2013 |
| | Anti-inflammatory; inhibition of (Th1 and Th17 cells | Kim et al. 2019a |
| | Antioxidant | Badawi et al. 2020 |
| Amyotrophic lateral sclerosis | Reduction of microglia marker expression | Yang et al. 2010 |
| | Increase in expression of α-synuclein nitration and phosphorylation; decrease of ubiquitinated protein; restoration of proteasomal activity | Yang et al. 2011 and Yang and Choi 2013 |

*(Continued)*

**Table 22.1** (Continued)

| Disease | Mechanism of Action | Reference |
|---|---|---|
| | Reduction of pro-inflammatory markers | Lee et al. 2014a and Lee et al. 2015 |
| | Reduction of TLR4, TNF-α, and CD14 | Cai et al. 2015 |
| Multiple sclerosis | Decrease of demyelination and inflammatory cell infiltration | Karimi et al. 2012 |
| | Activation of microglial cells | Mohammadi-Rad et al. 2019 |
| Skin diseases | Reduction of pro-inflammatory markers | Kim et al. 2013 |
| | Reduction of IL-1β, TNF-α, TLR2, and CD14 expression | An et al. 2014 |
| | Reduction IκB and IKK expression; reduction of pro-inflammatory markers | Lee et al. 2014b |
| | Reduction of TNF-α, IL-8, IL-1β, and IFN-γ expression | Kim et al. 2015 |
| | Reduction of macrophages and IgE levels | Shin et al. 2018 |
| | Regulation of CD55 via ERK pathway | Kim et al. 2019b |
| Cancer | Apoptotic cancer cell death | Wang et al. 2009 |
| | Antitumor effects | Huh et al. 2010 |
| | Inhibition of NF-κB and AP-1 | Park et al. 2010 |
| | Downregulation of cIAP-2, cPLA2, iNOS, Bcl-2, and COX-2; activation of caspase-3 and caspase-9 | Park et al. 2011 |
| | Activation of DR3, DR4, and DR6; inhibition of STAT3 | Jo et al. 2012 |
| | Enhanced DR3 expression and NF-κB pathway inhibition | Choi et al. 2014 |
| | Apoptotic cancer cell death | Jung et al. 2018 |
| | Antimelanoma effects | Lim et al. 2019 |
| Arthritis | Reduction of IL 1β, NF-κB, and TNF-α levels | El-Tedawy et al. 2020 |
| | Antioxidant effects | Tekeoğlu et al. 2020 |
| | Reduction of TGs, total cholesterol, and LDL; upregulation of HDL | Mohammed and Hassan 2019a, 2019b |
| | Anti-inflammatory | Lee et al. 2020 |
| Atherosclerosis | Increase in Treg cells | Kang et al. 2020 |
| Acute kidney injury | Anti-inflammatory; antioxidant effects | Kim et al. 2020 |
| Acute gastric ulceration | Reduction of pro-inflammatory cytokines, MDA, Bax levels, and caspase-3 expression | Mohamed et al. 2019 |

AP-1, activator protein 1; cIAP-2, cellular inhibitor of protein 2; COX, cyclooxygenase; cPLA2, cytosolic phospholipases; DR, death receptor; ERK, extracellular signal–regulated kinase; HDL, high-density lipoprotein; IFN, interferon; Ig, immunoglobulin; IKK, IκB kinase; IL, interleukin; iNOS, inducible nitric oxide synthase; LDL, low-density lipoprotein; MDA, malondialdehyde; NF-κB, nuclear factor κB; STAT3, signal transducer and activator of transcription 3; TG, triglyceride; Th, T helper; TLR4, Toll-like receptor 4; TNF-α, tumor necrosis factor-α; Treg, T regulatory.

## Medicinal Benefits of Bee Venom

### Neurological Disorders

#### Alzheimer's Disease

Alzheimer's disease (AD) is a fatal neurodegenerative disorder with no cure. It typically develops among older adults before the age of 65 years (early onset) or after the age of 65 years (late onset) (Giri et al. 2016). AD is characterized by the clinical features of progressive dementia followed by a gradual loss of cognitive abilities and disruption of verbal as well as motor functions (Kent et al. 2020). Pathologically, the accumulation of intraneuronal neurofibrillary tangles of *tau* protein and plaques of extracellular amyloid-beta (Aβ) protein is predominantly considered the prime hallmark associated with a series of inflammatory responses in the development of AD (Busche and Hyman 2020). A plethora of current treatments using natural products have been shown to possess a wide range of medicinal benefits for patients with AD. One such promising example is bee venom (Panda and Jhanji 2020).

Interestingly, bee venom phospholipase A2 (bvPLA2) showed its potential to attenuate the progression of AD under an in vivo mice model of AD (Ye et al. 2016). The administration of bvPLA2 (0.2 and 1.0 mg/kg) intraperitoneally was reported to enhance cognitive function in mice by reducing the aggregation of Aβ plaques, increasing glucose metabolism in the brain, and inhibiting hippocampal neuroinflammatory responses, in turn limiting AD progression. Another recent in vivo study by Baek et al. (2018) suggested a novel and combined treatment of bvPLA2 with Aβ vaccination could be an alternative therapy for AD in terms of preventing the adverse inflammatory response via the modulation of regulatory T cells (T regulatory cells [Tregs]; $CD4^+CD25^+Foxp3^+$ expressing cells).

#### Parkinson's Disease

Parkinson's disease (PD) is known as a neurodegenerative movement disorder clinically characterized by nonmotor symptoms, such as fatigue, memory and learning dysfunction, sleep disorders, and psychiatric disturbances, and motor symptoms, including muscle rigidity, tremor, hypokinesia, bradykinesia, and akinesia (Blauwendraat et al. 2020). Accumulation of α-synuclein in Lewy bodies along with the progressive loss of dopaminergic neurons in the brain – the substantia nigra – is the main contributing factor in the development of PD (Raza and Anjum 2019). According to the literature, bee venom treatment exerts some neuroprotective effects against PD under in vivo models via the protection of dopaminergic neurons, activation of Tregs, and restoration of neuroinflammatory and apoptotic markers.

A study has demonstrated that bee venom acupuncture (0.02 mL of bee venom [1:2000 w/v]; acupoint: GB34) could protect the degeneration of dopaminergic neurons against 1-methyl-4-phenyl-1,2,3,6-tetrahydropyridine (MPTP)–induced neurotoxicity in a PD mouse model via attenuation of phospho-c-Jun (c-Jun) immunoreactivity and restoration of tyrosine hydroxylase immunoreactivity (Doo et al. 2010). In this context, another study found that the suppression of dopaminergic neuronal cell loss is attributed to the anti-inflammatory activities of bee venom (Kim et al. 2011). The finding indicated that bee venom acupuncture (0.6 mg/kg; acupoint: ST36) exerted its therapeutic benefits against PD models through increased inducible nitric oxide synthase (iNOS) and macrophage-1 antigen (Mac-1) expression followed by microglial deactivation along the pathway of pro-inflammatory markers suppression. In a similar line, a preclinical bee venom acupuncture trial confirmed the neuroprotective ability against neuroinflammation, oxidative, and apoptosis induced by rotenone in the PD in vivo model (Tanner et al. 2011). The expression of apoptotic markers (Bcl-2 [B-cell lymphoma 2]-associated X protein [Bax]) was inhibited; DNA damage was repressed; and in turn, dopamine depletion was prevented. Also, the locomotor activity of bee venom–treated mice was restored. A subsequent study also suggested that bee venom acupuncture (0.02 mL of bee venom [1:2000 w/v]; acupoint: GB34) restored all neuroinflammatory and apoptotic markers (Khalil et al. 2015).

Moreover, the protective benefits of bee venom were also associated with increased CD4 T-cell infiltration, particularly an increased proportion of Tregs, against neuroinflammation in the MPTP-induced PD model (Chung et al. 2012). The protective potential of bee venom acupuncture (low dose, 12 μg/kg; high dose, 120 μg/kg) intraperitoneally against PD in vivo models also supported with the findings that bee venom offered substantial protection to dopaminergic neurons against MPTP-induced toxicity (Alvarez-Fischer et al. 2013). In tune with the previous literature, in a study by Kim et al. (2019a), bvPLA2 treatment (0.5 mg/kg subcutaneously) exhibited potent pharmacological activities against MPTP-induced PD in a mouse model. As reported, the behavioral abnormalities of bvPLA20-administrated mice were ameliorated. The result is assigned to the anti-inflammatory mechanism of bvPLA2 via the inhibition of inflammatory T helper 1 (Th1) and Th17 cells, along with the activation of Tregs. Additionally, bee venom was confirmed to potently inhibit rotenone-induced oxidative or nitrosative stress with evidence of reduced nitric oxide (NO) and malondialdehyde (MDA) along with increased total antioxidant capacity (Badawi et al. 2020).

#### Amyotrophic Lateral Sclerosis

Amyotrophic lateral sclerosis (ALS) is a fatal, adult-onset neurodegenerative disorder consisting of severe muscle denervation, weakness, hyper-reflexivity of the limbs, and fasciculations caused by a selective motor neuron degeneration in the spinal cord and brain motor cortex (Burk and Pasterkamp 2019). Accordingly, abnormal aggregation of mutant superoxide dismutase type 1 gene (mt*SOD1*) is said to be a significant hallmark of ALS (Boillée et al. 2006). An in vivo study was performed in mutant $SOD1_{G93A}$ transgenic mice to investigate the anti-inflammatory effects of bee venom acupuncture against ALS. The finding implicated that treatment of bee venom (0.1 μg/g; acupoint: ST36) subcutaneously significantly improved motor ability and prolonged the life expectancy of mice accompanied by a reduction of microglia marker expression in the biochemical analysis (Yang et al. 2010). Subsequent studies conducted in a similar ALS mice model proved that bee venom (0.1 μg/g; acupoint: ST36) subcutaneously or melittin (0.1 μg/g; acupoint: ST36) subcutaneously diminished the increased expression of α-synuclein nitration and phosphorylation (Yang and Choi 2013; Yang et al. 2011). Additionally, a decrease of ubiquitinated protein and restoration of proteasomal activity were observed in both bee venom– and melittin-treated mice groups.

Motivated by previous concerns, another experimental study was conducted using a similar ALS model to investigate the protective potential of melittin on ALS-induced inflammation in the lungs and spleen (Lee et al. 2014a). The activation of monocytes and T-cell numbers connected to disease progression is triggered by ALS immunocytes (Zhang et al. 2005, 2006). Furthermore, Zhang et al. (2005) have shown that the level of monocyte/macrophage activation was correlated with an increase in abnormally immunologic-sporadic ALS activation. It proposed that inflammation-involved markers may provide useful support for tracking ALS treatment and diseases of immune dysfunction. Remarkably, the result presented that melittin treatment (0.1 μg/g; acupoint: ST36) subcutaneously alleviated organ inflammation via the reduction of a cluster of differentiation 14 (CD14) and ionized calcium-binding adapter molecule 1 (Iba-1) expression in the lungs and COX-2 and CD14 in the spleen. These findings are supported by the observation that bee venom acupuncture was believed to enhance the immune response of mice against ALS and reduce these pro-inflammatory factors in the spleen, liver, and kidney (Lee et al. 2015). Also, the study highlighted that higher efficacy of bee venom treatment via a selective acupoint than an intraperitoneal route. To gain insight into the neuroprotective mechanism of bee venom acupuncture, Cai et al. (2015) demonstrated that this therapy improved motor function by decreasing death of motor neurons via an endogenous modulation of the immune system and inhibition of neuroinflammation with reduced levels of Toll-like receptor 4 (TLR4), tumor necrosis factor-α (TNF-α), and CD14.

#### Multiple Sclerosis

Multiple sclerosis (MS) is an autoimmune, neuroinflammatory disorder correlated with two major pathological hallmarks, inflammatory demyelination (early stage) and gliosis with neurodegeneration (chronic stage) in the CNS. These factors typically lead to nontraumatic neurological abnormality among young adults (Hauser and Cree 2020). In defiance of the present immunosuppressive and immunomodulatory therapies, MS is not fully controlled. Fascinatingly, a preclinical in vivo study concluded that bee venom potentially reduced the clinical symptoms and side effects from a rat model of experimental allergic encephalomyelitis (Karimi et al. 2012). Based on the observation, bee venom–treated rats showed decreased levels of nitrates, glutamate, and TNF-α; increased levels of γ-aminobutyric acid; and suppressed pathological changes, such as decreased demyelination and inflammatory cell infiltration, suggesting the pharmacological benefits of bee venom against MS by virtue of immunomodulatory, antioxidant, and anti-inflammatory activities.

Owing to the concerns regarding possible detrimental risks of severe allergic reactions of patients with MS caused by bee venom treatment, a preliminary clinical trial was performed by Castro et al. (2005) to access the safety of bee venom treatment on nine participants with MS. Although the safety of bee venom was suggested with certain levels of improvement in the symptoms in 60% of patients with MS because of the low sample size of the study group (n = 9), no definite conclusion was made. Following on this, another clinical study was evoked by Wesselius et al. (2005) with the goal of identifying the efficacy of bee venom therapy in 26 patients with MS (relapsing-remitting MS or relapsing secondary progressive MS). However, bee venom therapy with a duration of 24 weeks did not show any improvement of MS-related symptoms, such as fatigue, disability, or quality of life. In contrast, a recent investigation revealed that co-treatment with apamin potently prevented the cuprizone-induced myelin destruction by increasing the myelin content along with the inhibition of microglia cells activation in an MS-induced in vivo model (Mohammadi-Rad et al. 2019). Despite different study reports on bee venom against MS symptoms, considerable efforts should be taken in further studies relevant to different experimental protocols, such as type of MS study model and administrated dose.

### Skin Diseases

Considerable literature suggests that bee venom possesses antibacterial and anti-inflammatory effects against certain inflammatory skin diseases, including atopic dermatitis and acne vulgaris. Atopic dermatitis or atopic eczema is a chronic, complex inflammatory skin disease with clinical manifestations of skin dryness and itchiness (Stefanovic et al. 2020). It has been postulated that skin barrier deficits (genetic factors) and immune system dysregulation (immune factors) could contribute to the pathophysiology of atopic dermatitis associated with the overexpression of pro-inflammatory factors and activation of antigen-presenting cells (Weidinger et al. 2018). Acne vulgaris is also a cutaneous inflammatory disease with complex acne pathophysiology caused by *Cutibacterium acnes* (previously *Propionibacterium acnes*), which involves changes in keratinization of pilosebaceous follicles, secretion of pro-inflammatory cytokines, and subsequent inflammation (Cong et al. 2019).

In 2013, an in vivo study reported that bee venom (0.0.1 and 0.1 mg/kg) intraperitoneally confers some anti-itchy effects by attenuating scratching behavior in a skin mouse model induced by compound 48/80 (Kim et al. 2013). By inhibiting mast cell degranulation and nuclear factor κB (NF-κB) activation, bee venom remarkably regulated pro-inflammatory cytokine expression. To evaluate the effects of bee venom on skin sensitization, a study was conducted on rats and guinea pigs (Han et al. 2013). The findings suggested bee venom used as the ingredients for skin cosmetics with no dermal irritation and was well tolerated in both in vivo models. However, negligible sensitization potential of bee venom was reported.

Based on the previous findings, a clinical study of the advantages of bee venom serum treatment on human skin aging was performed (Han et al. 2015). Interestingly, it was observed that facial wrinkles were improved along with a decreased average depth of the wrinkles, total count, and total area of wrinkles, suggesting the possible efficacy of bee venom serum on aging skin.

By using a *C. acnes*–induced skin disease model, decreased ear thickness of mice was observed after bee venom treatment (1, 10, and 100 μg/ear), which was given on the skin surface of the ear (An et al. 2014). The finding is supported by a biochemical estimation in which the levels of interleukin (IL)-1β and TNF-α and the expression of TLR2 and CD14 were reduced, indicating the antibacterial activity of bee venom. To gain insight into the mechanism of melittin, both in vitro and in vivo studies were conducted by Lee et al. (2014b) in heat-killed *C. acnes*–treated HaCaT cells (human epidermal keratinocyte cells) and a *C. acnes*–induced mice model, respectively. Melittin (0.1, 0.5, and 1 μg/mL) was shown to ameliorate the side effects of *C. acnes* by reducing the expression of IκB and IκB kinase (IKK) in HaCaT cells via NF-κB signaling pathway. Also, the inhibition of pro-inflammatory cytokine expression attributed to melittin can suppress the phosphorylated p38 mitogen-activated protein kinase (MAPK) signaling pathway.

The medicinal benefits of melittin were also exhibited in a *C. acne*–induced mice model in which markable reduced granulomatous and swelling were observed in melittin-applied mice ears. Likewise, another study reported that bee venom (1, 10 and 100 ng/mL) was able to reduce heat-killed *C. acnes*–induced increased pro-inflammatory cytokine (TNF-α, IL-8, IL-1β, and interferon-γ) secretion in keratinocytes and monocytes as well as inhibit the expression of TLR2 (Kim et al. 2015).

A recent study demonstrated that bvPLA2 treatment (80 ng/ear) greatly attenuated the atopic dermatitis–like skin lesions and inflammation triggered by 2,4-dinitrochlorobenzene (DNCB) and dermatophagoides farinae extract via interaction with CD26 (Shin et al. 2018). Application of bvPLA2 also inhibited mast cell infiltration and reduced the levels of macrophages and serum IgE in the ears of an atopic dermatitis–induced mice model. On the contrary, Kim et al. (2019b) suggested the possible mechanism of bee venom in the attenuation of symptoms caused by atopic dermatitis involved the mediation of CD55 through the extracellular signal–regulated kinase (ERK) signaling pathway in both in vitro and in vivo models.

## Cancer

Bee venom, particularly its major component, melittin, has been widely investigated in different in vitro cancer cell models, such as liver, renal, prostate, ovarian, lung, and melanoma cancer cells, owing to their antitumor, apoptotic, antibacterial, and antimelanoma activities. In this context, the anticancer mechanism of melittin was studied in an in vitro hepatocellular carcinoma cells (HCC) cell line (Wang et al. 2009). The results confirmed the synergistic mechanism of melittin with tumor necrosis factor–related apoptosis-inducing ligand (TRAIL) to induce the cellular apoptosis of HCC via the activation of calmodulin-dependent protein kinase (CAMKII)–transforming growth factor-β–activated kinase 1 (TAK1)–MAPK (JNK/p38) signaling pathway along with the inhibition of the IKK–NF-κB pathway. Other evidence of bee venom emerged regarding its antitumor effects on tumor angiogenesis and pulmonary metastasis via the inhibition of the vascular endothelial growth factor (VEGF) and VEGF receptor 2 (VEGFR2) pathways in both an in vitro Lewis lung carcinoma (LLC) cell line and an in vivo mice model (Huh et al. 2010). Inhibition of LLC cells' viability without showing cytotoxicity was also induced by bee venom.

Both bee venom and melittin were also found to exhibit their antitumor properties on phorbol 12-myristate 13-acetate (PMA)–induced matrix metalloproteinase-9 (MMP-9) expression in an in vitro human renal carcinoma cell line (Park et al. 2010). MMP-9 is responsible for cancer cell invasion and metastasis. Bee venom and melittin showed an inhibitory effect on PMA-induced cancer cell invasion. The inhibition was ascertained to the melittin-suppressed expression of MMP-9 by blocking the activation of NF-κB and activator protein-1 (AP-1) via the ERK and c-Jun N-terminal kinase (JNK) pathways. On the contrary, the anticancer properties of both bee venom and melittin were supported by evidence of suppressing the growth of prostate cancer cells under an in vitro model (Park et al. 2011). Melittin was demonstrated to induce apoptotic cancer cell death by (1) downregulating the antiapoptotic and proliferative gene products, including cellular inhibitor of apoptosis 2 (cIAP-2), cytosolic phospholipases A2 (cPLA2), iNOS, Bcl-2, and COX-2 via inhibition of NF-κB signaling pathway, and (2) activating caspase-3 and caspase-9. Jo et al. (2012) also suggested bee venom (1–5 μg/mL) and melittin (0.5–2 μg/mL) induced apoptosis in ovarian cancer cells via the activation of a series of death receptors (DR3, DR4, and DR6) and the inhibition of signal transducer and activator of transcription 3 (STAT3) pathway.

Furthermore, the inhibitory effect of bee venom toward cancer cell growth was evaluated in human lung cancer cell lines (Choi et al. 2014). The findings showed that the bee venom treatment (1–5 μg/mL) dose dependently induced cell death by apoptosis in lung cancer cells mainly via the enhanced DR3 expression and NF-κB pathway inhibition. The data are in agreement with Park et al. (2011)'s finding that melittin reduced expression of Bcl-2 and increased expression of Bax, cleaved caspase-3, and cleaved caspase-9.

The apoptotic properties of bee venom have also been demonstrated in in vitro human breast cancer cells (MDA-MB-231), which could be evidenced by the degradation and denaturation of protein and the defragmentation of DNA observed from Raman spectroscopy (Jung et al. 2018). In a study by Lim et al. (2019), both bee venom and melittin exhibited antimelanoma activity by suppressing the invasion, migration, and growth of melanoma cells via the downregulation of MARK and phosphoinositide 3-kinase–protein kinase B–mammalian target of rapamycin signaling pathway. A new approach was suggested that combined melittin with temozolomide potently increased antimelanoma effects compared with melittin or temozolomide alone.

## Arthritis

Arthritis is a chronic inflammatory disease mainly characterized by inflammation, swelling, stiffness, and pain of one or more joints. Rheumatoid arthritis (RA), osteoarthritis, and gout (Spel and Martinon 2020) are the common arthritic diseases. Over the years, the medicinal benefits of bee venom have been investigated in different kinds of arthritic diseases. Considerable literature has confirmed that bee venom acupuncture was developed as an alternative therapy to control inflammation and pain in the treatment of people with arthritis (Krylov et al. 2007; Son et al. 2007).

In recent years, enormous efforts have been devoted to determine the possible underlying mechanisms of bee venom in the treatment of patients with arthritic diseases under in vitro or in vivo models. El-Tedawy et al. (2020) highlighted the potential of high-dose systemic bee venom administration (60 mg/kg) on the improvement of signs and symptoms of RA under a complete Freund's adjuvant-induced arthritic rat model. Bee venom was shown to reduce serum levels of IL-1β, NF-κB, and TNF-α in the joints, indicating anti-inflammatory, analgesic, and anti-arthritic properties. Using a similar model, another study revealed that bee venom treatment (20 μg/kg) significantly improved oxidative stress index and total oxidant level (Tekeoğlu et al. 2020). On the other hand, Mohammed and Hassan (2019a, 2019b) believed that the attenuation of arthritis symptoms by bee venom (1 mg/kg of body weight) was attributed to the improvement in certain blood and biochemical markers, which can be evidenced by decreased levels of triglycerides, total cholesterol, and low-density lipoprotein along with increased high-density protein. Also, it was also demonstrated that bee venom (0.5 and 1 mg/kg of body weight) and apamin (0.5 and 1 mg/kg of body weight) treatment exerted anti-inflammatory effects by inhibiting the formation of inflammasome and pro-inflammatory cytokine secretion in a monosodium urate–induced gouty arthritis model (Lee et al. 2020). Motivated by previous concerns, a clinical trial was conducted on patients with osteoarthritis (Conrad et al. 2019). A set of 15 bee venom dermal injection (100 μg) was given at prespecified acupoints. The findings implicated that bee venom biotherapy potently improved patients' knee pain and enhanced their physical function.

## Other Diseases

Of major interest, the medicinal benefits of bee venom have also been confirmed in the treatment of other disease models, such as atherosclerosis, acute kidney injury, and acute gastric ulceration. In a study by Kang et al. (2020), it was shown that bvPLA2 treatment (0.2 mg/kg of body weight) could be an alternative treatment for atherosclerosis. bvPLA2 potently increased Tregs and decreased lipid accumulation in the aortic valve and aorta in an atherosclerosis mice model. On the contrary, the therapeutic effects of bee venom were evaluated on lipopolysaccharide-induced acute kidney injury using a mouse model (Kim et al. 2020). The results showed that bee venom administration (100 μg/kg of body weight) remarkably reduced inflammatory cytokines, lipid peroxidation markers, tubular injury markers, and infiltration of immune cells and attenuated tubular cell apoptosis, subsequently ameliorating injury of kidney structure and renal dysfunction. The findings confirmed the anti-inflammatory and antioxidant properties of bee venom. Apart from this, bee venom (2 mg/kg of body weight) was also proved to possess healing effects on acute gastric ulceration by decreasing the levels of pro-inflammatory cytokines, MDA, Bax, and caspase-3 accompanied by an increase in heat shock protein 70 and endogenous antioxidant enzymes in an acetylsalicylic acid–induced acute gastric ulceration model (Mohamed et al. 2019).

## Mechanism of Action of Bee Venom on Diverse Diseases

The effects of bee venom on diverse diseases, including neurodegenerative diseases (AD, PD, ALS), autoimmune diseases (arthritis, MS), inflammatory skin diseases, and cancer, are depicted in Figure 22.2, where crude bee venom's therapeutic interests are addressed. The anti-inflammatory properties influence the signaling pathways of primary inflammation and cause suppression of the expression of pro-inflammatory genes. Among neurodegenerative diseases, including PD, AD, and ALS, bee venom has a neuroprotective capacity by substantially enhancing cognitive function and blocking disease progression in animal models. However, no wide-ranging mechanism-related studies have been performed on bee venom in patients with different diseases. There is also incomplete reporting on bee venom's application in clinical practise. Therefore, comprehensive studies on bee venom are required after a thorough study of their toxicity to improve their medicinal benefits and for further drug development processes. In connection with its pharmacokinetics, pharmacodynamics, and clinical trials, the authors hope that this chapter will encourage further

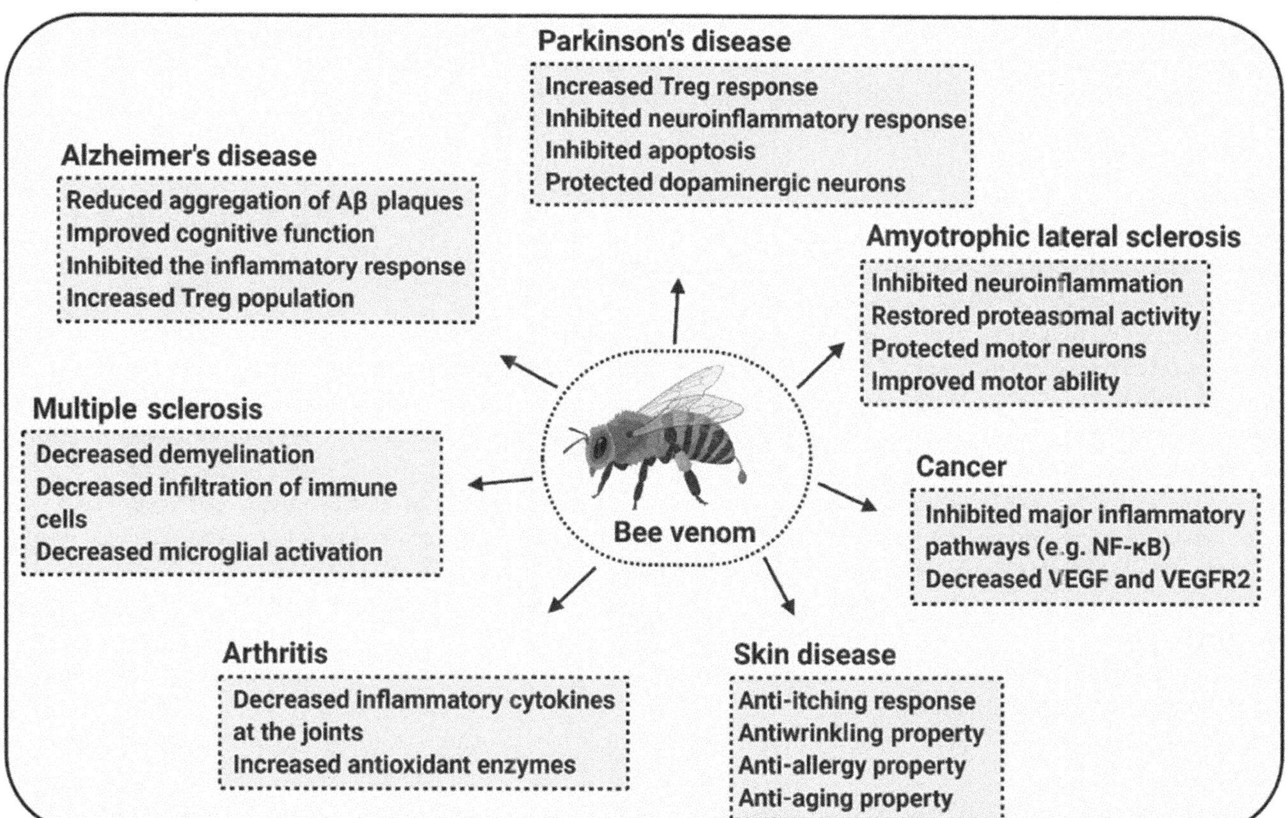


**Figure 22.2** Mechanisms of action of bee venom on diverse diseases. NF-kB, nuclear factor κB; Treg, T regulatory; VEGF, vascular endothelial growth factor.

investigations on bee venom. In the near future, all these additional parameters will speed up the production and use of bee venom as a promising medicinal agent.

## Conclusion and Future Perspectives

Bee venom has gained global attention for its various medicinal benefits, which appear to act primarily via its antioxidant and anti-inflammatory mechanisms. Research indicates that bee venom is primarily intended to help in the management of oxidative and inflammatory conditions, cancer, arthritis, neurodegenerative diseases, and skin diseases. Each component of bee venom may have various pharmacological activities based on the molecular target. Hence, for bee venom–based drug discovery, it is important to understand the biological role of each component. Thus, more studies on bee venom and its active components are warranted in the future, especially at cellular and molecular levels. Finally, the application of bee venom therapy in clinical settings is still a long way off. However, researchers believe that the ongoing work on this aspect will ultimately allow bee venom and its active components to be considered in the coming years as definitive candidates in various treatments.

## Acknowledgment

The authors also thank Universiti Kuala Lumpur Royal College of Medicine Perak, Malaysia, for providing necessary facilities and resources to complete this study for publication. The figures were created with the support of https://biorender.com under a paid subscription.

## Conflict of Interest

We wish to confirm that there are no known conflicts of interest associated with this publication, and there has been no significant financial support for this work that could have influenced its outcome.

## References

Alvarez-Fischer, D., Noelker, C., Vulinović, F., et al. (2013). Bee venom and its component apamin as neuroprotective agents in a Parkinson disease mouse model. *PLoS One* 8 (4): e61700.

An, H.-J., Lee, W.-R., Kim, K.-H., et al. (2014). Inhibitory effects of bee venom on Propionibacterium acnes-induced inflammatory skin disease in an animal model. *International Journal of Molecular Medicine* 34 (5): 1341–1348.

Badawi, H., Abdel-Salam, R., Abdel-Salam, O., et al. (2020). Bee venom attenuates neurodegeneration and motor impairment and modulates the response to L-dopa or rasagiline in a mice model of Parkinson's disease. *Iranian Journal of Basic Medical Sciences* 23 (12): 1628–1638.

Baek, H., Lee, C.-J., Choi, D.B., et al. (2018). Bee venom phospholipase A2 ameliorates Alzheimer's disease pathology in Aβ vaccination treatment without inducing neuro-inflammation in a 3xTg-AD mouse model. *Scientific Reports* 8 (1): 1–11.

Bellik, Y. (2015). Bee venom: its potential use in alternative medicine. *Anti-infective Agents* 13 (1): 3–16.

Blauwendraat, C., Nalls, M.A., and Singleton, A.B. (2020). The genetic architecture of Parkinson's disease. *The Lancet Neurology* 19 (2): 170–178.

Boillée, S., Yamanaka, K., Lobsiger, C.S., et al. (2006). Onset and progression in inherited ALS determined by motor neurons and microglia. *Science* 312 (5778): 1389–1392.

Burk, K. and Pasterkamp, R.J. (2019). Disrupted neuronal trafficking in amyotrophic lateral sclerosis. *Acta Neuropathologica* 137 (6): 859–877

Busche, M.A. and Hyman, B.T. (2020). Synergy between amyloid-β and tau in Alzheimer's disease. *Nature Neuroscience* 23 (10): 1183–1193.

Cai, M., Choi, S.-M., and Yang, E.J. (2015). The effects of bee venom acupuncture on the central nervous system and muscle in an animal hSOD1G93A mutant. *Toxins* 7 (3): 846–858.

Castro, H.J., Mendez-Inocencio, J.I., Omidvar, B., et al. (2005). A phase I study of the safety of honeybee venom extract as a possible treatment for patients with progressive forms of multiple sclerosis. *Allergy & Asthma Proceedings* 26: 470–476.

Chen, J., Guan, S.-M., Sun, W., and Fu, H. (2016). Melittin, the major pain-producing substance of bee venom. *Neuroscience Bulletin* 32 (3): 265–272.

Cherniack, E.P. and Govorushko, S. (2018). To bee or not to bee: the potential efficacy and safety of bee venom acupuncture in humans. *Toxicon* 154: 74–78.

Choi, K.E., Hwang, C.J., Gu, S.M., et al. (2014). Cancer cell growth inhibitory effect of bee venom via increase of death receptor 3 expression and inactivation of NF-kappa B in NSCLC cells. *Toxins* 6 (8): 2210–2228.

Chung, E.S., Kim, H., Lee, G., et al. (2012). Neuro-protective effects of bee venom by suppression of neuroinflammatory responses in a mouse model of Parkinson's disease: role of regulatory T cells. *Brain, Behavior, and Immunity* 26 (8): 1322–1330.

Chung, E.S., Lee, G., Lee, C., et al. (2015). Bee venom phospholipase A2, a novel Foxp3+ regulatory T cell inducer, protects dopaminergic neurons by modulating neuroinflammatory responses in a mouse model of Parkinson's disease. *Journal of Immunology* 195 (10): 4853–4860.

Cong, T.-X., Hao, D., Wen, X., et al. (2019). From pathogenesis of acne vulgaris to anti-acne agents. *Archives of Dermatological Research* 311 (5): 337–349.

Conrad, V.J., Hazan, L.L., Latorre, A.J., et al. (2019). Efficacy and safety of honey bee venom (Apis mellifera) dermal injections to treat osteoarthritis knee pain and physical disability: a randomized controlled trial. *The Journal of Alternative and Complementary Medicine* 25 (8): 845–855.

Doo, A.-R., Kim, S.-T., Kim, S.-N., et al. (2010). Neuroprotective effects of bee venom pharmaceutical acupuncture in acute 1-methyl-4-phenyl-1, 2, 3, 6-tetrahydropyridine-induced mouse model of Parkinson's disease. *Neurological Research* 32 (sup1): 88–91.

El-Tedawy, D.M., Abd-Alhaseeb, M.M., Helmy, M.W., and Ghoneim, A.I. (2020). Systemic bee venom exerts anti-arthritic and anti-inflammatory properties in a rat model of arthritis. *Biomedical Reports* 13 (4): 1-1.

Giri, M., Zhang, M., and Lü, Y. (2016). Genes associated with Alzheimer's disease: an overview and current status. *Clinical Interventions in Aging* 11: 665.

Han, S.M., Hong, I.P., Woo, S.O., et al. (2015). The beneficial effects of honeybee-venom serum on facial wrinkles in humans. *Clinical Interventions in Aging* 10: 1587.

Han, S.M., Lee, K.G., Park, K.K., and Pak, S.C. (2013). Skin sensitization study of bee venom (Apis mellifera L.) in guinea pigs and rats. *Cutaneous and Ocular Toxicology* 32 (1): 27–30.

Hanson, J.M., Morley, J., and Soria-Herrera, C. (1974). Anti-inflammatory property of 401 (MCD-peptide), a peptide from the venom of the bee Apis mellifera (L.). *British Journal of Pharmacology* 50 (3): 383.

Hauser, S.L. and Cree, B.A. (2020). Treatment of multiple sclerosis: a review. *The American Journal of Medicine* 133 (12): 1380–1390.

Hossen, M., Shapla, U.M., Gan, S.H., and Khalil, M. (2017). Impact of bee venom enzymes on diseases and immune responses. *Molecules* 22 (1): 25.

Huh, J.-E., Baek, Y.-H., Lee, M.-H., et al. (2010). Bee venom inhibits tumor angiogenesis and metastasis by inhibiting tyrosine phosphorylation of VEGFR-2 in LLC-tumor-bearing mice. *Cancer Letters* 292 (1): 98–110.

Jo, M., Park, M.H., Kollipara, P.S., et al. (2012). Anti-cancer effect of bee venom toxin and melittin in ovarian cancer cells through induction of death receptors and inhibition of JAK2/STAT3 pathway. *Toxicology and Applied Pharmacology* 258 (1): 72–81.

Jung, G.B., Huh, J.-E., Lee, H.-J., et al. (2018). Anti-cancer effect of bee venom on human MDA-MB-231 breast cancer cells using Raman spectroscopy. *Biomedical Optics Express* 9 (11): 5703–5718.

Kang, G.-H., Lee, S., Choi, D.B., et al. (2020). Bee Venom phospholipase A2 ameliorates atherosclerosis by modulating regulatory T cells. *Toxins* 12 (10): 609.

Karimi, A., Ahmadi, F., Parivar, K., et al. (2012). Effect of honey bee venom on Lewis rats with experimental allergic encephalomyelitis, a model for multiple sclerosis. *Iranian Journal of Pharmaceutical Research* 11 (2): 671.

Kent, S.A., Spires-Jones, T.L., and Durrant, C.S. (2020). The physiological roles of tau and Aβ: implications for Alzheimer's disease pathology and therapeutics. *Acta Neuropathologica* 140 (4): 447.

Khalil, W.K., Assaf, N., ElShebiney, S.A., and Salem, N.A. (2015). Neuroprotective effects of bee venom acupuncture therapy against rotenone-induced oxidative stress and apoptosis. *Neurochemistry International* 80: 79–86.

Kim, J.-I., Yang, E.J., Lee, M.S., et al. (2011). Bee venom reduces neuroinflammation in the MPTP-induced model of Parkinson's disease. *International Journal of Neuroscience* 121 (4): 209–217.

Kim, J.-Y., Lee, S.-J., Maeng, Y.-I., et al. (2020). Protective effects of bee venom against endotoxemia-related acute kidney injury in mice. *Biology* 9 (7): 154.

Kim, J.-Y., Lee, W.-R., Kim, K.-H., et al. (2015). Effects of bee venom against Propionibacterium acnes-induced inflammation in human keratinocytes and monocytes. *International Journal of Molecular Medicine* 35 (6): 1651–1656.

Kim, K.H., Kim, M., Lee, J., et al. (2019a). Comparison of the protective effects of bee venom extracts with varying pla2 compositions in a mouse model of Parkinson's disease. *Toxins* 11 (6): 358.

Kim, K.-H., Lee, W.-R., An, H.-J., et al. (2013). Bee venom ameliorates compound 48/80-induced atopic dermatitis-related symptoms. *International Journal of Clinical and Experimental Pathology* 6 (12): 2896.

Kim, Y., Lee, Y.-W., Kim, H., and Chung, D.K. (2019b). Bee venom alleviates atopic dermatitis symptoms through the upregulation of decay-accelerating factor (DAF/CD55). *Toxins* 11 (5): 239.

Krylov, V., Agafonov, A., Krivtsov, N., et al. (2007). *Theory and Agents of Apitherapy*. Moscow: Comme it faut Publishers.

Lee, S.-H., Choi, S.-M., and Yang, E.J. (2014a). Melittin ameliorates the inflammation of organs in an amyotrophic lateral sclerosis animal model. *Experimental Neurobiology* 23 (1): 86.

Lee, S.-H., Choi, S.-M., and Yang, E.J. (2015). Bee venom acupuncture augments anti-inflammation in the peripheral organs of hSOD1G93A transgenic mice. *Toxins* 7 (8): 2835–2844.

Lee, W.-R., Kim, K.-H., An, H.-J., et al. (2014b). The protective effects of Melittin on Propionibacterium acnes–induced inflammatory responses in vitro and in vivo. *Journal of Investigative Dermatology* 134 (7): 1922–1930.

Lee, Y.M., Cho, S.-N., Son, E., et al. (2020). Apamin from bee venom suppresses inflammation in a murine model of gouty arthritis. *Journal of Ethnopharmacology* 257: 112860.

Lim, H.N., Baek, S.B., and Jung, H.J. (2019). Bee venom and its peptide component melittin suppress growth and migration of Melanoma Cells via inhibition of PI3K/AKT/mTOR and MAPK pathways. *Molecules* 24 (5): 929.

Mohamed, W.A., Abd-Elhakim, Y.M., and Ismail, S.A. (2019). Involvement of the anti-inflammatory, anti-apoptotic, and anti-secretory activity of bee venom in its therapeutic effects on acetylsalicylic acid-induced gastric ulceration in rats. *Toxicology* 419: 11–23.

Mohammadi-Rad, M., Ghasemi, N., and Aliomrani, M. (2019). Evaluation of apamin effects on myelination process in C57BL/6 mice model of multiple sclerosis. *Research in Pharmaceutical Sciences* 14 (5): 424.

Mohammed, Z.I. and Hassan, A.J. (2019a). Effect of Bee venom on gene expression of HSP70 and IL-1β in male rats induced by arthritis in comparison with Prednisolone drug. *Research Journal of Pharmacy and Technology* 12 (11): 5269–5274.

Mohammed, Z.I. and Hassan, A.J. (2019b). Effect of bee venom on some blood and biochemical parameters in formaldehyde induced arthritis male rats in comparison with prednisolone drug. *Journal of Physics: Conference Series* 1234: 012066.

Panda, S.S. and Jhanji, N. (2020). Natural products as potential anti-Alzheimer agents. *Current Medicinal Chemistry* 27 (31): 5887–5917.

Park, J.H., Jeong, Y.-J., Park, K.-K., et al. (2010). Melittin suppresses PMA-induced tumor cell invasion by inhibiting NF-κB and AP-1-dependent MMP-9 expression. *Molecules and Cells* 29 (2): 209–215.

Park, M.H., Choi, M.S., Kwak, D.H., et al. (2011). Anti-cancer effect of bee venom in prostate cancer cells through activation of caspase pathway via inactivation of NF-κB. *The Prostate* 71 (8): 801–812.

Raza, C. and Anjum, R. (2019). Parkinson's disease: mechanisms, translational models and management strategies. *Life Sciences* 226: 77–90.

Shin, D., Choi, W., and Bae, H. (2018). Bee venom phospholipase A2 alleviate house dust mite-induced atopic dermatitis-like skin lesions by the CD206 mannose receptor. *Toxins* 10 (4): 146.

Son, D.J., Lee, J.W., Lee, Y.H., et al. (2007). Therapeutic application of anti-arthritis, pain-releasing, and anti-cancer effects of bee venom and its constituent compounds. *Pharmacology & Therapeutics* 115 (2): 246–270.

Spel, L. and Martinon, F. (2020). Inflammasomes contributing to inflammation in arthritis. *Immunological Reviews* 294 (1): 48–62.

Stefanovic, N., Flohr, C., and Irvine, A.D. (2020). The exposome in atopic dermatitis. *Allergy* 75 (1): 63–74.

Tanner, C.M., Kamel, F., Ross, G.W., et al. (2011). Rotenone, paraquat, and Parkinson's disease. *Environmental Health Perspectives* 119 (6): 866–872.

Tekeoğlu, İ., Akdoğan, M., and Çelik, İ. (2020). Investigation of anti-inflammatory effects of bee venom in experimentally induced adjuvant arthritis. *Reumatologia* 58 (5): 265–271.

Wang, C., Chen, T., Zhang, N., et al. (2009). Melittin, a major component of bee venom, sensitizes human hepatocellular carcinoma cells to tumor necrosis factor-related apoptosis-inducing ligand (TRAIL)-induced apoptosis by activating CaMKII-TAK1-JNK/p38 and inhibiting IκBα kinase-NFκB. *Journal of Biological Chemistry* 284 (6): 3804–3813.

Wehbe, R., Frangieh, J., Rima, M., et al. (2019). Bee venom: overview of main compounds and bioactivities for therapeutic interests. *Molecules* 24 (16): 2997.

Weidinger, S., Beck, L.A., Bieber, T., et al. (2018). Atopic dermatitis. *Nature Reviews Disease Primers* 4 (1): 1.

Wesselius, T., Heersema, D., Mostert, J., et al. (2005). A randomized crossover study of bee sting therapy for multiple sclerosis. *Neurology* 65 (11): 1764–1768.

Yang, E.J. and Choi, S.-M. (2013). α-Synuclein modification in an ALS animal model. *Evidence-Based Complementary and Alternative Medicine* 2013: 1–7.

Yang, E.J., Jiang, J.H., Lee, S.M., et al. (2010). Bee venom attenuates neuroinflammatory events and extends survival in amyotrophic lateral sclerosis models. *Journal of Neuroinflammation* 7 (1): 1–12.

Yang, E.J., Kim, S.H., Yang, S.C., et al. (2011). Melittin restores proteasome function in an animal model of ALS. *Journal of Neuroinflammation* 8 (1): 69.

Ye, M., Chung, H.-S., Lee, C., et al. (2016). Neuroprotective effects of bee venom phospholipase A2 in the 3xTg AD mouse model of Alzheimer's disease. *Journal of Neuroinflammation* 13 (1): 1–12.

Zhang, R., Gascon, R., Miller, R.G., et al. (2005). Evidence for systemic immune system alterations in sporadic amyotrophic lateral sclerosis (sALS). *Journal of Neuroimmunology* 159 (1–2): 215–224.

Zhang, R., Gascon, R., Miller, R.G., et al. (2006). MCP-1 chemokine receptor CCR2 is decreased on circulating monocytes in sporadic amyotrophic lateral sclerosis (sALS). *Journal of Neuroimmunology* 179 (1–2): 87–93.

# 23

# Medicinal Properties of Stingless Bee Honey

*Mahendran Sekar[1,2,*], Ahmad Yasser Hamdi Nor Azlan[3], Nur Najihah Izzati Mat Rani[3], and Siew Hua Gan[1]*

[1] *School of Pharmacy, Monash University Malaysia, Bandar Sunway, Subang Jaya 47500, Selangor, Malaysia*
[2] *Department of Pharmaceutical Chemistry, Faculty of Pharmacy and Health Sciences, Royal College of Medicine Perak, Universiti Kuala Lumpur, Ipoh 30450, Perak, Malaysia*
[3] *Faculty of Pharmacy and Health Sciences, Royal College of Medicine Perak, Universiti Kuala Lumpur, Ipoh 30450, Perak, Malaysia*
* *Corresponding author*

## Introduction

Since ancient times, honey has been an important natural food source because of its nutritional and medicinal properties. Honey is produced from plant exudate liquids, which have been gathered, modified, and stored by bees (Jalil et al. 2017). Honey can be produced by two bees, namely, European bees (also known as Western bees) and stingless bees. It is estimated that stingless bees (scientifically known as *Meliponini* spp.) have approximately 500 species on earth, which makes them the largest group of eusocial bees (Rosli et al. 2020). Forty species of bees are reported to have potential as honey producers (Al-Hatamleh et al. 2020). Al-Hatamleh et al. (2020) also stated that these species are distributed in the tropical and subtropical regions, with approximately 391 species in the neotropical region of South America followed by in the Indo-Malayan region (60), paleotropical region of Africa (50), and Australasia region of Australia (10). Stingless bees can be classified into two genera, namely, *Melipona* and *Trigona*; both have their own roles in the process of pollination. (Al-Hatamleh et al. 2020; Jalil et al. 2017). According to research done by the Malaysian Agricultural Research and Development Institute, the specialty of stingless bees is the ability to pollinate small-sized flowers because of their diminutive size, which cannot be achieved by relatively large honeybees. Also, stingless bees are not choosy in building their colony hives (The Sun Daily 2021). As a result, it is easier to build an artificial hive to manipulate the colony and increase honey production. Stingless bee honey (SBH) is amber brown and sour in nature and contains various phytochemical constituent that have a variety of medicinal properties. Honey has also been suggested as an effective immunomodulatory modulator because of its anti-inflammatory characteristics and multiple mechanisms of action in suppression and stimulation of the immune system (Figure 23.1).

Honey contains approximately 200 compounds, such as vitamins, enzymes, amino acids, and minerals, with the major content being water and sugars. Sugars make up approximately 95%–99% of honey's dry matter. Of the sugars in honey, fructose is the most prevalent, making up approximately 32%–38% of its total sugar (Rao et al. 2016). In addition, various vitamins, including ascorbic acid (vitamin C), thiamine (vitamin $B_1$), riboflavin (vitamin $B_2$), nicotinic acid (vitamin $B_3$), pantothenic acid (vitamin $B_5$), pyridoxine (vitamin $B_6$), biotin (vitamin $B_8$), folic acid (vitamin $B_9$), and cyanocobalamin (vitamin $B_{12}$), are present (Ciulu et al. 2011; Rao et al. 2016). Other phytochemicals available in honey are such as phenolic acids (caffeic, ellagic, ferulic acids), flavonoids (apigenin, chrysin, galangin, hesperetin, kaempferol, pinocembrin, and quercetin), and antioxidants (tocopherols, ascorbic acid, superoxide dismutase, catalase, and reduced glutathione) (Rao et al. 2016). SBH has been reported by several studies to possess countless medicinal properties, which are antibacterial activity, antioxidant, wound-healing activity, anti-inflammatory, antidiabetic, and anticancer, as well as effects on ocular disease, fertility disease, and hypertension (Figure 23.2) (Al-Hatamleh et al. 2020; Amin et al. 2018; Rosli et al. 2020). This chapter summarizes the medicinal properties of SBH that have been reported by researchers (in vitro and in vivo). Furthermore, the limitations and challenges affecting the phytochemical constituents and physicochemical characteristics of SBH as remedies are also highlighted in this chapter.

*Honey: Composition and Health Benefits*, First Edition. Edited by Md. Ibrahim Khalil, Gan Siew Hua, and Bey Hing Goh.

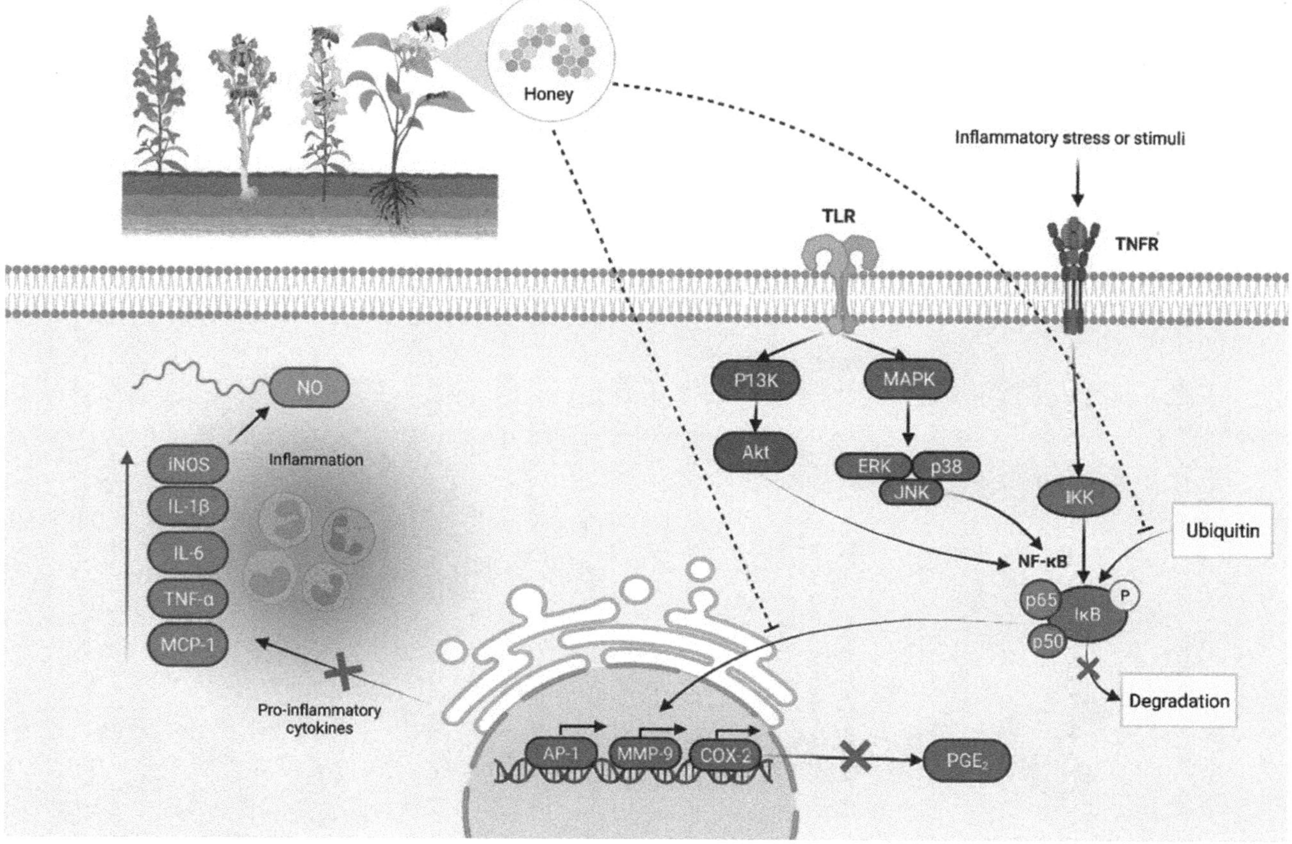


**Figure 23.1** The role of stingless bee honey in inflammation. Honey's immunomodulatory properties are thought to be achieved via reducing the generation of pro-inflammatory cytokines and downregulating inflammatory transcription factors. It can also exert anti-inflammatory activities by inducing the synthesis of inflammatory mediators, including prostaglandin E2 (PGE2) and cyclooxygenase-2 (COX-2). Akt, Ak strain transforming; AP-1, activated protein-1; ERK, extracellular signal-regulated kinase; IκB, inhibitor of κB; IKK, IκB kinase; IL, interleukin; iNOS, inducible nitric oxide synthase; JNK, c-Jun N-terminal kinase; MAPK, mitogen-activated protein kinase; MCP-1, monocyte chemoattractant protein-1; MMP-9, matrix metalloproteinase 9; NO, nitric oxide; P13K, phosphatidylinositol 3-kinase; TLR, Toll-like receptor; TNF-α, tumor necrosis factor-α; TNFR, tumor necrosis factor receptor.

## Medicinal Properties of Stingless Bee Honey

### Antioxidant Properties

Recently, there has been an escalating in interest in food rich in antioxidants such as honey, significantly enhancing interest in acquiring a healthy diet that can prevent chronic degenerative diseases in relation to oxidative stress (Moo-Huchin et al. 2016). Jalil et al. (2017) stated that SBH from several species from several countries – *Meliponini* tribe (Australia), *Tetragonula carbonaria* (Europe), and *Melipona fasciculate* (South America), and *Melipona* spp. (Malaysia) – possess high flavonoids and phenolic acids (e.g. protocatechuic acid and 4-hydroxyphenylacetic acid), which makes them possess antioxidants effects. The antioxidant level of SBH (*Trigona carbonaria* in Australia) has been determined previously via total antioxidant activity (TAA) (via 2,2-azinobis-[3-ethylbenzothiazoline-6-sulfonic acid] [ABTS] radical cation [$ABTS^+$] assays) and random scavenging activity (RSA) (via 1,1-diphenyl-2-picrylhydrazyl [DPPH] assay). The result revealed that the TAA values of honey produced by *T. carbonaria* was 233.96 μM Trolox equivalents, which was higher than Czech honeys in which TAA varied between 121.56 and 125.89 μM Trolox equivalents with antioxidant contents in floral honey and honeydew honey, respectively. The average RSA of honey produced by *T. carbonaria* is equivalent to 48% ascorbic acid, which is similar to that of Spanish honeys (a mix of floral and honeydew origins) (Oddo et al. 2008). A study conducted in Brazil determined the SBH samples produced by *Melipona (Michmelia) seminigra merrillae* collected in

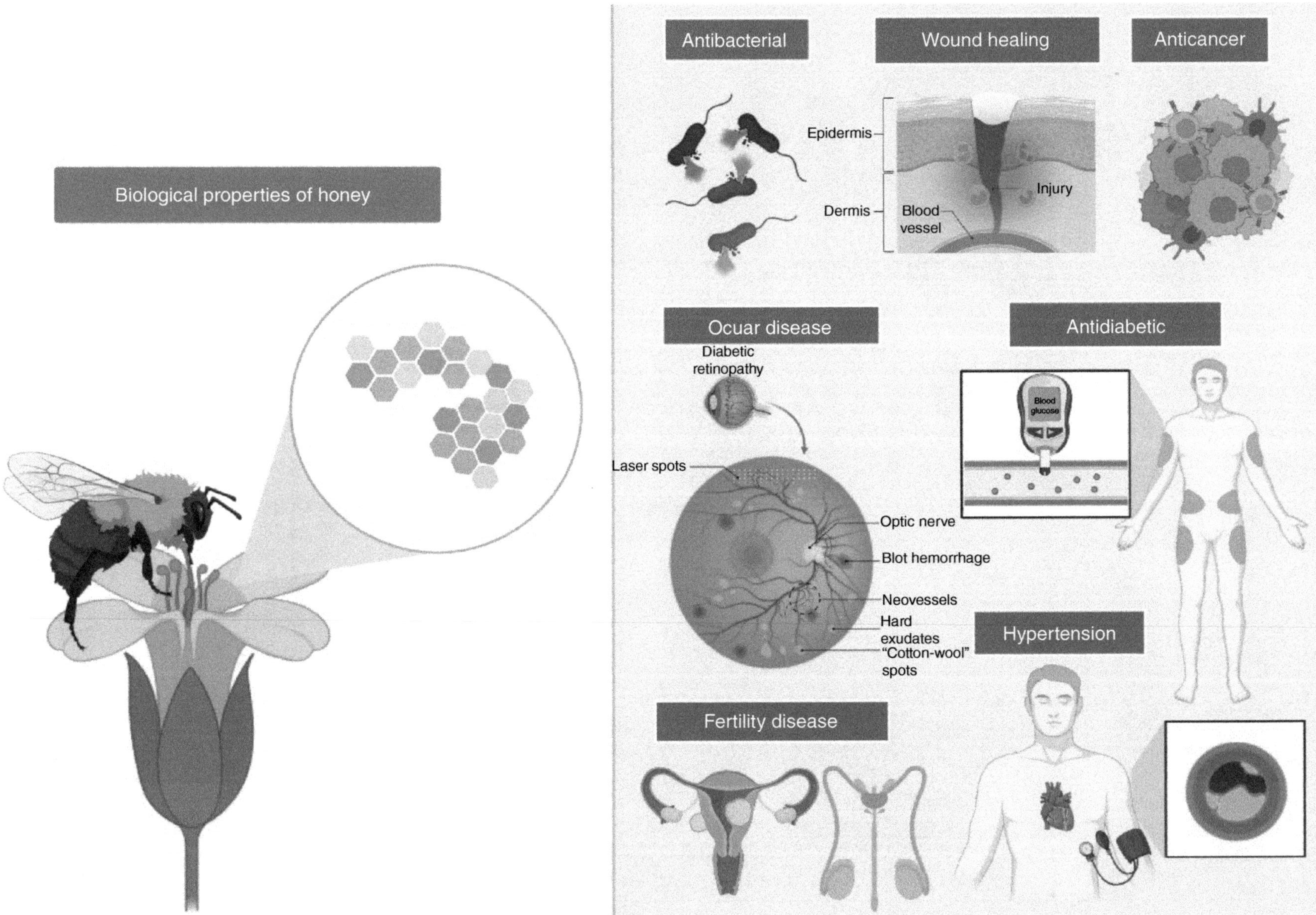


**Figure 23.2** Medicinal properties of stingless bee honey.

seven counties (distributed in the central and southern region of Amazonas state in Brazil). The antioxidant level were studied via $ABTS^{\cdot+}$ cation radical–scavenging activity, and the results revealed that the SBH collected from Maués, Lábrea, and Rio Preto da Eva showed the highest antioxidant capacity (210 ± 0.25 to 337 ± 3.17 mg/mL) because of a higher level of total phenolic content detected (43 ± 0.02 to 66 ± 0.05 mg GAE/g) (da Silva et al. 2013). The study also reported that there is a correlation between compounds with the capacity to scavenge free radicals such as phenolic compounds to enhance antioxidant activity.

In Malaysia alone, a study reported that the SBH produced by *Heterotrigona itama* collected in Jasin exhibited the highest levels of antioxidants activity via DPPH assay (97.30% ± 0.84%) compared with *H. itama* collected in Peserai (96.40% ± 0.08%) (Bakar et al. 2017).

Another study conducted by Tuksitha et al. (2018) reported that the SBH from Malaysia produced by *Geniotrigona thoracica* contained high amount of phenolics and flavonoid compounds (99.04 ± 5.14 mg/mL and 17.67 ± 0.75 mg/mL, respectively), meaning it possess greater antioxidant species than the SBH produced by *H. itama* and *Heterotrigona erythrogastra*. It is reported that the amount of antioxidant activity in SBH depends on several factors such as floral availability; origins of the botanical, season, and environment; method of processing; and the presence of pigments (e.g. flavonoids and phenolic compounds) (Bakar et al. 2017), which makes the amount of antioxidant activity detected different in SBH produced by different species in different countries. Moreover, several studies also reported that the antioxidant activity of SBH (produced from *Melipona beecheii* and *Trigona* spp.) contains higher level of antioxidants than Western bee honey (*Apis mellifera*) because of the presence of higher phenolic and flavonoid compounds (Alvarez-Suarez et al. 2018; Avila et al. 2018; Kek et al. 2014; Martinello and Mutinelli 2021).

### Antibacterial Activity

The continuously rising cases of antimicrobial resistance have become a global threat to our ability to treat infections. Hence, it is an urgent matter to develop new antibiotics that are more potent in inhibiting or killing bacteria. According to Yang et al. (2017), gram-positive bacteria such as *Staphylococcus aureus* and gram-negative bacteria such as *Pseudomonas aeruginosa* recently have developed an increase in resistance toward conventional antibiotics. Hence, currently, many researchers are developing antibiotics using natural resources such as plants, bacteria, and fungi. It has been reported that SBH has activity against several pathogens, including *Bacillus cereus, Candida albicans, Citrobacter freundii, Enterobacter cloacae, Escherichia coli, Klebsiella pneumoniae, Listeria monocytogenes, P. aeruginosa, Salmonella enterica, Serratia marcescens, Shigella sonnei, S. aureus, Streptococcus pyogenes*, and *Yersinia enterocolitica* (Domingos et al. 2021). Hence, studies on SBH as an antimicrobial agent have been investigated by several researchers to rule out biological activities involved in the inhibition or death of microorganisms. Additionally, the effectiveness of SBH as an antibacterial agent has been reported (Figure 23.3) to depend on the season, flower pollens, environment, and surroundings as well as species (Lim et al. 2019; Majid et al. 2019).

In 2013, de Queiroz et al. (2013) studied and compared the antibacterial activity of SBH of *Melipona compressipes manosensis* produced during dry and rainy seasons. The results revealed that SBH produced during the dry season is more effective because it can inhibit both gram-positive (*S. aureus*) and gram-negative (*E. coli, Shigella sonnei, Proteus vulgaris*, and *Klebsiella* spp.) organisms at different dilution rates. On the other hand, SBH collected during the rainy season had lower activity in which only two microorganisms could be inhibited in an undiluted condition. These results clearly indicate the influence of seasonality over the antibacterial activity of the honey obtained from *M. compressipes manaosensis*; the study reported that the honey produced during dry season presented with higher antimicrobial activity because of the phytochemical components from the nectar associated with bacterial growth inhibition (de Queiroz et al. 2013).

A study reported that the antibacterial activity of honey produced by *G. thoracica* was affected by the usage of different flower pollen in honey production. The studied classified honey produced by *G. thoracica* from multifloral (GTM) and unifloral (GTU) sources. The multifloral flowers in GTM were wildflowers from the nearby mountainous forest (5°44′47.0″N 101°52′01.4″E). In GTU, the unifloral honey was identified from the flowers of *Melastoma malabathricum L.* (locally known as Senduduk) by pollen analysis (5°44′45.8″N 101°52′01.7″E). The results showed that the antibacterial activities of both GTM and GTU were concentration dependent in which GTM was the most susceptible against all of the bacteria tested. The result demonstrated that *S. pyogenes* showed the highest sensitivity toward GTM compared with GTU in which the zone of inhibitions (ZOIs) were 24.3 ± 0.6 mm and 21.3 ± 1.5 mm, respectively, followed by methicillin-resistant *S. aureus* (MRSA) (ZOI, 21.3 ± 1.5 mm and 18.0 ± 1.0 mm) and *S. aureus* (ZOI, 15.7 ± 0.6 mm and 14.7 ± 0.6 mm).

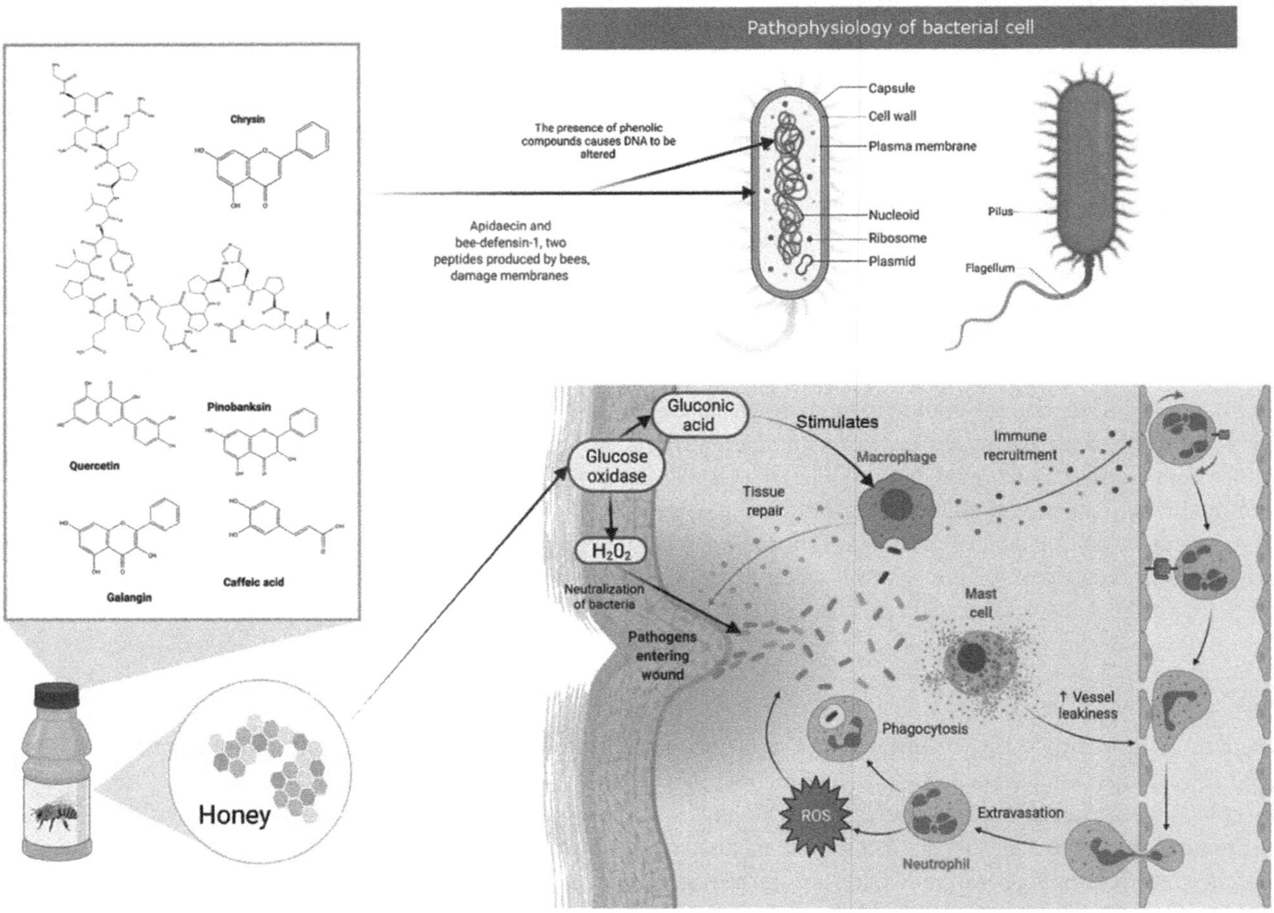


**Figure 23.3** Antibacterial activity of stingless bee honey. Numerous phenolic compounds and flavonoids, including apidaecin, bee-defensin 1, and chrysin, may act directly on bacteria, breaking their membranes and altering their DNA, ultimately resulting in bacterial death. Honey is a potent immunomodulator capable of enhancing the immune system by releasing enzymes such as gluconic acid and hydrogen peroxide that can trigger many cascades to promote an immunologic response to bacterial infection. ROS, reactive oxygen species.

Meanwhile, when tested against gram-negative bacteria, *P. aeruginosa* was the most susceptible toward GTM compared with GTU in which the ZOIs were 18.0 ± 1.0 mm and 16.7 ± 0.6 mm, respectively, followed by *E. coli* (ZOI, 12.3 ± 1.5 mm and 10.7 ± 0.6 mm) and *K. pneumoniae* (ZOI, 9.7 ± 1.1 mm and 14.3 ± 0.6 mm). The results obtained from this study clearly indicate that the antibacterial efficacy of SBH is more effective if the honey is produced from multifloral pollens (Omar et al. 2019).

As mentioned earlier, the surroundings, environment, and species also affect the efficacy of SBH as an antibacterial agent. A previous study conducted in Brazil investigated the antibacterial activity of SBH collected from Rio Branco-Acre (*Melipona eburnean*, *Melipona grandis*, and *Melipona seminigra*) and Xapuri-Acre (*Melipona flavolineata*). The results revealed that all types of SBH are more sensitive against gram-positive bacteria (*Enterococcus faecalis*, American Type Culture Collection [ATCC] 29212, *S. aureus* ATCC 25923, and *Streptococcus pneumoniae* ATCC 11733) in which honey produced by *M. flavolineata* and *M. grandis* able to inhibit all of the gram-positive bacteria with a minimum inhibitory concentration (MIC) of 3.12% (v/v). Only honey produced by *M. flavolineata* was able to inhibit gram-negative bacteria (*K. pneumoniae* ATCC 4352, *P. aeruginosa* ATCC 15442) with a MIC of 1.56% (v/v). Nevertheless, none of the honey was able to inhibit *E. coli* ATCC 25922 (Domingos et al. 2021).

Another study conducted by Rosli et al. (2020) also determined the antibacterial activity of different concentrations (50%, 25%, and 12.5% v/v) of SBH produced from eight different species (*H. itama*, *H. erythrogastra*, *Tetrigona apicalis*, *Lepidotrigona terminata*, *Tetrigona melanoleuca*, *Tetrigona bingami*, *G. thoracica*, and *Homotrigona fimbriata*) tested against five different bacteria (*S. aureus, Alcaligenes faecalis, Serratia marcescens, Bacillus subtilis*, and *E. coli*). Based on the results, the ability of SBH to inhibit the bacteria was concentration dependent. At the highest concentration (50% v/v), all honey were effective against both gram-positive and gram-negative bacteria except for honey produced by *H. erythrogastra*, which only exhibits ZOI when against *B. subtilis* (9 ± 1 mm). At a concentration of 25%, honey produced from *T. apicalis, Tetrigona binghami, G. thoracica*, and *H. erythrogastra* did not exhibit any inhibitory activity toward the five bacteria. Meanwhile, at 12.5% concentration, only three honey samples showed an inhibitory effect against one or more bacteria at the range of 10–16 mm in diameter of the inhibition zone. Overall, it is reported that *H. fimbriata* had the highest antimicrobial activity compared with other honey samples with inhibitory zone diameters up to 28 mm in size, and *H. erythrogastra* showed the lowest antibacterial effect (Rosli et al. 2020).

## Wound-Healing Activity

Coagulation, inflammation, tissue formation, and tissue remodeling are the steps involved in the wound-healing process. The wounds are differentiated into acute and chronic depending on the severity and duration of the healing process. Usually, wounds that are not healed for more than 90 days are classified as chronic (Guo and DiPietro 2010). Before, passive wound dressings were used in the management of wound care with the usage of non-occlusive gauzes and tulle dressings for covering the wound, removing the excess of exudates, preventing infection, and restoring function underneath (Dhivya et al. 2015; Moura et al. 2013). However, nowadays, a moist dressing is known to facilitate faster wound healing because it enhances the proliferation and migration of fibroblasts and keratinocytes as well as collagen synthesis, reducing the formation of scar. Also, a wound dressing should provide thermal insulation and gaseous exchange and facilitate drainage as well as debris removal for promoting tissue reconstruction processes (Moura et al. 2013). Oddo et al. (2008) reported that SBH has a high moisture content, which is relatively higher than that in Western bee honey. Naturally, honey has good moisturizing properties because of the presence of hydroxyl groups. Plus, the other basic components of honey such as sugars, proteins, and lactic acid also can act as moisturizers (Ediriweera and Premarathna 2012). Furthermore, several researchers reported that the addition of glycerin, propylene glycol, and sorbitol as solvents in cosmetic products can enhanced the moisturizing properties of honey (Bikowski 2001; Draelos 2010).

Previously, a study by Nordin et al. (2018) to evaluate the effect of freeze-dried SBH (produced by *Trigona* spp.) provided evidence that the honey has a potential wound-healing effect. The study reported on the cell viability study, cell morphological analysis, and cell-cycle progression of human dermal fibroblasts under exposure to selected concentrations of honey. Through 24-hour cell viability assays, the increased dose of SBH (up to 0.024 μg/mL) improved the viability of fibroblast cells before viability starts to decline. When the cells were reintroduced at 12.5 μg/mL, the cell viability starts to ameliorate up to 200 μg/mL before a decline in cell viability is seen. The third improvement of viability was reobserved at 1600–6400 μg/mL. Hence, the study was further tested on cell morphology in which the dermal fibroblast cell was observed at selected concentrations (0.024, 200, and 6400 μg/mL). The cells exhibited a normal, multipolar, and elongated shapes just before being exposed to honey, which indicates it is nontoxic. Additionally, the study reported that the selected dose of SBH also improved the proliferation of human dermal fibroblasts and did not alter the normal cell cycle (Nordin et al. 2018). The positive effect shown by SBH toward cell viability, morphology, and proliferation of human dermal fibroblasts suggested the ability of SBH to be developed as a therapy for wound healing.

Another study reported on the effect of SBH (produced by *Trigona* spp.) on preventing the formation of keloid scars, which are caused by an overproduction of the epithelial-to-mesenchymal transition (EMT) by transforming growth factor-β (TGF-β) in keratinocytes. The TGF-β–induced keratinocytes revealed a faster wound closure time and individual migration velocity in the in vitro scratch assays compared with the untreated control group. Furthermore, the circularity of TGF-β–induced keratinocytes was reduced and displayed classic EMT protein expression. However, keratinocytes treated with 0.0015% (v/v) of SBH reversed these effects, resulting in slower initial wound closure and individual migration velocity, which later sped up, resulting in greater wound closure at the final time point. Moreover, keratinocytes treated with SBH also led to greater directional migration compared with the control, showed reduced circularity in keratinocytes, and displayed a partial EMT protein expression, proving that SBH is a potential therapy in preventing keloid scar formation by attenuating TGF-β–induced EMT (Figure 23.4) (Nordin et al. 2020).

Another study reported on the development of the nanofibrous composite membrane using SBH and curcumin for wound-recuperating applications (via its antibacterial and antioxidant properties). The results revealed that the antioxidant activity tested in SBH and curcumin alone was lower (scavenging activity, 49%–56% and 57%–99%, respectively)

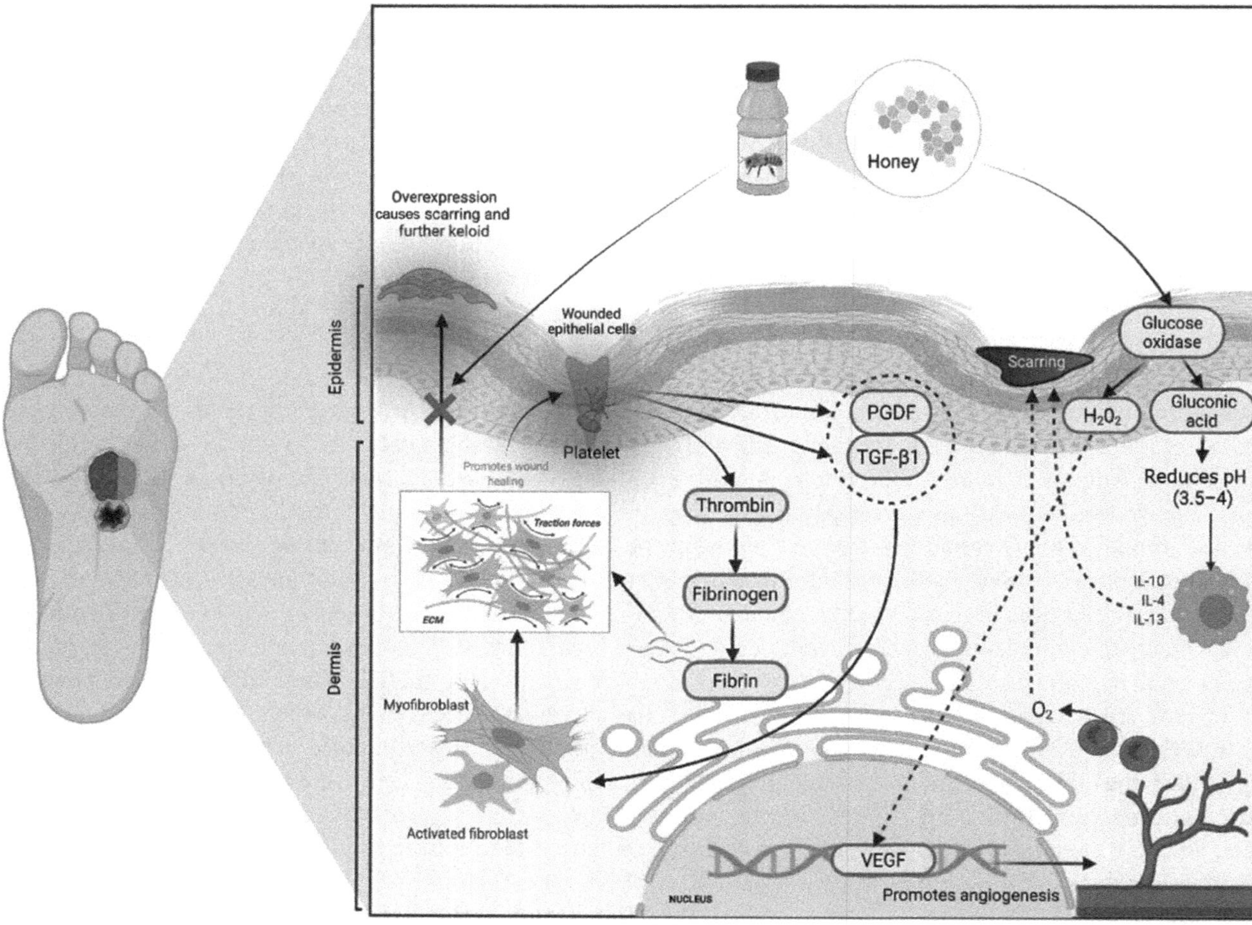


**Figure 23.4** Mechanism of action of stingless bee honey after an inflammatory response in restoring tissue homeostasis. This shows honey's mechanism of action in restoring tissue homeostasis after an inflammatory response. Honey's production of hydrogen peroxide ($H_2O_2$) stimulates the growth of fibroblasts and epithelial cells to repair inflammatory damage, and gluconic acid lowers the pH, making it suitable for tissue repair, elevation in oxygen release, reduction in protease activity, and stimulation of fibroblasts. Another mechanism involves platelets, which aid in the stimulation of fibroblasts for efficient wound healing by secreting transforming β1 and platelet-derived growth factors (PDGFs). ECM, extracellular matrix; IL, interleukin; VEGF, vascular endothelial growth factor.

compared with their combination (scavenging activity, 78%–100%) via DPPH assays. The findings based on the in vitro scratch assays were tested on L929 mouse fibroblast cells exposed to different concentration of nanofibrous composite membranes (10, 20, 30, 40, and 50 μL). The result obtained was that all the nanofibrous composites membranes were able to close the wound within 18 hours according to the positive control (cells treated with media only), which indicates that these nanocomposites were not toxic toward the fibroblasts cells and were able to close the wounds under normal conditions. Hence, these nanocomposite membranes containing SBH and curcumin have the potential for healing wounds and warrant further studies in specific wound models (Samraj et al. 2021).

## Ocular Therapeutic Properties

Vit et al. (1998) reported that SBH is a natural remedy for treatment of the eyes, and its marketed value is 20 times greater than that of *A. mellifera* honey. The effectiveness of SBH for eye therapy has been shown to be less than the effectiveness of other therapies. Traditionally, the Mayans used SBH (produced by *Hymenoptera* spp., Apidae spp., and *Meliponinae* spp.) as an eye drop to cure cataracts (Vit and Jacob 2008). One study conducted a survey using an open-ended structured questionnaire to investigate the uses of SBH produced by *L. arcifera* in the ethnomedicine of the Napali community in West Bengal, India. SBH was used by the Napali community to treat conjunctivitis; they usually applied two drops of SBH around the eyes two times daily until the condition was cured. Moreover, in treating eye infection, they insert pure filtered SBH into the eyes after washing the hands and eyes with boiled clean water (Biswa et al. 2017). In the in vitro studies conducted by Vit and Jacob (2008), the anticataract properties of flavonoid present in SBH were investigated. Briefly, cataracts were induced by incubating ovine lenses in 45% hypotonic HEPES buffered solution for 24 hours to test the anticataract action of 20 synthetic flavonoids at a concentration of $10^{-5}$ M. The results revealed that luteolin tetramethyl ether, luteolin 4'-glucoside, luteo in 3'-7-diglucoside, and orientin significantly inhibited cataracts induced in ovine lenses incubated in 45% hypotonic HBS within 24 hours, confirming that SBH has anticataract activity.

SBH was also effective in treating bacterial conjunctivitis in Hartley guinea pig as reported by Ilechie et al. (2012). Bacterial conjunctivitis was induced by *S. aureus* or *P. aeruginosa*. The results revealed that using of one drop (70 μL) of crude SBH (produced from *Meliponula* spp.) twice daily shortened inflammatory signs, the duration of infection, and the time for complete resolution. Moreover, the study also stated that the effectiveness of SBH as eye therapy was comparable to that of gentamicin (Ilechie et al. 2012). Collectively, SBH might be a potential natural remedy to be developed to treat people with ocular diseases.

## Antidiabetic Properties

The International Diabetes Federation reported that that 8.3% (382 million) of adults worldwide had diabetes in 2013 (Dagenais et al. 2016). In 2016, the World Health Organization estimated that around 422 million adults lived with diabetes in 2014 globally compared with the year 1980, when only 108 million adults lived with diabetes. This demonstrates that the global prevalence of diabetes has increased from 4.7% in 1980 to 8.5% in 2014 (World Health Organization 2016). Because of the increased number of patients with diabetes, this situation has triggered researchers to search for a potential natural resource that has antidiabetic effects as an alternative treatment. Honey (not only produced by stingless bee species) has been well known to reduce fasting blood glucose (FBG) levels, increase fasting C-peptide levels, and lower the glycemic index (GI) (Bobiş et al. 2018). A previous study stated that carbohydrates with high GIs (sucrose with a GI of 68) cause higher increases in blood sugar levels than those with lower GIs (fructose, with a G1 of 19) (Elliott et al. 2002). Fructose has been reported by Fasanmade and Alabi (2008) to be the major carbohydrate found in honey; hence, the intake of honey by patients with diabetes does not cause an increase in sugar level. Amin et al. (2018) also stated that α-amylase and α-glucosidase are the two main enzymes involved in elevation of blood glucose, and several studies also have been reported that the mechanism of action of the antidiabetic properties of SBH was via inhibition of α-glucosidase and α-amylase (Rahmawati et al. 2019; Sahlan et al. 2020). The inhibition of these two enzymes indicates a good antidiabetic effect because it helps to reduce blood glucose levels (Amin et al. 2018).

Another study was conducted to investigate the antidiabetic properties of SBH collected from different areas in peninsular Malaysia, mainly from Kulim, Kedah (honey 1); Tanjung Malim, Perak (honey 2);, and Kuala Selangor, Selangor (honey 3), via determination of an α-amylase inhibition assay. The results showed that all SBH samples exhibited antidiabetic activity in which the highest inhibition of 50% cell viability ($IC_{50}$) value was recorded by honey 1 (21.84 ± 5.88 μg/mL) followed by honey 2 (16.30 ± 2.98 μg/mL) and honey 3 (15.80 ± 1.02 μg/mL). Also, the $IC_{50}$ values of all the honey types was higher compared with the positive control, acarbose (14.56 ± 0.76 μg/mL). The study also reported that the phytochemical constituent in SBH that is responsible for the antidiabetic properties was phenolic acid because of the positive correlation observed between gallic acid content and $IC_{50}$ value. Moreover, the study also stated that the presence of antioxidants in SBH also contributed to antidiabetic activity by inducing the release of insulin and thus lowering blood glucose level (Zakaria et al. 2020).

Aziz et al. (2017) reported on the ability of SBH produced by *G. thoracica* in an in vivo rat diabetic model. The study was conducted for 28 days. The result obtained was that the diabetic rats that received 2 g/kg of SBH did not increase their fluid or food intake; the result obtained was similar to that of diabetic rats treated with glibenclamide (positive control). The FBG level shown by the diabetic group treated with SBH was in a normal range, probably because of the presence of L-phenylalanine (an agonist that potentiates glucose-dependent insulin secretion in the pancreatic β-cells) detected via liquid chromatography mass spectrophotometry (LC-MS). Also, the study reported on an improvement in lipid profiles in diabetic rats receiving 2 g/kg of SBH as a treatment, indicating that SBH can help to reduce risk for atherosclerosis development via an increase in of high-density lipoprotein cholesterol level (Aziz et al. 2017). Meanwhile, the diabetic groups treated with 1 g/kg did not show any changes on FBG level but decreased in body weight. The results obtained by Aziz et al. (2017) were similar to those of the study conducted by Sahlan et al. (2020) in which 1 g/kg of SBH (produced by *Tetragonula biroi*) given to male streptozotocin-induced diabetic Sprague Dawley rats did not cause any changes in FBG; however, the body weight of the rats decreased after 35 days of the therapy. Sahlan et al. (2020) also stated that SBH has the ability to break down fats and proteins and turn them into the energy, which makes the diabetic rats treated with 1 g/kg lose weight.

On the contrary, a study reported that the SBH tested did not possess any antidiabetic properties (Rashid et al. 2019). The researchers allowed consumption of SBH in an experimental group of 30 participants (18 years old and older and confirmed to be diagnosed with impaired fasting glucose [fasting plasma glucose values in range of 6.1–6.9 mmol/L]) and compared them with another 30 participants (similar inclusion criteria) as the control group. All participants in the experimental group received 30 g/day of SBH; the participants in the control group did not receive any honey. At the end of the intervention period, the researchers observed that there was no significant difference between the groups in FBG (Rashid et al. 2019), indicating that a bigger sample size may be needed to give stronger power to the study.

### Anticancer Activity

Honey has been studied by researchers throughout the world and has proven to be a chemopreventive agent. "Chemopreventive" can be defined as a natural or synthetic agent that has the potential to decrease the risk of cancer development (Kuppusamy et al. 2014). Amin et al. (2018) stated that the anticancer effects in honey are due to the presence of polyphenols such as quercetins, apigenin, chrysin, and luteolin, and their mechanism of action is via inhibition of cell proliferation, modulation of cancer signaling pathways, and induction of tumor cell apoptosis (Figure 23.5). Plus, SBH has been proven to have strong antioxidant activities; hence, it could be a potential chemopreventive agent from natural resources (Amin et al. 2018; Jalil et al. 2017). Several species of SBH have been reported to have anticancer properties, including *H. fimbriata, H. itama, Hetrotrigona bakeri, Tetragonula sarawakensis, Tetragonula testaceitarsis, Tetragonula fuscobalteata, Tetragonula laeviceps, Trigona incisa, T. apicalis, Trigona fusco-balteata*, and *Trigona fuscibasis* (Arung et al. 2021; Kustiawan et al. 2014). A previous study reported on the ability of SBH as a chemopreventive agent against colon cancer induced by azoxymethane in Sprague Dawley rats. The result obtained was that the cancer group treated with SBH managed to reduce the aberrant crypt foci and aberrant crypt levels, indicating the effectiveness of SBH as a chemopreventive agent. The study also stated that SBH contains caffeic acid phenethyl ester (CAPE), which is responsible for suppressing the transcriptional activity of β-catenin and hence can become an agent against colon cancer. Moreover, CAPE is a phenolic antioxidant that has biological and pharmacological functions that include immunoregulation, anti-inflammatory, antiviral, antibacterial, and antitumor activities (Saiful et al. 2016).

In 2019, a study was conducted in vitro to investigate the anticancer properties of different concentrations of SBH produced from *H. itama* against malignant glioma (U-87 MG) cells. The study was assessed via qualitative (morphological analysis and apoptosis analysis using acridine orange/propidium iodide [AO/PI] staining) and quantitative analysis (cytotoxicity assay). The result of cytotoxicity revealed that the SBH produced from *H. itama* spp. can exert cell proliferation inhibitory effect in U-87 MG cells in a time- and dose-dependent manner in which the $IC_{50}$ for U-87 MG cells was obtained at a concentration of 10% (observed at 72 hours after treatment). The results obtained are in line with qualitative analysis in which the cells exposed at 10% concentration started to shrink and appeared rodlike in shape and later to round up and lose adhesion after 72 hours. The highest amount of cell death was observed (detected as shown by the orange to red colored cells via AO/PI staining) in U-87 MG cells treated with 10% honey. Overall, it is confirmed that the SBH produced from *H. itama* can induce apoptosis in malignant glioma cells, which makes it a potential candidate for the development as an anticancer agent (Ahmad et al. 2019). A summary of the medicinal properties of SBH from different species and countries of origin is shown in Table 23.1.

## Limitations and Challenges of Stingless Bee Honey

### Factors Affecting the Phytochemical Constituent in Stingless Bee Honey

Despite having various medicinal properties that may be beneficial in the development of health therapy, there are a lot of problems and challenges that need to be taken into consideration before SBH can be used as a remedy. Many researchers have highlighted that the different amounts of phytochemical constituent content in SBH are affected by ecological factors, floral resources, and species (Braghini et al. 2021). The ecological factors usually affect resource collection (floral) whereby examples of ecological factors are climate, available flora, and competition (Leonhardt 2017). Furthermore, several researchers have stated that plant uptake from the soil and the environment from where the bees gather the nectar in the process of producing honey are the main factors that determine the mineral elements in honey itself (Alqarni et al. 2014; Alves et al. 2013; Maringgal et al. 2021). For instance, Kek et al. (2017) reported that Malaysian raw honey harvested in forests and jungles contains a higher level mineral element than honey collected from farm plantations.

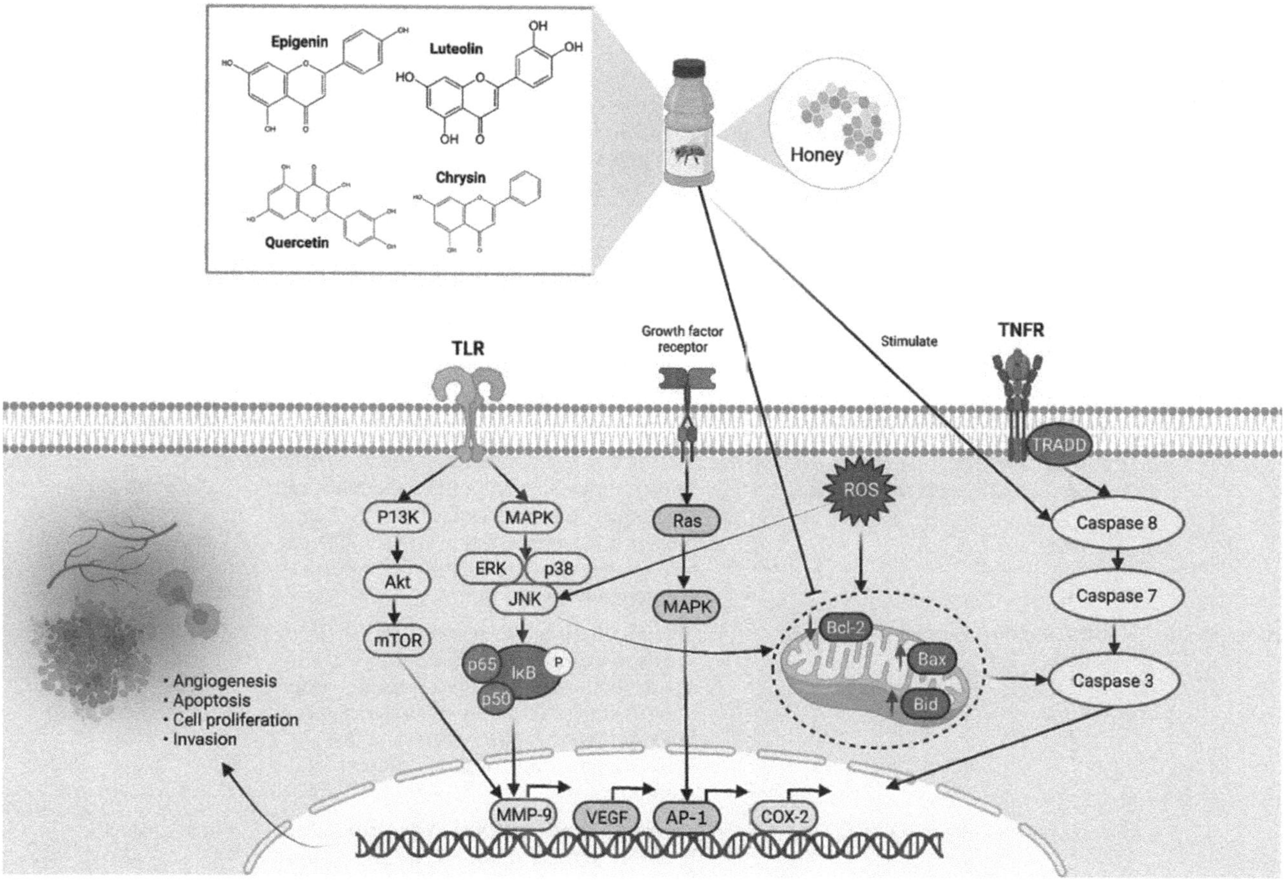


**Figure 23.5** Chemopreventive effect of honey. Honey contains polyphenols, which act on the apoptotic pathway of cancer cells by activating caspase-8; mitogen-activated protein kinase (MAPK), p-38 c-Jun N-terminal kinase (JNK), and activated protein-1 (AP-1); increasing reactive oxygen species (ROS); inhibiting angiogenesis; and halting cell proliferation and invasion. Akt, Ak strain transforming; Bax, Bcl-2-associated X protein; Bcl-2, B-cell lymphoma 2; Bid, BH3 interacting-domain death agonist; COX, cyclooxygenase; ERK, extracellular signal-regulated kinase; IκB, inhibitor of κB; MMP, matrix metalloproteinase; mTOR, mammalian target of rapamycin; P13K, phosphatidylinositol 3-kinase; Ras, Rat sarcoma virus; TLR, Toll-like receptor; TNFR, tumor necrosis factor receptor; VEGF, vascular endothelial growth factor.

Different species of stingless bees also have their own preferences in collecting different types of nectar. For instance, Leonhardt (2017) reported that all species of stingless bees preferred sugar-rich over sugar-poor artificial nectar, and the nectar must contain at least a 40%–60% sugar concentration. Biesmeijer et al. (1999) also reported that two *Melipona* species prefer specific sugars (sucrose over glucose and fructose). Another study also reported the same in which *Frieseomelitta nigra* avoids sugar solutions that contain potassium found in avocado (*Persea americana*) nectar (Afik et al. 2014). Also, factors such as morphology, bee size, bee behavior(Kajobe 2008), and microorganisms (Braghini et al. 2021) present in the hive (helping in the maturation and fermentation of the product) influence the phytochemical constituents of SBH. Therefore, it is important for researchers to analyze the phytochemical constituents of SBH (based on the targeted species) before conducting an experiment.

## The Effects of Certain Storage Conditions on the Physicochemical Properties of Stingless Bee Honey

There are standard conditions for maintaining the quality of honey, as stated by Souza et al. (2006) whereby Western bee honey quality was affected by moisture, ash, pH, acidity, electrical conductivity, diastase, 5-hydroxymethylfurfural (HMF), and carbohydrates. Nevertheless, up until now, few studies have researched the standard parameters needed for maintaining the quality of SBH. A previous study reported on the long-term storage of SBH produced from *T. laeviceps-pagdeni* from Thailand in which the physicochemical parameters of honey were determine and compared based on the storage period (stored at 4°, 30°, and 45°C for 6 and 12 months). The results obtained showed that moisture, ash, electrical conductivity,

**Table 23.1** Summary of medicinal properties of stingless bee honey from different species and origin countries.

| Medicinal Property | Stingless Bee Species and Country or Continent of Origin | Effects | Reference |
|---|---|---|---|
| Antioxidant | *Meliponini* tribe (Australia), *Tetragonula carbonaria* (Europe) *Melipona fasciculate* (South Africa), and *Melipona* spp. (Malaysia) | High flavonoids and phenolic acids (e.g. PCA and 4-hydroxyphenylacetic acid) makes them possess antioxidants effects. | Jalil et al. 2017 |
| | *T. carbonaria* (Australia) | TAA values of honey produced by *T. carbonaria* had equivalent antioxidant contents in floral honey and honeydew honey. Also, the average RSA of honey produced by *T. carbonaria* is equivalent to 48% ascorbic acid. | Oddo et al. 2008 |
| | *Melipona (Michmelia) seminigra merrillae* (Brazil) | The presence of phenolic compounds enhances the antioxidant activity of SBH. | da Silva et al. 2013 |
| | *Geniotrigona thoracica* (Malaysia) | This contains high amount of phenolics and flavonoid compounds, which makes them possess greater antioxidant species than the SBH produced by *Heterotrigona itama* and *Heterotrigona erythrogastra*. | Bakar et al. 2017 |
| Antibacterial | *Melipona compressipes manosensis* (Brazil) | SBH produced by *M. compressipes manosensis* during the dry season is effective against both gram-positive (*Staphylococcus aureus*) and gram-negative (*Escherichia coli*, *Shigella sonnei*, *Proteus vulgaris*, and *Klebsiella* spp.) organisms at different dilution rates. | de Queiroz et al. 2013 |
| | *G. thoracica* (Malaysia) | The results demonstrated that SBH produced by GTM from multifloral is effective against both gram-positive (*Streptococcus pyogenes*, MRSA, and *S. aureus*) and gram-negative (*Pseudomonas aeruginosa*, *E. coli*, and *K. pneumoniae*) bacteria. | Omar et al. 2019 |
| | *Melipona eburnean*, *Melipona grandis*, *M. seminigra*, and *Melipona flavolineata* (Brazil) | All SBH types are more sensitive against gram-positive bacteria (*Enterococcus faecalis*, *S. aureus* and *Streptococcus pneumoniae*). | Domingos et al. 2021 |
| | *H. itama*, *H. erythrogastra*, *Tetrigona apicalis*, *Lepidotrigona terminata*, *Tetrigona melanoleuca*, *Tetrigona bingami*, *G. thoracica*, and *Homotrigona fimbriata* (Malaysia) | The ability of SBH (produced from all species) to inhibit both gram-positive and gram-negative bacteria was concentration dependent. | Rosli et al. 2020 |
| Wound healing | *Trigona itama* and *Trigona thorasica* (Malaysia) | The positive effect shown by SBH (at concentrations 0.024, 200, and 6400 μg/mL) toward cell viability and morphology and proliferation of human dermal fibroblast suggested the ability of SBH to be developed as a therapy for wound healing. | Nordin et al. 2018 |
| | *T. itama* and *T. thorasica* (Malaysia) | SBH (at 0.015%, v/v) can prevent the formation of keloids caused by overproduction of EMT by TGF-β in keratinocytes; however, these effects resulted in slower initial wound closure and individual migration velocity, which later then sped up, resulting in greater wound closure at the final time point. | Nordin et al. 2020 |

**Table 23.1** (Continued)

| Medicinal Property | Stingless Bee Species and Country or Continent of Origin | Effects | Reference |
|---|---|---|---|
| Ocular therapeutic properties | *L. arciferal* (India) | Two drops of SBH placed around the eyes two times daily for treating conjunctivitis. For eye infection, pure filtered SBH was inserted into the eyes after washing the hands and eyes with boiled clean water. | Biswa et al. 2017 |
| | *Melipona* spp. (Venezuela) | The presence of luteolin tetra-methyl ether, luteolin 4'-glucoside, luteo in 3'-7-diglucoside, and orientin significantly inhibited cataracts induced in ovine lenses incubated in 45% hypotonic HBS within 24 hours, proving SBH's ability as an anticataract agent. | Vit and Jacob 2008 |
| | *Meliponula* spp. (Ghana) | One drop (70 μL) of crude SBH (produced from *Meliponula* spp.) twice daily shortened the inflammatory signs, duration of infection, and time for complete resolution. | Ilechie et al. 2012 |
| Antidiabetic | *Trigona* spp. (Malaysia) | The results shown that all SBH samples exhibited antidiabetic activity in which the highest $IC_{50}$ value recorded by all honey was higher than that of the positive control, acarbose. | Zakaria et al. 2020 |
| | *G. thoracica* (Malaysia) | The FBG level shown by the diabetic group treated with SBH was in a normal range, which indicates the ability of SBH as an antidiabetic agent. Plus, the antidiabetic activity was attributed to the presence of L-phenylalanine (an agonist that potentiates glucose-dependent insulin secretion in the pancreatic β-cells). | Aziz et al. 2017 |
| Anticancer | *Trigona* spp. (Malaysia) | SBH contains CAPE, which is responsible for suppressing the transcriptional activity of β-catenin and hence can become an agent against colon cancer. | Saiful et al. 2016 |
| | *H. itama* (Malaysia) | Through cytotoxicity test, the SBH produced from *H. itama* spp. proved to exert a cell proliferation inhibitory effect of 50% cell viability for U-87 MG cells at concentration of 10% (observed at 72 hours after treatment). The results obtained are also in line with qualitative analysis via morphological and AO/PI analysis. | Ahmad et al. 2019 |

AO/PI, acridine orange/propidium iodide; CAPE, caffeic acid phenethyl ester; EMT, epithelial-to-mesenchymal transition; FBG, fasting blood glucose; GTM, *Geniotrigona thoracica* from multifloral; HBS, hypotonic HEPES buffered solution; $IC_{50}$, inhibition of 50% cell viability; MG, malignant glioma; MRSA, methicillin-resistant *Staphylococcus aureus*; PCA, protocatechuic acid; RSA, random scavenging activity; SBH, stingless been honey; TAA, total antioxidant activity; TGF-β, transforming growth factor β.

and storage temperature slightly changed over time. Acidity was increased as the time of storage increased, and pH, diastase, and HMF demonstrated statistically significant changes for both time and temperature storage. Moreover, the level of carbohydrates (fructose, glucose, and maltose) decreased during time and temperature storage.

Longer SBH storage times and higher storage temperatures cause the greatest changes in all parameters (Chuttong et al. 2015). Similar results were obtained by Braghini et al. (2020) in which the storage of SBH at 40°C was not recommended because honey stored at this temperature shows the greatest reduction in terms of moisture, acidity, HMF, and diastase activity, as well as bioactive properties (antioxidant potential and individual phenolic compounds). Furthermore, Martínez et al. (2017) reported on important storage period and temperature of SBH produced by *Tetragonisca fiebrigi* in which the best conditions to maintain the properties of honey are at a cool temperature to prevent fermentation of the product and the crystallization process. The study also stated that to maintain the antimicrobial properties of honey, it is important to store it at room temperature. There has also been a study about transforming the SBH produced by *H. itama* into a dry form (by using thermal dehydration) and testing for its properties. The result of the study revealed that the dry form of honey contains higher phenolic compounds such as rosmarinic acid and quercetin that enhance the antioxidant activity in dehydrated honeys. However, the antibacterial activity of dehydrated honey was reduced significantly against *S. aureus* and *E. coli* (Chen et al. 2021). Overall, the storage period and storage temperature have a great influenced on the properties of SBH and affect its stability; hence, it is important to ensure proper storage conditions when dealing with SBH.

## Future Prospects

The development of therapies from natural remedies such as honey is very important despite the advancements in technologies that we have seen. Considering the beneficial effects of SBH reported by several researchers via in vitro and in vivo studies, honey can potentially be developed into therapy for combating infections and to be used as an eye therapy and for wound healing, diabetes, and even cancer. Plus, SBH is an excellent source and choice to be developed for therapeutic purposes because of the rising demand in the pharmaceutical and food industries for remedies made from natural resources that are both safe and cost-effective. SBH can also potentially be combined with other natural resources and biomaterials as a combination therapy. The findings from exploring the medicinal benefits of SBH can also facilitate growth in countries that can supply this resource.

## Conclusion

In conclusion, this chapter has summarized and highlighted the importance of the medicinal benefits of SBH produced by different species. SBH possesses superior medicinal properties such as antioxidant activity, antibacterial activity, wound-healing activity, antidiabetic properties, and anticancer activity (as a chemopreventive agent). The medicinal properties of SBH are attributed to its phytochemical constituents, especially phenolic compound and flavonoids. The limitations and challenges that affect the phytochemical constituents and physicochemical characteristics of SBH are also being highlighted, and these factors must be considered and monitored first before developing SBH into a therapy. Taken together, SBH is a beneficial natural resource that must be explored because it has the potential to be developed into remedies for various medical conditions.

## Acknowledgment

The authors thank Universiti Kuala Lumpur Royal College of Medicine Perak, Malaysia, for providing necessary facilities and resources to complete this study for publication. The figures were created with the support of https://biorender.com under a paid subscription.

## Conflict of Interest

We wish to confirm that there are no known conflicts of interest associated with this publication, and there has been no significant financial support for this work that could have influenced its outcome.

## References

Afik, O., Delaplane, K.S., Shafir, S., et al. (2014). Nectar minerals as regulators of flower visitation in stingless bees and nectar hoarding wasps. *Journal of Chemical Ecology* 40 (5): 476–483.

Ahmad, F., Seerangan, P., Mustafa, M.Z., et al. (2019). Anti-cancer properties of Heterotrigona itama sp. honey via induction of apoptosis in malignant glioma cells. *The Malaysian Journal of Medical Sciences* 26 (2): 30.

Al-Hatamleh, M.A., Boer, J.C., Wilson, K.L., et al. (2020). Antioxidant-based medicinal properties of stingless bee products: recent progress and future directions. *Biomolecules* 10 (6): 923.

Alqarni, A.S., Owayss, A.A., Mahmoud, A.A., and Hannan, M.A. (2014). Mineral content and physical properties of local and imported honeys in Saudi Arabia. *Journal of Saudi Chemical Society* 18 (5): 618–625.

Alvarez-Suarez, J.M., Giampieri, F., Brenciani, A., et al. (2018). Apis mellifera vs Melipona beecheii Cuban polifloral honeys: a comparison based on their physicochemical parameters, chemical composition and biological properties. *LWT* 87: 272–279.

Alves, A., Ramos, A., Gonçalves, M.M., et al. (2013). Antioxidant activity, quality parameters and mineral content of Portuguese monofloral honeys. *Journal of Food Composition and Analysis* 30 (2): 130–138.

Amin, F.A.Z., Sabri, S., Mohammad, S.M., et al. (2018). Therapeutic properties of stingless bee honey in comparison with European bee honey. *Advances in Pharmacological Sciences* 2018.

Arung, E.T., Ramadhan, R., Khairunnisa, B., et al. (2021). Cytotoxicity effect of honey, bee pollen, and propolis from seven stingless bees in some cancer cell lines. *Saudi Journal of Biological Sciences* 28: 7182–7189.

Avila, S., Beux, M.R., Ribani, R.H., and Zambiazi, R.C. (2018). Stingless bee honey: quality parameters, bioactive compounds, health-promotion properties and modification detection strategies. *Trends in Food Science & Technology* 81: 37–50.

Aziz, M.S.A., Giribabu, N., Rao, P.V., and Salleh, N. (2017). Pancreatoprotective effects of Geniotrigona thoracica stingless bee honey in streptozotocin-nicotinamide-induced male diabetic rats. *Biomedicine & Pharmacotherapy* 89: 135–145.

Bakar, M.A., Sanusi, S.B., Bakar, F.A., et al. (2017). Physicochemical and antioxidant potential of raw unprocessed honey from Malaysian stingless bees. *Pakistan Journal of Nutrition* 16 (11): 888–894.

Biesmeijer, J.C., Richter, J.A., Smeets, M.A., and Sommeijer, M.J. (1999). Niche differentiation in nectar-collecting stingless bees: the influence of morphology, floral choice and interference competition. *Ecological Entomology* 24 (4): 380–388.

Bikowski, J. (2001). The use of therapeutic moisturizers in various dermatologic disorders. *Cutis* 68 (5 Suppl): 3–11.

Biswa, R., Sarkar, A., and Khewa, S.S. (2017). Ethnomedicinal uses of honey of stingless bee by Nepali community of Darjeeling foothills of West Bengal, India.

Bobiş, O., Dezmirean, D.S., and Moise, A.R. (2018). Honey and diabetes: the importance of natural simple sugars in diet for preventing and treating different type of diabetes. *Oxidative Medicine and Cellular Longevity* 2018.

Braghini, F., Biluca, F.C., Ottequir, F., et al. (2020). Effect of different storage conditions on physicochemical and bioactive characteristics of thermally processed stingless bee honeys. *LWT* 131: 109724.

Braghini, F., Biluca, F.C., Schulz, M., et al. (2021). Stingless bee honey: a precious but unregulated product-reality and expectations. *Food Reviews International* 1–30.

Chen, Y.H., Chuah, W.C., and Chye, F.Y. (2021). Effect of drying on physicochemical and functional properties of stingless bee honey. *Journal of Food Processing and Preservation* 45 (4): e15328.

Chuttong, B., Chanbang, Y., Sringarm, K., and Burgett, M. (2015). Effects of long term storage on stingless bee (Hymenoptera: Apidae: Meliponini) honey. *Journal of Apicultural Research* 54 (5): 441–451.

Ciulu, M., Solinas, S., Floris, I., et al. (2011). RP-HPLC determination of water-soluble vitamins in honey. *Talanta* 83 (3): 924–929.

da Silva, I.A.A., da Silva, T.M.S., Camara, C.A., et al. (2013). Phenolic profile, antioxidant activity and palynological analysis of stingless bee honey from Amazonas, Northern Brazil. *Food Chemistry* 141 (4): 3552–3558.

Dagenais, G.R., Gerstein, H.C., Zhang, X., et al. (2016). Variations in diabetes prevalence in low-, middle-, and high-income countries: results from the prospective urban and rural epidemiological study. *Diabetes Care* 39 (5): 780–787.

de Queiroz, R.B.P., da Costa, C.A., Albuquerque, P.M., and Junior, S.D. (2013). Antimicrobial activity and rutin identification of honey produced by the stingless bee Melipona compressipes manaosensis and commercial honey. *BMC Complementary and Alternative Medicine* 13 (1): 1–14.

Dhivya, S., Padma, V., and Santhini, E. (2015). Wound dressings-A review. *BioMedicine (Netherlands)* 5 (4): 24–28.

Domingos, S.C.B., Clebis, V.H., Nakazato, G., et al. (2021). Antibacterial activity of honeys from Amazonian stingless bees of Melipona spp. and its effects on bacterial cell morphology. *Journal of the Science of Food and Agriculture* 101 (5): 2072–2077.

Draelos, Z.D. (2010). Active agents in common skin care products. *Plastic and Reconstructive Surgery* 125 (2): 719–724.

Ediriweera, E. and Premarathna, N. (2012). Medicinal and cosmetic uses of Bee's Honey–a review. *Ayu* 33 (2): 178.

Elliott, S.S., Keim, N.L., Stern, J.S., et al. (2002). Fructose, weight gain, and the insulin resistance syndrome. *The American Journal of Clinical Nutrition* 76 (5): 911–922.

Fasanmade, A.A. and Alabi, O.T. (2008). Differential effect of honey on selected variables in alloxan-induced and fructose-induced diabetic rats. *African Journal of Biomedical Research* 11 (2): 191–196.

Guo, S.A. and DiPietro, L.A. (2010). Factors affecting wound healing. *Journal of Dental Research* 89 (3): 219–229.

Ilechie, A.A., Kwapong, P.K., Mate-Kole, E., et al. (2012). The efficacy of stingless bee honey for the treatment of bacteria-induced conjunctivitis in guinea pigs. *Journal of Experimental Pharmacology* 4: 63.

Jalil, M.A.A., Kasmuri, A.R., and Hadi, H. (2017). Stingless bee honey, the natural wound healer: a review. *Skin Pharmacology and Physiology* 30 (2): 66–75.

Kajobe, R. (2008). Foraging behaviour of equatorial Afrotropical stingless bees: habitat selection and competition for resources. Doctoral dissertation, Utrecht University Repository, Netherlands.

Kek, S.P., Chin, N.L., Yusof, Y.A., et al. (2014). Total phenolic contents and colour intensity of Malaysian honeys from the Apis spp. and Trigona spp. bees. *Agriculture and Agricultural Science Procedia* 2: 150–155.

Kek, S.P., Chin, N.L., Yusof, Y.A., et al. (2017). Classification of entomological origin of honey based on its physicochemical and antioxidant properties. *International Journal of Food Properties* 20 (sup3): S2723–S2738.

Kuppusamy, P., Yusoff, M.M., Maniam, G.P., et al. (2014). Nutraceuticals as potential therapeutic agents for colon cancer: a review. *Acta Pharmaceutica Sinica B* 4 (3): 173–181.

Kustiawan, P.M., Puthong, S., Arung, E.T., and Chanchao, C. (2014). In vitro cytotoxicity of Indonesian stingless bee products against human cancer cell lines. *Asian Pacific Journal of Tropical Biomedicine* 4 (7): 549–556.

Leonhardt, S. (2017). Chemical ecology of stingless bees. *Journal of Chemical Ecology* 43 (4): 385–402.

Lim, D.C.C., Bakar, M.F.A., and Majid, M. (2019). Nutritional composition of stingless bee honey from different botanical origins. *IOP Conference Series: Earth and Environmental Science* 269: 012025.

Majid, M., Bakar, M.F.A., Mian, Z., et al. (2019). Variations of physicochemical properties of stingless bee honey from different botanical origin in state of Johor, Malaysia. *IOP Conference Series: Earth and Environmental Science* 269: 012028.

Maringgal, B., Hashim, N., Tawakkal, I., et al. (2021). Phytochemical content, antioxidant activity and mineral elements of honey produced by four different species of Malaysian stingless bees. *Food Research* 5: 39–46.

Martinello, M. and Mutinelli, F. (2021). Antioxidant activity in bee products: a review. *Antioxidants* 10: 71.

Martínez, R.A., Schvezov, N., Brumovsky, L.A., and Román, A.B.P. (2017). Influence of temperature and packaging type on quality parameters and antimicrobial properties during Yateí honey storage. *Food Science and Technology* 38: 196–202.

Moo-Huchin, V.M., Sauri-Duch, E., Moo-Huchin, M., et al. (2016). Antioxidant activity of stingless bee honey. *In: Stingless Bee's Honey from Yucatan: Culture, Traditional Uses and Nutraceutical Potential*. New York: Nova Science Publishers

Moura, L.I., Dias, A.M., Carvalho, E., and de Sousa, H.C. (2013). Recent advances on the development of wound dressings for diabetic foot ulcer treatment—a review. *Acta Biomaterialia* 9 (7): 7093–7114.

Nordin, A., Chowdhury, S.R., Saim, A.B., and Bt Hj Idrus, R. (2020). Effect of Kelulut honey on the cellular dynamics of TGFβ-induced epithelial to mesenchymal transition in primary human keratinocytes. *International Journal of Environmental Research and Public Health* 17 (9): 3229.

Nordin, A., Omar, N., Sainik, N.Q.A.V., et al. (2018). Low dose stingless bee honey increases viability of human dermal fibroblasts that could potentially promote wound healing. *Wound Medicine* 23: 22–27.

Oddo, L.P., Heard, T.A., Rodríguez-Malaver, A., et al. (2008). Composition and antioxidant activity of Trigona carbonaria honey from Australia. *Journal of Medicinal Food* 11 (4): 789–794.

Omar, S., Mat-Kamir, N.F., and Sanny, M. (2019). Antibacterial activity of Malaysian produced stingless-bee honey on wound pathogens. *Journal of Sustainability Science and Management* 14: 67–79.

Rahmawati, O., Pratami, D.K., Raffiudin, R., and Sahlan, M. (2019). Alpha-glucosidase inhibitory activity of stingless bee honey from Tetragonula biroi and Tetragonula laeviceps. *AIP Conference Proceedings* 2092: 030001.

Rao, P.V., Krishnan, K.T., Salleh, N., and Gan, S.H. (2016). Biological and therapeutic effects of honey produced by honey bees and stingless bees: a comparative review. *Revista Brasileira de Farmacognosia* 26: 657–664.

Rashid, M.R., Nor Aripin, K.N., Syed Mohideen, F.B., et al. (2019). The effect of Kelulut honey on fasting blood glucose and metabolic parameters in patients with impaired fasting glucose. *Journal of Nutrition and Metabolism* 2019.

Rosli, F.N., Hazemi, M.H.F., Akbar, M.A., et al. (2020). Stingless bee honey: evaluating its antibacterial activity and bacterial diversity. *Insects* 11 (8): 500.

Sahlan, M., Rahmawati, O., Pratami, D.K., et al. (2020). The effects of stingless bee (Tetragonula biroi) honey on streptozotocin-induced diabetes mellitus in rats. *Saudi Journal of Biological Sciences* 27 (8): 2025–2030.

Saiful, L.Y., Muhamad Zali, M.F.S., Mohd Ali, R., et al. (2016). Chemopreventive properties and toxicity of Kelulut honey in Sprague Dawley rats induced with Azoxymethane. *BioMed Research International* 2016.

Samraj, S.M.D., Kirupha, S.D., Elango, S., and Vadodaria, K. (2021). Fabrication of nanofibrous membrane using stingless bee honey and curcumin for wound healing applications. *Journal of Drug Delivery Science and Technology* 63: 102271.

Souza, B.A., Roubik, D.W., Barth, O.M., et al. (2006). Composition of stingless bee honey: setting quality standards. *Interciencia* 31 (12): 867–875.

The Sun Daily (2021). Weeding out unsustainable Kelulut honey harvesting practices [Online]. https://www.thesundaily.my/home/weeding-out-unsustainable-kelulut-honey-harvesting-practices-GK7927954 (accessed 18 September 2021).

Tuksitha, L., Chen, Y.-L.S., Chen, Y.-L., et al. (2018). Antioxidant and antibacterial capacity of stingless bee honey from Borneo (Sarawak). *Journal of Asia-Pacific Entomology* 21 (2): 563–570.

Vit, P. and Jacob, T.J. (2008). Putative anticataract properties of honey studied by the action of flavonoids on a lens culture model. *Journal of Health Science* 54 (2): 196–202.

Vit, P., Oddo, L.P., Marano, M.L., and de Mejias, E.S. (1998). Venezuelan stingless bee honeys characterized by multivariate analysis of physicochemical properties. *Apidologie* 29 (5): 377–389.

World Health Organization (2016). Global report on diabetes.

Yang, Q., Xu, Y.-C., Kiratisin, P., and Dowzicky, M.J. (2017). Antimicrobial activity among gram-positive and gram-negative organisms collected from the Asia-Pacific region as part of the Tigecycline Evaluation and Surveillance Trial: comparison of 2015 results with previous years. *Diagnostic Microbiology and Infectious Disease* 89 (4): 314–323.

Zakaria, N., Jaafar, N.M., and Mohamad, A.Z. (2020). Antioxidant, antibacterial and anti-diabetic activities of stingless bee honey from selected areas in Peninsular Malaysia. *IOP Conference Series: Earth and Environmental Science* 596: 012093.

# 24

# Economic Benefits of Honey and Honey Products

*Sridevi I. Puranik, Aimen A. Akbar, and Shridhar C. Ghagane*

## Introduction

The rearing of honeybees is called beekeeping or apiculture. Beekeeping is one of the modern agroindustries (Bansal et al. 2005). It involves honeybees and flowering plants that provide food to the bees and to the beekeepers who rear bees according to the environment and flora available for their commercial benefits. Apiculture generates employment and income and provides economic security (Rossi and Marrazzo 2021). India has had a good history of connection between humans and honeybees, which is witnessed by the oldest records of honey production in the form of paintings by ancient people in rock shelters (Seanbualuang 2013).

Honey remains the first sweet food consumed by the ancient Indians living in the forests. They hunted beehives for this natural product (Degrandi-Hoffman et al. 2019). The honeybee species *Apis mellifera* is common in India. Collection of honey from wild beehives is the traditional profession of tribal people and forest dwellers in different parts of India (Dafar 2018). The methods of gathering honey and beeswax from wild hives have changed only to some extent over the years (Sain 2017). The protection of beehive nests in marked tree cavities and periodic harvesting of honey were the first steps in apiculture. The next step might have been attracting bees into hives in protected bee gardens. Initially, bees were reared for production of honey (Ediriweera and Premarathna 2012).

Today the apiculture industry is not only restricted to honey alone but also includes production of hive products such as beeswax, bee collected pollen, royal jelly, bee venom, and propolis (Data et al. 2021). Honeybees are also involved in pollination; hence, they increase the yield of agricultural and horticultural crops. Honey is one of the most valued and appreciated natural substances known to humankind since prehistoric times. It is one of the finest products of sugar-rich nectar from flowers and honeybees with time-proven, universally accepted medicinal, cosmetic, and pharmaceutical properties (Koetz 2013). Honey is considered a natural and raw ingredient that is used as a sweetener and as medicine because of its therapeutic effects on human health (Bee and Project 2021). Apart from its nutritional and beneficial uses, honey has lot of implications on human life, thus making it of great ecologic and economic importance (Apergis 2017).

Consumption of honey has rapidly and remarkably become common for several years all around the world. Because there has been an increase in the number of honey consumers observed in recent years, it has become important to note the purity of types of honey and their products (Production quantity of honey (natural) in 2019 2020). Honeybee culture is an important human activity and a significant part of the bioeconomy. Honey products are classified primarily into honey and its products, pollination, regeneration (which is known as apitourism cultural and educational services), and aesthetic value (Shahbandeh 2019). The products of beekeeping are used in food and pharmacy and have commercial benefits for beekeepers. The improved mass and diversity of pollinators have a direct influence on crop production and can support small-scale farmers to increase their production by a global average of 24% (Partners and News 2021).

Honey production has significant social influences and a substantial impact on biodiversity. Economically and ecologically, it produces important profits to society. About 90% of honeybees' benefits comes from their pollination capacity; the other 10% comes from bee products such as honey, propolis, and wax (Grossman 2013). Beekeeping for production of honey in large quantities is a means of employment (Kamble et al. 2015). In this way, honey production generates extra income and aids in the development of a nation's economy. Besides the scope of honey, there's also opportunity for production and marketing of other honeybee products, thereby increasing the profitability of many beekeeping processes. This

*Honey: Composition and Health Benefits*, First Edition. Edited by Md. Ibrahim Khalil, Gan Siew Hua, and Bey Hing Goh.

chapter focuses on the international trade of honey, products of honeybees, segment analysis, regional industries, and their applications in the pharmaceutical industry and food market.

## Economics of Beekeeping

The practice of rearing and breeding honeybees for the production of honey, along with some secondary products such as propolis, bee wax, flower pollen, royal jelly, and bee pollen, is called apiculture. Only two honeybee species from the *Apis* genus, namely, *A. mellifera* L. (Western honeybee) and *Apiscerana indica* F. (Indian honeybee), are widely known for honey production (Mintel 2016). They are also distinguished by storing their honey and building perennial colonial hives from wax (Mintel 2016). Initial investments in beekeeping include the purchase of hives, apiculture equipment, bees, and queens. It is suggested that one should begin with a minimum of 100 bee colonies. The cost analysis is made on 10 bee colonies; the approximate values are listed in Table 24.1 for reference. These values may differ based on the location and availability. Furthermore, there could be some extra investment for maintenance of colonies during different seasons, which has not been included into this estimation but definitely has to be considered for practical implementation of honeybee farming (Worldwide Demand of Honey Market Share Calculated to Reach Over USD 10,840 Mn by 2026: Report by Facts & Factors 2021).

Landless and poor farmers are actively involved in beekeeping (apiculture) as an essential means of livelihood. This commercial farming of bees is deemed a beneficial action that gives additional income to an outsized group of people living in hills and tribal and rural people and for agriculturists, horticulturists, and hobbyists (García 2018; Oddo et al. 2004; Worldwide Demand of Honey Market Share Calculated to Reach Over USD 10,840 Mn by 2026: Report by Facts & Factors 2021). As such, high qualifications or a degree is not needed to work in the beekeeping industry. Beekeepers can be skilled on how to collect honey, rear honeybee colonies of required species in stated and standard boxes, install boxes in suitable places, and manage an ideal number of colonies throughout the year (Oddo et al. 2004). Also, the benefits related to honeybee rearing are cost-effective and low maintenance. It can also deliver healthier food and balanced nutrition to the farmers (García 2018). In addition, honeybees carry pollination of horticulture crops, produce hybrid seeds, and improve yields through cross-pollination. *A. mellifera* L. and *A. cerana indica* F. honeybee species are often incriminated and shifted from yield to yield for cross-pollination of crops and plants, thereby adding charges for the pollination services. A major component of commercial beekeeping that differs from smaller operations is transportation (Persistence Market Research n.d.). Currently used modern methodologies of moving beehives have led to revisiting the traditional role of beekeeping entrepreneurs, favoring commercial production at a large scale. Table 24.2 presents the costs involved and net profit of honey production (Persistence Market Research n.d.).

## Global Production and Trade

Honey is produced in almost all parts of the world with distinctive types of honey coming from different regions. Data about honey production in every country are published every year by the Food and Agriculture Organization (FAO) of the United Nations. Worldwide production of honey was estimated as 1.72 million metric tons with China being the largest contributor (Table 24.3) to total honey production followed by other major producers such as Turkey and Canada

**Table 24.1** Cost analysis of honeybee farming.

| Details of Items | Amount (INR) |
|---|---|
| Cost of 10 beehives per box | $2000 \times 80 = 1{,}60{,}000$ |
| Cost of 80 bee boxes | $80 \times 400 = 32{,}000$ |
| Cost of apiary equipment | 8000 |
| Cost of honey unit and uncapping tray | 15.500 |
| Bee wax sheet (1 kg) | 400 |

INR, Indian National Rupees.

**Table 24.2** Economics of honey production and returns.

| Details of Items | Amount (INR) |
|---|---|
| Honey production 80 Rs × 40 kg | 3200 Rs/box |
| For 80 boxes = 80 × 3200 | 2,56,000 |
| Labor (skilled) cost 5000 × 12 | 60,000 |
| Unskilled labor cost 3500 × 12 | 42,000 |
| Migration charges | 20,000 |
| Feed charges annual | 10,000 |
| Total | 1,32,000 |
| New 25 box × 700 each | 17,500 |
| Net profit | 1,49,500 |

INR, Indian National Rupees.

**Table 24.3** Production of natural honey in 2018–2019.

| Rank | Country | Share in Production (%) | Production Quantity (metric tons) |
|---|---|---|---|
| 1 | China | 25.81 | 444.100 |
| 2 | Turkey | 6.35 | 109.330 |
| 3 | Canada | 4.67 | 80.345 |
| 4 | Argentina | 4.59 | 78.927 |
| 5 | Iran | 4.39 | 75.463 |
| 6 | United States | 4.14 | 71.179 |
| 7 | Ukraine | 4.06 | 69.940 |
| 8 | India | 3.9 | 67.141 |
| 9 | World | | 1.72 million |

(Kantar 2016). India contributes around 67,141 tons of honey according to the FAO with the highest number of beehives totaling around 12.25 million because of India's vast forest area, facilitating the presence of nectar and pollen in its grasslands. Currently, there are 1.8 million bee colonies in India, and the production of honey is about 10 million tons. The production of beeswax is about 10,000 tons (Krishna 2005). The average cost of producing honey worldwide was estimated approximately USD$689,000 per metric ton in 2019.

Despite the demand and growing interest for honeybee by-products, there has been a decreasing trend with some peaks noticed for the past three decades in hive populations in some countries (Fatimah et al. 2013). A total of 1.93 million metric tons of honey was produced worldwide in 2016 and 2017, which was the highest rate until then and later slowly reduced to about 1.72 million metric tons. Figure 24.1 presents honey production data from 2014 to 2019 (Alam et al. 2014). Besides ranking the highest in production, China is also regarded as the top exporter of honey, contributing around 11.73% followed by New Zealand (11.47%), Argentina (7.35%), and Germany (7.35%), which have favorable production environments and extremely advanced colony management. The major consumer and importer globally is the Unites States, accounting for 21.33%, followed by Germany with 11.37% and Japan with 7.14%. Table 24.4 shows the top exporters and importers throughout the world based on ranking (Hananeh et al. 2015).

**Table 24.4** Global export and import of honey.

| Major Exporters | | Major Importers | |
|---|---|---|---|
| **Country** | **Export Share (%)** | **Country** | **Import Share (%)** |
| China | 11.73 | United States | 21.33 |
| New Zealand | 11.47 | Germany | 11.37 |
| Argentina | 7.35 | Japan | 7.14 |
| Germany | 6.85 | France | 5.86 |
| Ukraine | 5.62 | United Kingdom | 5.48 |
| India | 4.99 | China | 4.19 |

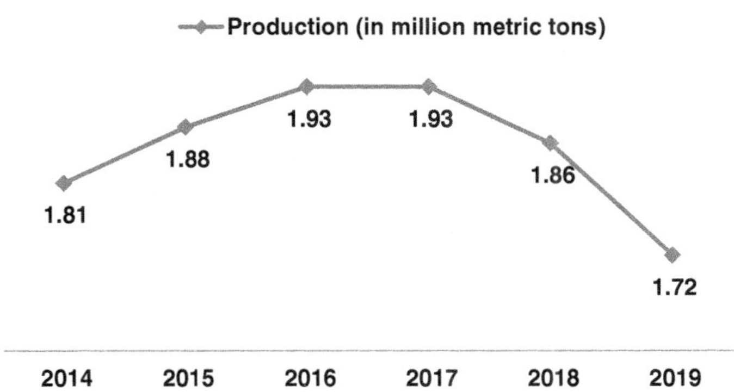


**Figure 24.1** Global honey production data by time series.

## Competitive Landscape

According to Alibaba, a global trading platform, the following companies are the top manufacturers and key players in the honey trade:

- Zhejiang Jiangshan Bee Enterprise Co., Ltd. (China)
- Beeyond the Hive (United States)
- Capilano Honey Ltd. (Australia)
- Caprilush International (Mexico)
- Bee Maid Honey (Canada)
- Groeb Farms, Inc. (United States)
- Duc Cuong Phat Company Limited (Vietnam)
- Dutch Golld Honey (United States)
- Navrang SL (Spain)
- Patanjali Ayurved (India)
- Nuxten Health Ltd. (New Zealand)
- New Zealand Honey Co. (New Zealand)
- Rowse Honey Ltd. (United Kingdom)
- Zanchetta Alimentos LTDA (Brazil)
- Dabur Ltd. (India)
- Oha Honey LP (New Zealand)

The global market is highly competitive because it depends on environmental conditions, and even the slightest change in the climate can drastically hamper production (Alam et al. 2014). The honey companies focus on driving market growth by introducing more research and development initiatives, invention and export operations, and commercialization. They also organize conferences and lectures and assist producers through mentoring and consulting services and industry-specific training and raise awareness about strategic business planning tools (Alam et al. 2014; Fatimah et al. 2013; Hananeh et al. 2015).

## Segment Analysis

The worldwide honey market remains to gain attraction due to the benefit of honey because of the current benefits of honey in health and the food and beverages industry, with increasing progress in coming days. Developments in food technology will further mark the importance of honey, prompting a substantial growth in business prospects for honey product manufacturers (Grossman 2013). Global demand for honey depends on the nature, application, packaging, and location of production (Figure 24.2).

## Apiculture as a Rural Small-Scale Industry

The forested type of apiculture is practiced in India and different southeast Asian countries. Various nectariferous plant species in moist and evergreen tropical and subtropical forests offer foraging (nectar and pollen) to honeybees (Hananeh et al. 2015). Hence, this environment provides the required raw materials available to beekeepers for honeybee culture. Bees' nests do not need additional space, nor do they struggle with agriculture for any input. Apiculture is therefore perfectly suited to agriculturalists as a part-time profession and can be taken as a rural small-scale industry (Eteraf-Oskouei and Najafi 2013). A bee box is the only basic and important requirement to begin in this industry. The outcome is relatively very high not only as a means of sustainable income generation but also as a novel food. It also aids a traditional medicine in the form of honey, royal jelly, protein-rich pollen, and brood to the village and tribal agriculturalists (Gibson et al. 2021). There has been good progress in apiculture using indigenous bees. Efforts were also made to introduce European bees, *A. mellifera L.*, in India successfully during the 1960s. Later, apiculture using European bees gained popularity.

## Global Honey Market

Maintaining the sweetness in foodstuffs and beverages has become a difficult job for industrialists because customers remain doubtful about the purity of these sweeteners. Also, large populations have diabetes globally; hence, food and beverage companies are advised to vigilantly include artificial sweeteners as well as natural sugars in their products (Grossman 2013; Kamble et al. 2015; Partners and News 2021). A speedy remedy for this global problem is the supplementation of pure honey in the manufacture of sweets, food products, and drinks. Honey is a naturally available organic sweet substance, which makes it one of the most readily available sweeteners in the world (Mintel 2016). Moreover, the carbohydrates present in honey offer its consumption for people with diabetes, promoting food and beverage manufactures to exchange any other sweetener for honey. Regular food preparations have included the use of honey, and customers purchase packaged honey to substitute sugars from their morning beverages such as tea or coffee (Gibson et al. 2021). The demand for honey has commercialized the production of nectar for the production of honey through naturally available nectar. Horticulture is raking in huge investments for growing commercial farming of bee culture. Inclusion of honey in top consumer retail goods such as sauces, mustards, and other seasoning flavors has created a surge in business opportunities for producers of honey products (Fakhrildin and Alsaadi 2014).

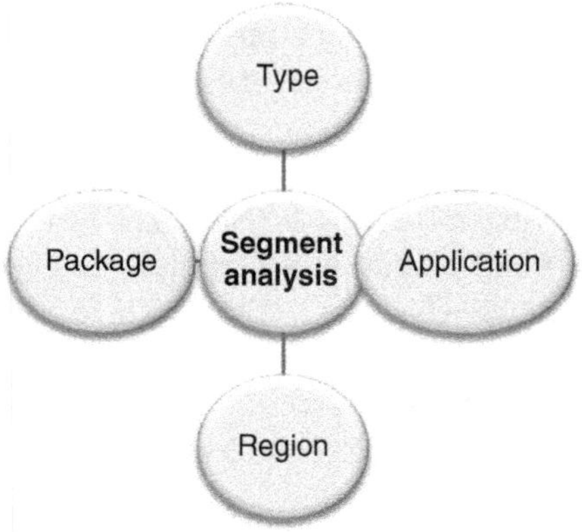


Figure 24.2 Segmentation of honey.

Developments in food-making technology will further increase the use of honey, prompting a significant increase in the business opportunities for honey product manufacturers (Abdelhafiz and Muhamad 2008). The international market for honey can be divided based on the of variety of honey.

The most commonly and commercially obtained honey varieties are date honey, flower honey, buckwheat honey, acacia honey, and linden honey, among others (Goharshenasan et al. 2016). Date honey is a basic means of energy, vitamins, minerals, and fiber. Acacia honey is preferred by people with diabetes because of the presence of low concentrations of sucrose and high concentrations of fructose in it (Popp et al. 2018).

Besides being a principal food additive in the world, honey is also a flourishing part of food and beverage production. The food and beverage industries now manufacture commercial beverages, honey mustard, and honey barbeque sauce as contemporary applications of honey (Alam et al. 2014). Honey production is also advanced in the drive toward fermentation for the production of sweet alcoholic drinks such as meads and honey wines. Honey manufacturing is playing a crucial role in the growth of the global honey market because of its healing benefits as a drug. Honey as a therapeutic agent is developing the market into drug production, medication, and production of medicinal disposables (Popp et al. 2018). The honey preservation period is relatively prolonged, allowing customers to store it even in the absence of freezers. However, the carbohydrate content of honey leads to consumer skepticism in the global market. Because honey is a natural sweetener, it is always being investigated by scientists and medical researchers (Food and Care 2021). The effect of intake of honey during treatment of open wounds leads to a major dislike among patients with diabetes (Food and Care 2021). Adulteration of honey by manufacturers leads to less intake by consumers, and impure products increase the health risks, which ultimately strengthens the honey industry in the global market. Botulism and health disorders caused by intake of impure and toxic honey have also affected the progress of the global honey market (Rao et al. 2016).

Pure honey production without adulteration includes critical monitoring systems that make sure that these products are free of adulterants (Ediriweera and Premarathna 2012). The incidence of bad practices such as mixing sugar syrups has contaminated the market presence of several honey production companies (Pasupuleti et al. 2017). Taking up new technologies and manufacturing equipment for a superior quality production of pure honey is now a prominent trend implemented by the market's contributors (Size et al. 2021). Some of the prominent firms in the global honey market are Lamex Foods UK Limited, Honey Sugar Product, Phondaghat Pharmacy, LITTLE BEE IMPEX, and Hi-Tech Natural Products (India) Ltd. (Ediriweera and Premarathna 2012).

## Product Types

Wildflower honey leads the honey market worldwide in terms of revenue, accounting for more than 25% of the total revenue share. The customer demand for this variety of honey is predicted to increase considering the presence of antioxidants in higher concentrations relative to other honey types. Other types of honey include clover, alfalfa, and buckwheat honeys. Buckwheat honey is widely popular in China, Russia, and several European countries (Ediriweera and Premarathna 2012).

## Application Insights

### Foods and Beverages

Honey is an invaluable natural substance that has been harvested by humans for centuries for food, both in raw and processed forms. The food and beverage sector has taken over one-third share of the total income generated by the global honey market and is expected to be in the top position in the coming years (Pasupuleti et al. 2017). It also dominates the U.S. honey market as the largest application with a share of more than 70% in 2020. This increased intake can contribute to the overall increase in living standards and a higher interest and demand for natural, nutritious health products. However, product adulteration is still one of the significant issues in the global market. On the market are some nonmedicinal honey-milk products such as pasteurized and homogenized milk sweetened with honey for long-term storage (Size et al. 2021).

### Personal Care and Cosmetics

Honey is used in the personal care and cosmetic industry in the form of face washes, moisturizers, soaps, lip balms, and other products. This segment is anticipated to grow at the fastest rate with a compound annual growth rate of more than 7% from 2021 to 2028 (Size et al. 2021). A main factor driving the manufacturing progress of honey in the cosmetic industry is the potential of the product in retaining skin texture, acting as an anti-aging agent, and inducing soothing and moisturizing effects on the skin (Size et al. 2021).

In recent years, there has been an enormous shift in the personal care industry and consumer behavior with consumers becoming extra aware of the ingredients used in their beauty products. According to the reports of Cosmetics Info (2016), honey is used in cosmetic products as a humectant to slow down the loss of moisture from products and is used as a fragrance and a skin-conditioning agent (Ediriweera and Premarathna 2012). Ingredients derived from superfoods, such as honey, oatmeal, and fruits, are becoming increasingly popular. Consumers want to be sure that ingredients have been sourced responsibly and that products are good for their skin. In fact, Mintel (2016) reported that around 60% of new product development is sourced from botanical and herbal ingredients in the personal care sector. In particular, honey is becoming a preferred component among consumers, with more than 75% of people being interested in purchasing products containing honey as a vital ingredient. This is likely because of the healing and medicinal properties of honey combined with the fact that it is a natural ingredient.

## Medicinal Uses

The therapeutic use of honey has been a matter of choice in modern medicine, exhibiting consistent and standardized antimicrobial activity. Scientists need to identify the floral species that give honey its antimicrobial characteristics (Pasupuleti et al. 2017; Size et al. 2021). Honeys contain low pH and high osmolarity combined through the enzymatic assembly of hydrogen peroxide, which exerts an antimicrobial result. Honey is most valued in traditional medicine and is used for the treatment of various ailments as a popular home remedy (Ediriweera and Premarathna 2012). It is excessively used in a conventional way in the treatment of diseases. It is especially used in the formulation of cough syrups and liquid dietary supplements. Because of its beneficial effects such as antiviral, antifungal, wound-healing, antioxidant, antibacterial, anti-inflammatory, and other health-enhancing properties, honey is receiving appreciation, demand, and broader acceptance in the pharmaceutical industry (Ministry of Agriculture & Farmers Welfare 2021).

Honey has been used extensively for various ailments, thus gaining popularity in modern medicine as a result of its antimicrobial activity. To mention a few uses of honey in medicine, cough has been a major problem for people of all ages and is a common frequent complaint presented to physicians. Cough is common among children and has multifactorial causes (Pasupuleti et al. 2017). The occurrence and the cause depend on the patient's age as well as geographical, environmental, weather, and epidemiological conditions. Thus, consumption of honey with milk is a routine remedy (Pasupuleti et al. 2017; Rao et al. 2016; Size et al. 2021).

The healing properties of honey are mainly attributed to its antibacterial activity, maintaining a moist wound condition, and its high viscosity, which help to provide a protective barrier to prevent infections (Ediriweera and Premarathna 2012). This particular medicinal significance of honey has been reported in the historical scientific literature. Honey has been identified for its outcomes on the healing process with multiple positive effects on wounds and is considered to enhance circulation and healing growth. In the research literature, honey has received high attention in the domain of wound healing, especially for burn wounds (Ediriweera and Premarathna 2012; Ministry of Agriculture & Farmers Welfare 2021; Size et al. 2021).

There has been a controversial debate on the impact of a high-carbohydrate diet on human health, particularly to understand the influence on changes in blood glucose levels. Presently, the significance of carbohydrate is frequently demonstrated as the glycemic index. Thus, honey still remains a controversial factor for patients with diabetes (Ediriweera and Premarathna 2012; Ministry of Agriculture & Farmers Welfare 2021; Pasupuleti et al. 2017; Rao et al. 2016; Size et al. 2021).

The ancient Egyptians used honey for fertility. Besides fertility, honey is consumed traditionally for enhancement of vitality among men. It has been known and demonstrated as a remedy for several causes of infertility (Ediriweera and Premarathna 2012; Food and Care 2021; Ministry of Agriculture & Farmers Welfare 2021; Pasupuleti et al. 2017; Ramachandra et al. 2012; Rao et al. 2016; Size et al. 2021). It has been described that there are several causes of infertility and potential remedies. As researched, honeybee pollen is thought to improve egg quality and general fertility and fecundity because of its rich content of iron, immunoenhancing properties, vitamins, other minerals, and amino acids. Men with impotency problems and problems related to infertility and women with erratic ovulation have been advised to ingest honey (Ediriweera and Premarathna 2012; Food and Care 2021; Ministry of Agriculture & Farmers Welfare 2021; Pasupuleti et al. 2017; Popp et al. 2018; Rao et al. 2016; Size et al. 2021).

Honey is rich in vitamins, particularly B vitamins, which are vital for the production of testosterone (Food and Care 2021). It has been thought that a higher concentration of nitric oxide (a chemical substance involved in vasodilatation present in honey) has resulted in improved erections in men diagnosed with dysfunctional erection or impotence. It has been researched that approximately 100 g of honey is required to increase nitric oxide levels about 50% in the blood (Food and Care 2021; Pasupuleti et al. 2017; Rao et al. 2016; Size et al. 2021).

It has been affirmed that honey is highly nutritional with promising antioxidant, anti-inflammatory, and antimicrobial properties as well as wound-healing and cough-reducing properties. Honey is also considered to improve serum testosterone concentrations, sperm counts, and fertility. The foremost concerns in the medicinal applications of honey in modern medicine are variations in its composition and the lack of clinical trials. Further investigation of honey in preclinical trials will pave a path to its usage in clinical settings for treating patients with various ailments at a large scale and improving health benefits with fewer side effects.

### Packaging

Honey may be packaged in glass jars, bottles, tubs, tubes, and so on. Bottle packaging is the most preferred globally (45.90%) because bottles are easy to store and transport, offer higher convenience compared with other types of packaging, and can be structured in different shapes and sizes, thus making them an ideal choice for consumers. Next is glass jars, with 24.33% of utilization. Mostly environment-friendly consumers highly prefer this type of product packaging (Hananeh et al. 2015).

## Regional Analysis

The worldwide honey market is divided into North America, Europe, Asia Pacific, and the rest of the world. By region, the Asia Pacific is deemed the leading region, accounting for the largest market share of revenue. This is projected to continue its regional demand in the estimated period of 2020–2027 (Eteraf-Oskouei and Najafi 2013). This regional dominance is due to major production and exporting countries such as China, India, Australia, and others. In addition, a few countries, namely, India and Australia, are boosting their production of honey to support the economy and market growth as the government is investing to promote beekeeping activity. For instance, the government of India allocated approximately USD$68 million toward apiculture under its Atmanirbhar Abhiyan scheme in May 2020 (Rao et al. 2016). Apiculture benefits consumers with different products (honey, wax, pollen, royal jelly, propolis, and so on) and amenities (pollination) to human beings and the environment. Throughout the world, honeybees support humankind and elevate the ecosystem. Apiculture is a significant source of income to rural communities related to farming and horticultural production. However, honey making is a smaller industry than expected in India's Uttara Kannada District, with 60% of its lands being forests and about 15% being farms (Pasupuleti et al. 2017). Case studies have highlighted that beekeeping could be elevated into a much more profitable venture requiring small capital investments and skilled labor for realizing high yields in comparison with other rural employment and poverty reduction programs (Ministry of Agriculture & Farmers Welfare 2021). Studies have also discussed the static performance of beekeeping societies, which had in their earlier days played vital role in spreading awareness on scientific beekeeping and its profitability, assisted people in setting up beekeeping units in their homes, and helped in marketing the products of bees (Ramachandra et al. 2012).

## Impact Analysis of COVID-19 in the Honey Market and Future Forecast

The honey industry has not only survived the crippling unprecedented impact of the COVID-19 (coronavirus) pandemic but has also witnessed a positive global demand shock across all regions (Global Honey Markets on the Rise in Spite of COVID-19 2021). People have increased their honey consumption to treat seasonal flu, throat infections, coughs, and cold caused by COVID-19. Furthermore, the global demand of immunity-boosting honey products is likely to spike sales and propel market growth even after the end of the pandemic. The worldwide honey market magnitude was valued about USD$7.19 billion in 2019 and exhibited a steady growth, reaching USD$9.21 billion in 2020 (Reportlinker 2019). In short, the consumption of honey has increased in developed markets as a direct result of the COVID-19 pandemic and is predicted to exponentially increase at a CAGR of around 4.8% between 2019 and 2025, generating approximately USD$10,336 million by 2025 (Global Apiculture Market Study 2020–2025: Honey Consumption on the Rise – ResearchAndMarkets.com 2020).

## Conclusion

The importance of honey goes beyond being a natural sweetener and is considered a promising off-farm enterprise that provides a regular source of income for rural households and a nation's economy in general. It is understood that honeybees should be taken care of because they are the only source for large-scale organic honey production. The honey industry also has positive potential spillovers through the facility of major contributions for the progress of associated industries, such as the pharmaceutical and brewery industries. In addition, it is substantially significant in environment conservation. Honey is a natural therapeutic that has a significant role in medicine. There is enormous evidence observed since ancient times endorsing the practice of using honey in disease management. Depending on the evidence, the use of honey in clinical wards is highly endorsed. Currently, there are many entrepreneurs in different regions of the world who have understood the importance of apiculture, which has a steady and rising market demand for honey as a nutritive food, as a therapeutic, for use as a sweetener, in the pharmaceutical industry, and so on.

## Conflict of Interest

The authors declare conflict of interest as none.

Compliance with Ethical Standards

## References


Abdelhafiz, A.T. and Muhamad, J.A. (2008 May 1). Midcycle pericoital intravaginal bee honey and royal jelly for male factor infertility. *International Journal of Gynecology & Obstetrics* 101 (2): 146–149.

Alam, F., Islam, M.D., Gan, S.H., and Khalil, M. (2014 January 1). Honey: a potential therapeutic agent for managing diabetic wounds. *Evidence-Based Complementary and Alternative Medicine* 2014: 1–17.

Apergis, N. (2017). The role of honey production in economic growth: evidence from a panel of major global producers. *International Journal of Agricultural Economics* 2 (5): 154.

Bansal, V., Medhi, B., and Pandhi, P. (2005 July 1). Honey–a remedy rediscovered and its therapeutic utility. *Kathmandu University Medical Journal (KUMJ)* 3 (3): 305–309.

Bee, H. and Project, F. (2021). Honey bee farming project report, cost, profits introduction to honey bee farming project report. Published online. 1–14.

Dafar, A. (2018). Review of economic and ecological importance of bee and bee products in Ethiopia. *Journal of Animal Husbandry and Dairy Science* 2 (2): 18–26.

Data, B.P., Feedback, G., Ton, M., and Nut, A. (2021). $6.89k. Published online. 2–3.

Degrandi-Hoffman, G., Graham, H., Ahumada, F., et al. (2019). The economics of honey bee (hymenoptera: apidae) management and overwintering strategies for colonies used to pollinate almonds. *Journal of Economic Entomology* 112 (6): 2524–2533.

Ediriweera, E.R. and Premarathna, N.Y. (2012 April). Medicinal and cosmetic uses of bee's honey a review. *Ayu* 33 (2): 178.

Ediriweera, E.R.H.S.S. and Premarathna, N.Y.S. (2012). Medicinal and cosmetic uses of bee's honey—a review. *AYU (An International Quarterly Journal of Research in Ayurveda)* 33 (2): 178. doi: 10.4103/0974-8520.105233.

Eteraf-Oskouei, T. and Najafi, M. (2013). Traditional and modern uses of natural honey in human diseases: a review. *Iranian Journal of Basic Medical Sciences* 16 (6): 731–742. doi: 10.22038/ijbms.2013.988.

Fakhrildin, M.B. and Alsaadi, R.A. (2014 March). Honey supplementation to semen-freezing medium improves human sperm parameters post-thawing. *Journal of Family & Reproductive Health* 8 (1): 27.

Fatimah, B., Abubakar, G., and Aliyu, S. (2013). Analysis of biochemical composition of honey samples from North-East Nigeria. *Biochemistry and Analytical Biochemistry* 2 (3): 1000139.

Food, A. and Care, P. (2021). Honey market size, share & COVID-19 impact analysis, by type (alfalfa, listen to audio version). Published online. 1–7.

García, N.L. (2018 July 3). The current situation on the international honey market. *Bee World* 95 (3): 89–94.

Gibson, L., Moffatt, C.J., Narahari, S.R., et al. (2021 October 1). Global knowledge gaps in equitable delivery of chronic edema care: a political economy case study analysis. *Lymphatic Research and Biology* 19 (5): 447–459. https://www.researchandmarkets.com/

(2020). Global apiculture market study 2020–2025: honey consumption on the rise – researchandmarkets.com. Published online.

(2021). Global honey markets on the rise in spite of COVID-19. Published online. 1–12.

Goharshenasan, P., Amini, S., Atria, A., et al. (2016). Topical application of honey on surgical wounds: a randomized clinical trial. *Complementary Medicine Research* 23 (1): 12–15.

Grossman E. (2013 March 5). Declining bee populations pose a threat to global agriculture. *Yale Environment*. 360.

Hananeh, W.M., Ismail, Z.B., Alshehabat, M.A., and Abeeleh, M.A. (2015 January 1). Effects of Sidr honey on second-intention healing of contaminated full-thickness skin wounds in healthy dogs. *Journal of Veterinary Research* 59 (3): 433–439.

Kamble, K.D., Pandit, R.S., and Rao, K.L. (2015). SMelittopalynological investigations of honey from Sunderban region, West Bengal, India. *Proceedings of the National Academy of Sciences, India Section B: Biological Sciences* 85 (1): 101–106.

Kantar, A. (2016 February). Update on pediatric cough. *Lung* 194 (1): 9–14.

Koetz, A.H. (2013 December). Ecology, behaviour and control of Apis cerana with a focus on relevance to the Australian incursion. *Insects* 4 (4): 558–592.

Krishna, R.S. (2005). Therapeutic uses of honey in ayurveda [Internet]. https://www.boloji.com/articles/1117/therapeutic-uses-of-honey-in-ayurveda (accessed 13 March 2022).

Ministry of Agriculture & Farmers Welfare. (2021). Government has allocated Rs. 500 crores towards beekeeping under AtmaNirbharAbhiyan: Shri Narendra Singh Tomar India among world's top five honey producers. Published online. 2–3.

Mintel. (2016 May). Category insight: shampoo & conditioner [Internet]. https://reports.mintel.com/display/860353 (accessed 14 March 2022).

Oddo, L.P., Piro, R., É, B., et al. (2004). Main European unifloral honeys: descriptive sheets. *Apidologie* 35 (Suppl. 1): S38–81.

Partners, T. and News, I. (2021). World' s top exports natural honey exports by country. Published online. 4–11.

Pasupuleti, V.R., Sammugam, L., Ramesh, N., and Gan, S.H. (2017). Honey, propolis, and royal jelly: a comprehensive review of their biological actions and health benefits. *Oxidative Medicine and Cellular Longevity* 2017: 1–22.

Persistence Market Research. (n.d.). Honey market: global industry analysis and forecast 2016–2022. https://www.persistencemarketresearch.com/market-research/honey-market.asp.

Popp, J., Kiss, A., Oláh, J., et al. (2018). Network analysis for the improvement of food safety in the international honey trade. *Amfiteatru Economic* 20 (47): 84–98.

Production quantity of honey (natural) in 2019. (2020). Livestock Primary/World Regions/Production Quantity from picklists. Food and Agriculture Organization of the United Nations. (Retrieved 7 February 2021).

Ramachandra, T.V., Chandran, M.S., Joshi, N.V., and Balachandran, C. (2012). Beekeeping: sustainable livelihood option in Uttara Kannada, central Western Ghats. *Environmental Information System Technical Report* (49): 1–49.

Rao, P.V., Krishnan, K.T., Salleh, N., and Gan, S.H. (2016). Biological and therapeutic effects of honey produced by honey bees and stingless bees: a comparative review. *RevistaBrasileira de Farmacognosia* 26 (5): 657–664.

Reportlinker. (2019). Global honey industry. *Report*. Published online. 2–5. https://www.prnewswire.com/news-releases/global-honey-industry-300790020.html.

Rossi, M. and Marrazzo, P. (2021). The potential of honeybee products for biomaterial applications. *Biomimetics* 6 (1): 1–14.

Sain, V. (2017). Economics and importance of beekeeping. *Biomedical Journal of Scientific & Technical Research* 1 (7): 1833–1834.

Seanbualuang, P. (2013 July 19). Basic knowledge of beekeeping. *Naresuan University Journal: Science and Technology (NUJST)* 20 (2): 93–100.

Shahbandeh, M.. (2019 August 23). Honey market worldwide and in the U.S. - statistics & facts.

Size, H.M., Care, P., Region, B., et al. (2021). Report Overview. 2028: 1–7.

Kool LH. Economic development on the periphery: a case study of East Galicia. *Temple University*; 1994.

# Index

Note: Page numbers followed by '*f*' refer to figures and '*t*' refer to tables.



*Honey: Composition and Health Benefits*, First Edition. Edited by Md. Ibrahim Khalil, Gan Siew Hua, and Bey Hing Goh.

***i***



***k***